MW01709540

Student Solutions Manual for Smith's

The Nature of Mathematics

Ninth Edition

Steven M. Watnik

Broward Community College

BROOKS/COLE

THOMSON LEARNING™

Australia • Canada • Mexico • Singapore • Spain • United Kingdom • United States

BROOKS/COLE

THOMSON LEARNING

Assistant Editor: *Rachael Sturgeon*
Marketing Team: *Leah Thomson, Samantha Cabaluna, Maria Salinas*
Editorial Assistant: *Jonathan Wegner*
Production Coordinator: *Dorothy Bell*
Cover Design: *Irene Morris*
Cover Photo: *Christopher Talbot Frank/ Panoramic Images*
Print Buyer: *Micky Lawler*
Printing and Binding: *Webcom Limited*

For more information about this or any other Brooks/Cole product, contact:
BROOKS/COLE
511 Forest Lodge Road
Pacific Grove, CA 93950 USA
www.brookscole.com
1-800-423-0563 (Thomson Learning Academic Resource Center)

Printed in Canada

10 9 8 7 6 5 4 3 2 1

ISBN 0-534-37985-0

Table Of Contents

Dedication

With infinite love and gratitude,
I dedicate this book to my folks,
Mimi and Joe Watnik,
exemplars of the selfless "nature of" parenting

Preface

Dear Reader,

I have endeavored to write a solution manual that is what its title purports: a manual (truly) for students. When I first contemplated this gargantuan project, I was motivated in part by my — and, more importantly, my students' — displeasure with many of the student manuals currently on the market. The class of manuals to which I refer tends to be written in a terse, formalistic style and often lacks the detailed explanation (and "novice friendly" language) that student readers desire. Many students find such resources more confusing than helpful; the more vocal among them complain of having been "ripped off." Often, I have felt disappointed that the publishers/authors squandered a seemingly valuable pedagogical opportunity and, in the process, engendered student frustration and anxiety (hence, diminishing the overall level of customer satisfaction). To the detriment of faculty and students alike, it is lamentable that many a solution manual is essentially (if not entirely) a reprint of the instructor's manual — except, of course, for the removal of 50% of the material (in particular, the even-numbered solutions).

With the support of Brooks/Cole, I have sought to author a manual that is a significant departure from many of its contemporaries. While remaining wholly true to the pedagogy, notation, terminology/usage, and spirit of Karl Smith's **Nature of Mathematics** *textbook, I have written a companion volume that complements Professor Smith's work and extends its ideas in meaningful ways. In addition, I have built into this manual several other distinguishing (in some cases, sacrilegious) features of which you should be apprised:*

- *The solutions are fully cross-referenced to the ninth edition of Smith's text, as well as to each other. This includes mention of page citations, textbook examples, and the similarities/differences among related problems. Generally speaking, I cite such references in the first paragraph of the solution, in the hope that the student who was formerly "stuck on" a problem will refer to the citation and actively reattempt the problem prior to (or in lieu of) passively reading the solution in its entirety. Moreover, for instructors (most notably, newcomers to Smith's* **Nature of Mathematics** *or neophytes to the subject matter), I am convinced this feature will greatly facilitate their use of Smith's text and enrich their awareness of its subtleties and themes.*

- *Multiple strategies/methods are provided for a given problem (or group of comparable problems). Sometimes I do this within the confines of a single solution. On other occasions, I vary the approach from problem to problem (among a series of related problems), usually with cross-references to preceding/succeeding solutions. Consequently, when you are checking the solution for a single problem, I recommend you peruse the solutions of similar-type problems.*

- *Each problem's* **final answer** *is prominently displayed (in an enlarged font and enclosed within a box), so as to visually distinguish it from the surrounding explanatory-type matter.*
- *Interspersed throughout the manuscript are hints, tips, and suggestions. Some pertain to study skills; others are more mathematical in nature.*

A final thought... I accepted this authorship assignment for the same reason daunting challenges in general, both professional and recreational, always have appealed to me: for the growth I hoped it would induce. Without brutally honest (no-holds-barred) feedback, growth is stunted. So, in this spirit, I enthusiastically invite all the criticisms, corrections, suggestions, and comments you wish to provide. If I can clarify a solution for you (especially via e-mail), I'll be glad to do so.

Steve Watnik
Broward Community College, South Campus
Department of Mathematics
7200 Pines Blvd., Bldg 69/Rm 236
Pembroke Pines, FL 33024
Phone: (954) 963-8935
E-Mail: swatnik@broward.cc.fl.us

All typesetting, layout design, and (original) graphics are solely the work of the author, Steve Watnik.

This manual was prepared with Microsoft ® Word 2000 and reproduced by Brooks/Cole from camera-ready copy supplied by the author. Many of the graphs for Chapters 11 and 12 were created in Maple6.0.

Acknowledgements

Thank you, Brooks/Cole Publishing, for this one-of-a-kind professional opportunity. Special thanks go to ancillary editor Rachael Sturgeon, for her professionalism, diligence, and instinctive knowledge of when to apply the firm riding crop and when to break out the cheerleader pompoms. I am also grateful to: Stephanie Schmidt, Rachael's predecessor; Karl Smith, for his blessing, oversight, meticulous proofreading, and numerous suggestions [thank you, Karl!]; JoAnne Drake, for her proofreading, error correction, and graciousness; and Diana Sulewski, my local Brooks/Cole sales rep and staunch advocate. Additionally, I feel compelled to seize this golden opportunity to express my affection for and indebtedness to a handful of the (many) people who indirectly have made this monumental accomplishment possible:

- *Without the strong (and abiding) influence of my five high school math teachers (and math team coaches) at Coconut Creek HS (Coconut Creek, FL), I never would have pursued a career in mathematics education — let alone written a book. They are Ruth Crane, Susan Greenberg, Betty Horn, Don Myerscough, and Had Oshry. May all kids be blessed with such dynamic, erudite, caring instruction.*

- *My mentor, Dr. M.J. DeLeon of Florida Atlantic University, a pervasive influence on countless mathematical and professional levels.*

- *The mathematics department at the University of Texas at Austin, for their rigorous graduate program and the many people who inspired me during my time there (most notably, my friend and treasured study-partner Mike Atchison).*

- *Prof. Joe Castillo, a cherished colleague, who for nine years has perpetually challenged me, encouraged me, and befriended me.*

- *Terrol Thompson, my first supervisor, and the recently-retired Jean Carpentier, both of Northeast High School (Oakland Park, FL) and both 30⁺-year classroom veterans, whose fatherly guidance and wise counsel have been indispensable from the time of my career's inception to the present day.*

- *My employer, Broward Community College, the seed from which so much of my professional opportunity and good fortune germinates. From my departmental colleagues to the institution's administration, I appreciate them all. I also would like to recognize BCC's new Pines Center (Jeff Allbritten, Myrna Bomser, Winston Chen, et al) for their hospitality and the use of their facility's computer equipment during much of this project.*

- *To my wonderful family and extraordinary longtime friends [you know who you are!], words cannot convey how eager I am to spend more time with each of you now that this all-consuming endeavor is behind us.*

- *And, finally, to my beloved Emily, my gifted partner. I thank you for your love, commitment, and tolerance during this entire episode.*

CHAPTER 1: The Nature of Problem Solving

Problem Set 1.1

1./3./5. **Answers vary**. Obviously, there is no one specific right — or, for that matter, completely wrong — answer to these three questions. Nonetheless, are you open to some friendly advice about these questions and related issues? Allow us to share two suggestions.

First, please do **not** avoid these questions, especially if you tend to be obsessed with grades (but not true learning) and are thinking "hey, these probably aren't gonna be on the test anyway." If this describes you, we urge you to stop thinking this way. Your attitude and approach to the subject matter is a crucial factor in determining your level of success. Said differently, if you devote your energies to learning (really learning!) as much as you can, rest assured the grade will take care of itself. However, if you focus your attention on grades, you are likely to be too distracted (due largely to self-induced stress) to fully focus on learning.

Second, here is one very healthy way to approach these three questions: try discussing — maybe even debating — these issues with others. Any of the following would be an appropriate person with whom to have this discussion/debate: a classmate (who may go on to become your study partner); your instructor (consider it a chance to begin your working relationship in a positive, non-threatening way); or a close friend or family member, preferably one who listens well, who asks thought-provoking questions, and is prone to challenging you. In short, find someone who can serve as your "sounding board." After having such a discussion, try to put your thoughts and views on paper, ideally as a first entry in your "journal" (as discussed on page 3 of your textbook).

7. Have you attempted to create "from scratch" Pascal's triangle? If not, before you read any further, please do so now. To paraphrase the logo located on the first page of Section 1.1 in your text, no passive spectators allowed!

Once you have become comfortable creating Pascal's triangle, all that is needed for you to answer this question is an ability to look at your created triangle from a variety of perspectives. Be creative! Trace the numbers, flip the triangle around, look at it diagonally, and so on. Eventually (or perhaps immediately), after you refer to the labeling system used in text Examples 1 and 2 (on pp. 8-10), you will notice that the **1st left diagonal and the 1st right diagonal** both consist of the sequence $1, 2, 3, 4, 5, \ldots$.

9. **Yes, the solution (namely, 120 paths) could have been obtained by going 3 blocks over and 7 blocks down.** After addressing the precise question asked in this exercise, it is important that you grasp the question's deeper significance. That is to say, do you think this question's sole purpose to help you find another way of arriving at an answer for Example 2? Indeed, it is not; rather Example

2 is intended to serve as a concrete, user-friendly "launching pad" for you to discover a far more important, more general pattern. In order to discover this pattern, you need to devise further "curiosity experiments" of your own to explore whether reversing the order of "down" and "over" works all the time or only occasionally. After experiencing firsthand that it works all the time, you should attempt to explain why. Your explanation should include mention the "symmetry" of Pascal's triangle and the central role it plays here. If the word "symmetry" is not in your vocabulary, please take a moment to look it up in a dictionary. It is a word (and a concept) you need to become familiar with.

11. This problem is directly related to textbook Examples 5 (located on pp. 12-13). Specifically, after experiencing (not just passively reading) this type of problem with smaller-sized families in Example 5, one discovers that examining the issue of a family with 6 children follows the same pattern and corresponds to the numbers in the 6th row of Pascal's triangle; namely, from left to right, $\{1, 6, 15, 20, 15, 6, 1\}$. Please remember that we are considering the '1' at the very top of the triangle to be the 0th row (not the 1st row).

From our experience with Example 5, we know the 1's in the sequence $\{1, 6, 15, 20, 15, 6, 1\}$ correspond to the number of ways of having all six children be of the same gender (i.e., either male or female). The 6's correspond to the number of ways of having five of the children be of one gender and the sixth be the other gender. The 15's reflect the number of ways of having 4 children of one gender and 2 of the other gender. Finally, the **'20'** is the number of ways of having three boys and three girls.

13. Based upon the ideas presented thus far (both in our solution to Problem 11 above and Example 5 in your text), your simplest approach is to find the 8th row of Pascal's triangle: $\{1, 8, 28, 56, 70, 56, 28, 8, 1\}$. You should again ask yourself what, in terms of the issue of children's genders, these numbers mean. For instance, what do the 1's on the ends signify? Like before, they should remind you that there is only one possible way to have eight children all of whom share the same gender. Work your way in from there. Be sure to ask at each stage what the numbers mean in terms of eight children's genders. After doing so, consider this question: on an intuitive level, does it make sense that the number of ways of having 3 boys and 5 girls would be identical to the number of ways of having 3 girls and 5 boys? If your answer is anything short of 100% certainty, take some more time to reflect upon this crucial issue. Finally, after you do understand this, you will find that the answer to this problem amounts to counting in three numbers (or five, if you prefer) from either the '1' on the left or right. The result is that the number of ways the parents can have 3 boys and 5 girls is **56**.

15. Let us consider two approaches:

I. ***Pascal's Triangle*** ⇒ This problem can be treated by making use of Pascal's triangle, as demonstrated in textbook Examples 1 & 2 (pp. 8-10). In other words, traveling from point A to B essentially amounts to a move of 3 "streets" over, followed by 2 up — or vice versa, if you prefer. In terms of Pascal's triangle, this means moving 2 down and 3 over (or, again, vice versa). Either way, one arrives at an answer of **10 paths**.

II. ***Numbered Vertices*** ⇒ The second tactic one could employ is the procedure demonstrated immediately prior to text Example 1. Refer to Figure 1.3 (on the top of page 8). Investigating this (preferably without blindly mimicking what is in your book) will help you obtain a deeper grasp of why and how Pascal's triangle works.

17. Once you adapt the same approach(es) described in our above solution to Problem 15, you will find the process of traveling "6 over and 4 up" can be performed along **210** different direct routes.

19. There are **35** possible paths one may traverse in moving (4 down and 3 over) from point A to point F. This result is obtained by making use of the same methods employed in solving Problems 14-17 (see above) and text Examples 1-2 (pp. 8-10).

21. There are **15** paths one could take in traveling from point A to point H, which is a movement of 2 up and 4 over.

23. **Answers vary**. Some will choose A for superficial reasons (i.e., both A and Problem 22 deal with penguins and bears). Better problem solvers will select C, which resembles Problem 22 in several regards; namely, there are two flower types (which is comparable to 2 body parts) and halving/doubling plays a role in both problems. Finally, others who are focused on the underlying structure of Problem 22 will find several reasons to select choice B.

25. If this problem gave you difficulty, did you attempt to construct a diagram? If not, please close this solution manual and go construct a diagram now.

Welcome back. Your diagram should be similar to the following:

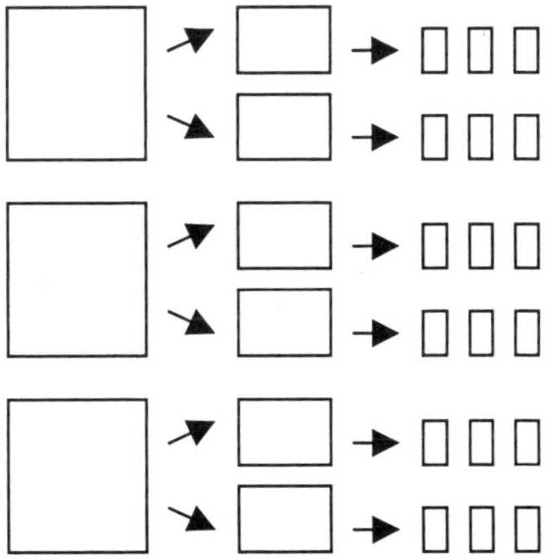

Thus, there is a total of **27 boxes**.

[Suggestion: At some point — either now or a day or so from now — you would be well advised to re-attempt this problem without the aid of a diagram.]

27. a) $1+1=\boxed{2}$

b) $1+2+1=\boxed{4}$

c) $1+3+3+1=\boxed{8}$

d) $1+4+6+4+1=\boxed{16}$

29. **37 different paths**. Since the path from point *A* to point *I* contains some intermediate paths (specifically, between Kearny Street and Stockton Street), we must make use of the "numbered paths" approach demonstrated by your text's author (see text Examples 1 and 3 for details). The diagram will appear as follows:

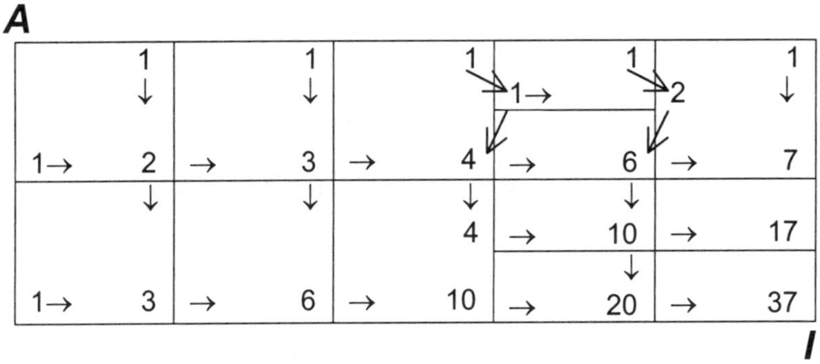

31. Based upon the diagram below, there are **11 possible paths**.

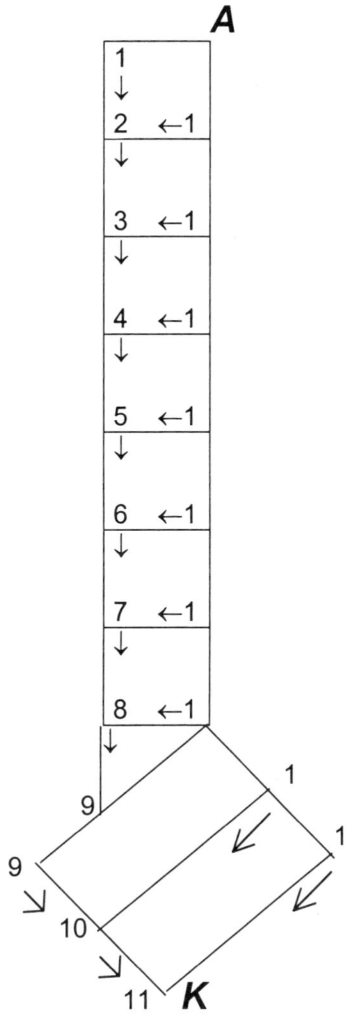

33. **There are various approaches** one may take. Among them is first grouping/pairing numbers in such a way that easy-to-manage subtotals are obtained (for example, subtotals ending in one or more zeros) and then summing these subtotals. Assorted other approaches exist. You may want to ask some older relatives and friends whose mathematics educations pre-date the use of electronic calculators. Ask them to share some of their "tricks of the [arithmetic] trade."

35. **Answers vary**. Included among the additional information you would be likely to consider are: (i) the number of items in the grocery carts of those not in the express lane, and (ii) the relative speeds of the cashiers.

37. Given that the boy and girl are 10 miles apart when they begin pedaling toward each other, we apply the basic formula $d = rt$ [distance = rate × time] and observe that

$$(6\text{ mi/hr})(1\text{ hr}) + (4\text{ mi/hr})(1\text{ hr}) = 10\text{ mi}.$$

This means it will take them one hour of cycling to reach each other. Consequently, the friendly fly has precisely one hour to buzz back and forth between the cyclists. Therefore, given its rate of 20 mph, the fly will cover a distance of **20 miles** in the one hour it takes the cyclists to meet.
Lastly, let us remind you of a powerful problem-solving strategy (in fact, it is part of the first step of Pólya's four-step problem solving model): *whenever a problem is not accompanied by an illustration, sketch your own*. So, if your textbook's author had not been as generous as he was by providing a picture of the cyclists and the fly, we hope your instinct would have been to create it yourself.

39. Sketching a simple diagram (see below) will reveal to you that this problem has **2 possible answers, 8 miles and 12 miles**. Since it is not known if Cal lives between Alex and Beverly or if Beverly lives between Alex and Cal, you must consider both possibilities:

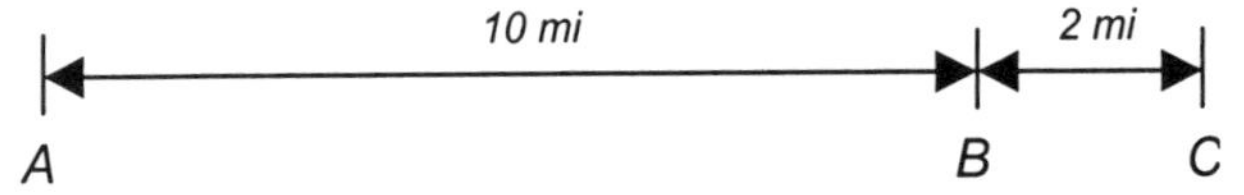

⇒ **12 miles**

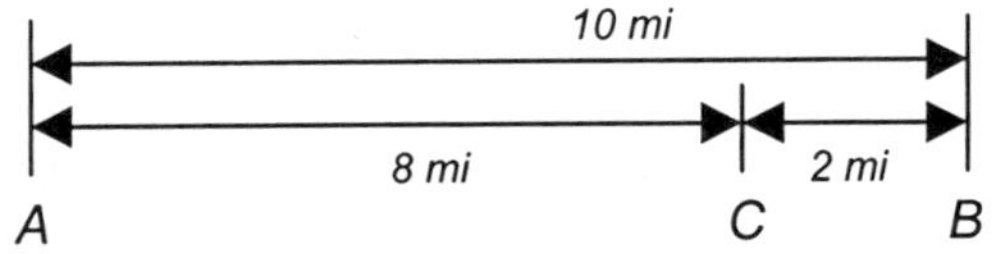

⇒ **8 miles**

41. **Yes, there is a pattern, but it does not work for all three-digit numbers**. The process described in the problem always yields the number 1089, except when the original three-digit number chosen has the same digit in both the hundreds place and the ones place (e.g., 474 or 919). Incidentally, there is a mathematical term for numbers that do not change when their digits are written in reverse order (e.g., 16561); such numbers are called ***palindromes***.

43. **8** (divisors of 1, 2, 4, 8); **10** (1, 2, 5, 10); **14** (1, 2, 7, 14); **15** (1, 3, 5, 15); and, as your text's author cites in the statement of the problem, **6**.

45. "A ton is a ton is a ton…." That is to say, a ton of anything weighs a ton. Therefore, no matter whether it is a ton of coal or a ton of feathers, **their weights are equal.**

47. First, for purposes of this explanation, please pretend you had personally recorded the 30 seconds with a stopwatch. You would have started the stopwatch at the sound of the grandfather clock's first stroke (which signifies one o'clock), and the stopwatch would have been stopped at the sound of the sixth stroke (which represents six o'clock). This means the thirty seconds you would have recorded were a measurement of the time elapsed between the 1st and 2nd strokes, the 2nd and 3rd strokes, the 3rd and 4th strokes, the 4th and 5th strokes, and the 5th and 6th strokes. In other words, the thirty-second interval you recorded is made up of five sub-intervals, each lasting 6 seconds; that is,

$$(30 \text{ seconds total}) \div (5 \text{ intervals}) = 6 \text{ seconds per interval.}$$

With this understood, we are ready to address the question posed in the problem; specifically, we need to determine how long it will take the same clock to strike noon (i.e., 12 strokes). All that remains for us to do is (i) figure out how many more sub-intervals there are between the sixth stroke and the twelfth stroke, and (ii) add this time to the thirty seconds that elapsed between the first and sixth strokes. You should find there to be 6 sub-intervals between the sixth and twelfth strokes. Since each sub-interval lasts six seconds, it will take 36 more seconds from the time of the sixth stroke to the twelfth. Hence, the total amount of time needed for the grandfather clock to strike noon is $(30 + 36) =$ **66 seconds**.

Please note that in this solution we have subtly introduced yet another valuable problem-solving strategy: we call it the process of "personalizing" the problem. In the case of solving this problem, it can help to pretend it is your grandfather clock, your stopwatch, and you who are doing the measuring. Feeling like you are in the middle of the action often aids one's grasp of the problem's details/intricacies. Additionally, a heightened level of emotional involvement will help you stay focused and motivated.

49. Did you read this one carefully? The key phrase is "one of these coins is not a nickel." Hence, the coins are **a quarter** (which, of course, is the "one" that is not a nickel) **and a nickel**.

51. The important word here is "full." It implies that the game ends after the bottom half of the ninth inning. Since each of the two teams is allowed three outs per inning, the full nine-inning game will consist of $9 \times 6 =$ **54** outs in all.

53. *[Before you read this solution, you may want to first consider the following hint: do not assume that once you get an item to the other side of the river that you can't (or won't want to) bring it back again.]*

Answers vary. Here are two possible solutions:

(i) Take the goose across; return and pick up the fox; drop off the fox and take the goose back to the first side; drop off the goose; pick up the corn and leave it with the fox; go back one last time and retrieve the goose.

(ii) Take the goose across; return and pick up the corn; drop off the corn and take the goose back to the first side; drop off the goose; pick up the fox and leave it with the corn; go back one last time and retrieve the goose.

55. *[As was provided in our solution to Problem 53, let us first provide you with a hint: can you find a way to "fill" an empty glass, hence turning it into a full glass? We hope you will ponder this for a while before proceeding to read the solution to this problem.]*
The solution is:

pick up the 5th glass (i.e., the middle full glass), pour its water content into the 2nd glass (i.e., the middle empty glass), and put the now-empty 5th glass back where it was originally positioned.

57. Using the "numbered paths" approach explained by your text's author (in Examples 1 and 3), we find:

a) **26 paths**

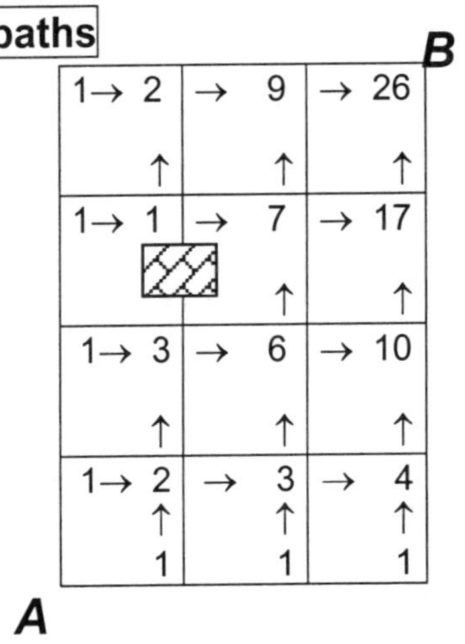

b) **23 paths**

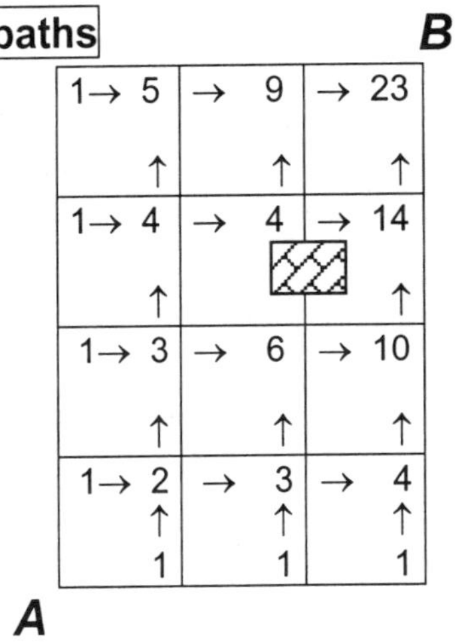

c) **27 paths**

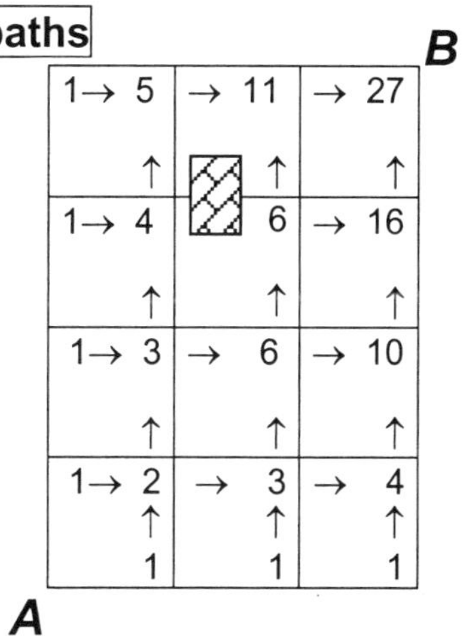

After comparing the unobstructed path that has 35 possible routes to the three obstructed paths above, we discover the general pattern: take the number of routes from the start to the destination (with the obstruction removed) and subtract from it the number of routes passing through the barricade. Let us illustrate/confirm this general pattern with our above solutions to parts (a), (b), and (c):

- In our diagram for part (a), take note of the number of paths prior to encountering the barricade. This number is 3. Then observe that there are three paths through the barricade.
  Hence, $35-(3+3+3)=26$ paths.
- Regarding part (b), look at the '6' prior to the barricade. Then notice that there are 2 routes through the barricade to be subtracted.
  Therefore, the number of paths is $35-(6+6)=23$.
- Finally, for part (c), take note of the '4' located immediately to the left of the barricade and the fact that there are two routes through the barricade.
  Accordingly, we have $35-(4+4)=27$ paths.

59. Since there are 52 cards in a standard deck, there are 26 cards in each pile. Remember that the hidden card is located 7 cards down in the first pile. From the second pile, three cards are removed (i.e., placed face up), leaving 23 cards in the second pile. To discover this pattern (or, for that matter, most any pattern), it is best to consider the possibilities in order:

Sum of the 3 face-up cards	*# of additional cards needed to get a sum of 30*	*# of cards remaining in 2nd pile (after additional cards needed to make sum of 30 have been removed)*	*# of cards needed to be counted down to find the hidden card (after the remaining cards in the 2nd pile have been placed on top of the 1st pile)*
30	0	23	30 = 23 + 7
29	1	22	29 = 22 + 7
28	2	21	28 = 21 + 7
⋮	⋮	⋮	⋮
8	22	1	8 = 1 + 7
7	23	0	7 = 0 + 7

A comparison of the columns on the far left and far right reveals the pattern. That is, if the three face-up cards total x, then we need to count $23-(30-x)+7=x$ cards down.

[Note: In the event the sum of the 3 face-up cards happened to be 6 (i.e., three cards with the number 2 on each), the trick would need to be altered a bit; namely, because there are only 23 cards in the 2nd pile yet 24 are necessary to get a total of 30 when the total of the 3 face-up cards is 6. So, before performing the trick for a crowd, you may wish to find your own way to "compensate" for this.]

Problem Set 1.2

1. **Answers vary**.

— Are you making an effort to discover your own unique "mathematical voice"?

It is critically important that you become knowledgeable about these two reasoning processes and how/when to apply them. They are indispensable tools in solving most any problem, mathematical or otherwise. ***Inductive reasoning*** involves drawing a general conclusion from repeated observations of specific (often simpler) examples and their behaviors (patterns). In contrast, ***deductive reasoning*** is characterized by applying previously established/accepted general principles to reach a conclusion regarding another circumstance.

3. Your text's author discusses the ***order of operations*** in the section. Refer to page 20 for the details. Once again, we encourage you to write your answer to this question in <u>your</u> own words — rather than copy the author's.

5. Again, you should be able to locate — and then rewrite in your own words — the explanation provided in your text. The ***substitution property*** is stated on p. 22 of your text.

7. a) $14+6\times3=14+18=\boxed{\mathbf{32}}$ b) $30\div5\times2=6\times2=\boxed{\mathbf{12}}$

9. a) $(8+6)\div2=14\div2=\boxed{\mathbf{7}}$ b) $8+6\div2=8+3=\boxed{\mathbf{11}}$

11. a) $450+550/10=450+55=\boxed{\mathbf{505}}$ b) $\dfrac{450+550}{10}=\dfrac{1000}{10}=\boxed{\mathbf{100}}$

13. a) $1+3\times2+4+3\times6=1+6+4+18=\boxed{\mathbf{29}}$

b) $3+6\times2+8+4\times3=3+12+8+12=\boxed{\mathbf{35}}$

15. a)
$$\begin{aligned}8+2(3+12)-5\times3&=8+2(15)-15\\&=8+30-15\\&=\boxed{\mathbf{23}}\end{aligned}$$

b)
$$\begin{aligned}25-4(12-2\times6)+3&=25-4(12-12)+3\\&=25-4(0)+3\\&=25-0+3\\&=\boxed{\mathbf{28}}\end{aligned}$$

17. a)
$$\begin{aligned}3+[(9\div3)\times2]+[(2\times6)\div3]&=3+[3\times2]+[12\div3]\\&=3+6+4\\&=\boxed{\mathbf{13}}\end{aligned}$$

b)
$$\begin{aligned}[(3+9)\div(3\times2)]+[(2\times6)\div3]&=[12\div6]+[12\div3]\\&=2+4\\&=\boxed{\mathbf{6}}\end{aligned}$$

19. This news clip illustrates **deductive reasoning. Explanations vary.** Because Sam's reasoning involved a conclusion based upon a set of premises, the form of reasoning he employed corresponds to the definition of deductive reasoning (provided on textbook p. 22). Please recall that inductive reasoning involves the use of particular facts/cases to reach a general conjecture, but Sam's reasoning was not such a process.

21. As shown below, the four pattern is **4, 8, 3, 7, 2, 6, 1, 5, 9**, 4, 8, 3 …

$$
\begin{aligned}
&4\times1=4 \Rightarrow 4\\
&4\times2=8 \Rightarrow 8\\
&4\times3=12 \Rightarrow 1+2=3\\
&4\times4=16 \Rightarrow 1+6=7\\
&4\times5=20 \Rightarrow 2+0=2\\
&4\times6=24 \Rightarrow 2+4=6\\
&4\times7=28 \Rightarrow 2+8=10 \Rightarrow 1+0=1\\
&4\times8=32 \Rightarrow 3+2=5\\
&4\times9=36 \Rightarrow 3+6=9\\
&4\times10=40 \Rightarrow 4+0=4\\
&4\times11=44 \Rightarrow 4+4=8\\
&4\times9=48 \Rightarrow 4+8=12 \Rightarrow 1+2=3\\
&\vdots
\end{aligned}
$$

23. By using the same approach as demonstrated in our solution to Problem 21 above (and, of course, also in textbook Example 1 on page 19), you will find the six pattern to be **6, 3, 9**, 6, 3, 9 …

25. **2500**. Here are <u>two approaches</u>:

(i) Refer to the inductive reasoning applied to Example 2 in your text. Using the patterns observed there, the sum of the first fifty consecutive odd numbers is:

$$\underbrace{1+3+5+\cdots+95+97+99}_{\text{the first 50 consecutive odd \#s}}=50^2=2500$$

(ii) Observe the pattern that unfolds when the numbers are reorganized, with the first number (namely, 1) being paired with the last (99), the second (3) with the second to last (97), and so on:

$$(1+99)+(3+97)+(5+95)+(7+93)+\cdots+(47+53)+(49+51)$$

Each pair's sum is 100. Since there are 50 numbers, there must be 25 pairs. Thus, 25 pairs of 100 amounts to $25\cdot100=2500$.

27. a) $\boxed{\mathbf{3+(2\cdot4)}}$ Please note that the parentheses here are optional. Make sure you understand why.

b) $\boxed{\mathbf{3(2+4)}}$ These parentheses are <u>not</u> optional. If you're not sure why, please review the order of operations.

29. a) $\boxed{\mathbf{8\cdot9+10}}$ b) $\boxed{\mathbf{8(9+10)}}$

31. a) $\boxed{3^2+2^3}$ b) $\boxed{3^3-2^2}$

33. a) $\boxed{4^2+9^2}$ b) $\boxed{(4+9)^2}$

35. Let n represent "the number."

a) $\boxed{3(n+4)=16}$

b) $\boxed{5(n+1)=5n+5}$

37. Let n represent the number.

a) $\boxed{n^2+8n=6}$

b) $\boxed{n^2+8n=n(8+n)}$

39. $\boxed{A=\frac{1}{2}bh}$ (or $A=\dfrac{bh}{2}$, $A=0.5bh$, etc.)

41. $\boxed{A=\frac{1}{2}h(a+b)}$ **43.** $\boxed{V=\ell wh}$ **45.** $\boxed{V=\pi r^2 h}$

47. $\boxed{9\times 54{,}321-1=488{,}888}$. Please remember: it is always a good idea to take the time to double check your answer (in this case, either by hand or with the aid of a calculator).

49. We begin with some important advice. Although there a number of possible approaches you could have used to make your prediction, we hope you attempted to find <u>several</u> approaches. Do not settle for just one. Mathematics is far more than "getting the right answer." Remember: all good problem solvers — no matter whether they are government leaders, corporate leaders, or in any other realm — by their very nature seek multiple approaches and perspectives when confronting problems.

With that preliminary advice stated, here are two approaches, an inefficient, not-so-desirable one and a more intelligent, truly inductive one:

I. *<u>The Inefficient Approach</u>*: Yes, we suppose you could have listed all the problems in the pattern between the last one given in Problem 47 and the one needed to answer this question. If you did this, you are to be congratulated for your energy and dedication. Keep it up! At the same time, however, we hope you followed up this commendable effort by seeking — and ultimately finding — a less time consuming strategy.

II. *<u>A More Efficient View (albeit not the only one)</u>*: We need to find a connection (i.e., pattern) between the number being multiplied by 9 in each problem and that problem's corresponding answer. There are two aspects to the answer that we need to address.

First, how many 8's will the answer contain? When 9 is multiplied by 21, the answer contains two 8's; when the multiplier of 9 is 321, the answer contains three 8's; when the multiplicand of 9 is 4321, the answer contains four 8's. Notice, for example in the case of 4321, that the leading digit (4) is identical to the number of digits in 4321. This pattern is consistent with the fact that the answer to Problem 48 contains nine 8's, which equals the number of digits in the number 987,654,321. Notice that the relevant number in this problem (10,987,654,321) should be viewed as a '10', followed by 9, 8, 7, 6, …, 1. Therefore, since our problem asks us to multiply 9 by 10,987,654,321, the answer will contain ten 8's.

The second issue to resolve is what the answer's leading digit will be. Again, we use Problem 47 to uncover this. When 9 was multiplied by 21, the answer's leading digit was 1; in the case of 321, the answer's leading digit was 2; as for 4321, the answer's leading digit was 3. Continuing this pattern, we found in Problem 48 (in which 9 was multiplied by 987,654,321) that the answer's leading digit was 8. So, in general, we should expect the answer's leading "digit" to be one less than the number of digits contained in the multiplier of 9. Since the number 10,987,654,321 contains 10 digits (that is, if we view '10' as a "digit," which formally speaking it is not), the leading digit in our answer will be one less than ten, namely 9.

So, the answer to the problem is **98,888,888,888**.

Please observe that this problem can be solved in a more efficient manner **if** one solved Problem 48 beforehand. Specifically, since Problem 48 revealed the fact that $9 \times 987{,}654{,}321 - 1$ equals 8,888,888,888, it follows that

$$\begin{aligned} 9\times 10{,}987{,}654{,}321-1 &= 9\times[10{,}000{,}000{,}000+987{,}654{,}321]-1 \\ &= 9\times 10{,}000{,}000{,}000+9\times 987{,}654{,}321-1 \\ &= [9\times 10{,}000{,}000{,}000]+[9\times 987{,}654{,}321-1] \\ &= 90{,}000{,}000{,}000+8{,}888{,}888{,}888 \\ &= 98{,}888{,}888{,}888. \end{aligned}$$

51. From Problem 50, recall that $123{,}456{,}789\times 9 = 1{,}111{,}111{,}101$. Hence,

$$\begin{aligned} 123{,}456{,}789\times 9{,}000 &= 123{,}456{,}789\times(9\times 1{,}000) \\ &= (123{,}456{,}789\times 9)\times 1{,}000 \\ &= 1{,}111{,}111{,}101\times 1{,}000 \\ &= \boxed{1{,}111{,}111{,}101{,}000} \end{aligned}$$

53. Did you take full advantage of the hint given by your text's author? If you could not determine the pattern, did you expand the hint to the next step, namely a 4-by-4 square? If not, please close this book for a moment and attempt do so now.

The 4-by-4 square 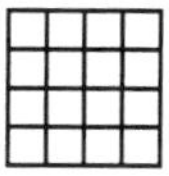 has:

1	4-by-4 square
4	3-by-3 squares
9	2-by-2 squares
16	1-by-1 squares

Are the numbers 1, 4, 9, 16 forming a pattern you recognize? They are all perfect squares; that is, $1^2, 2^2, 3^2, 4^2 \ldots$

Consequently, in the 6-by-6 square given in Figure 1.8, we should expect to find

$$\underbrace{1}_{\#\text{ of }6\times6\text{ squares}} + \underbrace{4}_{\#\text{ of }5\times5\text{ squares}} + \underbrace{9}_{\#\text{ of }4\times4\text{ squares}} + \underbrace{16}_{\#\text{ of }3\times3\text{ squares}} + \underbrace{25}_{\#\text{ of }2\times2\text{ squares}} + \underbrace{36}_{\#\text{ of }1\times1\text{ squares}} = \boxed{\textbf{91 squares}} \text{ in all.}$$

55. First, we must determine how many smaller one-inch cubes make up the overall four-inch cube. By using inductive reasoning or by drawing a picture (like the one below), we find there are <u>64 one-inch cubes</u> ($4 \times 4 \times 4$ or 4^3).

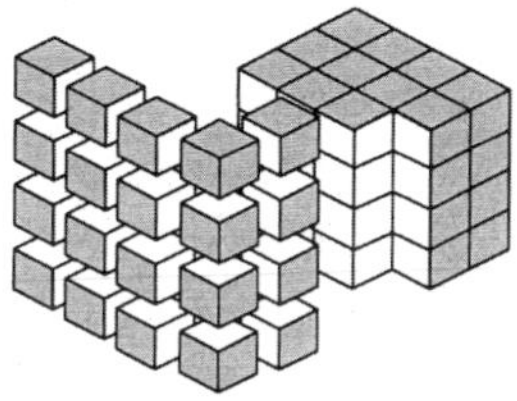

As for the number of one-inch cubes with paint on exactly one of their sides, we notice (when we look at any one face of the four-inch cube) that the one-inch cubes positioned on the edges (or corners) will be painted on more than one side. The only one-inch cubes that will have paint on <u>exactly</u> one of their sides are the four that are positioned in the middle of each face of the four-inch cube. Since the original four-inch cube has six faces and since there are 4 (one-inch) cubes per face with paint on exactly one side, there are $6 \times 4 =$ <u>24 one-inch cubes</u> containing paint on exactly one side.

Therefore, the fraction of one-inch cubes painted on exactly one side is

$$\frac{24}{64} = \boxed{\frac{3}{8}}.$$

57. ***Step 1:*** Place three of the coins on one pan, three on the other, and set the last three aside.

<u>Analysis</u>: If the scale balances, then the counterfeit (heavier) coin is among the three you set aside. If the scale is not in balance, the counterfeit coin is among the three on the heavier pan. In either case, you have narrowed the counterfeit down to three of the original nine coins.

***Step 2**:* From the group of three containing the counterfeit coin, place one coin on one pan, one on the other, and set aside the remaining coin.
Analysis: If the scale is in balance, the counterfeit coin is the one set aside. If the scale is not in balance, the counterfeit coin is on the heavier pan.

59. *[In the event you are reading this solution because you do not know what a "magic square" is, please refer to Problem 58 for the definition. In the future, when you cannot locate unfamiliar terms and concepts, remember to use your text's index and glossary.]*

To prove that Harry Nelson's answer is indeed a magic square, one must confirm that the respective sums of each vertical column, each horizontal row, and each of the two diagonals are equal. In the case of Nelson's magic square, the common sum is 4,440,084,513.

Problem Set 1.3

1. **Answers vary**. Refer to textbook page 28 for the details. Make sure you thoroughly understand and know how to use appropriately the associated terms (e.g., base, power, exponential) and corresponding notation.

3. **Answers vary**. Refer to both the user's manual for your calculator and your text's discussion (pp. 30-33) of arithmetic logic, algebraic logic, and RPN logic.

5. **Answers vary**. Among the considerations to keep in mind are the respective sizes (more specifically, dimensions and volumes) of the classrooms and of the balloons. To take full advantage of this problem-solving opportunity, you may wish to make up some reasonable room dimensions and typical balloon shapes/sizes.

7. **Answers vary**; for example, every individual living in the United States would have to contribute in excess of $3,600 for one trillion dollars to be accumulated *(based on an August 2000 U.S. Census Bureau estimate of 275 million)*.

	Scientific Notation	*Floating-Point Notation*	
9. a)	2.379×10^{1}	2.379 01	
b)	10^{-6} or 1×10^{-6}	1 −06	
c)	3.5×10^{10}	3.5 10	
11. a)	10^{100} or 1×10^{100}	1 100	*[This number is too large for most calculators.]*
b)	1.2003×10^{6}	1.2003 06	
c)	1.23×10^{-6}	1.23 −06	

13. a) **64** b) **0.0021** c) **40,700**

15. a) $\mathbf{0.02\overline{7}}$ b) **32,170,000** c) **0.00000 00000 889**

17. $\mathbf{8.6\times 10^{1}}$; $\mathbf{6.1\times 10^{3}}$ **19.** $\mathbf{6\times 10^{12}}$; $\mathbf{1.86\times 10^{5}}$

21. **906** **23.** **38,000,000,000,000,000,000,000,000**

25. <u>*Estimation*</u>: Making approximately 2000 copies of a roughly 500-page manuscript requires $\approx$ **1,000,000 pages**. *[Disclaimer: Estimates may vary.]*

<u>*Exact Calculation*</u>: (1,850 copies) $\times$ (487 pages per copy) = **900,950 pages**

27. <u>*Estimation*</u>: \$1500×12 mos = \$1500×10 + \$1500×2 **or** \$1000×12 + \$500×12 = **\$18,000**. *[Disclaimer: Estimates may vary.]*

<u>*Exact Calculation*</u>: (\$1543 per month) × (12 mo/yr) = **\$18,516**

29. <u>*Estimation*</u>: (1,000,000 people) ÷ (10,000 people/year) = **100 years**
[Disclaimer: Estimates may vary.]

<u>*Calculation*</u>: (1,000,000 people) ÷ (13,688 people/year) = $\mathbf{73\frac{97}{1{,}711}}$ **years** or

$\approx$ **73 years 21 days** or

$\approx$ **73 years**

In many real-world situations, one would report 73 years as the calculated answer. Additionally, be advised that many calculators have built-in fraction tools, including automatic reducing of fractions. On many calculators, the primary key for fractions looks like $a\,{}^{b}\!/\!{}_{c}$. This key and its related functions can be valuable timesavers. In fact, this is how we chose to determine the $\frac{97}{1711}$ reported above.

31. <u>*Estimation*</u>: (300 miles) ÷ (10 gallons) = 30 mi/gal *[Disclaimer: Estimates may vary.]*

<u>*Calculation*</u>: (280 miles) ÷ (10.2 gallons) = $27\frac{23}{51}$ mi/gal. $\approx$ **27.451 mi/gal**

33. We recommend you do these computations both by hand (as shown below) and with the aid of your calculator. Refer to Example 6 in your textbook (p. 33) for some calculator insights.

a) $$\frac{(5\times10^4)(8\times10^5)}{4\times10^6}=\frac{(5\times8)(10^4\times10^5)}{4\times10^6}$$
$$=\frac{40\times10^9}{4\times10^6}$$
$$=\left(\frac{40}{4}\right)\times10^{9-6}$$
$$=10\times10^3$$
$$=\boxed{\mathbf{10^4\text{ or }1\times10^4}}$$

b) $$\frac{(6\times10^{-3})(7\times10^8)}{3\times10^7}=\frac{(6\times7)(10^{-3}\times10^8)}{3\times10^7}$$
$$=\frac{42\times10^5}{3\times10^7}$$
$$=\left(\tfrac{42}{3}\right)\times\left(\tfrac{10^5}{10^7}\right)$$
$$=14\times10^{-2}$$
$$=\boxed{\mathbf{1.4\times10^{-1}}}$$

35. a) $$\frac{(2xy^{-2})(2^{-1}x^{-1}y^4)}{x^{-2}y^2}=\left(2\cdot2^{-1}\right)\left(\frac{x^1\cdot x^{-1}}{x^{-2}}\right)\left(\frac{y^{-2}\cdot y^4}{y^2}\right)$$
$$=\left(2^0\right)x^{[1+-1-(-2)]}y^{[-2+4-2]}$$
$$=2^0x^2y^0$$
$$=1\cdot x^2\cdot1$$
$$=\boxed{\boldsymbol{x^2}}$$

b) $$\frac{x^2y(2x^3y^{-5})}{2^{-2}x^4y^{-8}}=\left(\frac{2^1}{2^{-2}}\right)\left(\frac{x^2\cdot x^3}{x^4}\right)\left(\frac{y^1\cdot y^{-5}}{y^{-8}}\right)$$
$$=2^{[1-(-2)]}x^{[2+3-4]}y^{[1+-5-(-8)]}$$
$$=2^3x^1y^4$$
$$=\boxed{\boldsymbol{8xy^4}}$$

37. a) $\dfrac{4{,}500{,}000{,}000{,}000\times 0.00001}{50\times 0.0003}=\dfrac{(4.5\times 10^{12})(10^{-5})}{(5\times 10^{1})(3\times 10^{-4})}$

$$=\frac{4.5\times 10^{7}}{15\times 10^{-3}}$$

$$=\frac{4.5\times 10^{7}}{1.5\times 10^{-2}}$$

$$=\frac{4.5}{1.5}\times\frac{10^{7}}{10^{-2}}$$

$$=\boxed{\mathbf{3\times 10^{9}}}$$

b) $\dfrac{0.0348\times 0.00000\,00000\,00002}{0.000058\times 0.03}=\dfrac{(3.48\times 10^{-2})(2\times 10^{-15})}{(5.8\times 10^{-5})(3\times 10^{-2})}$

$$=\frac{6.96\times 10^{-17}}{17.4\times 10^{-7}}$$

$$=\frac{6.96\times 10^{-17}}{1.74\times 10^{-6}}$$

$$=\frac{6.96}{1.74}\times\frac{10^{-17}}{10^{-6}}$$

$$=\boxed{\mathbf{4\times 10^{-11}}}$$

39. **Answers vary; ≈ 30 miles** Remember that it makes a difference whether one is asked to find the distance along the indicated paths/roads, as opposed to the direct (linear) distance between the two points. For your information, in arriving at our estimate, we considered the paths/roads.

41. *[We begin with a hint: measure the photo and divide it into smaller square pieces — square centimeters, for instance, would work well. Then estimate how many people are typically present per square centimeter, and proceed from there.]*

The dimensions of the photo are 4 cm by 3 cm. We estimate there to be approximately 15 persons per square centimeter. Thus, we estimate the photo includes $(15$ persons/cm$^2) \times (12$ cm$^2) =$ **180 people**. *[Disclaimer: Estimates may vary.]*

43. In case you missed it, please refer to the hint provided for Problem 41 above. We estimate there to be roughly 8 penguins/cm^2 and, hence, approximately **300 penguins** in all. *[Disclaimer: Estimates may vary.]*

45. 5 quadrillion $= 5 \times 10^{15}$ 157 trillion $= 157 \times 10^{12} = 0.157 \times 10^{15}$

Thus, 5 quadrillion 157 trillion tons $\Rightarrow$ 5 quadrillion tons + 157 trillion tons

$\Rightarrow (5 \times 10^{15}$ tons$) + (0.157 \times 10^{15}$ tons$)$

$\Rightarrow$ **5.157×10^{15} tons**

Observe how well it served us to <u>not</u> write 157 trillion in standard scientific notation. Writing it in terms of 10^{15} made it easier for us to add it to the 5 quadrillion.

47. (1 million bills) ÷ (233 bills/in.) **≈ 4,292 in. or 358 ft. or 119 yds.**

49. Several options are available to you here. You can choose to convert the 2,260,000 grains/lb to grains/ton and then multiply by the 30,000,000 tons produced annually; or, alternatively, you can convert the 30,000,000 tons to pounds and then multiply by the 2,260,000 grains/lb. In either event, this problem requires two steps. What follows is the first of the two described approaches:

$$(2.26 \times 10^6 \text{ grains/lb}) \cdot (2 \times 10^3 \text{ lbs/ton}) \cdot (3 \times 10^7 \text{ tons})$$

$$(2.26)(2)(3)(10^6 \times 10^3 \times 10^7) \text{ grains}$$

$$(2.26)(2)(3)(10^{16}) \text{ grains}$$

≈ 1.356×10^{17} or 135,600,000,000,000,000 grains

51. If you directly entered 8^{16} on your calculator, you obtained an answer in floating-point form, equivalent in scientific notation to $2.814749767 \times 10^{14}$. In the statement of the problem, your text's author asks you to "use your calculator to *<u>help you</u>* find this number." We hope you took note of his italicized words "help you." That is to say, one still must do some paper-and-pencil calculations in order to find 8^{16} exactly (i.e., not in rounded scientific notation). There are a number of ways to obtain the desired result. Among them is:

$$8^{16} = 8^{11} \times 8^5 = 8{,}589{,}934{,}592 \times 32{,}768 = \mathbf{281{,}474{,}976{,}710{,}656}$$

53. *<u>Estimation</u>*: 238,000 mi ÷ 4 mi/min ≈ 240,000 mi ÷ 4 mi/min = 60,000 min

60,000 min ÷ 60 min/hr = 1,000 hrs

1,000 hrs ÷ 24 hrs/day ≈ 1,000 hrs ÷ 20 hrs/day = **50 days**

[<u>Disclaimer</u>: Estimates may vary.]

<u>Calculation</u>: (238,000 mi)÷(4 mi/min)÷(60 min/hr)÷(24 hrs/day) = $41.319\overline{4}$ days

41.319444... days $\Rightarrow$ 41 days, $(0.319\overline{4} \times 24)$ hours

41 days, 7.6666... hours $\Rightarrow$ 41 days, 7 hours, $(0.\overline{6} \times 60)$ minutes

$\Rightarrow$ **41 days, 7 hours, 40 minutes**

55. Here is one approach. First, determine the total number of digits to be written. For the numbers 1 to 1000, notice that there are 9 one-digit numbers (namely, the numbers 1 through 9), 90 two-digit numbers (i.e., 10 through 99), 900 three-digit numbers (100-999), and 1 four-digit number (1000). Pause to confirm that the sum of 9, 90, 900, and 1 is 1000, as you should have expected. Thus, the total number of digits to be written is $(9 \times 1) + (90 \times 2) + (900 \times 3) + (1 \times 4) = 2893$.

Since it takes one second per digit, the total time required is 2893 seconds, which is more clearly expressed as **48 minutes and 13 seconds**.

Two comments are worthy of your consideration:

I. A common misconception is to think, for example, that there are 89 numbers in the set $\{10, 11, 12, \ldots, 98, 99\}$, based on the flawed reasoning of $99 - 10 = 89$. If you fell into this trap, we encourage you to perform several inductive-reasoning experiments. For example, you might ask yourself how many numbers are in the set $\{10, 11, 12\}$. Surely, you would not say $12 - 10 = 2$. Conducting experiments like this will help you avoid such reasoning errors.

II. Instead of breaking up the set 1 to 1000 in the manner illustrated in our solution, you may prefer a different approach. One alternative is to tabulate how many numbers possess a units digit, how many have a tens digit, how many contain a hundreds digit, and how many possess a thousands digit. We encourage you to give this alternative approach a try.

57. This problem presents you with yet another pattern to uncover. It is our hope that when you need to uncover patterns you immediately think of the process of inductive reasoning and seek ways to apply it.

There are several ways to view this problem. One approach is to divide the numbers into the following categories:

- ***#s with exactly 1 zero (located in units digit):*** The 9 numbers $\{10, 20, \ldots, 90\}$ are contained in this category. This pattern is replicated with each of the three-digit numbers; for instance, $\{110, 120, \ldots, 190\}$ and $\{910, 920, \ldots, 990\}$. Thus, there are $9 \times 10 = 90$ such numbers, and accordingly, there are **90** zeros.
- ***#s with exactly 1 zero (located in tens digit):*** This list begins with the group $\{101, 102, \ldots, 109\}$ and ends with $\{901, 902, \ldots, 909\}$. Hence, we have $9 \times 9 = 81$ of these numbers, which equates to **81** more zeros.
- ***#s with exactly two zeros:*** This category consists of nine numbers, the set $\{100, 200, \ldots, 900\}$. So, we add $9 \times 2 = $ **18** more zeros to our tabulation.
- ***#s with exactly three zeros:*** The only number satisfying this requirement is 1000, which means we add **3** more zeros to our total.

We, therefore, have a grand total of $90 + 81 + 18 + 3 =$ **192 zeros**.

59. a) **Choice D**. Choice (A) is not worthy of mention — let alone consideration. Choice (B) is also ridiculously small; can you imagine being confined to 8 ft^2, which is a space smaller than 1 yd^2 (i.e., 9 ft^2)? A small mattress consumes more space. As for the astronomically large choice (E), it is as absurd as choice (A). Finally, choice (C) may be tempting at first, but one quickly realizes that it is less reasonable than choice (D), since 80 ft^2 is equivalent to each person being restricted to an 8 ft $\times$ 10 ft space, the size of a small bedroom. In light of the huge size of California, choice (D) is more reasonable.

b) We first need to recall how many square feet are in one square mile. Because 1 mi $= 5{,}280$ ft, we know 1 mi^2 = $(5{,}280)^2$ ft^2 = $27{,}878{,}400$ ft^2. Thus,
$(158{,}600$ mi$^2) \times (27{,}878{,}400$ ft^2/mi$^2) \div (5.5 \times 10^9$ people$) \approx$ **804 ft^2/person**.

Problem Set 1.4

1. **Answers vary**. Refer to textbook pages 42-44 for details.

We strongly encourage you to spend some time connecting (relating) the mathematical definitions of the terms used in your text to their non-mathematical usages. Said differently, you should be striving to find synonyms for each mathematical term. Doing so will benefit you greatly.

Permit us to illustrate this powerful learning strategy. Consider this question: in what other contexts have you heard the word ***union***? Do uses like "labor union" and "the union of holy matrimony" have relevance to determining the union of two sets? Don't they, too, suggest a "joining together" (a "uniting") of sorts?

Now, please take a moment to relate the set operations of ***intersection*** and ***complementation*** to a non-mathematical use of each word (or another grammatical form of the word). Refer to a dictionary as necessary.

3. **Answers vary**. As is proven in Example 7 (p. 51), DeMorgan's Laws establish the fact that $\overline{X} \cup \overline{Y} = \overline{X \cap Y}$ and $\overline{X} \cap \overline{Y} = \overline{X \cup Y}$ for any sets X and Y.

We suggest you attempt to translate these laws into your own written words. Let us illustrate this process with the statement $\overline{X} \cup \overline{Y} = \overline{X \cap Y}$. First, $\overline{X \cap Y}$ might be translated into the words "the elements that are not in the intersection of sets X and Y" or "the elements that are not in both set X and set Y." Pause for a moment to contemplate the meaning of these words. After that, $\overline{X} \cup \overline{Y}$ could be read as "the joining together (union) of the elements that are not in X with those not in Y." Finally, the all-important last step would be to make sure the translations of $\overline{X \cap Y}$ and $\overline{X} \cup \overline{Y}$ communicate the same set. We leave these remaining details to you and <u>your</u> unique mathematical voice.

5. **Answers vary**. One example would be the set consisting of all students in the U.S. currently enrolled in a mathematics course.

7. a) If you have spent some time relating the concept of set union to nonmathematical meanings of the word "union" (as recommended in our solution above to Problem 1), then you know that this problem asks you to "join" (or "unite" or "marry") the elements of $\{2, 6, 8\}$ with $\{6, 8, 10\}$. Thus, we obtain: $\{2, 6, 8\} \cup \{6, 8, 10\} =$ **{2, 6, 8, 10}**. Please remember that (i) we do not list the repeated elements (in this case, 6 and 8) twice, and (ii) the order in which you write the elements of the set is unimportant.

b) We hope you have found some synonyms for "intersection" that help you retain what this set operation asks you to do. For instance, if you think about the intersection of two roads (or two lines, for that matter), you will realize that the operation $\cap$ asks one to find the "overlap" (or "common ground") of two sets. Hence, we have $\{2, 6, 8\} \cap \{6, 8, 10\} =$ **{6, 8}**.

9. a) Recalling the meaning of "intersection" (as explained in our solution to part b of Problem 7 above), we find $\{1, 2, 3, 4, 5\} \cap \{3, 4, 5, 6, 7\} =$ **{3, 4, 5}**.

b) In this problem, the overbar is defined as the "complement" of {2, 8, 9}. This requires us to refer back to the given universal set U and form a set consisting of all the elements in U <u>except</u> for the elements 2, 8, and 9. Therefore, since U is defined as $\{1, 2, 3, 4, 5, 6, 7, 8, 9, 10\}$ for Problems 7-10, $\overline{\{2, 8, 9\}} =$ **{1, 3, 4, 5, 6, 7, 10}**.

11. a) $A \cup B = \{1,2,3,4\} \cup \{1,2,5,6\} =$ **{1, 2, 3, 4, 5, 6}**

Remember: set union implies uniting (joining, marrying, etc.) the two sets.

b) $A \cap B = \{1,2,3,4\} \cap \{1,2,5,6\} =$ **{1, 2}**

Finding the intersection requires finding the overlap (common ground, shared elements, etc.) of the two sets.

c) $\overline{A} = \overline{\{1,2,3,4\}} =$ **{5, 6, 7}**

The complement of set A is defined as the set of all elements in the universe that do not belong to set A.

13. a) $B \cap C = \{1,2,5,6\} \cap \{3,5,7\} =$ **{5}**

b) $B \cup C = \{1,2,5,6\} \cup \{3,5,7\} =$ **{1, 2, 3, 5, 6, 7}**

c) $\overline{B} = \overline{\{1,2,5,6\}} =$ **{3, 4, 7}**

15.

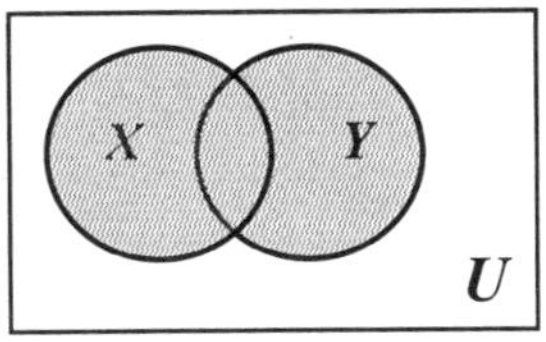

17.

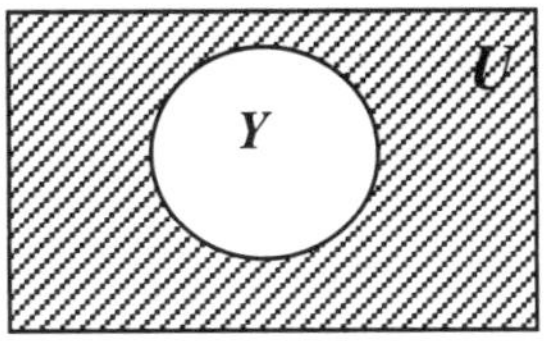

19. In addition to the very fine approach demonstrated in your textbook (see Examples 7 and 8 on pp. 51-52), there are other strategies you may wish to consider:

<u>Alternative Approach I</u>: If these problems are difficult for you, one approach (among several) which you may find helpful is to begin by labeling (e.g., lettering or numbering) each of the regions of the Venn diagram in some manner. For example, let $U = \{1, 2, 3, 4\}$, $A = \{1, 2\}$, and $B = \{2, 3\}$. That is,

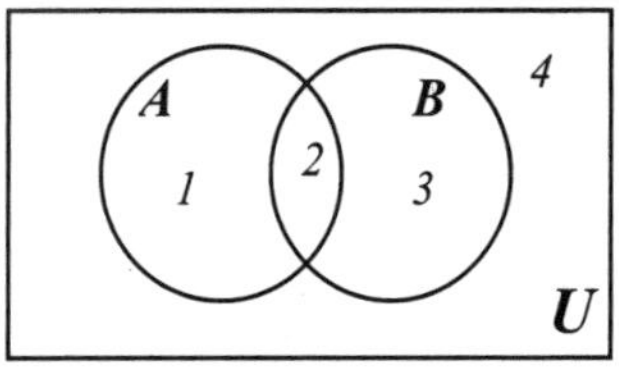

This implies $A \cap \overline{B} = \{1,2\} \cap \overline{\{2,3\}} = \{1,2\} \cap \{1,4\} = \{1\}$, which indicates that the region containing a '1' is to be shaded. Thus, our solution is:

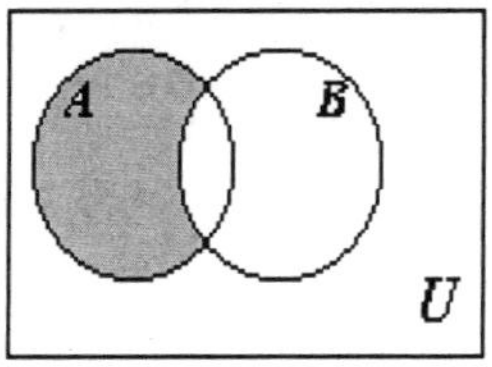

<u>Alternative Approach II</u>: Although it may take a while to get accustomed to, you may find it helpful to "translate" the set notation into words. For example, some would read the expression $A \cap \overline{B}$ as "those elements that are in set A and not in set B." Check whether this translation is equivalent to our shaded Venn diagram above. For further details, we refer you to your textbook (between Examples 2 and 3) where the author associates words like "and," "or", "both," "any" with their corresponding set operations.

21. As necessary, refer to the two alternative approaches demonstrated above in the solution to Problem 19. When confronted with three subsets of a universal set, realize that your generalized Venn diagram will consist of 8 regions.

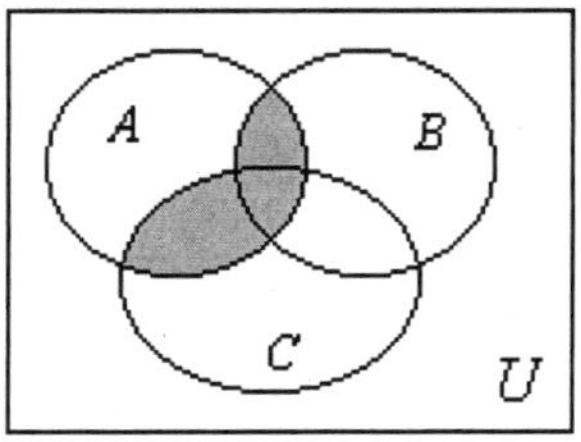

For Problems 23-26: Use the same approaches as demonstrated in your text (see Examples 7 & 8, pp. 51-53) and the two alternative approaches detailed in this manual's solution to Problem 19 (see above).

23.

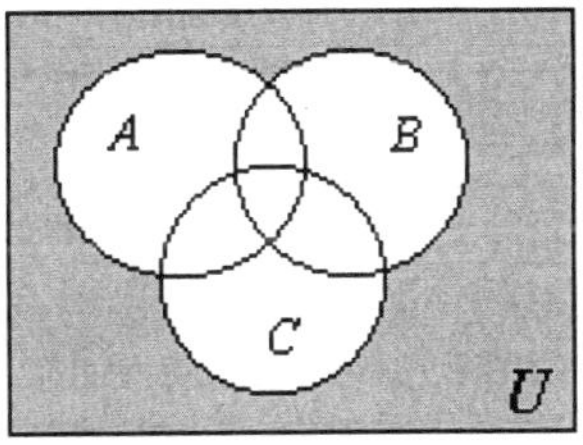

25.

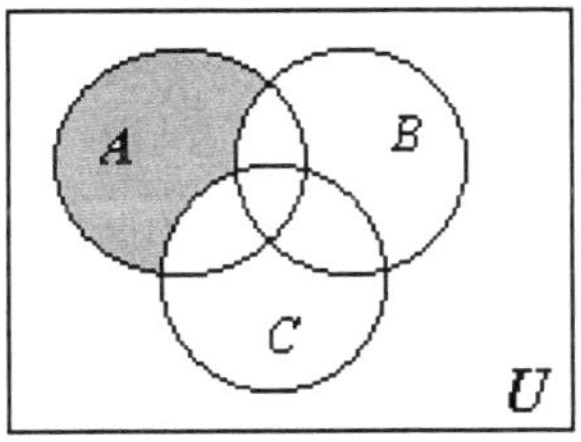

27. Prior to sketching a diagram, there are two inferences you must make: (i) the set of cats and the set of dogs are disjoint; and (ii) both sets are subsets of the (universal) set of all animals. In light of these two considerations, the following would be an appropriate Venn diagram:

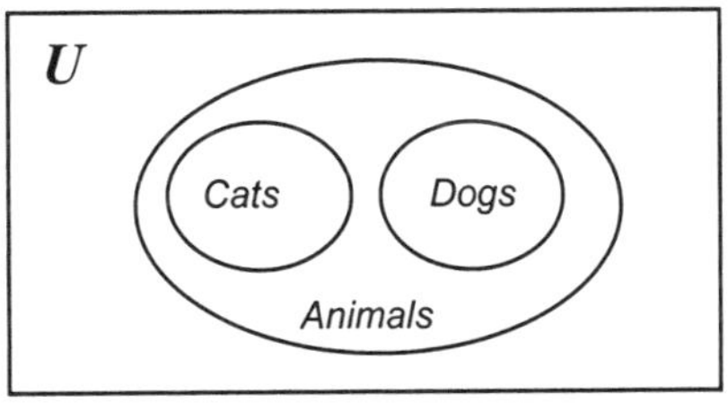

Disclaimer: In contrast to the Venn diagram above, you may have chosen to omit the outside rectangle and instead define your universal set U as the set of all animals. Doing so is acceptable. In case you are wondering what U might represent in our diagram above, it can be anything for which all animals are a subset; for example, U may be the set of all life forms on Earth.

29. Suppose we let U represent the set of all people. Then, U is composed of two parts (subsets): (i) people who are 30 or under and (ii) those who are over 30. Certainly, some of the members of these two subsets are people who drive. Hence, with this perspective, an appropriate Venn diagram would be:

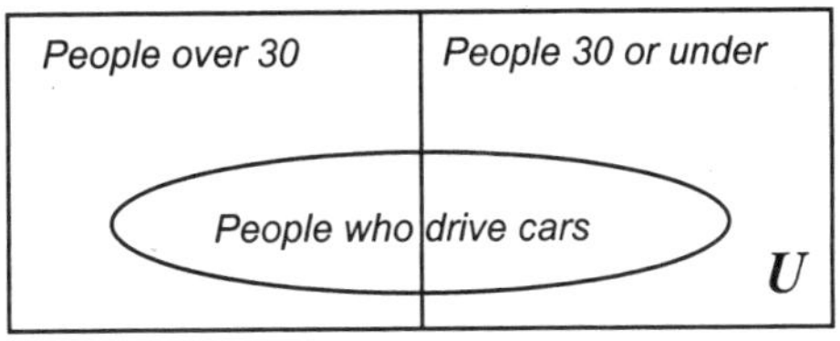

31. Our Venn diagram must be one in which the set of Chevrolets is a subset of the set of all cars. That is,

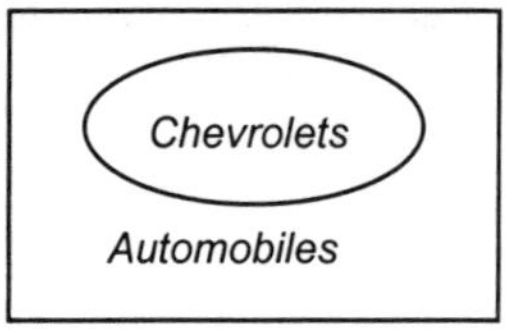

33. We begin by defining three sets. Let

U = the set of all Santa Rosa Junior College enrollees in the fall of 1999,
F = the set of all females in U, and
A = the set of all those over the age of 25 in U.

Please note that the selection (or lack thereof) of a set M which would signify all males in U is purely optional, since $M = \overline{F}$ (or, if preferred, $F = \overline{M}$). In any event, the given information leads to the following cardinalities:

$$|U| = 29{,}000$$

$$|F| = 29{,}000\,(0.58) = 16{,}820$$

$$|M| = |\overline{F}| = \begin{cases} |U| - |F| = 29{,}000 - 16{,}820 = 12{,}180 \\ \text{or} \\ 29{,}000(0.42) = 12{,}180 \end{cases}$$

$$|A| = 29{,}000\,(0.62) = 17{,}980$$

$$|M \cap A| = |\overline{F} \cap A| = 17{,}980\,(0.4) = 7{,}192$$

$$|F \cap A| = |\overline{M} \cap A| = \begin{cases} 17{,}980\,(0.6) = 10{,}788 \\ \text{or} \\ |A| - |M \cap A| = 17{,}980 - 7{,}192 = 10{,}788 \end{cases}$$

The following Venn diagram more clearly illustrates these relationships:

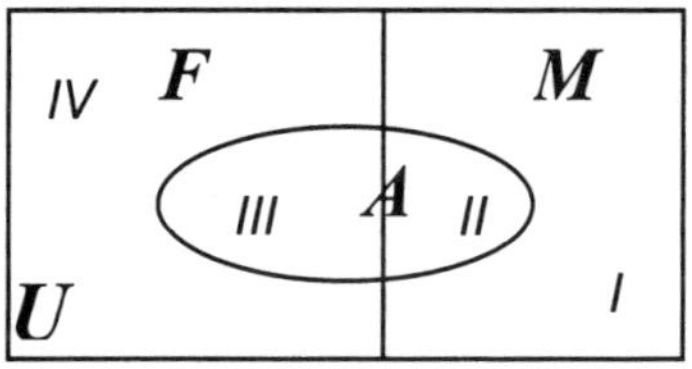

I: $M \cap \overline{A}$ = {male students who are 25 and under}

$$\Rightarrow |M \cap \overline{A}| = 12{,}180 - 7{,}192 = \boxed{\mathbf{4{,}988}}$$

II: $M \cap A = \{$male students who are over 25$\}$

$$\Rightarrow |M \cap A| = \boxed{7{,}192}$$

III: $F \cap A = \{$female students who are over 25$\}$

$$\Rightarrow |F \cap A| = \boxed{10{,}788}$$

IV: $F \cap \overline{A} = \{$female students who are 25 and under$\}$

$$\Rightarrow |F \cap \overline{A}| = 16{,}820 - 10{,}788 = \boxed{6{,}032}$$

Finally, permit us to remind you of a critically important, yet often overlooked, part of solving multi-step problems like this one. We speak of the fourth step of Pólya's problem solving guidelines ("Look Back"), as discussed in Section 1.1 of your text. In the case of this particular problem, one item (among several) you should check is the sum of the elements in regions I through IV. Does this sum equal the number of 1999 Santa Rosa Junior College enrollees?

> ***Strategy tip for Problems 35-38***: *You may find it helpful to begin by describing the shaded region(s) in your own words (perhaps using words such as "and", "or", "everything except", etc.) and then translating those words into their corresponding set notation. We demonstrate this tactic in our next solution.*

35. $\boxed{\textbf{Answers vary}}$; included among them are $\boxed{\overline{A \cup B}}$ and $\boxed{\overline{A} \cap \overline{B}}$, which we previously have encountered in Example 7 of your text (a la DeMorgan's Laws).

37. $\boxed{\textbf{Answers vary}}$.

Three possibilities (among many) are $\boxed{(A \cap B) \cup (A \cap C)}$, $\boxed{A \cap (B \cup C)}$, and $\boxed{(A \cap B \cap C) \cup (A \cap B \cap \overline{C}) \cup (A \cap \overline{B} \cap C)}$.

Let us take a moment to demonstrate the strategy tip mentioned in our solution to Problem 35. We begin by arbitrarily numbering the regions in the Venn diagram:

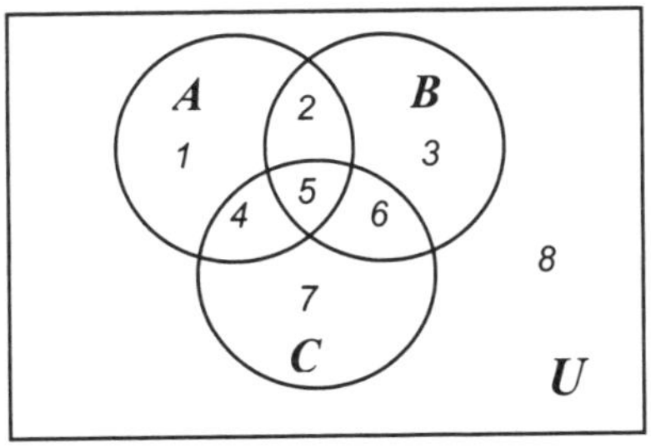

Next, we need to find language that adequately and accurately describes the relevant regions (in this case, regions 2, 4, and 5). Here are three perspectives:

I. Regions 2 and 5 are common to both sets A "and" B, which translates notationally to $A \cap B$. Regions 4 and 5 are common to sets A and C (i.e., $A \cap C$). "Joining" these two together yields regions 2, 4, and 5, and the word "joining" suggests, of course, the set operation "union." Therefore, we write $(A \cap B) \cup (A \cap C)$.

II. Here is another way to view regions 2, 4, and 5: they are the portions of set A that "overlap" (i.e., intersect with) the union of sets B and C. These words translate to $A \cap (B \cup C)$.

III. Lastly, we can choose to view regions 2, 4, and 5 individually, and then, once we have expressed each of them with proper notation, we can perform a union of these three expressions. Region 2 consists of those elements that are common to set A, set B, but not set C. This can be written as $A \cap B \cap \overline{C}$. Region 4 is made up of elements common to sets A and C, but not C, which we formally express as $A \cap \overline{B} \cap C$. As for region 5, its elements are common to A, B, and C (i.e., $A \cap B \cap C$). Lastly, in any order desired, we perform a union of these three expressions, thus writing $(A \cap B \cap C) \cup (A \cap B \cap \overline{C}) \cup (A \cap \overline{B} \cap C)$.

Note: Please do not be concerned if the words you (would) have chosen differ from ours. The specific words themselves are relatively unimportant; what is important is how appropriately and consistently you translate the words into formal set notation.

39. Using Venn diagrams as the directions state (and in the manner your text's author illustrates with Example 7 on p. 51), we arrive at the following result:

$\overline{A \cup B}$ $\quad \neq \quad$ $\overline{A} \cup \overline{B}$

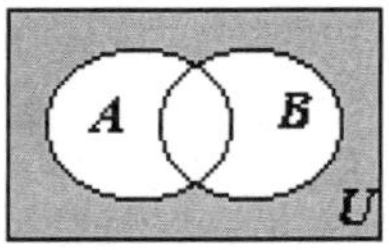

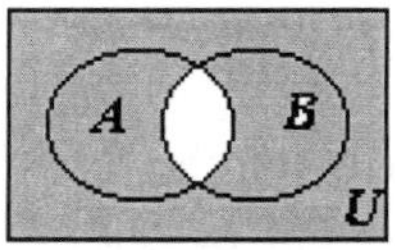

Now, let us take a moment to share an alternative, but similar, approach to the shaded-region tactic demonstrated in your text. Please note that we first illustrated this alternative approach in this manual's solution to Problem 19; feel free to refer to it, as needed. We begin by constructing the Venn diagram and arbitrarily labeling the four regions in some manner:

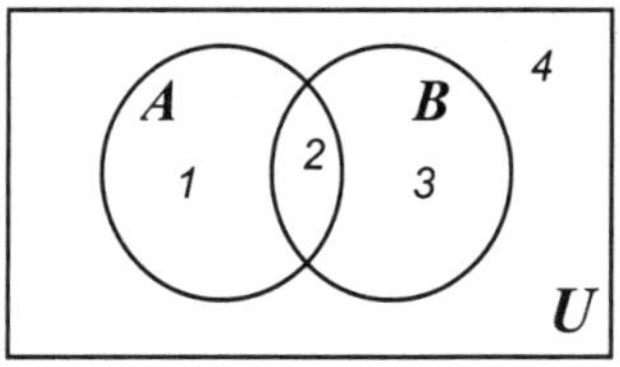

Based on our chosen labeling, we have $A = \{1,2\}$, $B = \{2,3\}$, and $U = \{1,2,3,4\}$ and accordingly obtain:

$$\overline{A \cup B} = \overline{\{1,2\} \cup \{2,3\}} = \overline{\{1,2,3\}} = \{4\}$$
$$\overline{A} \cup \overline{B} = \overline{\{1,2\}} \cup \overline{\{2,3\}} = \{3,4\} \cup \{1,4\} = \{1,3,4\}$$

Since $\{4\} \neq \{1,3,4\}$ as depicted in our shaded Venn diagrams above, we have disproved the statement $\overline{A \cup B} = \overline{A} \cup \overline{B}$. Therefore, $\overline{A \cup B} \neq \overline{A} \cup \overline{B}$.

41. There are at least three approaches one may choose to employ: (i) the shaded-region approach (used in your text with Example 6-8, pp. 50-53); (ii) the labeled-region strategy (as illustrated in this manual's solution to Problem 39 above and Problem 43 below), or (iii) the language-translation tactic (introduced as "Alternative Approach II" in this manual's solution to Problem 19). In any case, we find that the two expressions given are indeed **equal**. Both yield identically shaded Venn diagrams, specifically the following:

$(A \cup B) \cup C \qquad = \qquad A \cup (B \cup C)$

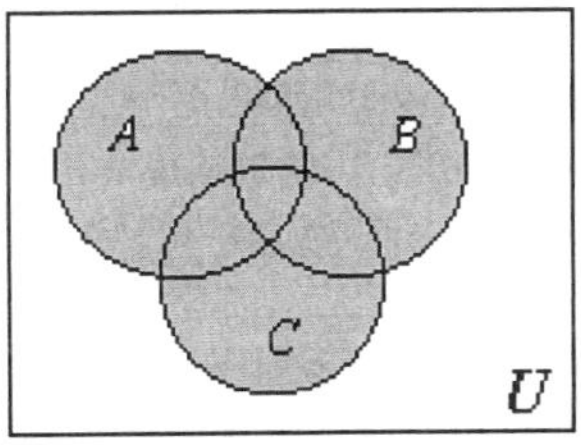

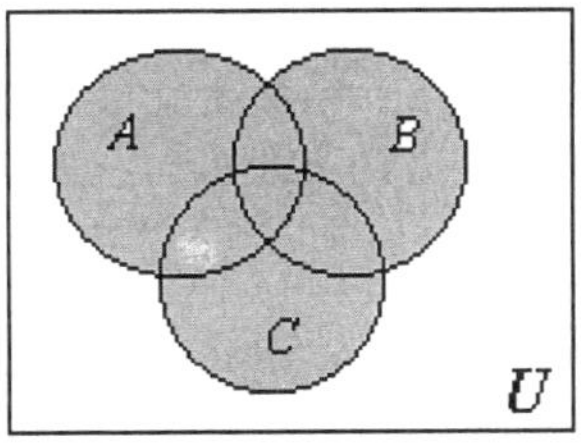

This proof establishes the following useful fact: <u>set union is an "associative" operation</u>, comparable to the associative property for addition and multiplication of real numbers (as mentioned in your text's glossary).
As an additional exercise, we encourage you to explore whether set <u>intersection</u> is also an associative operation.

43. Note that there are at least three approaches one could apply to this problem. Let us use this problem as another opportunity to reinforce the labeled-region strategy (previously employed in this manual's solutions to Problems 19 and 39). We begin by arbitrarily labeling the Venn diagram's regions:

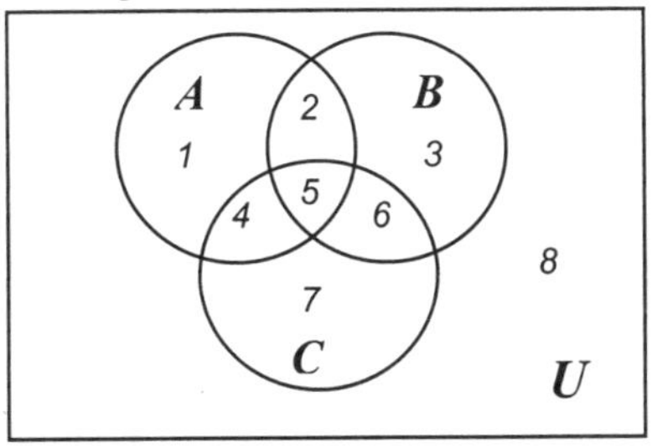

Based on labeling, we have $A = \{1,2,4,5\}$, $B = \{2,3,5,6\}$, and $C = \{4,5,6,7\}$. Hence,

$$
\begin{aligned}
A\cap(B\cup C) &= \{1,2,4,5\}\cap(\{2,3,5,6\}\cup\{4,5,6,7\}) \\
&= \{1,2,4,5\}\cap\{2,3,4,5,6,7\} \\
&= \{2,4,5\} \\
(A\cap B)\cap(A\cap C) &= \left(\{1,2,4,5\}\cap\{2,3,5,6\}\right)\cap\left(\{1,2,4,5\}\cap\{4,5,6,7\}\right) \\
&= \{2,5\}\cap\{4,5\} \\
&= \{5\}
\end{aligned}
$$

Therefore, upon shading the regions corresponding to these labels, we conclude:

$$A\cap(B\cup C) \quad \boxed{\neq} \quad (A\cap B)\cap(A\cap C)$$

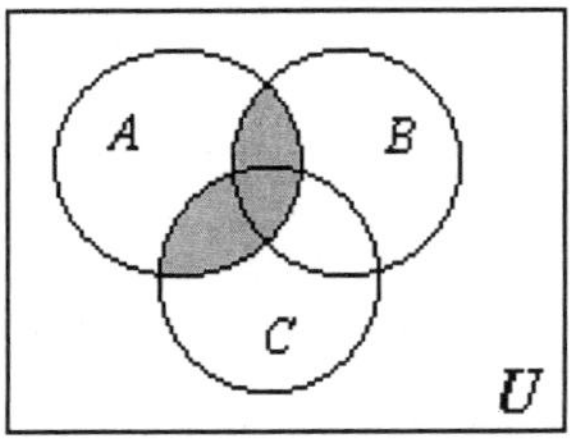

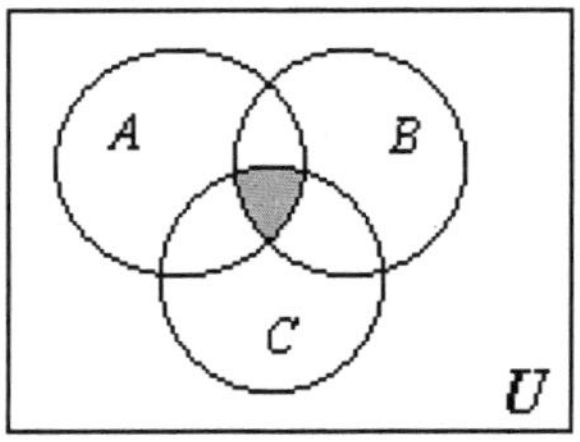

45.

<u>I (left-handed male non-citizens of U.S.)</u>:	**none**
<u>II (right-handed female non-citizens of U.S.)</u>:	**Steffi Graf, Evonne Goolagong, Virginia Wade**
<u>III (left-handed female U.S. citizens)</u>:	**Martina Navratilova**
<u>IV (left-handed male U.S. citizens)</u>:	**John McEnroe**
<u>V (right-handed male non-citizens of U.S.)</u>:	**Michael Stich, Stefan Edberg, Boris Becker, Pat Cash**
<u>VI (right-handed female U.S. citizens)</u>:	**Chris Evert Lloyd**
	<u>VII (right-handed male U.S. citizens)</u>: none
<u>VIII (left-handed female non-citizens of U.S.)</u>:	**none**

47. <u>*IMPORTANT WARNING*</u>: With Problems 46-54, a common student error is to misinterpret phrases such as "75 students owned stereos." Make sure you do <u>not</u> insert, mentally or otherwise, the word "only" into such phrases. That is to say, the given statement "75 students owned stereos" is <u>not</u> to be interpreted as saying "75 students owned <u>only</u> stereos." The given statement makes no such restrictions pertaining to what else, if anything, these 75 students owned.

a) Of the 75 students who owned stereos, 35 of these students also owned cars. Therefore, 40 (i.e., $75-35$) of them owned stereos but not cars. Similarly, since 35 of the 45 car owners also owned stereos, 10 owned cars but not stereos. You may find it easier to make such determinations by making use of a Venn diagram, which would ultimately look like this:

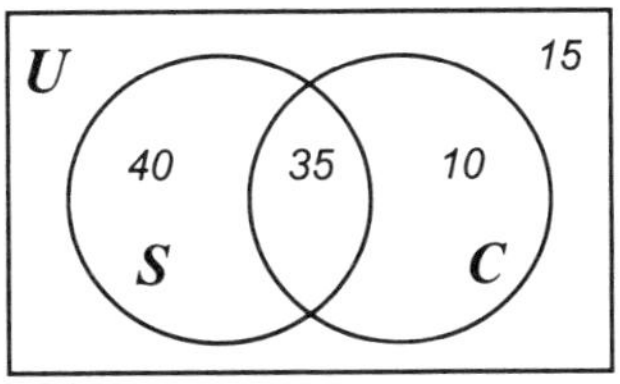

where S and C represent stereo owners and car owners, respectively.

Thus, $40 + 10 = \boxed{50}$ students own either a car or a stereo.

b) Since the Venn diagram above accounts for 85 (i.e., 45+35+10) of the 100 college students surveyed, the remaining $100 - 85 = \boxed{15}$ own neither a car nor a stereo.

49. The two most common approaches to this problem are (i) completing a Venn diagram and (ii) applying the formula for the cardinality of the union of two sets (as explained by your text's author above Example 4). If F and B respectively signify football and basketball players, the associated Venn diagram would appear as follows:

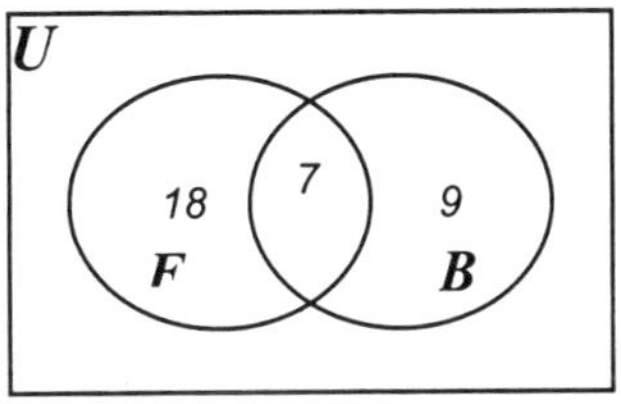

Therefore, there is a total of $18 + 7 + 9 = \boxed{34}$ different people playing these sports.

Alternatively, and perhaps more efficiently, we could have obtained this result by using the cardinality formula mentioned above. In formal notation, $|F \cup B|$ represents the number of students playing football or basketball. We then apply the formula shown on textbook p. 47:

$$\begin{aligned} |F \cup B| &= |F| + |B| - |F \cap B| \\ &= 25 + 16 - 7 \\ &= 34 \end{aligned}$$

51.

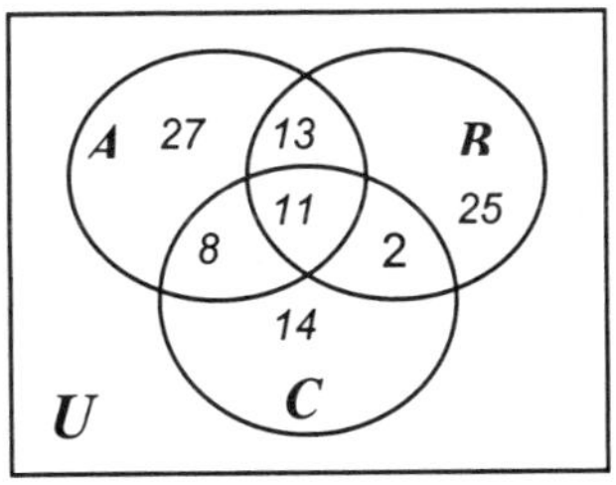

This problem is very similar to textbook Example 5 (pp. 48-49), in which the author suggested you "fill in the innermost region first, and then work [your] way outward."

In the event you are having difficulty, consider these two suggestions:

I. Heed the "important warning" provided above in our solution to Problem 47. That is, do not infer incorrectly, for example, that the words "59 use shampoo A" mean these 59 people "only" use shampoo A. The sentence "59 use shampoo A' means these fifty-nine people use "at least" shampoo A.

II. In the spirit of your textbook's motto "Math is not a spectator sport" (prominently displayed at the beginning of Section 1.1), use Example 5 as a practice problem. That is to say, first work it out on your own. Then refer to the author's solution to check your work, repair any errors, and discover alternative approaches.

53. $\boxed{20}$. While resembling its predecessors (namely, Problems 51-52 and textbook Example 5), this problem requires you, the problem solver, to interpret and then categorize two pieces of information. These are: (i) "used buses" suggests use of a form of public transportation, and (ii) "used their own cars" is intended to mean the same as "drove alone." The latter is made clear by the given information "10 used both a carpool and sometimes their own cars."

To answer the question, we again construct/label a Venn diagram and work outward. Letting A be the set of students who drive alone, P the set of car-poolers, and T public transportation riders, here are the steps:

I. Since 4 students used all three methods, we place a '4' in the innermost region (i.e., the intersection of all three sets).

II. Recalling our warning in the solution to Problem 47, the "6 [who] used buses as well as their own cars" include the 4 who used all three methods. Hence, $6 - 4 = 2$ belong in $A \cap T$ but "outside of" P (i.e., $A \cap T \cap \overline{P}$).

III. Similarly, the "10 [who] used both a carpool and sometimes their own cars" include the 4 who use all three methods. Consequently, $10 - 4 = 6$ are in $A \cap P$ but not T.

IV. Like steps II and III, the 9 who used both carpools and public transportation leads us to knowing that $9 - 4 = 5$ are in $P \cap T$ but not A.

V. The 31 who rode public transportation include those from step I (the 4 who used all three methods), step II (the 2 who rode public transportation and sometimes drove alone but don't carpool), and step IV (the 5 who used both carpools and public transportation but didn't drive alone). This leads us to concluding that $31 - 4 - 2 - 5 = 20$ used only public transportation but not carpools or their own cars. Said differently, $\left|T \cap \overline{A} \cap \overline{P}\right| = 20$.

VI. Applying step V's reasoning, we observe that the 28 who rode in a carpool include the four (from step I), six (from step III), and five (from step IV). This implies $28 - 4 - 6 - 5 = 13$ used carpools as their sole source of transportation to campus.

VII. Like steps V and VI, we know that the 42 who drove alone include the four, two, and six from steps I, II, and III, respectively. As a result, $42-4-2-6=30$ always got to campus by driving alone.

VIII. Finally, remember that 100 students were polled. The sum of the findings in steps I through VII is equal to $4+2+6+5+20+13+30=80$. Thus, there are 20 remaining students who used none of the three specified means of transportation.

The Venn diagram resulting from these eight steps looks like this:

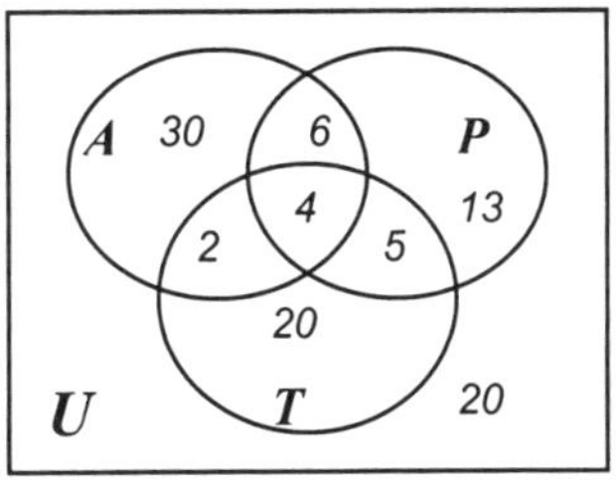

55. a) Let $U=$ {people who live in the United States}
$C=$ {people who use computers}.
$B=$ {black people}
$H=$ {hispanic people}
$O=$ {people who are classified as "other" than black, hispanic, or white}
$W=$ {white people}

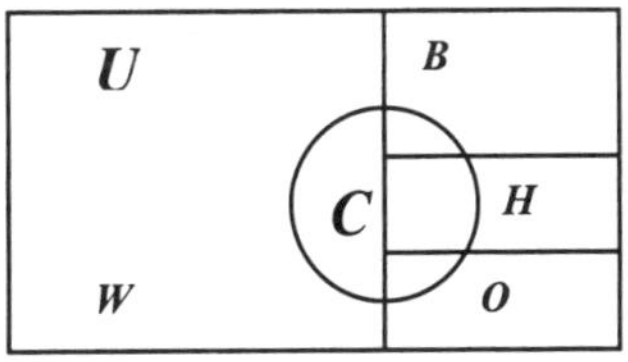

In the interest of clarity, we feel compelled to note that the use of the word "whites" in our Venn diagram — and in Mr. Brokaw's broadcast — means non-hispanic whites. Similarly, assume in this problem that the term "blacks" refers to non-hispanic blacks. In other words, please assume B, H, O, and W have no elements in common (i.e., their intersection equals $\varnothing$). Obviously, in our very diverse society, such distinctions are not so clear-cut; that is, since the label "hispanic" is a classification of ethnicity (as opposed to "white" and "black" which are racial classifications), there are black hispanics, white hispanics, asian hispanics, etc.

b) To demonstrate the error in Brokaw's comment, we make use of the data provided in Problem 34; namely, in millions, we have:

$|U|=263$ $\qquad$ $|W|=0.8|U|=210.4$

$|B|=0.12|U|=31.56$ $\qquad$ $|H|=0.09|U|=23.67$

Based on NBC's 1995 news report, 27% of whites use computers, 14% of blacks, and 13% of hispanics. Hence,

$$0.27|W| = 0.27(210.4) \approx 57 \text{ million whites use computers}$$

$$0.14|B| = 0.14(31.56) \approx 4.4 \text{ million blacks use computers}$$

$$0.13|H| = 0.13(23.67) \approx 3 \text{ million hispanics use computers}$$

Consequently, among blacks and hispanics, a total of approximately 7.4 million people use computers. Since this total does not even come close to the 57 million whites who use computers, it is clear that Tom Brokaw's addition of percentages lead to a seriously flawed analysis of the population of computer users.

57. This problem provides us with another opportunity to make use of the language-translation strategy first explained in this manual with our solution to Problem 37.

a) Region 1 can be described in several ways. First, one might view the elements of this region as "not in A, and not in B, and not in C, and not in D, but in E." Since the words "and" and "but" suggest set intersection and the word "not" indicates set complement, one could translate these words into the following set notation:

$$\boxed{\overline{A} \cap \overline{B} \cap \overline{C} \cap \overline{D} \cap E}$$

Alternatively, region 1 could be described as "the set of elements who are not in A or B or C or D but belong to E." Since, the word "or" suggests set union, this phrasing would be translated as follows:

$$\boxed{\overline{(A \cup B \cup C \cup D)} \cap E}$$

Both answers are equivalent (via DeMorgan's laws).

b) The elements of region 11 can be thought of in several ways. First, they are in A and E but not in B, C, or D. Second, they are in A and E and not B and not C and not D. Therefore, in a manner comparable to part (a), these two verbal descriptions translate respectively to the following solutions:

$$\boxed{(A \cap E) \cap \overline{(B \cup C \cup D)}} \text{ or } \boxed{A \cap E \cap \overline{B} \cap \overline{C} \cap \overline{D}}$$

c) The elements of region 21 belong to sets B, D, and E, but not A or C. Thus, we can describe this region as follows:

$$\boxed{(B \cap D \cap E) \cap \overline{(A \cup C)}} \text{ or } \boxed{\overline{A} \cap B \cap \overline{C} \cap D \cap E}$$

d) Region 31 is the set of elements "common to" (i.e., in the intersection of) all five sets. Hence, we write:

$$\boxed{A \cap B \cap C \cap D \cap E}$$

e) The elements of region 16 belong to sets A, B, and E, but not C or D. In a manner comparable to part (b), we obtain:

$$\boxed{(A \cap B \cap E) \cap (\overline{C \cup D})} \text{ or } \boxed{A \cap B \cap \overline{C} \cap \overline{D} \cap E}$$

59.

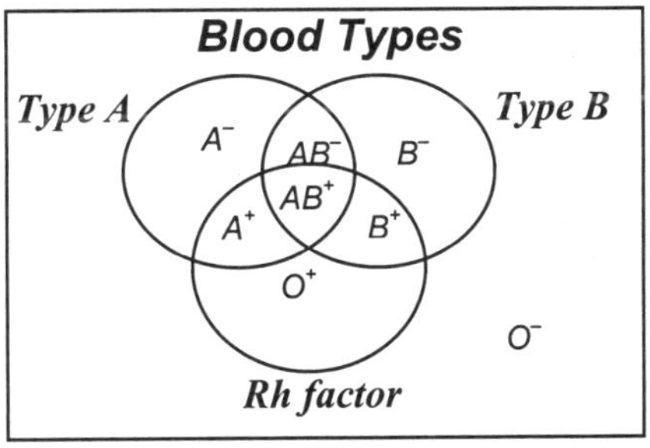

Chapter 1 Review Questions

1. Refer to textbook Section 1.1 (page 7) for details.

2. Counting 5 "blocks" down and 4 over (or 4 down and 5 over) on Pascal's triangle tells us there are **126 possible paths**. To review this concept, see textbook Examples 1 and 2 in Section 1.1 (pp. 8-10).

3. This problem should strike you as being very much the same as its predecessor, Problem 2. In terms of computing possible paths, does it make a difference whether a rook is moving horizontally and vertically on a chessboard or a person is walking north and west? Hopefully, your answer is a resounding "no." Traveling 4 over and 7 down along Pascal's triangle yields the solution: **330 possible routes**.

4. Immediately after you read this question, the powerful problem-solving technique known as ***inductive reasoning*** (discussed extensively in Section 1.2) should have popped into your mind. Did it? Let us explore smaller related problems and seek to discover a pattern:

$$1 \times 1 = 1$$

$$11 \times 11 = 121$$

$$111 \times 111 = 12{,}321$$

$$\vdots$$

Thus, we expect the product of the 9-digit number $111{,}111{,}111{,}111$ and itself to equal **12,345,678,987,654,321**.

5. A discussion of the concepts ***order of operations*** and ***scientific notation*** can be found in Sections 1.2 and 1.3, respectively — as well as in your textbook's glossary (Appendix G). Refer specifically to pages 20 and 29.

6. a) $\{1, 3, 5, 7, 9\} \cup \{2, 4, 6, 9, 10\} = \boxed{\{1, 2, 3, 4, 5, 6, 7, 9, 10\}}$

b) $\{1, 3, 5, 7, 9\} \cap \{2, 4, 6, 9, 10\} = \boxed{\{9\}}$

c) $\overline{\{2, 4, 6, 9, 10\}} = \boxed{\{1, 3, 5, 7, 8\}}$

d) Making use of the answer to part (a), we have:

$$\begin{aligned}\overline{A} \cap (B \cup A) &= \overline{\{1,3,5,7,9\}} \cap \{1,2,3,4,5,6,7,9,10\} \\ &= \{2,4,6,8,10\} \cap \{1,2,3,4,5,6,7,9,10\} \\ &= \boxed{\{2,4,6,10\}}\end{aligned}$$

7. a)

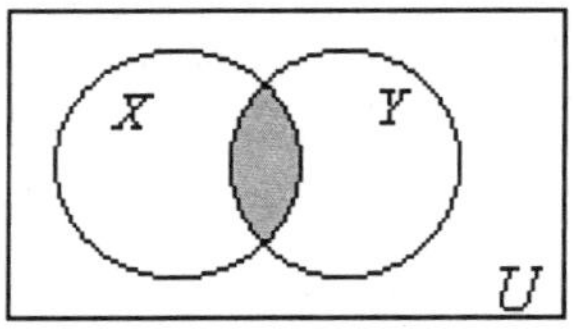

b)

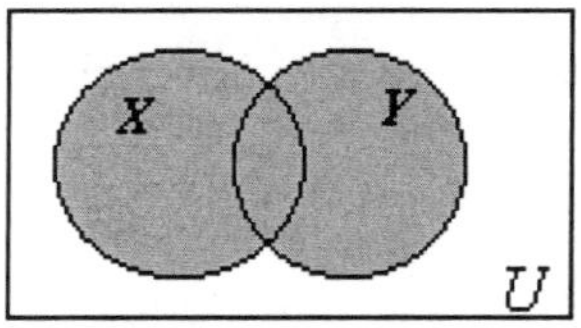

c)

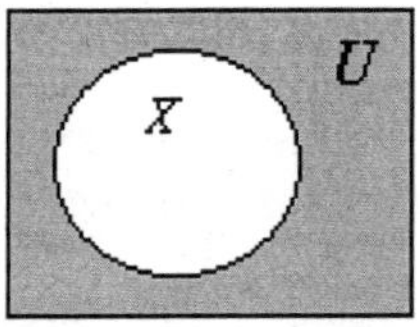

d)

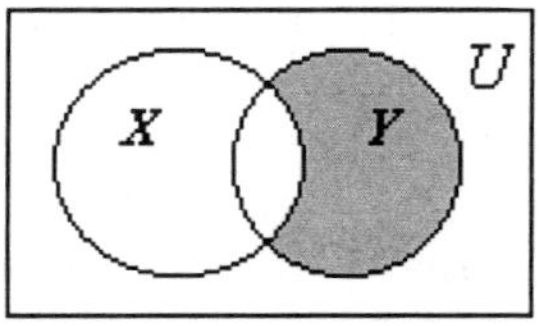

In the event you need a review of strategies for this type of problem (especially part d), we refer you to your text's Example 6 & 8 (pp. 50-52) <u>and</u> this manual's solution for Problem 19 (in Section 1.4), which is a very similar problem to this one and one for which this we include a detailed explanation.

8. Refer to this manual's solution to Problem 43 in Section 1.4 for a detailed illustration of one approach (among several) to this type of problem.

a) **Disproved/Unequal**; $A \cup (B \cap C) \quad \neq \quad (A \cup B) \cap C$

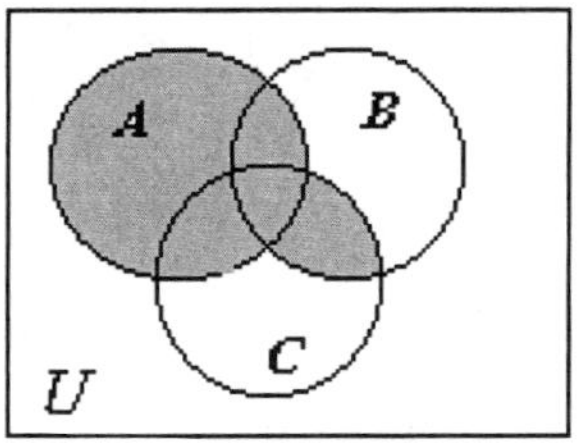

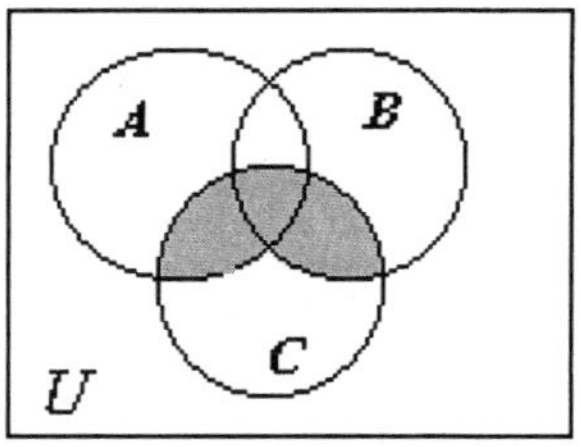

b) **Proved/Equal**; $(A \cup B) \cap C \quad = \quad (A \cap C) \cup (B \cap C)$

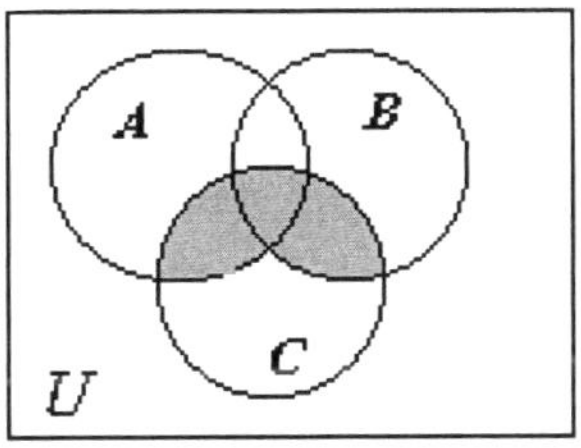

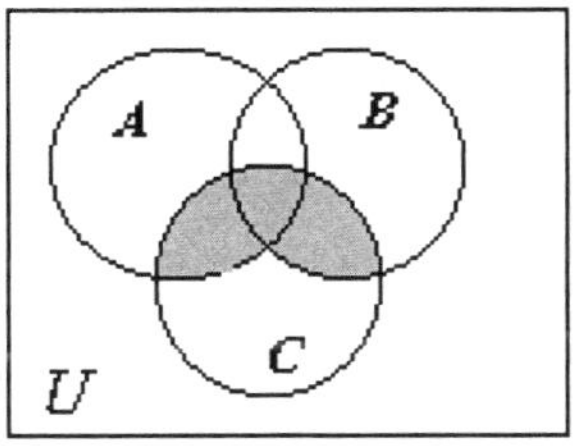

9. The story illustrates **inductive reasoning** because the answer to the third question ("Three baseball teams!") is obtained by utilizing the pattern of the two previous questions. Recall from textbook p. 21 that inductive reasoning "involves reasoning from particular facts or individual cases to a general conjecture."

10. **Answers vary**, but here are some of the more common calculator keystrokes:

a) *Keystrokes:* [2] [x^y] [63] [=] or [2] [^] [63] [=]

Result: **9.223372037 18** or **9.223372037E18**

b) *Keystrokes:* [9.22] [EXP] [18] [÷] [6.34] [EXP] [6] [=]

Result: **1.454258675 12** or **1.454258675E12**

11. **Choice C**

Choice A: As mentioned in Problem 17 (p. 59) and Problem 34 (p. 54), we need to remember that there are between 250 and 300 million people in the United States. Using (for purposes of this problem) a population estimate of 281.9 million people reveals that we would have to spend approximately $\$1{,}000$ per person; that is,

$(\$281.9$ billion$) \div (281.9$ million people$) = \$1000$ per person. Obviously, since the price of one steak dinner falls far short of costing $\$1{,}000$, choice A won't even come close to liquidating the windfall.

Choice B: The product $\$1 \times 60$ sec $\times$ 60 min $\times$ 24 hr $\times$ 365.25 days $\times$ 100 years approximately equals $\$3.16$ billion. Like choice A, this is well short of the available $\$281.9$ billion.

Choice C: If we estimate populations of 1 million San Franciscans and 1 billion Chinese, we would expend

$(\$80{,}000 \times 1$ million people$) + (\$200 \times 1$ billion people$)$, which equals $\$280$ billion. This expenditure virtually liquidates the budget windfall.

12. $\boxed{C = \frac{\ell w}{15}}$, where C represents the boat's capacity, and ℓ and w are its respective length and width.

13. Forty-five minutes each day equals ¾ of an hour, which, in turn, suggests approximately 274 hours in a one-year (365-day) period. Since $(274 \text{ hrs}) \div (100 \text{ books})$ is approximately equal to $2\,¾$ hrs per book, each book would have to consume, on average, $2¾$ hrs of listening time. Obviously, it is not possible for a "fully" transcribed (i.e., unabridged) book to be completed in such a short duration — especially if it is among the so-called "100 greatest books of all time."

14. $\boxed{10}$. This problem is very similar to Example 5 (pp. 48-49) and Problems 51-54 in Section 1.4 (see text and this solution manual for additional details).

<u>Step 1:</u> Construct a Venn diagram and label the relevant sets accordingly.

<u>Step 2:</u> Begin with the innermost region; namely, make use of the statement "15 had all three."

<u>Step 3:</u> Use the information pertaining to the 17, 35, and 25 to determine how many individuals had ***exactly*** two of the three items. As mentioned earlier in this solution manual (i.e., in our explanations for Problems 47, 51, and 53 in Section 1.4), make sure you read phrases like "25 had a TV and a bicycle" as meaning "25 had **[at least]** a TV and a bicycle." That is to say, included among these 25 students are the 15 who had all three.

<u>Step 4:</u> Use the information regarding the 42, 50, and 30 to determine how many individuals had ***exactly*** one of the three items. To accomplish this, you will need to perform a series of subtractions.

<u>Step 5:</u> Sum the seven numbers thus far placed in your diagram and subtract this total from the number of students surveyed (70).

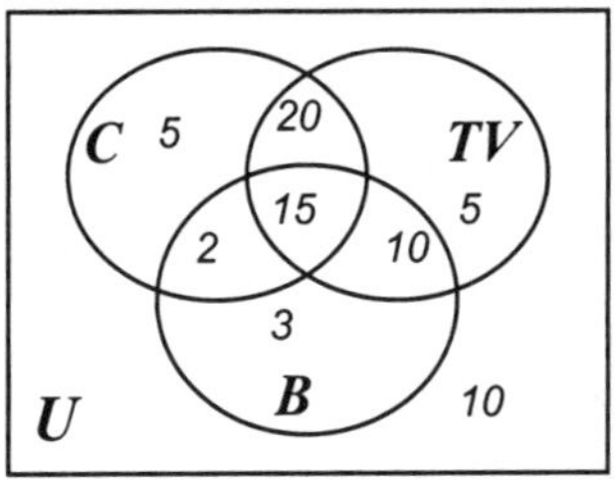

15. One way (among several) to accomplish this is to arrange the cards as:

$$\boxed{\begin{bmatrix} 2 & 7 & 6 \\ 9 & 5 & 1 \\ 4 & 3 & 8 \end{bmatrix}}.$$

Realize that other comparable (equally correct) arrangements exist; for example,

$$\boxed{\begin{bmatrix} 8 & 3 & 4 \\ 1 & 5 & 9 \\ 6 & 7 & 2 \end{bmatrix}}.$$

16. **Answers vary**. The answer given in the back of your textbook (p. 935) contains one interesting perspective. Let us share another.

Since a typical ice cube has a volume of about 1 in^3, it follows that 7×10^{16} ice cubes amounts to a total volume of approximately 7×10^{16} in.^{3}. According to *The 2000 World Almanac and Book of Facts*, the area of the United States (including all 50 states and the District of Columbia) is 3,717,796 square miles. This equates to:

$$3{,}717{,}796 \text{ mi}^2 \times \left(\frac{5{,}280^2 \text{ ft}^2}{1 \text{ mi}^2}\right) \times \left(\frac{12^2 \text{ in.}^2}{1 \text{ ft}^2}\right) \approx 1.49250533769 \times 10^{16} \text{in.}^2,$$

which suggests that covering the U.S. with a single layer of inch-high ice cubes would amount to a volume of $1.49250533769 \times 10^{16} \text{ in}^3$.

Therefore, since $(7 \times 10^{16}) \div (1.492503633769 \times 10^{16}) \approx 4.69$, a total volume of 7×10^{16} in.^{3} would amount to covering the entire U.S. with almost five layers of inch-high ice cubes. Said differently, if this iceberg's volume were redistributed it would cover the entire U.S. in roughly 4 ¾ inches of ice.

17. $\$5{,}670{,}000{,}000{,}000 = \$5.67 \times 10^{12}$

Each person's share of the national debt is $\dfrac{\$5.67 \times 10^{12}}{2.74 \times 10^8} \approx$ **$20,693**.

18. If a stack of 233 bills is 1 in. tall, then one bill is $\frac{1}{233}$ in. tall and consumes a volume of approximately $(6 \text{ in.}) \times (2.5 \text{ in.}) \times (\frac{1}{233} \text{ in.}) \approx .0643776824 \text{ in.}^3$. Therefore, since a 20' × 30' × 10' classroom has a volume of $6{,}000 \text{ ft}^3$ (or, more appropriately, $6{,}000 \text{ ft}^3 \times \left(12^3 \text{ in.}^3/\text{ft}^3\right) = 10{,}368{,}000 \text{ in.}^3$), one classroom will hold

$$\frac{10{,}368{,}000 \text{ in.}^3}{0.064377824 \text{ in.}^3} \approx 161{,}049{,}600 \text{ bills}$$

Using dollar bills, this translates to $\$161{,}049{,}600$ per classroom, which means the national debt would fill up more than $35{,}200$ classrooms; that is, $\$5.67 \times 10^{12}$ divided by $\$161{,}049{,}600$. However, filling this many classrooms is still hard to envision — not to mention the fact that a larger denomination than $\$1$ bills would be more likely and sensible. For example, **if $100 bills are used, the national debt would fill roughly 352 classrooms** (each classroom would hold $\$161{,}049{,}600{,}000$). Can you imagine 352 of the classrooms in your area filled with $\$100$ bills?

19. **Answers vary**; 22 is unhappy because $2^2 + 2^2 = 8$, which is unhappy. The number 23 is happy because $2^2 + 3^2 = 13$, which is happy.

20. Does it sound like it is again time to apply inductive reasoning? We begin by seeking a pattern for the last digit (also known as the "units digit" and printed below in boldfaced/underlined type):

$7^1 = \underline{\mathbf{7}}$	$7^2 = 4\underline{\mathbf{9}}$	$7^3 = 34\underline{\mathbf{3}}$	$7^4 = 240\underline{\mathbf{1}}$
$7^5 = 1680\underline{\mathbf{7}}$	$7^6 = 11764\underline{\mathbf{9}}$	$7^7 = 82354\underline{\mathbf{3}}$	$7^8 = 576480\underline{\mathbf{1}}$
$\vdots$	$\vdots$	$\vdots$	$\vdots$

Notice the repeating pattern $\{7, 9, 3, 1\}$. Since every fourth power of seven yields a units digit of '1' and since 1000 is a multiple of four, it follows that the last digit of 7^{1000} must be **1**.

CHAPTER 2: The Nature of Logic

Problem Set 2.1

1. The term ***operator*** — also known as a <u>*connective*</u> — is defined on page 65 of your text.

3. For details pertaining to the definition of a ***disjunction***, please refer to textbook pp. 66-67 (see Table 2.2).

5. See textbook pp. 69-70 for the details pertaining to (as well as examples of) this procedure.

7. Sentences **(a), (b), and (d)** are statements because each of them is a sentence that is either always true or always false. In contrast to these statements, sentence (c) is an opinion, and as such, its truth value cannot be determined. Please refer, if necessary, to the bottom half of textbook page 64 for additional information.

9. Only sentences **(a) and (b)** are statements. Sentence (c) is a command, and (d) is a question. As the author of your text explains on p. 64, commands and questions cannot be classified as true or false.

11. If you are unsure of what the term "negation" means, first go to pp. 67-69 in your text before you read this solution. Since, by definition, a statement and its negation must have opposite truth values, we can choose to find the negation by doing either of two things: (a) we can assume (or, if you will, "pretend") the given statement is true and ask ourselves what must be false; or, alternatively, (b) assume the given statement is false and ask what must be true. Although you are encouraged to experiment with both approaches, we have found most students prefer option (b). Here is, in a step-by-step — albeit somewhat dramatized, we admit — manner, how strategy (b) could be applied to the statement "all dogs have fleas":

I. <u>*Visualize (even personalize) the statement*</u>. Rephrasing it several times will help you grasp its full meaning. In the case of "all dogs have fleas," envision every single dog on the planet having one or more fleas on its body.

II. <u>*Pretend that the statement is false*</u>. You may find it helpful to dramatize this step; for instance, pretend you are a private investigator who has been hired to show that the claim "all dogs have fleas" is a lie.

III. <u>*Determine what "must be" true*</u>. We hope the all-important words "must be" caught your attention in both the preceding sentence and in the paragraph immediately above this three-step procedure. It is essential to identify what "must be" true, as contrasted from what "could be" or "might be" true. That is, if the statement "all dogs have fleas" is known to be false, what can we conclude must be true? Said differently, what must you, Ms. or Mr. Private Investigator, do to provide your client with sufficient evidence to show the given statement is a lie? Your answer, we hope, is go out and locate at least one dog on the planet that has absolutely no fleas.

Recall (from textbook p. 68) that the word "some" means "at least one." Thus, the proper negation is: **Some dogs do not have fleas**.

13. We apply the three-step strategy explained in our solution to Problem 11. First, make sure you grasp the full meaning of "some people do not pay taxes." This statement means there is at least one person who does not pay taxes. Second, assume this statement is false. Third, and finally, determine what must be true if you assume the original statement is false. From the perspective of the private investigator (as described in our solution to Problem 11 above), you, the investigator, must devise a way to disprove the statement "at least one person does not pay taxes." This means you must show that no one does not pay taxes. Said more clearly (i.e., without a double negative), the negation is:

All people pay taxes.

15. Again using the approach shown in Problems 11 and 13 above, we "pretend" the statement "no triangles are squares" is a falsehood and ask what then must be true. The answer is at least one triangle must be a square, which we would write (using the phrasings shown in your text) as: **Some triangles are squares**.

17. **Some counting numbers are not divisible by 1**. This statement possesses the same structure as Problem 11 (i.e. "all___are ___"), and as such, we refer to the approach shown above and on the preceding page of this solution manual.

19. Once again, we utilize the approach explained in our solution to Problems 11, 13, and 15 above. First, rephrase the statement "some integers are not odd." A good way is "at least one integer is not odd." Second, assume/pretend this statement is false. Lastly, determine what this assumption forces to be true. It should be noted that this last step might require several rephrasings on your part before you write the final negation. For example, pretending that the statement "at least one integer is not odd" is false may have lead you to conclude "not at least one integer is not odd." This, in turn, can be rewritten (overlooking the less-than-desirable grammar for the moment) as "no integer is not odd." Finally, repairing the grammatically-incorrect double negative, we arrive at the correct negation: **All integers are odd**.

21. **(a) and (c) are true**. To assume that prices rise and also taxes rise means we are beginning with the premise that both statement p and statement q are true. Compound statement (a) in Problem 20 is a disjunction (specifically, $p \vee \sim q$), and recall that a disjunction is true when <u>at least one</u> of its two component statements is true. Since at least one component of statement (a) is true (namely, p), compound statement (a) is true.

If this reasoning is hard for you to grasp, you may prefer (at least for the time being) the following alternative approach:

$$p \vee \sim q$$
$$\text{T} \vee \sim\text{T}$$
$$\text{T} \vee \text{F}$$
$$\textbf{T}$$

Thus, the statement $p \vee \sim q$ is true.

In contrast to statement (a), notice that compound statements (b), (c), and (d) are all conjunctions. Recall that a conjunction is true only when both of its two component statements are true. Out of conjunctions (b), (c), and (d), only statement (c) satisfies this requirement. Of course, as was shown above with statement (a), you may prefer to use the alternative (more symbolic) approach:

Statement (b)	*Statement (c)*	*Statement (d)*
$p \wedge \sim q$	$p \wedge q$	$\sim p \wedge q$
$\text{T} \wedge \sim\text{T}$	$\text{T} \wedge \text{T}$	$\sim\text{T} \wedge \text{T}$
$\text{T} \wedge \text{F}$	**T**	$\text{F} \wedge \text{T}$
F		**F**

Please refer in your textbook to Example 3 and especially Example 5 (pp. 69-70) for additional reinforcement.

23. *[Note: Specific word selection may vary.]*

a) **Prices or taxes will rise.**

b) **Prices will not rise, and taxes will rise.**

c) **Prices will rise, or taxes will not rise.**

d) **Prices will not rise, or taxes will not rise.** [Equivalently, you may write, "prices or taxes will not rise."]

25. Assuming that Paul like to read mathematics textbooks means we assume statement q in Problem 24 is true. Assuming Paul is not peculiar implies we shall assume statement p is false. Either of two approaches may be employed:

I. As was demonstrated on p. 70 (Example 5) in your text and in our solution to Problem 21 above, proceed to replace the p's and q's of your answers to Problem 24 with corresponding F's and T's and simplify accordingly.

II. Recognize that statements (a), (c), and (d) are conjunctions and recall that conjunctions are true only when all of their respective component statements are true. Similarly, note that (b) is a disjunction and remember that disjunctions are true when one or more of their component statements is/are true.

Both approaches will establish that **only (b) and (c) are true**.

27. *[Note: Specific wording may vary.]*

a) **Paul is peculiar and likes to read math textbooks.**

b) **Paul is not peculiar and likes to read math textbooks.**

c) **Paul is peculiar or does not like to read math textbooks.**

d) **Paul is not peculiar or does not like to read math textbooks.**

29. **Only (a) and (c) are true**. This problem is virtually identical to text Example 5 (see p. 70). For a review of the details, refer to this example and, if necessary, this manual's solution to Problem 21 above.

Alternatively, you may prefer to solve this problem by employing the less symbolic approach previously explained in our solutions to Problems 21 and 25. That is to say, notice that statements (a) and (b) are conjunctions, while (c) and (d) are disjunctions. Since conjunctions are true only when both of their simpler component statements are true and since we are to assume p and q are both true, it follows that (a) is true and (b) is false. Similarly, because a disjunction is true when at least one of its component statements is true, (c) is true (namely, because p is true), and (d) is false (since both of its component statements, $\sim p$ and $\sim q$, are false).

31. Please study text Example 4 (p. 69) before reading this solution.

a) $(r \vee s) \vee t$

$(\mathrm{F} \vee \mathrm{T}) \vee \mathrm{T}$	*Replace r, s, and t with their respective truth values.*
$\mathrm{T} \vee \mathrm{T}$	*Definition of disjunction*
T	*Definition of disjunction*

Notice that the truth values of statements r and s are inconsequential here. Since statement t is true and a disjunction is true when at least one of its component statements is true, we can conclude immediately (regardless of the outcome of $r \vee s$) that $(r \vee s) \vee t$ is true.

b) $(r \wedge s) \wedge \sim t$

$(\mathrm{F} \wedge \mathrm{T}) \wedge \sim\mathrm{T}$	*Replace r, s, and t with their respective truth values.*
$\mathrm{F} \wedge \mathrm{F}$	*Simplify (F∧T is equivalent to F, and ~T is equivalent to F).*
F	*Definition of conjunction*

Again, notice that the truth value of this statement could have been determined without considering $(r \wedge s)$. Since a conjunction is true only when all of its component statements are true, the conjunction $(r \wedge s) \wedge \sim t$ must be false because $\sim t$ is false.

33. Please study text Example 4 (p. 69) before reading this solution.

a) $(p \vee q) \wedge r$

(T ∨ T) ∧ F	*Replace p, q, and r with their respective truth values.*
T ∧ F	*Definition of disjunction*
F	*Definition of conjunction*

It should be noted that the truth values of p and q are irrelevant to this compound statement's truth value, since r is false and a conjunction is the last connective to be simplified.

b) $(p \wedge q) \wedge \sim r$

(T ∧ T) ∧ ~F	*Replace p, q, and r with their respective truth values.*
T ∧ T	*Simplify (T∧T is equivalent to T, and ~F to T).*
T	*Definition of conjunction*

35. a) $(p \vee q) \vee (r \wedge \sim q)$

(T ∨ T) ∨ (T ∧ ~T)	*Replace p, q, and r with their respective truth values.*
T ∨ (T ∧ F)	*Simplify (T∨T is equivalent to T, and ~T to F).*
T ∨ F	*Definition of conjunction*
T	*Definition of disjunction*

Note: From the truth value of $p \vee q$ alone, the final truth value can be determined. That is, for this particular problem, the outcome of $(r \wedge \sim q)$ was inconsequential.

b) $\sim(\sim p) \vee (p \wedge q)$

~ (~T) ∨ (T ∧ T)	*Replace p and q with their respective truth values.*
T ∨ T	*Simplify (~(~T)⇔~(F)⇔T and T∧T⇔T).*
T	*Definition of disjunction*

Again, it should be observed that the truth value of this particular disjunction could have been determined solely based upon the outcome of one or the other of its two component parts [i.e., either $\sim(\sim p)$ or $(p \wedge q)$].

37. a) $\sim(p \wedge q)$

~ (T ∧ F)

~F

T

b) $\sim p \wedge q$

~T ∧ F

F ∧ F

F

39. **e ∧ d ∧ t** where e: W.C. Fields is eating

d: W.C. Fields is drinking

t: W.C. Fields is having a good time

41. $\boxed{\sim t \wedge \sim m}$ where t: Jack will go tonight
m: Rosamond will go tomorrow

43. $\boxed{(j \vee i) \wedge \sim p}$ where j: The decision will depend on judgement
i: The decision will depend on intuition
p: The decision will depend on who paid the most

45. $\boxed{d \vee e}$ where d: The winner must have an A.A. degree
e: The winner must have three years of professional experience

47. Let s: Martha finished the sign
t: Martha finished the table
c: Martha finished a pair of chairs

a) $\boxed{(s \wedge t) \vee c}$ b) $\boxed{s \wedge (t \vee c)}$

49. $p \vee (q \wedge r)$
T ∨ (F ∧ F)
T ∨ F
$\boxed{\text{T}}$

[Here's a shortcut: Because the truth value of p is true, the disjunction is immediately known to be true. The outcome of the conjunction $q \wedge r$ is irrelevant.]

51. $(p \wedge \sim q) \vee (r \wedge \sim q)$
(T ∧ ~F) ∨ (F ∧ ~F)
(T ∧ T) ∨ (F ∧ T)
T ∨ F
$\boxed{\text{T}}$

[Shortcut: Because the truth value of $p \wedge \sim q$ is true, the disjunction is immediately known to be true. The outcome of the conjunction $r \wedge \sim q$ is irrelevant.]

53. $(r \wedge p) \vee (q \wedge r)$
(F ∧ T) ∨ (F ∧ F)
F ∨ F
$\boxed{\text{F}}$

55. $(p \wedge q) \vee (p \wedge r)$
(T ∧ F) ∨ (T ∧ F)
F ∨ F
$\boxed{\text{F}}$

57. $(q \vee \sim q) \wedge [(p \wedge \sim q) \vee (\sim r \wedge r)]$

$(F \vee \sim F) \wedge [(T \wedge \sim F) \vee (\sim F \wedge F)]$

$(F \vee T) \wedge [(T \wedge T) \vee (T \wedge F)]$

$T \wedge [T \vee F]$

$T \wedge T$

$\boxed{T}$

59. Within Melissa's note, there are four phrases that become negations when translated into symbols. They are: "I was really joking…"; "I didn't mean…"; "reconsidering my decision…"; and "not to…." With this in mind, we symbolically represent the statement "I was really joking when I told you that I didn't mean what I said about reconsidering my decision not to change my mind" as follows:

$\sim[\sim (\sim (\sim p))]$ where p denotes "I changed my mind"

Simplifying this — preferably by twice applying the law of double negation (see Problem 58) — reveals that $\sim[\sim (\sim (\sim p))]$ is equivalent to p.

Therefore, **Melissa did change her mind**.

Problem Set 2.2

1. A ***truth table*** is a device used in logic that summarizes all possible combinations of truth values for a given compound statement. In your text, a thorough discussion of truth tables begins on page 72 and is continued throughout much of the rest of the chapter.

3. The ***law of double negation*** may be found on textbook p. 74. Also, take a look at Problem 58 of Section 2.1 (p. 72).

5.

p	q	~p	~p ∨ q
T	T	F	**T**
T	F	F	**F**
F	T	T	**T**
F	F	T	**T**

7.

p	q	p ∧ q	~(p ∧ q)
T	T	T	**F**
T	F	F	**T**
F	T	F	**T**
F	F	F	**T**

9.

r	~r	~(~r)
T	F	**T**
F	T	**F**

11.

p	q	~q	p ∧ ~q
T	T	F	**F**
T	F	T	**T**
F	T	F	**F**
F	F	T	**F**

13.

p	q	~p	~p ∧ q	~q	(~p ∧ q) ∨ ~q
T	T	F	F	F	**F**
T	F	F	F	T	**T**
F	T	T	T	F	**T**
F	F	T	F	T	**T**

15.

p	q	p→q	p ∨ (p→q)
T	T	T	**T**
T	F	F	**T**
F	T	T	**T**
F	F	T	**T**

17.

p	q	p ∨ q	p ∧ (p ∨ q)	[p ∧ (p ∨ q)]→p
T	T	T	T	**T**
T	F	T	T	**T**
F	T	T	F	**T**
F	F	F	F	**T**

19.

p	q	r	p ∨ q	(p ∨ q) ∨ r
T	T	T	T	**T**
T	T	F	T	**T**
T	F	T	T	**T**
T	F	F	T	**T**
F	T	T	T	**T**
F	T	F	T	**T**
F	F	T	F	**T**
F	F	F	F	**F**

21.

p	q	r	p ∨ q	~r	(p ∨ q) ∧ ~r	[(p ∨ q) ∧ ~r] ∧ r
T	T	T	T	F	F	**F**
T	T	F	T	T	T	**F**
T	F	T	T	F	F	**F**
T	F	F	T	T	T	**F**
F	T	T	T	F	F	**F**
F	T	F	T	T	T	**F**
F	F	T	F	F	F	**F**
F	F	F	F	T	F	**F**

23. The definitions of converse, inverse, and contrapositive are provided on textbook p. 78. To obtain the ***converse*** of a conditional, interchange the antecedent and the consequent. To obtain the ***inverse***, negate both the antecedent and the consequent. To obtain the ***contrapositive***, interchange and negate the antecedent and consequent.

Statement: $\sim p \rightarrow \sim q$

Converse: ~q→~p Inverse: p→q Contrapositive: q→p

25. Statement: $\sim t \rightarrow \sim s$

Converse: ~s→~t Inverse: t→s Contrapositive: s→t

27. Let p: I get paid and q: I will go Saturday.
First rewrite the statement using the if-then connective, and then apply the definitions of converse, inverse, and contrapositive.

Statement ($p \rightarrow q$): If I get paid, then I will go Saturday.

Converse ($q \rightarrow p$): **If I go Saturday, then I will get paid.**

Inverse ($\sim p \rightarrow \sim q$): **If I do not get paid, then I will not go Saturday.**

Contrapositive ($\sim q \rightarrow \sim p$): **If I do not go Saturday, then I will not get paid.**

29. Translations for the conditional are located on the bottom of p. 77 in your text.

If it is a triangle, then it is a polygon.

31. Here are two possible phrasings (among several):

(i) **If one is a good person, then he/she will go to heaven.**

(ii) **If you are a good person, then you will go to heaven.**

33. **If we make proper use of those means which the God of Nature has placed in our power, then we are not weak.**

35. **If it is work, then it is noble.**

37. Both the antecedent, $5 + 10 = 16$, and the consequent, $15 - 10 = 3$, are false. Since F→F is true, this conditional statement is **true**.
Note: The truth-value of this statement's consequent is irrelevant, since any time a conditional's antecedent is false the conditional itself is automatically true. That is, F→F and F→T are both true.

39. This problem provides us with an opportunity to make use of the note provided in the solution to Problem 37 above. Since the antecedent $(1 + 1 = 10)$ is false, the conditional is **true** — regardless of the truth value of the consequent. Of course, if you prefer to assess the truth value of "the moon is made of green cheese," this conditional would be a case of F→F, which is a true compound statement.

41. First, determine the truth values of p and q. In this problem, they are both true. Then, apply the same procedure as was used in Section 2.1 with Problems 31-38 and Problems 48-57.

a) $\sim p \vee q$

$\sim T \vee T$

$F \vee T$

$\boxed{T}$

b) $\sim p \wedge \sim q$

$\sim T \wedge \sim T$

$F \wedge F$

$\boxed{F}$

c) $\sim (p \wedge q)$

$\sim (T \wedge T)$

$\sim T$

$\boxed{F}$

43. With p false and q true, we find:

a) $(p \wedge \sim q) \wedge p$

$(F \wedge \sim T) \wedge F$

$(F \wedge F) \wedge F$

$F \wedge F$

$\boxed{F}$

b) $p \vee (p \to q)$

$F \vee (F \to T)$

$F \vee T$

$\boxed{T}$

c) $(p \wedge q) \to p$

$(F \wedge T) \to F$

$F \to F$

$\boxed{T}$

45. $\boxed{(q \wedge \sim d) \to n}$ where q: The qualifying person is a child

d: The child is your dependent

n: You enter your child's name

Please take special note of how we phrased statement n. We avoided the words "enter this child's name" because such wording would communicate a command, and recall (from Section 2.1) that commands are not statements since they cannot be classified as true or false.

47. $\boxed{(b \vee c) \to n}$ where b: The amount on line 32 is \$86,025

c: The amount on line 32 is less than \$\$86,025

n: You multiply the total number of exemptions claimed on line 6e by \$2,500

Two comments are worthy of mention. First, similar to our note in the solution to Problem 45 above, please observe how we avoided phrasing statement n as a command. Second, some may choose to view the antecedent "... less than or equal to..." as a simple statement, but strictly speaking, the word "or" suggests the need for a compound antecedent (specifically, a disjunction).

49. $\boxed{(s \vee d) \to i}$ where s: You are a student

d: You are a disabled person

i: You see line 6 of the instructions.

Once again, as was explained in the solutions to Problems 45 and 47 above, we remind you to avoid the use of command-type phrasings.

51. Referring back to Example 5 (textbook pp. 76-77), assume statement w is false, while d, c, and b are each true. The compound statement's truth value is:

$$(\sim d \wedge c) \rightarrow (w \wedge b)$$
$$(\sim \text{T} \wedge \text{T}) \rightarrow (\text{F} \wedge \text{T})$$
$$(\text{F} \wedge \text{T}) \rightarrow \text{F}$$
$$\text{F} \rightarrow \text{F}$$
$$\boxed{\text{T}}$$

53. First, spend a moment interpreting the given statement; make sure you know who would (and wouldn't) be qualified for the position. After you have done this, rephrase the statement — without, of course, altering its meaning — to suit the symbolic forms provided in each of parts (a), (b), and (c):

a) "$a \rightarrow$ __" suggests the need to rephrase the statement as "If you are an applicant for the position, then you have a two-year college degree or five years of experience in the field." Therefore, we write $\boxed{\mathbf{a \rightarrow (e \vee f)}}$.

b) "$(a \wedge e) \rightarrow$ _" suggests a rephrasing of "If you are an applicant for the position and you have a two-year college degree in drafting, then you are qualified for the position." Consequently, we write $\boxed{\mathbf{(a \wedge e) \rightarrow q}}$.

c) "$(a \wedge f) \rightarrow$ __" leads us to a rephrasing of "If you are an applicant for the position and you have five years of experience in the field, then you are qualified for the position." Thus, we write $\boxed{\mathbf{(a \wedge f) \rightarrow q}}$.

55. Since the problem calls for the symbolic form "_$\rightarrow q$," we first need to rewrite the compound statement using the if-then connective with the consequent being "you qualify for the special fare." Both of the following phrasings accomplish this:

(i) "If you fly on Monday, Tuesday, Wednesday, or Thursday and you stay over a Saturday evening, then you qualify for the special rate."

(ii) "If you stay over a Saturday evening and you fly on Monday, Tuesday, Wednesday, or Thursday, then you qualify for the special rate."

Their symbolic representations are respectively:

$$\boxed{\mathbf{[(m \vee t \vee w \vee h) \wedge s] \rightarrow q}} \text{ or } \boxed{\mathbf{[s \wedge (m \vee t \vee w \vee h)] \rightarrow q}}$$

57. Since the problem calls for the symbolic form "$t \rightarrow$__," we first rewrite the compound statement using the if-then connective with the antecedent "you are a tenant." The following phrasing accomplishes this:

"If you are a tenant, then you lease the premises for 12 months beginning on September 1 and at a monthly rental charge of $800."

Hopefully, the consequent suggests to you a tenant must do <u>all</u> three of the following: (i) lease the premises for 12 months, (ii) begin on September 1, <u>and</u> (iii) pay $800 a month. Therefore, the consequent needs to be a three-part conjunction, and we write the conditional statement as:

$$\boxed{t \rightarrow (m \wedge s \wedge p)}$$

Please note that the order of the simple statements m, s, and p within the parentheses may vary.

59. This question is self-contradicting. We call such self-contradictions ***paradoxes***. That is, if Apollo can do anything, then he is capable of making anything, including an object he cannot lift. At the same time, if he can do anything, he can lift anything — even any object he makes. Therefore, this question is unanswerable since it gives rise to a declarative sentence that is neither true nor false.

Problem Set 2.3

1. **Answers vary**. The biconditional is defined on page 81 of your text.

3. **Answers vary**. The symbol ↔ denotes the biconditional, whereas the symbol ⇔ represents a logical equivalence (i.e., a biconditional that is also a tautology). The author of your textbook begins his presentation of this topic on p. 82.

5. **Answers vary**.

7.

p	q	p ∨ q	~p	q→~p	(p ∨ q) ∧ (q→~p)
T	T	T	F	F	**F**
T	F	T	F	T	**T**
F	T	T	T	T	**T**
F	F	F	T	T	**F**

Since the final column of the truth table does not consist entirely of T's, the statement is **not a tautology**.

9.

p	q	~q	~q→p	(~q→p)→q
T	T	F	T	**T**
T	F	T	T	**F**
F	T	F	T	**T**
F	F	T	F	**T**

Since the final column of the truth table does not consist entirely of T's, the statement is **not a tautology**.

11. The truth table below establishes the fact that $p \to q$ is logically equivalent to the disjunction $\sim p \vee q$. Hence, since the final column of the truth table consists entirely of T's, we answer: **yes, the statement is a tautology**.

p	q	p→q	~p	~p ∨ q	(p→q) ↔ (~p ∨ q)
T	T	T	F	T	**T**
T	F	F	F	F	**T**
F	T	T	T	T	**T**
F	F	T	T	T	**T**

Note: You would be well served to compare this problem with Example 3 (on textbook p. 82) and the role De Morgan's laws plays in the relationship between them.

13.

p	q	~p	~q	p ∨ ~q	~p ∨ q	(p ∨ ~q)↔(~p ∨ q)
T	T	F	F	T	T	**T**
T	F	F	T	T	F	**F**
F	T	T	F	F	T	**F**
F	F	T	T	T	T	**T**

No, not a tautology because the final column of the truth table does not consist entirely of T's.

15. In Problems 14-17, the task of "verifying" definitions consists of constructing a truth table and checking its final truth values with those provided in textbook Table 2.9 (p. 85). Additionally, we wish to take this opportunity to further enrich your understanding of the operator "neither p nor q." First, though, here is the truth table:

p	q	p ∨ q	~(p ∨ q)
T	T	T	**F**
T	F	T	**F**
F	T	T	**F**
F	F	F	**T**

This truth table agrees with the truth values provided in Table 2.9. Beyond checking this truth table (which, we realize, is all this problem's directions require), it is our hope that you have determined why each of the operators in Table 2.9 is defined in the manner it is; for instance, does it make sense to you to represent the statement "neither p nor q" by $\sim(p \vee q)$? If not, please keep reading. The statement "neither p nor q" is equivalent to "not p and not q," which translates to $\sim p \wedge \sim q$. From our previous work with DeMorgan's Laws (textbook page 83), we know that $\sim p \wedge \sim q$ is equivalent to $\sim(p \vee q)$. Therefore, one can choose to define "neither p nor q" as either one of these two symbolic statements. We encourage you to explore similarly the other statements included in Table 2.9.

17. Since the following truth table agrees with the truth values (in Table 2.9, page 85) for "no p is q," this operator's definition is verified.

p	q	~q	p→~q
T	T	F	**F**
T	F	T	**T**
F	T	F	**T**
F	F	T	**T**

In a manner similar to what we demonstrated in our solution to Problem 15, we invite you to spend some time exploring why the words "no p is q" are defined to be equivalent to the symbolic statement $p \to {\sim}q$.

19. **$\sim p \to \sim h$** where h: I will buy a new house

p: All provisions of the sale are clearly understood

Please recall, from textbook Table 2.9 (p. 85), the statement "a unless b" translates to ${\sim}b \to a$.

21. Using the definition provided in textbook Table 2.9, we write:

$(r \wedge s) \wedge (s \to r)$ where r: I am obligated to pay the rent

s: I signed the contract

23. **$m \to \sim i$** where m: It is a man

i: It is an island

Please observe that the pronoun "it" is a wise choice here, since it identifies an island equally as well as it does a man. Other pronouns (e.g., "he") tend to contribute to awkward sentence constructions.

25. First, recall that the contracted word " 'cause " is an informal use of the word "because." Second, in translating this statement into symbols, we need to be careful that each of our simple statements is a proper "statement" (i.e., a sentence that is either true or false). In particular, the words "be nice to people on your way up" convey a command, which we know (from our work in Section 2.1) is not a statement. Therefore, we rephrase this Jimmy Durante quotation and then translate it into symbols as follows:

"You are nice to people on your way up because you'll meet them on your way down."

$(n \wedge m) \wedge (m \to n)$ where n: You are nice to people on your way up

m: You will meet people on your way down

27. **$f \to r$** where f: The majority, by mere force of numbers, deprives a minority of a clearly written constitutional right

r: Revolution is justified

29. Let p: The cherries have turned red

q: The cherries are ready to be picked

The given if-then statement translates into the symbols $p \to q$, and its equivalent symbolic form $\sim p \vee q$ translates into the words:

The cherries have not turned red or they are ready to be picked.

31. The disjunction "Melissa will not watch Jay Leno or the NBC late-night orchestra" translates into the following symbols:

$\sim j \vee o$ where j: Melissa watches Jay Leno

o: Melissa watches the NBC late-night orchestra

The statement $\sim j \vee o$ is equivalent to $j \to o$, which translates into the words:

If Melissa watches Jay Leno, then she watches the NBC late-night orchestra.

33. The disjunction "The money is available or I will not take my vacation" translates into the following symbols:

$m \vee \sim v$ where m: The money is available

v: I will take my vacation

The statement $m \vee \sim v$ is equivalent to $\sim m \to \sim v$, which translates into the words:

If the money is not available, then I will not take my vacation.

35. The tautology shown in the truth table below establishes that $p \Leftrightarrow \sim(\sim p)$.

p	~p	~(~p)	p ↔ ~(~p)
T	F	T	**T**
F	T	F	**T**

37.

p	q	p∨q	~(p∨q)	~p	~q	~p∧~q	~(p∨q) ↔ (~p∧~q)
T	T	T	F	F	F	F	**T**
T	F	T	F	F	T	F	**T**
F	T	T	F	T	F	F	**T**
F	F	F	T	T	T	T	**T**

Thus, since $\sim(p \vee q) \leftrightarrow \sim p \wedge \sim q$ is a tautology, $\sim(p \vee q) \Leftrightarrow \sim p \wedge \sim q$.

39. The following truth table proves that $(p \to q) \Leftrightarrow (\sim p \vee q)$.

p	q	p→q	~p	~p∨q	(p→q) ↔ (~p∨q)
T	T	T	F	T	**T**
T	F	F	F	F	**T**
F	T	T	T	T	**T**
F	F	T	T	T	**T**

41. **$p \wedge q$** Here are two approaches:

I. On textbook p. 84 of your textbook, the negation of $b \to c$ is shown to be $b \wedge \sim c$; that is, $\sim(b \to c) \Leftrightarrow b \wedge \sim c$. Substituting p for b and $\sim q$ for c, the negation of $p \to \sim q$ must then be $p \wedge \sim(\sim q)$, which simplifies to $p \wedge q$.

II. If you are having difficulty remembering the negation of a conditional statement, you may find it helpful to make use of the approach demonstrated earlier in this solution manual for finding negations (see the solution to Problem 11 in Section 2.1). Specifically, assume the conditional $b \to c$ is false, and determine what must be true. We know that a conditional statement is false only in the case T→F. Therefore, if $b \to c$ is false, we know "b is true and c is false." The words enclosed in quotation marks translate into the symbols $b \wedge \sim c$. Therefore, the negation of $b \to c$ is $b \wedge \sim c$.

It is not difficult to adapt this strategy to the statement $p \to \sim q$. For this statement to be false, p must be true and $\sim q$ must be false. If $\sim q$ is false, then q is true. Therefore, for $p \to \sim q$ to be false, p must be true and q must be true. Thus, the negation of $p \to \sim q$ is $p \wedge q$.

[Suggestion: If you still encounter difficulty, you may find it helpful to work first with word statements and then gradually transition back to symbols.]

43. Using either of the two approaches shown above solution to Problem 41, we obtain $\sim(\sim p \to \sim q) \Leftrightarrow$ **$(\sim p \wedge q)$**.

45. Here are two strategies:

I. The given statement translates into the symbols $b \vee s$, where b and s respectively represent "Jane went to the basketball game" and "Jane went to the soccer game." Finding its negation amounts to finding $\sim(b \vee s)$. From our previous work with DeMorgan's laws (p. 83 in text), we know $\sim(b \vee s) \Leftrightarrow \sim b \wedge \sim s$. Translating $\sim b \wedge \sim s$ back into words, we obtain the negation: **Jane did not go to the basketball game and she did not go to the soccer game.**

II. Apply the tactic previously explained in this solution manual. Refer to the solutions to Problem 11 in Section 2.1 and to Problem 41 (in Section 2.3) above. First, assume the given statement "Jane went to the basketball game or to the soccer game" is false. Second, ask yourself what you can conclude from this assumption. In other words, when is an "or" statement (a disjunction) false? We know it is false only when both of its component statements are false. Therefore, the negation is "Jane did not go to the basketball game and she did not go to the soccer game."

47. This solution to this problem is very similar to that of Problem 45. Once again, consider these two approaches:

I. Apply DeMorgan's laws, namely $\sim(p \wedge q) \Leftrightarrow \sim p \vee \sim q$.

II. First, since a statement and its negation must have opposite truth values, assume the given statement is false and then determine what you can conclude, with certainty, to be true.

Both of these approaches yield the following negation:

Sally is on time or she did not miss the boat.

49. Once again, here are two strategies:

I. Translate the statement into symbols and apply the logical equivalence $\sim(p \rightarrow q) \Leftrightarrow p \wedge \sim q$.

II. Since a statement and its negation must have opposite truth values, assume the given conditional is false and then determine what must be true. Recall that the only case in which a conditional is false is when its antecedent is true and its consequent is false (i.e., $p \wedge \sim q$).

Both of these approaches yield this negation:

You're out of Schlitz and you have beer.

[Note: Specific word choice may vary.]

51. Using either of the two approaches demonstrated in each of the solutions above, we obtain a negation of: **x – 5 = 4 and x ≠ 1**

53. First, we rewrite the statement, using proper if-then form: If $x = 1$ and $y = 2$, then $2x + 3y = 8$. Recalling $\sim(p \rightarrow q) \Leftrightarrow p \wedge \sim q$, we know the negation of a conditional is a conjunction consisting of the conditional's antecedent ($x = 1$ and $y = 2$) and the negation of its consequent ($2x + 3y \neq 8$). Therefore, the negation is:

x = 1 and y = 2, and 2x + 3y ≠ 8

55. To qualify for this 40% discount, a customer must satisfy all five of the following requirements:

(i) Purchase the ticket between January 5 and February 15;
(ii) Fly round trip between February 20 and May 3;
(iii) Depart on a Monday, Tuesday, or Wednesday;
(iv) Return on a Tuesday, Wednesday, or Thursday;
(v) Stay over a Saturday night.

In other words, "if" all five of these conditions are met, "then" the 40% discount will be awarded. Therefore, we need to translate this statement into a conditional in which the antecedent is a five-part conjunction and the consequent is the 40% discount. Additionally, note that requirements (iii) and (iv) above include the word "or," which suggests the need for disjunctions.

$$[d \wedge f \wedge (m \vee t \vee w) \wedge (h \vee i \vee j) \wedge s] \rightarrow e$$

where d: You purchase your ticket between January 5 and February 15;

f: You fly round trip between February 20 and May 3;
m: You depart on Monday
t: You depart on Tuesday
w: You depart on Wednesday;
h: You return on Tuesday;
i: You return on Wednesday;
j: You return on Thursday;
s: You stay over a Saturday night;
e: You obtain 40% off a regular fare

57. Recall from textbook p. 85 that the author defined the statement "p unless q" as $\sim q \to p$. In this problem, q is the statement "written permission from the landlord is obtained." The symbolic translation $\sim q \to p$ indicates the need to rewrite the original contractual language in a manner comparable to "If written permission from the landlord is not obtained, then the tenant shall not let and shall not sublet the whole or any portion of the premises to anyone for any purpose whatsoever." In symbols, this translates to:

$$\boxed{\sim p \to [\sim(l \lor m) \land \sim(s \lor t)]}$$

where p: Permission is obtained;
l: The tenant lets the premises;
m: The tenant lets a portion of the premises;
s: The tenant sublets the premises;
t: The tenant sublets a portion of the premises.

It should be noted, by either applying DeMorgan's laws or phrasing the if-then contract language above differently, the statement also could have been written as:

$$\boxed{\sim p \to [(\sim l \land \sim m) \land (\sim s \land \sim t)]}$$
or
$$\boxed{\sim p \to \sim[(l \lor m) \lor (s \lor t)]}$$

59. Apply the definition of the statement "either p or q" given in textbook Table 2.9 (see p. 85).

Let a: Alfie is afraid to go
b: Bogie lied
c: Clyde lied

This translates to: Either $\sim a$ or $(b \land c)$

The symbolic form of the statement is: $\boxed{[(\sim a) \lor (b \land c)] \land \sim[(\sim a) \land (b \land c)]}$

Problem Set 2.4

1. Also known as modus ponens, the law of detachment, and assuming the antecedent, ***direct reasoning*** is a type of argument that consists of two premises and a conclusion presented in the argument form $[(p \to q) \land p] \to q$. Premises — sometimes called hypotheses — are statements assumed to be true. Refer to textbook pages 87-88 for more information (along with accompanying examples) pertaining to direct reasoning.

3. As explained on textbook p. 89, ***transitive reasoning*** is an argument type consisting of two premises and a conclusion in the form $[(p \to q) \land (q \to r)] \to (p \to r)$. Your understanding of transitive reasoning may be enhanced by (cautiously) relating it to the transitive property of equality (namely, if $a = b$ and $b = c$, then $a = c$), which is encountered in algebra classes. However, please heed this warning: if you choose to associate transitive reasoning with the transitive property of equality, you need to realize there is one very important difference. Order in the case of equality does not matter (i.e., $a = b$ is equivalent to $b = a$), but the order of a conditional's antecedent and consequent very much matters (i.e., $p \to q$ is not equivalent to its converse $q \to p$). The fallacy of the false chain pattern (found on textbook p. 93) illustrates the importance of this warning.

5. The term ***syllogism*** is defined on p. 87 in your text. Your textbook considers three types of syllogisms: direct reasoning, indirect reasoning, and transitive reasoning.

7. To prove that $[(p \to q) \land (p \to r)] \to (q \to r)$ is an invalid argument, one needs to show that this statement fails to be a tautology. The final column (specifically, the sixth row) of the following truth table reveals this:

p	q	r	p→q	p→r	(p→q) ∧(p→r)	q→r	[(p→q)∧(p→r)]→(q→r)
T	T	T	T	T	T	T	**T**
T	T	F	T	F	F	F	**T**
T	F	T	F	T	F	T	**T**
T	F	F	F	F	F	T	**T**
F	T	T	T	T	T	T	**T**
F	T	F	T	T	T	F	**F**
F	F	T	T	T	T	T	**T**
F	F	F	T	T	T	T	**T**

As identified on textbook p. 93, this fallacy is known as the **false chain pattern**.

9. First, recall that Pólya's problem-solving method can be found on p. 7 of your text. Second, remember that this problem asks you to rewrite ("in your own words") the given detective problem-solving steps using the language of Pólya's problem-solving method. With this in mind, what follows below is one attempt at performing this rewrite; of course, **wording may vary**.

DETECTIVE LANGUAGE	***PROBLEM-SOLVING LANGUAGE***
Understand the case *What are you looking for?*	***Understand the problem*** ***What is the unknown?***
Investigate the case *Have you solved a similar case?* *What are the facts?*	***Devise a plan*** ***Do you know a related problem?*** ***Can you simplify the problem?***
Analyze the facts/data *What information is important?* *What information is not important?* *What pieces of information do not seem to fit together logically?* *What data are inconsistent with the given information?*	***Carry out the plan*** ***What information is important?*** ***What information is not important?*** ***What pieces of information fit together logically?*** ***Which information is consistent with the given information?***
Reexamine the facts *Do the facts support the solution?* *Can we obtain a conviction?*	***Look back*** ***Examine the solution obtained.*** ***Does it make sense?***

11. a) **valid**; by constructing truth table and showing that $[(p\vee q)\wedge\sim p]\rightarrow q$ is a tautology. This form of reasoning is sometimes called **the law of the excluded middle**.

b) **valid; law of the excluded middle**

13. a) **valid; indirect reasoning**

b) **valid; direct reasoning** (also known as modus ponens, law of detachment, and assuming the antecedent)

15. **valid; indirect reasoning**

If we let p represent the statement "a^2 is even" and q represent "a is even," then the given argument translates as follows:

$p\rightarrow q$	*"If a^2 is even, then a must be even"*
$\underline{\sim q}$	*<u>"a is odd" (i.e., a is not even)</u>*
$\therefore \sim p$	$\therefore$ *"a^2 is odd" (i.e., a^2 is not even)*

This demonstrates clearly that the argument is of the indirect-reasoning type (textbook p. 88) and, accordingly, must be valid.

17. **valid; indirect reasoning** As was mentioned on textbook p. 77 and demonstrated in Example 5 (p. 91), please recall that the statement "all cats are animals" is equivalent to the proper conditional "if it is a cat, then it is an animal." Rewriting this premise in if-then form should make it easier for you to recognize this argument as a correct application of indirect reasoning. If not, go one step further by translating the argument into symbols (as was demonstrated above in our solution to Problem 15) and then examine its form.

19. **valid; direct reasoning**

21. **invalid; fallacy of the inverse**

23. **invalid; fallacy of the inverse**

25. **valid; indirect reasoning**

27. **invalid; fallacy of the converse**

29. **valid; direct reasoning**

31. **invalid; a conditional statement is not equivalent to its converse**. Recall that a conditional is equivalent to its contrapositive, but not its converse nor its inverse.

33. **valid; indirect reasoning**

35. **valid; indirect reasoning** In a manner similar to our solution for Problem 17 in this section, you may find it helpful to translate the premise "no students are enthusiastic" into if-then form: "If a person is a student, then that person is not enthusiastic." Doing this should enable you to recognize the argument as a proper use of indirect reasoning.

37. **If you learn mathematics, then you understand human nature.** This conclusion is based upon **transitive reasoning** (as introduced on textbook page 89).

39. **We do not go to the concert *(indirect reasoning)***

41. **$b = 0$ *(law of the excluded middle)*** For your reference, the law of the excluded middle is introduced on page 65 of your text. Please recall that we always assume in these arguments that all of the premises are true. Therefore, the statements "$a = 0$ or $b = 0$" and "$a \neq 0$" must both be true. For the disjunction to be true, at least one of its two component statements must be true. Since we are told the first component is not

true (namely, $a \neq 0$), it must be the case that the second component is true. Thus, b must equal zero.

43. **We do not interfere with the publication of false information** ***(indirect reasoning)***

45. **I will not eat that piece of pie** ***(indirect reasoning)***

47. **We will go to Europe**. There are two ways to approach this argument, and therefore, there are two reasons for this conclusion:

I. *Law of Transitivity and Direct Reasoning* – Please recall that we can rearrange the order of the three original premises as our needs require; for example, the first two premises may be viewed as follows:

If we are ingenious, we will win first prize
If we win first prize, we will go to Europe
∴ If we are ingenious, we will go to Europe *(transitive reasoning)*

We then consider this conclusion "if we are ingenious, we will go to Europe" and the given premise "we are ingenious." Applying direct reasoning to these two statements yields the conclusion "we will go to Europe."

II. *Direct Reasoning Applied Twice* – Given the premises "if we are ingenious, we will win first prize" and "we are ingenious," we can conclude (via direct reasoning) "we will win first prize." Then we apply direct reasoning again. Specifically, we apply it to the conclusion "we will win first prize" and the third of our three original premises "if we win first prize, then we will go to Europe." This reasoning enables us to conclude we will go to Europe.

Alternatively, you may find it easier to first translate the statements into symbols, then apply the types of reasoning previously discussed, and lastly, convert your symbolically-expressed conclusion back into words. We will demonstrate this approach in our next solution (Problem 49).

49. First, we rewrite the three given premises, wording them in if-then form. This makes it easier for us to translate them into symbols as follows:

If this is a baby, then it is an illogical creature ($b \rightarrow l$)

If it is a despised creature, then it is not one who can manage a crocodile ($d \rightarrow \sim c$)

If it is an illogical creature, then it is despised ($l \rightarrow d$)

where b: This is a baby
l: This is an illogical creature
d: This is a despised creature
c: This is one who can manage a crocodile

Then we obtain:

$$\begin{array}{ll} b \to l & \\ \underline{l \to d} & \\ \therefore b \to d & \textit{(transitive reasoning)} \\ \underline{d \to \sim c} & \\ \therefore \boldsymbol{b \to \sim c} & \textit{(transitive reasoning)} \end{array}$$

Therefore, the conclusion is: If it is a baby, then it cannot manage a crocodile. Stated without if-then form, we conclude that **babies cannot manage crocodiles**.

51. Like Problem 49 above, we first express the premises in if-then form and proceed to translate them into symbols.

If it is a duck, then it cannot waltz ($d \to \sim w$)
If it is an officer, then it waltzes ($f \to w$)
If it is my poultry, then it is a duck ($p \to d$)

where d: It is a duck
w: It waltzes
p: It is my poultry

This leads us to:

$$\begin{array}{ll} p \to d & \\ \underline{d \to \sim w} & \\ \therefore p \to \sim w & \textit{(transitive reasoning)} \\ \underline{\sim w \to \sim f} & \textit{(contrapositive of the premise } f \to w\textit{)} \\ \therefore \boldsymbol{p \to \sim f} & \textit{(transitive reasoning)} \end{array}$$

Thus, our conclusion is: If it is my poultry, then it is not an officer. Stated in the language of the original premises, this conclusion may be worded in several ways. Here are two possible phrasings: **None of my poultry are officers**

or **All my poultry are not officers**

Two additional comments should be made. First, recall (from the directions for Problems 37-54) that our conclusions for these arguments are to be based upon "all" the premises. Consequently, if your conclusion is different from that which is stated above, check whether you utilized all three premises. Second, by choosing to piece together the premises (and their contrapositives) differently than was shown above, your conclusion in symbols may have been $f \to \sim p$. This, of course, is equivalent to our $p \to \sim f$, since they are contrapositives of one another.

53. **Airsecond Aircraft Company suffers financial setbacks**

In our solutions to the preceding two problems (namely, 49 and 51), we chose to translate the premises into symbols. Doing so made it easier to determine a conclusion. While once again we could demonstrate this approach with this particular problem (which, by the way, we encourage you to attempt on your own), let us instead illustrate a different strategy, one that resembles our solution for Problem 47.

First, please number the three premises in the order in which they were presented in the text. Notice that the first two premises are conditional statements and the third is a simple statement. Applying ***direct reasoning*** to the first and third premises leads to the conclusion that Senator Firstair stands to earn a great deal of money. Next, we make use of this conclusion and the second premise (i.e., "If Airsecond Aircraft Company does not…"). Either of two approaches may then be used:

I. By ***indirect reasoning***, we can conclude Airsecond Aircraft Company suffers financial setbacks.

II. By first replacing this second premise with its contrapositive and then applying direct reasoning to both the resulting contrapositive and the conclusion regarding Senator Firstair standing to earn a great deal of money, we again conclude Airsecond Aircraft Company suffers financial setbacks.

55. **Because the building fuses for the entire building were blown, the janitor could not have taken the elevator.**

57. We first rewrite (in if-then form) two of the three premises and translate all three into symbols:

I. $b \to s$: If one really appreciates Beethoven, then one keeps silent while the *Moonlight Sonata* is being played

II. h: Guinea pigs are hopelessly ignorant of music

III. $h \to {\sim}s$: If one is hopelessly ignorant of music, then one does not keep silent while the *Moonlight Sonata* is being played

where b: One really appreciates Beethoven
s: One keeps silent while the *Moonlight Sonata* is being played
h: One is hopelessly ignorant of music

This yields:

$$\begin{array}{ll} h & \\ \underline{h \to {\sim}s} & \\ \therefore {\sim}s & \textit{(direct reasoning)} \\ \quad \underline{b \to s} & \\ \quad {\sim}b & \textit{(indirect reasoning)} \end{array}$$

Lastly, we translate this final conclusion, ${\sim}b$, back into words — making sure, of course, that it is applied to the creatures in question (namely, guinea pigs). Therefore, we conclude: **Guinea pigs do not appreciate Beethoven**.

59. We begin by numbering the five premises, carefully rephrasing (and simplifying) them, and translating each into symbols. You may find Table 2.9 (page 85) in your textbook to be of some use in this process.

(1) ${\sim}g \to u$: If I do not grumble, then I understand the problem

(2) ${\sim}o$: These problems are not arranged in regular order

(3) $e \to \sim h$: If a problem is easy, then it does not give me a headache

(4) $\sim o \to \sim u$: If a problem is not arranged in regular order, then I do not understand it

(5) $\sim h \to \sim g$: If a problem does not give me a headache, then I do not grumble

We then use premises (2) and (4) to reach one conclusion and separately use (3), (5), and (1) respectively to reach another:

(2) $\sim o$
(4) $\underline{\sim o \to \sim u}$
$\therefore \sim u$ *(direct reasoning)*

(3) $e \to \sim h$
(5) $\underline{\sim h \to \sim g}$
$\therefore e \to \sim g$ *(transitive reasoning)*
(1) $\underline{\sim g \to u}$
$\therefore e \to u$ *(transitive reasoning)*

Putting these two statements together gives us our final conclusion:

$\sim u$
$\underline{e \to u}$
$\therefore \sim e$ *(indirect reasoning)*

Hence, **these problems are not easy**.

Problem Set 2.5

1. First, since all three have raised their hands, each of them immediately would have concluded at least two of the three hats are black. Now, from the original statement of the problem, it is important to take note of the words "after a few moments." That is, if it had been the case that Moe's hat were white, both Harry and Larry would have immediately (that is, without a momentary delay) deduced their respective hats as having been black. However, since Harry and Larry fail to immediately reach a solution, Moe knows his hat also must be black.

You may find it useful to construct a truth table consisting of all the possible hat-color arrangements. You will find there to be eight possible cases, of which those with two or more white hats immediately will be discarded as impossibilities.

3. **White**. Since the only way for the bear to be both due south of the fox and simultaneously southeast of the fox is for them to be at the North Pole. Therefore, it is a polar bear.

5. No matter which two flowers are selected, it is a certainty that at least one of the two will be red. So, there can be only **one yellow** flower, and consequently the remaining **49 are red**.

7. Before you read this solution, you may find it helpful to construct a table in which you keep track of each suspect's statements and, as we deduce them, each statement's truth value. Such a table might start out like this:

	Statement I	*Statement II*	*Statement III*	*Statement IV*
Otto				
Curly				
Slim				
Mickey				
The Kid				

To solve the crime, we begin by eliminating Otto and The Kid as suspects:

- *Otto:* Assume for the moment that Otto did it. If this were so, it would follow that the first three of his four statements (at the very least) were false. But this can't happen since we know each suspect made only one false statement — not three or four. Therefore, Otto could not have killed Kiltraine.
- *The Kid:* Assume for the moment that The Kid did it. If this were so, it would follow that his first statement ("I did not kill Kiltraine") and his fourth statement ("Otto lied when he said I am guilty") were false. However, this would contradict the premise that only one of The Kid's four statements is false. Thus, The Kid couldn't have done it.

Now, since The Kid is not the murderer, Otto's third statement implicating The Kid must be false. Therefore, since each suspect is limited to only one false statement, each of Otto's other three statements must have been true, including his having said "Mickey and I are pals." The fact that Mickey and Otto are pals leads us to conclude that Mickey's one false statement must have been the one in which he said "I never saw Otto before." Since that is a lie, Mickey's other three statements must be true, including the words "I did not kill Kiltraine." Thus, we can eliminate Mickey as a suspect.

Then, we turn our attention to trying to determine which of Slim's four statements is false. From the statement of the problem, we know his second statement (citing the date of the murder) is true. Also, Slim's fourth statement regarding Otto's whereabouts must be true because Otto himself said he was in Chicago and we have already concluded that Otto's one false statement pertained to The Kid being guilty. This leaves us with needing to determine which of Slim's remaining two statements, his first or third, is false. To make this determination, we will make certain assumptions and check to see what contradictions (if any) arise. So, let us start by assuming for the moment that Slim's third statement ("One of us is guilty") is false. This leads us to two conclusions: (i) he and Curly, who is the only other remaining suspect, together must have killed Kiltraine; and because only one statement per suspect can be false, (ii) Slim's first statement ("Curly lied when he said he never owned a revolver") must be true. However, these two conclusions lead to a contradiction. Namely, conclusion (i) implies that Curly's first statement ("I did not kill Kiltraine") must be false, and conclusion (ii) forces Curly's second statement ("I never owned a revolver in my life") to also be false. Since Curly cannot have two false statements, our assumption regarding the Slim's third statement ("One of us is guilty") must have been wrong. Therefore, Slim's third statement must be true. By process of elimination, we now know that Slim's one false statement must have been his first ("Curly lied when he said he never owned a revolver"). Note that, for the moment, Slim still remains a viable suspect.

Let us turn our attention to Curly. At this stage, we know either he or Slim (but not both) killed Kiltraine. As for Curly's four statements, we know (from Slim's lie regarding Curly) that Curly's second statement ("I never owned a revolver") must be true. Also, we previously concluded that Mickey's fourth statement is true. Accordingly, Curly's fourth statement ("I was in Detroit the night of March 3rd") must be true. Now, we examine Curly's third statement ("The Kid knows me"). Since Mickey's one false statement pertained to Otto, we know his second statement ("The Kid has never been in Pontiac") is true. Thus, the The Kid's second statement ("I have never been in Pontiac") is true. This conclusion (and our previous ones) means that The Kid's one false statement must have been "I never saw Curly before." Since this is false, Curly's third statement ("The Kid knows me") must be true. Then, since Curly's second, third, and fourth statements are all true, his first statement ("I did not kill Kiltraine") must be false. Therefore, **Curly committed the murder.**

Please note that your completed table should look like this:

	Statement I	*Statement II*	*Statement III*	*Statement IV*
Otto	T	T	F	T
Curly	F	T	T	T
Slim	F	T	T	T
Mickey	T	T	F	T
The Kid	T	T	F	T

9. Only one question is necessary: ask Connie, who falsely claimed to have the mixed bag, to pull out one fruit from her bag. This leads to two possible scenarios:

I. Suppose Connie pulls out a peach. This means, she must have the bag containing two peaches. Since all three have lied, we can then conclude that Alice (who falsely claimed to have two peaches) must have two plums, which leaves Betty with the mixed bag.

II. Suppose Connie pulls out a plum. Then, she must have the bag containing two plums. Consequently, Betty must have the two-peach bag, and Alice has the mixed bag.

Problem Set 2.6

1. The light is on when the circuit is "complete" (i.e., when there is no break in it).

3. These circuit symbols are introduced on textbook page 104.

a) **battery (or other power source)**

b) **light**

c) **switch**

5.

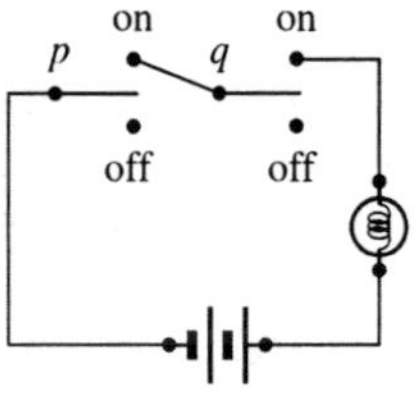

7.

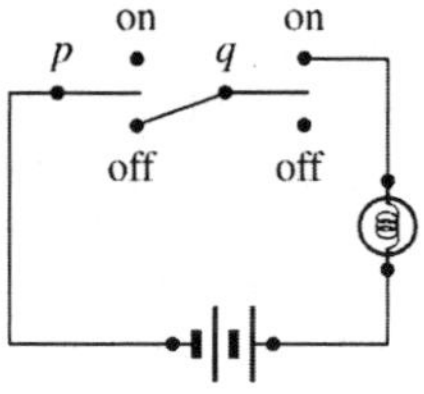

9.

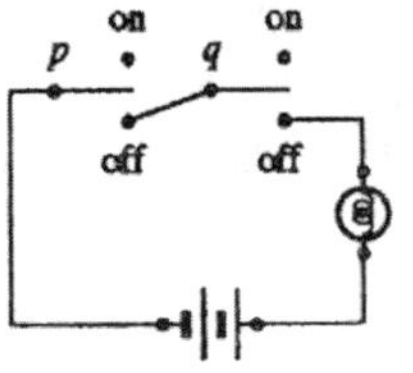

11.

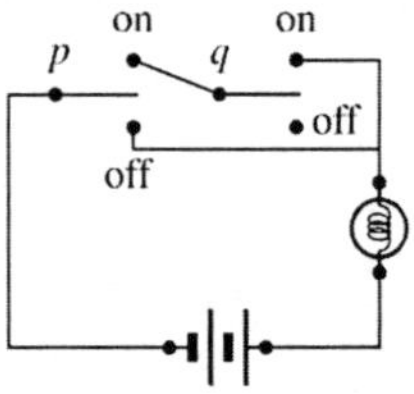

13.

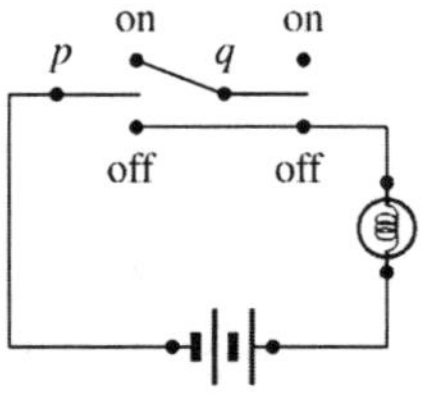

15.

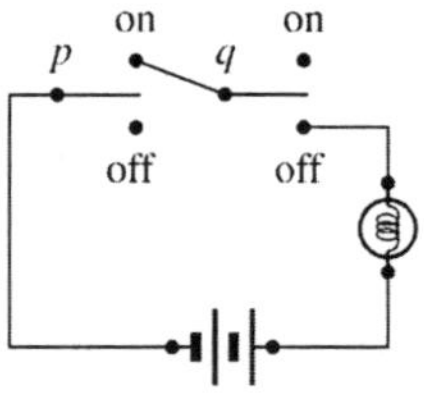

17.

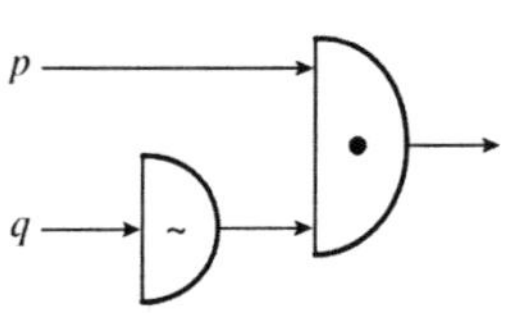

19.

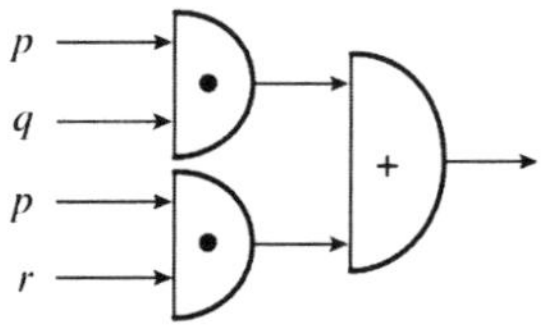

21.

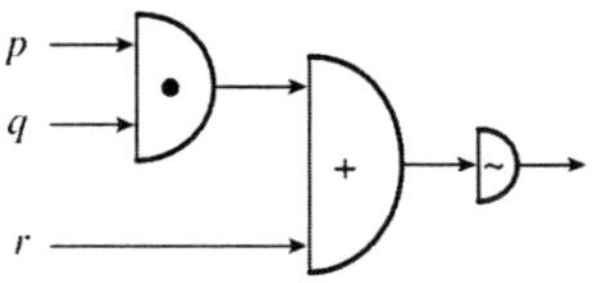

23.

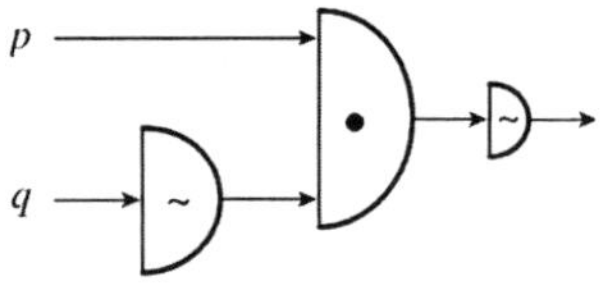

25. In solving Problem 24, we learned that $(p \to q) \Leftrightarrow (\sim p \vee q)$. So, it follows that $(\sim p \to q)$ is equivalent to $(p \vee q)$:

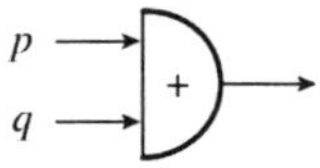

27. Recall that a conditional is equivalent to its contrapositive, so $(\sim q \to \sim p) \Leftrightarrow (p \to q)$. From Problem 24, we know $(p \to q) \Leftrightarrow (\sim p \vee q)$. Hence, $(\sim q \to \sim p) \Leftrightarrow (\sim p \vee q)$:

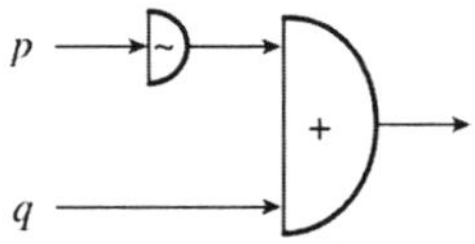

29. Let committee members Alfie, Bogie, and Clyde be represented by a, b, and c, respectively. The following circuit turns the light on for a yea majority and keeps it off otherwise.

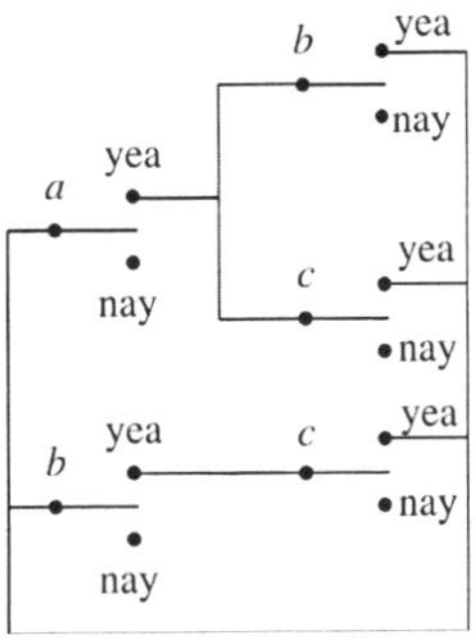

Chapter 2 Review Questions

1. a) This term is defined in Section 2.1 (page 64). A ***logical statement*** is a declarative statement that is either true or false.

b) Refer to the top of textbook page 82. A ***tautology*** is a statement in which the conclusion is equivalent to its premise. In terms of a truth table, a compound statement whose truth value is always true (regardless of the truth values of its component statements) is a tautology.

c) The author of your text introduces the ***law of contraposition*** on page 78. It is restated in symbols on page 83. In essence, this law establishes the fact that a conditional statement (i.e., $p \to q$) is logically equivalent to its contrapositive ($\sim q \to \sim p$).

2.

p	q	~p	p ∧ q	p ∨ q	p → q	p ↔ q
T	T	F	T	T	T	T
T	F	F	F	T	F	F
F	T	T	F	T	T	F
F	F	T	F	F	T	T

3.

p	q	p ∧ q	~(p ∧ q)
T	T	T	**F**
T	F	F	**T**
F	T	F	**T**
F	F	F	**T**

4.

p	q	~q	p ∨ ~q	~p	(p ∨ ~q) ∧ ~p	[(p ∨ ~q) ∧ ~p] → ~q
T	T	F	T	F	F	**T**
T	F	T	T	F	F	**T**
F	T	F	F	T	F	**T**
F	F	T	T	T	T	**T**

5.

p	q	r	p ∧ q	(p ∧ q) ∧ r	[(p ∧ q) ∧ r] → p
T	T	T	T	T	**T**
T	T	F	T	F	**T**
T	F	T	F	F	**T**
T	F	F	F	F	**T**
F	T	T	F	F	**T**
F	T	F	F	F	**T**
F	F	T	F	F	**T**
F	F	F	F	F	**T**

6.

p	q	p ∧ q	~(p ∧ q)	~p	~q	~p ∨ ~q	~(p ∧ q) ↔ (~p ∨ ~q)
T	T	T	F	F	F	F	**T**
T	F	F	T	F	T	T	**T**
F	T	F	T	T	F	T	**T**
F	F	F	T	T	T	T	**T**

The final outcome of this truth table should come as no surprise since the statement $\sim(p \wedge q) \Leftrightarrow \sim p \vee \sim q$ is a tautology, one with which you should be well familiar by now; it is one of De Morgan's laws.

7. The ***principle of direct reasoning*** is discussed on textbook pp. 87-88. In symbols, it can be expressed as follows:

$$\begin{array}{l} p \to q \\ \underline{p \quad\quad} \\ \therefore q \end{array}$$

To prove this principle, we need to show that the statement $[(p \to q) \wedge p] \to q$ is always true. The following truth table accomplishes this:

p	q	p →q	(p→q) ∧ p	[(p→q) ∧ p] → q
T	T	T	T	**T**
T	F	F	F	**T**
F	T	T	F	**T**
F	F	T	F	**T**

8. **Answers vary**. ***Indirect reasoning*** is defined, explained, and illustrated on textbook pp. 88-89. Here is one example:

If she is a strong reader, then the principle of indirect reasoning makes sense.
The principle of indirect reasoning does not make sense.
Therefore, she is not a strong reader.

9. **Answers vary**. Three of the more common fallacies are discussed in your text on pp. 92-93. Here is an example (of the fallacy of the converse):

If she is a strong reader, then the principle of indirect reasoning makes sense.
The principle of indirect reasoning makes sense.
Therefore, she is a strong reader.

10. **Yes, it is the principle of indirect reasoning**. In a manner comparable to our proof of the principle of direct reasoning (see the solution to Problem 7 above), we can formally prove the validity of this syllogism by constructing a truth table.

11. In the event you are still having difficulty with negations, you may find the three-step method previously explained in this solutions manual to be helpful. We first introduced this method in our solution to Problem 11 (Section 2.1) and further illustrated its application in our solutions to Problems 13, 15, 17, and 19. Here, once again, is our approach:

a) First, take a moment to visualize the statement "all birds have feathers." Envision every single bird on the planet having feathers (i.e., no feather-less birds exist). Second, pretend that this statement is false. Third, assuming that the claim "all birds have feathers" is a lie, what you can conclude must be true? If not all birds have feathers, then we can conclude there must be at least one bird on the planet who does not have feathers. Recall that the word "some" means "at least one." Thus, the negation of "all birds have feathers" is: **Some birds do not have feathers.**

b) If we assume to be false the statement "some apples are rotten," then we can conclude the following with certainty: **No apples are rotten.** If this is unsettling, you may find it helpful to double-check this answer by beginning with the statement "no apples are rotten" and asking yourself what its negation would be. Your answer should be "some apples are rotten." Remember that these statements must be negations of each other. So, your answer is correct only if it is reversible.

c) **Some cars have two wheels.**

d) **All smart people attend college.** Two notes are in order. First, the given statement "not all smart people attend college" is equivalent to "some smart people do not attend college." Second, please be aware of the importance of the placement of the word "not." That is, "not all smart people attend college" is a very different statement from "all smart people do not attend college."

e) Let's demonstrate two strategies:
- Recall (from textbook page 84) that the negation of the conditional $p \to q$ is $p \wedge \sim q$. Said differently, $\sim(p \to q) \Leftrightarrow p \wedge \sim q$. In this particular case, the statement "you go on Tuesday" is represented by p, and "you cannot win the lottery" can be expressed as q.
- In the event you find it difficult to remember the fact that $p \to q$ and $p \wedge \sim q$ are negations of each other, consider applying the three-step approach previously demonstrated in this manual. Specifically, since a statement and its negation must have opposite truth values, ask yourself what you would know to be true if you assume that the given conditional is false. Since the only case in which a conditional is false is when its antecedent is true and its consequent is false (i.e., $p \wedge \sim q$), we know that if the conditional is false then the statement "you go on Tuesday" is true while "you cannot win the lottery" must be false.

Both of these strategies yield the following negation:

"You go on Tuesday and you can win the lottery."

12. As we explain below, **all of the following statements are true**. Before reading our explanation, though, please make sure you know the answer to this very basic (but vital) question: when is a conditional statement false? If you are not sure, please review textbook pp. 74-76.

a) Because its antecedent is false (i.e., $111 + 1 \neq 1,000$), this conditional is true. Of course, the consequent is also false, but assessing its truth value is not essential.

b) Since the antecedent is a conjunction in which one of its component statements is false (namely, $6 + 2 = 10$), the antecedent is false. Therefore, regardless of the consequent's truth value, the conditional is true. Alternatively, you can view this in terms of symbols:

$$(\text{F} \wedge \text{T}) \to \text{T}$$
$$\text{F} \to \text{T}$$
$$\text{T}$$

c) This conditional's antecedent is a disjunction. Since at least one of its component statements is true (namely, $5 + 2 = 7$), the antecedent is true. Then, unlike statements (a) and (b), we must assess the truth value of the consequent. Of course, $7 + 2$ is equal to 9. Therefore, since both the antecedent and consequent are true, this conditional statement is true. In terms of symbols, this translates to:

$$(\text{F} \vee \text{T}) \to \text{T}$$
$$\text{T} \to \text{T}$$
$$\text{T}$$

d) This conditional statement has an antecedent that is false. Thus, regardless of the consequent's truth value, the conditional itself is true. We can express this in symbols as follows:

$$(T \wedge F) \rightarrow T$$
$$F \rightarrow T$$
$$T$$

e) Since the moon is not made of green cheese, the conditional is true (because its antecedent is false). That is, F → F is a true statement.

13. a) **If P is a prime number, then $P + 2$ is a prime number.**

b) Before translating this statement into its final verbal form, we first take a moment to assess the meaning of $(p \vee q) \wedge [\sim(p \wedge q)]$. One's initial reading of it might be something like this: "p or q and not p and q." This is equivalent to saying "p or q but not both p and q." However, in English, there is even a more concise way to say that: "either p or q." Therefore, our translation is:

Either P is a prime number or $P + 2$ is a prime number.

[Note: The translations for part (b) and other common statements are in textbook Table 2.9, p. 85.]

14. From your previous work with conditional statements (see the bottom of textbook p. 77), we know that the statement "all computers are incapable of self-direction" is equivalent to saying something like "if it is a computer, then it is not capable of self-direction." Of course, your specific word selection may differ from ours. Regardless, our symbolic forms should agree:

a) **p → ~q** where p: It is a computer
q: It is capable of self-direction

b) The contrapositive of $p \rightarrow \sim q$ is $\sim(\sim q) \rightarrow \sim p$ or **q → ~p**.

15. Let p: There are a finite number of primes
q: There is some natural number, greater than 1, that is not divisible by any prime

In terms of this choice of symbols, the argument translates as follows:

$$\begin{array}{l} p \rightarrow q \\ \underline{\sim q} \\ \therefore \sim p \end{array}$$

This valid argument is an illustration of **indirect reasoning**.

16. a)

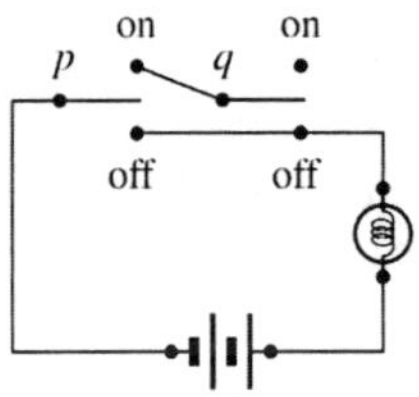

b)

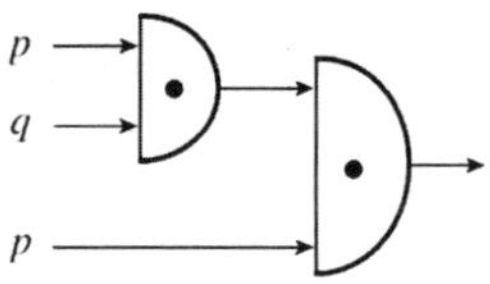

17. We begin by translating the three premises into symbols.

Let d: I attend to my duties

r: I am rewarded

z: I am lazy

We are given:

1st Premise: $d \to r$

2nd Premise: $z \to \sim r$

3rd Premise: z

By applying direct reasoning to the second and third premises, we can conclude:

$$\begin{array}{l} z \to \sim r \\ \underline{z \quad\quad} \\ \therefore \sim r \end{array}$$

(i.e., I am not rewarded)

Using this conclusion and the first premise, we then apply indirect reasoning:

$$\begin{array}{l} d \to r \\ \underline{\sim r \quad\quad} \\ \therefore \sim d \end{array}$$

Thus, having used all the statements in the problem (as specified in the directions for Problems 17-19), the final conclusion ($\sim d$) is:

I do not attend to my duties.

18. First, rewrite the three given premises in if-then form. Doing so makes it easier to translate them into symbols.

1st Premise: If this is organic food, then this is healthy ($o \to h$)

2nd Premise: If this is an artificial sweetener, then this is not healthy ($s \to \sim h$)

3rd Premise: If this is a prune, then it is organic food ($p \to o$)

Upon observing that all three premises are conditional statements, you should expect transitive reasoning and the law of contraposition to play key roles here. We first apply transitive reasoning with the 3rd and 1st premises, respectively:

$$\begin{array}{l} p \to o \\ \underline{o \to h} \\ \therefore p \to h \end{array}$$

Since transitive reasoning cannot be applied to this conclusion and the 2nd premise, we replace this 2nd premise with its contrapositive and then apply transitive reasoning:

$$p \to h$$
$$\underline{h \to \sim s} \quad \text{(contrapositive of 2nd premise)}$$
$$\therefore p \to \sim s$$

Written in if-then form, this conclusion translates to: "If this is a prune, then it is not an artificial sweetener." Finally, in words more closely resembling the problem's original phrasing, our final conclusion can be stated as follows:

No prune is an artificial sweetener.

19. As was done with Problem 18, we begin by rewriting the three premises in if-then form and translating these conditional statements into symbols:

1st Premise: If it is a square, then it is a rectangle ($s \to r$)
2nd Premise: If it is a rectangle, then it is a quadrilateral ($r \to q$)
3rd Premise: If it is a quadrilateral, then it is a polygon ($q \to p$)

We then apply transitive reasoning twice. First, from the 1st and 2nd premises, we can conclude:

$$s \to r$$
$$\underline{r \to q}$$
$$\therefore s \to q \quad \text{(transitive reasoning)}$$

Then, this conclusion and the 3rd premise yield our final conclusion:

$$s \to q$$
$$\underline{q \to p}$$
$$\therefore s \to p \quad \text{(transitive reasoning)}$$

Finally, we translate our final conclusion $s \to p$ ("if it is a square, then it is a polygon") back into words that are consistent with the problem's original phrasing:

All squares are polygons.

20. The fact that (i) Josie's hobby is gardening and (ii) she sits on Terry's left allows us to begin by seating Josie and Terry in some manner that is consistent with these two statements; for example:

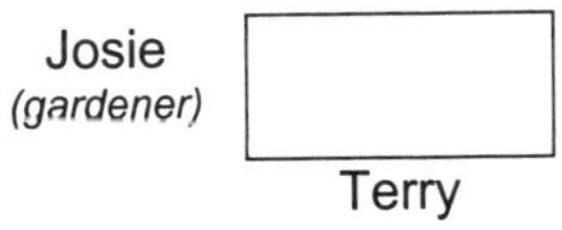

The last given statement ("Warren, who faces Terry, is not the baker") enables us to seat Warren and, in turn, Maureen. So, at this stage, we have the following seating:

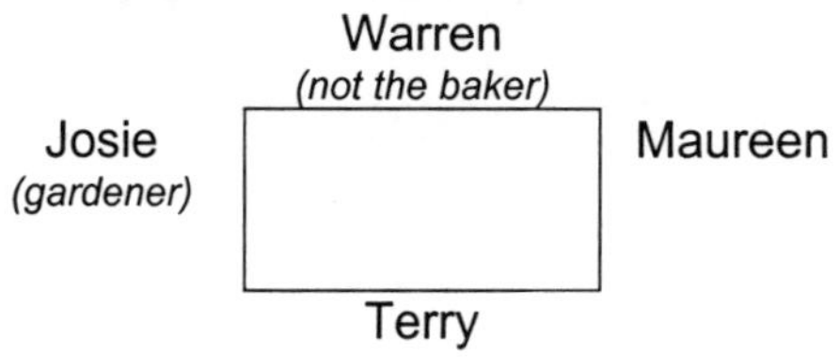

We then use our final clue: "Maureen sits at the hiker's right." In other words, Terry must be the hiker. Also, by process of elimination, since Warren is not the baker, Maureen must be the baker. Hence, Warren is the surfer.

Thus, our final drawing of this seating arrangement is:

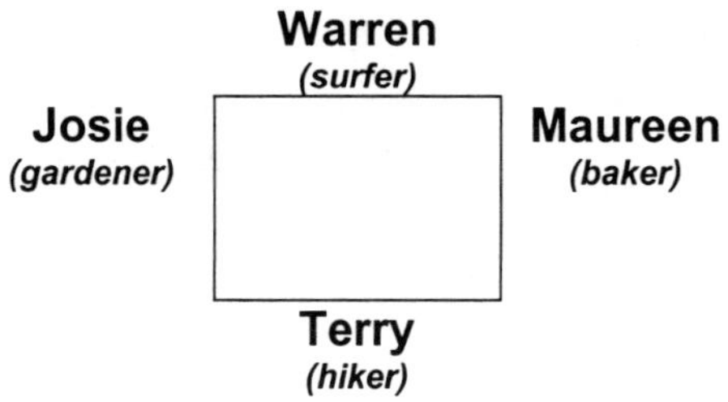

CHAPTER 3: The Nature of Calculation

Problem Set 3.1

1. **Examples vary**. A ***number*** is a concept, whereas a ***numeral*** is a symbol used to represent a number. To quote the author of your text (page 116), "The concept or idea of 'seventeenness' is called a *number*; the symbol used to represent the concept is called a *numeral*."

3. **Answers vary**. Among the Egyptian numeration system's shortcomings are its lack of a zero symbol and the difficulty it presented when writing large numbers. As for its contribution, it was a system based on simple grouping (specifically in powers of ten), which is a feature we make use of today — with our base-ten numeration system and with the metric system of measurement.

5. **Answers vary**. One of the Babylonian numeration system's shortcomings was the fact that it consisted of only two symbols, making it difficult to utilize. Additionally, as is mentioned on textbook p. 121, the Babylonian system, at least for most of its history, lacked a zero symbol. However, it did have significant advantages, mainly in terms of being a positional system, which is a benefit we retain to the present day.

7. a) **Roman, Egyptian**
   b) **Roman, Babylonian**. The Roman numeration system is a positional system because, for example, IX and XI represent different numbers.
   c) **Egyptian, Roman, Babylonian**
   d) **Egyptian, Roman, Babylonian**
   e) **Roman, Babylonian**
   f) **Roman**

9. $\boxed{\frac{1}{200}}$. As is explained on textbook pp. 117-118, in the Egyptian system an oval over a numeral indicates that the numeral is the denominator of a fraction. For additional details, revisit either textbook Table 3.1 (p. 116) or Table 3.4 (p. 122).

11. **100,010**. In the Egyptian system, the polliwog represents 100,000 and the heel bone 10. See textbook Table 3.1 (p. 116).

13. $\boxed{\frac{1}{12}}$. This problem is very similar to Problem 9 (see comments above).

15. $\boxed{\frac{1}{100}}$. This problem is comparable to Problems 9 and 13 (see solutions above).

17. **1,000,001**. Use textbook Table 3.1 (p. 116) as needed. The astonished man represents $1{,}000{,}000$ and the stroke 1.

19. **1,997**

1 Lotus Flower:	$1 \times 1{,}000$	$= 1000$
9 Scrolls:	9×100	$= 900$
9 Heel Bones:	9×10	$= 90$
7 Strokes:	7×1	$= + \ 7$
		$1{,}997$

21. **709** $\quad D = 500,\ CC = 200,\ IX = 9$

23. **2,001** $\quad MM = 2000,\ I = 1$

25. Remember: a bar placed over a Roman numeral (or a group of numerals) means $1{,}000$ times the value of the Roman numerals themselves.
So, because $CD = 500 - 100 = 400$, we find:

$$\overline{CD} = 400 \times 1{,}000 = \boxed{\mathbf{400{,}000}}.$$

27. This problem is similar to text Example 3d (p. 119).
Since $IX = 9$, $D = 500$, $CC = 200$, $X = 10$, and $II = 2$, it follows that:

$$\overline{IX}DCCXII = (9 \times 1000) + 500 + 200 + 10 + 2 = \boxed{\mathbf{9{,}712}}.$$

29. **24**. The Babylonian numerals are summarized in textbook Table 3.3 (p. 120).

31. $(11 \times 60) + 11 = \boxed{\mathbf{671}}$

33. $(3 \times 10) - 2 = \boxed{\mathbf{28}}$. [Note: The Babylonian's subtractive symbol, , is introduced on the bottom of textbook p. 120. Example 6a (p. 121) demonstrates its use.]

35. $(2 \times 10) + 5 = \boxed{\mathbf{25}}$

37. **One**; I am the only one known to be going to St. Ives. The others are those with whom I have had contact along the way; no information is stated about their destinations/whereabouts.

39. a) $75 = (7 \times 10) + (5 \times 1) \Rightarrow$ 

b) $75 = 50 + (2 \times 10) + 5 \Rightarrow$ **LXXV**

c) $75 = 60 + 10 + (5 \times 1) \Rightarrow$ 

41. a) $521 = (5 \times 100) + (2 \times 10) + 1 \Rightarrow$ 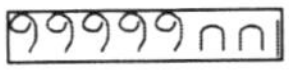

b) $521 = 500 + (2 \times 10) + 1 \Rightarrow$ **DXXI**

c) $521 = (8 \times 60) + (4 \times 10) + 1 \Rightarrow$ 

43. a) $2{,}001 = (2 \times 1{,}000) + 1 \Rightarrow$ 

b) $2{,}001 = (2 \times 1{,}000) + 1 \Rightarrow$ **MMI**

c) $2{,}001 = (33 \times 60) + (2 \times 10) + 1 \Rightarrow$ 

45. $45 + 34 + 54 =$ **133** bottles of water in all

47. Problems 46-51 are patterned after text Example 2 (p. 117).

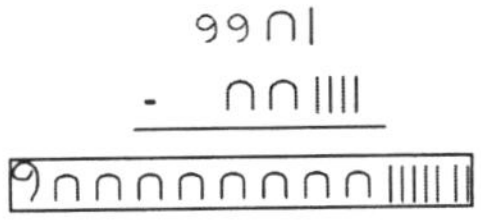

Check: $211 - 24 = 187$ ✓

49. We know $34 + 16 = 50$. So, in Babylonian numerals, we write:

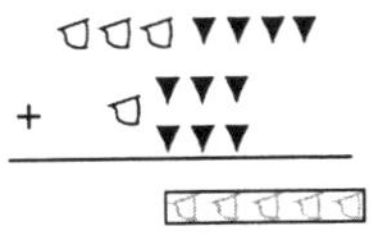

51. The difference $72 - 47$ is 25, which the following Babylonian arithmetic illustrates:

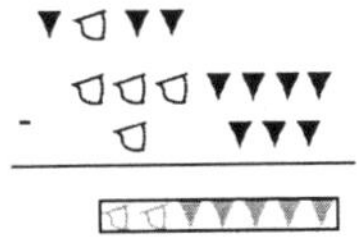

53. a) **MDCLXVI** $= (1000 + 500 + 100 + 50 + 10 + 5 + 1) = 1{,}666$

Note: The bar symbol (i.e., $\overline{\text{MDCLXVI}}$) cannot be used here. The bar is an operation symbol and not among the basic set of numerals.

b) **MCDXLIV** $= 1000 + 400 + 40 + 4 = 1{,}444$

55. a) $9 =$ **IX** b) $8 =$ **VIII** c) $99 =$ **XCIX**

d) $89 =$ **LXXXIX** e) $999 =$ **CMXCIX**

57. Let x and $(x + 10)$ represent the width and length, respectively.

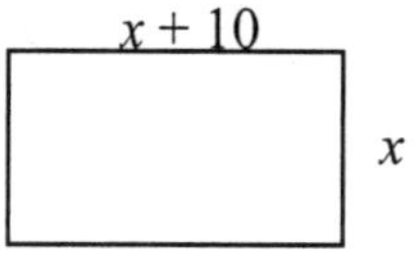

$$L \cdot W = \text{Area}$$
$$(x+10)x = 600$$
$$x^2 + 10x - 600 = 0$$
$$(x+30)(x-20) = 0$$
$$x + 30 = 0 \quad \text{or} \quad x - 20 = 0$$
$$x = -30 \qquad x = 20$$

↑ **Discard** *(because lengths cannot be negative)*

Therefore, the width is $x = 20$ units, and the length is $x + 10 = 30$ units. In terms of Babylonian notation, the dimensions (30 by 20) are expressed as:

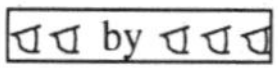

59. **Answers may vary** (namely, the writing of the Chinese symbols may vary). You may find it helpful to begin by constructing Pascal's triangle alongside the Chinese triangle (Figure 3.2). If you have forgotten Pascal's triangle, please revisit textbook Section 1.1.

– represents 1; = represents 2; ≡ represents 3; ≣ represents 4; ≣ represents 5; ├ represents 6; ° represents 10; and so forth…

Problem Set 3.2

1. 123 means we have 1 group of 100, 2 groups of 10, and 3 extra. This can be depicted with the following symbols:

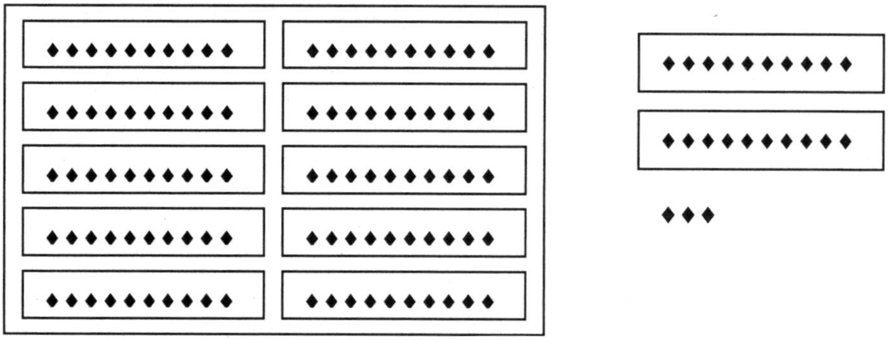

3. First, we must devise a way to represent 1,000.

Since 1,000 is ten 100's, we let **X** represent ◆◆◆◆◆◆◆◆◆◆ of the larger groups; that is, ten of the group enclosed in the *large* rectangles. We then note that 1,134 means 1 group of 1000, 1 group of 100, 3 groups of 10, and 4 extra. Pictorially, this may be represented as:

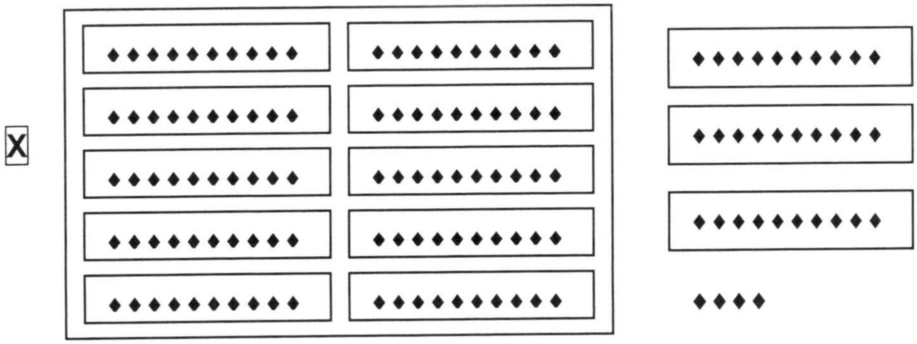

5. We define b^n (where b is a nonzero number and n is a counting number) as

$$\boxed{b^n = \underbrace{b \cdot b \cdot \cdots \cdot b}_{n \text{ factors of } b}}.$$

7. The 5 in 805 means **5 units**.

9. The 5 in 0.00567 represents **5 thousandths**; that is, $\frac{5}{1000}$.

11. a) **100,000** b) **1,000**

13. a) $10^{-4} = \frac{1}{10,000} = \boxed{0.0001}$ b) $10^{-3} = \frac{1}{1000} = \boxed{0.001}$

15. a) **5,000** b) **500**

17. a) $6 \times \frac{1}{10^2} = \frac{6}{100} = \boxed{0.06}$ b) $9 \times \frac{1}{10^5} = \frac{9}{10,000} = \boxed{0.00009}$

19. $10,000 + 200 + 30 + 4 = \boxed{10,234}$

21. $500,000 + 20,000 + 1,000 + 600 + 50 + 8 = \boxed{521,658}$

23. $7,000,000 + 0.03 = \boxed{7,000,000.03}$

25. $500,000 + 400 + 50 + 7 + 0.3 + 0.04 = \boxed{500,457.34}$

27. $20{,}000 + 600 + 0.4 + 0.007 + 0.0006 + 0.00009 = \boxed{\mathbf{20{,}600.40769}}$

29. $\boxed{\mathbf{7\times10^5+2\times10^4+8\times10^3+4\times10^2+7\times10^0}}$

31. $\boxed{\mathbf{2\times10^1+7\times10^0+5\times10^{-1}+7\times10^{-2}+2\times10^{-3}}}$

33. $\boxed{\mathbf{5\times10^2+2\times10^1+1\times10^0}}$

35. $\boxed{\mathbf{2\times10^6+3\times10^5+5\times10^3+6\times10^2+8\times10^1+1\times10^0}}$

37. $\boxed{\mathbf{5\times10^3+2\times10^2+4\times10^1+5\times10^0+5\times10^{-1}}}$

39. $\boxed{\mathbf{1\times10^5+1\times10^{-3}}}$

41. $\boxed{\mathbf{8\times10^0+5\times10^{-5}}}$

43. $\boxed{\begin{aligned}&\mathbf{5\times10^4+7\times10^3+2\times10^2+8\times10^1+5\times10^0+9\times10^{-1}+3\times10^{-2}}\\&\mathbf{+6\times10^{-3}+1\times10^{-4}}\end{aligned}}$

45. $3\times10^3+2\times10^2+1\times10^0=\boxed{\mathbf{3{,}201}}$

47. $5\times10^6+1\times10^3+5\times10^0=\boxed{\mathbf{5{,}001{,}005}}$

49. $(5+3)\times10^6+(5+4)\times10^3+2\times10^1+(5+1)\times10^0=\boxed{\mathbf{8{,}009{,}026}}$

51. $$\begin{aligned}849 &= 8\times10^2+4\times10^1+9\times10^0\\ &= (5+3)\times10^2+4\times10^1+(5+4)\times10^0\end{aligned}$$

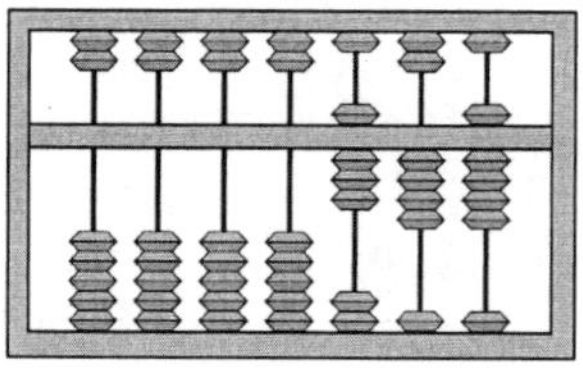

53. $$\begin{aligned}9{,}387 &= 9\times10^3+3\times10^2+8\times10^1+7\times10^0\\ &= (5+4)\times10^3+3\times10^2+(5+3)\times10^1+(5+2)\times10^0\end{aligned}$$

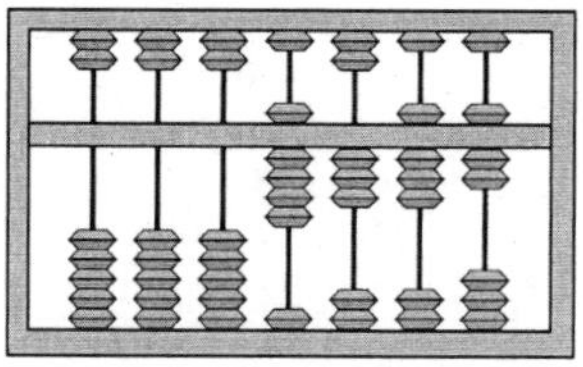

55. $2{,}001 = 2\times10^3 + 1\times10^0$

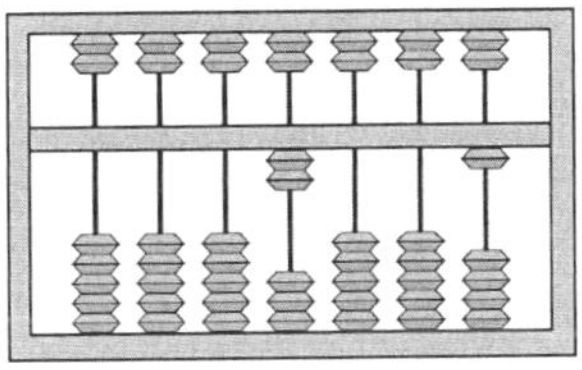

57. $8{,}007{,}009 = 8\times10^6 + 7\times10^3 + 9\times10^0$

$= (5+3)\times10^6 + (5+2)\times10^3 + (5+4)\times10^0$

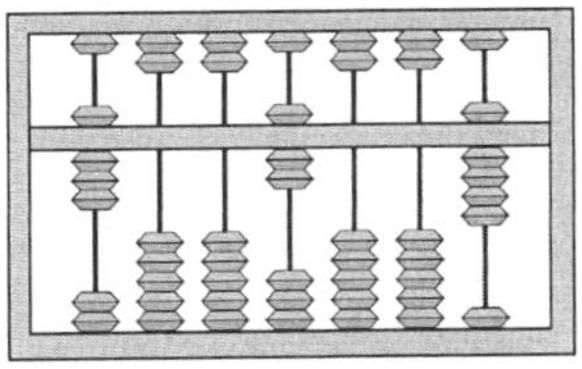

59. Notice that the only symbols (i.e., digits) being used are 0, 1, 2, 3, and 4. As your text's author explains thoroughly in the next section, this pattern is a demonstration of how one would count in a base-five numeration system. To help you comprehend this pattern without having to first read Section 3.3, we suggest you examine the mechanics of how counting is done in our decimal (Hindu-Arabic, base-ten) system. For example, think about how 100 follows 99: we roll the two 9's back to zero and elevate the next higher place value to a 1. In the case of a base-five system, this "elevation process" would occur, for example, at 44 (instead of 99), with 100 following 44. Hence, the pattern (of counting in base-five) is:

0, 1, 2, 3, 4, 10, 11, 12, 13, 14, 20, 21, **22, 23, 24, 30, 31, 32, 33, 34, 40, 41, 42, 43, 44, 100, 101, 102, 103, 104, 110, 111, 112, 113, 114, 120, ...**

Problem Set 3.3

1. As is summarized on textbook page 133 (immediately above the statement of this problem), **write the base-eight number in expanded form, and then perform the arithmetic**. Refer to Example 1 (pp. 130-131) to see a specific illustration of this.

3. Once again, as is summarized on p. 133 (and illustrated on pp. 131-132 in Examples 2 and 3), converting from base ten to base eight is a two-step process:

(i) **Perform repeated division by 8.**

(ii) **Read the remainders in top-down order.**

5. To convert from base two to base eight, there are two different approaches:

I. **Change from base two to base ten and then from base ten to base eight**;

or, alternatively,

II. **Use the simpler method provided in Section 3.4 (explained prior to Problem 49 and used for Problems 49-59)**. This requires first grouping the digits of the base-two number into groups of three, and then converting each group of three base-two numbers into a base-eight number using the correspondence provided in Table 3.9 (p. 139).

7. There are 13 people in all. In terms of each of the following bases, this number would be expressed as:

a) Base ten: 1 group of ten and 3 extra ⇒ $\boxed{13}$

b) Base five: 2 groups of five and 3 extra ⇒ $\boxed{23_{five}}$

c) Base thirteen: 1 group of thirteen and 0 extra ⇒ $\boxed{10_{thirteen}}$

d) Base eight: 1 group of eight and 5 extra ⇒ $\boxed{15_{eight}}$

e) Base two: 1 group of eight (2^3), 1 group of 4 (2^2), and 1 extra ⇒ $\boxed{1101_{two}}$

f) Base twelve: 1 group of twelve and 1 extra ⇒ $\boxed{11_{twelve}}$

9. a) $\boxed{6\times 8^2+4\times 8^1+3\times 8^0}$

b) $\boxed{5\times 12^3+3\times 12^2+8\times 12^1+7\times 12^0+9\times 12^{-1}}$

11. a) $\boxed{1\times 2^5+1\times 2^4+1\times 2^2+1\times 2^1+1\times 2^0+1\times 2^{-1}+1\times 2^{-4}}$

b) $\boxed{5\times 6^3+4\times 6^2+1\times 6^1+1\times 6^0+1\times 6^{-1}+2\times 6^{-3}+3\times 6^{-4}}$

13. a) $\boxed{3\times 4^5+2\times 4^4+3\times 4^3+2\times 4^{-1}}$

b) $\boxed{2\times 5^5+3\times 5^4+4\times 5^3}$

15. $527_{eight}=5\times 8^2+2\times 8^1+7\times 8^0=320+16+7=\boxed{343}$

17. $1101.11_{two}=1\times 2^3+1\times 2^2+1\times 2^0+1\times 2^{-1}+1\times 2^{-2}=8+4+1+\frac{1}{2}+\frac{1}{4}=\boxed{13.75}$

19. $431_{five}=4\times 5^2+3\times 5^1+1\times 5^0=100+15+1=\boxed{116}$

21. $1011.101_{two}=1\times 2^3+1\times 2^1+1\times 2^0+1\times 2^{-1}+1\times 2^{-3}=\boxed{11.625}$

23. $573_{twelve} = 5\times12^2 + 7\times12^1 + 3\times12^0 = \boxed{807}$

25. $2110_{three} = 2\times3^3 + 1\times3^2 + 1\times3^1 = \boxed{66}$

27. $537.1_{eight} = 5\times8^2 + 3\times8^1 + 7\times8^0 + 1\times8^{-1} = \boxed{351.125}$

29. $5742_{eight} = 5\times8^3 + 7\times8^2 + 4\times8^1 + 2\times8^0 = \boxed{3{,}042}$

31. In a manner identical to Examples 2 and 3 in your text, we begin with $4^4 = 256$ and demonstrate two approaches:

$$\begin{aligned} 628 &= \underline{2}\times4^4 + 116 \\ 116 &= \underline{1}\times4^3 + 52 \\ 52 &= \underline{3}\times4^2 + 4 \\ 4 &= \underline{1}\times4^1 + 0 \\ 0 &= \underline{0}\times4^0 \end{aligned}$$

or

$$\begin{array}{rl} 4\overline{)2} = 0 & \text{r.}2 \\ 4\overline{)9} & \text{r.}1 \\ 4\overline{)39} & \text{r.}3 \\ 4\overline{)157} & \text{r.}1 \\ 4\overline{)628} & \text{r.}0 \end{array}$$

Thus, $628 = \boxed{21310_{four}}$.

33. Here are two strategies:

I.
$$\begin{aligned} 427 &= \underline{2}\times12^2 + 139 \\ 139 &= \underline{11}\times12^1 + 7 \\ 7 &= \underline{7}\times12^0 + 0 \end{aligned}$$

Recall that $11 = E_{twelve}$. Thus, $427 = \boxed{2E7_{twelve}}$

II.
$$\begin{aligned} 427 \div 12 &= 35 \quad \text{r.}7 \\ 35 \div 12 &= 2 \quad \text{r.}11 \\ 2 \div 12 &= 0 \quad \text{r.}2 \end{aligned}$$

Collecting these remainders in reverse order and representing 11 as a base-twelve digit yields $2E7_{twelve}$. Of course, this approach is equivalent to the repeated division (and collection-of-remainders process) shown above in our solution to Problem 31 (see right side). We recommend that you, too, experiment with different division "layouts" in order to find the one that best suits you.

35.

$$615 = \underline{1}\times 8^3 + 103$$
$$103 = \underline{1}\times 8^2 + 39$$
$$39 = \underline{4}\times 8^1 + 7$$
$$7 = \underline{7}\times 8^0 + 0$$

Thus, $615 = \boxed{\mathbf{1147}_{eight}}$

37.

$$615 = \underline{1}\times 2^9 + 103$$
$$103 = \underline{1}\times 2^6 + 39$$
$$39 = \underline{1}\times 2^5 + 7$$
$$7 = \underline{1}\times 2^2 + 3$$
$$3 = \underline{1}\times 2^1 + 1$$
$$1 = \underline{1}\times 2^0 + 0$$

Therefore, $615 = \boxed{\mathbf{1001100111}_{two}}$

39. In the interest of exposing you to as wide a variety of approaches as possible, we momentarily step away from the tactic employed in our solutions to the two previous problems (35 and 37). Instead, let us demonstrate another method, one that we employed as our second approach to Problem 33 above. We begin by dividing 795 by 7 and then continue this process, repeatedly dividing each subsequent quotient by 7. We stop when a quotient of zero is obtained.

$$795 \div 7 = 113 \quad \text{r.}4$$
$$113 \div 7 = 16 \quad \text{r.}1$$
$$16 \div 7 = 2 \quad \text{r.}2$$
$$2 \div 7 = 0 \quad \text{r.}2$$

Finally, we collect the remainders and write them in bottom-to-top order.
Hence, $795 = \boxed{\mathbf{2214}_{seven}}$

41. For this problem, we shall demonstrate both the approach used in Problem 39 (which we will call Approach I) and the method employed in Problems 35 and 37 (Approach II).

Approach I:

$$4{,}731 = \underline{2}\times 12^3 + 1275$$
$$1275 = \underline{8}\times 12^2 + 123$$
$$123 = \underline{10}\times 12^1 + 3$$
$$3 = \underline{3}\times 12^0 + 0$$

Since $T = 10_{twelve}$, $4{,}731 = \boxed{\mathbf{28T3}_{twelve}}$

Approach II:

$$\begin{aligned} 4{,}731 \div 12 &= 394 \quad \text{r.}3 \\ 394 \div 12 &= 32 \quad \text{r.}10 \\ 32 \div 12 &= 2 \quad \text{r.}8 \\ 2 \div 12 &= 0 \quad \text{r.}2 \end{aligned}$$

After writing the remainders in bottom-to-top order (2, 8, 10, and 3) and writing 10 in its proper base-twelve representation (i.e., T), we have the answer:

$$4{,}731 = 28T3_{twelve}.$$

43.

$$\left.\begin{aligned} 76 &= \underline{1}\times 4^3 + 12 \\ 12 &= \underline{0}\times 4^2 + 12 \\ 12 &= \underline{3}\times 4^1 + 0 \\ 0 &= \underline{0}\times 4^0 + 0 \end{aligned}\right\} \quad \text{Therefore, } 76 = \boxed{\mathbf{1030}_{four}}$$

45. By using any of the methods illustrated above, one finds $2{,}000 = \boxed{\mathbf{3720}_{eight}}$.

47. Since $52 \div 7 = 7$ r.3, $52 = 73_{seven}$. Thus, 52 days equals **7 weeks, 3 days**.

49. Because there are 12 inches in a foot and $55 \div 12 = 4$ r.7, we know $55 = 47_{twelve}$. Therefore, 55 inches is equivalent to **4 ft, 7 in**.

51. This problem is comparable to —although it is easier than — textbook Example 4 (p.133). Since a gross consists of 144 (or 12^2) units and a dozen is comprised of 12 units, this problem essentially requires converting 500 to a base-twelve number. Making use of one of the methods employed in Problems 31-46, we obtain:

$$\begin{aligned} 500 &= \underline{3}\times 12^2 + 68 \\ 68 &= \underline{5}\times 12^1 + 8 \\ 8 &= \underline{8}\times 12^0 + 0 \end{aligned}$$

Since $500 = 358_{twelve}$, 500 units consists of **3 gross, 5 dozen, and 8 units**.

53. Note that the process of converting $\$0.84$ to quarters, nickels, and pennies is mathematically identical to the process of converting the base-ten number 84 to a base-five number. Thus, since $84 = 314_{five}$, the minimum number of coins needed to make $\$0.84$ is 3 quarters, 1 nickel, and 4 pennies, which is a total of **8 coins**.

55. Once again, since one gross is 144 (or 12^2) units, 9 gross, 5 dozen, and 4 pencils would be written in base-twelve as $\boxed{\mathbf{954}_{twelve}}$.
Since $954_{twelve} = 9 \times 12^2 + 5 \times 12 + 4$, this base-twelve number is equal (in base-ten) to $\boxed{\mathbf{1360}}$.

57. Since $44 = 62_{seven}$, 44 days = $\boxed{\textbf{6 weeks, 2 days}}$.

59. Because $29 = 15_{twenty\text{-}four}$, 29 hours is equivalent to $\boxed{\textbf{1 day and 5 hours}}$.

Problem Set 3.4

1. $\boxed{\textbf{Answers/wording varies.}}$ Figure 3.4 (p. 135) is an example of a two-state device, which is a visually effective way to illustrate how computers process information. The light being on represents a "1"; the light being off represents a "0".

3. We begin by converting the light display to a binary number. In this case, off-off-on-off-off-on-on-on is represented by 00100111_{two}, which is more simply expressed as 100111_{two}. Converting this to base-ten, we have:

$$1 \times 2^5 + 1 \times 2^2 + 1 \times 2^1 + 1 \times 2^0 = 32 + 4 + 2 + 1 = \boxed{\mathbf{39}}$$

5. Applying the approach used in our solution to Problem 3 above, we find:

$$10100111_{two} = 1 \times 2^7 + 1 \times 2^5 + 1 \times 2^2 + 1 \times 2^1 + 1 \times 2^0 = 128 + 32 + 4 + 2 + 1 = \boxed{\mathbf{167}}$$

7. $1101_{two} = 1 \times 2^3 + 1 \times 2^2 + 1 \times 2^0 = 8 + 4 + 1 = \boxed{\mathbf{13}}$

9. $1011_{two} = 1 \times 2^3 + 1 \times 2^1 + 1 \times 2^0 = 8 + 2 + 1 = \boxed{\mathbf{11}}$

11. $11101_{two} = 1 \times 2^4 + 1 \times 2^3 + 1 \times 2^2 + 1 \times 2^0 = 16 + 8 + 4 + 1 = \boxed{\mathbf{29}}$

13. $11011_{two} = 1 \times 2^4 + 1 \times 2^3 + 1 \times 2^1 + 1 \times 2^0 = 16 + 8 + 2 + 1 = \boxed{\mathbf{27}}$

15. $1100011_{two} = 1 \times 2^6 + 1 \times 2^5 + 1 \times 2^1 + 1 \times 2^0 = 64 + 32 + 2 + 1 = \boxed{\mathbf{99}}$

17. $10111000_{two} = 1 \times 2^7 + 1 \times 2^5 + 1 \times 2^4 + 1 \times 2^3 = 128 + 32 + 16 + 8 = \boxed{\mathbf{184}}$

19. Converting a base-ten number to a binary numeral can be done using any of the approaches employed in the previous section (3.3) of this text. In the event you need to review these approaches, refer to Problems 31-46 and, as needed, their solutions.

With that said, let us approach this particular problem by using repeated division by 2:

$$13 \div 2 = 6 \quad \text{r.}1$$
$$6 \div 2 = 3 \quad \text{r.}0$$
$$3 \div 2 = 1 \quad \text{r.}1$$
$$1 \div 2 = 0 \quad \text{r.}1$$

By collecting these remainders in bottom-to-top order, we know $13 = \boxed{\mathbf{1101}_{two}}$.

21. For this problem, we shall demonstrate two approaches. The first is identical to that used in Problem 19 immediately above. The second is another one of the methods used in Section 3.3 to convert decimal numbers to other bases.

Approach I:

$$\begin{aligned} 35 \div 2 &= 17 \quad \text{r.}1 \\ 17 \div 2 &= 8 \quad \text{r.}1 \\ 8 \div 2 &= 4 \quad \text{r.}0 \\ 4 \div 2 &= 2 \quad \text{r.}0 \\ 2 \div 2 &= 1 \quad \text{r.}0 \\ 1 \div 2 &= 0 \quad \text{r.}1 \end{aligned}$$

We write the remainders in bottom-to-top order and accordingly conclude that 35 equals $\boxed{\mathbf{100011}_{two}}$.

Approach II:

$$\left.\begin{aligned} 35 &= \underline{1} \times 2^5 + 3 \\ 3 &= \underline{0} \times 2^4 + 3 \\ 3 &= \underline{0} \times 2^3 + 3 \\ 3 &= \underline{0} \times 2^2 + 3 \\ 3 &= \underline{1} \times 2^1 + 1 \\ 1 &= \underline{1} \times 2^0 + 0 \end{aligned}\right\} \quad \text{Therefore, } 35 = \boxed{\mathbf{100011}_{two}}.$$

23. As you do more of these problems, it is our hope (and expectation) that you are observing patterns and trends which enable you to convert decimal numbers to binary with increasing speed and efficiency. Consistent with this expectation, let us demonstrate another approach to this problem, an approach that is slightly different from one employed in Problems 19 and 21. We begin by recalling our powers of 2 (namely, $1, 2, 4, 8, 16, 32, 64, \ldots$) and attempt to write 51 as a sum of these powers:

$$51 = 32 + 16 + 2 + 1$$
$$= 2^5 + 2^4 + 2^1 + 2^0$$
$$= \boxed{\mathbf{110011}_{two}}$$

If you are not sufficiently comfortable with the approaches previously shown, we encourage you to continue to practice those methods with both this problem and those that follow.

25. Once again, the approaches employed in previous problems could be employed here, but it should be noted that 64 is a power of 2. Specifically, 64 is equal to 2^6. Thus, $64 = \boxed{\mathbf{1000000}_{two}}$.

27. Like its predecessors (Problems 25-26), the decimal number used in this problem is a power of 2, namely $128 = 2^7$. Therefore, $128 = \boxed{\mathbf{10000000}_{two}}$.

29. For this final solution, let us demonstrate one of the approaches demonstrated above, and we encourage you to practice both it and the others demonstrated previously:

$$\left.\begin{aligned}
795 &= \underline{1} \times 2^9 + 283 \\
283 &= \underline{1} \times 2^8 + 27 \\
27 &= \underline{0} \times 2^7 + 27 \\
27 &= \underline{0} \times 2^6 + 27 \\
27 &= \underline{0} \times 2^5 + 27 \\
27 &= \underline{1} \times 2^4 + 11 \\
11 &= \underline{1} \times 2^3 + 3 \\
3 &= \underline{0} \times 2^2 + 3 \\
3 &= \underline{1} \times 2^1 + 1 \\
1 &= \underline{1} \times 2^0 + 0
\end{aligned}\right\} \quad \text{Thus, } 795 = \boxed{\mathbf{1100011011}_{two}}$$

31. Making use of Table 3.8 (p. 137), we find the ASCII code of the capital letters DO to be: **68 79**.

33. END written in ASCII code is **69 78 68**.

35. Using Table 3.8 to convert ASCII codes back to the respective letters each represents, we find that 72 65 86 69 represents **HAVE**.

37. The ASCII code 83 84 85 68 89 represents the word **STUDY**.

39. As the author of your text points out (on p. 128), you must remember when adding $1 + 1$ in binary that you put down a 0 and "carry" the 1. Similarly, when adding $1 + 1 + 1$ in binary, you put down a 1 and "carry" a 1 because 3 in binary is expressed as 11_{two}. Therefore,

$$\begin{array}{r} 11_{two} \\ +10_{two} \\ \hline \boxed{\mathbf{101}_{two}} \end{array}$$

Although this method is the preferred approach, it is worthy of mention that an alternative (albeit less efficient) approach would be to convert the original two numbers to decimal numerals, add them together, and then convert the sum back to binary.

41. We demonstrate this arithmetic in a step-by-step manner:

(i) Since in binary $0 + 1 = 1$, we first write:
$$\begin{array}{r} 110_{two} \\ +111_{two} \\ \hline 1_{two} \end{array}$$

(ii) Since $1 + 1 = 10$ in binary, we write a zero and carry the one:
$$\begin{array}{r} {}^{1}110_{two} \\ +111_{two} \\ \hline 01_{two} \end{array}$$

(iii) Since $1 + 1 + 1 = 11$ in binary, we write 11 to conclude this addition process:
$$\begin{array}{r} {}^{1}110_{two} \\ +111_{two} \\ \hline 1101_{two} \end{array}$$

Thus, the answer is $\boxed{\mathbf{1101}_{two}}$.

43. $\boxed{\mathbf{10}_{two}}$. We consider two different perspectives:

I. Think about how "borrowing" in binary should be done. Since borrowing in subtraction is the opposite of "carrying" in addition, borrowing a 1 means we must allot two 1's to the place value to its immediate right. That is,

$$\begin{array}{r} 101_{two} \\ -\ 11_{two} \\ \hline \end{array} \quad \Rightarrow \quad \begin{array}{r} {}^{\ 11}\\ 001_{two} \\ -\ 11_{two} \\ \hline 10_{two} \end{array}$$

II. Alternatively, it should be fairly easy for you to convert these two binary numbers to decimal, subtract them, and then convert the resulting difference back to binary. In this particular problem, $101_{two} = 5$ and $11_{two} = 3$. Thus, since $5 - 3 = 2$ and we know $2 = 10_{two}$, our final answer is 10_{two}.

45.

$$\begin{array}{r} 111_{two} \\ \times 101_{two} \\ \hline 111 \\ 000 \\ +111 \\ \hline \end{array}$$

Then, by using our binary addition skills (i.e., the notion of "carrying" acquired in Problems 39-42 and Example 4), we obtain a final answer of $\boxed{\mathbf{100011}_{two}}$. Please remember that you can check this answer by (i) converting the given binary numbers to decimal numerals, (ii) multiplying the decimal numerals, and (iii) validating that the resulting product is equal to our binary answer above. For this particular problem, you will find $7 \times 5 = 35$, which indeed is equivalent to 100011_{two}.

47. The computer is correct because the question asked essentially was: "Will there be war OR peace?" The important word here is the conjunction "or." As you learned in your study of logic (Chapter 2), an or-statement (known as a disjunction) is true when **at least one** of its component parts is true. If we let p represent the statement "there will be peace," it then follows that the negation of p (written $\sim p$) is "there will not be peace." We then can translate $\sim p$ to "there will be war." The four-star general's question amounts to assessing the truth table of the disjunction $p \vee \sim p$. As the truth table below reveals, this statement is always true (known as a "tautology" in logic). Thus, the computer's answer of "yes" is correct.

p	~p	p ∨ ~p
T	F	**T**
F	T	**T**

49. Table 3.9 (on p. 139) provides the answer to this two-part question.

a) $5_{eight} = \boxed{\mathbf{101}_{two}}$ b) $6_{eight} = \boxed{\mathbf{110}_{two}}$

51. We use the method explained by your text's author.

a) Since $1_{eight} = 001_{two}$ and $6_{eight} = 110_{two}$ and $7_{eight} = 111_{two}$, it follows that 167_{eight} equals $\boxed{\mathbf{001\ 110\ 111}_{two}}$, which also may be written as 1110111_{two}.

b) Similarly, $624_{eight} = \boxed{\mathbf{110\ 010\ 100}_{two}}$.

53. a) $\boxed{\mathbf{101\ 111\ 000\ 000}_{two}}$ b) $\boxed{\mathbf{000\ 100\ 011\ 010\ 000}_{two}}$

55. Table 3.9 (p. 139) provides the answer to this two-part question.

a) $011_{two} = \boxed{\mathbf{3}_{eight}}$ b) $001_{two} = \boxed{\mathbf{1}_{eight}}$

57. As was explained in your text, we perform this base conversion by replacing each three-digit binary grouping with its octal equivalent (found in Table 3.9). This yields:

$$100\ 000\ 000\ 101\ 110\ 111_{two} = \boxed{\mathbf{400567}_{eight}}$$

59. 453127_{eight}

Problem Set 3.5

1. **Answers vary**. Among the computing devices that predate the invention of the electronic computer are fingers, the abacus, Napier's rods, slide rule, Pascal's calculator, Leibniz' reckoning machine, and Babbage's calculating machines.

3. **Answers vary**. The central processing unit (CPU) controls the functions of the computer and coordinates the processing of information among its basic components.

5. **Answers vary**. Read-only memory (ROM) and random-access memory (RAM) are compared and contrasted in your text. Refer to textbook p. 148 for details regarding these two fundamental types of information storage. It is worthy of mention that one is likely to encounter assorted forms of RAM and ROM when shopping for a computer; for example, two forms of RAM found in many current computers are SDRAM and RDRAM.

7. **Answers vary**. Among the common uses of home computers are word processing, database management, spreadsheeting, e-mail, internet/online activity, use of home finance software, among others.

9. **Answers vary**. You may find it valuable to write two separate responses to this question — one defending the computer as a societal benefit, the other explaining its detriments.

11. a) **27**. With your hands held up and palms facing away from yourself (like on p. 141), the middle finger of your left hand should be bent, separating the first two fingers (denoting 20) from the remaining seven fingers (denoting 7).
    b) **63**. The seventh finger from the left should be bent, separating six fingers from the three remaining.
    c) **54**. The thumb on your right hand should be bent, revealing the five fingers on your left hand (denoting 50) and the remaining four on your right hand (indicating 4).

13. Holding both hands up with palms facing away, we use the method explained by your text's author (page 141).
    a) **243**. To compute 27×9, first group the first two fingers of your left hand (since 27 contains two tens), separating these fingers from the rest of your left hand. Then, because 27 has 7 units, bend the seventh finger from the left. The

first two fingers (on your left hand) indicate the product contains two 100's; the next four fingers (namely, the remaining three on your left hand and the your right hand's thumb) indicate four 10's; and, finally, the last three fingers on your right hand (after the bent finger) reveal three units.

b) In a manner similar to part (a), we multiply 48 and 9 by separating the first four fingers of our left hand away from the thumb and bend the eighth finger from the left (i.e., the right hand's middle finger). The first four fingers of the left hand reveal 400, the next three fingers indicate 30, and the two fingers to the right of the bent middle finger suggest 2 units. Therefore, 48×9 equals **432**.

c) In computing the product of 56 and 9, we separate our left hand away from the right (because of the 5 in 56) and bend the thumb of our right hand (due to the 6 in 56). First, the five fingers of the left hand tell us the product begins with 500. Second, since there are no other fingers to the left of the bent thumb, the product contains zero 10's. Finally, the four fingers to the right of the bent thumb reveal four units. Hence, the product of 56 and 9 is **504**.

15. Arranged in order of each computer's first appearance, we have:
ENIAC (1942-46), UNIVAC (1951), Cray(1958), Altair (1975), Apple (1976)

Note: Answers to Problems 16-33 vary.

17. Aristophanes **devised a finger-calculating system** in about 500 B.C.

19. Charles Babbage (1792-1871) **developed and designed two calculating "engines,"** neither one of which could be built because the technical knowledge needed to construct them was not far enough advanced.

21. Clifford Berry **helped John Vincent Atanasoff build the first electronic digital computer**, although they could not get it to work properly.

23. Seymour Cray, in 1958, **developed the first supercomputer**.

25. Douglas Engelbart **invented the computer mouse**.

27. Steven Jobs **co-founded Apple Computer, Inc. and co-designed the Apple II** computer.

29. Gottfried Leibniz **designed a mechanical calculator** (circa 1672). Finally built in 1695, it improved upon Pascal's calculators, although it, like Pascal's calculators, proved to be unreliable.

31. John Napier, in 1617, **invented a calculating device for multiplication** (commonly known as Napier's rods or Napier's bones).

33. Stephen Wozniak, along with Steven Jobs, **codesigned the Apple**, the first commercially successful personal computer.

35. **Yes; speed, complicated computations**

37. **Yes; speed, repetition**

39. **Yes; repetition**

41. **Yes; ability to readily make corrections**

43. **Yes (but not completely); it can help with some of the technical aspects**

45. **Yes; speed, complicated computations, repetition**

47. Choice E

49. Choice D

51. Answers vary.

53. Answers vary. In addressing a small part of the textbook author's larger question(s), you should recall that π is an irrational number and accordingly consider the computer's limitations in expressing such non-repeating, non-terminating decimal numbers.

55. Answers vary.

57. Answers vary.

59. Answers vary.

Chapter 3 Review Questions

1. Answers vary. A ***positional system*** is one in which the positioning (or ordering) of the individual digits is relevant. Refer to Section 3.1 (pp. 120-122) for examples of positional numeration systems.

2. Answers vary. Addition is easier in a simple grouping system; examples vary. Review Section 3.1 as necessary.

3. Answers vary. The ***Hindu-Arabic numeration system*** (also known as the decimal system) uses ten symbols; it is positional; it has a place-holder symbol (0); and it uses 10 as its basic unit for grouping. See Section 3.2 for additional information.

4. Answers vary. Your response should include mention of finger calculating, Napier's rods, Pascal's calculator, Lieniz' reckoning machine, Babbage's difference and analytic engines, ENIAC, UNIVAC, Atanasoff's and Eckert and Mauchly's computers, and the dispute they had in establishing their rightful places in the history of computers. You also should allude to the role and impact of the Apple and Macintosh computers, as well as supercomputers (like the Cray). Refer to Section 3.5 for details.

5. **Answers vary**. This topic is discussed in Section 3.5 (p. 150). Your response should include examples of illegal uses (e.g., breaking into another's computer; adding, modifying, or destroying information; and copying programs without authorization) and abuses caused by ignorance (e.g., assuming that output is correct and not using software for its intended purpose).

6. **Answers vary**
   a) ***hardware*** – the physical components (mechanical, magnetic, electronic) of a computer system
   b) ***software*** – programs and associated documentation in a computer system that allow the user to carry out particular tasks
   c) ***word processing*** – the process of creating, modifying, deleting, and formatting text and materials
   d) ***modem*** – a peripheral device that allows the computer to communicate with other computers across electronic cables or phone lines
   e) ***e-mail*** – electronic mail — that is, messages transmitted via computer modems (or other communication-type peripheral devices)
   f) ***RAM*** – Random-Access Memory is where programs and data are temporarily stored while the computer is processing the information or completing a task
   g) ***computer bulletin board*** – an electronic "meeting place" to exchange information with others, usually on a particular topic

7. 1 billion $= 1{,}000{,}000{,}000 = \boxed{10^9}$

8. $436.20001 = \boxed{4\times10^2+3\times10^1+6\times10^0+2\times10^{-1}+1\times10^{-5}}$

9. $523_{eight} = \boxed{5\times8^2+2\times8^1+3\times8^0}$

10. $1001110_{two} = \boxed{1\times2^6+1\times2^3+1\times2^2+1\times2^1}$

11. **4,020,005.62**

12. $11101_{two} = 1\times2^4+1\times2^3+1\times2^2+0\times2^1+1\times2^0$
$= 16+8+4+1$
$= \boxed{29}$

13. $1111011_{two} = 1\times2^6+1\times2^5+1\times2^4+1\times2^3+1\times2^1+1\times2^0$
$= 64+32+16+8+2+1$
$= \boxed{123}$

14. $123_{three} = 1 \times 3^2 + 2 \times 3^1 + 2 \times 3^0$
$= 9 + 6 + 2$
$= \boxed{17}$

15. $821_{twelve} = 8 \times 12^2 + 2 \times 12^1 + 1 \times 12^0$
$= 1152 + 24 + 1$
$= \boxed{1{,}177}$

16. In the back of your text (page 941), the author demonstrates the approach of repeated division and collection of remainders in bottom-to-top order. You should recall, from your work in Sections 3.3 and 3.4, that this is just one method among several shown previously in both your text and this solution manual.

Another perspective previously discussed involves recalling powers of two (1, 2, 4, 8, 16, ...) and expressing the given decimal numeral as a sum of such powers. In the case of 12, we write:

$$12 = 8 + 4 = 2^3 + 2^2 = 1 \times 2^3 + 1 \times 2^2 + 0 \times 2^1 + 0 \times 2^0 = \boxed{1100_{two}}$$

17. Like Problem 16, there is an assortment of approaches one may choose to use for this problem. The author of your text demonstrates one (provided with the answers in Appendix H, p. 941); please permit us to show you two others, both of which we previously used in our solutions to the problems of Sections 3.3 and 3.4:

Approach I:

$$\begin{aligned} 52 &= 32 + 16 + 4 \\ &= 2^5 + 2^4 + 2^2 \\ &= 1 \times 2^5 + 1 \times 2^4 + 0 \times 2^3 + 1 \times 2^2 + 0 \times 2^1 + 0 \times 2^0 \\ &= \boxed{110100_{two}} \end{aligned}$$

Approach II:

$$\left.\begin{aligned} 52 &= \underline{1} \times 2^5 + 20 \\ 20 &= \underline{1} \times 2^4 + 4 \\ 4 &= \underline{0} \times 2^3 + 4 \\ 4 &= \underline{1} \times 2^2 + 0 \\ 0 &= \underline{0} \times 2^1 + 0 \\ 0 &= \underline{0} \times 2^0 + 0 \end{aligned}\right\} \quad \text{Therefore, } 52 = 110100_{two}$$

18. **11111010011_{two}** In the interest of reminding you of the various methods previously shown in both your text and this manual, we again (as was done with the solutions to Problems 16-17) use this "review" opportunity to illustrate another approach, one that is different from what is shown in textbook Appendix H (p. 941). However, we do encourage you to make use of the method demonstrated in Appendix H (along with others shown elsewhere).

$$
\begin{aligned}
2003 &= \underline{1}\times 2^{10} + 979 \\
979 &= \underline{1}\times 2^{9} + 467 \\
467 &= \underline{1}\times 2^{8} + 211 \\
211 &= \underline{1}\times 2^{7} + 83 \\
83 &= \underline{1}\times 2^{6} + 19 \\
19 &= \underline{0}\times 2^{5} + 19 \\
19 &= \underline{1}\times 2^{4} + 3 \\
3 &= \underline{0}\times 2^{3} + 3 \\
3 &= \underline{0}\times 2^{2} + 3 \\
3 &= \underline{1}\times 2^{1} + 1 \\
1 &= \underline{1}\times 2^{0} + 0
\end{aligned}
$$

Thus, $2{,}003 = 11111010011_{two}$.

19. a) $1{,}331 \div 12 = 110$ r.11
$110 \div 12 = 9$ r.2
$9 \div 12 = 0$ r.9

Recall that 11 is represented in base twelve by the letter E. So, by grouping the remainders in bottom-to-top order, we know $1{,}331 =$ **$92E_{twelve}$**.

b) This problem can be done quickly by considering the powers of 5 $(1, 5, 25, 125, \ldots)$ and their relationship to 100. That is,

$$100 = 4 \times 25 = 4 \times 5^2 + 0 \times 5^1 + 0 \times 5^2 = \mathbf{400_{five}}$$

20. $1{,}000{,}000 = 2^{19} + 2^{18} + 2^{17} + 2^{16} + 2^{14} + 2^{9} + 2^{6} =$ **$11110100001001000000_{two}$**

CHAPTER 4: The Nature of Numbers

Problem Set 4.1

1. The set of ***natural numbers*** (also known as the counting numbers) consists of **{1, 2, 3, 4, 5, …}**.

3. As is explained on textbook page 163, ***subtraction*** is defined in terms of addition:

$a - b = x$ **means** $a = b + x$.

5. In terms of the specific operations of addition and multiplication, the ***commutative property*** is discussed on pp. 159-160. In more general terms, the commutative property may be stated as follows:

> **Given a set S and operation $\circ$, we say S is commutative for the operation $\circ$ provided:**
> **$a \circ b = b \circ a$ for any elements a and b in S.**

As the author of your text points out (in the margin on p. 159), making use of the root word "commute" can be a great aid in helping you remember the commutative property and avoid confusing it with others, such as the associative property. See this manual's explanation below for Problem 9.

7. ***Distributivity*** (i.e., the distributive property) is explained (for multiplication over addition) on p. 162. More generally, the distributive property may be stated as follows:

> **Given a set S and operations $\circ$ and $\otimes$, we say S is distributive for $\otimes$ over $\circ$ provided:**
> **$a \otimes (b \circ c) = (a \otimes b) \circ (a \otimes c)$ for any elements a, b, and c in S.**

9. **Answers vary**. Picking up on the comments made in the margin of textbook p. 159, we too recommend you make use of your prior (non-mathematical) understanding of these words and other forms of them (such as "commute" and "associate"). "Commute" can mean to switch places (like urban commuters do every day), while "associate" can mean to connect or group together (as you and those with whom you personally "associate" do).

11. Using the definition provided on p. 160, we find:

a) $3 \cdot 4$ means to add 4 three times, namely **4 + 4 + 4**.

b) $4 \cdot 3$ means to add 3 four times, namely **3 + 3 + 3 + 3**.

13. Use the definition provided on textbook p. 160.

a) Two terms of 184: $\boxed{\underbrace{184+184}_{\text{Note: 2 addends}}}$

b) A total of 184 terms of 2: $\boxed{\underbrace{2+2+2+\cdots+2}_{\text{184 addends}}}$

15. Again, we apply the definition provided on textbook p. 160.

a) A total of x terms of y: $\boxed{\underbrace{y+y+y+\cdots+y}_{x\text{ addends}}}$

b) A total of y terms of x: $\boxed{\underbrace{x+x+x+\cdots+x}_{y\text{ addends}}}$

17. **Commutative property**. Notice that 3 and 5 have "switched places."

19. **Associative property**. Observe that no switching of places ("commuting") has occurred; only the grouping ("associating") symbols have changed.

21. **Commutative property**. The parentheses (grouping symbols) have not moved; instead, the order of 2 and 3 has been "commuted."

23. **Commutative property** (for addition)

25. **Commutative property**

27. **Commutative property**, because the order of the words "sweet" and "dear" has been reversed.

29. **None are associative**

31. **Answers vary**. You will find dictionary definitions to be circular; that is, the definition of the word addition makes use of words like "add" and "sum," while the definitions of "add" and "sum" include the word "addition" (among others).

33. a) $\boxed{-i}$ b) $\boxed{-1}$ c) $\boxed{1}$ d) $\boxed{-i}$ e) $\boxed{i}$ f) $\boxed{-1}$

35. **Yes**, **reasons vary**. Since the product of any two elements in set F is an element of F, set F is closed for the operation $\times$.

37. a) **Yes**, **reasons vary**. Set F is associative for the operation $\times$ because it can be shown that $(a\times b)\times c = a\times(b\times c)$ for any elements a, b, c in F.

b) **Yes**, **reasons vary**. As the table provided in Problem 33 clearly demonstrates, $a \times b = b \times a$ for any a and b in set F.

39. **Yes**, **reasons vary**. The definition of the operation $\leftarrow$ guarantees that any set upon which this operation is used will be closed. That is to say, since $c \leftarrow d = c$ and since c is an element of the original set for which the operation $\leftarrow$ is defined, the original set (in this case, the natural numbers) is closed for the operation $\leftarrow$.

41. a) **The answer is □**. Notice that □ • Δ = Δ • □.

b) **The answer is O**. In contrast to part (a), notice that Δ • O ≠ O • Δ.

c) **Yes**, O • □ = □ • O since both are equal to Δ.
However, the answer to the second question is **no**. That is, as our mention in part (b) establishes, Δ • O ≠ O • Δ, so the set is not commutative for •.

d) **Yes**, (O • Δ) • Δ = □ • Δ = □ and O • (Δ • Δ) = O • Δ = □.

43. To answer this question, we must determine whether $a \downarrow (b \rightarrow c)$ equals $(a \downarrow b) \rightarrow (a \downarrow c)$ for any natural numbers a, b, and c.
For example, consider $a = 7$, $b = 1$, and $c = 4$:

$$7 \downarrow (1 \rightarrow 4) = 7 \downarrow 4 = 4 \text{ and } (7 \downarrow 1) \rightarrow (7 \downarrow 4) = 1 \rightarrow 4 = 4.$$

After trying several other examples on your own, you will become convinced that $a \downarrow (b \rightarrow c)$ is always equal to $(a \downarrow b) \rightarrow (a \downarrow c)$.
More generally, please remember this: to establish that a property does not hold, all one needs to do is find one counterexample. As can be more formally (and systematically) proven, no counterexamples exist for this problem, and, therefore, our answer to the question is: **Yes**, the distributive property holds.

45. Respectively, each of these should be mentally pictured as follows:

a) $5 \times (100 - 1) = 500 - 5 =$ **495**

b) $4 \times (90 - 2) = 360 - 8 =$ **352**

c) $8 \times (50 + 2) = 400 + 16 =$ **416**

47. **Yes**, $X \cap Y = Y \cap X$ for any sets X and Y. As was discussed back in textbook Section 1.4, Venn diagrams can be used to validate this. As shown on the next page, since both $X \cap Y$ and $Y \cap X$ yield the same Venn diagram, the operation of intersection with sets is commutative.

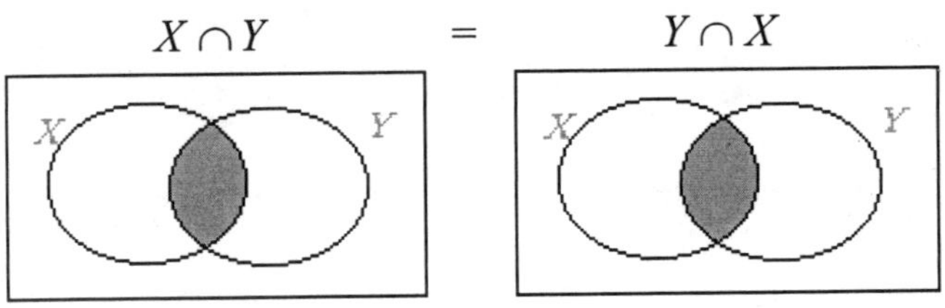

49. **Yes**, because the sum of any two even natural numbers is itself an even natural number.

51. **Yes**, because the product of odd natural numbers is always an odd natural number.

53. Simplifying the given definition of this operation is the vital step here. That is to say, recognizing that $a \odot b = 0 \cdot a + 1 \cdot b$ equals b immediately reveals the answer: **Yes**, since $0 \cdot a + 1 \cdot b = b$ and since b is an element of S, set S is closed for the operation of $\odot$.

55. It is **commutative** but **not associative**. Regarding the commutative property, you may find it helpful to begin with several numerical examples; for instance, $3 \otimes 4 = 14$ and $4 \otimes 3 = 14$. With each numerical example, you should be trying to figure out why the commutative property holds true. Eventually, you will notice since addition is a commutative operation for real numbers, it follows that:

$$a \otimes b = 2(a+b) = 2(b+a) = b \otimes a$$

Thus, because $a \otimes b = b \otimes a$ for any a and b in S, the operation is commutative for S.

To determine whether the associative property holds, you again should make up and evaluate several examples; for instance,

$$(2 \otimes 3) \otimes 4 = 10 \otimes 4 = 28 \text{ but } 2 \otimes (3 \otimes 4) = 2 \otimes 14 = 32.$$

Since we found a counterexample, the associative property does not hold. We should mention that you also could have shown that the associative property fails by evaluating the generalized expressions $(a \otimes b) \otimes c$ and $a \otimes (b \otimes c)$. While we leave the details in your capable hands, you ultimately will find:

$$(a \otimes b) \otimes c = 2[2(a+b)+c] = \underline{\underline{4a+4b+2c}}$$

yet

$$a \otimes (b \otimes c) = 2[a+2(b+c)] = \underline{\underline{2a+4b+4c}}$$

57. **Yes**. If you correctly completed the table in Problem 56, you found that the performance of this operation (i.e., ★) with any two elements of H always yields an element of H. Hence, H is closed for the operation of ★.

59. **Answers vary**. Here is one solution: $0+1+2-3+4-5-6+7-8+9$. Your text's author provides another (see p. 941 in Appendix H). Can you find a third correct solution?

Problem Set 4.2

1. **Wording varies**. As is explained in your textbook (pages 169-170), a ***prime number*** is a counting number that has exactly two divisors (namely, itself and 1).

3. **Wording varies**. The ***canonical representation of a number*** is the representation of that number as a product of prime factors with the factors written in ascending order. For additional details, refer to textbook p. 172.

5. **Answers vary**; ***l.c.m.*** stands for "least common multiple." It is the smallest number that all of the given numbers divide into evenly. The procedure for finding the l.c.m. is very clearly explained on textbook p. 174 (most notably, Figure 4.3 and Example 7). We leave it to you to translate this procedure into your own words.

7. a) **prime**. This may be determined by either (i) using the Sieve of Eratosthenes provided on p. 170 (Table 4.2); or (ii) confirming 59 is not divisible by any primes less than $\sqrt{59} \approx 7.7$ (namely, 2, 3, 5, and 7).

b) **not prime**; $57 = 3 \cdot 19$.

c) **not prime**. The number 1 only has one divisor, but a prime number, by definition, must have exactly two divisors.

d) **prime**. Since 1,997 is not divisible by any primes less than 45 (i.e., $\sqrt{1997} \approx 44.7$), it must be prime. Refer to pp. 170-171 for a more detailed explanation of this procedure.

9. a) **prime**; either (i) use the Sieve of Eratosthenes, or (ii) check to see whether any of the three primes that are less than $\sqrt{43} \approx 6.6$ (specifically, 2, 3, and 5) is a factor of 43.

b) **prime**; again, either use the sieve or confirm that 97 is not divisible by any prime less than or equal to 9 (i.e., $\sqrt{97} \approx 9.8$).

c) **not prime**; $171 = 3^2 \cdot 19$.

d) **prime**; because no prime less than $\sqrt{2001} \approx 44.7$ is a factor of 2001.

11. For assistance, refer to textbook p. 167.

a) **true** b) **false** c) **true** d) **false**

13. For assistance, refer to textbook p. 167. Additionally, for parts (b) through (d), we hope you avoided doing the actual division; instead, please make use of the appropriate divisibility rules (see the text's Table 4.1, p. 169).

a) **false** b) **true** c) **true** d) **true**

15. Since $\sqrt{300} \approx 17.3$, all you need to do is use the sieve-of-Eratosthenes technique (shown for 1 to 100 in textbook Table 4.2, p. 170). That is, you need to cross out all multiples (up to 300) of each prime (up to 17). The list of primes is:

2, 3, 5, 7, 11, 13, 17, 19, 23, 29, 31, 37, 41, 43, 47, 53, 59, 61, 67, 71, 73, 79, 83, 89, 97, 101, 103, 107, 109, 113, 127, 131, 137, 139, 149, 151, 157, 163, 167, 173, 179, 181, 191, 193, 197, 199, 211, 223, 227, 229, 233, 239, 241, 251, 257, 263, 269, 271, 277, 281, 283, and 293

17. Remember: ***Factor trees*** are a good way to determine prime factorizations. To review how this is done, see text Examples 4-5 (p. 172). Be sure, though, to write all final answers in canonical form.

a) $2^3 \cdot 3$ b) $2 \cdot 3 \cdot 5$ c) $2^2 \cdot 3 \cdot 5^2$ d) $2^4 \cdot 3^2$

19. Remember: As illustrated in text Examples 4-5 (p. 172), ***factor trees*** can be a helpful way to find prime factorizations. Be sure to write your final answers in canonical form.

a) $2^3 \cdot 3 \cdot 5$ b) $2 \cdot 3^2 \cdot 5$ c) $3 \cdot 5^2$ d) $3 \cdot 5^2 \cdot 13$

21. 83 is **prime**

23. 127 is **prime**

25. $377 = 13 \cdot 29$

27. $105 = 3 \cdot 5 \cdot 7$ Note: Since $3 \mid (1 + 0 + 5)$, we know $3 \mid 105$.

29. 67 is **prime**

31. $315 = 3^2 \cdot 5 \cdot 7$ Note: Since $9 \mid (3 + 1 + 5)$, it follows that $9 \mid 315$.

33. $567 = 3^4 \cdot 7$ Note: Since the sum $5 + 6 + 7$ is divisible by 9, it follows that $9 \mid 567$. We can then express 567 as the product of 9 and 63, from which the prime factorization and canonical representation readily follow.

35. $2{,}869 = \boxed{\mathbf{19 \cdot 151}}$

37. Since $60 = 2^2 \cdot 3 \cdot 5$ and $72 = 2^3 \cdot 3^2$, the g.c.f. $= 2^2 \cdot 3 = 4 \cdot 3 = \boxed{\mathbf{12}}$ and the l.c.m. $= 2^3 \cdot 3^2 \cdot 5 = 8 \cdot 9 \cdot 5 = \boxed{\mathbf{360}}$. For additional details, see text Examples 6 & 7.

39. $12 = 2^2 \cdot 3$, $54 = 2 \cdot 3^3$, $171 = 3^2 \cdot 19$

g.c.f. $= \boxed{\mathbf{3}}$ l.c.m. $= 2^2 \cdot 3^3 \cdot 19 = 4 \cdot 27 \cdot 19 = \boxed{\mathbf{2{,}052}}$

41. $9 = 3^2$, $12 = 2^2 \cdot 3$, $14 = 2 \cdot 7$

g.c.f. $= \boxed{\mathbf{1}}$ (i.e., the numbers 9, 12, and 14 are relatively prime)

l.c.m. $= 2^2 \cdot 3^2 \cdot 7 = 4 \cdot 9 \cdot 7 = \boxed{\mathbf{252}}$

43. $90 = 2 \cdot 3^2 \cdot 5$, $75 = 3 \cdot 5^2$, $120 = 2^3 \cdot 3 \cdot 5$

g.c.f. $= 3 \cdot 5 = \boxed{\mathbf{15}}$

l.c.m. $= 2^3 \cdot 3^2 \cdot 5^2 = 8 \cdot 9 \cdot 25 = \boxed{\mathbf{1{,}800}}$

45. You need to find the first night off that Bill and Sue have in common. The notion of "first common" should suggest to you the "least common multiple" (l.c.m.), which for 6 and 8 is 24. Therefore, given that Bill and Sue are both off tonight, it will be **24 nights (or 3 weeks and 3 days)** before they simultaneously are off again.

47. **Answers vary**. One pattern worthy of note is that every prime (with the exceptions of 2 and 3) is either one more or one less than a multiple of 6. For example, 42 is a multiple of 6, and the prime 41 is one less than 42 while the next prime 43 is one more than 42.

49. Before you unnecessarily spend time writing all the numbers from 1 to 100, first take a close look at Table 4.4 and make some observations. For instance, notice that the only numbers remaining after the first two steps of the "crossing out" process have been executed are numbers in the first and third columns. Consequently, there is no need to list all the numbers from 1 to 100. Instead, you can obtain the rest of the numbers in the first column by repeatedly adding six to 43 (i.e., 49, 55, 61…). Then, the numbers in the third column (i.e., 51, 57, 63…) can be obtained either in similar fashion or by adding two to each number in the first column.

We then turn our attention to the remaining steps of this "crossing out" process. You will find that crossing out every 7th number removes from the list 19, 39, 61, and 81. Next, we remove every 9th number, which deletes 27, 57, and 91. After that, removing every 13th number eliminates 45, and 97. Then, crossing out every 15th number gets rid of 55. Finally, crossing out every 21st number removes 85. The process ceases at this point because the next step calls for the removal of every 25th

number yet the remaining list only contains 23 numbers. Thus, the lucky numbers are: **1, 3, 7, 9, 13, 15, 21, 25, 31, 33, 37, 43, 49, 51, 63, 67, 69, 73, 75, 79, 87, 93, 99**.

51. Once we move beyond 7, any group of three consecutive odd numbers is certain to include one number that is a multiple of 3 (and, therefore, a composite number); for example, 37, 39, 41 includes the composite number 39. Expressed in more formal mathematical terms, we could write the following:
Let $2n+1$, $2n+3$, and $2n+5$ be any three consecutive odd numbers. If $n=1$, we have the prime triplets 3, 5, and 7. If $n>1$, then one of the three numbers must be divisible by 3. Suppose the first is divisible by 3, then the three numbers are not prime triplets. If the first is not divisible by 3, then dividing it by 3 leaves a remainder of either 1 or 2. If it leaves a remainder of 1, then the middle number is divisible by 3. If it leaves a remainder of 2, then the last number is divisible by 3. In all cases, the numbers will fail to be prime triplets.

53. a) **Yes**, the set is closed for the given operation because the l.c.m. of any pair of numbers in set S is itself an element of set S. This is so because set S consists of all the divisors of 30, and the l.c.m. of any pair of divisors of 30 will itself be a divisor of 30.

b) **Yes**. When finding the l.c.m. of a trio of numbers, the way in which you choose to associate the numbers makes no difference.

c) **Yes**, $a \,\mathbb{M}\, b = b \,\mathbb{M}\, a$.

55. One example is $\mathbf{2\cdot 3\cdot 5\cdot 7\cdot 11\cdot 13+1=30{,}031}$, which is not prime because $30{,}031 = 59\cdot 509$.

57. **Answers vary**. Some possibilities are $1^2+1=\mathbf{2}$, $4^2+1=\mathbf{17}$, $6^2+1=\mathbf{37}$, $10^2+1=\mathbf{101}$, and so on. A computer could be used to help answer this question. Refer to the answers in the back of your text for additional possibilities.

59. If $N=5$, then $2^{N-1}(2^N-1)=2^4(2^5-1)=16(31)=496$. The canonical representation of 496 is $2^4\cdot 31$. This leads us to the proper divisors of 496, which are 1, 2, 4, 8, 16, 31, 62, 124, and 248. Because the sum of its divisors is 496, the number 496 is said to be a ***perfect number***.

Problem Set 4.3

1. **Answers vary**. Refer to textbook pages 180-181. As always, we encourage you to express the provided information in <u>your</u> own words.

3. **Answers vary**. Express (in your own words) the information provided on pp. 182-183.

5. **Wording varies**. The expression $0 \div 5$ means "0 divided by 5"; that is, $\frac{0}{5}$ or $5\overline{)0}$. This expression asks, "What number multiplied by 5 yields a product of zero?" The answer, of course, is 0.

In contrast, $5 \div 0$ means "5 divided by 0," which can be expressed as $\frac{5}{0}$ or $0\overline{)5}$. It asks, "What number multiplied by 0 equals a product of 5?" Since there is no such number, $5 \div 0$ is said to be "undefined."

7. a) $\boxed{30}$ b) $\boxed{30}$ c) $\boxed{0}$ d) $\boxed{-30}$ e) $8 - 8 = \boxed{0}$

9. a) $\boxed{8}$ b) $\boxed{-2}$

11. a) $\boxed{10}$ b) $\boxed{-4}$

13. a) $\boxed{-7}$ b) $|-14+2| = |-12| = \boxed{12}$

15. a) $\boxed{3}$ b) $\boxed{-3}$

17. a) $\boxed{-18}$ b) $\boxed{-20}$

19. a) $\boxed{-70}$ b) $\boxed{70}$

21. a) $\boxed{-3}$ b) $\boxed{7}$

23. a) $\boxed{132}$ b) $\boxed{-1}$

25. a) $\boxed{-56}$ b) $\boxed{-75}$

27. a) $-2^2 = -(2^2) = -(2 \cdot 2) = -(4) = \boxed{-4}$

b) $(-2)^2 = (-2) \cdot (-2) = \boxed{4}$

29. a) $\boxed{150}$ b) $-12+[(-4)+(-3)]=-12+(-7)=\boxed{-19}$

31. a) $|-5-(-10)|=|-5+10|=|5|=\boxed{5}$

b) $|5-(-5)|=|5+5|=|10|=\boxed{10}$

33. a) $-4-8=-4+(-8)=\boxed{-12}$

b) $\boxed{15}$

35. a) $|-8|+(-8)=8+(-8)=\boxed{0}$

b) This time (in the interest of exposing you to a variety of mathematical presentation styles) we present the solution in a vertical format:

$$\begin{aligned}-9-(4-5)&=-9-[4+(-5)]\\&=-9-(-1)\\&=-9+1\\&=\boxed{-8}\end{aligned}$$

37. a) $-6-(-6)=-6+6=\boxed{0}$

b) $-18-5=-18+(-5)=\boxed{-23}$

39. a) $6+(-8)-5=6+(-8)+(-5)=\boxed{-7}$

b) $-2(-3)+(-1)(6)=6+(-6)=\boxed{0}$

41. a) $6-(-2)=6+2=\boxed{8}$

b) $-5-(-3)=-5+3=\boxed{-2}$

43. a) $\frac{-32}{-8}-5-(-7)=4+(-5)+7=\boxed{6}$

b) $\frac{-15}{-5}-4-(-8)=3+(-4)+8=\boxed{7}$

45. a) $|-3|-[-(-2)]=|-3|-[2]=3+(-2)=\boxed{1}$

b) $15-(-7)=15+7=\boxed{22}$

47. a) $[48\div(-6)]\div(-2)=-8\div(-2)=\boxed{4}$

b) $48\div[(-6)\div(-2)]=48\div 3=\boxed{16}$

49. a) $-5(2)+(-3)(-4)-6(-7)=-10+12+42=\boxed{44}$

b) $-8(3)-6(-4)-2(-8)=-24+24+16=\boxed{16}$

51. a) $\boxed{1}$ b) $\boxed{-1}$ c) $\boxed{-1}$

d) For any natural number k, the product $2k$ is an <u>even</u> natural number. So, it follows that the result of raising -1 to an even power is 1. That is, $(-1)^{2k}=\boxed{1}$.

e) For any natural number k, the quantity $2k+1$ will equal an <u>odd</u> natural number. [*If this fact is not readily apparent to you, we encourage you to take the time to experiment with this quantity; that is, choose several values for k — both even and odd — and evaluate $2k+1$.]* Thus, since $2k+1$ is odd and since the result of raising -1 to odd powers is -1, $(-1)^{2k+1}=\boxed{-1}$.

53. Recall that every natural number is also a whole number and every whole number is also an integer. The following Venn diagram express this relationship:

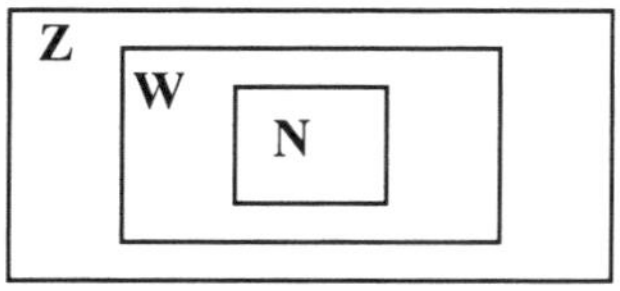

Expressed in terms of "subset" notation (as was done in textbook Section 1.4), we write: $\mathbf{N}\subseteq\mathbf{W}\subseteq\mathbf{Z}$,
which means $\mathbf{N}$ is a subset of $\mathbf{W}$, and $\mathbf{W}$ is in turn a subset of $\mathbf{Z}$.

55. a) $\boxed{\text{Wording varies}}$. Given a set S and operation $\circ$, we say S is ***commutative*** for the operation $\circ$ if $a\circ b=b\circ a$ for any elements a and b in S.

b) $\boxed{\text{Yes}}$, addition is commutative for the set of integers. Try several examples.

c) $\boxed{\text{No}}$, subtraction is not commutative for the integers because, for example, $7-5=2$ but $5-7\neq 2$.

d) $\boxed{\text{Yes}}$, multiplication is commutative for the integers. Since multiplication is commutative for the set of whole numbers and since the determination of signs (as shown on textbook p. 183) is commutative, multiplication is commutative for the integers.

e) $\boxed{\text{No}}$, division is not commutative for the set of integers, as illustrated by the fact that $8\div 4=2$ while $4\div 8\neq 2$.

57. To establish that the set of integers is closed for subtraction, use the facts that (i) the integers are closed for addition and (ii) the opposite of any integer is also an integer. From these two facts and the definition of subtraction (p. 184), the desired result follows.

59. a) **123,456,687,765,433**. Please note that the displays of some calculator models may not display enough digits to see this large number in its entirety. Other models will display the result in scientific notation, in which case you may still be able to (perhaps with some resourceful calculator usage on your part) determine the product.. Incidentally, as an alternative to the calculator, you may find it instructive to enter this product in a spreadsheet program (e.g., Microsoft Excel) and adjust the display format so that you can see this 15-digit product.

b) We make use of the inductive reasoning approaches (i.e., "seeking patterns") explored in Section 1.2 of your text. In this particular case, we begin with the following products:

$$1 \times 9 = 9$$

$$12 \times 99 = 1{,}188$$

$$123 \times 999 = 122{,}877$$

$$1234 \times 9999 = 12{,}338{,}766$$

An assortment of patterns exist here, so many in fact that it is likely you will notice some we neglect to mention in this limited space. First, note that the number of digits in the product is equal to the combined number of digits of the two numbers being multiplied (also known as the "multiplicands"). Second, with the exception of 1×9, it appears that the product begins with a '1' each time, then counts its way upward (2, 3, ...) and stops one short of the number of digits in each multiplicand. Then, the highest digit reached is repeated. An '8' appears. It is followed by a count downward as many digits as was previously counted upward, and again the last digit is repeated. Additionally, notice how the sum of the repeated digits is equal to 9. Thus, based on these patterns (and perhaps others you discovered), the result is:

$$1{,}234{,}567 \times 9{,}999{,}999 = \boxed{123{,}456{,}687{,}765{,}433}.$$

Problem Set 4.4

1. **Answers vary**. A fraction is said to be ***reduced*** when its numerator and denominator have no common factor other than 1 (or -1). Refer to textbook pages 188-189 for additional details.

3. **Answers vary**. Said simply,

$$\frac{a}{b} \div \frac{c}{d} = \frac{a}{b} \cdot \frac{d}{c} = \frac{ad}{bc}.$$

If you do not understand "why" this is so, please take a few moments to do two things:

(i) read Example 3 (pp. 190-191), in which your text's author very thoroughly explains the basis for division of rational numbers; and

(ii) create several simple examples (e.g., $\frac{1}{2} \div 2$) that will enable you to make sense of what division truly means and why we change the division operation to multiplication and the divisor to its reciprocal. In creating your examples, you may find it helpful to relate the fractions to something familiar (money, a measuring cup, tool sizes, pizza pies, etc.).

5. $\boxed{\frac{a}{b}-\frac{c}{d}=\frac{a}{b}+(-1)\frac{c}{d}=\frac{a}{b}+\frac{-c}{d}=\frac{a}{b}\cdot\frac{d}{d}+\frac{-c}{d}\cdot\frac{b}{b}=\frac{ad}{bd}+\frac{-bc}{bd}=\frac{ad}{bd}-\frac{bc}{bd}=\frac{ad-bc}{bd}}$

7. a) Since the g.c.f. of 2 and 10 is 2, $\frac{2}{10}=\frac{2\cdot 1}{2\cdot 5}=\boxed{\frac{1}{5}}$.

b) The g.c.f. is 3. Therefore, $\frac{3}{12}=\frac{3\cdot 1}{3\cdot 2^2}=\frac{1}{2^2}=\boxed{\frac{1}{4}}$.

9. a) $\boxed{2}$ b) $\boxed{2}$

11. a) $\boxed{3}$ b) $\boxed{\frac{2}{3}}$

13. a) $\boxed{\frac{1}{8}}$ c) $\boxed{\frac{1}{3}}$

15. a) Since $78=2\cdot 3\cdot 13$ and $455=5\cdot 7\cdot 13$, the g.c.f. of 78 and 455 is 13. Thus, $\frac{78}{455}=\frac{2\cdot 3\cdot 13}{5\cdot 7\cdot 13}=\frac{2\cdot 3}{5\cdot 7}=\boxed{\frac{6}{35}}$.

b) $\boxed{\frac{3}{20}}$

17. a) Using the skills acquired back in Section 4.2, we find $2431=11\cdot 13\cdot 17$ and $3003=3\cdot 7\cdot 11\cdot 13$. So, their g.c.f. is $11\cdot 13=143$. Therefore,

$$\frac{2{,}431}{3{,}003}=\frac{11\cdot 13\cdot 17}{11\cdot 13\cdot 3\cdot 7}=\frac{17}{3\cdot 7}=\boxed{\frac{17}{21}}.$$

b) Since $47957=7\cdot 13\cdot 17\cdot 31$ and $54808=2^3\cdot 13\cdot 17\cdot 31$, their g.c.f. is $13\cdot 17\cdot 31=6851$. Thus,

$$\frac{47{,}957}{54{,}808}=\frac{7}{2^3}=\boxed{\frac{7}{8}}.$$

19. a) Since $35 = 5 \cdot 7$ and $15 = 3 \cdot 5$, their least common multiple (l.c.m.) is $3 \cdot 5 \cdot 7 = 105$, which leads to the following:

$$\frac{-12}{35} - \frac{8}{15} = \frac{-12}{35} + \frac{-8}{15} = \frac{-12}{35} \cdot \frac{3}{3} + \frac{-8}{15} \cdot \frac{7}{7} = \frac{-36+(-56)}{105} = \boxed{-\frac{92}{105}}$$

b) $2^{-1} + 3^{-1} = \frac{1}{2} + \frac{1}{3} = \frac{1}{2} \cdot \frac{3}{3} + \frac{1}{3} \cdot \frac{2}{2} = \frac{3}{6} + \frac{2}{6} = \boxed{\frac{5}{6}}$

21. a) $\frac{7}{9} - \frac{2}{3} = \frac{7}{9} + \frac{-2}{3} = \frac{7}{9} + \frac{-2}{3} \cdot \frac{3}{3} = \frac{7}{9} + \frac{-6}{9} = \boxed{\frac{1}{9}}$

b) $\frac{4}{7} - \frac{-5}{9} = \frac{4}{7} + \frac{5}{9} = \frac{4}{7} \cdot \frac{9}{9} + \frac{5}{9} \cdot \frac{7}{7} = \frac{36}{63} + \frac{35}{63} = \boxed{\frac{71}{63}}$

23. a) $\frac{5}{3} \div \frac{7}{12} = \frac{5}{3} \cdot \frac{12}{7} = \frac{5}{1} \cdot \frac{4}{7} = \boxed{\frac{20}{7}}$ b) $\frac{-2}{9} \div \frac{6}{7} = \frac{\overset{1}{\cancel{-2}}}{9} \cdot \frac{7}{\underset{3}{\cancel{6}}} = \boxed{-\frac{7}{27}}$

25. a) $3^{-1} + 5^{-1} = \frac{1}{3} + \frac{1}{5} = \frac{1}{3} \cdot \frac{5}{5} + \frac{1}{5} \cdot \frac{3}{3} = \frac{5}{15} + \frac{3}{15} = \boxed{\frac{8}{15}}$

b) $2^{-1} + 5^{-1} = \frac{1}{2} + \frac{1}{5} = \frac{1}{2} \cdot \frac{5}{5} + \frac{1}{5} \cdot \frac{2}{2} = \frac{5}{10} + \frac{2}{10} = \boxed{\frac{7}{10}}$

27. a) $\frac{2}{3} \times \frac{5}{7} = \boxed{\frac{10}{21}}$ b) $\frac{-1}{\underset{4}{\cancel{8}}} \times \frac{\overset{1}{\cancel{2}}}{-5} = \frac{-1}{-20} = \boxed{\frac{1}{20}}$

29. a) $7 \times 7^{-1} = 7 \times \frac{1}{7} = \boxed{1}$

b) $-14 \times 14^{-1} = -14 \times \frac{1}{14} = \frac{-14}{1} \times \frac{1}{14} = \boxed{-1}$

Part (b) of this problem illustrates the huge — and hopefully obvious — difference between a negative number (as in -14) and a number raised to a negative exponent (as in 14^{-1}).

31. a) $6 \div 6^{-1} = 6 \div \frac{1}{6} = 6 \cdot 6 = \boxed{36}$

b) $-5 \div 5^{-1} = -5 \div \frac{1}{5} = -5 \cdot 5 = \boxed{-25}$

33. a) $\left(\frac{3}{7} \cdot \frac{3}{5}\right) \div \frac{1}{2} = \left(\frac{9}{35}\right) \cdot \frac{2}{1} = \boxed{\frac{18}{35}}$

b) $\frac{2}{-3} \cdot \frac{-2}{15} = \frac{-4}{-45} = \boxed{\frac{4}{45}}$

35. a) $\frac{2}{3} - \frac{7}{12} = \frac{2}{3} \cdot \frac{4}{4} + \frac{-7}{12} = \frac{8}{12} + \frac{-7}{12} = \boxed{\frac{1}{12}}$

b) $\frac{-7}{24} + \frac{-13}{16} = \frac{-7}{24} \cdot \frac{2}{2} + \frac{-13}{16} \cdot \frac{3}{3} = \frac{-14}{48} + \frac{-39}{48} = \boxed{-\frac{53}{48}}$

37. a) $-5+(-5)^2=-5+25=\boxed{\mathbf{20}}$

b) $-3+(-3)^2=-3+9=\boxed{\mathbf{6}}$

39. a) $\boxed{-\frac{5}{43}}$ Here are two methods:

Method I: Treat this problem in three steps. First, perform the indicated addition in the numerator, namely $\frac{1}{2}+\frac{-2}{3}$. Second, perform the subtraction problem posed in the denominator (i.e., $\frac{5}{6}-\frac{-3}{5}$). Lastly, divide the resulting fraction in the numerator (from the first step) by the resulting fraction in the denominator (from the second step).

In succession, this three-step process appears as follows:

$$\frac{\left[\frac{1}{2}+\frac{-2}{3}\right]}{\left[\frac{5}{6}-\frac{-3}{5}\right]}=\frac{\left[\frac{1}{2}+\frac{-2}{3}\right]}{\left[\frac{5}{6}+\frac{3}{5}\right]}$$

$$=\frac{\left[\frac{1}{2}\cdot\frac{3}{3}+\frac{-2}{3}\cdot\frac{2}{2}\right]}{\left[\frac{5}{6}\cdot\frac{5}{5}+\frac{3}{5}\cdot\frac{6}{6}\right]}$$

$$=\frac{\left[\frac{3}{6}+\frac{-4}{6}\right]}{\left[\frac{25}{30}+\frac{18}{30}\right]}$$

$$=\frac{\left(\frac{-1}{6}\right)}{\left(\frac{43}{30}\right)}$$

$$=\left(\frac{-1}{6}\right)\div\left(\frac{43}{30}\right)$$

$$=\frac{-1}{6}\cdot\frac{30}{43}$$

$$=\frac{-1}{1}\cdot\frac{5}{43}$$

$$=\frac{-5}{43}$$

Method II: First determine the l.c.m. of all four "little" denominators in this problem; that is, the lcm of 2, 3, 5, and 6 is 30. Then multiply both the numerator $\left(\frac{1}{2}+\frac{-2}{3}\right)$ and the denominator $\left(\frac{5}{6}-\frac{-3}{5}\right)$ by 30, which require use of the distributive property. Multiplying by the l.c.m. in this manner, as you will see in a moment, gets rid of the "little" denominators and makes it easier to simplify the expression. As always, of course, it is essential that you know "why" any process is mathematically valid. In this particular case, multiplying "both" the numerator and denominator by 30 means we are multiplying the given expression by $\frac{30}{30}$, which is a different form of the number 1. Since multiplying an expression by 1 does not change the expression's original value, this process is fundamentally sound.

Here is the execution of this method:

$$\frac{\frac{1}{2}+\frac{-2}{3}}{\frac{5}{6}-\frac{-3}{5}}=\frac{\frac{1}{2}+\frac{-2}{3}}{\frac{5}{6}+\frac{3}{5}}$$

$$=\frac{30}{30}\cdot\frac{\left[\frac{1}{2}+\frac{-2}{3}\right]}{\left[\frac{5}{6}+\frac{3}{5}\right]}$$

$$=\frac{\left[30\cdot\frac{1}{2}+30\cdot\frac{-2}{3}\right]}{\left[30\cdot\frac{5}{6}+30\cdot\frac{3}{5}\right]}$$

$$=\frac{15+10(-2)}{5(5)+6(3)}$$

$$=\frac{15+(-20)}{25+18}$$

$$=\frac{-5}{43}$$

b) $\boxed{\frac{28}{33}}$ Again, we perform the two procedures described above.

<u>**Method I**</u>:

$$\frac{\left[\frac{1}{3}-\frac{-1}{4}\right]}{\left[\frac{7}{8}-\frac{3}{16}\right]}=\frac{\left[\frac{1}{3}+\frac{1}{4}\right]}{\left[\frac{7}{8}+\frac{-3}{16}\right]}$$

$$=\frac{\left[\frac{1}{3}\cdot\frac{4}{4}+\frac{1}{4}\cdot\frac{3}{3}\right]}{\left[\frac{7}{8}\cdot\frac{2}{2}+\frac{-3}{16}\right]}$$

$$=\frac{\left[\frac{4}{12}+\frac{3}{12}\right]}{\left[\frac{14}{16}+\frac{-3}{16}\right]}$$

$$=\frac{\left(\frac{7}{12}\right)}{\left(\frac{11}{16}\right)}$$

$$=\left(\frac{7}{12}\right)\div\left(\frac{11}{16}\right)$$

$$=\frac{7}{12}\cdot\frac{16}{11}$$

$$=\frac{7}{3}\cdot\frac{4}{11}$$

$$=\frac{28}{33}$$

Method II: Note that the l.c.m. of 3, 4, 8, and 16 is 48.

$$\frac{\frac{1}{3}-\frac{-1}{4}}{\frac{7}{8}-\frac{3}{16}}=\frac{\frac{1}{3}+\frac{1}{4}}{\frac{7}{8}+\frac{-3}{16}}$$

$$=\frac{48}{48}\cdot\frac{\left[\frac{1}{3}+\frac{1}{4}\right]}{\left[\frac{7}{8}+\frac{-3}{16}\right]}$$

$$=\frac{\left[48\cdot\frac{1}{3}+48\cdot\frac{1}{4}\right]}{\left[48\cdot\frac{7}{8}+48\cdot\frac{-3}{16}\right]}$$

$$=\frac{16+12}{6(7)+3(-3)}$$

$$=\frac{28}{33}$$

41. a) $\frac{-3}{4}\cdot\frac{119}{200}+\frac{-3}{4}\cdot\frac{81}{200}=\frac{-3}{4}\left(\frac{119}{200}+\frac{81}{200}\right)=\frac{-3}{4}\left(\frac{200}{200}\right)=\frac{-3}{4}(1)=\boxed{-\frac{3}{4}}$

b) $\frac{-3}{4}\left(\frac{119}{200}+\frac{81}{200}\right)=\frac{-3}{4}\left(\frac{200}{200}\right)=\frac{-3}{4}(1)=\boxed{-\frac{3}{4}}$

43. Since $144=2^4\cdot3^2$, $300=2^2\cdot3\cdot5^2$, and $108=2^2\cdot3^3$, their l.c.m. is $2^4\cdot3^3\cdot5^2=10{,}800$. Thus,

$$\frac{11}{144}+\frac{17}{300}+\frac{7}{108}$$

$$\frac{11}{144}\left(\frac{3\cdot5^2}{3\cdot5^2}\right)+\frac{17}{300}\left(\frac{2^2\cdot3^2}{2^2\cdot3^2}\right)+\frac{7}{108}\left(\frac{2^2\cdot5^2}{2^2\cdot5^2}\right)$$

$$\frac{11}{144}\left(\frac{75}{75}\right)+\frac{17}{300}\left(\frac{36}{36}\right)+\frac{7}{108}\left(\frac{100}{100}\right)$$

$$\frac{825}{10{,}800}+\frac{612}{10{,}800}+\frac{700}{10{,}800}$$

$$\boxed{\frac{2{,}137}{10{,}800}}$$

45. We first reduce $\frac{21}{51}$ to lowest terms (i.e., $\frac{7}{17}$). In determining the l.c.m. of 60, 90, and 17, first observe that 17 and 60 are relatively prime — as are 17 and 90. This observation suggests a simple two-step approach to determining the l.c.m. of 17, 60, and 90. First, we determine the l.c.m. of 60 and 90, which happens to be 180. Then, because 17 and 180 must be relatively prime, it follows that the l.c.m. of 17 and 180 equals the product of 17 and 180; that is, $17\cdot180=3{,}060$. Therefore,

$$\frac{7}{60}-\frac{19}{90}+\frac{7}{17}=\frac{7}{60}\left(\frac{51}{51}\right)-\frac{19}{90}\left(\frac{34}{34}\right)+\frac{7}{17}\left(\frac{180}{180}\right)$$

$$=\frac{357}{3{,}060}-\frac{646}{3{,}060}+\frac{1260}{3{,}060}$$

$$\boxed{\frac{971}{3{,}060}}$$

47. We begin by determining the least common denominator:

$$\left.\begin{aligned} 210 &= 2\cdot 3\cdot 5\cdot 7 \\ 124 &= 2^2\cdot 31 \\ 1085 &= 5\cdot 7\cdot 31 \end{aligned}\right\} \Rightarrow \text{l.c.m.} = 2^2\cdot 3\cdot 5\cdot 7\cdot 31 = 13{,}020$$

Thus, $\frac{143}{210}+\frac{15}{124}+\frac{11}{1{,}085}=\frac{143}{210}\left(\frac{2\cdot 31}{2\cdot 31}\right)+\frac{15}{124}\left(\frac{3\cdot 5\cdot 7}{3\cdot 5\cdot 7}\right)+\frac{11}{1{,}085}\left(\frac{2^2\cdot 3}{2^2\cdot 3}\right)=\frac{8{,}866+1{,}575+132}{13{,}020}=\boxed{\frac{10{,}573}{13{,}020}}$.

49. To show that the set of rational numbers is closed for subtraction, we must prove that, for any two rational numbers $\frac{a}{b}$ and $\frac{c}{d}$, their difference $\frac{a}{b}-\frac{c}{d}$ is also rational.

We know the difference $\frac{a}{b}-\frac{c}{d}$ equals $\frac{ad-bc}{bd}$. By the definition of a rational number (p. 188), we know a, b, c, and d are integers, with $b\neq 0$ and $d\neq 0$. Since the set of integers is closed for multiplication, the products ad, bc, and bd are integers. We also know bd is a nonzero integer because both b and d are nonzero. Finally, since ad and bc are integers and the set of integers is closed for subtraction, it follows that $ad-bc$ is an integer. So, by the definition of a rational number, $\frac{ad-bc}{bd}$ is a rational number, and thus we have shown that the set of rational numbers is closed for subtraction.

51. The set of integers is **closed for addition, subtraction, and multiplication**, but the set of integers is **not closed for division** (as illustrated by the fact that $4\div 5=\frac{4}{5}=0.8$).

53. **Yes**, the set of rationals is commutative for addition.

55. **Answers vary**; two such answers are $\frac{3}{4}=\frac{1}{2}+\frac{1}{4}$ and $\frac{3}{4}=\frac{1}{3}+\frac{1}{4}+\frac{1}{8}+\frac{1}{24}$.

57. **Answers vary**; two such answers are $\frac{67}{120}=\frac{1}{3}+\frac{1}{8}+\frac{1}{10}$ and $\frac{67}{120}=\frac{1}{2}+\frac{1}{20}+\frac{1}{120}$.

59. This problem requires two steps: first solve the problem and then confirm that the answer given on the papyrus is equivalent to the solution.

The problem as posed (i.e., "a quantity and its two-thirds and its half and its one-seventh together make 33") lends itself to writing an equation. Letting x represent the unknown quantity, the problem translates into the following equation, which we then proceed to solve:

$$x + \tfrac{2}{3}x + \tfrac{1}{2}x + \tfrac{1}{7}x = 33$$
$$\left(1 + \tfrac{2}{3} + \tfrac{1}{2} + \tfrac{1}{7}\right)x = 33$$
$$\tfrac{97}{42}x = 33$$
$$x = 33\left(\tfrac{42}{97}\right)$$
$$x = \tfrac{1386}{97} \text{ or } 14\tfrac{28}{97}$$

Simplify — either by hand or with a calculator — the answer on the papyrus. Doing so confirms that $14 + \frac{1}{4} + \frac{1}{56} + \frac{1}{97} + \frac{1}{194} + \frac{1}{388} + \frac{1}{679} + \frac{1}{776}$ equals $14\frac{28}{97}$. Hence,

the answer given on the papyrus is correct.

Problem Set 4.5

1. **Wording (and variable choice) varies**. The ***Pythagorean theorem*** states: Given a right triangle with sides (legs) of length x and y and hypotenuse of length z, $x^2 + y^2 = z^2$. Also, the converse of the preceding sentence holds; that is, if $x^2 + y^2 = z^2$ for a triangle with sides x, y, and z, then the triangle must be a right triangle.

3. **Answers vary**. The ***exact value*** of an irrational number cannot be represented by a terminating or repeating decimal. That is to say, when an irrational number is written as a decimal number, the decimal number is only an approximation of the exact value. For instance, the symbol $\sqrt{2}$ is considered an exact representation for the number square root of 2, whereas the number 1.414213562 (perhaps displayed on a calculator screen) is considered an ***approximation*** of the square root of 2.

5. **Answers vary**. By definition, the square root of 2 is a number which, when squared, equals 2. Looking at the last digit of the given computer approximation, we see that it is 4. Thus, squaring this computer approximation will result in a number whose last two digits are 16 (i.e., $4 \cdot 4$). Obviously, the number 2 does not end with digits of '1' and '6'. Therefore, this decimal approximation is not exactly equal to $\sqrt{2}$.

7. Recall that a dozen consists of 12, a gross 144, and a score 20. We check the truth/falsity of the limerick by expressing it in terms of mathematical notation:

$$\frac{12 + 144 + 20 + 3\sqrt{4}}{7} + 5 \cdot 11 = 9^2$$

Because both sides of the equality equal 81, the limerick is **true**.

9. If necessary, refer to the definition of square root and Example 1, both of which are provided on textbook page 197.

a) **30** b) **36** c) **807** d) **169**

11. For the problems given in this section, remember that we are to assume all variables represent nonnegative numbers. Your text's author stated this in his definition of square root (p. 197). Please know, however, if this requirement had not been imposed, our answers (and the way in which they were obtained) would differ from the following:

a) $\boxed{a}$ b) $\boxed{xy}$

c) $2\sqrt{b}\times 2\sqrt{b}=(2\times 2)\times(\sqrt{b}\times\sqrt{b})=\boxed{\mathbf{4b}}$ d) $\boxed{\mathbf{40}w}$

13. a) **Irrational; 3.162** b) **Irrational; 5.477**
c) **Irrational; 9.870** d) **Irrational; 7.389**

15. a) **Rational; 13** b) **Rational; 20**
c) **Irrational; 23.141** d) **Irrational; 22.459**

17. a) **Rational; 32** b) **Rational; 44**
c) **Irrational; 1.772** d) **Irrational; 1.253**

19. There are various approaches one could take with each of these problems. In order for you to become sufficiently proficient with radicals (which are so frequently encountered in mathematics), we recommend you try to work each problem in several ways. The more methods you know the better off you will be. In this spirit, let us provide you with two perspectives/approaches for each of the following problems. The first approach taken for each problem is based on the methods demonstrated in your text; the second is a variation — albeit at times a very slight one — of the first.

a) Approach I: $\sqrt{1{,}000}=\sqrt{10^3}=\sqrt{10^2\cdot 10}=\sqrt{10^2}\cdot\sqrt{10}=\boxed{\mathbf{10\sqrt{10}}}$

Approach II: $\sqrt{1000}=\sqrt{100\cdot 10}=\sqrt{100}\cdot\sqrt{10}=10\sqrt{10}$

b) Approach I:

$$\begin{aligned}\sqrt{2{,}800}&=\sqrt{2^4\cdot 5^2\cdot 7}\\&=\sqrt{2^2\cdot 2^2\cdot 5^2\cdot 7}\\&=\sqrt{2^2}\cdot\sqrt{2^2}\cdot\sqrt{5^2}\cdot\sqrt{7}\\&=2\cdot 2\cdot 5\cdot\sqrt{7}\\&=\boxed{\mathbf{20\sqrt{7}}}\end{aligned}$$

Approach II: $\sqrt{2{,}800} = \sqrt{28 \cdot 100}$

$= \sqrt{4 \cdot 7 \cdot 100}$

$= \sqrt{4} \cdot \sqrt{7} \cdot \sqrt{100}$

$= 2 \cdot \sqrt{7} \cdot 10$

$= 20\sqrt{7}$

c) Approach I: $\sqrt{2{,}240} = \sqrt{2^6 \cdot 5 \cdot 7}$

$= \sqrt{2^2 \cdot 2^2 \cdot 2^2 \cdot 5 \cdot 7}$

$= \sqrt{2^2} \cdot \sqrt{2^2} \cdot \sqrt{2^2} \cdot \sqrt{35}$

$= 2 \cdot 2 \cdot 2\sqrt{35}$

$= \boxed{8\sqrt{35}}$

Approach II: $\sqrt{2{,}240} = \sqrt{64 \cdot 35} = \sqrt{64} \cdot \sqrt{35} = 8\sqrt{35}$

d) Approach I: $\sqrt{4{,}410} = \sqrt{3^2 \cdot 7^2 \cdot 2 \cdot 5} = 3 \cdot 7\sqrt{2 \cdot 5} = \boxed{21\sqrt{10}}$

Approach II: $\sqrt{4{,}410} = \sqrt{441 \cdot 10} = \sqrt{441} \cdot \sqrt{10} = 21\sqrt{10}$

21. a) $\sqrt{\frac{1}{2}} = \frac{1}{\sqrt{2}} = \frac{1}{\sqrt{2}} \cdot \frac{\sqrt{2}}{\sqrt{2}} = \boxed{\frac{\sqrt{2}}{2}}$

b) $\sqrt{\frac{1}{3}} = \frac{1}{\sqrt{3}} = \frac{1}{\sqrt{3}} \cdot \frac{\sqrt{3}}{\sqrt{3}} = \boxed{\frac{\sqrt{3}}{3}}$

c) $\sqrt{\frac{3}{5}} = \frac{\sqrt{3}}{\sqrt{5}} = \frac{\sqrt{3}}{\sqrt{5}} \cdot \frac{\sqrt{5}}{\sqrt{5}} = \boxed{\frac{\sqrt{15}}{5}}$

d) $\sqrt{\frac{3}{7}} = \frac{\sqrt{3}}{\sqrt{7}} = \frac{\sqrt{3}}{\sqrt{7}} \cdot \frac{\sqrt{7}}{\sqrt{7}} = \boxed{\frac{\sqrt{21}}{7}}$

23. a) $\frac{1}{\sqrt{2}} \cdot \frac{\sqrt{2}}{\sqrt{2}} = \boxed{\frac{\sqrt{2}}{2}}$

b) $\frac{-1}{\sqrt{3}} \cdot \frac{\sqrt{3}}{\sqrt{3}} = \boxed{-\frac{\sqrt{3}}{3}}$

c) $\frac{2}{\sqrt{5}} \cdot \frac{\sqrt{5}}{\sqrt{5}} = \frac{2\sqrt{5}}{5}$ or $\boxed{\frac{2}{5}\sqrt{5}}$

d) $\frac{5}{\sqrt{10}} \cdot \frac{\sqrt{10}}{\sqrt{10}} = \frac{5\sqrt{10}}{10} = \boxed{\frac{1}{2}\sqrt{10}}$ or $\frac{\sqrt{10}}{2}$

25. a) $\boxed{\sqrt{x^2 + 4}}$ This radical expression, as given, is in its simplest form.

b) $\sqrt{(x+2)^2} = \boxed{x+2}$

27. a) $\sqrt{10^2 - 4(5)(-5)} = \sqrt{100 + 100} = \sqrt{200} = \sqrt{100 \cdot 2} = \boxed{10\sqrt{2}}$

b) $\sqrt{12^2 - 4(3)(12)} = \sqrt{144 - 144} = \sqrt{0} = \boxed{0}$

29. a) Here are two methods, both of which we recommend you become familiar with:

I. $\dfrac{6+2\sqrt{5}}{2}=\dfrac{2(3+\sqrt{5})}{2}=\dfrac{\not{2}(3+\sqrt{5})}{\not{2}}=\boxed{\mathbf{3+\sqrt{5}}}$

II. $\dfrac{6+2\sqrt{5}}{2}=\dfrac{6}{2}+\dfrac{2\sqrt{5}}{2}=3+\sqrt{5}$

b) Once again, here are two approaches:

I. $\dfrac{8-4\sqrt{3}}{4}=\dfrac{4(2-\sqrt{3})}{4}=\dfrac{\not{4}(2-\sqrt{3})}{\not{4}}=\boxed{\mathbf{2-\sqrt{3}}}$

II. $\dfrac{8-4\sqrt{3}}{4}=\dfrac{8}{4}-\dfrac{4\sqrt{3}}{4}=2-\sqrt{3}$

31. Again, we share two approaches to each problem:

a) Approach I: $\dfrac{3-9\sqrt{x}}{3}=\dfrac{\not{3}(1-3\sqrt{x})}{\not{3}}=\boxed{\mathbf{1-3\sqrt{x}}}$

Approach II: $\dfrac{3-9\sqrt{x}}{3}=\dfrac{3}{3}-\dfrac{9\sqrt{x}}{3}=1-3\sqrt{x}$

b) Approach I: $\dfrac{9+3\sqrt{x}}{-3}=\dfrac{\not{3}(3+\sqrt{x})}{-\not{3}}=\dfrac{3+\sqrt{x}}{-1}=-(3+\sqrt{x})=\boxed{\mathbf{-3-\sqrt{x}}}$

Approach II $\dfrac{9+3\sqrt{x}}{-3}=\dfrac{9}{-3}+\dfrac{3\sqrt{x}}{-3}=-3-\sqrt{x}$

33. a) $\sqrt{\dfrac{4x^2}{25y}}=\dfrac{2x}{5\sqrt{y}}=\dfrac{2x}{5\sqrt{y}}\cdot\dfrac{\sqrt{y}}{\sqrt{y}}=\dfrac{2x\sqrt{y}}{5y}$ or $\boxed{\mathbf{\dfrac{2x}{5y}\sqrt{y}}}$

b) $\dfrac{\sqrt{5y}}{\sqrt{16x}}=\dfrac{\sqrt{5y}}{4\sqrt{x}}=\dfrac{\sqrt{5y}}{4\sqrt{x}}\cdot\dfrac{\sqrt{x}}{\sqrt{x}}=\dfrac{\sqrt{5xy}}{4x}$ or $\boxed{\mathbf{\dfrac{1}{4x}\sqrt{5xy}}}$

35. $\dfrac{-(-2)-\sqrt{(-2)^2-4(6)(-3)}}{2(6)}=\dfrac{2-\sqrt{4+4(18)}}{12}=\dfrac{2-\sqrt{76}}{12}=\dfrac{2-\sqrt{4\cdot 19}}{12}$

$=\dfrac{2-2\sqrt{19}}{12}=\dfrac{2(1-\sqrt{19})}{12}=\boxed{\mathbf{\dfrac{1-\sqrt{19}}{6}}}$ or $\frac{1}{6}-\frac{1}{6}\sqrt{19}$

37. $\dfrac{-(-12)+\sqrt{(-12)^2-4(1)(-1)}}{2(1)}=\dfrac{12+\sqrt{144+4}}{2}=\dfrac{12+\sqrt{148}}{2}=\dfrac{12+\sqrt{4\cdot 37}}{2}$

$=\dfrac{12+2\sqrt{37}}{2}=\dfrac{2\left(6+\sqrt{37}\right)}{2}=\boxed{6+\sqrt{37}}$

39. First, sketch a picture. The 26-ft ladder is the hypotenuse of the right triangle; the vertical leg is 10 ft long; and the horizontal leg is the length we need to find. Letting $b=10$ and $c=26$, we obtain:

$$a^2+b^2=c^2 \quad \Rightarrow \quad a=\sqrt{c^2-b^2}=\sqrt{26^2-10^2}=\sqrt{576}=\boxed{\textbf{24 ft.}}$$

41. Saying that the carpenter "wants to be sure that the corner … is square" means she wants the sides to meet at a right angle, and, therefore, she wants the lengths of the sides to satisfy the Pythagorean theorem. So, with the legs of the right triangle being 6 ft. and 8 ft., the length of the diagonal (i.e., the hypotenuse) must be:

$$c=\sqrt{a^2+b^2}=\sqrt{6^2+8^2}=\sqrt{100}=\boxed{\textbf{10 ft.}}$$

43. In the statement of the problem, take note of the word "exact." If you are unsure about its relevance, revisit p. 198 of your text and this manual's solution to Problem 3. With that said, the exact length of the hypotenuse is:

$$c=\sqrt{a^2+b^2}=\sqrt{2^2+2^2}=\boxed{\sqrt{8}\textbf{ in. or }2\sqrt{2}\textbf{ in.}}$$

45. First, walking diagonally across the lot suggests the need for us to find the length of the hypotenuse:

$$c=\sqrt{a^2+b^2}=\sqrt{400^2+300^2}=\sqrt{2500}=500\text{ ft.}$$

Thus, since walking the length and width of the lot is a 700 ft walk (300 ft + 400 ft), one saves 700 ft – 500 ft = $\boxed{\textbf{200 ft}}$ by walking diagonally across the lot.

47. A simple sketch of this right-triangle situation may help you. The wind speed of 3 ft per second represents the horizontal leg of the triangle; the rate of ascent (4 ft per second) represents the vertical leg; and the actual path of the balloon (specifically, its speed) can be depicted by the hypotenuse. By way of the Pythagorean theorem, we find that the balloon is traveling at a rate of:

$$c=\sqrt{a^2+b^2}=\sqrt{(3\text{ ft/sec})^2+(4\text{ ft/sec})^2}=\sqrt{25\text{ ft}^2/\text{sec}^2}=5\text{ ft/sec.}$$

Hence, in three seconds, the balloon will have traveled a distance of $3\times 5=\boxed{\textbf{15 ft}}$.

49. $\boxed{\textbf{Answers vary}}$. Since $1=\sqrt{1}$ and $3=\sqrt{9}$, it follows that included among the infinitely many irrational numbers between 1 and 3 are $\sqrt{2}$, $\sqrt{3}$, $\sqrt{5}$, $\sqrt{6}$, $\sqrt{7}$ and

$\sqrt{8}$. Additionally, the number e (discussed on textbook p. 198) is another such number.

51. **Answers vary**. Upon observing that $\frac{1}{11}=\frac{1}{\sqrt{121}}$ and $\frac{1}{10}=\frac{1}{\sqrt{100}}$, we realize that among the infinitely many irrational numbers between $\frac{1}{11}$ and $\frac{1}{10}$ are $\frac{1}{\sqrt{101}}$, $\frac{1}{\sqrt{102}}$, $\frac{1}{\sqrt{103}}, \ldots, \frac{1}{\sqrt{120}}$.

53. Since all three plates are of "uniform" thickness, our choice (of whether to take the two smaller squares or the one large square) need not be based on the total volume of gold. Instead, we can base our decision on which option yields the greater total area. For a square with side of length s, recall that its area is $s \times s = s^2$.

a) Since the two smaller squares have sides 1 in. in length, each of their areas is $(1 \text{ in.})^2$, or 1 square inch. Thus, their combined area is 2 sq. in. Meanwhile, the larger square of length 2 in. on a side has an area of $(2 \text{ in.})^2 = 4$ sq. in. Since the area of the larger square (4 in^2) is greater than the combined area of the two smaller squares (2 in^2), we **choose the one large square**.

Now, before proceeding with parts (b)-(e), let us "look back" at the approach in part (a). That is to say, let us do what all good problem solvers do, as is advised by the 4th step of Pólya's problem-solving model, which has been an ongoing theme of your text (since the very first chapter). In this problem, by "look back" we mean seek ways to obtain the above solution more efficiently. *To decide which gold square(s) to select, we need to obtain the sum of the areas of the two smaller squares and compare this sum to the area of the one larger square*. With that established, let us continue:

b) The sum of the areas of the two smaller squares is $(3 \text{ in.})^2 + (4 \text{ in.})^2 = 25 \text{ in.}^2$, while the area of the large square is also $(5 \text{ in.})^2 = 25 \text{ in.}^2$. Since both choices consist of the same amount of gold, **choose either option**.

c) Total area of the 2 smaller squares $= (4 \text{ in.})^2 + (5 \text{ in.})^2 = 39 \text{ in.}^2$
Area of the 1 large square = $(7 \text{ in.})^2 = 49 \text{ in.}^2$
Since $49 \text{ in}^2 > 36 \text{ in}^2$, **choose the one large square**.

d) Total area of the 2 smaller squares $= (5 \text{ in.})^2 + (7 \text{ in.})^2 = 74 \text{ in}^2$
Area of the 1 large square $= (9 \text{ in.})^2 = 81 \text{ in}^2$
Since $81 \text{ in}^2 > 74 \text{ in}^2$, **choose the one large square**.

e) Total area of the 2 smaller squares $= (10 \text{ in.})^2 + (11 \text{ in.})^2 = 221 \text{ in}^2$
Area of the 1 large square $= (15 \text{ in.})^2 = 225 \text{ in}^2$
Since $225 \text{ in}^2 > 221 \text{ in}^2$, **choose the one large square**.

55.

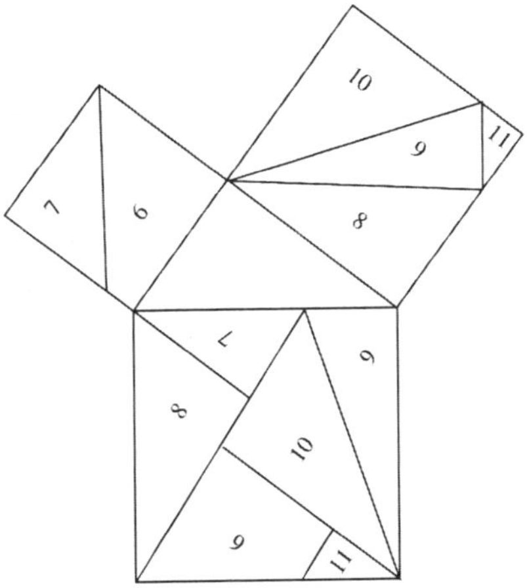

57. **Answers vary**. What Gauss meant was any natural number can be written as the sum of three (or less) triangular numbers. Recall that the triangular numbers consist of $1, 3, 6, 10, 15, 21, \ldots$. Pick any natural number and you will find no more than three triangular numbers added together will equal the number you picked.

59. Consider the set formed by collecting (in ascending order) those numbers located on the second diagonal of Pascal's triangle; that is, $\{1, 3, 6, 10, 15, 21, \ldots\}$. If we take the sum of any 2 adjacent numbers from this set (i.e., $1+3$, $3+6$, $6+10$, ...), we obtain the square numbers: $4, 9, 16, 25, \ldots$.
[*Remember*: we start counting with the 0th diagonal.]

Problem Set 4.6

1. **Answers vary**. When rational numbers are expressed as decimals, they either terminate or repeat. Irrational numbers, in contrast, have decimal representations that are nonterminating and nonrepeating.

You may find it helpful to associate the mathematical terms "rational" and "irrational" with their non-mathematical meanings. For example, if a person you know behaves "rationally," you might characterize their behavior as predictable, perhaps even systematic. Rational numbers behave in this way (i.e., their decimal representations either terminate or repeat). On the other hand, an irrationally behaving person is unpredictable (even erratic perhaps), which is an apt description for the decimal representations of irrational numbers.

3. **Answers vary**. First, refer to your author's explanation of the ***identity properties*** for addition and multiplication (pages 209-210). Then, refer to the generalized definition of an identity provided within the definition of a group (p. 211).

Finally, in a manner similar to what was demonstrated in our solution to Problem 1 above, can you form some connections between the mathematical term "identity" and its nonmathematical meanings?

5. For the definition of a ***group***, refer to the definition provided on textbook p. 211 and restated in Appendix H (the text's selected answers section).

7. a) **N, Z, Q, R** b) **Q, R** c) **Q′, R**
d) **Q′, R** e) **Q, R** f) **Q, R**
g) **Q, R** h) **N, Z, Q, R** i) **Z, Q, R**

9. a) $3 \div 2 = \boxed{1.5}$ b) $7 \div 10 = \boxed{0.7}$ c) $3 \div 5 = \boxed{0.6}$ d) $27 \div 15 = \boxed{1.8}$

11. a) $\boxed{0.\overline{6}}$ b) $\boxed{2.\overline{153846}}$ c) $\boxed{5}$ d) $\boxed{1.\overline{09}}$

13. a) $0.5 = \frac{5}{10} = \boxed{\frac{1}{2}}$ b) $0.8 = \frac{8}{10} = \boxed{\frac{4}{5}}$

15. a) $0.45 = \frac{45}{100} = \boxed{\frac{9}{20}}$ b) $0.234 = \frac{234}{1000} = \boxed{\frac{117}{500}}$

17. a) $98.7 = 98 + \frac{7}{10} = \frac{980}{10} + \frac{7}{10} = \boxed{\frac{987}{10}}$ b) $0.63 = \boxed{\frac{63}{100}}$

19. a) $15.3 = 15 + \frac{3}{10} = \frac{150}{10} + \frac{3}{10} = \boxed{\frac{153}{10}}$ b) $6.95 = \frac{600}{100} + \frac{95}{100} = \frac{695}{100} = \boxed{\frac{139}{20}}$

21. a) $\boxed{3.179}$ b) $\boxed{-5.504}$
c) $\boxed{1.901}$ d) $\boxed{-6.31}$

23. a) $\boxed{-0.13112}$ b) $\boxed{-65.415}$
c) $\boxed{12.4}$ d) $\boxed{0.85}$

25. a) $\boxed{7.46}$ b) $\boxed{-5.15}$
c) $\boxed{72.6}$ d) $\boxed{45.96}$

27. **Commutative property for multiplication**

29. **Distributive property**

31. **Commutative property**. Notice that the order of a and $(10+b)$ has been reversed, but the quantity enclosed by the parentheses has not been changed. If you are confused by this, we encourage you to take a few moments to compare this problem with Problem 30.

33. **Associative property**. Notice that the parentheses have been moved, but the order of the three colors (red, blue, yellow) remained unchanged.

35. **Distributive property** for set intersection ($\cap$) over union ($\cup$).

37. **No, not a group**. The set of natural numbers for subtraction fails to satisfy the closure property; for example, $3-7=-4$ but -4 is not a natural number.

39. **No, not a group**. The inverse property is not satisfied because the multiplicative inverses (i.e., the reciprocals) of $\{2, 3, 4, \ldots\}$ are not elements of the set of whole numbers. Additionally, take note of the fact that zero has no multiplicative inverse, and the definition of a group (see the bottom of p. 211) requires that each (i.e., every) element of the given set have an inverse.

41. **No, not a group** — for the same reason as was cited in Problem 39.

43. **No, not a group** because there is no inverse for the rational number 0 (zero). Be careful here not to confuse definitions. It is important to distinguish among the inverse property for multiplication (stated on the top part of textbook p. 211), the inverse property for a group (given on the bottom of p. 211), and the multiplicative inverse property for a field (p. 213). Notice that the inverse property for a group requires that "each $x \in S$" (i.e., every element of the given set) have an inverse, whereas the inverse property for multiplication and the multiplicative inverse property for a field both exclude zero. Therefore, when determining whether a set forms a group for a given operation, we must be certain that *all* the set's elements have an inverse.
Note: Although the set of rational numbers is not a group for multiplication, the set of nonzero rational numbers and the set of positive rational numbers are groups for the operation of multiplication.

45. a)

×	*1*	*2*	*3*	*4*
1	1	2	3	4
2	2	4	6	8
3	3	6	9	12
4	4	8	12	16

b)

*	1	2	3	4
1	2	2	2	2
2	4	4	4	4
3	6	6	6	6
4	8	8	8	8

c) **Closure**: Both operations, × and *, are not closed since, for example, $3 \times 2 = 6$ and $3 * 2 = 6$ but $6 \notin \{1, 2, 3, 4\}$.

Associative: The operation × is associative since multiplication of real numbers is associative. The operation * is not associative as the following example demonstrates:

$$(1 * 2) * 3 = 2 * 3 = 4 \text{ but } 1 * (2 * 3) = 1 * 4 = 2.$$

Identity: The identity for × is 1, but * has no identity.

Inverse: Neither operation satisfies the inverse property. For our given set $\{1, 2, 3, 4\}$ to satisfy this property with the operation ×, the set would need to contain the reciprocals of 2, 3, and 4. As for *, one cannot even begin to consider the inverse property since * has no identity element.

Commutative: We know that multiplication (×) is commutative. However, * is not commutative as illustrated by the following: $2 * 3 = 4$ but $3 * 2 = 6$.

47. **Answers vary**. There are infinitely many rational and irrational numbers between 2 and 3. Two such rational numbers are 2.5 and $\frac{17}{7} = 2.\overline{428571}$. As for irrational numbers between 2 and 3, we first note that $2 = \sqrt{4}$ and $3 = \sqrt{9}$. Consequently, among the infinitely many irrational numbers between 2 and 3 is $\sqrt{5}$, $\sqrt{6}$, $\sqrt{7}$ and $\sqrt{8}$. Additionally, the number e (introduced on textbook p. 198) and $\frac{3\pi}{4}$ (≈ 2.356) are two other irrational numbers — among infinitely many — between 2 and 3.

49. **Answers vary**. Once again, there are infinitely many rational and irrational numbers between 4.5 and 4.6. One such rational number is $\frac{41}{9} = 4.\overline{5}$, and one such irrational number is $\sqrt{21}$.

51. In order to be certain not to overlook any rational numbers between 0 and 1, we devise a systematic counting method. We begin with the smallest denominator possible (i.e., 2), and, as shown in the table below, we progressively increase the denominator:

Denominator	Rational #s between 0 and 1	Comments
2	$\frac{1}{2}$	
3	$\frac{1}{3}, \frac{2}{3}$	
4	$\frac{1}{4}, \frac{3}{4}$	$\frac{2}{4}$ is redundant with $\frac{1}{2}$ (already included above).
5	$\frac{1}{5}, \frac{2}{5}, \frac{3}{5}, \frac{4}{5}$	
6	$\frac{1}{6}, \frac{5}{6}$	$\frac{2}{6}, \frac{3}{6}, \frac{4}{6}$ were already included above.
⋮	⋮	

Therefore, the (infinite) set of rational numbers between 0 and 1 is as follows:

$$\boxed{\left\{\frac{1}{2}, \frac{1}{3}, \frac{2}{3}, \frac{1}{4}, \frac{3}{4}, \frac{1}{5}, \frac{2}{5}, \frac{3}{5}, \frac{4}{5}, \frac{1}{6}, \frac{5}{6}, \frac{1}{7}, \frac{2}{7}, \ldots\right\}}$$

53. $\boxed{a \circ b = a \cdot b}$

55. $\boxed{a \circ b = a + b + 1}$

57. $\boxed{a \circ b = 2(a + b)}$

59. $\boxed{a \circ b = a^2 + 1}$

Problem Set 4.7

1. **Answers vary**.

3. **Answers vary**.

5. a) $9 + 6 = \boxed{3}$. To adjust to 12-hour clock arithmetic, you may need to begin by thinking of $9 + 6$ as asking a question; for example, "if it's 9:00AM now, what time will it be 6 hours from now?" After you understand this (and the numbers in Table 4.6), you should begin to apply your knowledge of a modulo 12 system to 12-hour clock arithmetic, which is what it essentially is — except for the fact that 12, in 12-hour clock arithmetic, is replaced with 0 in modulo 12. Hence, $9 + 6 = 15 \equiv 3, (\text{mod} 12)$.

   b) $5 - 7 = \boxed{10}$ We provide two explanations.
   First, by using the definition of subtraction provided in your text (located immediately above Example 2), we know $5 - 7$ means $5 = 7 + x$. That is, in our 12-hour clock system, we need to find what number we add to 7 to obtain 5. That number is 10.

Secondly, we again remind you that your knowledge of modulo 12 arithmetic may be applied to 12-hour clock system problems. That is, since $5 \equiv 17, (\text{mod} 12)$, we may replace 5 with 17 and obtain the following:

$$5-7 \equiv 17-7 = 10, (\text{mod} 12)$$

7. a) Reinforcing the above solution to Problem 5a, we provide you with two perspectives. First, you may think of $7 + 10$ as asking you to determine the time 10 hours after 7 o'clock (or, if you prefer, 7 hours after 10 o'clock). Secondly, you can view this problem as asking you to determine $7+10, (\text{mod} 12)$. In either case, $7 + 10 = \boxed{5}$.

b) Again, we share two approaches. First, from the definition of subtraction in your text, we know $7 - 9$ means $7 = 9 + x$. In terms of 12-hour clock arithmetic, this means we need to find what number we add to 9 to obtain 7. That number is $\boxed{10}$. Alternatively, we again remind you that a 12-hour clock system is very similar to a modulo 12 system. In this instance, since $7 \equiv 19, (\text{mod} 12)$, we may replace 7 with 19 and obtain the following:

$$7-9 \equiv 19-9 = 10, (\text{mod} 12)$$

9. Using the approaches demonstrated in our solutions to Problems 7 and 9, we find:

a) $5 + 7 = \boxed{12}$. Please note, if you chose to restate this problem in terms of modulo 12, you would have found $5+7=12 \equiv 0, (\text{mod} 12)$. However, in the 12-hour clock system, zero is not included in the set $\{1, 2, \ldots, 12\}$; it is instead represented by the number 12.

b) $4 - 8 = \boxed{8}$.

11. a) The definition of multiplication (located immediately above Example 2 in your text) tells us that $4 \times 8 = 8+8+8+8$. From Table 4.6 (among other means), we know that in the 12-hour clock system $8+8=4$ and $4+4=8$, and from this it follows that:

$$\begin{aligned} 4 \times 8 &= 8+8+8+8 \\ &= (8+8)+(8+8) \\ &= (4)+(4) \\ &= \boxed{8} \end{aligned}$$

b) Since $2 \times 3 = 3+3$ (from the definition of multiplication) and $3+3=6$ (in 12-hour clock arithmetic), $2 \times 3 = \boxed{6}$.

13. a) In 12-hour clock arithmetic, $3 \times 5 = 5+5+5 = 3$. Thus, $(3 \times 5) - 7 = 3 - 7$. From the definition of subtraction, we know $3-7$ means $3 = 7 + x$. From Table 4.6, we see $x = 8$. Thus, $3 \times 5 - 7 = \boxed{8}$. *[Note: As was shown in some* *of our previous solutions, you may find it more efficient to apply modulo 12 arithmetic to this problem.]*

b) In a fashion similar to part (a), we find:

$$\begin{aligned}7+(3\times 2) &= 7+(2+2+2)\\ &= 7+6\\ &= \boxed{1}\end{aligned}$$

15. For additional details/reinforcement, please refer in your text to the definition of congruence mod m and Example 3 (p. 219).

a) Since $5+8=13$ and since 13 and 1 differ by 12 (which is a multiple of 6), the statement $5+8\equiv 1,(\text{mod}\,6)$ is **true**.

b) Since 9 and 1 differ by 8 (which is not a multiple of 7), it is the case that $4+5\not\equiv 1,(\text{mod}\,7)$. Thus, the given statement is **false**.

17. a) Since 47 and 2 differ by 45 (which is a multiple of 5), the statement is **true**.

b) Since 108 and 12 differ by 96 (which is a multiple of 8), the statement is **true**.

19. a) Since 1996 and 0 differ by 1996, this statement is **true**. It is worthy of mention that this particular problem reveals the following (more general) truth:

$$n\equiv 0,(\text{mod}\,n)$$

b) Since 246 and 150 differ by 96 (which is divisible by 6), the statement is **true**.

21. a) Noting that 15 is divisible by 5, $9+6=15\equiv$ **0, (mod 5)**.

b) **8, (mod 12)** Since $7\equiv 19,(\text{mod}\,12)$, we can replace 7 with 19, which yields:

$$\begin{aligned}7-11,(\text{mod}\,12) &\equiv 19-11,(\text{mod}\,12)\\ &\equiv 8,(\text{mod}\,12)\end{aligned}$$

23. a) Since the remainder of $7\div 4$ is 3, we obtain $5+2=7\equiv$ **3, (mod 4)**

b) Since $2\equiv 7$, (mod 5), we replace 2 with 7 and obtain:

$$2-4,(\text{mod}\,5)\equiv 7-4,(\text{mod}\,5)\equiv \boxed{3,\ (\text{mod}\ 5)}$$

25. a) $4+3=7\equiv$ **2, (mod 5)** because the remainder of $7\div 5$ is 2.

b) Since $6\equiv 14$, (mod 8), we replace 6 with 14 as follows:

$$6-12,(\text{mod}\,8)\equiv 14-12,(\text{mod}\,8)\equiv \boxed{2,\ (\text{mod}\ 8)}$$

27. a) $7+41=48\equiv$ **3, (mod 5)** because the remainder of $48\div 5$ is 3.

b) Since $4\equiv 15$, (mod 11), we can replace 4 with 15, which yields:

$$4-5,(\text{mod}\,11) \equiv 15-5,(\text{mod}\,11) \equiv \boxed{\mathbf{10, (mod\ 11)}}$$

29. a) Since $7 \equiv 0,(\text{mod}\,7)$, we can let $x+3=7$ or, more generally, we can let $x+3$ equal any multiple of 7 (i.e., $x+3=7k$). The only values of x that will satisfy the equation $x+3 \equiv 0,(\text{mod}\,7)$ are numbers which are three less than multiples of 7 (e.g., 4, 11, 18, 25). Thus, $\boxed{\mathbf{x = 4, (mod\ 7)}}$.

b) In modulo 5, there are five possible values of x; namely, 0, 1, 2, 3, and 4. We solve this equation by checking each of these values:

$x=0$: $4(0)=0,(\text{mod}\,5)$

$x=1$: $4(1)=4,(\text{mod}\,5)$

$x=2$: $4(2)=8 \equiv 3,(\text{mod}\,5)$

$x=3$: $4(3)=12 \equiv 2,(\text{mod}\,5)$

$x=4$: $4(4)=16 \equiv 1,(\text{mod}\,5)$

Hence, the only value of the variable that satisfies the equation $4x \equiv 1,(\text{mod}\,5)$ is $\boxed{\mathbf{x = 4, (mod\ 5)}}$.

31. a) There are several approaches you can take here. First, of course, is the approach of evaluating the equation for each possible value of x; in a modulo 6 system, that consists of x = 0, 1, 2, 3, 4, and 5. Alternatively (and more efficiently), we could choose to add 2 to both sides of the equation, from which we would obtain:

$$x-2 \equiv 3,(\text{mod}\,6)$$
$$x-2+\underline{2} \equiv 3+\underline{2},(\text{mod}\,6)$$
$$x \equiv 5,(\text{mod}\,6)$$

Therefore, this equation's solution set is $\boxed{\mathbf{5, (mod\ 6)}}$.

b) In modulo 7, there are seven possible values of x; namely, 0, 1, 2, 3, …, 6. We solve this equation by checking each of these values:

$x=0$: $3(0)=0,(\text{mod}\,7)$

$x=1$: $3(1)=3,(\text{mod}\,7)$

$x=2$: $3(2)=6,(\text{mod}\,7)$

$x=3$: $3(3)=9 \equiv 2,(\text{mod}\,7)$

$x=4$: $3(4)=12 \equiv 5,(\text{mod}\,7)$

$x=5$: $3(5)=15 \equiv 1,(\text{mod}\,7)$

$x=6$: $3(6)=18 \equiv 4,(\text{mod}\,7)$

Hence, the only value of the variable that satisfies the equation $3x \equiv 2,(\text{mod}\,7)$ is $\boxed{\mathbf{x = 3, (mod\ 7)}}$.

33. a) We solve this equation by evaluating $x^2,(\text{mod}\,4)$ for all possible modulo 4 values of x:

$x = 0$: $(0)^2 = 0 \not\equiv 1,(\text{mod}\,4)$

$x = 1$: $(1)^2 = 1,(\text{mod}\,4)$

$x = 2$: $(2)^2 = 4 \equiv 0 \not\equiv 1,(\text{mod}\,4)$

$x = 3$: $(3)^2 = 9 \equiv 1,(\text{mod}\,4)$

Thus, our solution set consists of **1,(mod 4) and 3,(mod 4)**.

b) Let us first review the meaning of division in a modulo arithmetic system. From your text, we know $x \div 4 = t$ means $x = 4t$. Said differently, $x \div 4$ asks one to find the number which when multiplied by 4 is congruent to x, $(\text{mod}\,9)$. So, to solve this equation, let us share two approaches, one that exhausts all the possible solutions (but relies on the approach previously demonstrated) and the other that is more efficient (and makes wise use of the definition of division in a modulo arithmetic system).

Our first approach requires us to consider all possible modulo 9 values of x:

$x = 0$: $0 \div 4 = 0$ because $0 = 4(0)$; $0 \div 4 \equiv 0 \not\equiv 5,(\text{mod}\,9)$

$x = 1$: $1 \div 4 = 7$ because $1 \equiv 4(7)$; $1 \div 4 \equiv 7 \not\equiv 5,(\text{mod}\,9)$

$x = 2$: $2 \div 4 \equiv 5$ because $2 \equiv 4(5)$; $2 \div 4 \equiv 5,(\text{mod}\,9)$

$x = 3$: $3 \div 4 \equiv 3$ because $3 \equiv 4(3)$; $3 \div 4 \equiv 3 \not\equiv 5,(\text{mod}\,9)$

$x = 4$: $4 \div 4 = 1$ because $4 = 4(1)$; $4 \div 4 \equiv 1 \not\equiv 5,(\text{mod}\,9)$

$x = 5$: $5 \div 4 \equiv 8$ because $5 \equiv 4(8)$; $5 \div 4 \equiv 8 \not\equiv 5,(\text{mod}\,9)$

$x = 6$: $6 \div 4 \equiv 6$ because $6 \equiv 4(6)$; $6 \div 4 \equiv 6 \not\equiv 5,(\text{mod}\,9)$

$x = 7$: $7 \div 4 \equiv 4$ because $7 \equiv 4(4)$; $7 \div 4 \equiv 4 \not\equiv 5,(\text{mod}\,9)$

$x = 8$: $8 \div 4 = 2$ because $8 = 4(2)$; $8 \div 4 \equiv 2 \not\equiv 5,(\text{mod}\,9)$

Thus, the solution is **2,(mod 9)**.

The alternative (and more time-efficient) approach relies on our applying the definition of division (see textbook p. 217):

$$\frac{x}{4} \equiv 5,(\text{mod}\,9) \text{ means } x = 4 \cdot 5 = 20 \equiv 2,(\text{mod}\,9).$$

35. a) Given that k represents any natural number, $4k$ represents 4, 8, 12, 16, and so on. Since all of these numbers are multiples of 4, it follows that $4k \equiv 0,(\text{mod}\,4)$. Thus, the solution to the equation $4k \equiv x,(\text{mod}\,4)$ is:

x = 0, (mod 4).

b) Notice that $4k+2$ consists of the numbers $\{6,\ 10,\ 14,\ 18,\ \ldots\}$ and each of these numbers is congruent to $2,(\text{mod}\,4)$. Therefore, it follows that $4k+2 \equiv 2,(\text{mod}\,4)$, which leads us to the conclusion that **x = 2, (mod 4)**.

37. In a modulo 2 system, there are only two possible values of x to check:

$$x=0: \quad 5(0)^3-3(0)^2+70=70\equiv 0,(\text{mod}\,2)$$

$$x=1: \quad 5(1)^3-3(1)^2+70=72\equiv 0,(\text{mod}\,2)$$

Thus, $5x^3-3x^2+70\equiv 0,(\text{mod}\,2)$ is satisfied by **all x's (mod 2)**.

39. a) Assuming that today is Friday and given that it is labeled as the 6th day of the week (i.e., Sunday is considered the 1st day of the week), thirty days from today will be day 36 (i.e., 6 + 30). Since $36\equiv 1,(\text{mod}\,7)$, day 36 is the first day of the week, which is **Sunday**.

b) 195 days from today (Friday, the 6th day) equates to day 201 (i.e., 6 + 195). Since $201\equiv 5,(\text{mod}\,7)$, day 201 is equivalent to the 5th day of the week, which (based on Sunday being the week's 1st day) is **Thursday**.

c) 390 days from Friday (the week's 6th day) is day 396. Because $396\equiv 4,(\text{mod}\,7)$, day 396 is the 4th day of the week, namely **Wednesday**.

d) Assuming no leap years, three years is equivalent to $3\times 365=1095$ days. 1095 days from Friday is the $1095+6=1101$st day. So, since $1101\equiv 2,(\text{mod}\,7)$, the 1101st day is a **Monday**.

41. This problem bears a strong resemble to Example 5 (page 220). We begin by letting x equal the number of miles between your house and the aunt's house. Since there are six round trips (12 one-way trips), the total mileage is $12x$. Because the units digit has increased by 8 miles, we know $12x\equiv 8,(\text{mod}\,10)$. We then process to solve this equation in the same manner demonstrated previously with Problems 29-37. That is, we consider all possible values of x modulo 10 (i.e., $0,\ 1,\ 2,\ \ldots,\ 9$). Upon doing so, we find the solutions to the equation to be $x=4,(\text{mod}\,9)$ and $x=9,(\text{mod}\,9)$. Therefore, the possible distances (in miles) are:

{4, 14, 24, 34, …} or {9, 19, 29, 39, …}

43. Given our conclusions in Problem 41 and the fact that the distance is between 10 and 15 miles, her house must be **14 miles** away.

45. Let A = $\{0,\ 1,\ 2,\ 3,\ 4,\ 5,\ 6\}$. To confirm whether set A is a group for addition modulo 7, we need to check whether all of the following requirements are satisfied:

- *Closure*: Since addition (modulo 7) of any two elements of set A yields a result that is also an element of A, the closure requirement is satisfied.

- <u>*Identity*</u>: There is an identity element (namely, 0). That is to say, the sum of zero and any element x in set A is equal to that element of A (i.e., $x+0 \equiv x$ for all $x \in A$).
- <u>*Associative*</u>: It is easy enough to show that $(a+b)+c \equiv a+(b+c)$ for any elements a, b, and c belonging to A. Therefore, this requirement is satisfied.
- <u>*Inverse*</u>: Each element in A has an inverse. That is, for any x in A, there exists a unique element x' such that $x+x' \equiv 0$. Specifically,
  $0+0 \equiv 0, (\text{mod } 7)$; $1+6 \equiv 0, (\text{mod } 7)$; $2+5 \equiv 0, (\text{mod } 7)$;
  $3+4 \equiv 0, (\text{mod } 7)$; $4+3 \equiv 0, (\text{mod } 7)$; $5+2 \equiv 0, (\text{mod } 7)$;
  $$6+1 \equiv 0, (\text{mod } 7)$$

Since all four requirements are satisfied, the answer to the question is **yes**, set A is a group for addition modulo 7. More generally, by the way, it is true that all modular addition systems are groups. However, not all modulo multiplication systems are groups (as you'll see in some of the next several problems).

47.

+	***0***	***1***	***2***	***3***	***4***	***5***	***6***	***7***	***8***	***9***	***10***
0	0	1	2	3	4	5	6	7	8	9	10
1	1	2	3	4	5	6	7	8	9	10	0
2	2	3	4	5	6	7	8	9	10	0	1
3	3	4	5	6	7	8	9	10	0	1	2
4	4	5	6	7	8	9	10	0	1	2	3
5	5	6	7	8	9	10	0	1	2	3	4
6	6	7	8	9	10	0	1	2	3	4	5
7	7	8	9	10	0	1	2	3	4	5	6
8	8	9	10	0	1	2	3	4	5	6	7
9	9	10	0	1	2	3	4	5	6	7	8
10	10	0	1	2	3	4	5	6	7	8	9

$\times$	***0***	***1***	***2***	***3***	***4***	***5***	***6***	***7***	***8***	***9***	***10***
0	0	0	0	0	0	0	0	0	0	0	0
1	0	1	2	3	4	5	6	7	8	9	10
2	0	2	4	6	8	10	1	3	5	7	9
3	0	3	6	9	1	4	7	10	2	5	8
4	0	4	8	1	5	9	2	6	10	3	7
5	0	5	10	4	9	3	8	2	7	1	6
6	0	6	1	7	2	8	3	9	4	10	5
7	0	7	3	10	6	2	9	5	1	8	4
8	0	8	5	2	10	7	4	1	9	6	3
9	0	9	7	5	3	1	10	8	6	4	2
10	0	10	9	8	7	6	5	4	3	2	1

49. **No**, this set is not a group for multiplication modulo 11 because 0 does not have a multiplicative inverse. However, if 0 were excluded from this set, it would be a group for multiplication modulo 11. Additionally, please note that the fact that 11 is a prime number is important. That is, if Problems 47-52 had involved a set in which 0 were excluded but a composite number had been used [e.g., the set $\{1, 2, 3, 4, 5, 6, 7, 8, 9\}$ with multiplication modulo 10] the set still would not be a group for this operation because, for instance, the set would not be closed [e.g., $5 \times 8 = 40 \equiv 0 \pmod{10}$].

51. **No**. The operation of multiplication for this set is indeed commutative. However, since the set fails to be a group (for the reason cited in our answer to Problem 49), it cannot be a commutative group.

53. a) We begin by computing the following:

$$10(0)+9(5)+8(3)+7(4)+6(1)+5(3)+4(7)+3(2)+2(8)$$

This equals 168, which is congruent to $3, (\text{mod}11)$. Therefore, the check digit x must satisfy the equation $3+x \equiv 0, (\text{mod}11)$. The one-digit solution to this equation is $x=8$. Thus, the check digit is **8**.

b) Once again, we begin with the necessary computation:

$$10(0)+9(6)+8(9)+7(1)+6(0)+5(2)+4(3)+3(5)+2(6)$$

which equals 182. Since $182 \equiv 6, (\text{mod}11)$, the check digit x must satisfy the equation $6+x \equiv 0, (\text{mod}11)$. Because the one-digit solution to this equation is $x=5$, we know the check digit is **5**.

55. The possible check digits for ISBNs are **0, 1, 2, 3, ..., 10**. As for the book with ISBN 0-13-330325-X, the check digit X stands for **10**. This determination is made by employing the exact same procedure as was demonstrated in our solution to Problem 53 above.

57. **Answers vary**. There are many possible ways to write such a schedule. However, since this section of your text has largely been about modular arithmetic, this scheduling problem presents us with a valuable opportunity to apply that knowledge. One way to do so is to begin by (i) assigning eleven of the twelve teams the numbers 1 to 11 and (ii) considering the last team (which we will call number 12) as a sort of "left over" team. On the first day of play, we let all teams whose assigned team numbers sum to 1 (mod 11) play each other. Day 2 pairs those whose numbers sum to 2 (mod 11). Day 3 pairs those whose sum is 3 (mod 11). And so on.... Additionally, so that each team plays everyday (i.e., no team is idle), each day's last team plays the "left over" team (i.e., number 12). With that said, the schedule follows:

Day 1:	1 vs. 11;	2 vs. 10;	3 vs. 9;	4 vs. 8;	5 vs. 7;	6 vs. 12
Day 2:	2 vs. 11;	3 vs. 10;	4 vs. 9;	5 vs. 8;	6 vs. 7;	1 vs. 12
Day 3:	3 vs. 11;	4 vs. 10;	5 vs. 10;	6 vs. 8;	1 vs. 2;	7 vs. 12
Day 4:	4 vs. 11;	5 vs. 10;	6 vs. 9;	7 vs. 8;	1 vs. 3;	2 vs. 12
Day 5:	5 vs. 11;	6 vs. 10;	7 vs. 9;	1 vs. 4;	2 vs. 3;	8 vs. 12
Day 6:	6 vs. 11;	7 vs. 10;	8 vs. 9;	1 vs. 5;	2 vs. 4;	3 vs. 12
Day 7:	7 vs. 11;	8 vs. 10;	1 vs. 6;	2 vs. 5;	3 vs. 4;	9 vs. 12
Day 8:	8 vs. 11;	9 vs. 10;	1 vs. 7;	2 vs. 6;	3 vs. 5;	4 vs. 12
Day 9:	9 vs. 11;	1 vs. 8;	2 vs. 7;	3 vs. 6;	4 vs. 5;	10 vs. 12
Day 10:	10 vs. 11;	1 vs. 9;	2 vs. 8;	3 vs. 7;	4 vs. 6;	5 vs. 12
Day 11:	1 vs. 10;	2 vs. 9;	3 vs. 8;	4 vs. 7;	5 vs. 6;	11 vs. 12

59. Let x represent the total number of coins (doubloons). The given information enables us to write three equations:

(i) Since x coins divided equally among 17 pirates resulted in a remainder of 3 coins, we know $x \equiv 3,(\text{mod}\,17)$;

(ii) After 6 pirates were killed in a fight, 11 pirates remained. Since dividing the x doubloons among these 11 remaining pirates resulted in a remainder of 4, it follows that $x \equiv 4,(\text{mod}\,11)$;

(iii) After the shipwreck, 5 more pirates perished, leaving 6 pirates in all. When the x doubloons were equally divided among these 6 pirates, the remainder was 5 coins for the cook. This information translates into the equation $x \equiv 5,(\text{mod}\,6)$.

Thus, the question posed in the original problem amounts to finding the smallest possible solution to the following system of congruence equations:

$$\begin{cases} x \equiv 3,(\text{mod}\,17) \\ x \equiv 4,(\text{mod}\,11) \\ x \equiv 5,(\text{mod}\,6) \end{cases}$$

There are methods for solving systems of equations such as this; they include the Chinese Remainder Theorem and Diophantine equations, both of which are typically learned in a first course in an area of mathematics known as number theory. These techniques, however, are beyond the scope of your text, and because they are, we can only solve this problem by resorting to a more crude “brute force” approach. In so doing, we encourage you to make use of computer software or, if you know how, write a computer program to solve this problem.

To demonstrate, we chose to use a very popular computer spreadsheet program (Microsoft ® Excel) to solve this problem. In the event you wish to explore this further, our spreadsheet began as follows:

	A	B	C	D	E
1	1	=MOD(A1,17)	=MOD(A1,11)	=MOD(A1,6)	=AND(B1=3,C1=4,D1=5)
2	=A1+1	=MOD(A2,17)	=MOD(A2,11)	=MOD(A2,6)	=AND(B2=3,C2=4,D2=5)

We then replicated the contents of cells A2 through E2 for row 3, row 4, and onward (for 1000+ rows). The output displayed looked like this:

	A	B	C	D	E
1	1	1	1	1	FALSE
2	2	2	2	2	FALSE
3	3	3	3	3	FALSE
	⋮	⋮	⋮	⋮	⋮
783	783	1	2	3	FALSE
784	784	2	3	4	FALSE
785	785	3	4	5	TRUE
786	786	4	5	0	FALSE
787	787	5	6	1	FALSE
788	788	6	7	2	FALSE
	⋮	⋮	⋮	⋮	⋮

Without going into any further detail, note that the first cell to contain the word "TRUE" is E785. Thus, after the deaths of all the pirates, the smallest number of coins the cook would have finally received is **785**. Upon further exploration of such a spreadsheet, one would find the next smallest solution to be 1907. The general solution to the above system of congruences turns out to be $x \equiv 785, (\text{mod} 1122)$, and notice that 1122 is equal to $17 \times 11 \times 6$.

Problem Set 4.8

1. To encode this message, we have to encode the three words themselves, the two spaces between the words, and the period at the end of the message. If desired, we can use Figure 4.21 (textbook page 228) to help us. The result is:

14-5-22-5-18-29-19-1-25-29-14-5-22-5-18-28

3. **25-15-21-29-2-5-20-29-25-15-21-18-29-12-9-6-5-28**

5. Problems 5-8 require reversing the process performed in solving Problems 1-4. Before doing these problems, you may wish to either make use of Figure 4.21 (p. 228) or, if preferred, construct your own decoding key. The resulting decoded message is:

ARE WE HAVING FUN YET

7. **FAILURE TEACHES SUCCESS**

9. As the author of your text states (on p. 229), "the decoding key is the inverse [function] of the encoding key." Therefore, since the encoding key is to multiply by 8, the decoding key is to **divide by 8**.

11. When we have a multi-step encoding key, the decoding key must consist of the inverse of each of these steps and must be performed in reverse order. In the event this does not make intuitive sense to you, please take a moment to test it. Pick a number, say 17. If we multiply it by 20 and add 2, the encoded result is 342. To decode 342, we must first subtract 2 (which yields 340) and then divide by 20. This process successfully brings us back to our original number (in this case, 17). Thus the decoding key for "times 20, add 2" is: **Subtract 2, divide by 20**.

13. As we explained in our solution to Problem 11, the decoding key for this three-step encoding key must consist of the inverse of each of these three steps and the order must be reversed. Therefore, the decoding key for "multiply by 4 and add 2, then double the result" is: **Divide by 2, subtract 2, then divide by 4**.

15. The process of encoding a message consists of four steps:

(i) replacing each letter, space, and punctuation mark with its numerical value;
(ii) applying the encoding key to these numerical values;
(iii) "modulating" (i.e., writing each of these answers modulo 29);
(iv) rewriting these modulated numbers in terms of letters/spaces/punctuation.

Note that Figure 4.21 (p. 228) can make it much easier (and faster) to perform this process' first, third, and fourth steps.
You may remember that we previously encountered this problem's message ("NEVER SAY NEVER.") in Problem 1 of this section. Therefore, we have previously performed the first of our four steps.

(i) N E V E R   S A Y   N E V E R .
14-5-22-5-18-29-19-1-25-29-14-5-22-5-18-28
(ii) We apply the encoding key (in this case, multiply by 3) to each of these values:
42-15-66-15-54-87-57-3-75-87-42-15-66-15-54-84
(iii) Using Figure 4.21 as an aid, we modulate the numbers from step (ii):
13-15-8-15-25-29-28-3-17-29-13-15-8-15-25-26
(iv) We obtain our coded message by replacing each of these modulated numbers with its assigned letter (or space or punctuation mark):

MOHOY .CQ MOHOYZ

17. Recall that this message was used earlier in this section (see Problem 3). To encode it, we apply the four-step process explained in our solution to Problem 15 above:

(i) Y O U   B E T   Y O U R   L I F E .
25-15-21-29-2-5-20-29-25-15-21-18-29-12-9-6-5-28
(ii) We apply this two-step encoding key (i.e., first multiply by 2 and then add 5):
55-35-47-63-9-15-45-63-55-35-47-41-63-29-23-17-15-61
(iii) We modulate these numbers (modulo 29) as follows:
26-6-18-5-9-15-16-5-26-6-18-12-5-29-23-17-15-3

(iv) By replacing each of these modulated numbers with its assigned letter (space, or punctuation mark), we obtain our encoded message:

ZFREIOPEZFRLE WQOC

19. To decode this message, we must reverse the previously described four-step process (as demonstrated in our solutions to Problems 15 and 17). Example 1 (on page 229 in your text) illustrates how this is done.

First, we use Figure 4.21 to write the encoded message in terms of numerical values:

X E J S B Q LE J S B Q , . C R CG G U E Q Z
24-5-10-19-2-17-29-12-5-10-19-2-17-29-27-28-29-3-29-18-3-7-7-29-21-5-17-26

Second, we determine the decoding key. Since the encoding key is to "multiply by 3," the decoding key is to divide by 3. Before doing this, however, we replace all the numbers above that are not divisible by 3 with (congruent modulo 29) numbers that are divisible by 3:

24-63-39-48-60-75-87-12-63-39-48-60-75-87-27-57-87-3-87-18-3-36-36-87-21-63-75-84

We now apply the decoding key (i.e., divide by 3):

8-21-13-16-20-25-29-4-21-13-16-20-25-29-9-19-29-1-29-6-1-12-12-29-7-21-25-28

Finally, we replace (with the assistance of textbook Figure 4.21) each of these numbers with its corresponding letter space, or punctuation mark. The resulting decoding message is:

HUMPTY DUMPTY IS A FALL GUY.

21. Just like we did in our solution to Problem 19, we begin by writing the encoded message in terms of numerical values:

S K A X B V I X A V X V G K Q D V . W S Y Q C L T
19-11-1-24-2-22-9-24-1-22-24-22-7-11-17-4-22-28-23-19-25-17-3-12-20

Second, since the encoding key is "multiply by 2 and subtract 7," the decoding key is add 7 and then divide by 2. Let us first add 7 to each numerical value:

26-18-8-31-9-29-16-31-8-29-31-29-14-18-24-11-29-35-30-26-32-24-10-19-27

Then, before we can divide by 2, we replace all the numbers above that are not divisible by 2 with (congruent modulo 29) numbers that are divisible by 2:

26-18-8-60-38-58-16-60-8-58-60-58-14-18-24-40-58-64-30-26-32-24-10-48-56
Now, we divide each number by two:

13-9-4-30-19-29-8-30-4-29-30-29-7-9-12-20-29-32-15-13-16-12-5-24-28

Lastly, we obtain the resulting decoded message by replacing (with the assistance of textbook Figure 4.21) each of these numbers with its corresponding letter space, or punctuation mark:

MIDAS HAD A GILT COMPLEX.

23. To encode this message, we apply the four-step procedure previously explained in our solutions to Problems 15 and 17:

(i) S H O W M E A D R O P O U T F R O M A
19-8-15-23-29-13-5-29-1-29-4-18-15-16-15-21-20-29-6-18-15-13-29-1-29-

DA T A P R O C E S S I N G S C H O O L
4-1-20-1-29-16-18-15-3-5-19-19-9-14-7-29-19-3-8-15-15-12-29-

AN D I W I L L S H O W Y O U A
1-14-4-29-9-29-23-9-12-12-29-19-8-15-23-29-25-15-21-29-1-29-

N I N C O M PU T E R .
14-9-14-3-15-13-16-21-20-5-18-28

(ii) To these numbers, we apply the two-step encoding key (i.e., first multiply by 3 and then add 5):

62-29-50-74-92-44-20-92-8-92-17-59-50-53-50-68-65-92-23-59-50-44-92-8-92-

17-8-65-8-92-53-59-50-14-20-62-62-32-47-26-92-62-14-29-50-50-41-92-

8-47-17-92-32-92-74-32-41-41-92-62-29-50-74-92-80-50-68-92-8-92-

47-32-47-14-50-44-53-68-65-20-59-89

(iii) We modulate these numbers (modulo 29) as follows:

4-29-21-16-5-15-20-5-8-5-17-1-21-24-21-10-7-5-23-1-21-15-5-8-5-

17-8-7-8-5-24-1-21-14-20-4-4-3-18-26-5-4-14-29-21-21-12-5-

8-18-17-5-3-5-16-3-12-12-5-4-29-21-16-5-22-21-10-5-8-5-

18-3-18-14-21-15-24-10-7-20-1-2

(iv) Replacing each of these modulated numbers with its assigned symbol (letter, space, or punctuation mark) yields our encoded message:

D UPEOTEHEQAUXUJGEWAUOEHEQHGHEXAUNTDDCRZEDN UULEHRQECEPCLLED UPEVUJEHERCRNUOXJGTAB

25. We first take advantage of the provided hint. From this, we conclude that the three-letter word GHV represents AND. Assuming that G is A, H is N, and V is D, we can fill in those letters of the puzzle:

AN N A A , A D, AND N
GHJWHY NDW LOGKL TGRLBK, UBLRGXV, GHV XYOZLD WH DZL DWR

D A A A A N N .
VWS ZL RXBOJ G UGH MWX GOO LYGLWHZHSL.

Take note of the two-letter word WH. Since H is N, W must be a vowel. It can't be A, though, because G is A. Consequently, W must be either I or O. Based on the first two words (GHJWHY and NDW), it seems more likely that the W is an O. We could be wrong, but let's try it:

AN ON O A A , A D, AND ON O
GHJWHY NDW LOGKL TGRLBK, UBLRGXV, GHV XYOZLD WH DZL DWR

DO A A O A A ON N .
VWS ZL RXBOJ G UGH MWX GOO LYGLWHZHSL.

A reasonable guess for the first word (GHJWHY) is ANYONE. If that is correct, the second word (NDW) is probably WHO. So, let us assume that J is Y, Y is E, N is W, and D is H. If this is correct, the coded message would be:

ANYONE WHO A A , A D, AND E H ON H HO
GHJWHY NDW LOGKL TGRLBK, UBLRGXV, GHV XYOZLD WH DZL DWR

DO YA AN O A EA ON N .
VWS ZL RXBOJ G UGH MWX GOO LYGLWHZHSL.

The two-letter word ZL suggests Z is one of the two remaining vowels, either I or U. If Z is U, then L must be P, but the three-letter word DZL (along with the high number of L's) makes us doubt the likelihood of this line of reasoning. So, it is more likely that Z is I, and L is S or T. After testing both possibilities, let's go with L being S, which means the cipher would read:

ANYONE WHO S AS A S , S A D, AND E ISH ON HIS HO
GHJWHY NDW LOGKL TGRLBK, UBLRGXV, GHV XYOZLD WH DZL DWR

DO IS A AN O A SEASON IN S.
VWS ZL RXBOJ G UGH MWX GOO LYGLWHZHSL.

It then becomes clear that S is G. Then R (from DWR) is probably T. Also, since the word GOO begins with an A and has two repeated letters, O must be L. All of this leads us to:

ANYONE WHO SLA S A S , S A D, AND ELISH ON HIS HOT
GHJWHY NDW LOGKL TGRLBK, UBLRGXV, GHV XYOZLD WH DZL DWR

DOG IS L A AN O ALL SEASO NINGS
VWS ZL RXBOJ G UGH MWX GOO LYGLWHZHSL.

Now, we make some trial-and-error guesses. In light of the context of the sentence (namely, hot dogs), X must be R, and it would be reasonable to assume that UBLRGXV is MUSTARD. Replacing X with R, U with M, B with U, R with T yields:

ANYONE WHO SLA S ATSU , MUSTARD, AND RELISH ON HIS HOT
GHJWHY NDW LOGKL TGRLBK, UBLRGXV, GHV XYOZLD WH DZL DWR

DOG IS TRULY A MAN OR ALL SEASONINGS.
VWS ZL RXBOJ G UGH MWX GOO LYGLWHZHSL.

We can now complete the deciphered message:

ANYONE WHO SLAPS CATSUP, MUSTARD, AND RELISH ON HIS HOT DOG IS TRULY A MAN FOR ALL SEASONINGS.

27. Via the same type of reasoning as demonstrated in this manual's solution to Problem 25, the following deciphered message is obtained:

CRYSTAL-CLEAR AIR OF ROCKY MOUNTAINS PROVIDES IDEAL ENVIRONMENT FOR WEATHER STATION.

29. Note: This answer is not unique.

$$\begin{array}{r} 3915 \\ 15 \\ +\,4826 \\ \hline 8756 \end{array}$$

Chapter 4 Review Questions

1. $-4+5(-3)=-4+(-15)=\boxed{-19}$

2. $\frac{4}{7}+\frac{5}{9}=\frac{4}{7}\cdot\frac{9}{9}+\frac{5}{9}\cdot\frac{7}{7}=\frac{36}{63}+\frac{35}{63}=\boxed{\frac{71}{63}}$

3. Since $30=2\cdot3\cdot5$, $42=2\cdot3\cdot7$, and $99=3^2\cdot11$, their least common multiple (l.c.m.) is $2\cdot3^2\cdot5\cdot7\cdot11=6{,}930$. Thus,

$$\begin{aligned}\tfrac{7}{30}+\tfrac{5}{42}+\tfrac{5}{99} &= \tfrac{7}{30}\left(\tfrac{3\cdot7\cdot11}{3\cdot7\cdot11}\right)+\tfrac{5}{42}\left(\tfrac{3\cdot5\cdot11}{3\cdot5\cdot11}\right)+\tfrac{5}{99}\left(\tfrac{2\cdot5\cdot7}{2\cdot5\cdot7}\right)\\ &= \tfrac{7}{30}\left(\tfrac{231}{231}\right)+\tfrac{5}{42}\left(\tfrac{165}{165}\right)+\tfrac{5}{99}\left(\tfrac{70}{70}\right)\\ &= \frac{1{,}617}{6{,}930}+\frac{825}{6{,}930}+\frac{350}{6{,}930}\\ &= \frac{2{,}792}{6{,}930}\\ &= \boxed{\frac{1{,}396}{3{,}465}}\end{aligned}$$

4. $-\sqrt{10}\cdot\sqrt{10}=-\sqrt{100}=\boxed{-10}$

5. This problem presents us with an ideal opportunity to take full advantage of the commutative and distributive properties:

$$\begin{aligned}\left(\tfrac{11}{12}+2\right)+\tfrac{-11}{12} &= \left(2+\tfrac{11}{12}\right)+\tfrac{-11}{12}\\ &= 2+\left(\tfrac{11}{12}+\tfrac{-11}{12}\right)\\ &= 2+0\\ &= \boxed{2}\end{aligned}$$

6. $$\frac{3^{-1}+4^{-1}}{6}=\frac{\frac{1}{3}+\frac{1}{4}}{6}=\frac{\frac{7}{12}}{6}=\tfrac{7}{12}\div 6=\tfrac{7}{12}\cdot\tfrac{1}{6}=\boxed{\frac{7}{72}}$$

7. The easiest (most efficient) way to do this problem begins with one having observed that both terms in this sum contain the factor $\frac{-7}{9}$, which prompts the following application of the distributive property:

$$\begin{aligned}\tfrac{-7}{9}\cdot\tfrac{99}{174}+\tfrac{-7}{9}\cdot\tfrac{75}{174} &= \tfrac{-7}{9}\left(\tfrac{99}{174}+\tfrac{75}{174}\right)\\ &= \tfrac{-7}{9}\left(\tfrac{174}{174}\right)\\ &= \tfrac{-7}{9}(1)\\ &= \boxed{-\frac{7}{9}}\end{aligned}$$

8. $$\frac{-3+\sqrt{3^2+4(2)(3)}}{2(2)}=\frac{-3+\sqrt{9+24}}{4}=\boxed{\frac{-3+\sqrt{33}}{4}}$$

9. $\boxed{\frac{8}{3}}$ is already expressed in reduced form.

10. $\frac{16}{18}=\frac{2\cdot8}{2\cdot9}=\boxed{\frac{8}{9}}$

11. $\frac{100}{825}=\frac{25\cdot4}{25\cdot33}=\boxed{\frac{4}{33}}$

12. $\frac{184}{207} = \frac{23 \cdot 8}{23 \cdot 9} = \boxed{\frac{8}{9}}$

13. $\frac{1{,}209}{2{,}821} = \frac{403 \cdot 3}{403 \cdot 7} = \boxed{\frac{3}{7}}$

14. $\frac{3}{8} = 3 \div 8 = \boxed{0.375}$

15. $\frac{7}{3} = 7 \div 3 = \boxed{2.\overline{3}}$

16. $\frac{3}{7} = 3 \div 7 = \boxed{0.\overline{428571}}$

17. $\frac{125}{20} = \frac{25}{4} = 25 \div 4 = \boxed{6.25}$

18. $\frac{3}{13} = 3 \div 13 = \boxed{0.\overline{230769}}$

Since each of these fractions is a ratio of integers, all are (by definition) rational numbers. See p. 188 for the definition of a rational number. Notice that each decimal representation either repeats or terminates; recall that all rational numbers do so.

19. Recall (from the top of textbook p. 171) that if 89 is composite, then one of its factors must be less than or equal to $\sqrt{89} \approx 9.4$. However, since no natural number less than 9 (most notably, 2, 3, 5, and 7) is a divisor of 89, we know 89 is $\boxed{\text{prime}}$.

20. Based on reasoning comparable to what we utilized in our solution to Problem 19, we conclude that 101 is $\boxed{\text{prime}}$.

21. Since 349 is not divisible by any prime number less than $\sqrt{349} \approx 18.7$ (i.e., 2, 3, 5, 7, 11, 13, and 17), 349 must be $\boxed{\text{prime}}$.

22. Because $\sqrt{1001} \approx 31.6$, we know at least one of the prime divisors (if indeed there are any at all) must be less than or equal to 31. With the aid of a calculator, we quickly divide $1{,}001$ by $2, 3, 5, 7, \ldots$. From this, we find: $1{,}001 = \boxed{7 \cdot 11 \cdot 13}$.

23. $6{,}825 = \boxed{3 \cdot 5^2 \cdot 7 \cdot 13}$. To determine this factorization, you may find it helpful to construct a ***factor tree***, as demonstrated on textbook p. 172 (in Section 4.2).

24. Recall (from textbook p. 217) that $\frac{x}{5} \equiv 2, (\text{mod}\, 8)$ means:

$$x = 5 \cdot 2 = 10 \equiv \boxed{2, (\text{mod}\, 8)}$$

25. In modulo 7, there are seven possible values of x; namely, $0, 1, 2, 3, \ldots, 6$. We solve this equation by checking each of these values:

$$\begin{aligned}
x=0: &\quad 2(0)=0,(\text{mod}\,7)\\
x=1: &\quad 2(1)=2,(\text{mod}\,7)\\
x=2: &\quad 2(2)=4,(\text{mod}\,7)\\
x=3: &\quad 2(3)=6,(\text{mod}\,7)\\
x=4: &\quad 2(4)=8\equiv 1,(\text{mod}\,7)\\
x=5: &\quad 2(5)=10\equiv 3,(\text{mod}\,7)\\
x=6: &\quad 2(6)=12\equiv 5,(\text{mod}\,7)
\end{aligned}$$

Thus, the only value of the variable that satisfies the equation $2x \equiv 3,(\text{mod}\,7)$ is

$$\boxed{x = 5, (\text{mod } 7)}.$$

26. We solve this equation by evaluating $2x^2+7x+1,(\text{mod}\,2)$ for all possible modulo 2 values of x:

$$\begin{aligned}
x=0: &\quad 2(0)^2+7(0)+1=1,(\text{mod}\,2)\\
x=1: &\quad 2(1)^2+7(1)+1=10\equiv 0,(\text{mod}\,2)
\end{aligned}$$

Hence, our solution set consists of $\boxed{1,(\text{mod } 2)}$.

27. Recall that the greatest common factor (g.c.f.) is the largest number that divides (evenly) into each of the given numbers. Since $49=7^2$, $1{,}001=7\cdot 11\cdot 13$, and $2{,}401=7^4$, the g.c.f. is $\boxed{7}$. To review this topic, refer to textbook page 173.

28. The least common multiple (l.c.m.) is defined as the smallest number that each of the given numbers divides into evenly. Using the factorizations shown in our solution to Problem 27, the least common multiple (l.c.m.) is $7^4\cdot 11\cdot 13$, which equals $\boxed{343{,}343}$. See p. 174 for additional reinforcement.

29. The parentheses in the expression $(1 \odot 2) \odot (3 \odot 4)$ suggest we need to first compute $(1 \odot 2)$ and $(3 \odot 4)$. Then, afterwards, we use these resulting values to compute the final answer. Therefore, we have the following:

$1 \odot 2 = 1\times 2+1+2 = 2+1+2 = \underline{\underline{5}}$

$3 \odot 4 = 3\times 4+3+4 = 12+3+4 = \underline{\underline{19}}$

So,

$5 \odot 19 = 5\times 19+5+19 = \boxed{119}$

30. We perform the operations in a step-by-step manner, beginning within each pair of parentheses, and obtain the following:

$$(1 \downarrow 2) \uparrow (2 \downarrow 3)$$
$$(1) \uparrow (2)$$
$$\boxed{2}$$

31. This definition is stated on textbook p. 160. For $a \neq 0$, ***multiplication*** is defined as: $a \times b$ means $\underbrace{b+b+b+\cdots+b}_{a \text{ addends}}$. If $a=0$, $0 \times b = 0$.

32. ***Subtraction*** is defined on textbook p. 163 (for natural numbers) and p. 184 (for integers). Please revisit those pages accordingly.

33. As defined on textbook p. 184, $a \div b = \dfrac{a}{b} = z$ means $a = bz$, where $b \neq 0$.

34. **Explanations vary**. On p. 185, the author of your text provides an excellent explanation.

35. **Answers vary**. Noting that $34 = \sqrt{1156}$ and $35 = \sqrt{1225}$, we know that among the infinitely many irrational numbers between 34 and 35 are $\sqrt{1157}$, $\sqrt{1158}$, ..., $\sqrt{1224}$. Additionally, using the numbers π and e (see textbook p. 198), 11π and $12.75e$ are two more irrational numbers between 34 and 35. On page 945, your text's author provides another answer ($34.1011011101111011\ldots$), one which clearly conveys the idea that rational numbers are numbers whose decimal representations do not terminate nor repeat.

36. The definition of a ***field*** can be found on textbook page 213. Beyond knowing its definition/properties, we suggest you spend some time expressing these concepts in your own words. Additionally, you should be familiar with several examples of fields as well as examples that fail to satisfy all the field properties.

37. To confirm whether the set of natural numbers is a group for the operation of g.c.f., we need to check whether all of the following requirements are satisfied:

- <u>Closure</u>: Since the product of natural numbers is itself a natural number (because the natural numbers are closed for multiplication) and since the g.c.f. is a product of factors that are natural numbers, the g.c.f. of any two natural numbers is itself a natural number . Therefore, the closure requirement is satisfied.
- <u>Associative</u>: $(a\mathbb{D}b)\mathbb{D}c = a\mathbb{D}(b\mathbb{D}c)$ because the g.c.f. is a product of common factors and multiplication is associative. So, this requirement is satisfied.

- *Identity*: We need to have a natural number i such that $x\mathbb{D}i = i\mathbb{D}x = x$ for every natural number x. However, no such number exists.
- *Inverse*: Since there is no identity element, this property cannot even be considered.

Since two of the requirements are not satisfied, the set of natural numbers is not a group for the operation of g.c.f. Since it is not a group, it obviously is not a commutative group — despite the fact that $a\mathbb{D}b = b\mathbb{D}a$ for all natural numbers a and b. Thus, our answer to the question is **NO**.

38. The formula for body mass index is: $\boxed{BMI = \dfrac{703w}{h^2}}$.

a) A height of 5 ft. 5 in. means $h = 65$. Hence, this person's body mass index is:

$$BMI = \frac{703(165)}{(65)^2} \approx \boxed{27.5}$$

b) A height of 6 ft. means $h = 72$. So, this individual's body mass index is:

$$BMI = \frac{703(185)}{(72)^2} \approx \boxed{25.1}$$

[Note: We would hope you expected, before performing the actual computation, that person (a) would have a higher BMI than person (b).]

c) Before performing the actual computations, you should attempt to predict (estimate) the answer based on your experiences with parts (a) and (b) of this problem. Since this 5-ft., 6-in. person has approximately the same height as the person in part (a) but has a lower BMI, one should anticipate this individual's weight to be lower than the 165-lb. person in part (a).

For a 5-ft., 6-in. tall person, we let $h = 66$. We are given the value of the BMI and, thus, need to solve the resulting equation for w.

$$\frac{703w}{h^2} = BMI$$

$$\frac{703w}{(66)^2} = 25$$

$$703w = 25(66)^2$$

$$w = \frac{25(66)^2}{703} \approx 154.9$$

Rounded to the nearest pound, this person's weight is **155 lbs**.

39. In the spirit of the first step of Pólya's problem-solving method (i.e., "Understand the Problem"), we recommend you begin by spending some time examining several specific doors. For example, which tenants affect (i.e. open or close) the 10th door? There are four tenants who affect it. The first tenant opens it, the second closes it, the fifth reopens it, and the tenth closes it for good. What about the 29th door? The only two tenants who interact with it are the first tenant (who opens it) and the twenty-ninth tenant (who closes it). Stated in more general terms, the only tenants who will open/close a given door are those whose number is a divisor of the door number.

This sort of analysis aids us in assessing the question asked in the article; namely, which doors ultimately will remain open? The only way for a door to remain open is for it to be affected by an odd number of tenants (e.g., open, close, and then open). If an even number of tenants affect a given door, then the door ultimately will be closed (e.g., the 10th and 29th doors illustrated above). Consequently, to determine which doors will ultimately remain open, we need to find those door numbers that have an odd number of divisors. For example, the 9th door will remain open because the first tenant opens it, the third closes it, and the ninth reopens it. More generally, the only such numbers that have an odd number of divisors are the perfect squares: 1, 4, 9, 16, 25, 36, ..., 841, 900, and 961. Noting that $961 = 31^2$, we, therefore, conclude that **31 doors are left open**.

40. Since the vertical rise is 8 ft (i.e., 96 in.) and the maximum rise for each step is 8 in., we need (96 in.) ÷ (8 in. per step) = **12 steps**.

The sum of the lengths of the vertical segments making up the steps must equal the total vertical rise of 8 ft. Similarly, the sum of the lengths of the horizontal segments must equal the total horizontal run of 12 ft. Therefore, the total length of the segments making up the stairs is the sum of the vertical and horizontal lengths; namely, 8 ft + 12 ft = **20 ft of segments**.

Finally, we notice that the diagonal is the hypotenuse of a right triangle. Thus, we compute its length by applying the Pythagorean theorem (introduced in Section 4.5):

$$\text{Length of diagonal} = \sqrt{12^2+8^2} = \sqrt{208} = \sqrt{16\cdot 13} = \boxed{4\sqrt{13}\ \text{ft}}$$

CHAPTER 5: The Nature of Algebra

Problem Set 5.1

1. According to your text, a ***polynomial*** is **a term or a sum of terms**. Please refer to the first page of this textbook section (page 238) and acquaint yourself with the definition of a term and other important polynomial-related words.

3. **Answers vary**. When adding and subtract polynomials, we combine so-called like (or similar) terms. For example, $5x^3 - 11x^3 = -6x^3$. The justification for combining terms in this manner is the distributive property and the order of operations (textbook p. 20). In the case of our example, we know the distributive property permits us to rewrite $5x^3 - 11x^3$ as $x^3(5-11)$ or $(5-11)x^3$, and then performing the arithmetic in the parentheses yields the result: $-6x^3$.

5. **Answers vary**. ***FOIL*** is an acronym standing for "First Outer Inner Last." It is not a new mathematical concept, rather its sole purpose is to make it easier (more efficient) for students to multiply binomials. As your textbook's author makes clear (pp. 240-241), the process of multiplying binomials is a direct consequence of the distributive property.

7. The ***binomial theorem*** is stated and explained on textbook pp. 243-244.

9. If necessary, refer to Example 2f (textbook p. 239) for further details. By combining like terms, we obtain:
$(2x-4)-(3x+4) = 2x-4-3x-4 = \boxed{-x-8}$; **[binomial, degree 1]**

11. $(x-y-z)+(2x-5y-3z) = (x+2x)+(-y-5y)+(-z-3z)$
$= \boxed{3x-6y-4z}$

Since each term is of degree 1, this **trinomial** is of **degree 1**.

13. $(2x^2-5x+4)+(3x^2-2x-11) = (2x^2+3x^2)+(-5x-2x)+(4-11)$
$= \boxed{5x^2-7x-7}$; **[trinomial, degree 2]**

15. $(x^2+4x-3)-(2x^2+9x-6) = x^2+4x-3-2x^2-9x+6$
$= (x^2-2x^2)+(4x-9x)+(-3+6)$
$= \boxed{-x^2-5x+3}$; **[trinomial, degree 2]**

17. $3(x-5)-2(x+8)=3(x-5)+(-2)(x+8)=3x-15-2x-16=\boxed{x-31}$;
[binomial, degree 1]

19. $\cdots=6x^2+15x-15+10x^2-6x+12=\boxed{16x^2+9x+3}$;
[trinomial, degree 2]

21. $\cdots=(2x+6)+(-3x^2+9x-3)+(4x-20)$
$=-3x^2+(2x+9x+4x)+(6-3-20)$
$=\boxed{-3x^2+15x-17}$; **[trinomial, degree 2]**

23. If you are not sure what your text's author means by "multiply mentally," refer to Example 5 (p. 241) for details.

a) $\boxed{x^2+5x+6}$ b) $\boxed{y^2+6y+5}$ c) $\boxed{z^2+4z-12}$ d) $\boxed{s^2+s-20}$

25. a) $\boxed{c^2-6c-7}$ b) $\boxed{z^2+2z-15}$ c) $\boxed{2x^2-x-1}$ d) $\boxed{2x^2-5x+3}$

27. a) $\boxed{x^2+2xy+y^2}$ b) $\boxed{x^2-2xy+y^2}$ c) $\boxed{x^2-y^2}$ d) $\boxed{a^2-b^2}$

29. "To square" a quantity means to multiply it by itself, so it follows that $(a+b)^2$ means $(a+b)(a+b)$ which, in turn, equals $a^2+2ab+b^2$. With this in mind, we elaborate on the "mental" multiplication behind part (a) of this problem and leave parts (b), (c), and (d) for you.

a) We begin by translating $(a+b)^2=a^2+2ab+b^2$ into words. One way to view this relationship is:

$$(a+b)^2=\underbrace{a^2}_{\text{"square the first term"}}+\underbrace{2ab}_{\text{"double the product of the two terms"}}+\underbrace{b^2}_{\text{"square the last term"}}$$

Then, in the case of $(x+4)^2$, we mentally perform these three steps:

- *"Square the first term"* $\Rightarrow$ x^2
- *"Double the product of the two terms"* $\Rightarrow$ $2(4x)=8x$
- *"Square the last term"* $\Rightarrow$ $4^2=16$

Thus, we obtain $\boxed{x^2+8x+16}$.

b) $\boxed{y^2-6y+9}$ c) $\boxed{s^2+2st+t^2}$ d) $\boxed{u^2-2uv+v^2}$

31. This problem is very much like textbook Example 3c. We begin by distributing the binomial $(2x-1)$ to each term of the trinomial $3x^2+2x-5$:

$$\begin{aligned}(2x-1)(3x^2+2x-5) &= 3x^2(2x-1)+2x(2x-1)-5(2x-1)\\ &= 6x^3-3x^2+4x^2-2x-10x+5\\ &= \boxed{\mathbf{6x^3+x^2-12x+5}}\end{aligned}$$

33. $$\begin{aligned}(5x+1)(x^3-2x^2+3x) &= x^3(5x+1)-2x^2(5x+1)+3x(5x+1)\\ &= 5x^4+x^3-10x^3-2x^2+15x^2+3x\\ &= \boxed{\mathbf{5x^4-9x^3+13x^2+3x}}\end{aligned}$$

35. $3(3x^2-5x+2)+(-4)(x^3-4x^2+x-4)$

$(9x^2-15x+6)+(-4x^3+16x^2+-4x+16)$

$\boxed{\mathbf{-4x^3+25x^2-19x+22}}$

37. We begin by performing "mental multiplication" (as was practiced in Problems 23-29) and then proceed to combine like terms:

$$(6x^2-5x-6)+(x^2+5x+6)$$

$$(6x^2+x^2)+(-5x+5x)+(-6+6)$$

$$\boxed{\mathbf{7x^2}}$$

39. <u>Algebraically</u>: $\boxed{\mathbf{x^2+5x+4}}$

<u>Geometrically</u>:

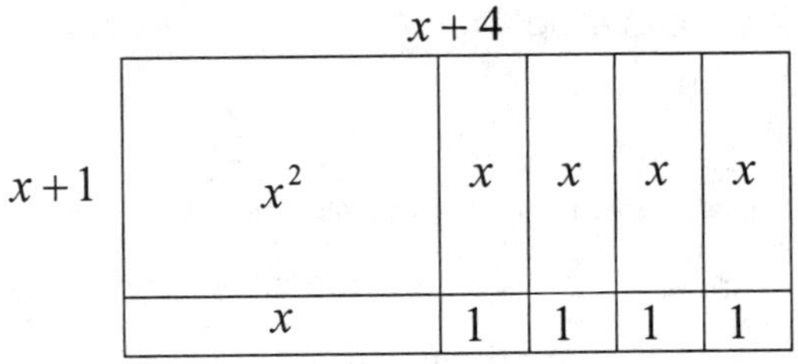

Since there is one square (x^2), five rectangles ($5x$), and four units (4), we conclude: $(x+1)(x+4)=x^2+5x+4$.

41. <u>**Algebraically**</u>: $\boxed{\mathbf{x^2 + 7x + 12}}$

<u>**Geometrically**</u>:

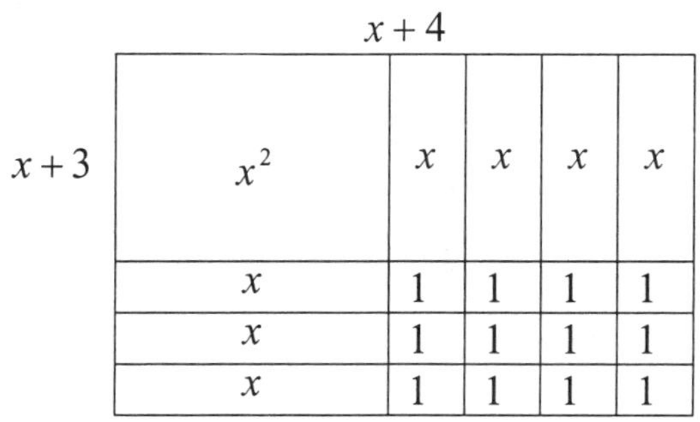

Since there is one square (x^2), seven rectangles ($7x$), and twelve units (12), it follows that $(x+3)(x+4) = x^2 + 7x + 12$.

43. <u>**Algebraically**</u>: $\boxed{\mathbf{4x^2 + 8x + 3}}$

<u>**Geometrically**</u>:

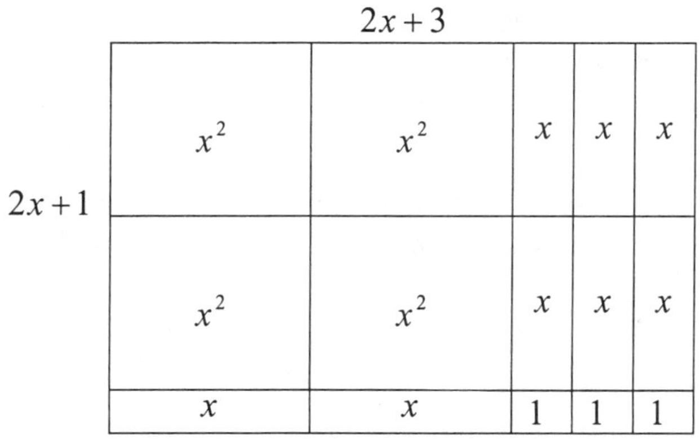

Having four squares ($4x^2$), eight rectangles ($8x$), and three units (3), we know that the product of $(2x+1)$ and $(2x+3)$ is $4x^2 + 8x + 3$.

45. For this first solution, let us proceed in a detailed manner by applying the binomial theorem (textbook p. 243) to $(a+b)^3$. In our solutions to the next several problems (47, 49, 51), we will forego this extra detail, which should become unnecessary as you become more skilled at applying the binomial theorem.

We know

$$(a+b)^3 = \binom{3}{0}a^3 + \binom{3}{1}a^2b + \binom{3}{2}ab^2 + \binom{3}{3}b^3$$
$$= 1a^3 + 3a^2b + 3ab^2 + 1b^3$$

Reading this polynomial from left to right, two observations are worthy of mention:

- the powers of a descend (3, 2, 1, 0), while the powers of b ascend (0, 1, 2, 3);
- the coefficients (1, 3, 3, 1) are obtained from the third row of Pascal's triangle (see textbook Figure 5.1, p. 243).

Finally, we finish the problem $(x-1)^3$ by letting $a = x$ and $b = -1$ and simplifying appropriately:

$$1(x)^3 + 3(x)^2(-1) + 3(x)(-1)^2 + 1(-1)^3$$
$$\boxed{\mathbf{x^3 - 3x^2 + 3x - 1}}$$

47. We begin by applying the observations cited in our solution to Problem 45. First, in left-to-right order, we expect the powers of x to descend (6, 5, ..., 0) and the powers of y to ascend (0, 1, ..., 6). Second, since $n = 6$, the respective coefficients of the expanded polynomial are 1-6-15-20-15-6-1 (as obtained from the 6th row of Pascal's triangle). Hence,

$$(x+y)^6 = \boxed{\mathbf{x^6 + 6x^5y + 15x^4y^2 + 20x^3y^3 + 15x^2y^4 + 6xy^5 + y^6}}.$$

49. First, note that $(x-y)^8 = [x+(-y)]^8$. We then perform three steps: (i) use as our coefficients the numbers found on the eighth row of Pascal's triangle; (ii) place (in left-to-right order) the powers of x and y in descending order and ascending order, respectively; and (iii) simplify accordingly. This process yields

$$[x+(-y)]^8$$
$$1x^8 + 8x^7(-y)^1 + 28x^6(-y)^2 + 56x^5(-y)^3 + 70x^4(-y)^4 + 56x^3(-y)^5 + 28x^2(-y)^6 + 8x^1(-y)^7 + 1(-y)^8$$
$$\boxed{\mathbf{x^8 - 8x^7y + 28x^6y^2 - 56x^5y^3 + 70x^4y^4 - 56x^3y^5 + 28x^2y^6 - 8xy^7 + y^8}}$$

51. Knowing that $(a+b)^4 = a^4 + 4a^3b + 6a^2b^2 + 4ab^3 + b^4$, we let $a = 2x$ and $b = -3y$. Then, upon simplifying appropriately, we obtain:

$$(2x+-3y)^4 = (2x)^4 + 4(2x)^3(-3y) + 6(2x)^2(-3y)^2 + 4(2x)(-3y)^3 + (-3y)^4$$
$$= 16x^4 + 4(8x^3)(-3y) + 6(4x^2)(9y^2) + 4(2x)(-27y^3) + 81y^4$$
$$= \boxed{\mathbf{16x^4 - 96x^3y + 216x^2y^2 - 216xy^3 + 81y^4}}$$

53. The last three terms in the expansion of $(x+y)^{14}$ are:

$$\binom{14}{12}x^2y^{12}+\binom{14}{13}xy^{13}+\binom{14}{14}y^{14}$$

Then, upon referring to the 12th, 13th, and 14th numbers on the 14th row of Pascal's triangle, we obtain the answer:

$$\boxed{91x^2y^{12}+14xy^{13}+y^{14}}$$

55. If you are reading this solution because you are having difficulty answering this question, we recommend you consider doing what all good problem solvers often do: make the question more concrete by first asking yourself a simpler (but still related) question. For example, if a classroom has 6 rows and 5 seats in each row, how many seats are in the classroom? We expect you would find the answer by multiplying 6×5 and would report there to be 30 seats in the classroom. Thus, in similar fashion, the answer to the original problem is

$\boxed{(6x+2)(51x-7) \text{ or } 306x^2+60x-14 \text{ seats}}$ in the auditorium.

57. If this problem is difficult for you, you may want to do what we recommended in our solution to Problem 55 above: begin by temporarily replacing the binomials with simple numbers and considering what you would do under those circumstances. That is to say, for example, if a boat is traveling at a rate of 35 miles per hour for a time of 2 hours, what is the distance traveled by the boat? We expect you would say 70 miles (i.e., 35×2). Hence, the distance traveled by the boat described in the original problem is $\boxed{(6b+15)(10-2b) \text{ or } -12b^2+30b+150 \text{ miles}}$.

[Disclaimer: In lieu of the admittedly lower-level approach just described, we actually prefer you apply the formula $d = rt$ (i.e., distance = rate × time).]

59. First, note that the given first number is $(10x+y)$ and the second is $(10x+z)$, with both having x as their tens digit and y and z as their respective units digits. We then proceed to assess their product:

$$\begin{aligned}(10x+y)(10x+z) &= 100x^2+10xy+10xz+yz\\ &= 100x^2+10x(y+z)+yz\end{aligned}$$

Recalling the given assumption $y+z=10$, we further simplify this product:

$$\begin{aligned}&= 100x^2+10x(10)+yz\\ &= 100x^2+100x+yz\\ &= 100(x^2+x)+yz\\ &= 100[x(x+1)]+yz\end{aligned}$$

Quoting directly from Problem 58, we find:

- *"...mentally multiply the units digit(s)"* $\Rightarrow$ yz
- *"...multiply the tens digit of the first number by one more than the tens digit of the second number"* $\Rightarrow$ $x(x+1)$

Hence, the mental calculation process described in Problem 58 "works."

Problem Set 5.2

1. **Answers vary**. (1) Look for a common factor, then either (2) factor a difference of squares or (3) use FOIL to factor if it is a trinomial.

3. This binomial has a greatest common factor (gcf) of $2x$. Hence, the binomial's factorization is: $10xy - 6x = \boxed{\mathbf{2x(5y-3)}}$

5. $8xy - 6x = \boxed{\mathbf{2x(4y-3)}}$

7. Using the approaches illustrated in textbook Examples 3 & 4 (pp. 248-249), we obtain the factorization of $x^2 - 4x + 3$: $\boxed{\mathbf{(x-1)(x-3)}}$

9. $\boxed{\mathbf{(x-3)(x-2)}}$

11. $\boxed{\mathbf{(x-4)(x-3)}}$

13. $\boxed{\mathbf{(x-6)(x+5)}}$

15. $\boxed{\mathbf{(x-7)(x+5)}}$

17. $\boxed{\mathbf{(3x+10)(x-1)}}$

19. $\boxed{\mathbf{(2x-1)(x-3)}}$

21. $\boxed{\mathbf{(3x+1)(x-2)}}$

23. $\boxed{\mathbf{(2x+1)(x+4)}}$

25. $\boxed{\mathbf{(3x-2)(x+1)}}$

27. $5x^3 + 7x^2 - 6x = x(5x^2 + 7x - 6) = \boxed{\mathbf{x(5x-3)(x+2)}}$

29. $7x^4 - 11x^3 - 6x^2 = x^2(7x^2 - 11x - 6) = \boxed{\mathbf{x^2(7x+3)(x-2)}}$

31. Refer, as necessary, to your textbook author's explanation of a difference of squares (pp. 250-251). $x^2 - 64 = \boxed{\mathbf{(x-8)(x+8)}}$

33. $\boxed{\mathbf{25(x^2+2)}}$

35. This problem requires factoring a difference of squares <u>twice</u>:

$$\begin{aligned} x^4-1 &= (x^2-1)(x^2+1) \\ &= \boxed{\mathbf{(x-1)(x+1)(x^2+1)}} \end{aligned}$$

37.

$x+3$ (top), $x+2$ (side)

x^2	x	x	x
x	1	1	1
x	1	1	1

Thus, $x^2+5x+6=\boxed{\mathbf{(x+2)(x+3)}}$.

39.

$x+3$ (top), $x+1$ (side)

x^2	x	x	x
x	1	1	1

Hence, $x^2+4x+3=\boxed{\mathbf{(x+1)(x+3)}}$.

41.

$x+2$ (top), $x+4$ (side)

x^2	x	x
x	1	1
x	1	1
x	1	1
x	1	1

So, $x^2+6x+8=\boxed{\mathbf{(x+4)(x+2)}}$.

43. We begin with a square (representing the x^2 of x^2-1). From this square, we remove one unit (representing the -1 of x^2-1), which is the shaded portion below:

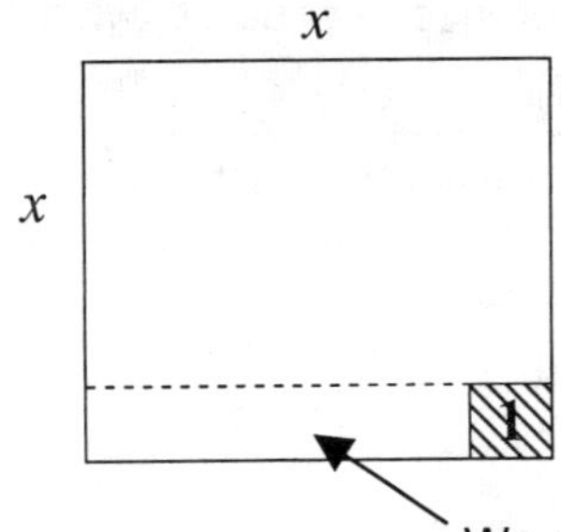

We then move this unshaded piece to the right side (as shown below) to form a single rectangular area.

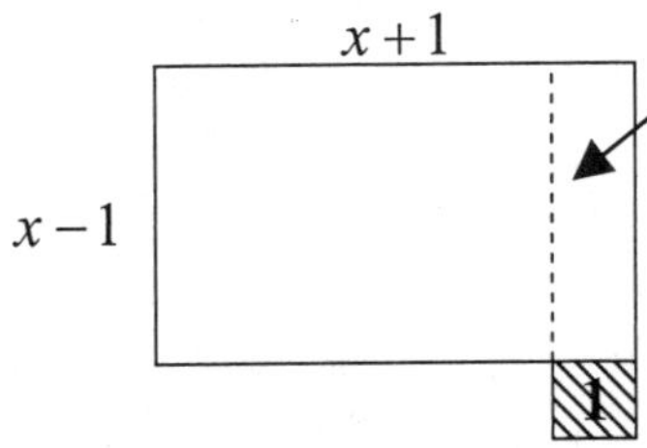

Hence, $x^2-1=\boxed{(x-1)(x+1)}$. Refer to textbook Example 2 (pp. 247-248) for additional details.

45. We begin by depicting the x^2+x portion of the trinomial x^2+x-2 with one large square (representing x^2) and one rectangular strip (representing x) attached to the right side of the large square. From this, we remove two 1-by-1 small squares (shaded in the picture below). This leaves us with the task of moving the pieces around to form a single rectangular figure, a task which we demonstrate below.

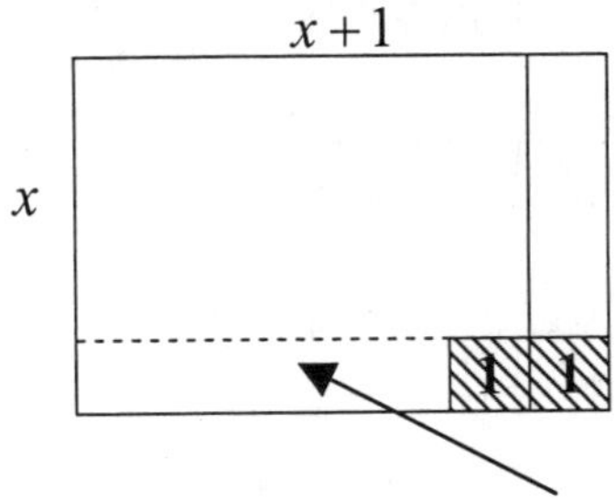

Move this unshaded portion to the right side (as shown on the next page).

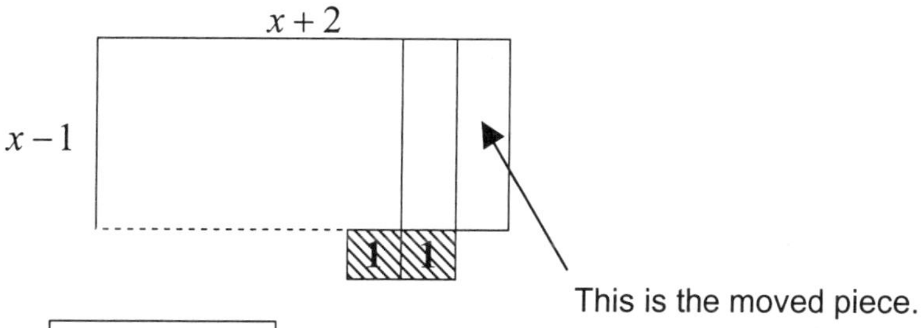

Thus, $x^2 + x - 2 = \boxed{(\boldsymbol{x-1})(\boldsymbol{x+2})}$.

47. We begin with one large square (depicting x^2). From this, we must remove the following: (i) one rectangular strip of area x (representing the $-x$ term of $x^2 - x - 2$); and (ii) two small 1-by-1 squares (representing the -2 term of $x^2 - x - 2$). The portions we remove from the large square are the shaded portions of the following picture:

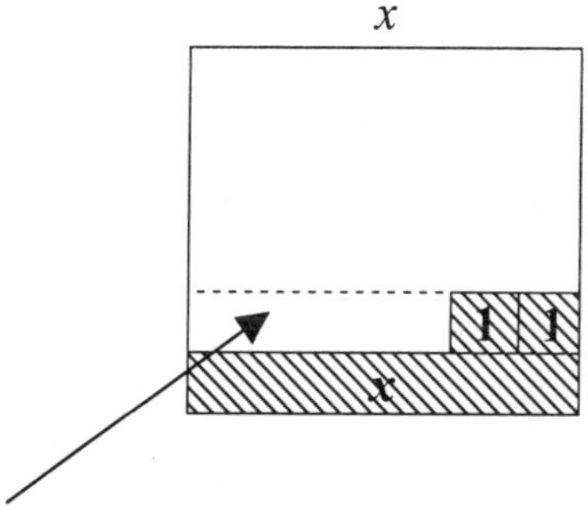

We then move this unshaded piece to the right side (as shown below) to obtain one rectangular figure.

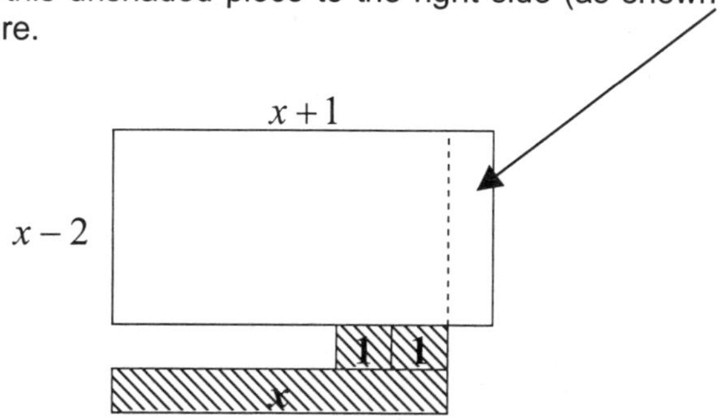

Thus, the factorization of $x^2 - x - 2$ is $\boxed{(\boldsymbol{x-2})(\boldsymbol{x+1})}$.

49. Recall that: (i) the word "dimensions" refers to the width and length of the rectangle; and (ii) the area of a rectangle equals the product of its width and length (i.e., $A = L \times W$).

Since the area of this rectangle is $x^2-2x-143$ square feet and this trinomial has a factorization of $(x-13)(x+11)$, it follows that the dimensions of this rectangle are

$$\boxed{(x-13)\text{ feet by }(x+11)\text{ feet}}.$$

51. Since $d=rt$ (i.e., distance = rate × time), we know $t=\frac{d}{r}$.

Given that $d=6x^2+5x-4$ and $r=3x+4$, the travel time is:

$$t=\frac{d}{r}=\frac{6x^2+5x-4}{3x+4}=\frac{(3x+4)(2x-1)}{(3x+4)}=\boxed{2x-1\text{ hours}}$$

53. We begin by multiplying the two pairs of binomials and combining like terms:

$$(x+2)(x+4)+(5x+6)(x-1)$$
$$x^2+6x+8+5x^2+x-6$$
$$6x^2+7x+2$$

The factorization of the resulting trinomial is $\boxed{(3x+2)(2x+1)}$.

55. This trinomial has a greatest common factor (gcf) of x^2. After factoring this gcf, we factor the remaining trinomial:

$$x^6-13x^4+36x^2$$
$$x^2(x^4-13x^2+36)$$
$$x^2(x^2-4)(x^2-9)$$

Notice that the two binomials (i.e., x^2-4 and x^2-9) are both differences of two perfect squares. Hence, the complete factorization of $x^6-13x^4+36x^2$ is:

$$\boxed{x^2(x-2)(x+2)(x-3)(x+3)}$$

57. Like the previous problem (Problem 55), this trinomial has a gcf of x^2. This leads to the following factorization:

$$20x^2y^2+17x^2yz-10x^2z^2$$
$$x^2(20y^2+17yz-10z^2)$$
$$\boxed{x^2(4y+5z)(5y-2z)}$$

59. Let $x-1$, x, and $x+1$ represent the three consecutive integers. The product of the first and third is $(x-1)(x+1)=x^2-1$, which is indeed equal to one less than the square of the middle integer x.

Problem Set 5.3

1. **Answers vary**. A ***variable*** is a symbol used to represent an unspecified member of some set (called the domain). Think of a variable as a "place holder" for the name of some member of the set.

3. **Answers vary**.

5. a) **+(2/3)∗A1∧2** b) **+5∗A1∧2 – 6∗A2∧2**

7. a) **+12∗ (A1∧2+4)** b) **+(15∗A1+7)/2**

9. a) **+(5–A1)∗(A1+3)∧2** b) **+6∗(A1+3)∗(2∗A1–7)∧2**

11. a) **+(1/4)∗A1∧2–(1/2)∗A1+12** b) **+(2/3)∗A1∧2+(1/3)∗A1–17**

13. a) $\boxed{4x+3}$ b) $\boxed{5x^2-3x+4}$

15. a) $\boxed{\frac{5}{4}x+14^2}$ b) $\boxed{\left(\frac{5}{4}x+14\right)^2}$

17. a) $\boxed{\frac{x}{y}(z)}$ b) $\boxed{\frac{x}{yz}}$

19. $A = x + z + 8 = (1) + (4) + 8 = \boxed{13}$

21. $C = 10 - w = 10 - (2) = \boxed{8}$

23. $E = 25 - y^2 = 25 - (2)^2 = 25 - 4 = \boxed{21}$

25. $G = 5x + 3z + 2 = 5(1) + 3(4) + 2 = 5 + 12 + 2 = \boxed{19}$

27. $I = 5y - 2z = 5(2) - 2(4) = 10 - 8 = \boxed{2}$

29. $K = wxy = (2)(1)(2) = \boxed{4}$

31. $M = (x + y)^2 = (1 + 2)^2 = (3)^2 = \boxed{9}$

33. $P = y^2 + z^2 = (2)^2 + (4)^2 = 4 + 16 = \boxed{20}$

35. $R = z^2 - y^2 - x^2 = (4)^2 - (2)^2 - (1)^2 = 16 - 4 - 1 = \boxed{11}$

37. $T = x^2 + y^2 z = (1)^2 + (2)^2(4) = 1 + (4)(4) = 1 + 16 = \boxed{17}$

39. $V = \dfrac{3wyz}{x} = \dfrac{3(2)(2)(4)}{1} = \dfrac{48}{1} = \boxed{48}$

41. $X = (x^2 z + x)^2 z = \left[(1)^2(4) + 1\right]^2(4) = [5]^2(4) = 25(4) = \boxed{100}$

43. In order to complete this problem, you first must solve every one of Problems 19-42 — both the even and odd exercises. After doing so, you will find the resulting secret message to be a statement that is virtually impossible to dispute:

ALGEBRA_IS__THE_GREATEST LABOR SAVING DEVICE_EVER_INVENTED.

45. a) The values of cells A1, B1, and C1 are 6, -4, and 2, respectively. Thus, the formula +A1–B1/C1 is numerically equivalent to $6 - \frac{(-4)}{2} = 6 + 2 = \boxed{8}$.

b) +(A1–B1)/C1 is equivalent to $\dfrac{6-(-4)}{2} = \frac{10}{2} = \boxed{5}$.

c) $\frac{6}{2} \cdot 3 = 3 \cdot 3 = \boxed{9}$

d) $(6 + -4 + 2 + 3) \div 2 = 7 \div 2 = \boxed{3.5}$

47. Recall, from textbook p. 261, that the # symbol (as in #A#1) holds constant the reference to a row and/or column, depending upon the placement of the symbol. In this problem, entering #A#1 forces the reference to cell A1 to remain fixed, regardless of the cell(s) in which it is replicated. As each of the formulas in parts (a) and (b) are entered in C1 and subsequently copied across row 1, the formula's reference to A1 will not change (due to the inclusion of #), but references to B1 will change (because they do not include the # symbol). With that said, we turn our attention to each problem:

a) After we enter the formula +#A#1+B1 in cell C1 and replicate it in cells D1 and E1, cells D1 and E1 respectively will contain +#A#1+C1 and +#A#1+D1. That is,

	A	B	C	D	E
1	1	3	+#A#1+B1	+#A#1+C1	+#A#1+D1
2					
3					
4					
5					

Upon determining the values of cells C1, D1, and E1, we arrive at the answer to the problem:

	A	B	C	D	E
1	1	3	4	5	6
2					
3					
4					
5					

b) In a manner comparable to part (a), the formulas in cells C1, D1, and E1 will appear as follows:

	A	B	C	D	E
1	1	3	+#A#1∗B1	+#A#1∗C1	+#A#1∗D1
2					
3					
4					
5					

This results in a spreadsheet with the following values:

	A	B	C	D	E
1	1	3	3	3	3
2					
3					
4					
5					

49. a) If we enter zero (0) in cell B1, the following chain of computations will occur:

- The formula in cell C3 takes this 0 from B1, adds 1 to it, and places the result (of 1) in C3;
- Cell A2 references cell C3 and adds 1 to C3's value, yielding a value of 2 in Cell A2;
- Cell A3 uses the value of 2 (from cell A2) to compute its value (2 + 1= 3);
- And so on… In other words, cells B2, C1, C2, A1, and B3 in order add 1 to each of their immediate predecessors.

This sequence of computations results in the following spreadsheet:

	A	B	C	D
1	7	0	5	
2	2	4	6	
3	3	8	1	
4				
5				

b) The major difference between this problem and part (a) is that cell B1 has a value of 100, instead of 0. This results in cell C3 becoming 101, A2 becoming 102, and so on. Thus, the values of these nine cells become:

	A	B	C	D
1	107	100	105	
2	102	104	106	
3	103	108	101	
4				
5				

51. Since profit is equal to the gross sales minus the cost, cells D2 through D21 require a formula that subtracts the corresponding contents of column C from those of column B. Cells E2 through E21 need to compute 8% of their corresponding entries in column D. This results in the following spreadsheet:

	A	B	C	D	E
1	NAME	SALES	COST	PROFIT	COMMISSION
2				+B2–C2	+D2∗0.08
3					
4	*Replicate Row 2 for Rows 2 through 21.*				
5					

53. This problem is very similar to Example 4 (p. 255) and the explanatory material that precedes it. With $p = 0.65$ and $q = 0.35$, we find the following percentages of the population with each genotype:

black (BB): $p^2 = (0.65)^2 = 0.4225 =$ **42.25%**

black w/ recessive brown (bB or Bb): $2pq = 2(0.65)(0.35) = 0.455 =$ **45.5%**

brown (bb): $q^2 = (0.35)^2 = 0.1225 =$ **12.25%**

As for the phenotype (which refers to the physical appearance), **87.75% have black fur** (i.e., BB, bB, and Bb), while **12.25% have brown fur**.

55. In this problem, the genotypes and phenotypes are the same.

Red flower (rr): $(0.2)^2 = 0.04 =$ **4%**

White flower (ww): $(0.8)^2 = 0.64 =$ **64%**

Hybrid pink flower (rw or wr): $2(0.2)(0.8) = 0.32 =$ **32%**

57. Recall that $(B+b)^2 = B^2 + 2Bb + b^2$. Since brown is the recessive gene and 25% of the population observed has brown fur, $b^2 = 25\% = 0.25$ suggests $b = 0.5$. And, since $B + b = 1$ and $b = 0.5$, it follows that $B = 0.5$, too. Thus, we expect the gene pool to be approximately evenly divided, **50% is B and 50% is b**.

59. **Answers vary**. For your information, in the general population, FF is 49%, Ff (a.k.a. fF) is 42%, and ff is 9%. That is, free hanging (FF, Ff, and fF) is 91% and attached (ff) is 9%. Said differently, F is 70% and f is 30%.

Problem Set 5.4

1. **Answers/wording vary**. We solve first-degree (linear) equations by using the equation properties (described on page 264 in your text). The objective is to isolate the variable on one side of the equation.

3. a)

$x-5=10$	*Given linear equation*
$x-5+5=10+5$	*Add 5 to both sides.*
$\boxed{\mathbf{x=15}}$	*Simplify.*

b)

$6=x-2$	*Given linear equation*
$6+2=x-2+2$	*Add 2 to both sides.*
$\boxed{\mathbf{8=x}}$	*Simplify.*

5. a)

$\frac{x}{4}=8$	*Given linear equation*
$4\left(\frac{x}{4}\right)=4(8)$	*Multiply both sides by 4.*
$\boxed{\mathbf{x=32}}$	*Simplify.*

b)

$\frac{x}{-4}=11$	*Given linear equation*
$-4\left(\frac{x}{-4}\right)=-4(11)$	*Multiply both sides by –4.*
$\boxed{\mathbf{x=-44}}$	*Simplify.*

7. a)

$13x=0$	*Given linear equation*
$\frac{13x}{13}=\frac{0}{13}$	*Divide both sides by 13.*
$\boxed{\mathbf{x=0}}$	*Simplify.*

b)

$-\frac{1}{2}x=0$	*Given linear equation*
$-2\left(-\frac{1}{2}x\right)=-2(0)$	*Multiply both sides by –2.*
$\boxed{\mathbf{x=0}}$	*Simplify.*

9. a) $-5X = -1$ *Given linear equation*

$\frac{-5X}{-5} = \frac{-1}{-5}$ *Divide both sides by –5.*

$\boxed{X = \frac{1}{5}}$ *Simplify.*

b) $6 = C - 4$ *Given linear equation*

$6 + 4 = C - 4 + 4$ *Add 4 to both sides.*

$\boxed{10 = C}$ *Simplify.*

11. a) $16F - 5 = 11$ *Given linear equation*

$16F - 5 + 5 = 11 + 5$ *Add 5 to both sides.*

$16F = 16$ *Simplify.*

$\frac{16F}{16} = \frac{16}{16}$ *Divide both sides by 16.*

$\boxed{x = 1}$ *Simplify.*

b) Since this problem is so similar to part (a), let us use it as an opportunity to demonstrate a more succinct approach; specifically, let's remove certain written details which can be calculated mentally. We strongly recommend you, too, begin to condense your equation-solving work in this manner. The more you practice, the more automatic these mental calculations will become.

$6 = 5G - 24$ *Given linear equation*

$30 = 5G$ *Add 24 to both sides.*

$\boxed{6 = G}$ *Divide both sides by 5.*

13. a) $\frac{J}{5} = 3$ *Given linear equation*

$\boxed{J = 15}$ *Multiply both sides by 5.*

b) $\frac{2K}{3} = 6$ *Given linear equation*

$\frac{3}{2}\left(\frac{2K}{3}\right) = \frac{3}{2}(6)$ *Multiply both sides by the reciprocal of $\frac{2}{3}$, namely $\frac{3}{2}$.*

$\boxed{K = 9}$ *Simplify.*

[Note: Naturally, we are assuming you know how to multiply fractions. Without such basic arithmetic skills, solving equations can be an excruciatingly difficult, mentally exhausting process.]

15. a) $7 = \frac{2N}{3} + 11$ — *Given linear equation*

$-4 = \frac{2N}{3}$ — *Subtract 11 from both sides.*

$\frac{3}{2}(-4) = \frac{3}{2}\left(\frac{2N}{3}\right)$ — *Multiply both sides by the reciprocal of $\frac{2}{3}$.*

$\boxed{\mathbf{-6 = N}}$ — *Simplify.*

b) $-5 = \frac{2P+1}{3}$ — *Given linear equation*

$-15 = 2P + 1$ — *Multiply both sides by 3.*

$-16 = 2P$ — *Subtract 1 from both sides.*

$\boxed{\mathbf{-8 = P}}$ — *Divide both sides by 2.*

17. $5(6S - 81) = -3(15 + 5S)$ — *Given linear equation*

$30S - 405 = -45 - 15S$ — *Simplify (i.e., apply the distributive property).*

$45S - 405 = -45$ — *Add 15S to both sides.*

$45S = 360$ — *Add 405 to both sides.*

$\boxed{\mathbf{S = 8}}$ — *Divide both sides by 45.*

19. $3(U - 3) - 2(U - 12) = 18$ — *Given linear equation*

$3(U - 3) + (-2)(U - 12) = 18$ — *Apply the definition of subtraction (which, beginning with our solution for Problem 21, will be performed mentally).*

$3U - 9 - 2U + 24 = 18$ — *Distribute [Be especially cautious when distributing the –2].*

$U + 15 = 18$ — *Simplify (i.e., combine like terms).*

$\boxed{\mathbf{U = 3}}$ — *Subtract 15 from both sides.*

21. $5(W + 3) - 6(W + 5) = 0$ — *Given linear equation*

$5W + 15 - 6W - 30 = 0$ — *Apply the distributive property (be careful with the signs).*

$-W - 15 = 0$ — *Simplify (i.e., combine like terms).*

$\boxed{\mathbf{-15 = W}}$ — *Add W to both sides.*

23. $5(Z - 2) - 3(Z + 3) = 9$ — *Given linear equation*

$5Z - 10 - 3Z - 9 = 9$ — *Simplify (distributive property).*

$2Z - 19 = 9$ — *Simplify (combine like terms).*

$2Z = 28$ — *Add 19 to both sides.*

$\boxed{\mathbf{Z = 14}}$ — *Divide both sides by 2.*

25. $x^2 = 10x$ — *Given quadratic equation*

$x^2 - 10x = 0$ — *Subtract 10x from both sides to rewrite in $ax^2+bx+c=0$ form.*

$x(x-10) = 0$ — *Factor, if possible.*

$x = 0$ or $x - 10 = 0$ — *Apply zero-product rule (see textbook p. 267).*

$\boxed{\mathbf{x = 0,\ x = 10}}$ — *Solve each of the resulting linear equations.*

27. $5x + 66 = x^2$ — *Given quadratic equation*

$0 = x^2 - 5x - 66$ — *Subtract 5x and 66 from both sides to rewrite in $ax^2+bx+c=0$ form.*

$0 = (x-11)(x+6)$ — *Factor, if possible.*

$x - 11 = 0$ or $x + 6 = 0$ — *Apply zero-product rule (see textbook p. 267).*

$\boxed{\mathbf{x = 11,\ x = -6}}$ — *Solve each of the resulting linear equations.*

29. $x^3 = 4x$ — *Given cubic equation*

$x^3 - 4x = 0$ — *Subtract 4x from both sides.*

$x(x^2 - 4) = 0$ — *First factor greatest common factor (g.c.f.) x.*

$x(x-2)(x+2) = 0$ — *Factor difference of squares.*

$x = 0$ or $x - 2 = 0$ or $x + 2 = 0$ — *Apply zero-product rule.*

$\boxed{\mathbf{x = 0,\ x = 2,\ x = -2}}$ — *Solve each of the resulting linear equations.*

31. $4x(x-9) = 9(1-4x)$ — *Given equation*

$4x^2 - 36x = 9 - 36x$ — *Simplify (by applying the distributive property).*

$4x^2 - 9 = 0$ — *Rewrite the quadratic equation in $ax^2+bx+c=0$ form.*

$(2x-3)(2x+3) = 0$ — *Factor difference of squares.*

$\boxed{\mathbf{x = \tfrac{3}{2},\ x = -\tfrac{3}{2}}}$ — *Set each factor equal to 0 and mentally solve.*

33. Notice that $x^2 + 7x + 2 = 0$ has a zero on one side and also that the left-hand side of the equation does <u>not</u> factor (by using integers). Therefore, we use the quadratic formula. First, we mentally identify $a = 1$, $b = 7$, and $c = 2$. Then, applying the quadratic formula (shown on page 267 of your text) and simplifying appropriately yields the following:

$$x = \frac{-(7) \pm \sqrt{(7)^2 - 4(1)(2)}}{2(1)} = \boxed{\frac{-7 \pm \sqrt{41}}{2}}$$

35. $x^2 - 5x - 3 = 0$ — *Given quadratic equation, one that is not factorable with integers.*

$$x = \frac{-(-5) \pm \sqrt{(-5)^2 - 4(1)(-3)}}{2(1)}$$ *Apply quadratic formula (with a = 1, b = –5, and c = –3).*

$$= \boxed{\frac{5 \pm \sqrt{37}}{2}}$$ *Simplify.*

37. $x^2 - 6x + 7 = 0$ — *Given quadratic equation*

$$x = \frac{-(-6) \pm \sqrt{(-6)^2 - 4(1)(7)}}{2(1)}$$ *Apply quadratic formula (with a = 1, b = –6, and c = 7).*

$$= \frac{6 \pm \sqrt{8}}{2}$$ *Simplify.*

$$= \frac{6 \pm 2\sqrt{2}}{2}$$ *Simplify radical.*

$$= \frac{6}{2} \pm \frac{2\sqrt{2}}{2} \quad \text{or} \quad \frac{2\left(3 \pm \sqrt{2}\right)}{2}$$ *<u>Either</u> rewrite as two fractions or factor 2 from numerator.*

This step is usually done mentally.

$$= \boxed{3 \pm \sqrt{2}}$$ *Reduce accordingly.*

39. $3x^2 + 5x - 4 = 0$ — *Given quadratic equation*

$$x = \frac{-(5) \pm \sqrt{(5)^2 - 4(3)(-4)}}{2(3)}$$ *Apply quadratic formula (with a = 3, b = 5, and c = –4).*

$$= \boxed{\frac{-5 \pm \sqrt{73}}{6}}$$ *Simplify.*

41. $4x^2 + 2x = -5$ — *Given quadratic equation*

$4x^2 + 2x + 5 = 0$ — *Add 5 to both sides to rewrite in $ax^2+bx+c=0$ form.*

$$x = \frac{-(2) \pm \sqrt{(2)^2 - 4(4)(5)}}{2(4)}$$ *Apply quadratic formula (with a = 3, b = 5, and c = –4).*

$$= \frac{-2 \pm \sqrt{-76}}{8}$$ *Simplify.*

Because we are working within the set of real numbers and square roots of negative numbers are not real numbers, we say there is **no solution**.

43. $3x^2 = 11x + 4$ — *Given quadratic equation*

$3x^2 - 11x - 4 = 0$ — *Subtract 11x and 4 from both sides to rewrite in $ax^2+bx+c=0$ form.*

$(3x - 4)(x + 1) = 0$ — *Factor.*

$\boxed{x = \frac{4}{3},\ x = -1}$ — *Solve by mentally applying the zero-product rule.*

Note: Even though this trinomial is relatively easy to factor, you still may wish to use an alternative method to solve this equation, specifically the quadratic formula. In the spirit of encouraging you to strive to become as skilled a problem solver as possible, we suggest you take advantage of this alternative method, whenever possible, because doing so will help you become more proficient with manipulating fractions and radicals. Realize, of course, that regardless of the method chosen, the final solutions should be identical.

45. $6x^2 = 5x$ — *Given quadratic equation*

$6x^2 - 5x = 0$ — *Subtract 5x from both sides to rewrite in $ax^2+bx+c=0$ form.*

$x(6x - 5) = 0$ — *Factor the greatest common factor (g.c.f.) x.*

$\boxed{x = 0,\ x = \frac{5}{6}}$ — *Solve by mentally applying the zero-product rule.*

47. $x^2 + 4 = 3\sqrt{2}x$ — *Given quadratic equation*

$x^2 - 3\sqrt{2}x + 4 = 0$ — *Rewrite in $ax^2+bx+c=0$ form.*

$$x = \frac{-(-3\sqrt{2}) \pm \sqrt{(-3\sqrt{2})^2 - 4(1)(4)}}{2(1)}$$ *Apply the quadratic formula.*

$$= \frac{3\sqrt{2} \pm \sqrt{18 - 16}}{2}$$ *Simplify.*

$$= \frac{3\sqrt{2} \pm \sqrt{2}}{2}$$

At this stage, you may choose to either (i) use your calculator with the above-simplified solution set or (ii) consider the solutions individually (call them, for example, x_1 and x_2) as we demonstrate below.

$$x_1 = \frac{3\sqrt{2} + \sqrt{2}}{2} = \frac{4\sqrt{2}}{2} = 2\sqrt{2} \approx \boxed{2.83}$$

$$x_2 = \frac{3\sqrt{2} - \sqrt{2}}{2} = \frac{2\sqrt{2}}{2} = \sqrt{2} \approx \boxed{1.41}$$

49. $\sqrt{2}x^2 + 2x - 3 = 0$ *Given quadratic equation*

$$x = \frac{-(2) \pm \sqrt{(2)^2 - 4(\sqrt{2})(-3)}}{2(\sqrt{2})}$$ *Apply the quadratic formula.*

$$= \frac{-2 \pm \sqrt{4 + 12\sqrt{2}}}{2\sqrt{2}}$$ *Simplify.*

$\approx$ **0.91, –2.33** *Use calculator to obtain approximated solution.*

Please note that (prior to using your calculator) you may wish to further simplify the radical expression above. Some students find that doing so makes it easier to use the calculator; others prefer to minimize the number of algebraic manipulations and instead like to use the calculator as early as possible. Regardless of your preference, the following illustrates the simplification process:

$$\frac{-2 \pm \sqrt{4 + 12\sqrt{2}}}{2\sqrt{2}} = \frac{-2 \pm \sqrt{4(1 + 3\sqrt{2})}}{2\sqrt{2}}$$

$$= \frac{-2 \pm 2\sqrt{1 + 3\sqrt{2}}}{2\sqrt{2}}$$

$$= \frac{\cancel{2}\left(-1 \pm \sqrt{1 + 3\sqrt{2}}\right)}{\cancel{2}\sqrt{2}}$$

$$= \frac{-1 \pm \sqrt{1 + 3\sqrt{2}}}{\sqrt{2}}$$

$$\approx 0.91, -2.33$$

51. $0.02x^2 + 0.831x + 0.0069 = 0$ *Given quadratic equation*

$$x = \frac{-(0.831) \pm \sqrt{(0.831)^2 - 4(0.02)(0.0069)}}{2(0.02)}$$ *Apply quadratic formula.*

$$= \frac{-0.831 \pm \sqrt{0.690009}}{0.04}$$ *Simplify (optional), but do not round.*

$\approx$ **–41.54, –0.01**

53. $x^2 - 11.001x + 24.098 = 0$ *Given quadratic equation*

$$x = \frac{-(-11.001) \pm \sqrt{(-11.001)^2 - 4(1)(24.098)}}{2(1)}$$ *Apply quadratic formula.*

$$= \frac{11.001 \pm \sqrt{24.630001}}{2}$$ *Simplify (optional), but do not round.*

$$\approx \boxed{7.98,\ 3.02}$$

55. a) Substituting the given values into the Young's Rule yields the result:

$$\begin{aligned}\text{Child's Dose} &= \frac{\text{Age of Child}}{\text{Age of Child} + 12} \times \text{Adult Dose} \\ &= \frac{(10)}{(10)+12} \times (100\text{ mg}) \\ &= \tfrac{10}{22} \times (100\text{ mg}) \\ &= 45.\overline{45}\text{ mg} \\ &\approx \boxed{45.5\text{ mg}}\end{aligned}$$

b) To obtain the adult's dose, we first substitute into Young's Rule the values given (for the child's age and dosage) and then solve the resulting equation for the adult dosage:

$$\begin{aligned}\text{Child's Dose} &= \frac{\text{Age of Child}}{\text{Age of Child} + 12} \times \text{Adult Dose} \\ 10\text{ mg} &= \frac{(12)}{(12)+12} \times \text{Adult Dose} \\ 10\text{ mg} &= \tfrac{12}{24} \times \text{Adult Dose} \\ 10\text{ mg} &= \tfrac{1}{2} \times \text{Adult Dose}\end{aligned}$$

Multiplying both sides by 2 yields the final answer: Adult Dose = $\boxed{20\text{ mg}}$.

Note: There is a second approach, one which may be preferable — especially for someone who regularly needs to convert children's dosages to adult dosages. This approach would consist of the same steps shown above but performed in reverse order; namely, we would first rewrite Young's Rule in terms of the adult dose (which would provide us with another "version" of Young's Rule which we could use thereafter) and then substitute the values given. For your information, this version of Young's Rule would be as follows:

$$\text{Adult Dose} = \frac{\text{Age of Child} + 12}{\text{Age of Child}} \times \text{Child's Dose}$$

57. There is an error in the last step, the step in which both sides are divided by $a+b-c$. Given the fact that $a+b=c$, it follows that $a+b-c$ equals zero. Consequently, dividing by $a+b-c$ in the last step means we are dividing by zero, but recall that division by zero cannot be done (i.e., division by zero is undefined).

59. a) $x^2 + 10x = 39$ *Given quadratic equation*

$x^2 + 10x - 39 = 0$ *Rewrite in $ax^2+bx+c=0$ form.*

$(x+13)(x-3) = 0$ *Factor*

$\boxed{x = -13,\ x = 3}$ *Solve by mentally applying the zero-product rule.*

b) These directions are a geometric proof (which, by the way, gives only one of the answers). Al-Khwârizmî begins with a square. If he lets its length be represented by x, it follows that its area is x^2 (as shown below).

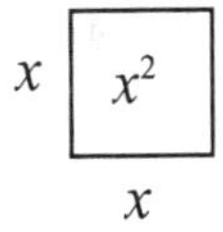

Then, al-Khwârizmî mentions that the square is to be "combined with ten of its roots" and this "will give a sum total of 39." So, to the square, he must add $10x$. This is done by adding four rectangles each of breadth $10/4$ the length x to the square. That is,

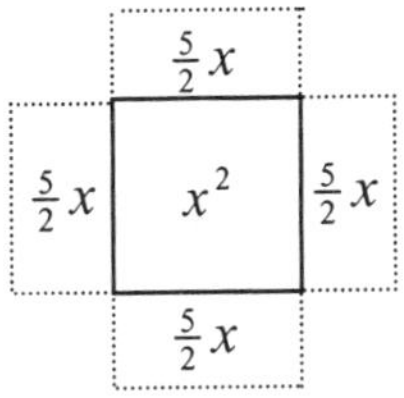

The figure now has area $x^2 + 10x$, which is equal to 39. Now, to complete the square, he adds four little squares each of area $(5/2)^2 = 25/4$.

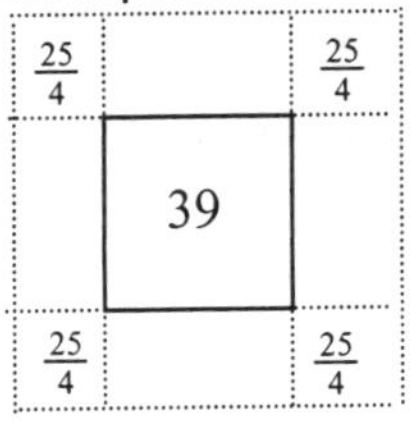

Thus, the newly formed (large) square has an area of

$$4(\tfrac{25}{4}) + 39 = 64.$$

Therefore, each side of the square is 8 units in length. At the same time, though, each side of the square is $(5/2) + x + (5/2) = x + 5$ units in length. So, $x + 5 = 8$, which gives $x = 3$.

Problem Set 5.5

1. **Wording varies**. The ***comparison property*** is stated on textbook page 272.

3.

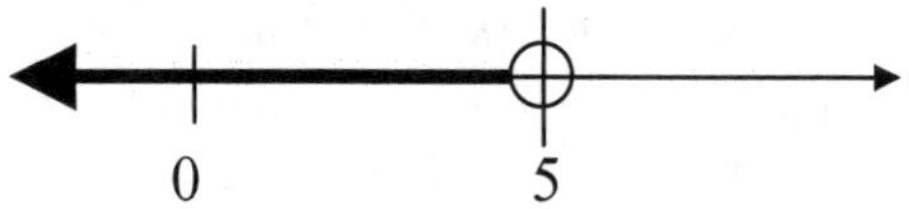

5.

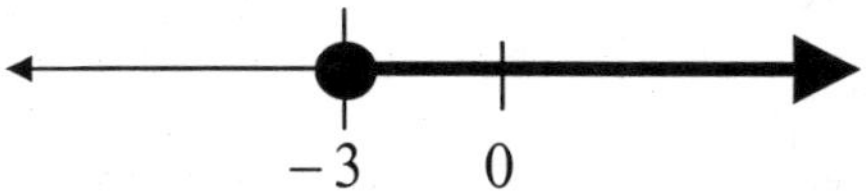

7. Do not read this inequality from left to right; you should read it from right to left. In general, always read an inequality in terms of its variable. So, in the case of $4 \geq x$, we say "x is less than or equal to 4" (i.e., $x \leq 4$).

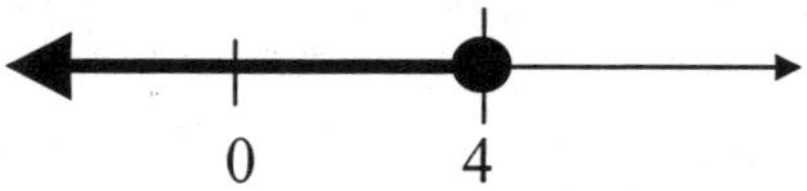

9. It is important to remember what these graphs represent: the set of possible values for the variable. Hence, we first must solve the given inequality $\frac{x}{2} > 3$ for its variable, x. Multiplying both sides by 2 gives us the statement $x > 6$, which leads us to the following graph:

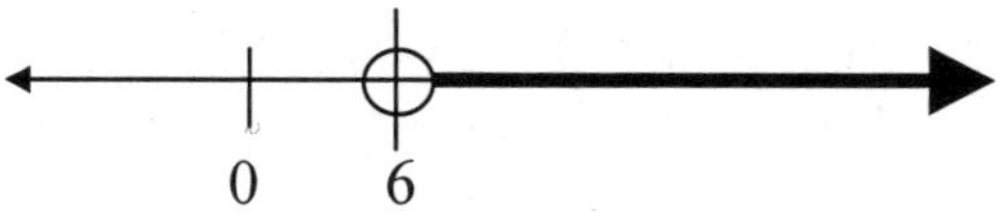

11. As explained in our solution to Problem 9 above, we must solve the inequality before we can graph it.

$-2 > \frac{x}{-4}$	*Given inequality*
$(-4) \cdot -2 < \frac{x}{-4} \cdot (-4)$	*Multiply both sides by –4 and reverse the inequality.*
$8 < x$	*Simplify.*

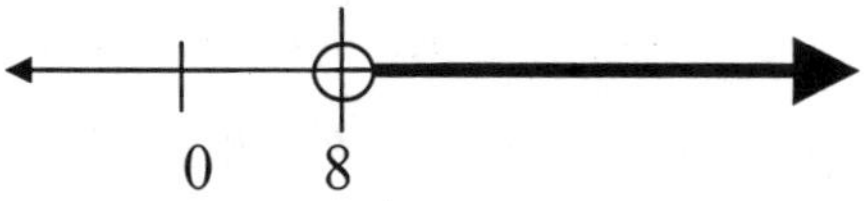

13.

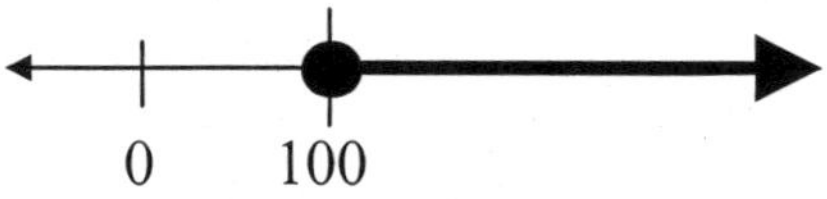

15.

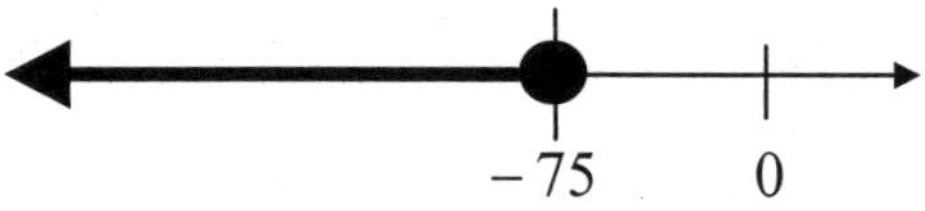

17. As previously noted (in our solution to Problem 7), inequalities should be read in terms of their variable — not necessarily from left to right. That is, in the case of $-40 < x$, it is to be read as "x is greater than -40" (i.e., $x > -40$).

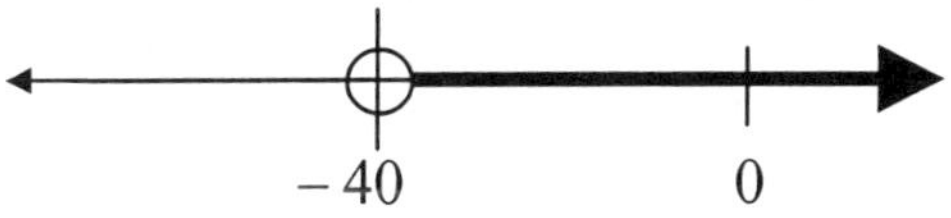

19. $x - 2 \le 5$ *Given inequality*

$\boxed{x \le 7}$ *Add 2 to both sides.*

21. $10 < 5 + y$ *Given inequality*

$5 < y$ *Subtract 5 from both sides.*

$\boxed{y > 5}$ *Rewrite solution in a more suitable, easy-to-read form.*

23. $-3 < 5 + y$ *Given inequality*

$-8 < y$ *Subtract 5 from both sides.*

$\boxed{y > -8}$ *Rewrite solution.*

25. $-t \le -3$ *Given inequality*

$\boxed{t \ge 3}$ *Multiply both sides by –1 and reverse the inequality.*

27. Since this problem and Problem 29 are so similar, permit us to illustrate one approach with this problem and then demonstrate another with Problem 29.

$5 \le 4 - y$ *Given inequality*

$1 \le -y$ *Subtract 4 from both sides.*

$1 \ge y$ *Multiply both sides by –1 and reverse the inequality.*

$\boxed{y \le 1}$ *Rewrite solution.*

29. Having illustrated one method with Problem 27 (as mentioned above), let us now demonstrate another:

$5 \ge 1 - w$ *Given inequality*

$5 + w \ge 1$ *Add w to both sides.*

$\boxed{w \ge -4}$ *Subtract 5 from both sides.*

31. $3y - 6 \ge 9$ *Given inequality*

$3y \ge 15$ *Add 6 to both sides.*

$\boxed{y \ge 5}$ *Divide both sides by 3.*

33. $\boxed{s > -3}$. Here are two approaches:

Approach I		**Approach II**	
$2 < 2s + 8$	*Given equation*	$2 < 2s + 8$	*Given equation*
$-6 < 2s$	*Subtract 8 from both sides.*	$1 < s + 4$	*Divide both sides by 2.*
$-3 < s$	*Divide both sides by 2.*	$-3 < s$	*Subtract 4 from both sides.*

35.

$3 > 2b - 13$	*Given inequality*
$16 > 2b$	*Add 13 to both sides.*
$8 > b$	*Divide both sides by 2.*
$\boxed{b < 8}$	*Rewrite solution.*

37.

$5t - 7 \geq 8$	*Given inequality*
$5t \geq 15$	*Add 7 to both sides.*
$\boxed{t \geq 3}$	*Divide both sides by 3.*

39.

$9 - 2v < 5$	*Given inequality*
$-2v < -4$	*Subtract 9 from both sides*
$\boxed{v > 2}$	*Divide both sides by –2 and reverse the inequality.*

41.

$2 - x \geq 3x + 10$	*Given inequality*
$-4x \geq 8$	*Subtract 3x and 2 from both sides.*
$\boxed{x \leq -2}$	*Divide both sides by –4 and reverse the inequality.*

43.

$B > 3(1 + B)$	*Given inequality*
$B > 3 + 3B$	*Apply the distributive property.*
$-2B > 3$	*Subtract 3B from both sides.*
$\boxed{B < -\frac{3}{2}}$	*Divide both sides by –2 and reverse the inequality.*

45.

$2(D + 7) > 2 - D$	*Given inequality*
$2D + 14 > 2 - D$	*Apply the distributive property.*
$3D > -12$	*From both sides, subtract 14 and add D.*
$\boxed{D > -4}$	*Divide both sides by 3.*

47.

$3(3F - 2) > 4F - 3$	*Given inequality*
$9F - 6 > 4F - 3$	*Apply the distributive property.*
$5F > 3$	*From both sides, add 6 and subtract 4F.*
$\boxed{F > \frac{3}{5}}$	*Divide both sides by 5.*

49.

$5(4+H)<3(H+1)$	*Given inequality*
$20+5H<3H+3$	*Apply the distributive property twice.*
$2H<-17$	*From both sides, subtract 20 and 3H.*
$\boxed{H<-\frac{17}{2}}$	*Divide both sides by 2.*

51.

$7(J-2)+5\le 3(2+J)$	*Given inequality*
$7J-14+5\le 6+3J$	*Apply the distributive property twice.*
$7J-9\le 6+3J$	*Simplify (combine like terms).*
$4J\le 15$	*From both sides, add 9 and subtract 3J.*
$\boxed{J\le \frac{15}{4}}$	*Divide both sides by 4.*

53. Let x equal the unknown number(s). We then translate the words of the problem into mathematical symbols:

Words	***Mathematical Symbols***
"the opposite of a number"	$-x$
"must be greater than"	$>$
"twice the number"	$2x$

Putting this together in the form of a mathematical statement yields the following inequality, which we then proceed to solve:

$-x>2x$	*Given inequality (i.e., the translation of the problem)*
$-3x>0$	*Subtract 2x from both sides.*
$x<0$	*Divide both sides by –3 and reverse the inequality.*

Finally, we express the mathematical statement $x<0$ in terms of the language of the original problem. Hence, the possible numbers are **any numbers less than 0**.

We, of course, strongly encourage you to **check this answer** by selecting some numbers and confirming that they indeed do satisfy the conditions of the original problem.

55. This problem is very similar to Problem 53 (see above solution). Letting x represent the unknown number(s) and translating the words "the opposite of a number must be less than 5" leads us to write the following inequality, which we, in turn, solve:

$-x<5$	*Given inequality (i.e., the translation of the problem)*
$x>-5$	*Multiply both sides by –1 and reverse the inequality.*

Thus, **any numbers greater than –5** satisfy the stated condition.

As was mentioned in our solution to Problem 53, we strongly encourage you to check this answer by selecting some numbers and confirming that they satisfy the original problem's conditions.

57. Let x represent the unknown number(s). We then translate the words of the problem into mathematical symbols:

Words	***Mathematical Symbols***
"a number"	x
"is"	$=$
"six less than…"	$\ldots - 6$
"twice its opposite"	$2(-x)$

Putting these pieces together yields the following equation, which we then solve:

$x = 2(-x) - 6$ *Given equation (i.e., the translation of the problem)*

$x = -2x - 6$ *Simplify.*

$3x = -6$ *Add 3x to both sides.*

$x = -2$ *Divide both sides by 2.*

Therefore, the only possible number is **−2**. We leave it to you to check this answer.

59. Let x represent the unknown number(s). In the same manner as demonstrated above (see solutions to Problems 53, 55, and 57), we translate the words "a number is less than six minus twice its opposite" and proceed to solve the resulting inequality.

$x < 6 - 2(-x)$ *Given inequality (i.e., the translation of the problem)*

$x < 6 + 2x$ *Simplify.*

$-x < 6$ *Subtract 2x from both sides.*

$x > -6$ *Multiply both sides by –1 and reverse the inequality.*

Thus, the possible numbers are **any numbers greater than –6**. As we have mentioned in assorted previous solutions, please take the time to check this answer.

Problem Set 5.6

1. **Answers/wording vary**. Your answer should include mention of Pólya's four-step problem-solving model, which initially was introduced in the very first section of the first chapter of your text (see page 7). Please note that your text's author does an especially outstanding job of rephrasing these steps on pp. 277-278. We recommend you study it.

***For Problems 3–25**: Karl Smith, your text's author, is asking you to "translate the problem." As he writes repeatedly in this section, "translate" means "write down a verbal description of the [main idea of the] problem using operation signs and an equal or inequality sign." For additional details, please refer to the second step of the procedure summarized on textbook p. 277.*

3. Translating the words "add seven to twice a number… is seventeen" means:

2(NUMBER) + 7 = 17

We hope it goes without saying that writing "7 + 2(number) = 17" is equivalent to the above translation.

5. **5(NUMBER) – (–10) = –30**

7. **INTEGER + NEXT INTEGER = 117**

9. **INTEGER + NEXT INTEGER + THIRD CONSECUTIVE INTEGER = 105**

11. Remember that this first ("translating") step of the problem-solving procedure only requires you to translate the problem's **main** idea. In this problem, the main idea is:

VALUE OF HOUSE + VALUE OF LOT = $212,400

Later, when you actually solve the problem (see Problem 34), you will then go on to translate any remaining ideas/conditions; for example, in reading this problem, you would have observed that the VALUE OF HOUSE = 5(VALUE OF LOT).

13. The "main" idea of the problem can be translated as follows:

AMT. PAID FOR CORRECT SOLNS. + AMT. FINED FOR INCORRECT = $0

Once again, as we noted in our solution to Problem 11, we consider the remaining details when we actually solve the problem (which is done in Problem 36).

15. Before you can translate the problem, you must first "understand the problem," which is the first-step of Pólya's problem-solving guidelines. Drawing a picture will greatly enhance your level of understanding. Before studying the one drawn below, you, of course, should have attempted to sketch your own.

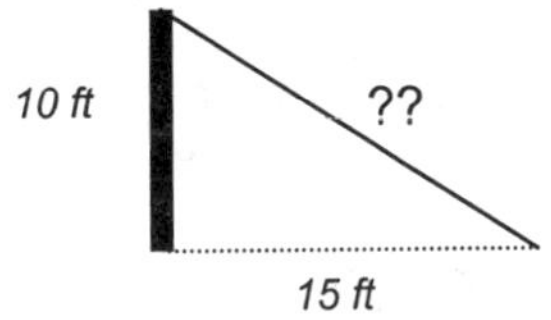

The thick vertical line depicts the 10-ft pole; the dotted horizontal line represents the ground. The problem's unknown is depicted by the thin solid line, which represents just one of the four guy wires connected to the top of the pole and the ground ("15 ft from the base of the pole").
Since the three sides form a right triangle, we can relate their lengths to one another by applying the Pythagorean theorem (see textbook p. 283):

$$10^2 + 15^2 = (\text{LENGTH OF GUY WIRE})^2$$

17. Just like we did with Problem 15, we recommend you begin by sketching a picture. If you have not already attempted to create such a sketch, we recommend you go do so now, before reading the rest of this solution.
From the wording of the problem, it is clear that the cities are located in the following order:

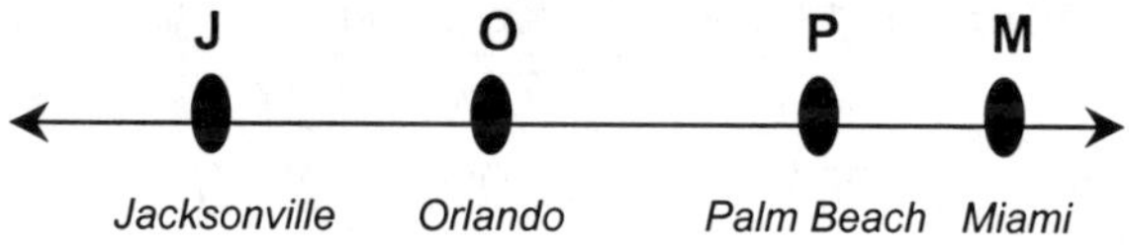

DIST FROM J TO O + DIST FROM O TO P + DIST FROM P TO M = TOTAL DISTANCE

Please note that the additional details given in the problem (for example, the fact that the total distance equals 370 miles) become included once we move beyond the "translate" problem-solving step to the "evolve" step. To see how this is done with a problem closely related to this one, please refer to textbook Example 3 (pp. 280-281).

19. This problem is very similar to Problem 17 (see solution above) and textbook Example 3 (pp. 280-281). We encourage you to begin by sketching a diagram. Letting S represent San Antonio, A signify Austin, W stand for Waco, and D connote Dallas, the problem translates to:

DIST FROM S TO A + DIST FROM A TO W + DIST FROM W TO D = TOTAL DISTANCE

21. This problem is similar to textbook Example 4 (pp. 281-283). Drawing a diagram may help you better understand the problem.

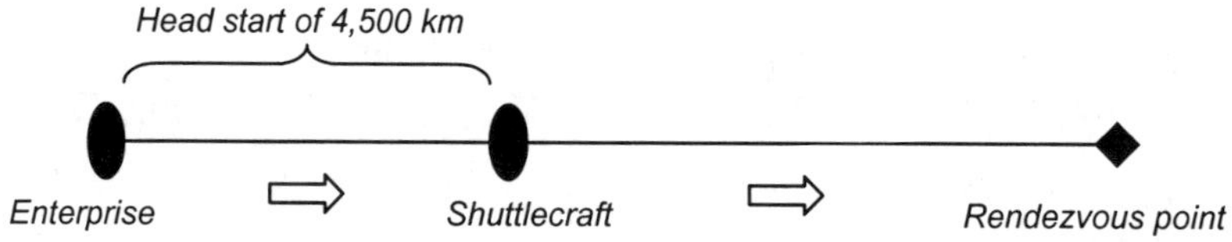

Based upon our diagram and letting S and E signify respectively shuttlecraft and enterprise, the main idea of the problem is:

DISTANCE S TRAVELS + HEAD START = DISTANCE E TRAVELS

23. This problem is very similar to Problem 21 (see solution above) and textbook Example 4 (pp. 281-283). The slower walker has a 1-mile head start on the faster walker, yet they finish together (i.e., the faster walker catches up at the rendezvous point). Upon sketching a diagram comparable to the one on textbook page 282 and the one in our solution to Problem 21, it becomes clear that this problem translates to:

SLOWER WALKER'S DISTANCE + HEAD START = FASTER WALKER'S DISTANCE

25. Once again, we must "understand the problem" before we can do anything else. A well-sketched diagram helps us accomplish this:

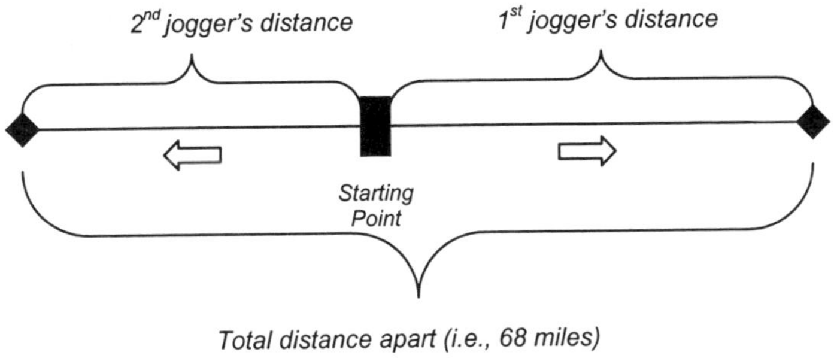

Please note that in our sketch we arbitrarily chose to let the first jogger represent the faster of the two. From this sketch, the relationship among the relevant distances becomes apparent and leads to the following translation of the problem:

Distance of 1st jogger + Distance of 2nd jogger = Total Distance

27. This problem builds upon the verbal translation written for Problem 4. We apply the text's problem-solving procedure.

Translate: 2(NUMBER) – 12 = 6

Evolve: Let x = the unknown NUMBER. We then substitute this variable into the verbal equation above and write: $2x-12=6$.

Solve:

$$2x-12=6$$
$$2x=18$$
$$x=9$$

Answer: We then state our solution in words: **The number is 9**.

Check: $2(9)-12=18-12=6$ ✓

29. This problem builds upon the verbal translation written for Problem 6. We apply the text's problem-solving procedure.

Translate: 3(NUMBER) – 6 = 2(NUMBER)

Evolve: Let x = the NUMBER. We then substitute this variable into the verbal equation above and write: $3x-6=2x$.

Solve:

$$3x-6=2x$$
$$-6=-x$$
$$6=x$$

Answer: We then state our solution in words: **The number is 6**.

Check: $3(6)-6=18-6=12$ ✓, which equals $2(6)$ ✓

31. This problem builds upon the verbal translation written for Problem 8. We apply the text's problem-solving procedure.

Translate: EVEN INTEGER + NEXT EVEN INTEGER = 94

Evolve: Please note that there are many ways one could choose variables for this problem. Here is just one approach: if we let $x =$ the first EVEN INTEGER, then $x+2 =$ the NEXT EVEN INTEGER. Substituting these into the verbal equation gives us the equation $x+(x+2)=94$.

Solve:
$$x+(x+2)=94$$
$$2x+2=94$$
$$2x=92$$
$$x=46$$

Answer: Since $x=46$, $x+2=48$. We then state our solution in words:

The two consecutive even integers are 46 and 48.

Check: $46+48=94$ ✓

33. This problem builds upon the verbal translation written for Problem 10. Also, please note that this problem is quite similar to textbook Example 2 (pp. 279-280). We once again apply the text's problem-solving procedure.

Translate: INTEGER + NEXT INTEGER + 3rd INTEGER + 4th INTEGER = 74

Evolve: There are many ways one could choose variables for this problem. One approach is to let $x =$ the first INTEGER, from which it follows that

$x+1 =$ NEXT INTEGER

$x+2 =$ 3rd INTEGER

$x+3 =$ 4th INTEGER

Substituting these into the verbal equation above gives us the equation
$$x+(x+1)+(x+2)+(x+3)=74.$$

Solve:
$$x+(x+1)+(x+2)+(x+3)=74$$
$$4x+6=74$$
$$4x=68$$
$$x=17$$

Answer: Since $x=17$, we know $x+1=18$, $x+2=19$, and $x+3=20$. We then state our solution in words:

The four consecutive integers are 17, 18, 19, and 20.

Check: $17+18+19+20=74$ ✓

35. This problem builds upon the verbal translation written for Problem 12. We again apply the text's problem-solving procedure.

Translate: From Problem 12, we know the main idea of this problem can be summarized by the following verbal equation:

PRICE OF 1st CABINET + PRICE OF 2nd CABINET = $4,150

Evolve: Notice that this verbal equation has two unknowns. Our goal is to "evolve" the equation into one with a single unknown. Since we know the cost of one type of cabinet is four times the cost of the other, we can write this second verbal equation:

PRICE OF 1st CABINET = 4(PRICE OF 2nd CABINET)

Please note that the labeling of "first" cabinet and "second" cabinet is a completely arbitrary selection.

Based on this second verbal equation, if we let

x = PRICE OF 2nd CABINET, then it follows that

$4x$ = PRICE OF 1st CABINET.

We now have a single unknown. Substituting into the main verbal equation gives us the one-variable equation $4x + x = 4{,}150$.

Solve:

$$4x + x = 4{,}150$$
$$5x = 4{,}150$$
$$x = 830$$

Answer: Since $x = 830$, we know $4x = 3{,}320$. We then state our solution in words: **The prices of the two cabinets are \$830 and \$3,320**.

Check: $\$3{,}320 + \$830 = \$4{,}150$ ✓ and $8(\$830) = \$3{,}320$ ✓

37. This problem builds upon the verbal translation written for Problem 14. We again apply your text's problem-solving procedure.

Translate: Let us review the "translation" process (i.e., the task of Problem 12). Notice that after collecting his winnings from the third race, the professional gambler has four times ("quadrupled") the amount of money he started with. With this quadrupled amount in hand, he proceeds to bet (and lose) \$72 on the fourth race, after which he is left with \$48. Hence, what transpired during the fourth race reveals the main idea of this problem, which can be summarized by the following verbal equation:

4(AMOUNT OF MONEY TO START) – 72 = 48.

Evolve: Letting x = AMOUNT OF MONEY TO START, we then write the equation $4x - 72 = 48$ and proceed to solve it.

Solve:

$$4x - 72 = 48$$
$$4x = 120$$
$$x = 30$$

Answer: After making sure we understand what x was chosen to represent, we are prepared to state our final solution in words:

The gambler started with \$30.

Check: $4(\$30) - \$72 = \$48$ ✓

39. This problem builds upon the verbal translation written for Problem 16. We again apply your text's problem-solving procedure.

Translate: Before you can "translate" this problem (i.e., perform the task of Problem 12), you must first "understand the problem." Drawing a picture is an especially good way to achieve the necessary level of understanding.

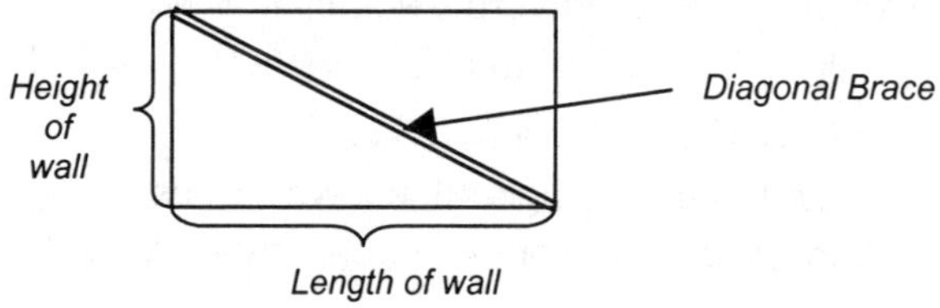

Notice that we have a right triangle, with the diagonal brace representing its hypotenuse. Based upon our knowledge of the Pythagorean theorem and the given dimensions of the wall (namely, 8 ft high and 14 ft long, we can translate the problem into a verbal equation:

$$8^2 + 14^2 = (\text{LENGTH OF BRACE})^2$$

Evolve: Letting x = LENGTH OF BRACE, we then write the algebraic equation $8^2 + 14^2 = x^2$ and proceed to solve it.

Solve: $x^2 = 8^2 + 14^2$

$x^2 = 260$

From the square root property, we know $x = \pm\sqrt{260}$.

Simplifying yields: $x = \pm\sqrt{260} = \pm\sqrt{4 \cdot 65} = \pm 2\sqrt{65}$.

However, because x represents a length (and lengths cannot be negative), we only consider the positive solution $x = 2\sqrt{65} \approx 16.12$.

Answer: We then state our final solution in words:

The length of the brace is exactly $2\sqrt{65}$ ft, or approximately 16 ft.

Check: Besides checking the arithmetic/algebra, it is important to ask whether the answer is "reasonable." We hope you will ask yourself this important question after every problem you do.

41. This problem builds upon the verbal translation written for Problem 18, which is almost identical to the verbal translations (explained in this manual) for Problems 17 and 19. Additionally, you may wish to refer to textbook Example 3 (pp.280-281).

Translate: From the wording of the problem, it is clear that the cities are located in the following order.

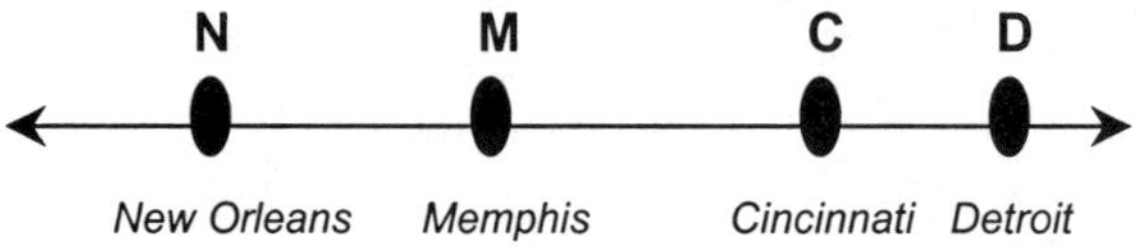

Hence, we translate the problem into the following verbal equation:

(DIST N TO M) + (DIST M TO C) + (DIST C TO D) = TOTAL DISTANCE

Evolve: We now translate the other pieces of information given in the statement of problem. First, we know the TOTAL DISTANCE equals 1,140 miles. Also, by adding to the smaller distance in each case, we can write:

(DIST N TO M) + 90 = (DIST M TO C)

(DIST N TO M) = (DIST C TO D) + 150

We can rewrite this second equation as:

(DIST N TO M) – 150 = (DIST C TO D)

We can use these verbal equation to substitute into the main equation:

(DIST N TO M) + **(DIST M TO C)** + **(DIST C TO D)** = TOTAL DISTANCE

↑ ↑

(DIST N TO M) + 90 ***(DIST N TO M) – 150***

Letting $d =$ DIST N TO M and substituting into our main verbal equation yields the equation

$$d+(d+90)+(d-150)=1{,}140.$$

Solve:

$$d+(d+90)+(d-150)=1{,}140$$
$$3d-60=1{,}140$$
$$3d=1{,}200$$
$$d=400$$

Answer: Although the equation is solved, we cannot automatically conclude that the answer to the problem is "$d = 400$." Remember that d represents the distance from New Orleans to Memphis, but the problem asks for the length of the Cincinnati-Detroit leg of the trip, which equals $d-150$ (i.e., $400 - 150$). Having a clear understanding of the question, we now can state our final answer in words:

The distance from Cincinnati to Detroit is 250 miles.

Check: $400+490+250=1140$ ✓

43. This problem builds upon the verbal translation written for Problem 20, which is almost identical to the verbal translations (explained in this manual) for Problems 21 and 23. Additionally, you may wish to refer to textbook Example 4 (pp.281-283).

Translate: Before we can translate any problem, we must first understand it. In this regard, sketching a diagram like the following often can be helpful:

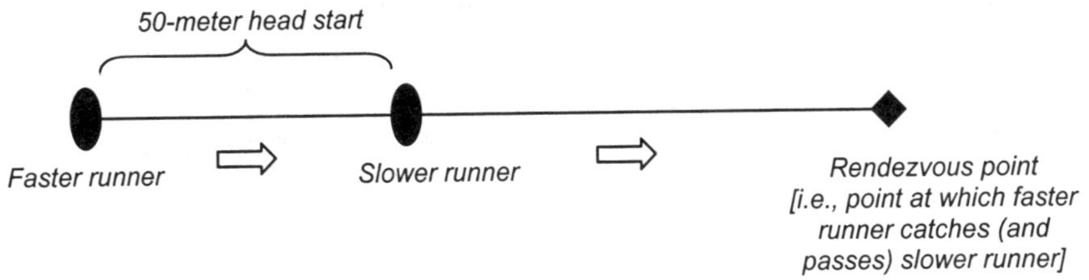

At the moment the faster runner catches up to the slower runner, the faster runner's distance traveled is exactly 50 meters more than the slower runner's. Hence, we translate the problem into the following verbal equation:

SLOWER RUNNER'S DIST + HEAD START = FASTER RUNNER 'S DIST

Evolve: We now translate the other pieces of information given in the statement of problem. First, we know the HEAD START equals 50 meters. Also, since both runners start running at the same instant, we know by using the distance-rate-time relationship (i.e., $d = rt$) that:

SLOWER RUNNER'S DIST = (SLOWER RUNNER'S RATE)(TIME TO RENDEZVOUS)

FASTER RUNNER'S DIST = (FASTER RUNNER'S RATE)(TIME TO RENDEZVOUS)

However, each runner's respective rate is given in the statement of the problem; the slower runner's rate is 6 meters per second, and the faster runner's rate is 10 meters per second. From this, it follows that

SLOWER RUNNER'S DIST = 6(TIME TO RENDEZVOUS)

FASTER RUNNER'S DIST = 10(TIME TO RENDEZVOUS)

Substituting these into the main equation gives us:

SLOWER RUNNER'S DIST	+	HEAD START	=	FASTER RUNNER 'S DIST
↑		↑		↑
6(TIME TO RENDEZVOUS)	**+**	***50***	**=**	***10(TIME TO RENDEZVOUS)***

Letting $t =$ TIME TO RENDEZVOUS (in seconds) and substituting into our main verbal equation yields the following one-variable equation:

$$6t + 50 = 10t$$

Solve:

$$\begin{aligned} 6t + 50 &= 10t \\ 50 &= 4t \\ \tfrac{50}{4} &= t \\ t &= 12.5 \end{aligned}$$

Answer: Remember that t represents the TIME TO RENDEZVOUS. With a clear understanding of the question being asked and what the chosen variable represents, we now can state our final answer in words:

It takes the faster runner 12.5 seconds to catch up.

Check: Since the faster runner's rate is 10 m/sec and the slower runner's rate is 6 m/sec, in 12.5 sec the faster runner will have run in 125 meters (i.e., 10×12.5), while the slower runner will have run 75 meters (6×12.5). Note that $75 + 50 = 125$. This confirms that it does indeed take the faster runner 12.5 sec to make up the additional 50 meters needed to catch the slower runner. ✓

45. This problem is almost identical to Problem 43 (see above solution), Problem 44, and textbook Example 4 (pp. 281-283). Furthermore, it builds upon the verbal translation written for Problem 22, which is almost identical to the verbal translations (explained in this manual) for Problems 21 and 23. Therefore, instead of providing you with a fully detailed solution (like we did for Problem 43), this time we have chosen to share a more concise explanation.

Recall the verbal equation obtained for Problem 22:

DIST OF SPEEDING CAR + HEAD START = DISTANCE OF POLICE CAR

With the word "rendezvous" meaning the point at which the police car catches the speeding car, we let $t =$ TIME TO RENDEZVOUS (in hours). We then apply the distance-rate-time formula:

DIST OF SPEEDING CAR = (RATE OF SPEEDING CAR)(TIME TO RENDEZVOUS)
= $80t$

DIST OF POLICE CAR = (RATE OF POLICE CAR)(TIME TO RENDEZVOUS)
= $100t$

Knowing that the speeding car has a 2-mile head start, the main verbal equation written above can be rewritten as:

$$80t + 2 = 100t$$

This leads to the following solution:

$$2 = 20t$$
$$\tfrac{2}{20} = t$$
$$t = \tfrac{1}{10}$$

In all word problems, it is important to remember what the variables represent, <u>including</u> their respective units. In this case, t represents the TIME TO RENDEZVOUS, with the units being hours. With that in mind, we may now state the final answer and, in the process, express it in terms of units appropriate to the situation:

It takes the police car 6 minutes (or 0.1 hr) to catch the speeding car.

47. Once again, we apply Pólya's problem-solving procedure.

Translate: In order to translate a problem, one first must understand it. Sketching a diagram like the following helps enhance our understanding:

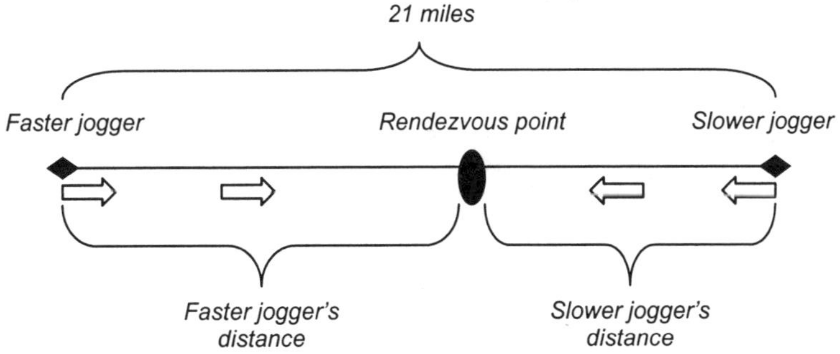

Hence, we translate the main idea of the problem into a verbal equation:

FASTER JOGGER'S DIST + SLOWER JOGGER'S DIST = TOTAL DISTANCE

Evolve: We now translate the other pieces of information given in the statement of problem. First, we know the TOTAL DISTANCE equals 21 miles. Also, since both joggers start running at the same instant, we know by using the distance-rate-time relationship (i.e., $d = rt$) that:

SLOWER JOGGER'S DIST = (SLOWER JOGGER'S RATE)(TIME TO RENDEZVOUS)

FASTER JOGGER'S DIST = (FASTER JOGGER'S RATE)(TIME TO RENDEZVOUS)

However, the TIME TO RENDEZVOUS is given in the statement of the problem; its value is 1.5 hrs. Additionally, we are given information that allows us to compare the rates:

FASTER JOGGER'S RATE = SLOWER JOGGER'S RATE + 2 mph

Letting $r =$ SLOWER JOGGER'S RATE (in miles/hour) and substituting into our previous verbal equations, we obtain the following:

FASTER JOGGER'S RATE = $r + 2$

SLOWER JOGGER'S DIST = $(r)(1.5) = 1.5r$

FASTER JOGGER'S DIST = $(r + 2)(1.5) = 1.5(r + 2)$

From this, we evolve our main verbal equation into:

FASTER JOGGER'S DIST + SLOWER JOGGER'S DIST = TOTAL DISTANCE

↑ ↑ ↑

1.5(r + 2) **+** ***1.5r*** **=** ***21***

Solve:

$$1.5r + 1.5(r + 2) = 21$$
$$1.5r + 1.5r + 3 = 21$$
$$3r = 18$$
$$r = 6$$

Answer: Remember that r represents the SLOWER JOGGER'S RATE. With a clear understanding of the question being asked and what the chosen variable represents, we are prepared to state our final answer in words:

One jogs at a rate of 6 mph; the other jogs 8 mph.

Check: Based on a rate of 8 mph and a jogging time of an hour and a half, we know the faster jogger will cover 12 miles. The slower jogger, whose rate is 6 mph, will cover 9 miles in that time span. Since $12 + 9 = 21$ and the joggers live 21 miles apart, they will meet after an hour and a half of jogging toward each other. ✓

49. Area $= \frac{1}{2}bh = 17.5\text{ cm}^2$. Letting $x =$ the length of the shorter leg, it follows that $x + 2 =$ the length of the longer leg. In no particular order, one of these two legs must be the base of the triangle, and the other must be the height. So,

$$\tfrac{1}{2}(x+2)x = 17.5$$

$$x^2 + 2x = 35$$ *Multiply both sides by 2 and then distribute x.*

$$x^2 + 2x - 35 = 0$$ *Rewrite the quadratic equation in $ax^2+bx+c=0$ form.*

$$(x+7)(x-5) = 0$$ *Factor.*

$$x = -7,\ x = 5$$ *Solve by mentally applying the zero-product rule.*

Since lengths cannot be negative and since x equals the length of the triangle's shortest side, we can conclude that **the length of the shortest side is 5 cm**.
[Reminder: As always, do not hesitate to begin by sketching (and labeling) a picture of the described right triangle. Drawing a picture is among the best ways to acquire the level of understanding necessary to solve a problem.]

51. Area $= \tfrac{1}{2}bh = 3$ ft^2. With h representing the height of the triangle and given that the triangle's base b is 2 ft longer than the height, it follows that $b = h + 2$. So,

$$\tfrac{1}{2}(h+2)h = 3$$

$$h^2 + 2h = 6$$ *Multiply both sides by 2 and then distribute h.*

$$h^2 + 2h - 6 = 0$$ *Rewrite the quadratic equation in $ah^2+bh+c=0$ form.*

Applying the quadratic formula,

$$h = \frac{-(2) \pm \sqrt{(2)^2 - 4(1)(-6)}}{2(1)} \approx 1.646,\ -3.646$$

However, since lengths cannot be negative, we disregard -3.646. Hence, the triangle's **height is ≈1.6 ft and its base ≈3.6 ft**.

53. If an investment of \$1 is "to double in 2 years," then obviously we have \$2 in 2 years. Hence, A equals \$2, and we substitute this value into the given formula and proceed to solve for r:

$$(1+r)^2 = 2$$

$$1 + r = \sqrt{2}$$ *Square root both sides of the equation; disregard negative value.*

$$r = \sqrt{2} - 1$$ *Subtract 1 from both sides.*

$$r \approx 0.4142$$ *Use calculator to obtain approximate solution.*

One, therefore, needs to obtain an annual rate of return of **approximately 41.4%** compounded annually.

55. This problem is almost identical to textbook Example 6 (pp. 285-286); the only difference between the two is the length of the expansion.

Since the expansion in this problem is ½ inch, one-half the arc (see Figure 5.8, p. 284) would have length 0.5 mile + 0.25 inch. By applying to this problem the equations and unit conversions previously established in Example 6 (p. 285), we obtain the following equation:

$$(0.5\text{ mile})^2 + (\text{HEIGHT})^2 = (0.5\text{ mile} + 0.25\text{ inch})^2$$

$$(31{,}680\text{ in.})^2 + h^2 = (31{,}680.25\text{ in.})^2$$

Solving for h, we obtain

$$h^2 = (31{,}680.25\text{ in.})^2 - (31{,}680\text{ in.})^2$$

$$h = \sqrt{(31{,}680.25\text{ in.})^2 - (31{,}680\text{ in.})^2}$$

$$\approx 125.86$$

This solution, 125.86 in., is approximately 10.5 ft. Therefore, for ½ inch of expansion, the 1-mile-long pipe would bow **approximately 10.5 ft** at the middle.

For additional details, please study text Example 6. Note that it is important to understand the author's comment [see "Devise A Plan" in the margin on p. 285] regarding why a right triangle can be used to model/approximate an arc length.

57. We begin by translating the problem's main idea into a verbal equation:

TOTAL NUM – (NUM TO MAHADEV + NUM TO HURI + NUM TO SUN + NUM TO DEVI) = NUM LEFT

Given that the number of lilies remaining is six, we know NUM LEFT equals 6. As for the respective number of lilies given to Mahadev, Huri, Sun, and Devi, each of these amounts is expressed (in the statement of the problem) in terms of the total number of lilies. Hence, it makes sense to choose a variable to represent the total number of lilies. Letting x represent TOTAL NUM, we "evolve" our verbal equation as follows:

TOTAL NUM – (NUM TO MAHADEV + NUM TO HURI + NUM TO SUN + NUM TO DEVI) = NUM LEFT

↑ ↑ ↑ ↑ ↑ ↑

$x \quad - \quad [\quad \frac{1}{3}x \quad + \quad \frac{1}{5}x \quad + \quad \frac{1}{6}x \quad + \quad \frac{1}{4}x \quad] = \quad 6$

We then proceed to solve this linear equation [note: the steps may vary]:

$$x - \left[\tfrac{1}{3}x + \tfrac{1}{5}x + \tfrac{1}{6}x + \tfrac{1}{4}x\right] = 6$$

$$60x - [20x + 12x + 10x + 15x] = 360 \quad \textit{Multiply both sides by 60 (the LCM of 3,4,5,6).}$$

$$60x - 57x = 360 \quad \textit{Combine like terms.}$$

$$3x = 360$$

$$x = 120$$

Therefore, **the total number of lilies is 120.**

Check: If there are 120 lilies in all, 40 (i.e., 1/3 of 120) must have gone to Mahadev, 24 to Huri , 20 to Sun, and 30 to Devi. Since 40 + 24 + 20 + 30 equals 114, there were indeed only 6 lilies remaining for the spiritual teacher. ✓

59. To better understand this problem, we heartily recommend you begin by drawing a picture.

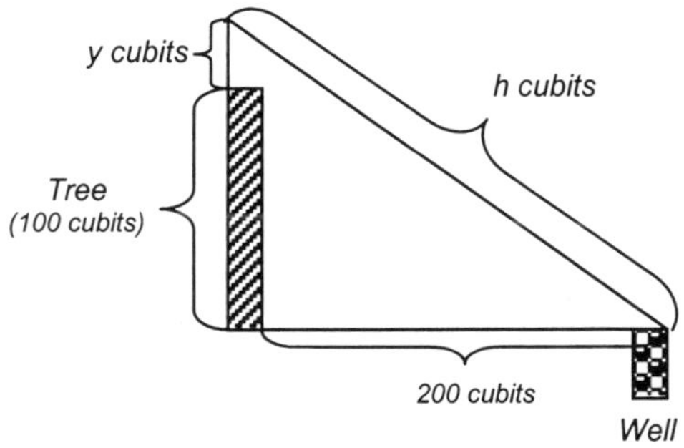

The main idea of the problem is found within the phrase "...both pass over an equal space." This can be translated into the following verbal equation:

FIRST MONKEY'S DISTANCE = SECOND MONKEY'S DISTANCE

Since the first monkey climbs 100 cubits down the tree and travels another 200 cubits directly to the well, the FIRST MONKEY'S DISTANCE totals 300 cubits. The second monkey first leaps up in the air (from the top of the tree) an additional y cubits above the top of the tree and then descends h cubits to arrive at the well. Consequently, the SECOND MONKEY'S DISTANCE is $(y+h)$ cubits. Thus, we can rewrite ("evolve") the above verbal equation as $300 = y + h$.

Then, having observed that our sketch forms a right triangle, we can apply the Pythagorean theorem to write:

$$200^2 + (100+y)^2 = h^2$$

This equation has two unknowns. However, if we express h in terms of y, we will have an equation whose only variable is y, which represents the height of the leap (and is precisely what the problem asks us to find). Since $300 = y + h$, $h = 300 - y$. Thus, we substitute for h in our previous equation and proceed to solve:

$$200^2 + (100+y)^2 = (300-y)^2$$
$$200^2 + 100^2 + 200y + y^2 = 300^2 - 600y + y^2$$
$$200^2 + 100^2 - 300^2 = -600y - 200y$$
$$40{,}000 + 10{,}000 - 90{,}000 = -800y$$
$$-40{,}000 = -800y$$
$$y = 50$$

Therefore, **the second monkey leapt 50 cubits into the air.**

Problem Set 5.7

1. **Answers vary**. A ***ratio*** is a quotient of two numbers or expressions. A ***proportion*** is a statement of equality between ratios.

3. **Answers vary**. Compared to our previously established methods for solving equations, the ***procedure for solving proportions*** (as defined on textbook page 294) is more efficient. It requires only one step, whereas previous equation-solving approaches generally required several.

5. The ratio of cement to water can be written (and reduced) as follows:

$$\frac{60\text{ lbs}}{3\text{ gal}}=\frac{20\text{ lbs}}{1\text{ gal}}$$

Thus, expressed in lowest terms, this is a **20 to 1** ratio. This reduced form certainly would be helpful to a person mixing cement.

7. $\frac{106\text{ males}}{100\text{ females}}=\frac{53\text{ gal}}{50\text{ females}}$. Therefore, in simplified form, this is a **53 to 50** ratio.

9. A ratio of 279 miles to $15\frac{1}{2}$ gallons of gasoline can be simplified as follows:

$$\begin{aligned}\frac{279\text{ miles}}{15\frac{1}{2}\text{ gallons}}&=279\div 15\tfrac{1}{2}\\&=\frac{279}{1}\div\frac{31}{2}\\&=\frac{279}{1}\cdot\frac{2}{31}\\&=\frac{9}{1}\cdot\frac{2}{1}\\&=\frac{18}{1}\end{aligned}$$

Thus, our simplified ratio is **18 to 1**. This simplified ratio tells us that this vehicle's gas mileage (for this particular trip) was 18 miles per gallon, which we often write as 18 mpg.

11. Let us begin by applying the "means-extremes" method shown in your text (see pages 290-291).

a) The product of the means (i.e., 1×21) equals 21, and the product of the extremes (i.e., 3×7) also equals 21. Since the product of the means equals the product of the extremes, we conclude **Yes**, the ratios form a proportion.

b) **Yes**, these ratios form a proportion because:

Product of means $=8\times 9=\underline{\underline{72}}$ *and* Product of extremes $=6\times 12=\underline{\underline{72}}$.

c) **Yes**, these ratios form a proportion because:

Product of means $= 6 \times 5 = \underline{\underline{30}}$ *and* Product of extremes $= 3 \times 10 = \underline{\underline{30}}$.

You should know that there are assorted other methods one can use to determine whether ratios form a proportion. Among them are the following three strategies:

(i) reduce both ratios to lowest terms and then check whether they are equal;

(ii) rewrite both ratios so that they share a common denominator and then compare them; and

(iii) express each ratio in decimal form and then compare the decimal numbers.

We encourage you to experiment with all three of these approaches.

13. Using the approach explained on pp. 290-291 of your text and demonstrated in our solution to Problem 11 above, we obtain the following answers:

a) **Yes**, product of means $= 4 \times 75 = 300$ and product of extremes $= 3 \times 100 = 300$.

b) **No**, the product of the means is $3 \times 67 = 201$, while the product of the extremes is $2 \times 100 = 200$. Since $201 \neq 200$, the ratios $\frac{2}{3}$ and $\frac{67}{100}$ do <u>not</u> form a proportion.

c) **No**, these two ratios do not form a proportion because the product of the means does not equal the product of the extremes. That is,

Means	***Extremes***
$3 \times 7\frac{1}{3}$	5×4
$\frac{3}{1} \times \frac{22}{3}$	20
22	

Since $22 \neq 20$, we can conclude $\frac{5}{3} \neq \frac{7\frac{1}{3}}{4}$.

15. For this problem, there are various methods you may use. They include:

(i) apply the means-extremes approach demonstrated in your text (see Example 4 on pp. 291-292);

(ii) rewrite both ratios so that they share a common denominator and then compare them;

(iii) rewrite each ratio as a decimal (or, if preferred, a percentage) and then compare them;

(iv) reduce the ratios to lowest terms first and then use any of the previous three approaches described; and

(v) try to picture the ratios in terms of a context you understand (for example, as pie slices or measuring cups or wrench sizes) and compare them accordingly.

Since ratios and proportions play such a prominent role in everyday life, we <u>strongly</u> encourage you to become as proficient with as many strategies as you can. For this problem, regardless of the approach used, your answers should be as follows:

a) $\frac{1}{6}$ **>** $\frac{1}{8}$ b) $\frac{1}{4}$ **<** $\frac{1}{3}$ c) $\frac{1}{5}$ **>** $\frac{1}{8}$

17. As the author of your text recommends (p. 292), it may help you to begin by writing "each decimal with the same number of places by affixing trailing zeros." Let us demonstrate:

a) Since $0.8 = 0.8000$ and since 8000 ten thousandths is smaller than 8001 ten thousands, we know $0.8 \boxed{<} 0.8001$.

b) Since $0.8 = 0.8000$ and $8000 > 7999$, it follows that $0.8 \boxed{>} 0.7999$.

c) Since 8.00 is larger than 2.81, we conclude $8 \boxed{>} 2.81$.

d) Since 2.80 is smaller than 2.88, it follows that $2.8 \boxed{<} 2.88$.

19. Using the approach demonstrated by your text's author (in Example 6 on pp. 292-293), we obtain the following solution (contained within the box):

$$\frac{5}{1} = \frac{A}{6}$$

PRODUCT OF MEANS = *PRODUCT OF EXTREMES*

$$1A = 5(6)$$

$$\boxed{A = \mathbf{30}}$$

Please note that Example 7 (p. 294) demonstrates a more efficient method, one that builds upon the experience acquired with Example 6. In our solutions below for Problems 23-41, we illustrate this improved approach.

21. $\dfrac{C}{2} = \dfrac{5}{1}$ *PRODUCT OF MEANS = PRODUCT OF EXTREMES*

$$2(5) = 1C$$

$$10 = C$$

$$\boxed{C = \mathbf{10}}$$

As mentioned in our solution to Problem 21, this approach can be refined. We demonstrate this in the solutions that follow.

23. In our previous two solutions, we demonstrated the approach introduced by your text's author on p. 293. Let us now show a more efficient method; namely, the procedure for solving proportions explained on textbook p. 294 and demonstrated in Example 7.

$$\frac{12}{18} = \frac{E}{12}$$ *Given equation*

$$E = \frac{12 \times 12}{18}$$ ← *Product of the extremes* / ← *Number diagonally opposite the unknown*

$$E = \frac{2 \times 12}{3}$$ *Reduce one of the 12's with 18 (by dividing each by 6)*

$$E = \frac{2 \times 4}{1}$$ *Reduce the 12 with 3 (by dividing each by 4)*

Therefore, $\boxed{\mathbf{E = 8}}$.

Several tips are worthy of mention:

- You may find it easier to reduce the $\frac{12}{18}$ in the given equation before applying the solving procedure;
- With the aid of a suitable calculator, all of the operations shown above (reducing, etc.) can be performed in one automatic sequence of keystrokes. Your text's author uses parts (a) and (b) of Example 7 (p. 294) to demonstrate this.

25. By applying the procedure for solving proportions (textbook p. 294) and then simplifying either by hand or with a calculator, we obtain the following:

$$\frac{G}{24} = \frac{14}{16}$$ *Given equation*

$$G = \frac{24 \times 14}{16}$$ ← *Product of the means* (numerator); ← *Number diagonally opposite the unknown* (denominator)

$$\boxed{\mathbf{G = 21}}$$

27. We again apply your text's procedure for solving proportions (see p. 294) and then simplify accordingly (either by hand or with a calculator).

$$\frac{2}{3} = \frac{I}{24}$$ *Given equation*

$$I = \frac{2 \times 24}{3}$$ ← *Product of the extremes* (numerator); ← *Number diagonally opposite the unknown* (denominator)

$$\boxed{\mathbf{I = 16}}$$

29.

$$\frac{3}{K} = \frac{2}{5}$$ *Given equation*

$$K = \frac{3 \times 5}{2}$$ ← *Product of the extremes* (numerator); ← *Number diagonally opposite the unknown* (denominator)

$$\boxed{\mathbf{K = \tfrac{15}{2} \text{ or } 7\tfrac{1}{2}}}$$

31.

$$\frac{7\frac{1}{5}}{9} = \frac{M}{5}$$ *Given equation*

$$M = \frac{7\frac{1}{5} \times 5}{9}$$ ← *Product of the extremes* (numerator); ← *Number diagonally opposite the unknown* (denominator)

$$M=\frac{\frac{36}{5}\times\frac{5}{1}}{9}$$ *Simplify either by hand (as shown here) or with a calculator.*

$$M=\frac{36}{9}$$

$$\boxed{M=4}$$

Once again, remember that (with the aid of an appropriately equipped calculator) the expression

$$\frac{7\frac{1}{5}\times 5}{9}$$

can be entered in one sequence of keystrokes, and the calculator will immediately produce the final simplified result.

33.

$$\frac{P}{4}=\frac{4\frac{1}{2}}{6}$$ *Given equation*

$$P=\frac{4\times 4\frac{1}{2}}{6}$$ ← *Product of the means* / ← *Number diagonally opposite the unknown*

$$P=\frac{\frac{4}{1}\times\frac{9}{2}}{6}$$ *Simplify either by hand (as shown here) or with a calculator.*

$$P=\frac{18}{6}$$

$$\boxed{P=3}$$

35.

$$\frac{5}{R}=\frac{7}{12\frac{3}{5}}$$ *Given equation*

$$R=\frac{5\times 12\frac{3}{5}}{7}$$ ← *Product of the extremes* / ← *Number diagonally opposite the unknown*

$$R=\frac{\frac{5}{1}\times\frac{63}{5}}{7}$$ *Simplify either by hand (as shown here) or with a calculator.*

$$R=\frac{63}{7}$$

$$\boxed{R=9}$$

37.

$$\frac{33}{2\frac{1}{5}}=\frac{3\frac{3}{4}}{T}$$ *Given equation*

$$T=\frac{2\frac{1}{5}\times 3\frac{3}{4}}{33}$$ ← *Product of the means* / ← *Number diagonally opposite the unknown*

$$T=\frac{\frac{11}{5}\times\frac{15}{4}}{33}$$ *Simplify either by hand (as shown here) or with a calculator.*

$$T=\frac{11\times\frac{3}{4}}{33}$$

$$T=\frac{1\times\frac{3}{4}}{3}$$

$$T=\tfrac{3}{4}\div 3$$

$$T=\tfrac{3}{4}\cdot\tfrac{1}{3}$$

$$\boxed{T=\tfrac{1}{4}}$$

39. $$\frac{\frac{1}{5}}{\frac{2}{3}}=\frac{\frac{3}{4}}{V}$$ *Given equation*

$$V=\frac{\frac{2}{3}\times\frac{3}{4}}{\frac{1}{5}}$$ ← *Product of the means* ← *Number diagonally opposite the unknown*

$$V=\frac{\frac{1}{2}}{\frac{1}{5}}$$ *Simplify either by hand (as shown here) or with a calculator.*

$$V=\tfrac{1}{2}\cdot\tfrac{5}{1}$$

$$\boxed{V=\tfrac{5}{2}\text{ or }2\tfrac{1}{2}}$$

41. $$\frac{9}{Y}=\frac{1\frac{1}{2}}{3\frac{2}{3}}$$ *Given equation*

$$Y=\frac{9\times 3\frac{2}{3}}{1\frac{1}{2}}$$ ← *Product of the extremes* ← *Number diagonally opposite the unknown*

$$Y=\frac{\frac{9}{1}\times\frac{11}{3}}{\frac{3}{2}}$$ *Simplify either by hand (as shown here) or with a calculator.*

$$Y=\frac{3\times 11}{\frac{3}{2}}$$

$$Y=33\div\tfrac{3}{2}$$

$$Y=\tfrac{33}{1}\cdot\tfrac{2}{3}$$

$$\boxed{Y=22}$$

43. This problem is almost identical to textbook Example 8 (p. 295). To quote the author of your textbook (see top of p. 295), "Whenever you are working an applied problem, you should estimate an answer so that you will know whether the result you obtain is reasonable." Since 4 melons cost \$0.52, it follows that 8 melons cost 2(\$0.52) = \$1.04. Hence, the cost of 7 melons must be somewhere between \$0.52 and \$1.04, closer to the latter amount.

Expressing its wording in terms of a proportion, the problem asks the following question: "4 melons is to \$0.52 as 7 melons is to what?"

This prompts us to write the following equation:

$$\frac{4 \text{ melons}}{0.52 \text{ dollars}} = \frac{7 \text{ melons}}{x \text{ dollars}}$$

Then, upon applying the text's procedure for solving proportions and simplifying either by hand or with a calculator, we find:

$$x = \frac{0.52 \times 7}{4} = 0.91$$

Therefore, 7 melons will cost **\$0.91**.

45. As the text's author writes (top of p. 295), "whenever you are working an applied problem, you should estimate an answer so that you will know whether the result you obtain is reasonable." Since this family uses $3\frac{1}{2}$ gallons of milk per week and since 4 days is slightly more than half a week, the amount of milk the family needs for four days should be slightly more than half of $3\frac{1}{2}$ gallons.

When setting up a proportion with units, make sure to use like units; for example, we cannot use days for one ratio and weeks for the other. Consequently, we need to view the problem as asking a question along the lines of:

"7 days is to $3\frac{1}{2}$ gallons as 4 days is to how many gallons?"

$$\frac{\overset{\text{days}}{\downarrow}\;7}{3\frac{1}{2}\;\underset{\text{gallons}}{\uparrow}} = \frac{\overset{\text{days}}{\downarrow}\;4}{x\;\underset{\text{gallons}}{\uparrow}}$$

So, $x = \dfrac{3\frac{1}{2} \times 4}{7} = \dfrac{\frac{7}{2} \times \frac{4}{1}}{7} = \dfrac{14}{7} = 2$.

Thus, for a four-day period, this family needs **2 gallons** of milk.

47. Expressing its wording in terms of a proportion, the problem asks the following question:

"3 miles is to 40 minutes as 2 miles is to how many minutes?"

<u>Remember</u>: Whenever you are working an applied problem, you first should estimate an answer so that you will know whether the result you obtain is reasonable. After making such an estimate, one would then proceed to write (and subsequently solve) a proportion similar to the following:

$$\underset{\text{minutes}}{\overset{\text{miles}}{\frac{3}{40}}} = \underset{\text{minutes}}{\overset{\text{miles}}{\frac{2}{x}}}$$

Upon applying the text's procedure for solving proportions and simplifying either by hand or with a calculator, we find:

$$x = \frac{40 \times 2}{3} = \frac{80}{3} = 26.\overline{6}$$

Therefore, to the nearest minute, it will take **27 minutes** for Jack to jog 2 miles.

49. Expressing its wording in terms of a proportion, the problem asks the following question:

"140 lbs is to 2,100 calories as 165 lbs is to how many calories?"

Remember: Whenever you are working an applied problem, you first should estimate an answer so that you will know whether the result you obtain is reasonable.

After making such an estimate, one would then write (and solve) a proportion similar to the following:

$$\underset{\text{calories}}{\overset{\text{lbs}}{\frac{140}{2{,}100}}} = \underset{\text{calories}}{\overset{\text{lbs}}{\frac{165}{x}}}$$

Then, upon applying the text's procedure for solving proportions and simplifying either by hand or with a calculator, we find:

$$x = \frac{2{,}100 \times 165}{140} = 2{,}475$$

Therefore, **2,475 calories** are necessary to maintain a moderately active 165-pound person.

51. Substituting the given values into the proportion $\frac{V}{T} = \frac{V'}{T'}$ and then applying the procedure for solving proportions (textbook p. 294) yields the following:

$$\frac{175}{300} = \frac{V'}{273} \quad \Longrightarrow \quad V' = \frac{175 \times 273}{300} = \boxed{159.25}$$

53. Since the span given is 24 feet, the half-span is 12 feet. Hence, the pitch is:

$$\frac{\text{rise}}{\text{half-span}} = \frac{8\text{ ft}}{12\text{ ft}} = \boxed{\frac{2}{3}}$$

55. For the original 3 in. by 5 in. photo, the ratio of its shorter side to its longer side is 3 to 5. Consequently, the ratio of the enlargement's shorter side to its longer side must also be 3 to 5. In other words, the respective ratios (shorter to longer) for the original photo and for the enlarged poster must be proportional. This notion can be represented by the following equation:

$$\frac{\overset{\substack{\text{photo's shorter side} \\ \text{(in inches)}}}{\downarrow}\;3}{\underset{\substack{\text{photo's longer side} \\ \text{(in inches)}}}{5\;\uparrow}} = \frac{\overset{\substack{\text{poster's shorter side} \\ \text{(in feet)}}}{\downarrow}\;2}{\underset{\substack{\text{poster's longer side} \\ \text{(in feet)}}}{x\;\uparrow}}$$

Then, upon applying the text's procedure for solving proportions and simplifying either by hand or with a calculator, we find:

$$x = \frac{5 \times 2}{3} = \frac{10}{3}$$

Therefore, the size of the poster's longer side should be **3 1/3 ft or 3 ft, 4 in.**

57. This problem is almost identical to textbook Example 10 (p. 296). Expressing its wording in terms of a proportion, the problem asks the following question:

"$180,000 is to $1,080 as $130,000 is to what?"

$$\frac{\overset{\text{value}}{\downarrow}\;180{,}000}{\underset{\text{tax}}{1{,}080\;\uparrow}} = \frac{\overset{\text{value}}{\downarrow}\;130{,}000}{\underset{\text{tax}}{x\;\uparrow}}$$

Then, upon applying the text's procedure for solving proportions and simplifying either by hand or with a calculator, we find:

$$x = \frac{1{,}080 \times 130{,}000}{180{,}000} = 780$$

Therefore, the tax is **$780**.

Remember: Whenever you are working an applied problem, you first should estimate an answer (before performing any computations) so that you will know whether the (computed) result you later obtain is reasonable.

59. Notice that (each day) she filled the soft-drink barrels with only pure water — never with any more pure soft drink. Consequently, if after four days only 1 pint of pure soft drink remained, she must have served all 32 gallons except for the 1 remaining pint. Hence, recalling that there are 4 quarts in a gallon and 2 pints in a quart, she served

32 gal – 1 pt	⟹	31 gal 4 qt – 1 pt	⟹	31 gal 3 qt 2 pt – 1 pt

31gal 3qt 1pt
of pure soft drink.

Problem Set 5.8

1. **Answers vary**. The author of your text addresses this issue on page 299.

3. Let x and $x+2$ represent the two consecutive positive even numbers. Given that their product is 440, we write and solve the following quadratic equation:

$$x(x+2)=440$$

$$x^2+2x-440=0 \quad \text{Rewrite the quadratic equation in } ax^2+bx+c=0 \text{ form.}$$

$$(x-20)(x+22)=0 \quad \text{Factor, if possible (or, if preferred, use quadratic formula).}$$

$$x=20 \text{ or } x=-22 \quad \text{Apply zero-product rule (see textbook p. 267).}$$

We discard $x=-22$ because the problem asks for two consecutive "positive" even numbers. Finally, since $x=20$, it follows that $x+2=22$.

Therefore, **the numbers are 20 and 22.**

[Recommendation: As always, we encourage you to (i) check the solution, and, more generally, (ii) perform Pólya's "look back" step.]

5. If we let x equal an integer, then the reciprocal of x is $1/x$. Hence, we translate the given information ("the sum of an integer and its reciprocal is $\frac{5}{2}$") as follows:

$$x+\frac{1}{x}=\frac{5}{2}$$

$$2x\left(x+\frac{1}{x}\right)=\left(\frac{5}{2}\right)2x \quad \text{Multiply both sides by the lcm of 2 and x.}$$

$$2x^2+2=5x \quad \text{Simplify.}$$

$$2x^2-5x+2=0 \quad \text{Rewrite the quadratic equation in } ax^2+bx+c=0 \text{ form.}$$

$$(2x-1)(x-2)=0 \quad \text{Factor, if possible (or, if preferred, use quadratic formula).}$$

$$x=\tfrac{1}{2} \text{ or } x=2 \quad \text{Apply zero-product rule (see textbook p. 267).}$$

Since we chose to let x represent an integer, we discard the solution $x=\frac{1}{2}$. Finally, after taking a moment to remind ourselves of the question asked ("What is the integer"), we write our final answer in words: **The integer is 2.**

Check: $2+\frac{1}{2}=2\frac{1}{2}=\frac{5}{2}$ ✓

7. Begin by translating the problem into a verbal equation, such as:

BUS/DRIVER COST + MEAL COST + ADMISSION FEE = TOTAL COST

All 20 people will eat lunch; namely, 10 children, 6 teens, 3 chaperones, and the bus driver. Given a cost of \$2.50 per meal, it follows that the total MEAL COST will equal \$50.00 (i.e., \$2.50/meal × 20 meals).
The camp needs to purchase admission tickets for everyone except the bus driver. Hence, the total ADMISSION FEE is

(\$5.00/child × 10 children) + (\$6.00/teen × 6 teens) + (\$7.00/adult × 3 chaperones),

which equals \$107. Please note that this 19-person group is not large enough to receive the \$20 discount.

Since the total cost is equal to the sum of \$150 (for the bus and driver), \$50 (in meals), and \$107 (to admit 19 people), the answer is **Choice A** (\$307).

9. Choice C seems absurdly large and choice B appears suspiciously small.
Since there were 48 games in all with half played at home and half on the road, there were 24 home games and 24 road games. Winning 2/3 of the 24 home games means the team won 16 games [i.e., $\frac{2}{3}\times 24$] at home. Similarly, winning 1/4 of their road games means the team had 6 road wins [i.e., $\frac{1}{4}\times 24$]. Therefore, the total number of games won is 22 (16 home wins + 6 road wins), which corresponds to **Choice A**.

11. We begin by writing a verbal equation that summarizes the problem's main idea:

(LBS OF KUMQUAT SOLD) × (PRICE/LB) = ¼ (INITIAL INVESTMENT)

The initial investment per tree is \$26; namely, the sum of \$15 (the tree cost), \$3.50 (soil amendments), and \$7.50 (irrigation supplies). Consequently, the total INITIAL INVESTMENT for 6 trees is \$156 [i.e., \$26/tree × 6].

Since 5/6 of the 24 pounds of fruit harvested is sold, the gardener sells 20 pounds of fruit. Hence, LBS OF KUMQUAT SOLD equals 20.

Thus, upon letting x represent the price per pound and substituting values appropriately, the verbal equation stated above can be rewritten as follows:

$$20x = \tfrac{1}{4}(156)$$

Solving this linear equation yields:

$$20x = 39$$
$$x = \tfrac{39}{20} = \$1.95$$

Therefore, we select **Choice A**.
Note: If you know anything about the price of kumquats (i.e., they are not a particularly expensive delicacy), you should have immediately eliminated choices B and D.

13. Noting that the question pertains only to puppies, we ignore the extraneous information regarding adult poodles. Since all 12 puppies (5 female, 7 male) get vaccinated (at a cost of \$16 each), the total cost of vaccination equals \$16(12) = \$192.

At a rate of \$30/male, the cost of getting 5 males neutered is \$150. As for the five female puppies, since all but one of them gets neutered, the total cost of neutering 4 females is \$180 [i.e., 4 × \$45/female].

So, the total cost of treatment for these puppies equals the sum of \$192 (for vaccinations) and \$330 (to neuter 5 males and 4 females), and we therefore select **Choice B** (\$522).

15. Recall that 16 oz = 1lb. An average yield of 6 oz of seed per sunflower suggests we should anticipate 1,200 sunflowers producing 7,200 oz of seeds, which is equivalent to 450 lbs. At a selling price of \$0.20 per lb, Melissa should expect to receive \$0.20/lb × 450 lb = \$90.00 in total revenue. Thus, we select **Choice D**.

17. If 1/3 of the 48-member class consists of freshmen, then there must be 16 freshmen (i.e., $\frac{1}{3} \times 48$). Similarly, we find there to be 12 sophomores (one-fourth of 48) and 6 juniors (one-eighth of 48). Therefore, the number of seniors equals

$$48 - (\text{FRESHMEN} + \text{SOPHOMORES} + \text{JUNIORS}) = 48 - (16 + 12 + 6) = \underline{\underline{14}},$$

which corresponds to **Choice C**.

Alternative Strategy: Knowing that 1/3 of the class are freshmen, 1/4 sophomores, and 1/8 juniors means $\frac{1}{3} + \frac{1}{4} + \frac{1}{8} = \frac{8}{24} + \frac{6}{24} + \frac{3}{24} = \underline{\underline{\frac{17}{24}}}$ of the class are not seniors. Hence, the remaining $\frac{7}{24}$ of the 48-member class are seniors (i.e, $1 - \frac{17}{24}$), and $\frac{7}{24} \times 48$ equals 14.

19. By simple inspection, you should immediately eliminate Choice D because Billie paid less (not more) than Abbie. Also, a basic knowledge of percentages should prompt you to eliminate Choice B (\$100) as being far too low a price; a \$25 savings would be far more than 8% of \$125. Similarly, notice that Choice C (\$117) is \$8 less than \$125, but we know \$8 is not equal to 8% of \$125. Thus, by process of elimination, one should select **Choice A**.

Let us now share two strategies for working the problem directly, <u>both</u> of which we hope you will utilize with future problems:

I. We begin with an efficient one-step approach. If Billie paid 8% less than the \$125 Abbie paid, Billie paid 92% of \$125, which equals 0.92(\$125) = \$115.

II. Alternatively, there is the common two-step procedure. We could first compute Billie's savings: $0.08(\$125) = \$10$. Then, since Billie paid $\$10$ less than the $\$125$ Abbie paid, Billie paid $\$125 - \$10 = \$115$.

21. Inspection (and careful reading of the question) prompts us to eliminate Choice B. There are various approaches one could apply to this problem. Let us share two.

I. We use our experiences with proportions (from Section 5.7) to translate the problem as "60% is to 40% as 240 is to what?" and, accordingly, write (and proceed to solve) the following proportion:

$$\frac{\overset{\text{voted against}}{60}}{\underset{\text{voted in favor}}{40}} = \frac{\overset{\text{voted against}}{240}}{\underset{\text{voted in favor}}{x}}$$

Then, upon applying the text's procedure for solving proportions (on p. 294) and simplifying either by hand or with a calculator, we find:

$$x = \frac{40 \times 240}{60} = 40 \times 4 = 160$$

Therefore, we select **Choice A**.

II. Alternatively, we can begin by writing verbal equations, such as:

0.60(NUMBER OF VOTERS) = NUMBER WHO VOTED AGAINST

0.40(NUMBER OF VOTERS) = NUMBER WHO VOTED IN FAVOR

Let x represent the NUMBER OF VOTERS. Knowing that 240 people voted against the referendum, we can rewrite and solve the first of our two verbal equations:

$$0.60x = 240$$
$$x = \tfrac{240}{0.60}$$
$$x = 400$$

Finally, since NUMBER OF VOTERS equals 400, we may use our second verbal equation to find:

NUMBER WHO VOTED IN FAVOR $= 0.40(400) = 160$.

23. By inspection, we immediately eliminate Choice C. We then may translate the given information as follows:

TOTAL EXPENSE = AUTO LOAN PAYMENT + AUTO INSURANCE PAYMENT

AUTO INSURANCE PAYMENT = 0.40(AUTO LOAN PAYMENT)

Letting x represent AUTO LOAN PAYMENT, we "evolve" (as the author of your text describes it) the second verbal equation as follows:

AUTO INSURANCE PAYMENT $=$ 0.40(AUTO LOAN PAYMENT)

$$120 = 0.4x$$
$$\tfrac{120}{0.4} = x$$
$$300 = x$$

Thus, since the monthly AUTO LOAN PAYMENT is $\$300$ and the monthly AUTO INSURANCE PAYMENT is $\$120$, the TOTAL EXPENSE is $\$300 + \$120 = \$420$, which corresponds to **Choice B**.

25. Only those who have no understanding of percentages would select Choice A, so we eliminate it. We must first compute Doug's salary in 2000 and then use it to determine his 2001 salary. To determine his salary in 2000, there are two approaches we could take. First, a 5% raise means Doug's new salary in 2000 is 105% of his 1999 salary; that is, $1.05(800) = \$840$ is Doug's weekly salary in 2000. Alternatively, we could have determined this $\$840$ amount by first computing the amount of Doug's raise in 2000, which is $0.05(800) = \$40$, and then adding this $\$40$ raise to his old 1999 salary of $\$800$. Finally, as for his 2001 weekly salary (which is 10% higher than his 2000 salary), we may either (i) compute 110% of $\$840$ [i.e., $1.10 \times \$840 = \$924$] or (ii) compute his 10% raise [$0.10 \times \$840 = \$84$] and add it to his old 2000 salary of $\$840$ [i.e., $\$840 + \$84 = \$924$]. Thus, we select **Choice D**.
[Caution: It would be incorrect to conclude that Doug's 2001 salary is 15% higher than his 1999 salary because the 5% raise is based upon his 1999 salary but the 10% raise applies to his 2000 salary.]

27. The fact that the quarterback "ran" with the ball 75 times is not relevant to this question, which pertains to "passing." Since 180 of his 450 passes were incomplete and 15 were intercepted, the remaining 255 were complete (i.e., $450–180–15$). Clearly, 255 is more than 50%, so we eliminate Choice A by inspection. As for his completion percentage, we compute $\frac{255}{450} \times 100$ and, hence, obtain an answer of approximately 56.7% and accordingly select **Choice B**.

29. If a whole number leaves a remainder of 2 when divided into 50, then the whole number must divide "evenly" into 48. The words "divide evenly" imply that the whole number must be a divisor of 48. The set of whole number divisors of 48 is $\{1, 2, 3, 4, 6, 8, 12, 16, 24, 48\}$. However, 1 and 2 do not leave a remainder of 2 when divided into 50, so we eliminate both of them from consideration.

Similarly, whole numbers who leave a remainder of 5 when divided into 65 must be divisors of 60, which consists of the set {1, 2, 3, 4, 5, 6, 10, 12, 15, 20, 30, 60}. This time, we eliminate $1, 2, 3, 4$, and 5 because none of them can leave a remainder of 5 when divided into 65.

Finally, we need to identify those whole numbers that satisfy both remainder conditions. Said differently, we need to determine the intersection of $\{3, 4, 6, 8, 12,$

16, 24, 48} and {6, 10, 12, 15, 20, 30, 60}. Hence, the only two positive whole numbers that leave a remainder of 2 when divided into 50 and leave a remainder of 5 when divided into 65 are 6 and 12. Therefore, we select **Choice B**.

31. This problem is similar in some respects to Problem 29. If a number leaves a remainder of 1 when divided into 46, then the number must be a divisor of 45; that is, 1, 3, 5, 9, 15, or 45. Among these six numbers, only 3, 9, 15, and 45 satisfy the condition that our number be a "multiple of 3." Finally, we divide each into 20 and identify those that leave a remainder of 5. The numbers 3, 9, and 45 fail to do so. Thus, the smallest positive multiple of 3 that satisfies the stated conditions is 15, which corresponds to **Choice C**.

33. If a positive number leaves a remainder of 5 when divided into 50, then that positive number must also be a divisor of 45. This leads us to the numbers 9, 15, and 45, all three of which are positive multiples of 3. Thus, we select **Choice D**.

35. **Answers vary**. It should be noted that we discussed the roles of inspection and estimating in our solutions above to Problems 9, 11, 19, 21, 23, 25, and 27. Please refer to each of those solutions for additional details.

37. Again, permit us to share two approaches:

I. If we were to declare that there are 92 musicians in all (i.e., 60 band members plus 32 orchestra members), we would be wrong because we would have "double counted" the 19 people who are members of both groups. Therefore, there are actually $92 - 19 = 73$ musicians in all.

II. Alternatively, using your knowledge of set theory (from Section 1.4), you may find a Venn diagram to be a helpful way to visualize this problem:

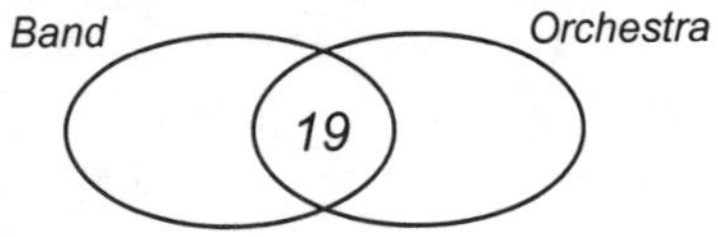

Because 19 of the 60-member band are also in the orchestra, the remaining 41 band members are not orchestra members. Similarly, since 19 of the 32-member orchestra are also in the band, the reamining 13 orchestra members are not band members. Thus, we complete our Venn diagram as follows:

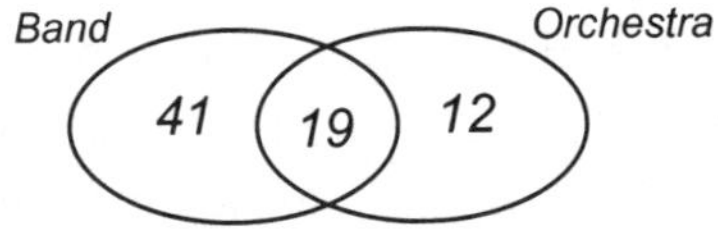

So, once again, we have $41 + 19 + 12 = 73$ musicians in all. Therefore,

Yes, the band and orchestra can travel in two 40-passenger buses.

39. Consider these verbal equations:

HEIGHT OF STANDARD OIL BLDG + HEIGHT OF SEARS BLDG = 2,590 ft

HEIGHT OF SEARS BLDG = 318 + HEIGHT OF STANDARD OIL BLDG

Let h represent the HEIGHT OF STANDARD OIL BLDG. It then follows that

HEIGHT OF SEARS BLDG = 318 + h

Substituting into our first verbal equation yields the following linear equation:

(HEIGHT OF STANDARD OIL BLDG) + (HEIGHT OF SEARS BLDG) = 2,590 ft

$$\underset{h}{\uparrow} \quad + \quad \underset{(318+h)}{\uparrow} \quad = 2{,}590$$

Solving this equation, we find

$$2h + 318 = 2{,}590$$
$$2h = 2{,}272$$
$$h = 1{,}136$$

Therefore, **the height of the Standard Oil building is 1,136 ft**, and **the height of the Sears tower is 1,454 ft** (i.e., $318 + h$).

41. The problem asks us to find the sum $2+4+6+8+\cdots+194+196+198+200$. Observe the pattern that unfolds when the numbers are reorganized so that the first number (namely, 2) is paired with the last (200), the second (4) with the second-to-last (198), and so on:

$$(2+200)+(4+198)+(6+198)+(8+194)+\cdots+(98+104)+(100+102)$$

Each pair's sum is 202. Since there are 100 numbers in all, there must be 50 pairs. Thus, 50 pairs of 202 amounts to $50 \cdot 202 = \boxed{10{,}100}$.

Alternatively, you may prefer this variation of the above explanation:

$$2+4+6+8+\cdots+194+196+198+200$$
$$2(1)+2(2)+2(3)+2(4)+\cdots+2(97)+2(98)+2(99)+2(100)$$

We factor the g.c.f. as follows:

$$2[1+2+3+4+\cdots+97+98+99+100]$$

Finally, we reorganize the terms and compute the result:

$$2[(1+100)+(2+99)+(3+98)+(4+97)+\cdots+(49+52)+(50+51)]$$
$$2[101+101+101+101+\cdots+101+101]$$
$$2[101(50)]$$
$$10{,}100$$

43. Remember that an MBA is a graduate-level (masters) degree (in business) that one earns after having acquired an undergraduate-level (bachelors) degree.

Of the 520 MBA degree holders, we know that 450 of them also hold (undergraduate) business degrees. Therefore, the remaining $\boxed{70}$ hold nonbusiness (undergraduate) degrees.

45. This problem is almost identical to text Example 3 (pp. 302-303).
When the drain pipe is closed, it takes 36 hours to fill the entire pool. Hence, it takes 18 hours (under these conditions) to fill half the pool.
It is essential to remember that the drain pipe is opened after half the pool has already been filled (i.e., after 18 hours have elapsed). This (second) half of the filling process can be described as follows:

WORK DONE BY INLET PIPE = (RATE OF INLET PIPE)(TIME TO FILL)
= $\frac{1}{36}$ (TIME)

WORK DONE BY DRAIN = (RATE OF DRAIN)(TIME TO DRAIN)
= $\frac{1}{40}$ (TIME)

WORK DONE BY INLET PIPE – WORK DONE BY DRAIN = ONE-HALF JOB COMPLETED

$$\frac{1}{36}\text{(TIME)} \qquad \frac{1}{40}\text{(TIME)} \quad = \quad \frac{1}{2}$$

Let t = TIME:

$$\frac{1}{36}t - \frac{1}{40}t = \frac{1}{2}$$

$$(360)\frac{1}{36}t - (360)\frac{1}{40}t = (360)\frac{1}{2}$$ *[Note: The lcm of 36, 40, and 2 is 360.]*

$$10t - 9t = 180$$

$$t = 180$$

This means it will take 180 hours to fill half the pool when the drain pipe is left open. Recall, however, that 18 hours had elapsed prior to the drain pipe being opened, and during this time, half the pool had been filled. Therefore, under the given conditions,

it will take 198 hours to fill the entire pool.

47. This problem is a continuation of Problem 46, which is based upon text Example 2 (pp. 300-302). We begin by recalling the equation that the author of your text developed in detail and eventually solved for Example 2:

$$\frac{x}{4} + \frac{3-x}{2} = 1$$

where x represented the length of time (in hours) that one should record on LP mode. Note that the '3' in the equation for Example 2 was based on the assumption that one wished to tape a 3-hour movie. Our problem, though, does not give us a specific movie length. Thus, we need to determine an appropriate (more generalized) replacement for the '3'; namely, an expression that represents the length of the movie (less the commercials) that we need to tape.

Given that the movie is ℓ hours long and it includes c minutes of commercials <u>per hour</u>, there will be a total of $(\ell \cdot c)$ minutes of commercials. Since there are 60 minutes in an hour, $(\ell \cdot c)$ min of commercials amounts to $(\frac{\ell \cdot c}{60})$ hr of commercials. Thus, the movie itself (without the commercials) is ℓ hours long <u>minus</u> $(\frac{\ell \cdot c}{60})$ hr of commercials. That is, we will need to tape for a total of $(\ell - \frac{\ell c}{60})$ hrs.

Now, we replace the '3' in the equation used to solve Example 2 with $(l - \frac{\ell c}{60})$:

$$\frac{x}{4} + \frac{(\ell - \frac{\ell c}{60}) - x}{2} = 1$$

We begin by solving for x by multiplying both sides by 4, and we proceed as follows:

$$4 \cdot \left[\frac{x}{4}\right] + 4 \cdot \left[\frac{(\ell - \frac{\ell c}{60}) - x}{2}\right] = 4 \cdot 1$$

$$x + 2[\ell - \frac{\ell c}{60} - x] = 4$$

$$x + 2\ell - \frac{\ell c}{30} - 2x = 4$$

$$-x = -2\ell + \frac{\ell c}{30} + 4$$

$$x = 2\ell - \frac{\ell c}{30} - 4$$

Remember that x represents the amount of time we will record in LP mode. Therefore, for an ℓ – hour movie that includes c minutes of commercials, we should **tape in LP mode for $(2\ell - \frac{1}{30}c\ell - 4)$ hours** and tape the rest in SP mode.

49. Applying the problem-solving strategy emphasized heavily in your text, we begin by translating the main idea of this problem into a verbal equation like:

SELLING PRICE – (COMMISSION + OTHER COSTS + MORTGAGE) = SETTLEMENT

We let p represent the SELLING PRICE and substitute into this verbal equation the appropriate values and expressions as follows:

SELLING PRICE –	(COMMISSION +	OTHER COSTS +	MORTGAGE) =	SETTLEMENT
↑	↑	↑	↑	↑
p –	($0.06p$ +	\$2,525 +	\$65,250) =	\$142,785

By solving this linear equation, we obtain the selling price:

$$1p-0.06p-2{,}525-65{,}250=142{,}785$$
$$0.94p-67{,}775=142{,}785$$
$$0.94p=210{,}560$$
$$p=\frac{210{,}560}{0.94}$$
$$p=224{,}000$$

Therefore, **the selling price was $224,000**.

51. Given that the box is 1 inch shorter and 3 inches longer than it is wide, the dimensions of the rectangular box can be described in terms of its width, namely:

Let $w=$ width,
$w+3=$ length,
$w-1=$ height.

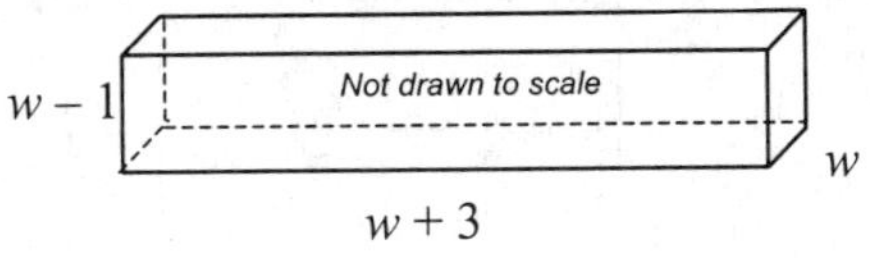

Let $d=$ the length of the diagonal on the bottom of the box. By viewing the bottom of the box, we discover a right-triangle relationship (and, accordingly, we can apply the Pythagorean theorem):

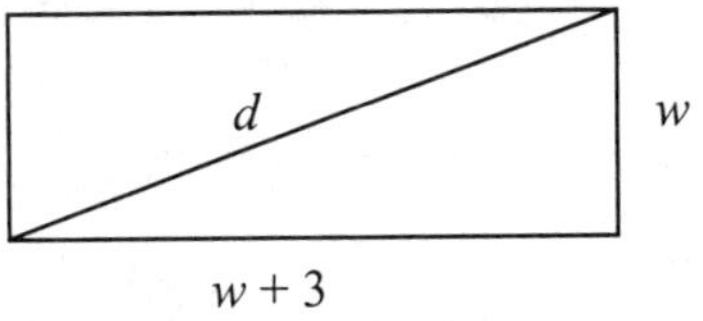

$$d^2=w^2+(w+3)^2$$

Now, let us view a cross section of the box, with the 17-in. bar spanning diagonally from one top corner to the diagonally-opposite bottom corner:

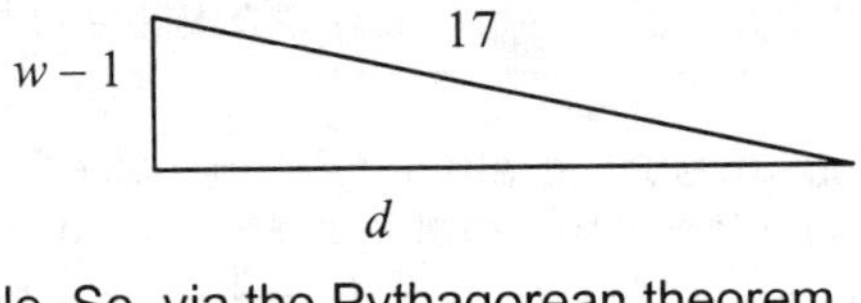

This, too, forms a right triangle. So, via the Pythagorean theorem, we write:

$$17^2=d^2+(w-1)^2.$$

Since $d^2=w^2+(w+3)^2$, we can substitute for d^2 in the preceding equation and then solve for w as follows:

$17^2=[w^2+(w+3)^2]+(w-1)^2$	*Substitution*
$289=w^2+(w^2+6w+9)+(w^2-2w+1)$	*Use "FOIL" to square the binomials.*
$289=3w^2+4w+10$	*Combine like terms.*
$0=3w^2+4w-279$	*Write in ax^2+bx+c form.*

$$0 = (3w+31)(w-9)$$ *Factor the trinomial.*

$$w = -\tfrac{10}{3} \text{ or } w = 9$$ *Mentally apply the zero-product rule.*

Because a width cannot be negative, we disregard $w = -\frac{10}{3}$ and conclude that the box's width is 9 in. Since $w = 9$, it follows that the box's length is $w+3 = 9+3 = 12$ in. and its height is $w-1 = 9-1 = 8$ in. Therefore,

the dimensions of the box are 8 in. × 9 in. × 12 in.

To find how much must be cut off the rod in order for it to lie flat on the bottom of the container, we must compute d, the length of the diagonal on the bottom of the box. Recalling that $d^2 = w^2 + (w+3)^2$ and $w = 9$, we find:

$$\begin{aligned} d^2 &= (9)^2 + [(9)+3]^2 \\ &= 81 + 144 \\ &= 225 \end{aligned}$$

So, $d = \sqrt{225} = 15$ in. Hence, in order for the 17-in. rod to fit along the 15-in. diagonal across the bottom of the box, one must **cut off 2 in.**

53. Let r and w represent respectively the speed of the plane in still air and the speed of the wind. If the plane is traveling "with" the wind (i.e., a tail wind), then it travels faster than it would in still air. Hence, its speed will be $(r+w)$ miles per hour. Conversely, if the plane is traveling "against" the wind (i.e., a head wind), then it is slowed by the wind, in which case its speed will be $(r-w)$ miles per hour.

From our previous experience with the distance-rate-time relationship, we know:

(SPEED WITH WIND)(TIME WITH WIND) = DISTANCE

and (SPEED AGAINST WIND)(TIME AGAINST WIND) = DISTANCE

We "evolve" these to write: $(r+w)(2.5) = 630$

$$(r-w)(3) = 630$$

By dividing both sides of the equations by their respective times, we find:

$r + w = 252$ mph and $r - w = 210$.

We then solve one of the equations for r (expressed in terms of w) and substitute into the other equation; for example, since $r = 252 - w$, we can write

$$\begin{aligned} (252-w) - w &= 210 \\ 252 - 2w &= 210 \\ -2w &= -42 \\ w &= 21 \end{aligned}$$

Therefore, **the speed of the wind is 21 mph.**

Check: If the speed of the wind is 21 mph, then the speed of the plane in still air is 231 mph. With the assistance of the tail wind, the plane will travel 252 mph, and we confirm that $252(2.5) = 630$ miles. When slowed down by the head wind, the plane travels $231–21=210$ mph, and we can see that $210(3) = 630$ miles. ✓

55. To better understand this problem, it may help to begin by expressing the problem in terms of a table that depicts the tour agency's income (also known as "revenue"). Then we can seek to discover a pattern:

# of participants ×	***Cost per person*** =	***Revenue***
100	\$2,500	\$250,000
101	\$2,495	\$251,995
102	\$2,490	\$253,980
103	\$2,485	\$255,955
⋮	⋮	⋮
100 + x	2,500 – 5x	(100 + x)(2,500 – 5x)

Note that the number of participants is increasing in increments of 1 and the cost per person is decreasing in decrements of \$5. If we let x represent the number of additional ("new") participants, then $(100+x)$ represents the total number of participants. Meanwhile, if x additional persons sign up, then the price (for everyone attending) will be decreased by \$($5x$). So, each person will pay $\$(2500-5x)$ for the tour. Hence, since the tour agency's income is made based on the number of participants and the price per participant, we can express its income as:

$$(100+x)(2{,}500-5x)$$

We then turn our attention to the agency's expenses. These consist of (i) the \$240,500 cost for the plane and (ii) incidental fees of \$500 per person. If $(100+x)$ people participate, then the incidental fees amount to $\$500(100+x)$. Thus, the agency's total expenses can be expressed as:

$$240{,}500+500(100+x)$$

The question asked in the problem is about the price per ticket that should be charged if income and expenses are equal. Therefore, we write and solve:

INCOME = EXPENSES

$$(100+x)(2{,}500-5x) = 240{,}500+500(100+x)$$

$$250{,}000+2{,}000x-5x^2 = 240{,}500+50{,}000+500x$$

$$-5x^2+1{,}500x-40{,}500 = 0$$

$$x^2-300x+8{,}100 = 0$$

$$(x-30)(x-270) = 0$$ *[If preferred, use the quadratic formula.]*

$$x = 30 \text{ or } x = 270$$

We disregard $x = 270$ because that would mean we would have $x + 100 = 270$ participants in all, but the airplane only seats 150 people. Thus, we will have 30 additional participants, or 130 participants in all. Since the price per person is $2{,}500 - 5x$ and $x = 30$, **each participant should be charged $2,350.**

57. This situation is no different than placing on a scale an aquarium containing a fish; regardless of whether the fish is swimming or chooses to rest motionless on the bottom of the aquarium, the reading on the scale remains unchanged. Just like the water surrounding the fish, the air surrounding the canary is a medium with weight. Therefore, **no**, the reading would not change.

59. a) **Answers vary**. Since there are infinitely many counting numbers, there are infinitely many "sacred" numbers. The first five of these infinitely many "sacred" numbers are:

$\boxed{2} = 1^2 + 1^2$; $\boxed{5} = 1^2 + 2^2$; $\boxed{8} = 2^2 + 2^2$; $\boxed{10} = 1^2 + 3^2$; $\boxed{13} = 2^2 + 3^2$

b) **No**. One counterexample would be $2 \cdot 8 = 16$. Although 2 and 8 are both sacred, 16 is not.
Note: To answer this question correctly, you must understand the use of the word "any" in the statement of the question. By using this word, the question is asking, "Will the product of two sacred numbers always be sacred?"

Problem Set 5.9

1. **Answers vary**. You may wish to refer to the glossary of your text (Appendix G, p. 906) for a definition of the term ***computer program***.

3. **Answers vary**. An ***editor command*** allows a programmer to make corrections to ("edit") the commands of a program. A ***system command*** is a command which causes a program to write, read, or be edited. A ***program command*** is a command, which is part of a program. For additional details, see textbook pages 309-310.

5. Line 10 will print the sum of 6 and 5, while line 20 will print their difference. As for line 30, the comma causes the product 6*5 and the quotient 6/3 to be displayed on the same line, with a separation between them. Thus, the output appears as follows:

```
11
1
30      2
```

7. Line 10 assigns to variable A the value 2. Similarly, line 30 assigns to variable B the value 3. Notice that Line 20 does not end with a semicolon. Consequently, after the word "Working" is printed, this command causes a line feed to occur. Line 40 prints the sum of A and B. The final output, therefore, appears as follows:

```
Working
5
```

9. Since line 10 ends in a semicolon, the letters T and H are printed side by side (i.e., without any line feed taking place). Line 20 also prevents (because of its semicolon) a line feed from occurring after the printing of the letter H. After that, the FOR-NEXT loop in lines 30-50 is performed twice — first when J = 1 and again when the value of J is elevated to 2. If necessary, refer to textbook pp. 316-317 to review the concept of a loop. It is important to notice two details in line 40: (i) the quotation marks include the letters I and S and also a blank space; (ii) the semicolon at the end of the line will prevent any line feeds. Therefore, lines 30-50 print twice "IS" followed by a space. Lines 60 and 70 are comparable to lines 10 and 20, except for the actual letters printed. Finally, lines 80-100 print the letter R twice (without any spaces and without line feeds), and line 110 prints "ECT" followed by a period. Thus, the output is:

```
THIS IS CORRECT.
```

11.
```
10 PRINT "THIS IS CORRECT."
20 END
```

13. Since M will be assigned the value 12 and C becomes 10, we know

$$(M/C)*100 = (12/10)*100 = 120.$$

Note that the INPUT command (used in lines 30 and 50) has the affect of pausing the program until the user types in a value and presses RETURN. Finally, recall that the command PRINT as used in Line 60 causes a blank line to be printed (i.e., a line feed).
The program produces the following output:

```
I will calculate IQ.
What is mental age? [input 12 here]
What is chronological age? [input 10 here]

IQ is 120
```

15.
```
I will calculate IQ.
What is mental age? [input 8 here]
What is chronological age? [input 10 here]

IQ is 80
```

17.
```
10 PRINT 5000*(1 + 0.06)^2
20 END
```

19. The proper placement of parentheses is especially important here. In order to have the computer divide the <u>entire</u> numerator by two, we need to enclose the numerator in parentheses. Otherwise, the operations will not be performed in the correct order.

```
10 PRINT (23.5^2 – 5*61.1)/2
20 END
```

21.

```
10 PRINT 12850*(1 + 0.09/12)^360
20 END
```

23. Since a variable is used (in this case, x), the program needs an INPUT statement. Such a statement enables the user of the program to specify the desired value(s) of x for which the expression $2x^2 + 3x - 4$ will be evaluated.

```
10 PRINT "WHAT IS X";
20 INPUT X
30 PRINT "2X^2 + 3X – 4 = "; 2*X^2 + 3*X – 4
40 END
```

Several reminders are in order:

(i) Recall that the INPUT command by itself prints out a question mark ("?"). That is why we do not need to include one in Line 10.

(ii) The way the final output — and, for that matter, the INPUT statement, too — is displayed is to some degree a matter of personal preference. For an illustration of this, take note of the program given in Example 3 (textbook pp. 313-314) and its output. Its output is a bit fancier than what the program above produces. So, feel free to enhance your programs as you (and those for whom you are writing) see fit.

25. This program is very similar to the one written for Problem 23. Accordingly, please take note of our comments in the solution above.

```
10 PRINT "WHAT IS X";
20 INPUT X
30 PRINT "5X^2 + 17X – 128 = "; 5*X^2 + 17*X – 128
40 END
```

27. This expression includes two variables. Consequently, our program needs to include either two INPUT statements (i.e., one per variable) or one INPUT statement in which the variables are separated by a comma (see p. 313 for details). Permit us to demonstrate both approaches:

```
10 PRINT "WHAT IS R";
20 INPUT R
30 PRINT "WHAT IS n";
40 INPUT N
50 PRINT "5000(1 + R/12)^(12n) = "; 5000*(1 + R/12)^(12*N)
60 END
```

or, alternatively,

```
10 PRINT "What are the respective values of R and n";
20 INPUT R, N
30 PRINT "5000(1 + R/12)^(12n) = "; 5000*(1 + R/12)^(12*N)
40 END
```

In order for the second program to be executed correctly, remember that the user must know to separate the two values by a comma and then press RETURN. If the program is likely to be run by users who do not know this, the first program (i.e., the one with two separate INPUT statements) would be more user friendly.

29. This problem is similar in many respects to Problem 27. Please refer, as necessary, to our solution to that problem.

```
10 PRINT "WHAT IS P";
20 INPUT P
30 PRINT "WHAT IS R";
40 INPUT R
50 PRINT "P(1 + R/12)^360 = "; P*(1 + R/12)^360
60 END
```

Once again, as was explained above in our solution to Problem 27, the details of your program may differ from ours; for example, you may have chosen to use one INPUT statement instead of two.

31. To reduce the number of steps to four, the most likely approach is to delete line number 30 and replace the variable A in the PRINT statement with X*Y. Upon doing so, the rewritten program would appear as follows:

```
10 LET X = 5
20 LET Y = 9
30 PRINT "The area is "; X*Y
40 END
```

There are several other ways, we suppose, one could reduce the number of steps to four. One approach would be to remove from the program mention of the variable X (i.e., delete line 10) and instead use its value (of 5) in line 30 (i.e., LET A = 5*Y).

33. Since line 20 defines M as (X + Y + Z)/3, it follows that:

$$M = \frac{X+Y+Z}{3} = \frac{(2)+(3)+(4)}{3} = \tfrac{9}{3} = \boxed{3}$$

35. Please note that the BASIC programming language will display at most eight places to the right of the decimal, unless special functions/commands are used. Consequently, the repeating decimal $3.\overline{3}$ is displayed as follows:

$$M = \frac{X+Y+Z}{3} = \frac{(-3)+(18)+(-5)}{3} = \tfrac{10}{3} = \boxed{3.33333333}$$

37.
```
10 PRINT "What are your sales for this week";
20 INPUT S
30 PRINT "Your salary for this week is "; 0.05*S + 100
40 END
```

39. Since the materials necessary to manufacture sticky widgets cost $2.68 per item and the labor costs $14 per item, the cost per item (materials and labor combined) is $16.68. Therefore, to compute the total monthly costs, multiply the number of items by the cost per item of $16.68, and add the result to the monthly fixed expenses of $6,000. The following program performs this process:

```
10 PRINT "How many items were manufactured for the month";
20 INPUT N
30 PRINT "The total monthly cost is "; 16.68*N + 6000
40 END
```

41. All of this program's output is produced by the PRINT statement in line 30, which prints the values of N and A. Since line 30 is contained within the FOR-NEXT loop, the PRINT command is executed as many times as the loop is performed (which, in this program, is ten times). To better understand the details, let us take a moment together to trace the program:

- Line 10 begins by assigning to the variable N a value of 1.
- Line 20 assigns to the variable A a value of 2*N + 3 = 2(1) + 3 = 5.
- Line 30 prints the values of N and A (i.e., 1 and 5, respectively).
- Line 40 raises the value of N by 1 (making it now 2) and takes us back to Line 10.
- Line 20 now assigns to A a value 7 [i.e., 2(2) + 3].
- Line 30 prints the current values of N and A (namely, 2 and 7).
- Line 40 again increases N by 1(making it 3) and loops us back to Line 10.
- And so on... This process continues until N reaches a value of 10.

Therefore, this program's output is:

```
 1    5
 2    7
 3    9
 4   11
 5   13
 6   15
 7   17
 8   19
 9   21
10   23
```

43. In a manner similar to what was shown in our solution above to Problem 41, we recommend you trace this program in a step-by-step manner. In so doing, you may find the format used by your text's author (see Example 6. pp. 317-318) to be more efficient than the somewhat "wordy" trace we performed in our solution to Problem 41.

Let us use this problem to demonstrate your author's approach. For this program, we need to create four columns: (i) a column entitled "Line Number," which tells us which step is being performed (and when); (ii) a column for the values assigned to variable A, (iii) another column for K; and, finally, (iv) a column for the output.

Line Number	*A*	*K*	*Output*
10	1		
20		1	
30	1		
40		2	
30	2		
40		3	
30	6		
40		4	
30	24		
40		5	
30	120		
40		6	
30	720		
50			720

Therefore, the output of this program is: **720**
[Incidentally, if you are already familiar with the material covered in Chapter 9 (p. 613) of your text, you may have noticed what this program, in effect, has accomplished: it computed the "factorial" of 6 (i.e., $6! = 6 \cdot 5 \cdot 4 \cdot 3 \cdot 2 \cdot 1$*).]*

45. Once again, as we demonstrated in our solutions above to Problems 41 and 43, you may trace this program in a step-by-step manner. Alternatively, as you become more familiar with the BASIC programming language, you will begin to find it fairly easy to look at a program and trace it mentally. Let us demonstrate.

Notice that line 10 prints "NUMBER" and "NUMBER SQUARED" exactly as they appear in the quotation marks. Because of the comma, line 10 prints these three words on the same line with some separation between "NUMBER" and "NUMBER SQUARED". Line 10 also gives the person tracing the program a big hint regarding the nature of the output.

From line 20 to line 50, a FOR-NEXT loop is performed — with the variable N assigned successively the values 1, 2, 3, ..., and, finally, 10. Meanwhile, line 30 squares each successive value of N (i.e., 1^2, 2^2, 3^2, ..., 10^2) and assigns the results (1, 4, 9, ..., 100) to the variable S. Finally, line 40 prints each respective value of N and S. Thus, when this program is executed, the following output is obtained:

NUMBER	**NUMBER SQUARED**
1	**1**
2	**4**
3	**9**
4	**16**
5	**25**
6	**36**
7	**49**
8	**64**
9	**81**
10	**100**

47. To understand this program, you need to be familiar with the concept of branching. To review this concept, please refer to textbook page 318-319.
Lines 30 through 70 of this program comprise an IF-ELSE branching command that performs as follows:

- When the value of N is less than or equal to 5 (see line 30), line 40 is executed, which prints the value of N. Thus, the first five times in which the FOR-NEXT loop is performed, the values of N are printed; namely, 1, 2, 3, 4, and 5.
- When N is not less than or equal to 5 (i.e., when N > 5), line 60 is executed, which prints the value of $N^2 - 36$. This occurs when N equals 6, 7, 8, 9, and 10. For these cases, line 60 prints the values of $6^2–36$, $7^2–36$, $8^2–36$, $9^2–36$, and $10^2 – 36$; that is, 0, 13, 28, 45, and 64, respectively.

Although there are five PRINT statements in this program, it is important to note that only two are contained within the FOR-NEXT loop; namely, those in lines 40 and 60, both of which are part of the IF-ELSE command.

Upon tracing the program and acquiring an understanding of both the FOR-NEXT loop and the IF-ELSE within the loop, we find the program's output to be:

```
   EVALUATE F
   1
   2
   3
   4
   5
   0
   13
   28
   45
   64
The values of F have been checked.
from 1 to 10.
```

49. This problem is simply an extension of textbook Example 6 (p. 317), which is a program that determines the sum of the first five counting numbers. A program similar to the one shown below extends this idea to the first 100 counting numbers:

```
10 LET SUM = 0
20 FOR J = 1 TO 100
30      LET SUM = SUM + J
40 NEXT J
50 PRINT SUM
60 END
```

Of course, the details/style of your program may differ from ours. For instance, you may have chosen other variable names, and you may have preferred to enhance line 50 (e.g., PRINT "The sum of the first 100 counting numbers is "; SUM).

51. This program is similar in structure to that of Example 6 (p. 317) and Problems 49 and 50. It is important, however, to understand the significant difference between this problem and Problem 50, its immediate predecessor. In this program, we need to square each of the first 100 counting numbers and then determine the sum of the squares. Upon grasping this essential detail, one's program would be along the lines of:

```
10 LET SUM = 0
20 FOR J = 1 TO 100
30      LET SUM = SUM + J^2
40 NEXT J
50 PRINT SUM
60 END
```

Needless to say, our previous comments regarding differences in programming details/style (see solution to Problem 49 above) once again apply.

53. Here is one program (among many) that will print the first one hundred counting numbers backwards:

```
10 FOR K = 0 TO 99
20      PRINT 100 – K
30 NEXT K
40 END
```

55. **Programs vary**. Here is one approach (among many):

```
10    PRINT "How many scores to average";
20    INPUT N
30    PRINT "What is first score";
40    INPUT S
50    FOR K = 1 TO N – 1
60          PRINT "What is next score";
70          INPUT T
80          LET S = S + T
90    NEXT K
100   IF N > 0 THEN
110         PRINT "The average is"; S/N
120   ELSE
130   PRINT "Error dividing by 0."
140   END
```

57. We begin by recalling that linear equations (i.e., $ax+b=0$ where $a \neq 0$) have only one root (solution), namely $x = -\dfrac{b}{a}$.

Our program needs to do three things: (i) allow the user to input the values of a and b; (ii) avoid a division-by-zero error in the event the user inputs a value of zero for a, and, of course, (iii) calculate and print the equation's root. The following program illustrates one way to accomplish this three-part task:

```
10   PRINT "I will solve any linear equation, ax + b =0."
20   PRINT "What is the value of a";
30   INPUT A
40   PRINT "What is the value of b";
50   INPUT B
60   IF A = 0 THEN
70        PRINT "a cannot be 0; try again."
80   ELSE
90        PRINT "The root is "; -1*B/A
100  ENDIF
110  END
```

Plenty of variations of this program are possible; for example, you might wish to refine the above program by consolidating lines 20 through 50 as one INPUT command.

59. a) Let us follow (and, in the process, translate) the steps in order:

- *"Think of a number"* ⇒ Let x represent the number
- *"Add 9"* ⇒ $x+9$
- *"Square"* ⇒ $(x+9)^2 = x^2+18x+81$
- *"Subtract the square of the original number"* ⇒ $(x^2+18x+81)-x^2=18x+81$
- *"Subtract 61"* ⇒ $18x+81-61=18x+20$
- *"Multiply by 2"* ⇒ $2(18x+20)=36x+40$
- *"Add 24"* ⇒ $36x+40+24=36x+64$
- *"Subtract 36 times the original number"* ⇒ $36x+64-36x=64$
- *"Take the square root"* ⇒ $\sqrt{64}=8$

Thus, regardless of the number chosen, the result is always **8**.

b) Here is a BASIC program that performs this arithmetic and prints the result after each step:

```
10   PRINT "Think of a number N.  What is this number";
20   INPUT N
30   LET X = N + 9
40   PRINT X
50   LET X = X^2
60   PRINT X
70   LET X = X – N^2
80   PRINT X
90   LET X = X – 61
100  PRINT X
110  LET X = 2*X
120  PRINT X
130  LET X = X + 24
```

```
140  PRINT X
150  LET X = X – 36*N
160  PRINT X
170  LET X = SQR(X)
180  PRINT X
190  END
```

Chapter 5 Review Questions

1. a) **Answers/wording vary**. As defined on textbook page 238, ***algebra*** refers to a structure, or a set of axioms, that forms the basis for what is accepted and what is not. Additionally, you may find the definition provided in the text's glossary (Appendix G) to be of interest.

b) The four main processes of algebra are: ***simplify*** [carry out all operations according to the order of operations agreement and write the result in a prescribed form], ***evaluate*** [replace the variable(s) with specified numbers and then simplify], ***factor*** [write the expression as a product], and ***solve*** [find the replacement(s) for the variable(s) that make the equation true].

2. a) As shown below, first apply the distributive property and then simplify (i.e., combine like terms):

$$\begin{aligned}(x-1)(x^2+2x+8) &= (x-1)(x^2)+(x-1)(2x)+(x-1)(8)\\ &= x^3-x^2+2x^2-2x+8x-8\\ &= \boxed{\mathbf{x^3+x^2+6x-8}}\end{aligned}$$

b)
$$\begin{aligned}x^2(x^2-y)-xy(x^2-1) &= x^2(x^2-y)+(-xy)(x^2-1)\\ &= x^4-x^2y+(-x^3y)+xy\\ &= \boxed{\mathbf{x^4-x^3y-x^2y+xy}}\end{aligned}$$

3. a) $x^3-y^3=(2)^3-(3)^3=8-27=\boxed{\mathbf{-19}}$

b)
$$\begin{aligned}(x-y)(x^2+xy+y^2) &= [(2)-(3)][(2)^2+(2)(3)+(3)^2]\\ &= [-1][19]\\ &= \boxed{\mathbf{-19}}\end{aligned}$$

4. a) First, factor the gcf (greatest common factor), which in this case is 3. Then, upon recognizing that x^2-9 is a difference of squares, factor it accordingly. Therefore, $3x^2-27=3(x^2-9)=\boxed{\mathbf{3(x-3)(x+3)}}$.

b) $x^2 - 5x - 6 = \boxed{(x-6)(x+1)}$

5. a) $8x - 12 = 0$ *Given equation (notice that it is a "linear" equation)*

$8x = 12$ *Add 12 to both sides.*

$x = \frac{12}{8}$ *Divide both sides by 8.*

$\boxed{x = \frac{3}{2}}$ *Simplify (reduce the fraction).*

b) $8x - 12 = -2x^2$ *Given equation (notice that it is a "quadratic" equation)*

$2x^2 + 8x - 12 = 0$ *Add $2x^2$ to both sides to rewrite in $ax^2+bx+c=0$ form.*

$x^2 + 4x - 6 = 0$ *Divide both sides by 2.*

Since the left-hand side of the equation does not factor easily, we will use the quadratic formula. We begin by mentally identifying $a = 1$, $b = 4$, and $c = -6$. Applying the quadratic formula (shown on textbook page 267) and simplifying appropriately yields the following solution set:

$$x = \frac{-(4) \pm \sqrt{(4)^2 - 4(1)(-6)}}{2(1)} = \frac{-4 \pm \sqrt{40}}{2}$$

$$= \frac{-4 \pm 2\sqrt{10}}{2} = \frac{2\left(-2 \pm \sqrt{10}\right)}{2}$$

Thus, $\boxed{x = -2 \pm \sqrt{10}}$.

6. To review this geometric approach to factoring polynomials, review Section 5.2 (textbook pp. 246-251) and our solution in that section to Problems 37-47.

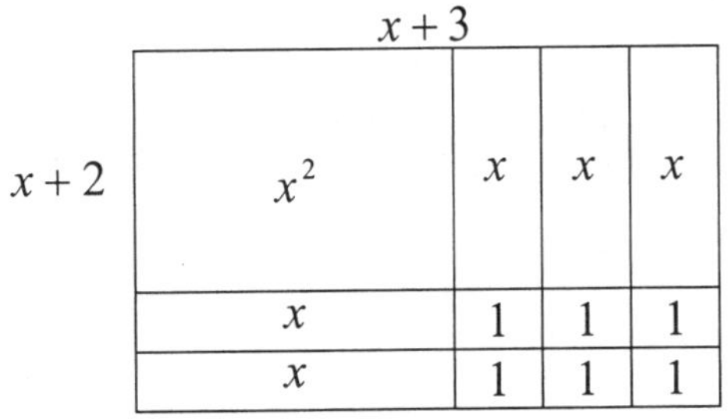

So, $x^2 + 5x + 6 = \boxed{(x+2)(x+3)}$.

7. $2x + 5 = 13$ *Given linear equation*

$2x = 8$ *Subtract 5 from both sides.*

$\boxed{x = 4}$ *Divide both sides by 2.*

8. $3x+1=7x$ *Given linear equation*

$1=4x$ *Subtract 3x from both sides.*

$\boxed{\frac{1}{4}=x}$ *Divide both sides by 4.*

9. $\frac{2x}{3}=6$ *Given linear equation*

$\frac{3}{2}\cdot\frac{2x}{3}=6\cdot\frac{3}{2}$ *Multiply both sides by the reciprocal of $\frac{2}{3}$ (namely, $\frac{3}{2}$)*

$\boxed{x=9}$ *Simplify.*

There are, of course, other approaches to this problem; for example, some students prefer to first multiply both sides by 3 and afterwards divide both sides by 2. Regardless of the method you prefer, we hope you remain open minded about — and have a sincere desire to understand — alternative strategies.

10. $2x-7=5x$ *Given linear equation*

$-7=3x$ *Subtract 2x from both sides.*

$\boxed{-\frac{7}{3}=x}$ *Divide both sides by 3.*

11. By applying the text's procedure for solving proportions (see page 294) and then simplifying either by hand or with a calculator, we obtain the following:

$\frac{P}{100}=\frac{3}{20}$ *Given equation*

$P=\frac{100\times 3}{20}$ ← *Product of the means* (numerator); ← *Number diagonally opposite the unknown* (denominator)

$\boxed{P=15}$

12. Like Problem 11, we again apply the procedure for solving proportions (see textbook p. 294) and then simplify (either by hand or with a calculator):

$\frac{25}{W}=\frac{80}{12}$ *Given equation*

$W=\frac{25\times 12}{80}$ ← *Product of the extremes* (numerator); ← *Number diagonally opposite the unknown* (denominator)

$\boxed{W=\frac{15}{4}\text{ or }3\frac{3}{4}\text{ or }3.75}$

13. $x^2=4x+5$ *Given quadratic equation*

$x^2-4x-5=0$ *Subtract 4x and 5 from both sides to rewrite in $ax^2+bx+c=0$ form.*

$(x-5)(x+1)=0$ *Factor, if possible. [Otherwise, use the quadratic formula.]*

$x-5=0$ or $x+1=0$ *Apply zero-product rule (see textbook p. 267).*

$\boxed{x=5,\ x=-1}$ *Solve each of the resulting linear equations.*

14. $4x^2+1=6x$ *Given quadratic equation*

$4x^2-6x+1=0$ *Subtract 6x from both sides to rewrite in $ax^2+bx+c=0$ form.*

Since the left-hand side of the equation does not factor easily, we will use the quadratic formula. We begin by mentally identifying $a=4$, $b=-6$, and $c=1$. Applying the quadratic formula (shown on textbook page 267) and simplifying appropriately yields the following solution set:

$$x=\frac{-(-6)\pm\sqrt{(-6)^2-4(4)(1)}}{2(4)}=\frac{6\pm\sqrt{20}}{8}=\frac{6\pm\sqrt{4\cdot 5}}{8}=\frac{6\pm 2\sqrt{5}}{8}=\frac{{}^{1}\cancel{2}\left(3\pm\sqrt{5}\right)}{\cancel{8}_{4}}$$

Therefore, $x=\boxed{\dfrac{3\pm\sqrt{5}}{4}}$.

15. $3<-x$ *Given inequality*

$-3>x$ *Multiply both sides by –1 and reverse the inequality.*

$\boxed{x<-3}$ *Rewrite solution in a more suitable, easy-to-read form.*

16. $2-x\geq 4$ *Given inequality*

$-x\geq 2$ *Subtract 2 from both sides.*

$\boxed{x\leq -2}$ *Multiply both sides by –1 and reverse the inequality.*

17. $3x+2\leq x+6$ *Given inequality*

$2x\leq 4$ *Subtract 2 and x from both sides.*

$\boxed{x\leq 2}$ *Divide both sides by 2.*

18. $14>5x-1$ *Given inequality*

$15>5x$ *Add 1 to both sides.*

$3>x$ *Divide both sides by 5.*

$\boxed{x<3}$ *Rewrite solution.*

19. This problem is comparable to Problems 15-18 in Section 5.7 (p. 297) and Examples 4-5 (pp. 291-292). For additional details, we refer you to both the explanation provided in this solution manual (for Problems 15 and 17) and your textbook's explanation provided in Section 5.7 (pp. 291-292).

a) $\frac{2}{3}\ \boxed{<}\ \frac{67}{100}$ b) $\frac{3}{4}\ \boxed{=}\ \frac{75}{100}$ c) $\frac{23}{27}\ \boxed{<}\ \frac{92}{107}$

d) $0.05\ \boxed{<}\ 0.10$ e) $0.990\ \boxed{>}\ 0.909$

20. In the words of your text's author (see top of page 295), "whenever you are working an applied problem, you should estimate an answer so that you will know whether the result you obtain is reasonable." In this case, since the ratio of water to flour is 1 to 5, we know 1 cup of water requires 5 cups of flour. Hence, 2 cups of water require 10 cups flour, 3 cups water implies 15 cups flour, and so on. Thus, since the problem calls for $2\frac{1}{2}$ cups of water and since $2\frac{1}{2}$ is between 2 and 3, the cups of flour required must be between 10 and 15.

Expressing its wording in terms of a proportion, the problem asks the following question: "1 cup water is to 5 cups flour as $2\frac{1}{2}$ cups water is to what?"
This prompts us to write the following equation:

$$\frac{\overset{\substack{\textit{water}\\ \downarrow}}{1}}{\underset{\substack{\uparrow\\ \textit{flour}}}{5}} = \frac{\overset{\substack{\textit{water}\\ \downarrow}}{2\frac{1}{2}}}{\underset{\substack{\uparrow\\ \textit{flour}}}{x}}$$

Then, upon applying the text's procedure for solving proportions (p. 294) and simplifying either by hand or with a calculator, we find:

$$x = \frac{5 \times 2\frac{1}{2}}{1} = \frac{\frac{5}{1} \times \frac{5}{2}}{1} = \frac{25}{2} = 12\tfrac{1}{2}$$

Thus, if $2\frac{1}{2}$ cups water is called for, **12 1/2 cups of flour** is needed. Please note that this answer is, as the problem's directions require, "to the nearest tenth." Since $\frac{1}{2} = \frac{5}{10}$, no rounding is necessary.

21. This problem is similar to both Problem 33 in Section 5.6 (for which this manual contains a detailed solution) and textbook Example 2 (pp. 279-280), among others. Instead of providing another fully detailed explanation (like we did for Problem 33) in which we illustrate Pólya's problem-solving procedure, this time we share a more concise solution.

Let x, $x+2$, $x+4$, and $x+6$ be the four consecutive even integers. Please note that this is just one way (among many) to choose the variables.

Since the sum of these four integers is 100, we write and solve the following equation:

$$\begin{aligned} x+(x+2)+(x+4)+(x+6) &= 100 \\ 4x+12 &= 100 \\ 4x &= 88 \\ x &= 22 \end{aligned}$$

Since $x = 22$, we know $x+2 = 24$, $x+4 = 26$, and $x+6 = 28$. We then state our solution in words:

The four consecutive even integers are 22, 24, 26, and 28.

Finally, remember to check your solution: $22 + 24 + 26 + 28 = 100$ ✓

22. This type of genetics application was first introduced on textbook pages 254-255. Example 4 (p. 255) and Problems 52-55 & 57-60 (p. 263) provide you with additional practice; refer to this manual's solutions (to the odd-numbered problems) for additional support.

Using the same notation as the author of your text (pp. 254-255), we begin by letting $p = 0.52$ and $q = 0.48$. We then find the fractions (i.e., percentages) of the population having each genotype:

<u>tall (TT)</u>: $p^2 = (0.52)^2 = 0.2704 \approx$ **27%**

<u>tall w/ recessive short (tT or Tt)</u>: $2pq = 2(0.52)(0.48) = 0.4992 \approx$ **50%**

<u>short (tt)</u>: $q^2 = (0.48)^2 = 0.2304 \approx$ **23%**

As for the <u>phenotype</u> (which refers to the outward <u>physical appearance</u>), approximately **77% are tall** (i.e., TT, tT, and Tt) and roughly **23% are short**.

23. This problem is almost identical to text Example 3 (pp. 280-281) and Problems 40-42 in Section 5.6. This time — instead of providing another fully detailed explanation (like we did for Problem 41 and your text's author did for Example 3) — we share an abbreviated solution.

The following verbal equation summarizes the main idea of the problem:

(<u>DIST NY TO CHICAGO</u>) + (DIST CHICAGO TO SF) + (<u>DIST SF TO HONOLULU</u>) = 4,980

↑ (DIST CHI TO SF – 1,140) ↑ (DIST CHI TO SF + 540)

Letting $x =$ DIST CHICAGO TO SF, we "evolve" this verbal equation into an algebraic equation and proceed to solve it:

$$(x - 1{,}140) + x + (x + 540) = 4{,}980$$

$$3x - 600 = 4{,}980$$

$$3x = 5{,}580$$

$$x = 1{,}860$$

Therefore, noting that $x = 1{,}860$, $x - 1{,}140 = 720$, and $x + 540 = 2{,}400$, we are prepared to state our final answer:

The distance from Chicago to San Francisco is 1,860 miles, from New York to Chicago is 720 mi, and from San Francisco to Honolulu is 2,400 mi.

24.

```
10 PRINT "WHAT IS X";
20 INPUT X
30 PRINT "18X^2 + 10 = "; 18*X^2 + 10
40 END
```

For additional details, please review Section 5.9; specifically, Example 3 (on pp. 313-314) and this manual's solutions to Problems 23, 25, 27, and 29 (from textbook page 320).

25. If the value of x is placed in cell A1, the formula below in cell B1 will compute the value of $18x^2 + 10$. Refer to Section 5.3 for additional details.

	A	B	C
1		+18∗A1^2+10	
2			
3			
4			
5			

CHAPTER 6: The Nature of Geometry

Problem Set 6.1

1. **Answers vary**. Depending upon the perspective from which you view it, the figure depicts both a young woman and an old woman.

3. **Answers vary**. A procedure for using a compass to construct a line segment congruent to a given segment is well described and nicely illustrated (in Figure 6.7) on the top half of textbook page 330.

5. **Wording varies**. The notion of an undefined term is discussed on p. 328 of your text. In order to properly define all words in a mathematical system, we must avoid using any words within our definitions whose meanings either are not given or lack clarity. Such nebulous (fuzzy) words are called ***undefined terms***.

7. **Wording varies**. As explained on textbook p. 329, an ***axiom*** (also called a *postulate*) is a statement that is accepted/assumed as fact without proof. In contrast, a ***theorem*** is a result/statement that is proved using defined terms, axioms, other previously proved theorems, and the laws of logic.

Please do not make the mistake of confusing the concept of a theorem with that of a theory; they are very different notions.

9. By using a compass and applying the procedure in your text (see the top half of textbook p. 330 and Figure 6.7), we obtain:

11. Once again, with the aid of a compass, we apply the procedure explained on textbook p. 330.

13. To review how to construct a circle with a given radius, refer to the step-by-step directions provided in textbook Figure 6.8 (bottom half of p. 330).

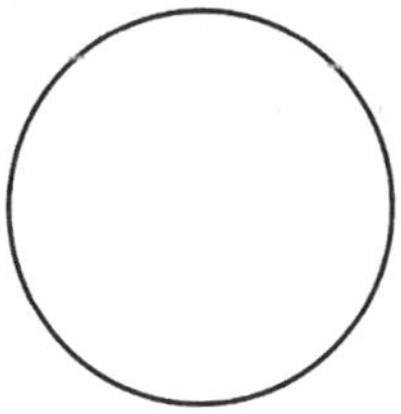

15. Once again, we apply the procedure explained in textbook Figure 6.8 (p. 330).

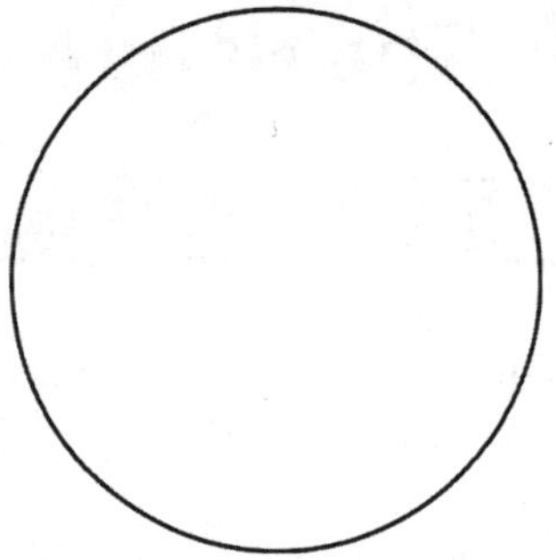

17. The procedure for constructing a line parallel to a given line through a given point is explained and illustrated on textbook pp. 330-331 (see Figures 6.9 and 6.10).

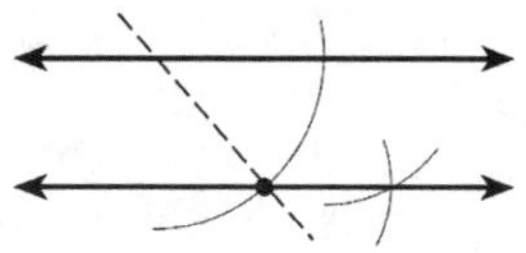

19. Again, apply the procedure explained on textbook pp. 330-331 (see Figures 6.9 and 6.10).

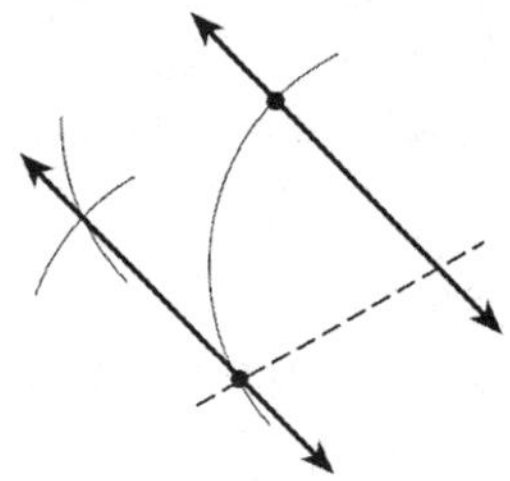

21. A **vertical line through the middle of the bell**, extending through the surrounding circle, is a line of symmetry for this corporate logo.

23. A **vertical line cutting in half each of the three phones** would be a line of symmetry for this illustration.

25. The Alabama state flag has **several lines of symmetry**; among them are an appropriately placed vertical line and a horizontal line. Note that diagonally positioned lines through the flag's corners will only be lines of symmetry if the flag's dimensions are square. To acquire a better understanding of this, consider spending a few moments folding in different ways a rectangular-shaped piece of notebook paper (contrast this with a square-shaped napkin, for instance).

27. Like the Alabama flag from Problem 25, the New Mexico flag has **several lines of symmetry**, a vertical line and a horizontal line.

29. From your text's previous work with sets in Section 1.4, recall that the symbol $\cup$ denotes set union. That is, $\overline{PQ} \cup \overline{QR}$ asks you to identify the set of all points that belong to segment PQ <u>or</u> segment QR <u>or</u> both.

Alternatively, as we explained in our solutions for Section 1.4's problems, you may find it helpful to make use of synonyms for the word "union." Consider using, for example, words and phrases like "joining together," "marrying," and "uniting."

In any event, the union of $\overline{PQ}$ and $\overline{QR}$ is $\boxed{\overline{PR}}$ (i.e., segment PR).

31. This problem asks you to determine the intersection ($\cap$) of two rays. In a manner similar to what we suggested in our solution above to Problem 29 (and in our previous work with set operations in Section 1.4), you might find it helpful to use synonyms for the word "intersection." A few synonyms that come to mind are "overlap," "common ground," and "shared elements."

A picture also might help clarify the situation.

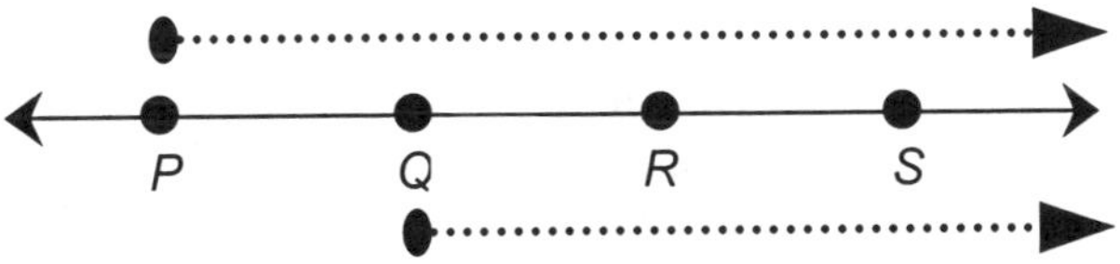

We have sketched ray $\overrightarrow{PQ}$ above the given line and ray $\overrightarrow{QS}$ below it. From this picture, it becomes clear that the points in the intersection of these two rays (i.e., the points that have been covered twice) consist of point Q and all points located to the right of Q.

Thus, our answer is the ray $\boxed{\overrightarrow{QS}}$ which could also be written as $\overrightarrow{QR}$.

33. If this set union (marriage, joining, etc.) is difficult for you to picture, consider drawing a sketch. With ray PQ sketched above the given line and ray SR below it, here is one such sketch:

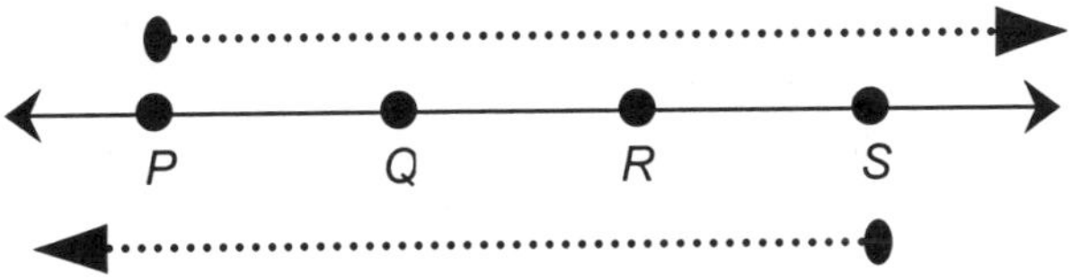

The concept of set union suggests we need to join these rays together. Doing so gives us the original line.

Hence, our answer is line $\boxed{\overleftrightarrow{PS}}$.

<u>Note</u>: It is equally correct to write this answer using any other pair of points (PQ, QR, QS, etc.) — <u>provided</u> the (double-arrowed) line symbol is placed over them.

35. The only point "shared by" (i.e., "in the intersection of") segment $\overline{PQ}$ and point Q is **point Q** itself.

37. The chambered nautilus has **no line symmetry**.

39. The *Winged Lion* has a **vertical line of symmetry**.

41. The human brain has **no line symmetry**. The different "halves" of the brain perform different functions and have different geometric attributes.

43. Yes, this restaurant **exhibits line symmetry**.

45. No, upon looking closely at its photograph, it is clear the Kaiser porcelain vase has **no line symmetry**.

47. This cartoon illustrates a **rotation** of the letters 'NNNNN' (or should we say 'ZZZZZ'?). To get better acquainted with the different types of geometric transformations, see textbook pp. 331-332 (including Figure 6.14 on p. 332).

> ***For Problems 48-53**: If you have difficulty with this type of spatial visualization, we recommend you either create a cube out of paper or look around your house for one (e.g., a child's letter blocks or a single die from a board game). Affix the letters C, U, B, and E on its faces in the manner shown in the directions (on textbook p. 337), and then spend some time exploring the letter patterns.*

49.

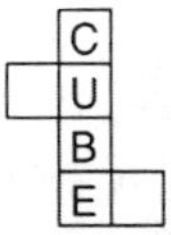

51.

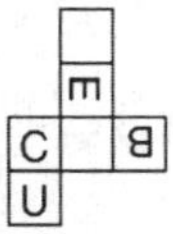

53.

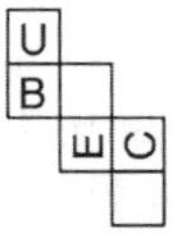

55. **Choice C**. The first rotation ($90°$ to the left) has the effect of moving the T to where the L is, the L is rotated out of view, and (from an overhead perspective) the V on the top of the cube is turned $90°$ clockwise. The second rotation ($90°$ downward) moves the T to the bottom of the cube and rotates the V into T's former location with the orientation shown in choice C.

57. **Choice C**. The first rotation ($90°$ upward) moves the yellow side (the one painted with a symbol resembling ∅) to the top, where the green lined side previously was positioned. The green side moves out of view. As for the N, it remains where it is, except for being rotated $90°$ clockwise. The second rotation ($90°$ leftward) moves the sideways N to the front left side. The green lined side is rotated back into view (in the front right position), and, finally, the ∅-labeled side remains on top but (when viewed from an overhead perspective) is rotated $90°$ clockwise. Choice C accurately depicts all of this.

59. Before you read the boxed solution below, first consider this **hint**:

Which of the four categories of letters exhibit a line of symmetry? For those categories that do have line symmetry, is the line vertical, horizontal, or both?

Your answer to these two questions should be:

Category (1) consists of letters with no symmetry;
Category (2)'s letters are symmetric around a horizontal line;
Category (3) is made up of letters with a vertical line of symmetry; and
Category (4)'s letters exhibit both vertical and horizontal line symmetry.

Problem Set 6.2

1. **Wording varies**. An ***angle*** consists of two rays that share a common endpoint. Refer to textbook page 339 for additional details.

3. **Wording varies**. Rays and half-lines are first introduced on textbook p. 329. According to your text, a ray includes the endpoint, but a half-line may (or may not) include the endpoint.
An open half-line looks like this: 

5. **Wording varies**. In a diagram on a printed page, any line that is parallel to the top and bottom edges of the page is considered ***horizontal***. Lines that are perpendicular to a horizontal line are considered to be ***vertical***. See textbook Figure 6.25 (p. 344) for an illustration of these two types of lines.

7. **Wording varies**. Consider two intersecting lines forming four angles. ***Adjacent angles*** are two angles which, when taken together, form a straight angle. ***Vertical angles*** are angles that are not adjacent. For additional clarification, refer to textbook p. 341.

As for corresponding angles, consider two parallel lines cut by a transversal. ***Corresponding angles*** are two nonadjacent angles whose interiors lie on the same side of the transversal such that one angle lies between the parallel lines and the other does not. Corresponding angles are congruent. Part (c) of Example 4 (textbook p. 343) provides you with an illustration of corresponding angles.

9. a) A four-sided polygon is known as a **quadrilateral**. Please note that the text's directions for Problems 9-14 ask you to classify the polygons according to "the number of sides." Consequently, you need not (and should not) in these problems concern yourself with terms like square, rectangle, etc.

b) A five-sided polygon is called a **pentagon**.

11. a) A three-sided polygon is known as a **triangle**.

b) A six-sided polygon is called a **hexagon**.

13. a) A four-sided polygon is known as a **quadrilateral**.

b) A seven-sided polygon is called a **heptagon**.

15. a) **True**. Based on the textbook's definition of a square (see p. 342), "a square is a rectangle" because squares possess all the characteristics of rectangles. That is to say, a square, like a rectangle, is a parallelogram that contains a right angle.

b) **True**. Because a square is a quadrilateral with opposite sides parallel, a square is a parallelogram. Said differently, since a square is a rectangle and a rectangle is a parallelogram, it follows that a square is a parallelogram.

17. a) **True**. Since a square is a 4-sided polygon, it is, by definition, a quadrilateral.

b) **False**. Although the converse is true (i.e., all rectangles are parallelograms), it is not true that all parallelograms are rectangles. Not all parallelograms contain a right angle.

19. a) **False**. Not all quadrilaterals have two parallel sides.

b) **False**. A trapezoid has <u>exactly</u> one pair of parallel sides, whereas a parallelogram is a quadrilateral that has two pairs of parallel sides.

21. a) **Yes** b) **Yes** c) **Yes** d) **Yes** e) **Yes**

23. a) **No** b) **No** c) **No** d) **No** e) **No**

25./27./29. To review this type of construction, refer to the bottom half of textbook p. 339 (and Figure 6.18 on p. 340).

31. a) $\angle BAC$ is $\angle 3$. Since $\angle 3$ and $\angle 1$ form a pair of vertical angles, they are congruent (i.e., $\angle 3 \simeq \angle 1$). Given that $m\angle 1 = 30°$, it follows that $m\angle 3 = 30°$. Therefore, $\angle BAC$ is **acute**.

b) $\angle ABC$ is $\angle 6$. Since $\angle 5$ and $\angle 6$ are supplementary angles (i.e., the sum of their measures is $180°$), $m\angle 6 = 180° - m\angle 5$. Given that $m\angle 5 = 90°$, it follows that $m\angle 6 = 180° - m\angle 5 = 180° - 90° = 90°$. Thus, $\angle ABC$ is **right**.

33. a) This problem is virtually identical to Problem 31b, whose solution is given above. Since $\angle 10$ and $\angle 11$ are supplementary angles and since $m\angle 11 = 90°$, the measure of $\angle 10$ must also be $90°$. Therefore, $\angle 10$ is **right**.

b) Since $\angle 15$ and $\angle 16$ are supplementary angles, $m\angle 16 = 180° - m\angle 15$. Given that $m\angle 15 = 30°$, it follows that

$$m\angle 16 = 180° - m\angle 15 = 180° - 30° = 150°.$$

Thus, because 150° is greater than 90°, $\angle 16$ is **obtuse**.

35. a) Once again, since $\angle 1$ and $\angle 2$ are supplementary angles, we can conclude that $m\angle 1 = 180° - m\angle 2 = 180° - 130° = 50°$.
Therefore, because $m\angle 1 = 50°$ and $50° < 90°$, $\angle 1$ is **acute**.

b) Upon recalling that lines ℓ_1 and ℓ_2 are parallel with transversal ℓ_3, we can see that $\angle 5$ and $\angle 15$ are a pair of alternate exterior angles and, hence, congruent. Therefore, since $\angle 5 \simeq \angle 15$ and $m\angle 15 = 88°$, $\angle 5$ is **acute**.

37. $\angle 13$ and $\angle 14$ are **adjacent and supplementary**.
[FYI: Based upon the directions for Problems 36-43, it would be sufficient to provide one of these two words as your final answer. Both words are desirable, but only one is necessary.]

39. $\angle 9$ and $\angle 10$ are **adjacent and supplementary**.
[Please take note of the "FYI" comment for Problem 37 (see above).]

41. $\angle 9$ and $\angle 11$ are a pair of **vertical** angles.

43. Note that lines ℓ_1 and ℓ_2 are parallel with transversal ℓ_3. $\angle 7$ and $\angle 13$ are **alternate interior angles**.

45. **False**. Judging from their appearance on textbook p. 346, lines b_1 and b_2 are horizontal lines.

47. **True**.

49. **False**. Actually, lines a_2 and a_3 are neither vertical nor parallel. The fact that they are not vertical is readily apparent from the diagram, and they are not parallel because $50° + 120° \neq 180°$.

51. $\angle 2$ and $\angle 4$ are vertical angles and, hence, congruent. Because $\angle 2$ and $\angle 6$ are corresponding angles, they are congruent. Since $\angle 2$ and $\angle 8$ are alternate interior angles, they are congruent. Therefore, $\mathbf{m\angle 2 = m\angle 4 = m\angle 6 = m\angle 8 = 65°}$.

Since $\angle 1$ and $\angle 2$ are supplementary, $m\angle 1 = 180° - 65° = 115°$. It is then easy to establish that $\angle 1$, $\angle 3$, $\angle 5$, and $\angle 7$ are supplementary to $\angle 2$, $\angle 4$, $\angle 6$, and $\angle 8$. Thus, $\mathbf{m\angle 1 = m\angle 3 = m\angle 5 = m\angle 7 = 115°}$.

53. $\mathbf{m\angle 1 = m\angle 3 = m\angle 5 = m\angle 7 = 153°}$

$\mathbf{m\angle 2 = m\angle 4 = m\angle 6 = m\angle 8 = 27°}$

55. $\mathbf{m\angle 1 = m\angle 3 = m\angle 5 = m\angle 7 = 163°}$

$\mathbf{m\angle 2 = m\angle 4 = m\angle 6 = m\angle 8 = 17°}$

57. Since important facts pertaining to triangles are not discussed in your text until Section 6.3, permit us to share a solution that relies solely upon concepts introduced in Section 6.2. However, you should be aware that this problem can be solved much more efficiently using several of the ideas discussed in Section 6.3 (e.g., isosceles triangles and the sum of the measures of a triangle's angles).

We begin by drawing a (dotted) line through point *C* parallel to the table/ground and label the angles as follows:

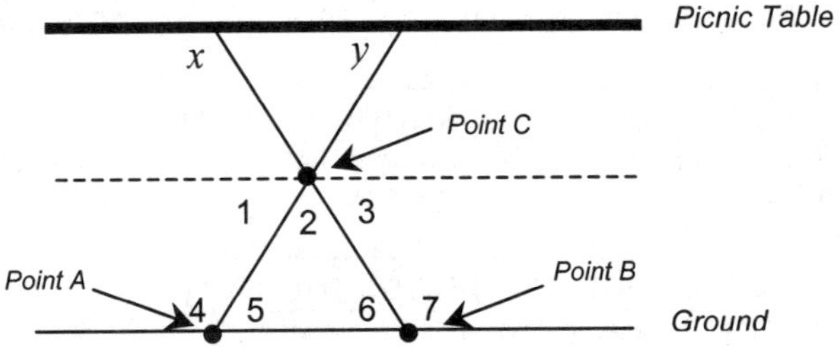

From the statement of the problem, we know $m\angle 2 = m\angle ACB = 85°$. We then take note of several relationships among the marked angles:

- $\angle 1$, $\angle 2$, and $\angle 3$ together form a straight angle (i.e., $m\angle 1 + m\angle 2 + m\angle 3 = 180°$);
- $\angle 1 \simeq \angle 5$ and $\angle 3 \simeq \angle 6$ because both form pairs of alternate interior angles;
- Since $\overline{AC}$ and $\overline{BC}$ are equal in length and the picnic table is parallel with the ground, it must follow that $m\angle 5 = m\angle 6$.

Since $m\angle 1 + m\angle 2 + m\angle 3 = 180°$ and since $\angle 1 \simeq \angle 5$ and $\angle 3 \simeq \angle 6$, we know that $m\angle 5 + m\angle 2 + m\angle 6 = 180°$. Replacing $m\angle 2$ with $85°$ and $m\angle 5 = m\angle 6$ with w, we can write and solve the following:

$$w + 85° + w = 180°$$
$$2w = 95°$$
$$w = 47.5°$$

Since $\angle 5$ and $\angle y$ are alternate interior angles and since $m\angle 5 = 47.5°$, we can conclude **$m\angle y = 47.5°$**. Then, since $\angle 7$ and $\angle 6$ are supplementary angles and since $\angle 7$ and $\angle x$ are alternate interior angles, $\angle x$ and $\angle 6$ are supplementary. Therefore, since $m\angle 6 = 47.5°$, **$m\angle x = 132.5°$**.

59.

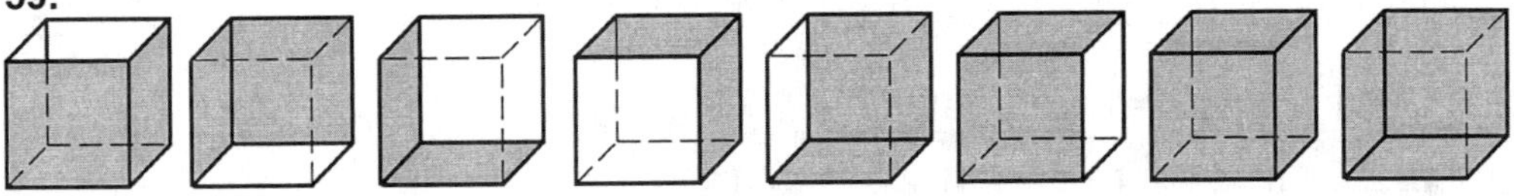

Problem Set 6.3

1. **Wording varies**. A ***triangle*** is a three-sided polygon.

3. **Wording varies**. As is explained on textbook pages 348-349, the given notation $\Delta ABC \simeq \Delta DEF$ means triangle ABC is congruent to triangle DEF. This means the two triangles have the same size and shape. Furthermore, based upon the ordering of the letters, we can conclude that $\angle A \simeq \angle D$, $\angle B \simeq \angle E$, $\angle C \simeq \angle F$, $\overline{AB} \simeq \overline{DE}$, $\overline{AC} \simeq \overline{DF}$, and $\overline{BC} \simeq \overline{EF}$.

5. $\overline{AB} \simeq \overline{ED}$; $\overline{AC} \simeq \overline{EF}$; $\overline{BC} \simeq \overline{DF}$; $\angle A \simeq \angle E$; $\angle B \simeq \angle D$; $\angle C \simeq \angle F$

7. $\overline{RS} \simeq \overline{TU}$; $\overline{RT} \simeq \overline{RT}$; $\overline{ST} \simeq \overline{UR}$; $\angle SRT \simeq \angle UTR$; $\angle S \simeq \angle U$; $\angle STR \simeq \angle URT$

9. $\overline{JL} \simeq \overline{PN}$; $\overline{LK} \simeq \overline{NM}$; $\overline{JK} \simeq \overline{PM}$; $\angle J \simeq \angle P$; $\angle K \simeq \angle M$; $\angle L \simeq \angle N$

11. $180° - (46° + 46°) = 180° - 92° = \boxed{88°}$. For additional details, see pp. 349-350 (including Example 2) in your text.

13. $180° - (20° + 15°) = 180° - 35° = \boxed{145°}$

15. $180° - (83° + 41°) = 180° - 124° = \boxed{56°}$

17. As your text's author proves (see pp. 350-351), the measure of the exterior angle of a triangle equals the sum of the measures of the two opposite interior angles. That is,

$$\overbrace{40+35}^{\text{sum of interior angles}} = \overbrace{x}^{\text{exterior angle}}$$
$$75 = x$$

Thus, the measure of the exterior angle is $\boxed{75°}$.

19. Angle x and the angle adjacent to it (marked with a measure of $100°$) together form a straight angle. In other words, they are supplementary angles. So,

$$x + 100 = 180$$
$$x = 80$$

Thus, the measure of the exterior angle is $\boxed{80°}$.

21. Here are two approaches:

I. Angle x and the $80°$ angle are supplementary. That is,

$$x + 80 = 180$$
$$x = 100$$

II. Via textbook p. 351, we know the measure of the exterior angle (in this case, labeled angle x) is equal to the sum of the measures of the two opposite interior angles (i.e., the $70°$ and $30°$ angles). That is,

$$\overbrace{70 + 30}^{\text{sum of interior angles}} = \overbrace{x}^{\text{exterior angle}}$$
$$100 = x$$

Therefore, the measure of the exterior angle is **100°**.

23./25./27. To review this type of construction, refer to the procedure explained on textbook p. 348 (accompanied by Figure 6.33).

29. **Wording varies**. An ***isosceles right triangle*** is a right triangle whose legs are equal in length.

31. Since the sum of the measures of the angles in any triangle is $180°$, we know

$$x + 3x + 5x = 180$$
$$9x = 180$$
$$x = \tfrac{180}{9}$$
$$x = \boxed{20°}$$

33.
$$x + 2(x + 12) + 3(x + 10) = 180$$
$$x + 2x + 24 + 3x + 30 = 180$$
$$6x + 54 = 180$$
$$6x = 126$$
$$x = \boxed{21°}$$

35. Since the measure of the exterior angle of a triangle equals the sum of the measures of the two opposite angles (see text's proof on pp. 350-351), we know

$$\overbrace{5x + 6x}^{\text{sum of opposite interior angles}} = \overbrace{4x + 42}^{\text{exterior angle}}$$
$$11x = 4x + 42$$
$$7x = 42$$
$$x = \boxed{6°}$$

37. In contrast to Problems 31-36, please note that this problem does <u>not</u> ask for the value of x; it instead asks for the measure of the triangle's angles.

Since the sum of the measures of the angles in any triangle is $180°$, we may write and solve the following equation:

$$x + (14 + 3x) + 3(x + 25) = 180$$
$$x + 14 + 3x + 3x + 75 = 180$$
$$7x + 89 = 180$$
$$7x = 91$$
$$x = 13$$

It then follows that $14 + 3x = 14 + 3(13) = 53$ and $3(x + 25) = 3(13 + 25) = 114$.

Therefore, the measures of this triangle's angles are **13°, 53°, and 114°**.

Of course, you always should check the solution to such a problem by making sure the sum of measures of the angles (i.e., $13 + 53 + 114$) equals $180°$.

<u>For Problems 39-46</u>: Answers may vary.

39. As explained on textbook p. 348, an ***acute scalene triangle*** is one in which all three angles are acute (i.e., have measures less than $90°$) and no two sides are equal in length. If no two sides are equal in length, then no two angles are equal (based on the Isosceles Triangle Property, p. 351). Here is an example of an acute scalene triangle:

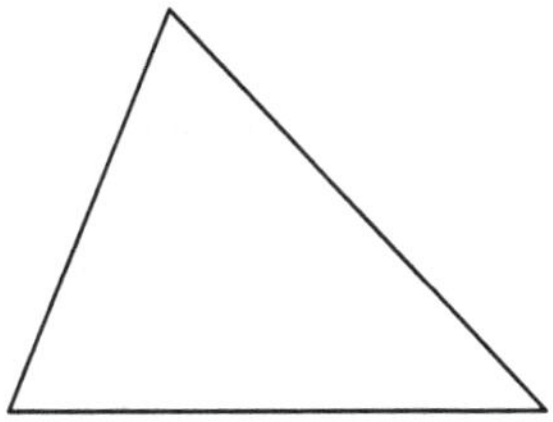

41. An ***equilateral triangle*** is one in which all three sides (and, hence, all three angles) are equal. Since the sum of the measures of the angles in any triangle is $180°$, the measure of each angle in an equilateral triangle must be $60°$. Therefore, all equilateral triangles are acute triangles. Here is an illustration:

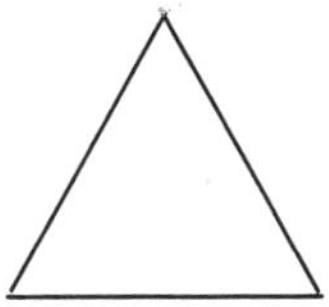

43. An ***isosceles right triangle*** is a right triangle whose legs are equal in length. Since the two legs are equal in length, the acute angles opposite them must be equal. Furthermore, since the sum of the three angles is $180°$ with one measuring $90°$, these two acute angles must each have a measure of $45°$.

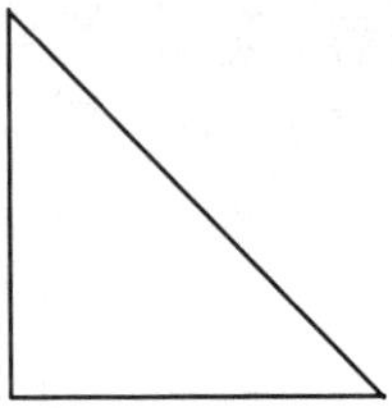

45. An ***obtuse scalene triangle*** is a triangle in which there is one obtuse angle and no equal sides.

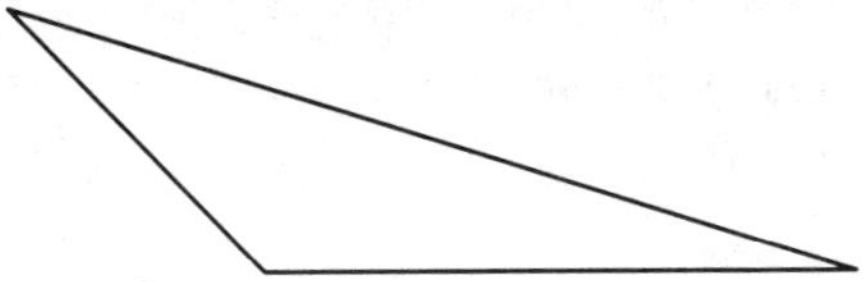

47. In the event you have forgotten how to construct congruent line segments (i.e., sides of a triangle), refer to Section 6.1 (textbook p. 330, including Figure 6.7). Like all other constructions, of course, your final sketch should be, if done correctly, an exact replica of the triangle given in Problem 25.

49. To reproduce the triangle by using ASA (angle-side-angle):

(i) Begin by choosing one of the sides of Problem 27's triangle — perhaps the vertically aligned side is easiest;

(ii) Construct a line segment congruent to the side you chose (in the same manner as shown in Section 6.1, textbook p. 330);

(iii) Move the compass to one of the two ends of the side chosen in step (i);

(iv) Viewing this endpoint as a vertex of one of the triangle's angles, construct an angle congruent to this angle (in the same manner as shown in Section 6.2, pp. 339-340);

(v) Move the compass to the other end of the side chosen in step (i);

(vi) View this endpoint as a vertex of an angle and construct an angle congruent to this angle.

Assuming that the rays constructed in steps (iv) and (vi) were drawn with adequate length, these rays should intersect, leaving you with a reproduction (via ASA) of the triangle shown in Problem 25 (p. 353).

51. **Arguments vary**. Suppose that $\angle x$ has two complements, one named $\angle y$ and the other $\angle z$. Since $\angle x$ and $\angle y$ are complements, we know (by definition of a complement, p. 341) that $m\angle x + m\angle y = 90°$. Similarly, since $\angle x$ and $\angle z$ are complements, $m\angle x + m\angle z = 90°$. Because both sums equal $90°$, they are equal to each other. That is,

$$m\angle x + m\angle y = m\angle x + m\angle z.$$

Subtracting $m\angle x$ from both sides of this equation yields

$$m\angle y = m\angle z$$

Therefore, complements of the same angle are equal.

53. a) In informal terms, $\angle COD \cup \angle BOC$ asks us to determine the result of joining together (marrying, uniting, putting together, ...) $\angle COD$ with $\angle BOC$. Together these two angles form a straight angle, namely $\boxed{\angle \mathbf{BOD}}$ (which may also be written as $\angle DOB$). *[Note: Strictly speaking, the union of ∠COD and ∠BOC includes ray OA.]*

b) Ray OB and ray OA share a common vertex (point O). Thus, the union of these two rays is $\boxed{\angle \mathbf{AOB}}$ (otherwise expressible as $\angle BOA$).

55. a) In informal terms, $\angle AOC \cap \angle AOB$ asks us to determine the intersection (overlap, common ground, shared points, ...) for these two angles. They share ray OA.

Therefore, $\angle AOC \cap \angle AOB = \boxed{\overrightarrow{OA}}$. Please recall that the order of the letters and the direction of the arrow are important when describing a ray.

b) The union of these three segments is a triangle, specifically $\boxed{\Delta \mathbf{AOB}}$. Remember, when describing a triangle, that the order of the letters is unimportant.

57. **Self-explanatory**. Although the particular triangles drawn may vary, the three-step procedure, if performed correctly, will show that the sum of the measures of the interior angles of **any** triangle is $180°$.

59. A pentagon can be separated into three non-overlapping triangles as follows:

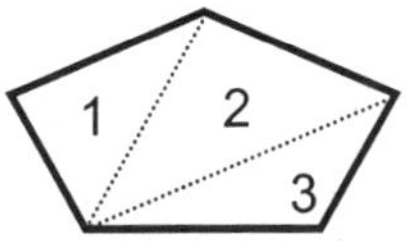

Since the sum of the measures of each triangle's angles is $180°$, the total measure of the three triangles' nine angles is $180° \times 3 = 540°$. Therefore, since the pentagon's five angles consist of the three triangles' nine angles, the sum of the measures of pentagon's five angles equals $\boxed{\mathbf{540°}}$.

Please note that, upon considering part (f) of Problem 58, one would discover that an n-sided polygon can be separated into $(n - 2)$ non-overlapping triangles. By applying inductive reasoning, we then obtain the general formula $S = 180°(n - 2)$, where S represents the sum of the measures of the angles in an n-sided polygon. So, for example, with a pentagon, we have $n = 5$ and, therefore,

$$S = 180°(5 - 2) = 180°(3) = 540°.$$

Problem Set 6.4

1. **Answers vary**. ***Congruent triangles*** have the same size and shape, whereas ***similar triangles*** have the same shape but not necessarily the same size. If two triangles are congruent, the lengths of their corresponding sides and the measures of their corresponding angles are equal. If two triangles are similar, the measures of their corresponding angles are equal and the measures of their corresponding sides are proportional. For additional information, refer to textbook page 355.

3. Since $\angle A \simeq \angle E$ and $\angle B \simeq \angle D$, the two triangles are **similar** (via the Similar Triangle Theorem, p. 355).

5. Since the sum of the measures of the angles in any triangle must equal $180°$, we know $m\angle B'' = 180° - (20° + 68°) = 92°$ and $m\angle D'' = 180° - (20° + 70°) = 90°$. Therefore, these two triangles have different shapes and, hence, are **not similar**.

7. Since $m\angle P = 30°$ and $m\angle S = 60°$, both ΔPMN and ΔRSQ are $30°$-$60°$-$90°$ triangles. Thus, they satisfy the conditions of the Similar Triangle Theorem and, accordingly, are **similar**.

9. Based upon the figures' marked angles, we know $\angle A \simeq \angle A'$ and $\angle B \simeq \angle B'$. Then, because the sum of the measures of the angles of any triangle is $180°$, we know $m\angle C = m\angle C' = 180° - (25° + 75°) = 80°$. We now summarize these findings:

m∠A = m∠A = 25°; m∠B = m∠B = 75°; m∠C = m∠C = 80°

11. **m∠A = m∠D = 38°; m∠B = m∠E = 68°; m∠C = m∠F = 74°**

13. **m∠A = m∠A = 54°; m∠B = m∠B = 36°; m∠C = m∠C = 90°**

15. The marked angles tell us that $m\angle A = m\angle A'$ and $m\angle B = m\angle B'$. That is, $\angle A$ corresponds with $\angle A'$, and $\angle B$ corresponds with $\angle B'$. By the Similar Triangle Theorem (p. 355), it follows that $\Delta ABC \sim \Delta A'B'C'$. So, $\angle C$ corresponds with $\angle C'$. Then, since $\angle A$ and $\angle A'$ are corresponding angles and since side $\overline{BC}$ is opposite $\angle A$ and side $\overline{B'C'}$ is opposite $\angle A'$, we know side $\overline{BC}$ corresponds with $\overline{B'C'}$. Since $\angle B$ and $\angle B'$ are corresponding angles and since side $\overline{AC}$ is opposite $\angle B$ and side $\overline{A'C'}$ is opposite $\angle B'$, side $\overline{AC}$ must correspond with $\overline{A'C'}$. Similarly, side $\overline{AB}$ corresponds with $\overline{A'B'}$. Since corresponding sides are proportional, we have the follow [note: other proportions are possible]:

$$\frac{|\overline{AC}|}{|\overline{A'C'}|} = \frac{|\overline{BC}|}{|\overline{B'C'}|} \qquad \frac{|\overline{AC}|}{|\overline{A'C'}|} = \frac{|\overline{AB}|}{|\overline{A'B'}|}$$

$$\frac{11}{22} = \frac{5}{|\overline{B'C'}|} \qquad \frac{11}{22} = \frac{11}{|\overline{A'B'}|}$$

$$|\overline{B'C'}| = \frac{22 \times 5}{11} = 10 \qquad |\overline{A'B'}| = 22$$

[FYI: Recall that we solved proportions like these in textbook Section 5.7. Review as necessary.]

In summary, we have:

$$\boxed{|\overline{\mathbf{BC}}| = \mathbf{5};\ |\overline{\mathbf{B'C'}}| = \mathbf{10};\ |\overline{\mathbf{AB}}| = |\overline{\mathbf{AC}}| = \mathbf{11};\ |\overline{\mathbf{A'B'}}| = |\overline{\mathbf{A'C'}}| = \mathbf{22}}$$

Note: Like many similar triangle problems, this problem can be solved more simply by observing the relationship between the triangles' corresponding sides. In this case, as revealed by the indicated lengths of sides $\overline{AC}$ and $\overline{A'C'}$, $\Delta A'B'C'$ is double the size of ΔABC.

17. By the Similar Triangle Theorem (p. 355), since $\angle D \simeq \angle G$ and $\angle E \simeq \angle H$, we can conclude that $\Delta DEF \sim \Delta GHI$. The corresponding sides and angles are:

- $\angle D$ and $\angle G$ are corresponding angles;
- $\angle E$ and $\angle H$ are corresponding angles;
- $\angle F$ and $\angle I$ are corresponding angles;
- $\overline{EF}$ (which is opposite $\angle D$) and $\overline{HI}$ (opposite $\angle G$) are corresponding sides;
- $\overline{DF}$ (which is opposite $\angle E$) and $\overline{GI}$ (opposite $\angle H$) are corresponding sides;
- $\overline{DE}$ (which is opposite $\angle F$) and $\overline{GH}$ (opposite $\angle I$) are corresponding sides;

Since corresponding sides are proportional, we have the following (although other proportions are possible):

$$\frac{|DE|}{14} = \frac{20}{16}, \text{ so } |DE| = \frac{14 \times 20}{16} = 17.5$$

$$\frac{|EF|}{12} = \frac{20}{16}, \text{ so } |EF| = \frac{12 \times 20}{16} = 15$$

In summary,

$$\boxed{|\overline{\mathbf{GH}}| = \mathbf{14};\ |\overline{\mathbf{HI}}| = \mathbf{12};\ |\overline{\mathbf{GI}}| = \mathbf{16};\ |\overline{\mathbf{DF}}| = \mathbf{20};\ |\overline{\mathbf{DE}}| = \mathbf{17.5};\ |\overline{\mathbf{EF}}| = \mathbf{15}}$$

19. By the same type of reasoning explained in our solutions to Problems 15 and 17, we know $\Delta ABC \sim \Delta A'B'C'$. Therefore, corresponding sides are proportional, and we have (specific proportions used may vary):

$$\frac{|\overline{A'B'}|}{|\overline{B'C'}|} = \frac{|\overline{AB}|}{|\overline{BC}|} \qquad \frac{|\overline{A'C'}|}{|\overline{B'C'}|} = \frac{|\overline{AC}|}{|\overline{BC}|}$$

$$\frac{|\overline{A'B'}|}{5} = \frac{10}{8} \qquad \frac{|\overline{A'C'}|}{5} = \frac{6}{8}$$

$$|\overline{A'B'}| = \frac{5 \times 10}{8} = 6.25 \qquad |\overline{A'C'}| = \frac{5 \times 6}{8} = 3.75$$

In summary, we have:

$$\boxed{|\overline{\mathbf{AC}}| = \mathbf{6};\ |\overline{\mathbf{BC}}| = \mathbf{8};\ |\overline{\mathbf{AB}}| = \mathbf{10};\ |\overline{\mathbf{A'C'}}| = \mathbf{3.75};\ |\overline{\mathbf{B'C'}}| = \mathbf{5};\ |\overline{\mathbf{A'B'}}| = \mathbf{6.25}}$$

21. By the Pythagorean theorem,

$$c = \sqrt{a^2 + b^2} = \sqrt{4^2 + 8^2} = \boxed{\sqrt{80} \text{ or } 4\sqrt{5}}$$

23. This problem is very similar to textbook Example 2 (p. 356).

$$\frac{a}{b} = \frac{a'}{b'}$$

$$\frac{4}{8} = \frac{2}{b'}$$

[Note: Other proportions are possible.]

$$b' = \frac{8 \times 2}{4}$$

$$b' = \boxed{4}$$

25. Once again, since the triangles are similar, their corresponding sides are proportional. Thus,

$$\frac{a}{c}=\frac{a'}{c'}$$

$$\frac{4}{6}=\frac{a'}{8}$$

$$a'=\frac{4\times 8}{6}$$

$$a'=\boxed{\frac{16}{3}\text{ or }5.\overline{3}}$$

27. Since $\frac{b}{c}=\frac{b'}{c'}$, we have: $\frac{b}{4}=\frac{8}{12}$. So, $b=\frac{4\times 8}{12}=\boxed{\frac{8}{3}\text{ or }2.\overline{6}}$.

29. It is important to make sure you <u>clearly</u> understand each triangle and its respective lengths. You may find it helpful to draw them separately:

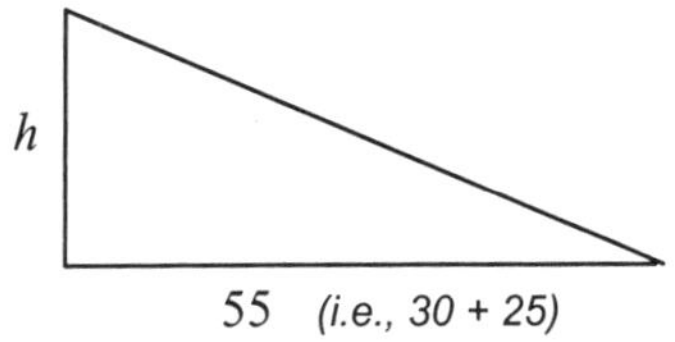

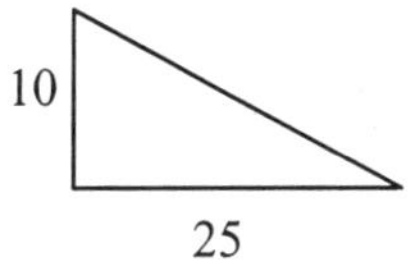

Since corresponding sides are proportional, we have

$$\frac{h}{55}=\frac{10}{25}\text{, so } h=\frac{55\times 10}{25}=\boxed{22}.$$

31. $\frac{x}{18}=\frac{10}{5+18} \Longrightarrow x=\frac{18\times 10}{23} \Longrightarrow \boxed{x\approx 7.8}$

33. We begin by drawing the two triangles separately and letting h represent the height of the larger triangle:

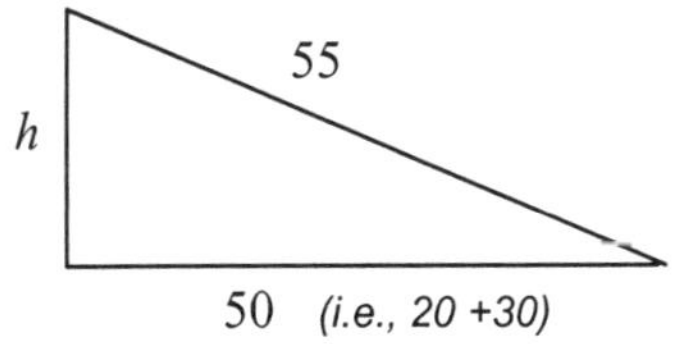

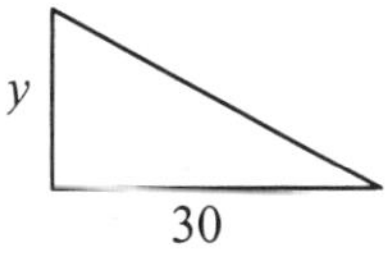

Since the triangles involved are right triangles, we can use the Pythagorean theorem to find h:

$$h^2+50^2=55^2\text{, so } h=\sqrt{55^2-50^2}=\sqrt{525}$$

Since the triangles are similar, their corresponding sides are proportional. Therefore,

$$\frac{y}{30}=\frac{\sqrt{525}}{50} \implies y=\frac{30\times\sqrt{525}}{50} \implies \boxed{y\approx 13.7}$$

35. Let us communicate our solution (proof) in two ways: (i) in terms of complete sentences and paragraphs and (ii) in terms of what we will call the two-column format (a statement in one column and its justification in the other). At this level of mathematics learning, we recommend you concentrate on the two-column format, unless your instructor deems otherwise.

(i) Paragraph format: Since $\angle BMA$ and $\angle BMC$ are both right angles (because $\overline{AC}\perp\overline{MB}$), we know $m\angle BMA = m\angle BMC$. Then, by the Isosceles Triangle Property (textbook p. 351), since ΔABC is equilateral, $m\angle A = m\angle C$.

Therefore, since two angles of ΔABM are equal to two angles of ΔCBM, we have established that $\Delta ABM \sim \Delta CBM$ (via the Similar Triangle Theorem, p. 355).

(ii) Two-Column format:

$m\angle BMA = m\angle BMC$	*Both are right angles because* $\overline{AC}\perp\overline{MB}$
$m\angle A = m\angle C$	*Isosceles Triangle Property because ΔABC is equilateral*
$\Delta ABM \sim \Delta CBM$	*Similar Triangle Theorem (p. 355)*

37. In our solution/proof to Problem 35, we communicated (for demonstration purposes) our reasoning in two formats. For Problem 37, let us instead reinforce the succinct two-column format (which is the presentation style recommended for most students), which consists of a sequence of mathematical statements and (in italics) their respective justifications:

$m\angle TCA = m\angle TCO$	*By definition of an angle bisector*
$m\angle CTA = m\angle CTO$	*By definition of an angle bisector*
$\Delta CAT \sim \Delta COT$	*Similar Triangle Theorem (p. 355)*

39. This problem is very similar to Problem 29, a problem for which this manual contains a detailed solution.

Let L = the length of the lake. Since corresponding sides of similar triangles are proportional, it follows that

$$\frac{L}{210+140}=\frac{50}{140}$$

$$L=\frac{350\times 50}{140}$$

$$L=125$$

Therefore, **the lake is 125 ft long.**

41. Let h = height of the house. So,

$$\frac{h}{6}=\frac{16}{4}$$ *[Note: Other proportions are possible.]*

$$\frac{h}{6}=4$$

$$h=24$$

Therefore, excluding the roof, **the height of the house is 24 ft.**

43. This problem is similar to textbook Example 3 (p. 357). With h representing the height of the building, it may help to begin with a diagram:

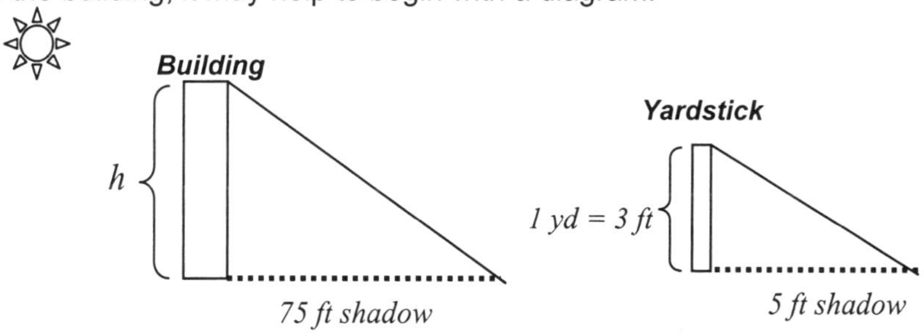

This sketch prompts us to write and solve the following proportion:

$$\frac{h}{75}=\frac{3}{5}$$ *[Note: Other proportions are possible.]*

$$h=\frac{75\times 3}{5}$$

$$h=45$$

Thus, **the building is 45 ft tall.**

45. This problem is similar to text Example 3 (p. 357) and the preceding two problems, namely Problems 43-44.

Let h = the height (in feet) of the bell tower. It is important in this problem to be cautious (and consistent) with the units chosen for the yardstick and the bell tower, respectively. Here is one approach:

bell tower ↓ *yardstick* ↓

$$\frac{h\text{ ft}}{45\text{ ft}}=\frac{36\text{ in.}}{23\text{ in.}}$$ *[Note: Other proportions are possible.]*

$$h=\frac{45\times 36}{23}$$

$$h\approx 70.43\text{ ft}$$

Therefore, to the nearest foot, **the bell tower is 70 ft tall.**

47. This problem is similar to its four predecessors (Problems 43-46) and text Example 3 (p. 357). Let t = the height (in feet) of the tree.

$$\underset{\text{tree}}{\downarrow}\qquad\underset{\text{person}}{\downarrow}$$

$$\frac{t}{10}=\frac{5}{3}$$

[Note: Other proportions are possible.]

$$t=\frac{10\times 5}{3}$$

$$t=16.\overline{6}\text{ ft}$$

Therefore, to the nearest foot, **the tree is 17 ft tall.**

49. First, we convert the given lengths to inches:

$$4\text{ ft }5\text{ in.} = 53\text{ in.};\quad 5\text{ ft }9\text{ in.} = 69\text{ in.};\quad 3\text{ ft }10\text{ in.} = 46\text{ in.}$$

Let t = the height of the tree.

$$\underset{\text{tree}}{\downarrow}\qquad\underset{\text{person}}{\downarrow}$$

$$\frac{t}{53}=\frac{69}{46}$$

[Note: Other proportions are possible.]

$$t=\frac{53\times 69}{46}$$

$$t=79.5\text{ in.}$$

$$t\approx 80\text{ in.}$$

Therefore, to the nearest inch, **the tree is 6 ft 8 in. tall.**

51. Since the triangle formed on the map and the triangle formed by the actual cities themselves are similar triangles, we use proportions to solve this problem. As noted in previous problems, remember that the actual proportions used may vary (i.e., other correct proportions are possible).

$$\frac{\overset{\text{N.O. - Denver Map}}{10.8}}{\underset{\text{N.O. - Denver Actual}}{x}}=\frac{\overset{\text{Chicago - N.O. Map}}{9.5}}{\underset{\text{Chicago - N.O. Actual}}{950}}\qquad\qquad\frac{\overset{\text{Chicago - Denver Map}}{10.2}}{\underset{\text{Chicago - Denver Actual}}{y}}=\frac{\overset{\text{Chicago - N.O. Map}}{9.5}}{\underset{\text{Chicago - N.O. Actual}}{950}}$$

$$x=\frac{10.8\times 950}{9.5}=1{,}080\qquad\qquad y=\frac{10.2\times 950}{9.5}=1{,}020$$

Therefore, the **distance between New Orleans and Denver is 1,080 mi** and the **distance between Chicago and Denver is 1,020 mi**.

53. a) We begin by identifying (labeling) two additional points. Let point *H* be the top of the 6-ft person's head. Let point *B* be the point at ground level (directly below point *T*) that the person is looking down toward. Our diagram then appears as follows:

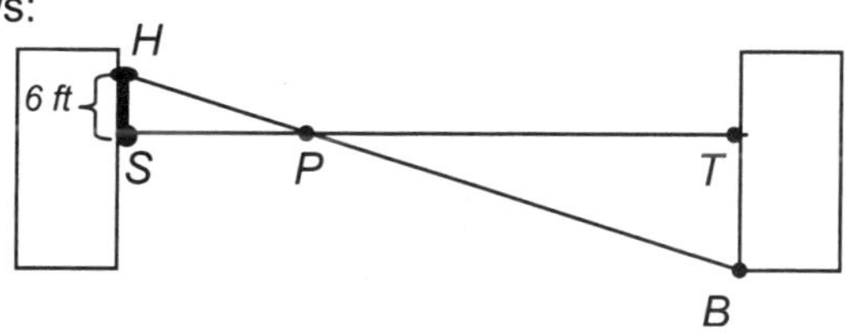

Based on this labeling, the two triangles are **ΔHSP and ΔBTP**.

b) **Yes**, the triangles are similar. $\angle PSH$ and $\angle PTB$ are both right angles, and, hence, they are equal. Since $\angle SPH$ and $\angle TPB$ are vertical angles, they, too, are equal. Therefore, via the Similar Triangle Theorem (textbook p. 355), we know the two triangles are similar.

c) In the above diagram, $\left|\overline{BT}\right|$ denotes the height of the bridge. Thus far, we have concluded that $\angle PSH$ and $\angle PTB$ are corresponding angles, and so are $\angle SPH$ and $\angle TPB$. Consequently, $\angle PHS$ must correspond to $\angle PBT$. Therefore, since $\overline{SP}$ and $\overline{PT}$ are opposite $\angle PHS$ and $\angle PBT$ respectively, $\overline{SP}$ corresponds to $\overline{PT}$. Similarly, since $\overline{HS}$ and $\overline{BT}$ are opposite vertical angles $\angle SPH$ and $\angle TPB$ respectively, $\overline{HS}$ corresponds to $\overline{BT}$. Since similar triangles' corresponding sides are proportional, we have:

$$\frac{\left|\overline{HS}\right|}{\left|\overline{SP}\right|} = \frac{\left|\overline{BT}\right|}{\left|\overline{PT}\right|}$$ *[Note: Other proportions are possible.]*

$$\frac{6}{10} = \frac{\left|\overline{BT}\right|}{35}$$

$$\left|\overline{BT}\right| = \frac{6 \times 35}{10} = 21$$

Thus, **the height of the footbridge is 21 ft.**

55. We begin by separating the diagram into three triangles:

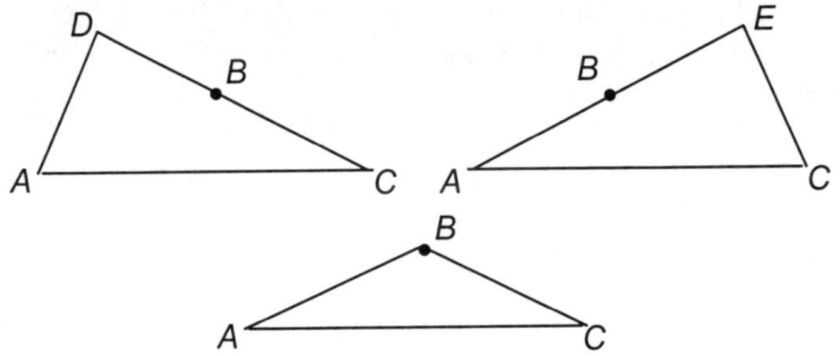

As mentioned in our solutions for Problems 35 and 37, there are typically two ways to present a formal mathematical proof: (i) in terms of complete sentences and paragraphs or (ii) in terms of the so-called two-column format (i.e., a statement in one column and its justification in the other). At this level of mathematics learning, we recommend you concentrate on the two-column format, unless your instructor deems otherwise. Here are both perspectives:

Paragraph format: Since $|\overline{AB}| = |\overline{BC}|$, ΔABC is, by definition, an isosceles triangle. Via the isosceles triangle property (textbook p. 351), we can conclude that $m\angle BAC = m\angle BCA$.

Referring to ΔCEA, we can see that $\angle BAC$ and $\angle EAC$ are the same angle. So, $m\angle BAC = m\angle EAC$. Similarly, referring to ΔADC, it is clear that $\angle BCA$ and $\angle DCA$ are the same and, hence, $m\angle BCA = m\angle DCA$. From the preceding paragraph we know $m\angle BAC = m\angle BCA$; therefore, $m\angle EAC = m\angle DCA$.

Finally, given that $m\angle D = m\angle E$ (i.e., $m\angle ADC = m\angle CEA$) and having concluded in the preceding paragraph that $m\angle EAC = m\angle DCA$, we can conclude (via the Similar Triangle Theorem, p. 355) that $\Delta ADC \sim \Delta CEA$.

Two-Column Format:

(1)	$\lvert\overline{AB}\rvert = \lvert\overline{BC}\rvert$	*Given*
(2)	$m\angle D = m\angle E$	*Given (i.e., $m\angle ADC = m\angle CEA$)*
(3)	ΔABC is isosceles	*Definition of an isosceles triangle*
(4)	$m\angle BAC = m\angle BCA$	*Isosceles Triangle Property (textbook p. 351)*
(5)	$m\angle BAC = m\angle EAC$	*Definition of an angle (i.e., $\angle BAC$ and $\angle EAC$ are the same angle)*
(6)	$m\angle BCA = m\angle DCA$	*Definition of an angle (i.e., $\angle BCA$ and $\angle DCA$ are the same angle)*
(7)	$m\angle EAC = m\angle DCA$	*Substitution Property (p. 22) [replacing line (4) with (5) and (6)]*
(8)	$\Delta ADC \sim \Delta CEA$	*Similar Triangle Theorem (p. 355)*

*[**Note**: In the Instructor's Manual, Karl Smith, the text's author, provides a different proof. If interested, ask your instructor for the details.]*

57. For purposes of this explanation, envision $\overline{CD}$ in Figure 6.49 as being the segment you are instructed to divide. Draw $\overline{AB}$ and line $\overleftrightarrow{PF}$ so that $|\overline{AF}| = 3$ units and $|\overline{BF}| = 7$ units (or, if preferred, vice versa). By having created the triangle and then having placed line $\overleftrightarrow{PF}$ so that $\overline{AB}$ is in a 3-to-7 ratio, you will have successfully divided $\overline{CD}$, the segment you intended to divide, in a 3-to-7 ratio.

59. It may help to begin with a diagram:

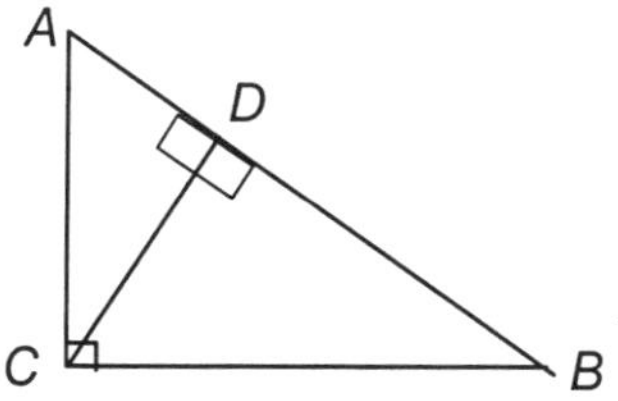

We need to show that ΔACD and ΔCBD are similar.

Since $\overline{CD} \perp \overline{AB}$, $\angle ADC$ and $\angle BDC$ are right angles; so, $m\angle ADC = m\angle BDC$. Because $\angle C$ is a right angle $m\angle ACD + m\angle BCD = 90°$. Also, since $\angle ADC$ is a right angle and since the sum of the measures of the angles in any triangle is $180°$, it follows that $m\angle ACD + m\angle A = 90°$. Hence,

$$m\angle ACD + m\angle A = 90° = m\angle ACD + m\angle BCD.$$

So, $m\angle ACD + m\angle A = m\angle ACD + m\angle BCD$. By subtracting $m\angle ACD$ from both sides of this equation, we obtain: $m\angle A = m\angle BCD$.

Therefore, via the Similar Triangle Theorem, since $m\angle ADC = m\angle BDC$ and $m\angle A = m\angle BCD$, we conclude that $\Delta ACD \sim \Delta CBD$.

[Note: You may find it easier to present your proof in two-column format (as explained and demonstrated in this manual's solutions for Problems 35, 37, and 55.]

Problem Set 6.5

1. **Wording varies**. Stated on textbook page 362, the ***Pythagorean theorem*** is a biconditional statement (a la Section 2.3) that says:

- If the two legs of a right triangle have lengths a and b, and the hypotenuse has length c, then $a^2 + b^2 = c^2$.
- Conversely, given a triangle with sides of length a, b, and c, where c is the length of the longest side, if $a^2 + b^2 = c^2$, then the triangle is a right triangle.

3. **Wording varies**. As explained on textbook page 363 (and also in our solution to Problem 13 below), for a given acute angle in a right triangle, its ***cosine*** is the ratio of the length of the side adjacent to this angle to the length of the hypotenuse.

5. **Rope A** would form a right triangle because the lengths of its sides satisfy the Pythagorean theorem (textbook p. 362); namely, since $5^2 + 12^2 = 13^2$, the triangle is a right triangle. However, in the case of Rope B, we find $2^2 + 3^2 = 13 \neq 4^2$; so Rope B does not form a right triangle.

7. **Side b** is opposite $\angle B$. As the figure below shows, angle B opens up toward side b.

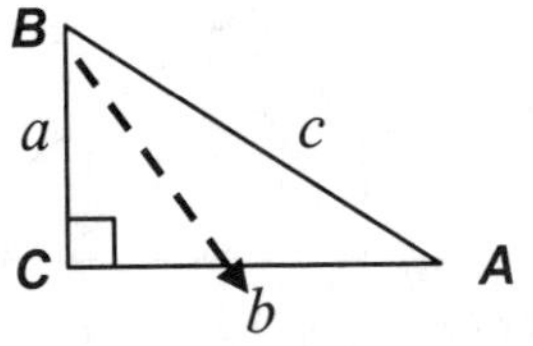

9. In ΔABC, $\angle B$ is made up of line segments $\overline{BC}$ (i.e., side a) and $\overline{BA}$ (side c). Although one who is unfamiliar with trigonometry might think of $\overline{BA}$ as being "adjacent" to $\angle B$, it is not; $\overline{BA}$ is the hypotenuse. We only consider legs of right triangles to be adjacent to particular angles.

As the figure below illustrates, **side a** is adjacent to $\angle B$.

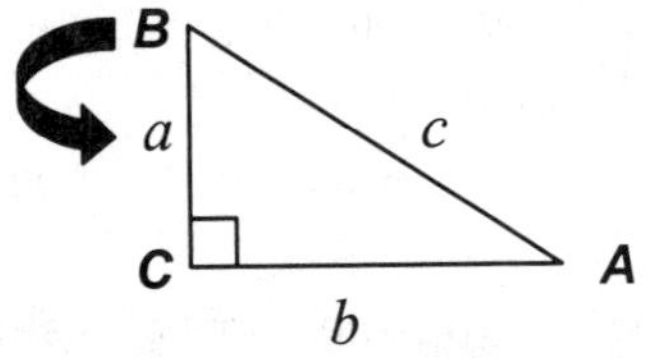

11. $\sin A = \dfrac{\text{the length of the side opposite angle } A}{\text{the length of the hypotenuse}} = \boxed{\dfrac{a}{c}}$

13. $\cos A = \dfrac{\text{the length of the side adjacent to angle } A}{\text{the length of the hypotenuse}} = \boxed{\dfrac{b}{c}}$

15. $\tan A = \dfrac{\text{the length of the side opposite angle } A}{\text{the length of the side adjacent to angle } A} = \boxed{\dfrac{a}{b}}$

Remember: (i) If you are using a calculator for Problems 17-37, make sure your calculator is set on DEGREE mode. For the details, refer to your calculator owner's manual.
(ii) If you are using a table, note that one is provided on textbook page 900 (Appendix E).

17. $\sin 56° \approx \boxed{\mathbf{0.8290}}$

19. $\sin 61° \approx \boxed{\mathbf{0.8746}}$

21. $\cos 54° \approx \boxed{\mathbf{0.5878}}$

23. $\cos 90° = \boxed{\mathbf{0}}$

25. $\tan 24° \approx \boxed{\mathbf{0.4452}}$

27. $\tan 75° \approx \boxed{\mathbf{3.7321}}$

29. $\sin^{-1} \frac{1}{2} = \boxed{\mathbf{30°}}$

31. $\tan^{-1} 1 = \boxed{\mathbf{45°}}$

33. $\tan^{-1} \sqrt{3} = \boxed{\mathbf{60°}}$

35. $\tan^{-1} 1.5 \approx 56.31° \approx \boxed{\mathbf{56°}}$

37. $\cos^{-1} 0.8 \approx 36.87° \approx \boxed{\mathbf{37°}}$

39. This problem and its successors (up through Problem 47) are similar to textbook Example 2 (p. 363).

Side $\overline{BC}$ is opposite angle A, and side $\overline{AC}$ is adjacent to angle A. By use of the Pythagorean theorem, we find the length of the hypotenuse:

$$\left|\overline{AB}\right| = \sqrt{5^2 + 12^2} = \sqrt{169} = 13$$

Recall that $\sin A = \frac{\text{Opp}}{\text{Hyp}}$, $\cos A = \frac{\text{Adj}}{\text{Hyp}}$, and $\tan A = \frac{\text{Opp}}{\text{Adj}}$.

Thus, we have: $\boxed{\sin A = \frac{12}{13};\ \cos A = \frac{5}{13};\ \tan A = \frac{12}{5}}$.

41. Side $\overline{BC}$ is opposite angle A, and side $\overline{AC}$ is adjacent to angle A. By use of the Pythagorean theorem, we find the length of leg $\overline{BC}$:

$$\left|\overline{BC}\right| = \sqrt{6^2 - 1^2} = \sqrt{35}$$

Thus, we have:

$$\boxed{\sin A = \tfrac{\sqrt{35}}{6} \approx \mathbf{0.9860};\ \cos A = \tfrac{1}{6} \approx \mathbf{0.1667};\ \tan A = \tfrac{\sqrt{35}}{1} = \sqrt{35} \approx \mathbf{5.9161}}.$$

43. Via the Pythagorean theorem, we find the length of the hypotenuse:

$$\left|\overline{AB}\right| = \sqrt{1^2 + 2^2} = \sqrt{5}$$

Hence,

$$\boxed{\sin A = \tfrac{1}{\sqrt{5}} = \tfrac{\sqrt{5}}{5} \approx \mathbf{0.4472};\ \cos A = \tfrac{2}{\sqrt{5}} = \tfrac{2\sqrt{5}}{5} \approx \mathbf{0.8944};\ \tan A = \tfrac{1}{2} = \mathbf{0.5000}}.$$

45. Via the Pythagorean theorem, we find the length of the hypotenuse:

$$|\overline{AB}| = \sqrt{1^2 + \left(\sqrt{3}\right)^2} = \sqrt{1+3} = 2$$

Thus, $\boxed{\sin A = \frac{1}{2} = \mathbf{0.5000}\text{ ; } \cos A = \frac{\sqrt{3}}{2} \approx \mathbf{0.8660}\text{ ; } \tan A = \frac{1}{\sqrt{3}} = \frac{\sqrt{3}}{3} \approx \mathbf{0.5774}}$.

47. $|\overline{AB}| = \sqrt{\left(\sqrt{2}\right)^2 + \left(\sqrt{7}\right)^2} = \sqrt{2+7} = 3$. We then find:

$$\boxed{\sin A = \frac{\sqrt{2}}{3} \approx \mathbf{0.4714}\text{ ; } \cos A = \frac{\sqrt{7}}{3} \approx \mathbf{0.8819}\text{ ; } \tan A = \frac{\sqrt{2}}{\sqrt{7}} = \frac{\sqrt{14}}{7} \approx \mathbf{0.5345}}$$

49. Letting h represent the height of the tower, we begin by creating a diagram:

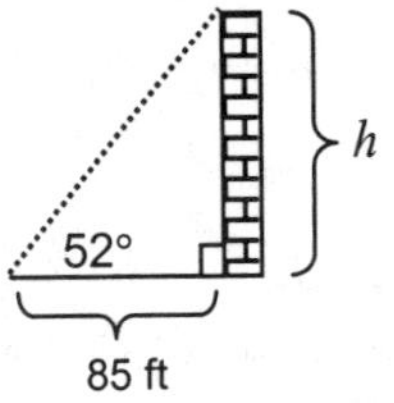

From this sketch, we see that: $\tan 52° = \frac{h}{85}$. Multiplying both sides of this equation by 85 yields: $h = 85 \tan 52°$. Finally, by using either a table (textbook Appendix E) or a calculator, we find $h \approx 108.8$ ft.

Therefore, rounded to the nearest foot, **the height of the tower is 109 ft.**

51. Recalling (from p. 366 in your text) that the angle of depression is the acute angle measured down from a <u>horizontal</u> line to the line of sight, we begin with a sketch:

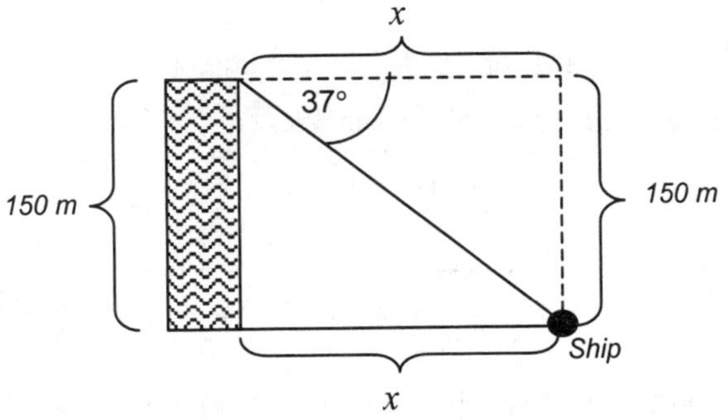

With x representing the distance from the ship to the point directly below the observer, we write the following trigonometric equation:

$$\tan 37° = \frac{150}{x}$$

Then, by using our proportion-solving skills (acquired in Section 5.7 and most recently practiced in Section 6.4), we obtain:

$$x = \frac{150}{\tan 37°}$$

Finally, we find $x \approx 199.0567$ (via calculator) or $x \approx 199.0446$ (with a table like Appendix E, p. 900). Regardless, to the nearest meter, **the distance is 199 m.**

53. Letting h represent the distance between the top of the ladder and the ground, we once again begin by drawing a picture:

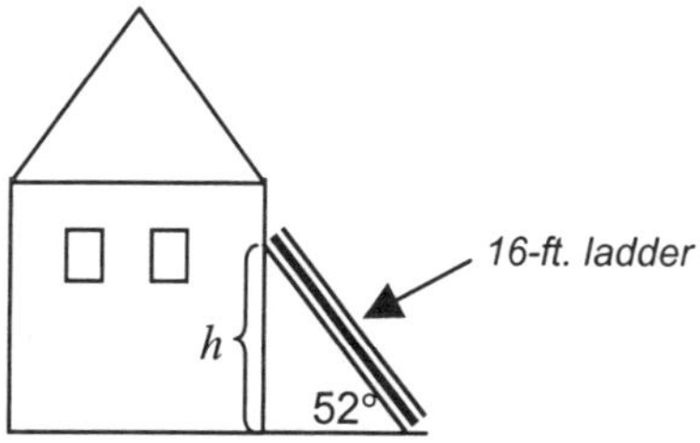

From the picture, we see that:

$$\sin 52° = \frac{h}{16} \quad \Longrightarrow \quad h = 16\sin 52° \approx 12.608\text{ ft}$$

It then follows that 12.608 ft = 12 ft + (0.608 ft × 12 in/ft) = 12 ft + 7.296 in. Therefore, to the nearest inch, **the top of the ladder is 12 ft 7 in. above the ground.**

55. Recalling (from p. 367 in your text) that the angle of elevation is measured from a horizontal line up to the line of sight, we again begin with a sketch:

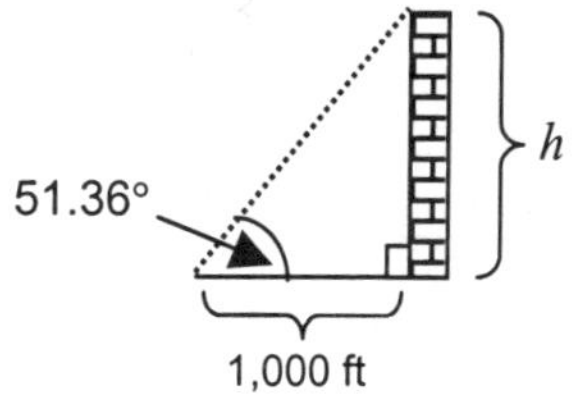

With h representing the height of the chimney stack, we write the following trigonometric equation:

$$\tan 51.36° = \frac{h}{1{,}000} \quad \Longrightarrow \quad h = 1{,}000\tan 51.36°$$

With the aid of a calculator, we find $h \approx 1{,}250.8863$. Therefore, to the nearest foot, **the chimney stack is 1,251 ft tall.**

57. Let x represent the length of the leg opposite the $30°$ angle. Given that "the length of the side adjacent to the $30°$ angle is $\sqrt{3}$ times the length of the side opposite the $30°$ angle," we have the following:

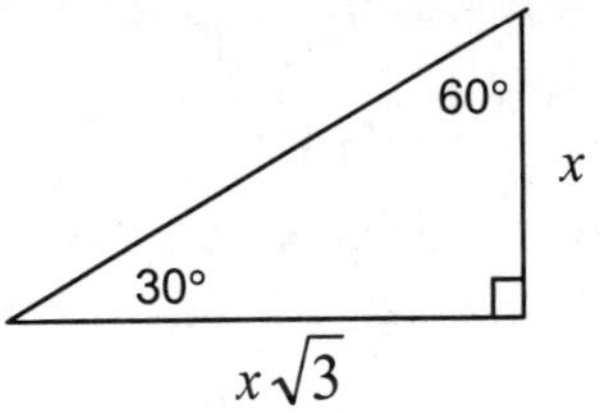

By applying the Pythagorean theorem, we can express the length of the hypotenuse (call it c) in terms of x:

$$c = \sqrt{a^2 + b^2} = \sqrt{x^2 + \left(x\sqrt{3}\right)^2} = \sqrt{x^2 + 3x^2} = \sqrt{4x^2} = 2x$$

So, the hypotenuse's length is $2x$; the length of the side adjacent to the $30°$ angle is $x\sqrt{3}$; and the length of the side opposite the $30°$ angle is x. Therefore, the exact cosine, sine, and tangent values of a $30°$ angle are as follows:

$$\cos 30° = \frac{\text{adj}}{\text{hyp}} = \frac{x\sqrt{3}}{2x} = \frac{\sqrt{3}}{2}$$

$$\sin 30° = \frac{\text{opp}}{\text{hyp}} = \frac{x}{2x} = \frac{1}{2}$$

$$\tan 30° = \frac{\text{opp}}{\text{adj}} = \frac{x}{x\sqrt{3}} = \frac{1}{\sqrt{3}} = \frac{\sqrt{3}}{3}$$

$\cos 30° = \frac{\sqrt{3}}{2}$; $\sin 30° = \frac{1}{2}$; $\tan 30° = \frac{\sqrt{3}}{3}$

59. We begin by sketching a right triangle whose hypotenuse is formed by connecting the center of the circle with a corner of the rectangle (i.e., the hypotenuse is the circle's radius):

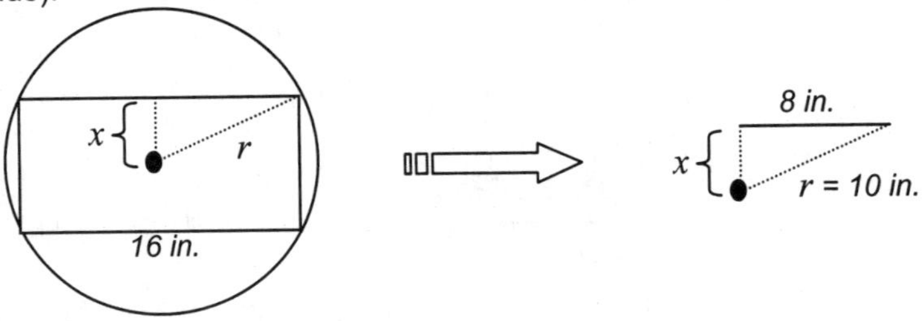

As indicated in our sketch, the length marked x (which happens to be the leg of a right triangle) represents **half** the width of the rectangle. By applying the Pythagorean theorem, we determine the value of x:

$$x = \sqrt{10^2 - 8^2} = \sqrt{36} = 6$$

Therefore, **the width of the rectangle is 12 in.**

Problem Set 6.6

1. **Wording varies**. As explained in your text (beginning on page 371), the ***golden ratio***, denoted by τ (the lower-case Greek letter tau), equals

$$\frac{1+\sqrt{5}}{2} \approx 1.618033989 .$$

With its origins dating back to the ancient Greeks, the golden ratio appears time and time again in art, architecture, nature, music, etc. Rectangles with such a length-to-width ratio have been thought by many to be the most visually pleasing, harmoniously proportioned rectangles.

3. **Wording varies**. Given a rectangle with height h and width w, the ***divine proportion*** is

$$\frac{h}{w} = \frac{w}{h+w} .$$

For additional details, refer in your text to pp. 370-371.

5. **Answers (sequences) vary**. This problem is an extension of the explanation provided on the bottom of textbook p. 371 (and continued on p. 372).

   a) Suppose we pick 6 and 8 as our two nonzero numbers. These yield the following Fibonacci-type sequence:

   $$6, 8, 14, 22, 36, 58, 94, \ldots$$

   b) We then obtain the following ratios of successive terms:

   $$\tfrac{8}{6} \approx 1.33 ;\ \tfrac{14}{8} = 1.75 ;\ \tfrac{22}{14} \approx 1.57 ;\ \tfrac{36}{22} \approx 1.64 ;\ \tfrac{58}{36} \approx 1.61 ;\ \tfrac{94}{58} \approx 1.62$$

   As you can see, these ratios oscillate around the golden ratio τ.

Please note that (although we chose two simple counting numbers, 6 and 8) you may choose — as your text's author writes — "any two nonzero numbers." This includes negative numbers, nonzero fractions, irrational numbers, and so on. We encourage you to further explore this. Be adventurous!

Consider, for example, $\sqrt{7}$ and $\sqrt{15}$. These nonzero numbers yield the following Fibonacci-type sequence: $\sqrt{7}$, $\sqrt{15}$, $\sqrt{7}+\sqrt{15}$, $\sqrt{7}+2\sqrt{15}$, $2\sqrt{7}+3\sqrt{15}$, $3\sqrt{7}+5\sqrt{15}$, and so on. With the aid of a calculator, you will find that the ratios of successive terms for this sequence also oscillate around the golden ratio. We leave the remaining details to you.

7. The first ten terms of the Lucas sequence are:

1, 3, 4, 7, 11, 18, 29, 47, 76, 123

The ratios of successive terms are:

$\frac{3}{1}=\mathbf{3}$; $\frac{4}{3}\approx\mathbf{1.33}$; $\frac{7}{4}=\mathbf{1.75}$; $\frac{11}{7}\approx\mathbf{1.57}$; $\frac{18}{11}\approx\mathbf{1.64}$;
$\frac{29}{18}\approx\mathbf{1.61}$; $\frac{47}{29}\approx\mathbf{1.62}$; $\frac{76}{47}\approx\mathbf{1.62}$; $\frac{123}{76}\approx\mathbf{1.62}$

These ratios are oscillating about (and progressively approaching the value of) the golden ratio τ.

9. The labeling of the text's accompanying illustration (located immediately below the statement of the problem) provides a valuable hint. Upon computing the various possible ratios, we find <u>two</u> that are very close to τ:

s to $\frac{1}{2}b$ $\Longrightarrow$ $\frac{612}{378}\approx 1.62$

b to h $\Longrightarrow$ $\frac{756}{481}\approx 1.57$

11. $\frac{300 \text{ yd}}{160 \text{ yd}}=$ **1.875. This is fairly close to τ.**

13. a) The dimensions (length by width) of a standard index card are **$5'' \times 3''$**.

b) $\frac{L}{W}=\frac{5}{3}=$ **1.6666... ≈ 1.67, which is very close to τ**.

15. a) The length and width of (the pages of) this textbook are **9 ¼ in. by 8 in.**.

b) $\frac{L}{W}=\frac{9.25}{8}=$ **1.15625 ≈ 1.16, which is not all that close to τ**.

17. **Answers vary** (according to one's body measurements).

19. a) There will be **2** branches in March.

b) There will be **8** branches in June.

c) Upon exploring the pattern month by month, you will find that the number of branches corresponds to the Fibonacci sequence (1, 1, 2, 3, 5, 8, ...). Therefore, after 12 months, there will be **144** branches (i.e., the 12th Fibonacci number).

d) The number of branches for each of the first 12 months respectively, from January to December, is 1, 1, 2, 3, 5, 8, 13, 21, 34, 55, 89, and 144. Hence, the sequence of ratios of successive terms is as follows:

$\frac{1}{1}=\mathbf{1}$; $\frac{2}{1}=\mathbf{2}$; $\frac{3}{2}=\mathbf{1.50}$; $\frac{5}{3}\approx\mathbf{1.667}$; $\frac{8}{5}=\mathbf{1.60}$; $\frac{13}{8}=\mathbf{1.625}$;
$\frac{21}{13}\approx\mathbf{1.615}$; $\frac{34}{21}\approx\mathbf{1.619}$; $\frac{55}{34}\approx\mathbf{1.618}$; $\frac{89}{55}\approx\mathbf{1.618}$; $\frac{144}{89}\approx\mathbf{1.618}$

As expected from both the text's explanation beginning on the bottom of page 371 and our prior experience with Problems 5 and 7, we can see that this sequence is oscillating about (and progressively converging upon the value of) the golden ratio τ.

21. In Problem 20, we discovered that the number of possible paths for n reflections is an increasing sequence that corresponds with the Fibonacci sequence. That is,

$$
\begin{aligned}
&0 \text{ reflections} \Rightarrow 1 \text{ path}\\
&1 \text{ reflection} \Rightarrow 2 \text{ paths}\\
&2 \text{ reflections} \Rightarrow 3 \text{ paths}\\
&3 \text{ reflections} \Rightarrow 5 \text{ paths}\\
&4 \text{ reflections} \Rightarrow 8 \text{ paths}\\
&\qquad\vdots\\
&n \text{ reflections} \Rightarrow \#\text{ of paths} = (n+2)^{\text{nd}} \text{ Fibonacci number}
\end{aligned}
$$

Thus, for the number of successive paths, we obtain the following sequence of ratios:

$$\boxed{\tfrac{2}{1} = \mathbf{2};\ \tfrac{3}{2} = \mathbf{1.50};\ \tfrac{5}{3} \approx \mathbf{1.67};\ \tfrac{8}{5} = \mathbf{1.60};\ \tfrac{13}{8} \approx \mathbf{1.63};\ \tfrac{21}{13} \approx \mathbf{1.62};\ \ldots}$$

Once again, we can see that these numbers are oscillating about (and progressively converging upon the value of) the golden ratio τ.

23. Unlike Problem 22, notice that it is not known whether the height is shorter than the width (or vice versa). Hence, we must consider both possibilities:

I. Assuming that the height is to be the window's <u>shorter</u> side, we have:

$$\frac{\text{Width}}{\text{Height}} = \tau$$

So, we let h represent the window's height and solve for this unknown:

$$\frac{5}{h} = \tau$$

$$h = \frac{5}{\tau} = \frac{5}{\frac{1+\sqrt{5}}{2}} = 5 \div \left(\frac{1+\sqrt{5}}{2}\right) = 5 \cdot \left(\frac{2}{1+\sqrt{5}}\right) = \frac{10}{1+\sqrt{5}} \approx \boxed{\mathbf{3.1\ ft}}$$

II. Assuming that the height is to be the window's <u>longer</u> side, we have:

$$\frac{\text{Height}}{\text{Width}} = \tau$$

Again, we let h represent the window's height and solve for this unknown:

$$\frac{h}{5} = \tau$$

$$h = 5\tau = 5\left(\frac{1+\sqrt{5}}{2}\right) = \frac{5+5\sqrt{5}}{2} \approx \boxed{\mathbf{8.1\ ft}}$$

25. Permit us to share <u>two</u> approaches.

First, recall (from textbook pp. 370-371) that the divine proportion is:

$$\frac{h}{w}=\frac{w}{h+w}$$

Given that the photograph is 9 cm high, we have

$$\frac{9}{w}=\frac{w}{9+w}$$

We then rewrite this proportion as a quadratic equation:

$$81+9w=w^2 \implies w^2-9w-81=0$$

Applying the quadratic formula, we solve for the unknown:

$$w=\frac{-(-9)\pm\sqrt{(-9)^2-4(1)(-81)}}{2(1)}=\frac{9\pm\sqrt{405}}{2}=\frac{9\pm\sqrt{81\cdot 5}}{2}=\frac{9\pm 9\sqrt{5}}{2}$$

Since w is a length, we disregard the negative value. Hence,

$$w=\frac{9+9\sqrt{5}}{2}\approx 14.56$$

Therefore, rounding (to the nearest centimeter) as directed in the statement of the problem, **the width of the photograph is 15 cm.**

Our second approach is more efficient and one that we recommend you utilize in future, related problems. On textbook p. 371, with the divine proportion, your text's author proved that when $h = 1$, we have $w = \tau$. Thus, based on our knowledge of ratios and proportions, if the divine proportion is a 1-to-τ ratio, it follows that when h equals 9 the divine proportion becomes a 9-to-9τ ratio. Therefore, given that $h = 9$ cm in this problem, it follows that:

$$w=9\tau=9\left(\frac{1+\sqrt{5}}{2}\right)=\frac{9+9\sqrt{5}}{2}\approx 14.56$$

27. Based upon the statement of the problem and its accompanying illustration, we begin by translating the librarian's ratio. In order for the length-to-width ratio to remain unchanged, we must have:

$$\frac{L}{W}=\frac{2W}{L}$$

$$\frac{L^2}{W^2}=2$$

$$\left(\frac{L}{W}\right)^2=2$$

$$\boxed{\frac{L}{W}=\sqrt{2}\approx 1.41421}$$

29. a) $x^2 - x - 1 = 0$ *Given quadratic equation*

$$x = \frac{-(-1) \pm \sqrt{(-1)^2 - 4(1)(-1)}}{2(1)}$$ *Apply quadratic formula (a = 1, b = -1, and c = –1).*

$$= \frac{1 \pm \sqrt{5}}{2}$$ *Simplify.*

b) These two numbers, τ and $\overline{\tau}$, are **negative reciprocals** of each other. That is,

$$\overline{\tau} = -\frac{1}{\tau} \text{ and } \tau = -\frac{1}{\overline{\tau}}$$

This can be supported in (at least) two ways. First, with a calculator, one could show that

$$\tau = \frac{1+\sqrt{5}}{2} \text{ equals } -\frac{1}{\overline{\tau}} = -\left(\frac{2}{1-\sqrt{5}}\right)$$

Second, we could validate (prove) this relationship algebraically:

$$-\frac{1}{\overline{\tau}} = -\left(\frac{2}{1-\sqrt{5}}\right)$$

$$= \frac{2}{\sqrt{5}-1}$$ *Distribute negative to denominator*

$$= \frac{2}{\sqrt{5}-1} \cdot \left(\frac{\sqrt{5}+1}{\sqrt{5}+1}\right)$$ *Multiply by the conjugate (see an algebra text for details).*

$$= \frac{2\left(\sqrt{5}+1\right)}{4}$$ *Multiply [Note: denominator is a difference of squares].*

$$= \frac{1+\sqrt{5}}{2}$$ *Reduce 2 and 4. Rewrite numerator (commutative prop.).*

$$= \tau$$

Problem Set 6.7

1. **Descriptions vary**. Refer to textbook pages 377-378 for details.

3. **Answers vary**. In terms of Pólya's problem-solving method, your text's author provides a fairly detailed explanation of the famed mathematician Leonhard Euler's solution. See pp. 380-381 and, as necessary, p. 378.

5. **Wording varies**. The difference between Euler circuits and Hamiltonian cycles has to do with the roles of — and, more specifically, the restrictions we place upon — vertices (also known as "points" or "nodes") and edges (paths or arcs).

For a network to be an ***Euler circuit***, it must be possible to begin at some vertex, travel on each edge exactly once, and then return to the starting vertex. The number of times each vertex is visited is inconsequential.

In contrast, to say that a network has a ***Hamiltonian cycle*** means that it is possible to visit each vertex exactly once and end at the original vertex. The number of times each edge is traversed is inconsequential.

7. No, **this network is not an Euler circuit** because it does not satisfy the conditions of Euler's Circuit Theorem (textbook page 381); namely, vertices A and D are odd.

Recall the following (from the bottom of p. 380 in your text):

> *"If there are two odd vertices, the network is traversable; one odd vertex must be a starting point and the other odd vertex must be the ending point."*

Thus, based upon vertices A and D, we know **this network is traversable.** There are four ways in which it can be traversed. Here is one of them:

$$\mathbf{D \to C \to B \to A \to D \to E \to F \to A}$$

Can you determine the other three?

9. **This network is neither an Euler circuit nor traversable.** It is not an Euler circuit because each of its four vertices has a degree of 3, and we know (from Euler's Circuit Theorem) that every vertex on a graph with an Euler circuit must have an even degree. As to why this network is not traversable, we refer you to the conclusions provided on the bottom of textbook page 380; namely,

> *"If there are more than two odd vertices, the network is not traversable."*

11. **This network is neither an Euler circuit nor traversable.** Of its 12 vertices, all but H and E are odd (specifically, with a degree of 3). Accordingly, the same reasoning as that provided in our solution to Problem 9 above also applies here.

<u>For Problems 12-17</u>: As stated on textbook page 384, determining whether a network has a Hamiltonian cycle is not as simple as determining whether it is an Euler circuit. As of this writing, Hamiltonian cycles remain one of the great unsolved problems of mathematics. So, to solve Problems 12-17, we will use trial-and-error approaches. Feel free, though, to visit www.mathnature.com if you are interested in learning more about Hamiltonian cycles and the traveling salesperson problem (TSP).

13. Yes, **this network has a Hamiltonian cycle.**

Here is one possible route (among many): $\mathbf{A \to B \to C \to D \to A}$.

15. Yes, **this network has a Hamiltonian cycle.**
Here is one possible route (among many):

$$\mathbf{A \to B \to C \to G \to H \to F \to E \to D \to A}.$$

17. Yes, **this network has a Hamiltonian cycle.**
Here is one possible route (among many): $\mathbf{A \to B \to C \to D \to A}$.

19. This problem (and, for that matter, all of Problems 18-23) are very similar to the text's Example 4 (pp. 382-383).

Viewing each room as a vertex and each door as an edge, we have 5 vertices. Three of which are odd; namely, Rooms 1, 2, and 4 have a degree of 5. The other two are even, each with a degree of 4. Since there are more than two odd vertices, we know (based on the conclusions stated on p. 380) that this floor plan is **not traversable**.

21. This floor plan consists of five rooms (i.e., vertices), the one in the middle and the four surrounding it. The middle room and the rightmost room are even (each with degree 2), while the other three rooms are odd (degree 3). Therefore, just like Problem 19, since there are more than two odd rooms, this floor plan is **not traversable**.

23. Like Problem 21, this floor plan consists of five rooms (vertices). This time, however, only two of the rooms (the leftmost and rightmost) are odd, each with a degree of 3. The remaining three rooms are all even (degree of 4).

Recall the following (from the bottom of textbook p. 380):

> *"If there are two odd vertices, the network is traversable; one odd vertex must be a starting point and the other odd vertex must be the ending point."*

Therefore, this floor plan is **traversable**. Here is one such path:

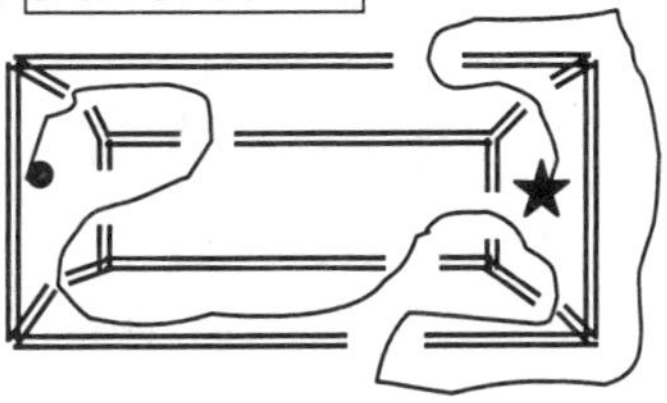

25. **This network is neither an Euler circuit nor traversable.** It is not an Euler circuit because four of its vertices (A, B, D, and E) are odd, and we know (from Euler's Circuit Theorem) that every vertex on a graph with an Euler circuit must have an even degree.

As to why this network is not traversable, we once again refer you to the conclusions provided on the bottom of textbook p. 380; namely,

> *"If there are more than two odd vertices, the network is not traversable."*

27. **This network is neither an Euler circuit nor traversable.** Since vertices A, C, E, F, G, and H are odd, the same justification as that given in our solution to Problem 25 (and other problems) applies.

29. **Not possible**. As one can confirm by exhausting all possible paths (via trial-and-error approaches), this network does not have a Hamiltonian cycle.

31. **Not possible**. Clearly, there is no way to visit vertex H without having visited vertex G at least twice — no matter whether H is the original starting/ending vertex or not.

33. **Answers may vary**. One such route is:
1→2→3→10→11→12→4→5→14→13→19→18→17→9→8→7→16→20→15→6→1

35. In a manner similar to what we did with the problems pertaining to floor plans, we begin by transforming the problem into a network. The streets represent the edges, and their intersections represent the vertices.

We then notice that there are more than two odd vertices; specifically, we have:

- the intersection of Pacific & Healdsburg is of degree 3;
- the intersection of Healdsburg, College, et al is of degree 5;
- the intersection of College and 4th Street is of degree 3; and
- the intersection of Mendocino and 4th Street is of degree 3.

Since any network with more than two odd vertices is not traversable (textbook page 380), it is **not possible** to choose a route so that all of the permitted streets are traveled exactly once.

37. **Yes**, it is possible. This "network" is traversable because it contains exactly two odd vertices, namely Manhattan and Queens. Consequently, based on the conclusions established on the bottom of textbook p. 380, we should begin at either Manhattan or Queens.

39. **Yes**, it is possible because the network formed by the streets marked in color contains exactly two odd vertices.

41. **No**, the edges of a cube are not traversable. Consider the cube pictured below:

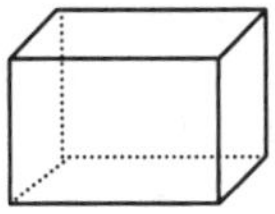

With its edges viewed as forming a three-dimensional network connecting the cube's corners, we see that all eight corners (i.e., vertices) are odd, each with a degree of 3. Since any network with more than two odd vertices is not traversable (textbook p. 380), we know the edges of the cube do not form a traversable network.

43. This problem is similar to textbook Example 6 (pp. 384-385).

a) The nearest-neighbor method requires traveling from each city to its nearest neighboring city. Given that we are starting in New York, we must first travel to Boston (a trip of 197 miles), followed by Washington, D.C. (471 miles), and then back to New York (235 miles). Therefore, the solution obtained with the nearest-neighbor method is **NYC → B → D.C. → NYC**, which has a total distance of **903 miles** (i.e., $197+471+235$).

Note that Cleveland is omitted because it is never the nearest neighbor if we start in New York City. Consequently, since this (potential) solution fails to visit all four cities, it ultimately would be disregarded.

[Correction: The distance between Cleveland and Washington, D.C. should appear in textbook Problem 42 as 375 mi. In some copies of the text, it was mistakenly printed as 712 mi.]

b) The sorted-edge method is a refinement of the nearest-neighbor method. As explained on textbook p. 384, "in the sorted-edge method, we sort the choices by selecting the nearest neighbor that does not form a loop." That is, if the nearest-neighbor method yields a route that excludes one or more cities (i.e., "a loop"), we select the next nearest neighbor that won't form a loop.

We begin by sorting the distances between the cities from smallest to largest: $197, 235, 375, 471, 481, 667$. As in part (a), we begin by traveling from NYC to Boston (197 miles) and then from Boston to Washington (471 miles). However, from Washington we ignore the shortest distance (235 miles to NYC) because it forms a loop that would exclude Cleveland. Therefore, we instead travel from Washington to Cleveland, and then finally back to NYC.

In summary, the solution obtained with the sorted-edge method is:

NYC → B → D.C. → C → NYC,

which has a total distance of **1,524 miles** (i.e., $197+471+375+481$).

45. Like Problems 42-44, this problem is similar to textbook Example 6 (pp. 384-385). The brute-force method requires listing every possible route, computing their respective total distances, and ultimately choosing the one with the shortest distance. In this case, with four cities, we only need to consider three routes:

$\text{D} \xrightarrow{1,344} \text{N.O.} \xrightarrow{677} \text{St.L.} \xrightarrow{1,844} \text{L.A.} \xrightarrow{1,062} \text{D}$ Total: 4,927 miles

$\text{D} \xrightarrow{879} \text{St.L.} \xrightarrow{677} \text{N.O.} \xrightarrow{2,009} \text{L.A.} \xrightarrow{1,062} \text{D}$ Total: 4,627 miles

$\text{D} \xrightarrow{879} \text{St.L.} \xrightarrow{1,844} \text{L.A.} \xrightarrow{2,009} \text{N.O.} \xrightarrow{1,344} \text{D}$ Total: 6,076 miles

Therefore, the best overall solution is:

$\textbf{D} \xrightarrow{879} \textbf{St.L.} \xrightarrow{677} \textbf{N.O.} \xrightarrow{2,009} \textbf{L.A.} \xrightarrow{1,062} \textbf{D}$ **Total: 4,627 miles**,

which is the same solution as was obtained with the sorted-edge method (in part b of Problem 44).

Please note that we did not need to calculate the following three possible routes:

$$D \to L.A. \to N.O. \to St.L. \to D$$
$$D \to L.A. \to St.L. \to N.O. \to D$$
$$D \to N.O. \to L.A. \to St.L. \to D$$

Except for being in reverse order, these are no different than the three routes we previously considered. Hence, their total distances were already known from our previous computations.

47. *[**Corrections**: In some copies of the text, errors appeared in the printing of this problem. First, this problem is an extension of Problem 42 (not Problem 44). Second, in the provided table, the distance between Boston and New York should appear as 197 mi (not 1,196), and the distance between Cleveland and New York should be 481 mi (not 471). Additionally, in Figure 6.73 (which accompanies Problem 42), the distance between Cleveland and Washington should appear as 375 mi (not 712).]*

By using the sorted-edge method, we find the shortest trip (starting in New York that visits each of these cities) to be:

$$\boxed{\textbf{NYC} \xrightarrow{197} \textbf{B} \xrightarrow{471} \textbf{DC} \xrightarrow{375} \textbf{C} \xrightarrow{780} \textbf{A} \xrightarrow{887} \textbf{NYC} \quad \textbf{Total: 2,710 miles}}$$

Recall that the sorted-edge method is a refinement of the nearest-neighbor method. First, sort the distances between the cities from smallest to largest: 197, 235, 375, 471, 481, 634, 667, 780, 887, $1{,}115$. Departing from New York, the shortest trip is to Boston (197 miles), so we go there. From Boston, we choose the next shortest destination (ignore "looping" to NYC, of course), which happens to be to Washington (471 miles). However, from Washington we ignore the shortest distance (235 miles to NYC) because it forms a loop that would exclude Cleveland and Atlanta. Therefore, we instead travel from Washington to Cleveland (since this distance is shorter than from Washington to Atlanta). Finally, we go from Cleveland to Atlanta, before returning back to NYC.

49. In the event you do not know what a tetrahedron is and are unable to locate one in your text, we recommend you open a dictionary before reading the rest of this solution.

A ***tetrahedron*** is a three-dimensional surface consisting of four triangular faces. Incidentally, a picture of one can be found on textbook p. 465 (within Problem 60 of Section 7.4).

Because each face is a triangle, the sum of the measures of the angles on each face is $180°$. Therefore, since there are four such faces, the sum of the measures of a tetrahedron's face angles is $4(180°) = \boxed{\mathbf{720°}}$.

51. It is important to keep in mind the restriction that Luke "always moves to an adjacent room and always in a southerly direction."

a) Let us address these four questions one at a time.

- For Luke, there is only $\boxed{\mathbf{1}}$ possible path to room 1.
- To room 2, there are $\boxed{\mathbf{2}}$ possible paths; specifically, (i) Luke can move directly to room 2 or (ii) he can move first to room 1 and then to room 2.

- To room 3, there are **3** possible paths. They are: (i) move to room 1 and then directly to room 3, (ii) move directly to room 2 and then to room 3, and (iii) move to room 1, then to room 2, and finally to room 3.
- To room 4, there are **5** possible paths. They are 1→2→3→4; 1→2→4; 1→3→4; 2→3→4; and 2→4.

b) If we continued the previous part's questions (with room 5, room 6, and so on), we would soon discover, by inductive reasoning, that the number of possible paths forms a sequence corresponding to the Fibonacci sequence. The only distinction is that the Fibonacci sequence begins with two 1's while this sequence of possible paths begins with only one '1'. In other words, the number of possible paths to room 1 can be viewed as equaling the 2nd number of the Fibonacci sequence; the number of possible paths to room 2 equals the 3rd number of the Fibonacci sequence; and so on. Hence, based upon this pattern, the number of possible paths to room 10 equals the 11th number in the Fibonacci sequence (1, 1, 2, 3, 5, 8, 13, 21, 34, 55, 89, 144, 233, 377, …). This number is **89**.

c) Based on the pattern explained in part (b), the number of possible paths to room 13 is equal to the 14th number in the Fibonacci sequence, namely **377**.

53. a) The secret is to **start at one of the two odd vertices**. One is located at the top of the pattern; the other is on the right side. All the other vertices are even.

b) Only two of this figure's 75+ vertices are even. Consequently, it is difficult to draw this figure without knowing something about networks because **the likelihood (probability, chance,…) that you will choose one of the two odd vertices at random is very small**.

55. The result is **a twisted band twice as long and half as wide as the band with which you began**.

57. The result is **two interlocking pieces, one twice as long and half as wide as the other**.

59. After giving the strip three half-twists and joining the ends together, you will find (by traversing it with your finger, a pencil, etc.) that it has **one edge and one side**.

After cutting it down the center, you will have **a single loop with a knot**.

Problem Set 6.8

1. **Answers vary**. ***Topology*** is a broad and fundamental branch of mathematics. It is concerned with discovering and analyzing essential similarities and differences among sets and figures. One important topological idea is called ***elastic motion***. For additional information, see textbook pages 390-392.

3. As explained on textbook p. 391, two geometric figures are said to be ***topologically equivalent*** if one figure can be elastically twisted, stretched, bent, shrunk, or straightened into the same shape as the other. However, one can cut the figure only if at some point the cut edges are "glued" back together again in such a way that the figure is the same as before.

Based upon this definition of "equivalent," this problem's eight figures fall into three topologically equivalent classes (with E and H in classes by themselves):

A/F; B/G; C/D

5. At the risk of sounding silly by stating the obvious, we begin with an important clarification: a ***simple closed curve*** (sometimes called a ***Jordan curve***) is a curve that is both simple and closed. A curve is said to be ***closed*** if it divides the two-dimensional plane into three disjoint subsets: (i) the set of points on the curve itself, (ii) the set of points *interior* to (i.e., "inside") the curve, and (iii) the set of points *exterior* to ("outside") the curve.

Thus, by definition, the only two simple closed curves in Problem 3 are **A and F**. In one way or another, each of the other figures fails to divide the plane into the three above-mentioned disjoint subsets.

7. One class is: **C, E, F, G, H, I, J, K, L, M, N, S, T, U, V, W, X, Y, Z**; **A, D, O, P, Q, R** form another class; **B** is in a third class.

One way to topologically classify each letter is according to its ***genus*** (i.e., its number of "holes"). This notion is explained on textbook p. 392. Notice that the letters A, D, O, P, Q, and R all have 1 hole (hence, genus 1), while the letter B has 2 holes (genus 2), and the rest of the alphabet have no holes (genus 0).

Please note that if the letter Q had a visible loop [like the one shown here: Q], then it would have belonged to the same class as the letter B.

9. **A/C/E/F; B/D; G** Like Problem 7 (see explanation above), we can classify these objects according to genus (i.e., number of holes). A drinking straw and a sewing needle both have one hole; so they are of genus 1. A funnel with a handle has 2 holes (genus 2). The remaining objects (bolt, horseshoe, brick, and pencil) have no holes (genus 0).

11. **B and C are inside** the given simple closed curve; **A is outside**.

13. Since this curve is rather complex, it may help to use a thin pointer of some sort (pencil, pen cap, etc.). Start by positioning the pointer in contact with the page somewhere outside the figure. Without ever crossing the curve itself and without ever

pulling the pointer off the page, begin to trace "through" the figure. Which of points A, B, or C are you eventually able to come in contact with? Such points are on the outside of the figure (since we started on the outside and never crossed through the curve); any points we are incapable of touching are on the inside.

Upon tracing the curve in this manner, you will find that **C is inside** the simple closed curve, while **A and B are outside**.

15. a) Regardless of where on the outside of the figure point X is positioned, line segment **$\overline{AX}$ will cross the curve an odd number of times** (up to a maximum of seven times).

$\overline{BX}$ will cross an even number of times (from zero times up to at most four).

$\overline{CX}$ will cross the curve an odd number of times.

b) **For $\overline{BX}$ and $\overline{CX}$, an odd number of times; for $\overline{AX}$, even.**

c) **For $\overline{AX}$ and $\overline{CX}$, an odd number of times; for $\overline{BX}$, even.**

d) Based on parts (a)-(c), a reasonable conjecture would be:

Given a point X exterior to a simple closed curve. If line segment $\overline{AX}$ crosses the curve an even number of times, then A is on the outside of (exterior to) the curve. If line segment $\overline{AX}$ crosses the curve an odd number of times, then A is on the inside of (interior to) the curve.

17. **Illustrations vary**. The map shown in Problem 16 is an example of a map in which each vertex is of even degree and, hence, can be colored with exactly two colors.

19. **Illustrations vary**.

21. **Answers vary**. For assistance in constructing this fractal curve, refer to textbook p. 394 and Problem 20 (p. 398).

23. **Constructions vary**. Tessellations are introduced on p. 395 in your text.

25. a) **Answers vary**. Please note that there are two possibilities, both of which the text's author, Karl Smith, illustrates in the textbook's Instructor's Manual. If interested, ask your instructor for the details.

b) **Answers vary**. There are 364 possible designs, one of which is shown in the Instructor's Manual.

27. **Mosaics vary**.

29. a) No, it will now fit the left hand.

b) Yes, they are topologically equivalent.

Problem Set 6.9

1. Wording varies. A ***non-Euclidean geometry*** is a geometry that does not assume Euclid's fifth postulate (also known as the parallel postulate). For additional information, refer to textbook pages 402-405, as well as the author's website (www.mathnature.com).

3. Answers vary. Euclidean geometry remains prevalent today because it accurately describes, depicts, and models so much of our real world.

5. Euclidean geometry. See Table 6.3 (textbook p. 405) and note that any formal proof of this statement requires the parallel postulate.

7. In a Euclidean geometry, the summit angles of a Saccheri quadrilateral are right angles. In all three of the geometries summarized in Table 6.3 (p. 405), the base angles of a Saccheri quadrilateral are, by Saccheri's definition, right angles.

9. Elliptic geometry, as explained on textbook p. 404.

11. In a hyperbolic geometry, the summit angles of a Saccheri quadrilateral are acute angles. In all three of the geometries summarized in Table 6.3 (p. 405), the base angles of a Saccheri quadrilateral are, by Saccheri's definition, right angles.

13. Yes, this is a Saccheri quadrilateral. It is identical to Figure 6.90 (p. 403), except for having been rotated 90° counterclockwise.

15. No, this is not a Saccheri quadrilateral. As noted on page 403 (see Figure 6.90), the sides of equal length in a Saccheri quadrilateral are not adjacent, whereas those marked in this figure's problem are.

17. Statement B. Refer to pp. 401-402 in your text.

19. Statement C. Statement A is absurd; despite the very high quality of this text (and, in particular, Figure 6.34), the book does not contain any groundbreaking (i.e., previously undiscovered) mathematical proofs. As for Statement B, anyone who chooses it has failed to comprehend the major ideas conveyed in textbook Section 6.9 and needs to reread this material.

21. Their sum is **greater than 180°**.

23. a) As defined on textbook p. 404, a ***great circle*** is a circle with a diameter equal to the diameter of the sphere.

b) In elliptic geometry (see p. 404 and Table 6.3 on p. 405), lines are defined as great circles. Therefore, in Figure 6.93, **line L is a line but M is not**.

c) **Yes**, it does.

25. As we stated in our solution to Problem 23, lines are defined (in elliptic geometry) as great circles. Consequently, what's wrong with this problem's contention is that M is not a great circle. So, **M is not a line**.

27. This is a **true** statement.

Let S equal the sum of the measures of the angles of a triangle in Lobachevskian geometry. The statement of the problem tells us $S < 180°$.

As previously discussed in your text (see Problem 58, p. 354), the sum of the measures of a quadrilateral can be determined by dividing the quadrilateral into two triangles and then determining the sum of the triangles' six angles. In other words, the sum of the angles in a quadrilateral equals $2S$. Since $S < 180°$, we know (by multiplying both sides of this inequality by 2) that $2S < 360°$. Therefore, the sum of the angles in a quadrilateral (i.e., $2S$) is less than $360°$.

29. Just down from the North Pole, there is a parallel that has a circumference of exactly one mile. If you begin anywhere on the circle that is a parallel one mile north of this, the conditions of Problem 26 will be satisfied.

Problem Set 6.10

1. **Wording varies**. LOGO is a computer language that is very useful in teaching concepts of geometry, especially to children.

3. **Type EDIT.** For additional information, see textbook pages 409-410.

5.

7.

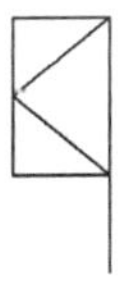

9.

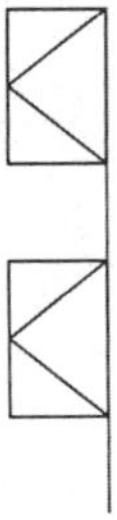

11.

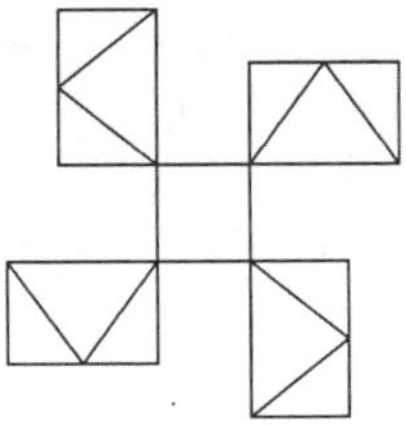

13.

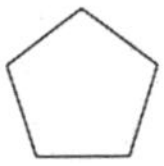

15.

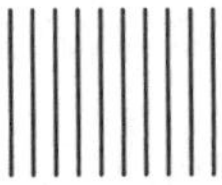

17.

```
TO SQUARE
SQUARER 100
END
```

19.

```
TO PROBLEM19
RT 30
TRIANGLER 10
TRIANGLER 20
TRIANGLER 30
TRIANGLER 40
TRIANGLER 50
END
```

21.

```
TO PROBLEM21
REPEAT 8[PROBLEM20 RT 45]
END
```

23.

```
TO PROBLEM23
REPEAT 4[PROBLEM22 LT 90]
END
```

25.

```
TO PROBLEM25
LT 45
REPEAT 4[PETAL 30 RT 90]
RT 45
REPEAT 4[PETAL 20 RT 90]
END
```

27.

```
TO PROBLEM27
CIRCLE 12
PROBLEM26 13
CIRCLE 25
PROBLEM26 30
CIRCLE 40
PROBLEM 26 53
END
```

29. **Answers vary**.

Chapter 6 Review Questions

1. Consider this depiction of the described cube:

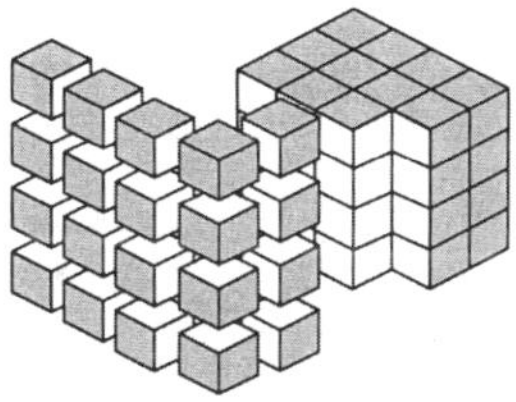

a) **No cubes** will be painted on four sides.

b) **8 cubes** will be painted on three sides, specifically those positioned on the eight corners of the original 4-cm cube.

c) Along each of the 12 edges of the original 4-cm cube, the two inner ones (i.e., those not on a corner) will be painted on two sides. Therefore, since there are 2 cubes per edge and 12 edges in all, **24 cubes** will be painted on two sides.

d) On each of the original 4-cm cube's six faces, the four middle cubes will be painted on one side. Thus, with 6 faces and 4 cubes per face painted on one side, there must be a total of **24 cubes** painted on one side.

e) Recall that there are 64 1-cm cubes in all. No one cube is painted on any more than three of its sides. Based upon our answers to parts (b) through (d), 56 (i.e., $24 + 24 + 8$) of the 1-cm cubes are painted in some way. Hence, the remaining **8 cubes** (i.e., $64 - 56$) must be painted on no sides. As the picture accompanying this solution reveals, these 8 cubes are located within the core of the original 4-cm cube.

2. a) **The flag itself (excluding the flagpole) is symmetric**, but the illustration (with the flagpole included) is not. The flag itself has two lines of symmetry, one horizontal and the other vertical. Of course, both must be drawn through the middle of the flag.

b) The flag of Nepal is **not symmetric**.

c) The Safeway logo is **not symmetric**.

d) The vair is a **symmetric** heraldic representation. Draw a vertical line through the middle of the figure.

3. Review, as necessary, textbook Section 6.2.

a) $\angle 1$ and $\angle 5$ are **corresponding angles**.

b) $\angle 5$ and $\angle 6$ are **adjacent/supplementary**.

c) $\angle 2$ and $\angle 4$ are a pair of **vertical angles**.

d) $\angle 2$ and $\angle 8$ are **alternate interior angles**.

e) $\angle 1$ and $\angle 7$ are **alternate exterior angles**.

f) $\angle 1$ and $\angle 3$ are **vertical angles**.

g) In the statement of the problem, we are given that lines **ℓ_1 and ℓ_2 are parallel**. With respect to the page on which it is printed, line **ℓ_3 is horizontal**.

4. By either using a calculator or reading a table like the one provided in Appendix E of your text, one obtains the following values (rounded to four decimal places):

a) $\sin 59° \approx$ **0.8572** b) $\tan 0° =$ **0**

c) $\cos 18° \approx$ **0.9511** d) $\tan 82° \approx$ **7.1154**

5. a) This network is **not traversable** because all (i.e., more than two) of its eight vertices are odd. Each has a degree of 3. From p. 380 of Section 6.7, you will recall that: *"If there are more than two odd vertices, the network is not traversable."*

b) Since exactly two of its vertices are odd (both with degrees of 5), this network is **traversable**. As to how to traverse it, remember the following (from p. 380): *"If there are two odd vertices, the network is traversable; one odd vertex must be a starting point and the other odd vertex must be the ending point."*

c) Two of the rooms (i.e., vertices) are odd (of degree 3); the other four rooms are even. Therefore, like part (b), these rooms are **traversable**. Start in one of the odd rooms and end in the other.

6. For additional practice with this type of problem, refer to Section 6.5 (specifically, Problems 48-56, 58-60 on pp. 369-370).

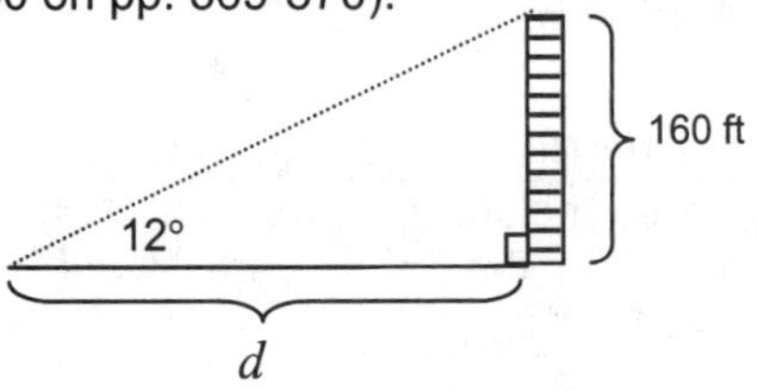

With d representing the distance from the boat to the base of the lighthouse, we apply the definition of the tangent ratio:

$$\tan 12° = \frac{160}{d}$$

Then, by using basic equation-solving skills (or the proportion-solving skills acquired in Section 5.7 and most recently practiced in Section 6.4), we obtain:

$$d = \frac{160}{\tan 12°}$$

Finally, we find $d \approx 752.7408$ (via calculator) or $d \approx 752.5870$ (with a table like Appendix E, p. 900). Regardless, to the nearest foot, **the distance is 753 ft.**

7. **Answers vary**.

a)

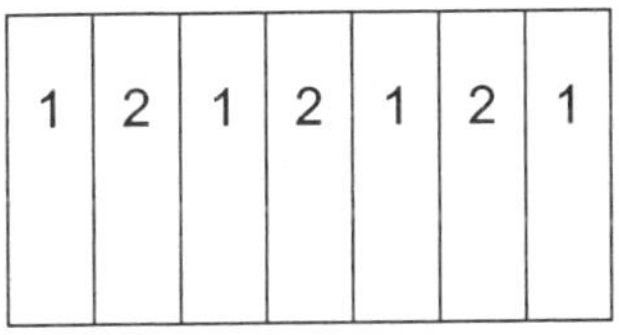

b)

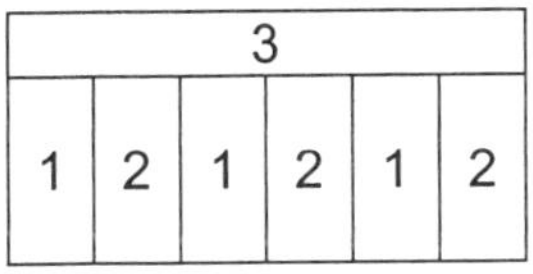

c)

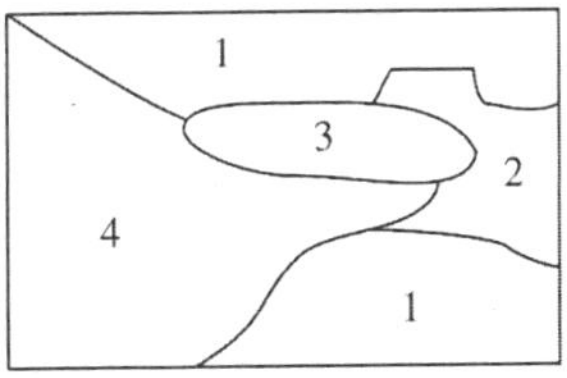

d) No such map exists.

8. After giving the strip four half-twists and joining its edges together, the resulting band has **two edges and two sides**.

By cutting the band down the center, we obtain **two interlocking loops**.

For additional related problems, go to Problems 54-59 (in Section 6.7, pp. 389-390).

9. As discussed in Section 6.9 and illustrated on p. 403 (Figure 6.90), a Saccheri quadrilateral is a quadrilateral $ABCD$ with base angles A and B right angles and with sides AC and BD with equal lengths. If summit angles C and D are right angles, then the resulting geometry is Euclidean. If the summit angles are acute, then the resulting geometric is hyperbolic. If the summit angles are obtuse, then the resulting geometry is elliptic.

10.

```
TO PENTAGON :SIDE
FD :SIDE
RT 72
FD :SIDE
RT 72
FD :SIDE
RT 72
FD :SIDE
RT 72
FD :SIDE
RT 72
END
```

TO PENTAGON: 50

CHAPTER 7: The Nature of Measurement and Growth

Problem Set 7.1

1. **Answers vary**. ***Precision*** refers to the actual act of measuring and the unit of measurement selected, whereas ***accuracy*** refers to the answer itself (in particular, the number of decimal places retained after performing any computations).

On page 424, your text's author specifies that the accuracy of an answer should not exceed the precision of the measurement. For example, suppose you desired to compute the area of a square and, in order to do so, had measured the length of one of its sides to the nearest tenth of an inch. If your measurement of the side had been 2.1 inches, you might have been tempted to report the area as $2.1^2 = 4.41$ square inches. However, since the author has stipulated that the accuracy (of the answer) is not to exceed the accuracy of the measurement (which, in this case, is one decimal place), we round the answer to one decimal place and report the square's area as 4.4 in.^{2}.

3. Please note that we expect you to "understand" these common-sense formulas. Do not blindly memorize them! With that disclaimer stated, the perimeter (P) formulas for each of the following polygons are respectively:

Square: $\boxed{P = 4s}$, where s is the length of a side

Rectangle: $\boxed{P = 2\ell + 2w \text{ or } P = 2(\ell + w)}$, where ℓ is the rectangle's length and w is its width

Equilateral Triangle: $\boxed{P = 3s}$, where s is the length of a side

Regular Pentagon: $\boxed{P = 5s}$, where s is the length of a side

Reminder: the adjectives "equilateral" and "regular" mean all sides are equal in length.

5. **Wording varies**. As is explained on textbook p. 429, the irrational number π (pi) is the ratio formed by dividing the circumference of a circle by the length of its diameter.

7. a) ______________________

b) ___

c) _____

9. **Choice A**. You may wish to use "process of elimination" here: choice B is clearly too small (since 9 cm is less than 5 in.), and choice C is too large.

11. **Choice B**. Please pay special attention to the use of the word "length" — in contrast to the word "width," which is used in Problem 12.

13. **Choice A**. Based on the answer to Problem 11, Choice B is much too small, and choice C is much too large.

15. **Choice C**. A typical "**new**" pencil is longer than 4 in. and shorter than 18 in.

17. **Choice B**. Since 1 meter is only a few inches longer than a yard, choice C is too small to be the distance around an automobile tire. Note that choice A is even smaller than choice C.

19. **Choice B**. If you chose A or C, do you think you would be able to fit a VISA credit card in a typical wallet — let alone your pants pocket? We think not; the card would be much too big.

21. **Choice C**. A "**console**" TV set suggests a larger screen size. Consequently, choices A and B are too small.

23. **Choice C**. We hope choices A and B immediately struck you as absurd. For those of you who live in more rural environments, we realize Choice C also may not be too realistic, although it still is clearly the best of the three choices.

25. **Choice C**. Again, we arrive at the answer by process of elimination. Choice A is much too large. Choice B is a reasonable height for a newborn infant but obviously not for an adult.

27. **Choice A**. Typically, ceiling heights are 8 ft or so. Choice A is closest to that height.

29. **Choice C**. If you have ever looked at a meter stick, you know it is approximately 39 inches in length, which is longer than a yard (which equals 36 in.). Therefore, since a yard is shorter than a meter, a 100-yard football field is equivalent to a length less than 100 meters.

31. **Choice A**. Continuing the above solution to Problem 29, we know a yard is shorter in length than a meter. So, 100 yards is shorter than 100 meters. Thus, a person who could run the 100-meter dash in exactly 10 seconds would be able to run 100 yards in less than 10 sec.

33. **Choice A**. The prefix milli- means one-thousandth (choice B), and centi- means one-hundredth (choice C). Refer to textbook Figure 7.3 (p. 427) for additional metric prefixes.

35. a) **3 cm** b) **3.4 cm** c) **1 in.** d) $\mathbf{1\frac{3}{8}}$ **in.**

37. There are two equivalent perspectives for determining the perimeter of a rectangle. First, using the formula $P = 2(\ell + w)$, all we need to do is add the length and width together and then double the resulting sum. That is, 4 in. + 5 in. equals 9 in.; 9 in. doubled equals 18 in. Therefore, the perimeter is **18 in.** Alternatively, you may use the equivalent formula $P = 2\ell + 2w$, which requires doubling both the length and the width and then adding the results. That is, $2(4 \text{ in.}) = 8 \text{ in.}$ and $2(5 \text{ in.}) = 10 \text{ in.}$, and the sum of 8 in. and 10 in. is 18 in.

39. $P = 3s = 3(3 \text{ dm}) =$ **9 dm**.

41. $C = 2\pi r = 2\pi(2.40) = 4.80\pi$ m. Then, by using the π-key on your calculator (or a suitable approximation if your calculator does not have a π-key), we obtain an approximate circumference of $4.80\pi \text{ m} \approx$ **15.08 m**.

43. First, we break this figure into three pieces. The original figure can be viewed as being made up of two semi-circles (on each end) and a rectangle (in the middle). The two semi-circles form one circle with diameter of length 50.00 ft. As for the rectangle, it is 120.00 ft long by 50.00 ft wide, but note that the 50-feet long sides of the rectangle are not part of the original figure. Therefore, the distance around the original figure is equal to the sum of 120.00 ft, another 120.00 ft, and the circumference of the circle [$C = \pi d = 50.00\pi \approx 157.08$ ft]. That is, $120.00 + 120.00 + 157.08 =$ **397.08 ft**.

45. The process of finding the perimeter of a parallelogram is identical to that of a rectangle. Here are two equivalent approaches:

(i) $P = 2(9 \text{ ft} + 14 \text{ ft}) = 2(23 \text{ ft}) =$ **46 ft**;

(ii) $P = 2(9 \text{ ft}) + 2(14 \text{ ft}) = 18 \text{ ft} + 28 \text{ ft} = 46 \text{ ft}$.

47. Remember: the directions for Problems 37-51 ask you to determine the "distance around" the figure. Consequently, the two circles removed from the inside of the figure's bottom section are not relevant to this question.

We begin with the top section of the figure. As one travels "around" the entire figure, the three sides of the top section are traversed; specifically, the top edge (with length 0.30 cm) and the two vertical sides (each of length 1.40 cm). Therefore, this top section contributes $0.30 \text{ cm} + 2(1.40 \text{ cm}) = \underline{3.10 \text{ cm}}$ to the overall perimeter.

As for the middle section, two vertical sides (each of 1.00 cm) and the two top edges (each of 0.30 cm) must be included. Hence, the middle section's contribution to the total perimeter is $2(1.00 \text{ cm}) + 2(0.30 \text{ cm}) = \underline{2.60 \text{ cm}}$.

Determining the bottom section's perimeter is a little more involved. We begin with the one part of this section whose dimensions are not labeled, namely the top edges. Notice that the very top of the overall figure is 0.30 cm long and the ends of the section below that are also 0.30 cm. Since the width of the very bottom section is 1.70 cm long, the top edges of this bottom section must have a combined length of $1.70 \text{ cm} - 0.30 \text{ cm} - 2(0.30 \text{ cm}) = 0.8$ cm. In other words, the length of each edge is half of 0.8 cm (i.e., 0.4 cm long). With that information obtained, we can computer the perimeter of the bottom section, which includes those two top edges (0.4 cm each), the two vertical sides (3.50 cm each), and the base of the figure (1.70 cm). Thus, the perimeter of the bottom section is $2(3.50 \text{ cm}) + 2(0.40 \text{ cm}) + 1.70$ cm, which equals $\underline{\underline{9.50 \text{ cm}}}$.

Finally, to obtain the total perimeter, we sum the perimeters of each section (which are double-underlined in the preceding paragraphs):

$$3.10 \text{ cm} + 2.60 \text{ cm} + 9.50 \text{ cm} = \boxed{\mathbf{15.20 \text{ cm}}}$$

49. This problem is similar to parts (d)-(e) of textbook Example 3 (p. 429).

As shown in the sketch below, this figure is one-fourth of a circle. The distance around the figure consists of an arc length (equivalent to one-fourth of the circle's circumference) and two 10-inch radii (plural for "radius").

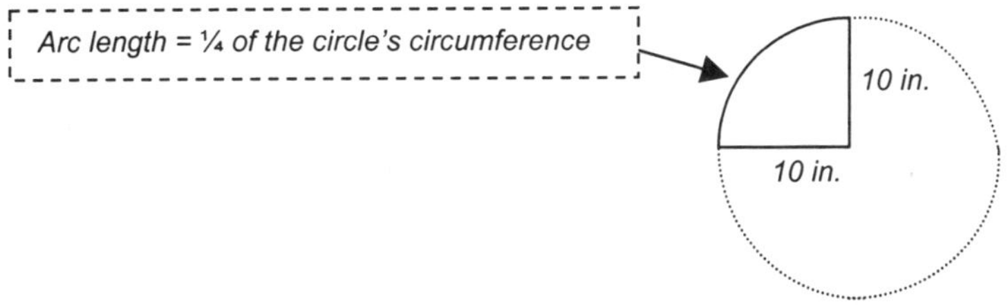

Since the circle's circumference C equals $2\pi r = 2\pi(10) = 20\pi$ inches, the length of the figure's arc equals $\frac{1}{4}C = \frac{1}{4}(20\pi) = 5\pi$ inches. So, the exact distance around the figure equals 10 in. + 10 in. + 5π in = $(20 + 5\pi)$ in. Therefore, rounded to two decimal places, the approximate distance around the figure is $\boxed{\mathbf{35.71 \text{ in.}}}$

51.

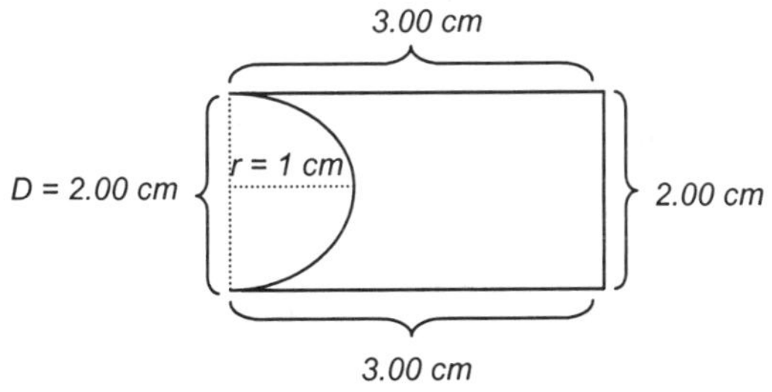

As the illustration above makes clear, the figure's curved portion is a semicircle whose diameter (d) is 2.00 cm in length. Of course, the distance around a semicircle is half of the full circle's circumference. So, since the circumference C of a circle with a diameter of d = 2.00 cm is $C = \pi d = \pi(2.00) = 2\pi$ cm, it follows that the distance around the figure's curved portion equals $\frac{1}{2}(2\pi) = \pi$ cm.

In total, the figure consists of (i) three sides of a rectangle with total length equal to 8.00 cm (i.e., 3.00 + 3.00 + 2.00) and (ii) the aforementioned semicircle. Therefore, the distance around the entire figure equals $(8.00 + \pi)$ cm, which is approximately (to the nearest hundredth) **11.14 cm**.

53. We know the perimeter formula for a rectangle is $P = 2\ell + 2w$. Given that $P = 750$ m and $w = 75$ m, we make the appropriate substitutions and solve for the length (ℓ) in the following equation:

$$\begin{aligned} 2\ell + 2w &= P \\ 2\ell + 2(75) &= 750 \\ 2\ell + 150 &= 750 \\ 2\ell &= 600 \\ \ell &= 300 \end{aligned}$$

Therefore, the length the rectangular lot is **300 meters**.

55. Perimeter $= P = 2x + 3x + 4x = 117$

$$\begin{aligned} 9x &= 117 \\ x &= 13 \end{aligned}$$

So, $|\text{BC}| = 2x = 26$; $|\text{AC}| = 3x = 39$; and $|\text{AB}| = 4x = 52$. Therefore, the lengths of the triangles sides are **26 in., 39 in., and 52 in.**

57. Answers vary. With the exception of those of us who are especially tall with unusually lengthy forearms and hands, most elbow-to-fingertip lengths are smaller than the Louvre's Egyptian cubit (52.5 cm $\approx$ 20.7 in.).

59. From Figure 7.5 (p. 432), we know that 1 cubit = 52.5 cm = 0.525 m. Thus, 300 cubits = 300 cubits $\times$ (0.525 m/cubit) = 157.50 m. Similarly, 50 cubits = 26.25 m and 30 cubits = 15.75 m. Therefore, to the nearest meter, the approximate dimensions (length by width by height) of Noah's ark are **158 m by 26 m by 16 m**.

Problem Set 7.2

1. Wording varies. As defined on textbook page 424, the number representing an object's linear dimension, as measured from end to end, is called its ***length***.

3. **Wording varies**. The area of a circle is determined by multiplying π by the square of its radius; that is, $A = \pi r^2$. Since a radius is a linear measure (measured in inches, feet, meters, miles, …) and since the computation of a circle's area requires squaring the radius, area is measured in square units (e.g., in.^{2}, ft^2, m^2, mi^2). In contrast, circumference is a linear measure determined either by multiplying π by twice the radius (i.e., $C = 2\pi r$) or by multiplying π by the diameter (i.e., $C = \pi d$).

For additional details, please refer to textbook pp. 429-430 and 437-438.

5. It is important to pay special attention to the grid upon which each of these figures is placed. The dimensions of each square within this underlying grid are 1 cm by 1 cm; that is to say, each square has an area of 1 sq cm.

a) If we consider the side positioned horizontally to be the base, it is evident (by looking at the underlying grid) that this triangle's base and height are each 3 cm. So, we have

$$A = \tfrac{1}{2}bh = \tfrac{1}{2} \times 3 \text{ cm} \times 3 \text{ cm} = 4.5 \text{ cm}^2.$$

Therefore, to the nearest square centimeter, this triangle's area is **5 cm^2**.

b) Begin by sketching this circle's horizontal (or vertical) diameter. What is the length of the diameter? Upon referring to the underlying grid, your answer should be 3 cm. So, the circle's radius must be 1.5 cm long, from which it follows that:

$$A = \pi r^2 = \pi (1.5)^2 = 2.25\pi \approx 7.0686$$

Therefore, to the nearest square centimeter, this circle's area is **7 cm^2**.

c) With the horizontally positioned side considered as the base, we see that this parallelogram's base is 4 cm in length and its height is 3 cm. So, the parallelogram's area is $b \times h = 4 \text{ cm} \times 3 \text{ cm} =$ **12 cm^2**.

7. As the following illustrates, the given figure consists of a trapezoid affixed to a triangle:

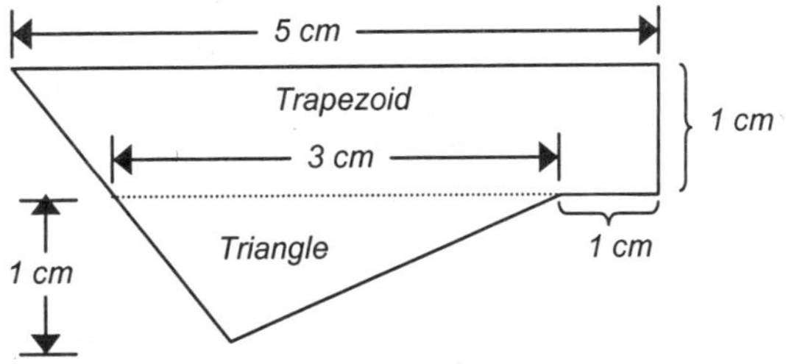

The underlying grid (provided in the text's figure) reveals the trapezoid's height (1 cm) and the lengths of its bases (5 cm and 4 cm). As for the triangle, it possesses a 3-cm base and a 1-cm height. So,

$$\begin{aligned} A_{\text{total}} &= A_{\text{trapezoid}} + A_{\text{triangle}} \\ &= \tfrac{1}{2}h(b+B) + \tfrac{1}{2}bh \\ &= \tfrac{1}{2}(1)(4+5) + \tfrac{1}{2}(3)(1) \\ &= 4.5 + 1.5 \\ &= 6.0 \end{aligned}$$

Therefore, to the nearest square centimeter, this figure's area is $\boxed{6 \text{ cm}^2}$.

9. **Choice C**. Choice A is immediately eliminated from consideration because area is never measured in centimeters; it is always measured in square units. In the event you were tempted to select choice B, take a moment to look back at the underlying grid given in your text for Problems 4c and 7. Their 5 cm by 2 cm grids each have an area of 10 cm^2; clearly, 10 cm^2 is much smaller than the area of a five-dollar bill.

11. **Choice B**. For the same reason as given above in our solution to Problem 9, we eliminate choice A because of its inappropriate units. As for Choice C, 8 cm^2 is an area that is significantly smaller than that of a credit card; Choice B is a more reasonable estimate.

13. **Choice A**. Do you know the dimensions of a typical (letter-size) sheet of notebook paper? We immediately should eliminate Choice B as being absurdly too small. Since letter-size paper is 8.5" $\times$ 11", we should estimate the area as being midway between 88 in.^2 (i.e., 8 $\times$ 11) and 99 in^2 (i.e., 9 $\times$ 11). Based upon this estimation, choice A is reasonable, while choice C is much too large.

[Note: If you are having difficulty acquiring "an intuitive feel" for square centimeters, take a look back at the underlying grids given in Problems 4 through 7 (textbook p. 440). The largest grids are shown in Problems 5c and 6, each of which has an area of 15 cm^2; the smallest grid is in Problem 4b, with an area of 4 cm^2. Needless to say, all of these grids are much smaller than a sheet of notebook paper.]

15. **Choice B**. Choice A is an area that is more reasonable for a small bedroom (e.g., a 10 ft by 10 ft room) than a classroom. Since an acre exceeds 40,000 ft^2, half an acre (which is choice C) is still much larger than the typical classroom — excluding massive university lecture auditoriums, of course.

17. **Choice C**. If you chose A or B, there are several tongue-in-cheek questions we are tempted to ask: (i) Isn't it impossible for you to find shoes — let alone socks — in your size?; (ii) You must frequently stub your toes, huh?; and (iii) Have you written the folks at the Guinness Book of World Records? We're sure they would love to talk with you.

19. $A = \ell w = (5 \text{ in.})(3 \text{ in.}) = \boxed{15 \text{ in.}^2}$, which alternatively may be written as 15 sq in. To review the concept of area (as it pertains to rectangles), please refer to textbook pages 433-434.

21. $A = \ell w = (23\text{ m})(52\text{ m}) = \boxed{\textbf{1,196 m}^2}$, which alternatively may be written as 1,196 sq m. Note that the concept of area (as it pertains to rectangles) is discussed on textbook pp. 433-434.

23. The area of this square is: $A = s^2 = (10\text{ mm})^2 = \boxed{\textbf{100 mm}^2}$. For additional information, refer to textbook pp. 434-434.

25. This problem is similar to textbook Example 3 (p. 435). The area of this particular parallelogram is: $A = bh = (120\text{ ft})(63\text{ ft}) = \boxed{\textbf{7,560 ft}^2}$.

27. As explained (and proven) on textbook pp. 435-436, the formula for the area of a triangle is $A = \frac{1}{2}bh$. So, this particular triangle's area is $\frac{1}{2}(21)(13) = \boxed{\textbf{136.5 dm}^2}$.

29. This figure is a trapezoid. From pp. 436-437 in your text, you will recall that the area formula for a trapezoid is: $A = \frac{1}{2}h(b+B)$. This particular trapezoid is such that its height (h) is 30 cm and the length of its bases (b and B) are 280 cm and 160 cm. So, this trapezoid's area is:

$$\begin{aligned} A &= \tfrac{1}{2}h(b+B) \\ &= \tfrac{1}{2}(30)(280+160) \\ &= 15(440) \\ &= \boxed{\textbf{6,600 cm}^2}. \end{aligned}$$

31. $A = \frac{1}{2}h(b+B) = \frac{1}{2}(4)(9+5) = 2(14) = \boxed{\textbf{28 in.}^2}$

33. Since its diameter has a length of 20 in., this circle's radius is 10 in. long. From textbook pp. 437-438, recall that the area formula for a circle is: $A = \pi r^2$. So, this circle's exact area is $\pi(10\text{ in.})^2 = 100\pi\text{ in.}^2$. To the nearest tenth of a square inch, this circle's approximate area is $\boxed{\textbf{314.2 in}^2}$.

35. This problem is very similar to its two predecessors, Problem 33 (see above solution) and Problem 34. Given that the length of this circle's diameter is 10 in., it follows that its radius is 5 in. long and, hence, its exact area is:

$$A = \pi r^2 = \pi(5\text{ in.})^2 = 25\pi\text{ in.}^2$$

So, to the nearest tenth of a square inch, this circle's approximate area is $\boxed{\textbf{78.5 in.}^2}$.

37. This problem is similar to your text's Example 6b (p. 438). Because only half the circle is shaded and its radius is 14 in. long, the area of this shaded region is:

$$A = \frac{\pi r^2}{2} = \frac{\pi(14)^2}{2} = \frac{196\pi}{2} = 98\pi \approx \boxed{\textbf{307.9 in.}^2}$$

39. As illustrated below, the shaded figure consists of a rectangle and an adjoining semicircle. The rectangle has a length of 3 cm and width of 2 cm. The semicircle's diameter is 2 cm long, so its radius is 1 cm long.

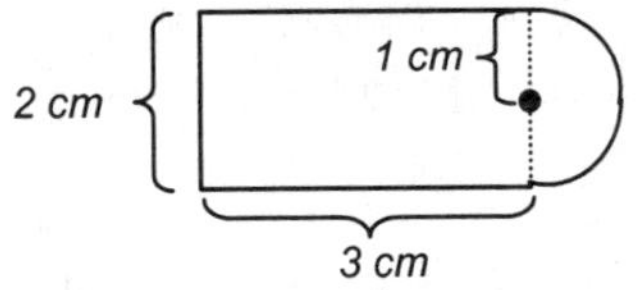

The total area of this figure equals the sum of the rectangle's area and the semicircle's area:

$$\begin{aligned} A_{\text{total}} &= A_{\text{rectangle}} + A_{\text{semicircle}} \\ &= \ell w \quad + \quad \tfrac{1}{2}\pi r^2 \\ &= (3)(2) + \tfrac{1}{2}\pi(1)^2 \\ &= 6 + \tfrac{\pi}{2} \\ &\approx 7.57079632679 \end{aligned}$$

Therefore, to the nearest tenth of a square centimeter, this figure's approximate area is **7.6 cm²**.

41. Triangular Lot A has an area of $\frac{1}{2}bh = \frac{1}{2}(135)(70) = 4{,}725$ ft². So, given a total cost of $\$13{,}500$, Lot A's cost per square foot is ($\$13{,}500 \div 4{,}725$ ft²) $\approx \$2.86$.

As for rectangular Lot B, its area is $\ell w = (160)(50) = 8{,}000$ ft². So, since its total cost is $\$25{,}500$, Lot B costs approximately $\$3.19$/ft² (i.e., $\$25{,}500 \div 8{,}000$ ft²).

Therefore, in terms of cost per square foot, **Lot A** costs less.

43. This problem is very similar to text Example 8 (p. 439). The area of this rectangular piece of property is $\ell w = (750)(1{,}290) = 967{,}500$ ft². Recalling that one acre equals $43{,}560$ ft², we convert from square feet to acres by dividing the square footage by $43{,}560$. Hence, to the nearest tenth, the acreage of this piece of property is $(967{,}500 \div 43{,}560) \approx$ **22.2 acres**.

45. Needless to say, television screens are rectangular in shape. Thus, this screen's area is $\ell w = (12)(18) =$ **216 in²**.

47. $A = \ell w = (8\tfrac{1}{2})(11) = (8.5)(11) = 93.5$ or **93 ½ in²**.

49. First, we compute the lawn's area. Since it is rectangular shaped, the lawn has an area of $\ell w = (100)(30) = 3{,}000$ ft².

Second, we determine how many pounds of seed are needed. Since 1 lb covers 150 ft^2, we need to purchase 20 lbs (i.e., 3,000 ft^2 $\div$ 150 ft^2/lb).

Finally, given that each pound of seed costs \$5.85, the 20 lbs we need to purchase costs **\$117** (i.e., 20 lbs $\times$ \$5.85/lb).

We encourage you to take a moment to consider how you would have estimated this cost if, for instance, you were standing in the garden department of a local store and did not have access to a calculator. First, knowing the dimensions of your lawn (100 ft by 30 ft), you should be quite capable of mentally multiplying to obtain the lawn's area (3,000 ft^2). Then, since you know 30 $\div$ 15, minimal effort should be required to compute 3,000 $\div$ 150, which reveals that 20 pounds of seed are needed. Finally, given that we are computing an estimate, most of us would think of a seed price of \$5.85 per pound as roughly equaling \$6; others might use \$5. So, needing 20 lbs at a cost slightly less than \$6/lb, we would anticipate spending slightly less than \$120 (i.e., 20 $\times$ 6), which is consistent with the \$117 exact cost computed above.

51. Notice that Heron's (or Hero's) formula requires that we first compute s, which is sometimes referred to as the semi-perimeter. In this case, we have

$$s = 0.5(a+b+c) = 0.5(5+8+10) = 0.5(23) = 11.5\text{ ft}$$

Then, with the aid of a calculator and Heron's formula, we obtain the area:

$$\begin{aligned} A &= \sqrt{s(s-a)(s-b)(s-c)} \\ &= \sqrt{11.5(11.5-5)(11.5-8)(11.5-10)} \\ &= \sqrt{11.5(6.5)(3.5)(1.5)} \\ &\approx 19.8100353357 \end{aligned}$$

So, to the nearest square foot, this triangle's area is **20 ft^2**.

53. This figure consists of a rectangle and an adjoining triangle. To compute this figure's total area, we need to add the area of the rectangle to the area of the triangle. The area of the rectangle is $\ell w = (240\text{ ft})(180\text{ ft}) = 43{,}200\text{ ft}^2$. To obtain the area of the triangle, we use Heron's formula:

$$\begin{aligned} s &= 0.5(a+b+c) = 0.5(130+150+180) = 0.5(460) = 230\text{ ft} \\ A &= \sqrt{s(s-a)(s-b)(s-c)} \\ &= \sqrt{230(230-130)(230-150)(230-180)} \\ &= \sqrt{230(100)(80)(50)} \\ &\approx 9{,}591.66304663 \end{aligned}$$

So, to the nearest square foot, the total area is $43{,}200 + 9{,}592 =$ **52,792 ft^2**.

55. There are several ways to view the pattern that exists between the area of a regular octagon and the length of its side. In this solution, we will share two views:

I. Take note of the representative triangle drawn within the octagon and observe that each octagon consists of 8 of these (identical) triangles. Said differently, the area of one of these triangles equals one-eighth the area of the octagon. So, perhaps we can discover the underlying pattern by considering the areas of these triangles:

Length of side	*Octagon's Area*	*Triangle's Area (Octagon's Area ÷ 8)*	*Pattern for Triangle's Area*
2	$8(1+\sqrt{2})$	$1(1+\sqrt{2})$	$1^2(1+\sqrt{2})$
4	$32(1+\sqrt{2})$	$4(1+\sqrt{2})$	$2^2(1+\sqrt{2})$
6	$72(1+\sqrt{2})$	$9(1+\sqrt{2})$	$3^2(1+\sqrt{2})$
⋮	⋮	⋮	⋮
s			$\left(\frac{s}{2}\right)^2(1+\sqrt{2})$

The third column from the left reveals the pattern. The numbers $1, 4,$ and 9 are numbers you should recognize: they are perfect squares (respectively, the squares of $1, 2,$ and 3). As to their relationship to the length of each octagon's side, we notice that $1, 2,$ and 3 are respectively half of $2, 4,$ and 6.

Based upon this pattern, we infer that a regular octagon 10 units long on a side contains eight triangles, with each triangle having area

$$\left(\frac{10}{2}\right)^2(1+\sqrt{2}) = (5)^2(1+\sqrt{2}) = 25(1+\sqrt{2}).$$

So, the area of the octagon must equal $8\times 25(1+\sqrt{2})$, or $200(1+\sqrt{2})$, which corresponds to **Choice C**.

II. By writing $8, 32,$ and 72 in terms of their prime factorizations, the pattern becomes apparent:

Length of side	*Octagon's Area*	*Pattern*
2	$8(1+\sqrt{2})$	$2\cdot 2\cdot 2\cdot \mathbf{1}\cdot\mathbf{1}(1+\sqrt{2})$
4	$32(1+\sqrt{2})$	$2\cdot 2\cdot 2\cdot \mathbf{2}\cdot\mathbf{2}(1+\sqrt{2})$
6	$72(1+\sqrt{2})$	$2\cdot 2\cdot 2\cdot \mathbf{3}\cdot\mathbf{3}(1+\sqrt{2})$
8	⋮	$2\cdot 2\cdot 2\cdot \mathbf{4}\cdot\mathbf{4}(1+\sqrt{2})$
10	⋮	$2\cdot 2\cdot 2\cdot \mathbf{5}\cdot\mathbf{5}(1+\sqrt{2})$

So, given an octagon with a side equal to 10, this pattern leads us to conjecture that its area is $2 \cdot 2 \cdot 2 \cdot 5 \cdot 5\,(1+\sqrt{2})$, or $200(1+\sqrt{2})$, which again corresponds to Choice C.

57. Just like its predecessors (Problems 54-56), there are several ways to view the pattern that exists between the area A and the distance W. In this solution, we share one perspective and hope you (and your classmates) will uncover others. By writing each value of A as a multiple of 3, we discover the following:

W	A	*Pattern*
3	3	$1(3) = \mathbf{1}^2(3)$
6	12	$4(3) = \mathbf{2}^2(3)$
9	27	$9(3) = \mathbf{3}^2(3)$
12		$\mathbf{4}^2(3)$

Based upon this pattern, we would infer that when $W = 12$, the area A of a figure similar to those depicted will equal $\underline{4}^2(3) = 16(3) = 48$ square units. So, we select **Choice D**.

59. Notice that the shaded region is made up of 4 identical shapes, each of which is a quarter circle. If these four quarter circles were spliced together, the resulting would be one complete circle. Since the square containing them is 10 inches on a side, the radius of this circle would be half that, namely 5 inches. Therefore, the area of the shaded region is $\pi r^2 = \pi(5)^2 = 25\pi \approx 78.5398$, which (to the nearest square inch) is approximately **79 in^2**.

Problem Set 7.3

1. **Answers vary**. ***Length*** is a one-dimensional measurement (expressed in linear units, such as in., ft, km, etc.) that provides an end-to-end measurement of an object. ***Area*** is a two-dimensional measurement (in.^{2}, ft^2, km^2, etc.) that quantifies the amount of space enclosed by a plane figure. ***Volume*** is a three-dimensional measurement (in.^{3}, ft^3, km^3, etc.) that expresses the number of cubic units enclosed within a solid figure.

3. **Answers vary**. ***Volume*** is the number of cubic units a solid object contains, whereas ***capacity*** the amount of liquid a container will hold.

5. **Answers vary**. As mentioned on textbook page 448 (see Figure 7.13), 1 liter is equivalent to 1.06 quarts (i.e., 1 qt equals 0.95 L). So, comparatively, a quart is a bit smaller than a liter; or, conversely, a liter is slightly larger than a quart.

7. This problem is somewhat similar to textbook Examples 1-2 (pp. 443-444). In this problem, we have a "topless" cube; so, there are five faces (a front, a back, two sides, and a bottom). Each face has an area of $(25 \text{ cm})^2 = 625 \text{ cm}^2$. Therefore, the cube's surface area is (5 faces) × (625 cm^2 per face) = **$3{,}125 \text{ cm}^2$**.

9. To find this rectangular box's surface area, we sum the areas of its six faces:

Top:	$10 \times 25 = 250 \text{ cm}^2$	
Bottom:	$10 \times 25 = 250 \text{ cm}^2$	*(same as top)*
Front:	$10 \times 4 = 40 \text{ cm}^2$	
Back:	$10 \times 4 = 40 \text{ cm}^2$	*(same as front)*
Side:	$25 \times 4 = 100 \text{ cm}^2$	
Side:	$25 \times 4 = 100 \text{ cm}^2$	*(obviously, the sides are the same size)*

Surface Area equals **780 cm^2**

In the spirit of Pólya's problem-solving procedure, we encourage you to take a moment to "look back." That is, can you think of any ways — or even formulas you could devise — that would help you solve this problem more efficiently? Please note that your text's author provides such formulas in Section 7.4.

11. This rectangular box (also known as a <u>*parallelepiped*</u>) does not have a top. The front and back faces of the box are both squares, each with an area of $2{,}500 \text{ cm}^2$ (i.e., 50×50). The box's bottom face and its two side faces each are 200 cm by 50 cm; consequently, each has an area of $10{,}000 \text{ cm}^2$ (i.e., 50×200). Therefore, the surface area of this box is:

$$\underbrace{(\text{Area of front \& back})}_{(2 \times 2{,}500)} + \underbrace{(\text{Area of 2 sides \& bottom})}_{(3 \times 10{,}000)}$$

which equals **$35{,}000 \text{ cm}^2$**.

13. This problem is very similar to textbook Example 2b (p. 444). In addition to referring to this example, we suggest you actually examine how tin cans are constructed. In the event you do not have access to a can at this moment, take a sheet of paper, roll it up into a cylinder and take some a moment to study its surface area. You will find that most cans have a seam along the side. After cutting along the seam and unrolling the can, it becomes apparent that the cylindrical-shaped part of the can is nothing more than a rolled-up rectangle whose top/bottom edges have a length equal to the circumference of the cylinder and whose side edges have a length equal to the height of the cylinder.

So, to determine the surface area of this particular cylindrical can (which has a height $h = 8$ cm and radius $r = 2$ cm), we need to add two areas together: (i) the area of the circle that serves as the bottom of the can, and (ii) the area of the "rolled up" rectangle that forms the sides of the can. That is,

Side:	$A = \ell w = (2\pi r)(h) = 2\pi(2)(8) = 32\pi$
Bottom:	$A = \pi r^2 = \pi(2)^2 = 4\pi$
Surface Area:	$32\pi + 4\pi = \boxed{36\pi \text{ cm}^2 \approx 113.1 \text{ cm}^2}$

15. For additional details, please refer to textbook Example 2b (p. 444) and this manual's solution to Problem 13 (see above). Unlike those two problems, though, this one has a "units disagreement"; namely, the radius is given in inches, while the height is in feet. So, we need consistency; here are two ways to achieve it:

I. Convert the radius from inches to feet (i.e., 6 in. $= 0.5$ ft).

Side:	$A = \ell w = (2\pi r)(h) = 2\pi(0.5)(1) = \pi$
Bottom:	$A = \pi r^2 = \pi(0.5)^2 = 0.25\pi$
Surface Area:	$\pi + 0.25\pi = \boxed{1.25\pi \text{ ft}^2 \approx 3.9 \text{ ft}^2}$

II. Convert the height from feet to inches (i.e., 1 ft $= 12$ in.).

Side:	$A = \ell w = (2\pi r)(h) = 2\pi(6)(12) = 144\pi$
Bottom:	$A = \pi r^2 = \pi(6)^2 = 36\pi$
Surface Area:	$144\pi + 36\pi = \boxed{180\pi \text{ in.}^2 \approx 565.5 \text{ in.}^2}$

17. This triangular prism consists of five faces. Its top and bottom are triangular, while its other three faces are rectangular. So, to find this prism's surface area, we sum the areas of its five faces:

Top Triangle:	$\frac{1}{2}bh = \frac{1}{2}(2 \text{ in.})(1.25 \text{ in.})$	$= 1.25 \text{ in.}^2$
Bottom Triangle:	$\frac{1}{2}bh = \frac{1}{2}(2 \text{ in.})(1.25 \text{ in.})$	$= 1.25 \text{ in.}^2$ *(same as top)*
Front Rectangle:	$\ell w = (2 \text{ in.})(1 \text{ in.})$	$= 2.00 \text{ in.}^2$
Back-Right Rectangle:	$\ell w = (1.35 \text{ in.})(1 \text{ in.})$	$= 1.35 \text{ in.}^2$
Back-Left Rectangle:	$\ell w = (1.95 \text{ in.})(1 \text{ in.})$	$= 1.95 \text{ in.}^2$

Surface Area equals $\boxed{7.80 \text{ in.}^2}$

19. Problems 19-20 are patterned after text Example 4 (p. 446).
Here are two perspectives:

I. There are $5 \times 3 = 15$ cubic centimeters on the bottom layer of cubes. Since there are 4 layers, the total number of cubes is $4 \times 15 = 60$. Hence, the volume is $\boxed{60 \text{ cm}^3}$.

II. Given a length of 5 cm, width of 3 cm, and height of 4 cm, it follows that the volume of this particular rectangular solid is

$$V = \ell \times w \times h = (5 \times 3 \times 4) \text{ cm}^3 = 60 \text{ cm}^3.$$

21. $V = s^3 = (5 \text{ ft})^3 = \boxed{125 \text{ ft}^3}$ (or 125 cu ft). To review the concept of volume (particularly as its applies to cubes), see textbook pp. 445-447.

23. $V = s^3 = (20 \text{ cm})^3 = \boxed{8{,}000 \text{ cm}^3}$ (or 8,000 cu cm)

25. $V = \ell wh = (3 \text{ ft})(2 \text{ ft})(4 \text{ ft}) = \boxed{24 \text{ ft}^3}$ (or 24 cu ft). The volume of a rectangular box (also known as a *parallelepiped*) is explained on textbook pp. 446-447.

27. $V = \ell wh = (40 \text{ cm})(30 \text{ cm})(80 \text{ cm}) = \boxed{96{,}000 \text{ cm}^3}$ (or 96,000 cu cm).

29. a) **2 c** b) **16 oz**

31. a) **13 oz** b) **380 ml**

33. a) **25 ml** b) **75 ml** c) **70 ml** *[Note: Each small mark represents 5 ml.]*

35. **Choice B**. Choice A is absurd; 5 ml would hardly be enough to paint this page, let alone some small bookshelves. Choice C is too much; 5 L is significantly more than a gallon. Most of us would probably buy a quart of paint for such a small-scale painting project.

37. **Choice A**. Choice C is ridiculously too large. Choice B might be a reasonable capacity for a container of cough medicine, but not for a single dose.

39. **Choice C**. We know a gallon equals four quarts, a quart approximately amounts to a liter, and a liter is equivalent to 1000 ml. Based upon this knowledge, we immediately disregard choice B, noting that 200 ml is less than a full cup. As for choice A, 15 L is approximately 4 gallons, and certainly automobile gas tanks hold more gas than that. Thus, we select choice C and note that 70 liters is approximately equivalent to 18 gallons.

41. **Choice B**. 50 kl equals 50,000 L. Since 4 liters is "roughly" equivalent to one gallon, 50,000 L significantly exceeds 10,000 gal. The capacity of water needed for a bath is substantially less than 10,000+ gallons; hence, we eliminate choice A. Similarly, we disregard choice C, because the potable water demands of a large city far exceed 10,000+ gallons.

43. **Choice B**. Please note that your text's author provides a user-friendly table summarizing metric prefixes (see Table 7.3 in Section 7.4, p. 462). You may find his memory aid helpful, or you may want to create your own acronym.

45. This problem is similar to textbook Example 8b (p. 451). We first compute this cube's volume: $V = s^3 = (10 \text{ in.})^3 = 1{,}000 \text{ in.}^3$. Then, since each 231 in.^3 in volume equates to a capacity of 1 gallon, we divide $1{,}000$ by 231 to obtain the cube's capacity, which (to the nearest tenth) equals **4.3 gal**.

47. For additional details, refer to textbook Example 8a (p. 451).

$$V = \ell wh = (10\text{ cm})(10\text{ cm})(25\text{ cm}) = 2{,}500\text{ cm}^3.$$

Since each $1{,}000\text{ cm}^3$ in volume equates to 1 liter of capacity, a volume of $2{,}500\text{ cm}^3$ is equivalent to a capacity of **2.5 L** (i.e., $2{,}500\text{ cm}^3 \div 1{,}000\text{ cm}^3/\text{L}$).

49. $V = \ell wh = (20\text{ cm})(15\text{ cm})(30\text{ cm}) = 9{,}000\text{ cm}^3$. Since each $1{,}000\text{ cm}^3$ in volume equates to 1 liter of capacity, this benzene container has a capacity of

$$9{,}000\text{ cm}^3 \div 1{,}000\text{ cm}^3/\text{L} = \mathbf{9\ L}.$$

51. $V = \ell wh = (50\text{ cm})(200\text{ cm})(50\text{ cm}) = 500{,}000\text{ cm}^3$. Because each $1{,}000\text{ cm}^3$ in volume is equivalent to 1 liter of capacity, this container has a capacity of

$$500{,}000\text{ cm}^3 \div 1{,}000\text{ cm}^3/\text{L} = \mathbf{500\ L}.$$

53. a) Permit us to share two approaches:

I. We first determine the refrigerator's volume in cubic inches, and then we convert cubic inches to cubic feet. That is,

$$V = \ell wh = (36\text{ in.})(33\text{ in.})(66\text{ in.}) = 78{,}408\text{ in.}^3.$$

Since $1\text{ ft}^3 = 1{,}728\text{ in}^3$ (see Figure 7.15, p. 450), the refrigerator's volume (in cubic feet) is

$$78{,}408\text{ in.}^3 \div 1{,}728\text{ in.}^3/\text{ft}^3 = \mathbf{45.375\ ft^3}.$$

II. Alternatively, we first convert each of the refrigerator's dimensions from inches to feet and then we determine its volume. That is, since 1 ft is equivalent to 12 in., it follows that

$$36\text{ in.} = 3\text{ ft}; 33\text{ in.} = 2.75\text{ ft.}; \text{and } 66\text{ in.} = 5.5\text{ ft.}$$

So, the refrigerator's volume is

$$V = \ell wh = (3\text{ ft})(2.75\text{ ft})(5.5\text{ ft}) = 45.375\text{ ft}^3.$$

b) If the <u>entire</u> refrigerator (motor, insulation, doors, storage space, etc.) has a volume of 45.375 ft^3 and if 19 ft^3 is the actual storage space <u>inside</u>, the difference between these two volumes reveals how much space is consumed by the motor, insulation, and so on: $45.375\text{ ft}^3 - 19\text{ ft}^3 =$ **26.375 ft³**.

55. A concrete sidewalk is a solid object. Therefore, to find how much concrete is required, we need to determine the sidewalk's <u>volume</u> (which is, quoting from textbook p. 445, "the amount of space inside a solid object").

Note: (i) two of the dimensions are given in feet and one in inches, and (ii) the directions ask for the volume "to the nearest ½ cubic yard." So, we need first to rectify this units conflict and then ultimately express our answer in terms of yd^3. There are several ways to do this, but let us share one.

Recalling that 1 yd = 3 ft = 36 in., we begin by converting each of the dimensions to yards:

$$50 \text{ ft} = 16\tfrac{2}{3} \text{ yd}; \; 4 \text{ ft} = 1\tfrac{1}{3} \text{ yd}; \text{ and } 4 \text{ in.} = \tfrac{4}{36} \text{ yd} = \tfrac{1}{9} \text{ yd}.$$

Hence, the sidewalk's volume is

$$V = \ell wh = (16\tfrac{2}{3} \text{ yd})(1\tfrac{1}{3} \text{ yd})(\tfrac{1}{9} \text{ yd}) = (\tfrac{50}{3} \times \tfrac{4}{3} \times \tfrac{1}{9}) \text{ yd}^3 = \tfrac{200}{81} \text{ yd}^3 \approx 2.469 \text{ yd}^3,$$

which equals (to the nearest ½ cubic yard) $\boxed{\textbf{2.5 yd}^3}$.

57. The question posed (i.e., "How much water…?"), indicates that we need to find the swimming pool's capacity. We know 1 m = 100 cm, from which it follows that

$$\mathbf{1\ m^3} = (1 \text{ m})^3 = (100 \text{ cm})^3 = 100^3 \text{ cm}^3 = \mathbf{1{,}000{,}000\ cm^3}.$$

The volume of a 7-m by 8-m by 2-m swimming pool is 112 m^3 (i.e., $V = \ell wh$). Since 1 m^3 equals 1,000,000 cm^3, 112 m^3 is equivalent to 112,000,000 cm^3. Then, by recalling that 1,000 cm^3 = 1 L and 1,000 L = 1 kl, we find:

$$112{,}000{,}000 \text{ cm}^3 = 112{,}000 \text{ L} = 112 \text{ kl}.$$

Therefore, this swimming pool can contain $\boxed{\textbf{112 kl}}$ of water.

59. a) People per sq mi = (Total Area in 1 sq mi) ÷ (Area Per Person)

$$= \left(\frac{2.8 \times 10^7 \text{ ft}^2}{1 \text{ mi}^2}\right) \div \left(\frac{50 \text{ ft}^2}{1 \text{ person}}\right)$$

$$= \left(\frac{2.8 \times 10^7 \text{ ft}^2}{1 \text{ mi}^2}\right) \times \left(\frac{1 \text{ person}}{50 \text{ ft}^2}\right)$$

$$= \boxed{\textbf{560,000 people/mi}^2}$$

b) $\left(\dfrac{1 \text{ mi}^2}{560{,}000 \text{ people}}\right) \times (5 \times 10^9 \text{ people}) \approx 8{,}928.57 \text{ mi}^2 \approx \boxed{\textbf{8,929 mi}^2}$

c) Acres per person = Total Acreage ÷ Population

$$= \underbrace{\left(5.2 \times 10^7 \text{ mi}^2 \times \frac{640 \text{ acres}}{1 \text{ mi}^2}\right)}_{\text{Total Acreage}} \div \underbrace{\left(5 \times 10^9 \text{ people}\right)}_{\text{Population}}$$

$$= \boxed{\textbf{6.656 acres/person}}$$

Problem Set 7.4

1. **Answers vary**. Your textbook's author does an excellent job of addressing many aspects of this issue. Read the bottom two lines of page 424, followed by page 425 in its entirety.

3. **Answers vary**. We convert from one metric unit to another by simply moving the decimal point. For details and examples, refer to textbook pp. 461-463.

5. a) **centimeter** b) **meter**

7. a) **liter** b) **kiloliter**

9. Celsius is the metric system's one common temperature unit.
   a) **Celsius** b) **Celsius**

11. **Choice C**. Keep in mind that 1 kg is approximately equivalent to 2.2 lbs; for example, your textbook weighs about 1 kg. Certainly, a typical can of grocery-store carrots would not weigh four times as much as your text (so, we eliminate B) — let alone forty times as much (hence, choice A is illogical).

13. **Choice C**. Because 795 mg = 0.795 kg and 1 kg $\approx$ 2.2 lb, we immediately eliminate choice A. Similarly, realizing that 4 kg is less than 10 lb, we know a 4-kg turkey would be insufficient for a group of 15, so we disregard choice B. To feed such a large group, one should purchase a turkey weighing certainly no less than 20 lb, and choice C (i.e., 12 kg $\approx$ 25 lb) is the only answer that comes even remotely close.

15. **Choice B**. At a temperature of 32°F, water freezes, but a day in which the outside temperature is 32°C would be considered rather summery and, hence, conducive to water skiing.

17. **Choice C**. An oven's broiling (not boiling) temperature is in excess of 500°F. In light of the fact that water boils at a temperature of 212°F (or 100°C) and steaks are broiled at a temperature roughly triple that, we eliminate choice A (because it is much too cool a temperature for broiling) and choice B (because it is much too hot).

19. **Choice A**. Recall that 1 kg $\approx$ 2.2 lb.

21. **Choice B**. Please review, as needed, textbook pp. 461-462. If you are having difficulty remembering the metric prefixes, you may find it helpful to either use the author's memory aid (see Table 7.3, p. 462) or create your own mnemonic device.

23. **Choice C** 25. **Choice B**

27. **Choice B** **29.** **Choice A**

31. Problems 30-37 are very similar to textbook Example 7 (pp. 462-463). Given a metric measurement of 63 ml, we write:

0.000063 kiloliter; 0.00063 hectoliter; 0.0063 dekaliter; 0.063 liter; 0.63 deciliter; 6.3 centiliter; 63 milliliter

33. Given a metric measurement of 3.5 L, we write:

0.0035 kiloliter; 0.035 hectoliter; 0.35 dekaliter; 3.5 liter; 35 deciliter; 350 centiliter; 3,500 milliliter

35. Given a metric measurement of 8 dkl, we write:

0.08 kiloliter; 0.8 hectoliter; 8 dekaliter; 80 liter; 800 deciliter; 8,000 centiliter; 80,000 milliliter

37. Given a metric measurement of 3.1 hl, we write:

0.31 kiloliter; 3.1 hectoliter; 31 dekaliter; 310 liter; 3,100 deciliter; 31,000 centiliter; 310,000 milliliter

39. Problems 38-42 are very similar to textbook Example 8 (p. 463). As needed, please review textbook pages 461-463.

a) From mm to m is a change of three decimal places; a larger prefix (m) suggests a smaller number, so we move the decimal point three places to the left:

$$1 \text{ mm} = \boxed{0.001 \text{ m}}$$

b) From mm to km is a change of six decimal places; a larger prefix (km) suggests a smaller number, so we move the decimal point six places to the left:

$$1 \text{ mm} = \boxed{0.000001 \text{ km}}$$

c) From m to mm is a change of three decimal places; a smaller prefix (mm) suggests a larger number, so we move the decimal point three places to the right:

$$1 \text{ m} = \boxed{1{,}000 \text{ mm}}$$

d) From m to cm is a change of two decimal places; a smaller prefix (cm) suggests a larger number, so we move the decimal point two places to the right:

$$1 \text{ m} = \boxed{100 \text{ cm}}$$

41. Once again, by using the pattern made evident by textbook Example 7 (p. 462) and subsequently demonstrated/explained in Example 8 (p. 463) and in this manual's solution to Problem 39 (see above), we obtain the following:

a) 1 L = **1,000 ml** b) 1 L = **10 dl**

c) 1 L = **0.001 kl** d) 1 kl = **1,000,000 ml**

43. In the interest of exposing you to as many perspectives as possible, let us share two ways to view this problem:

I. This "topless" box consists of a 3.5 cm by 3.5 cm square base and four identical sides, each of which is a 3.5 cm by 4.2 cm rectangle. So, the box's surface area is:

$$\underbrace{\text{(Area of square base)}} + \underbrace{\text{(Area of 4 rectangles)}}$$

$$s^2 \qquad + \qquad 4(\ell w)$$

$$(3.5)^2 \qquad + \qquad 4(3.5)(4.2)$$

which equals 71.05 cm^2. Therefore, rounded to the nearest square centimeter, this box's surface area is **71 cm^2**.

II. From Figure 7.19 in your text (p. 456), we know the surface area formula for right rectangular prisms (i.e., "boxes") is $S = 2\ell w + 2wh + 2\ell h$. However, since this particular box does not have a top, we slightly modify (the first term of) this formula and proceed to compute the surface area:

$$\begin{aligned} S &= \ell w + 2wh + 2\ell h \\ &= (3.5)(3.5) + 2(3.5)(4.2) + 2(3.5)(4.2) \\ &= 12.25 + 29.4 + 29.4 \\ &= 71.05 \\ &\approx 71 \text{ cm}^2 \end{aligned}$$

45. We use the formula provided in textbook Figure 7.19 (p. 456):

$$S = 4\pi r^2 = 4\pi\,(15)^2 = 900\pi \approx 2{,}827.4333 \approx \boxed{2{,}827 \text{ cm}^2}$$

47. We refer you to the formula given in textbook Figure 7.19 (p. 456). This figure is a right rectangular prism, so we use the formula $V = Bh$, where B represents the area of the prism's rectangular base and h signifies its height. It should be obvious that this formula is equivalent to $V = \ell wh$ (from Section 7.3) because, for a rectangular base, $B = \ell w$. Hence,

$$V = Bh = \ell wh = (5 \text{ ft})(3 \text{ ft})(2 \text{ ft}) = \boxed{30 \text{ ft}^3} \text{ (or 30 cu ft).}$$

For your information, the volume of boxes (also known as <u>*rectangular parallelepipeds*</u>) is discussed on textbook pp. 446-447.

49. Like textbook Example 2 (p. 457), this figure is a triangular prism. If necessary, imagine the figure standing on its side so that its "base" is triangular. The formula for a prism's volume is $V = Bh$, where B is the area of the prism's triangular base and h is its height. Since the base is triangular, $B = \frac{1}{2}ba$, where b is the triangle's base (in this case, $b = 10$ cm) and a is the height/altitude of the triangle (here, $a = 6$ cm). So, the volume of this prism is

$$V = Bh = \tfrac{1}{2}bah = \tfrac{1}{2}(10\text{ cm})(6\text{ cm})(20\text{ cm}) = \boxed{\textbf{600 cm}^3}.$$

51. Via the formulas given on textbook p. 456 (Figure 7.19), we find:

Surface Area $= S = 4\pi r^2 = 4\pi(6)^2 = 144\pi \approx 452.389 \approx \boxed{\textbf{452 in.}^2}$ (or 452 sq in.)

Volume $= V = \frac{4}{3}\pi r^3 = \frac{4}{3}\pi(6)^3 = 288\pi \approx 904.779 \approx \boxed{\textbf{905 in.}^3}$ (or 905 cu in.)

53. Suppose the original rectangle has length ℓ and width w. Then it has area $A = \ell w$. If we then create a new (enlarged) rectangle whose length is double the size the original rectangle's length, the new rectangle's length is 2ℓ. Similarly, if the new rectangle's width is triple the original rectangle's width, the new rectangle's width is $3w$. So, with dimensions 2ℓ by $3w$, the new rectangle's area is

$$A = (2\ell)(3w) = 6\ell w.$$

Therefore, since the original rectangle has area ℓw and the enlarged rectangle has area $6\ell w$, we can see that **the area is increased six-fold.**

[Suggestion: If you found this problem to be too abstract, consider making it more concrete by replacing ℓ and w with your own made-up numbers. Then, perform the "enlargement process" and compare the two areas.]

55. Suppose we begin with a sphere with radius r. From textbook Figure 7.19 (p. 456), we know the surface area of this sphere is $S = 4\pi r^2$.

Then, if we create a new (enlarged) sphere whose radius is triple the size of the original sphere, we have a sphere with radius $3r$ and, hence, surface area equal to $S = 4\pi(3r)^2 = 4\pi(9r^2) = 36\pi r^2$.

Upon comparing the two spheres' surface areas ($4\pi r^2$ versus $36\pi r^2$), it becomes evident that that **the area is increased nine-fold.**

[As suggested in our solution to Problem 53, if you find problems like this to be too abstract, consider making them more concrete by choosing your own made-up radius. Then, perform the "enlargement process" and compare the two areas.]

57. We begin with a circle having diameter d, radius r, and, hence, area $A = \pi r^2$.[or, if preferred $A = \pi(d/2)^2 = \frac{1}{4}\pi d^2$]. Then, suppose we create a new (enlarged) circle by doubling the original circle's diameter (i.e., $2d$). Since doubling a circle's diameter has the effect of doubling its radius, this new circle has a radius of $2r$ and, therefore, an area of

$$A = \pi\,(2r)^2 = \pi\,(4r^2) = 4\pi\,r^2 \quad \underline{\text{or}} \quad A = \pi\left(2d/2\right)^2 = \pi d^2$$

Upon comparing the original circle's area ($\pi\,r^2 = \frac{1}{4}\pi d^2$) to the enlarged circle's area ($4\pi\,r^2 = \pi d^2$), we conclude that the result of doubling the diameter is a **four-fold increase** in the area.

59. Reminder: Textbook Section 5.3 provides a student-friendly introduction to spreadsheets. The following spreadsheet will perform the desired task:

	A	B
1	TEMPERATURE	CONVERSIONS
2		
3	Celsius	Fahrenheit
4	0	+(9/5)∗A4+32
5	+A4+5	+(9/5)∗A5+32
6		
7	*Replicate Row 5 for*	
8	*Rows 6 through 12.*	

Here is the resulting output:

	A	B
1	TEMPERATURE	CONVERSIONS
2		
3	Celsius	Fahrenheit
4	0	32
5	5	41
6	10	50
7	15	59
8	20	68
9	25	77
10	30	86
11	35	95
12	40	104

Problem Set 7.5

1. **Wording varies**. A ***logarithm*** is the exponent to which a fixed number (called the *base*) must be raised in order to produce a given number (called the *argument*). Said differently, when we evaluate a logarithmic expression (such as $\log_b x$), we are determining an exponent (in particular, the exponent to which we raise the base b to obtain the argument x). *[We urge you to spend ample time studying the definition of logarithm introduced on textbook page 467.]*

3. **Wording varies**. Introduced on p. 468 of your text, a ***natural logarithm*** (written $\ln x$) is a logarithm whose base is the irrational number e; that is, $\ln x = \log_e x$. The number e (which was originally introduced in textbook Section 4.5, p. 198) is sometimes referred to as the Euler number (also Euler constant or natural base). It has numerous applications, many of which you will encounter in Chapter 8. Remember that $e \approx 2.71828182846$.

5. **Answers vary (depending on calculator model)**.
   a) On most calculators, the common logarithm $\log_{10} N$ is evaluated by pressing either **log** **N** or **N** **log**. Additionally, you may need to press the **=** (or **ENTER** key).
   b) The natural logarithm $\ln N$ is evaluated by pressing **ln** **N** (or perhaps **N** **log**). Once again, you may need to press the **=**-key (or its equivalent).
   c) To evaluate a base-b logarithm (obviously, for which $b \neq 10$ and $b \neq e$), we use the change-of-base theorem (see textbook pp. 469-470). Specifically, to evaluate $\log_b N$ with a calculator, we have two options:

   log **N** **÷** **log** **b** or **ln** **N** **÷** **ln** **b** *(followed by* **=** *or its equivalent).*

7. **Wording varies**. As classified on the bottom of textbook p. 470, the three types of exponential equations are $10^x = N$, $e^x = N$, and $b^x = N$.

9. This problem is very similar to text Example 2 (p. 468).
   a) For the exponential equation $64 = 2^6$, 2 serves as the base and 6 is the exponent. Remember that ***a logarithm is equal to an exponent***. Consequently, we translate the given exponential equation into the following logarithmic equation:

   $$\boxed{6 = \log_2 64}$$

   b) With $100 = 10^2$, 10 is the base and 2 is the exponent. So, $\boxed{2 = \log 100}$. Please recall that we call a logarithm with a base of 10 a ***common logarithm*** and agree to write it without explicit mention of the base (i.e., $\log x = \log_{10} x$).
   c) $m = n^p \Rightarrow \boxed{p = \log_n m}$

11. This problem is very similar to its predecessors. For additional details, please refer to the text's Example 2 (see p. 468) and this manual's solution to Problem 9 (see above).
   a) $\frac{1}{10} = 10^{-1} \Rightarrow \boxed{-1 = \log \frac{1}{10}}$
   b) $36 = 6^2 \Rightarrow \boxed{2 = \log_6 36}$
   c) $s = t^n \Rightarrow \boxed{n = \log_t s}$

13. This problem is akin to text Example 3 (p. 468).

Important Note: Before you can successfully simplify expressions like these, you **must** (must!) acquire a **true** understanding (i.e., no blind memorization) of the definition of logarithm. Said differently, have you created for yourself a phrase/question that makes clear to you what the notation $\log_b x$ means? In this solution, we share such a phrase and recommend you find one that best suits you:

$\log_b x$ asks us to "find the exponent to which we raise b to get x" (i.e., $b^{?} = x$).

a) In the case of the logarithmic expression $\log_{10} 10$, it implicitly asks for the exponent to which 10 must be raised to get 10. Clearly, 10^1 equals 10, so $\log_{10} 10 = \boxed{1}$.

b) $\log_{10} 1{,}000$ asks us to determine the exponent to which we raise 10 to get $1{,}000$. We know $10^3 = 1{,}000$. So, it follows that $\log_{10} 1{,}000 = \boxed{3}$.

c) $\log_{10} 10^{-5}$ implicitly asks a circular (self-answering) question. In particular, $\log_{10} 10^{-5}$ asks, "To what exponent do we raise 10 to get 10^{-5}?" Obviously, the answer is $\boxed{-5}$. *[If reading that question did not put a smile on your face, please read it again. Doesn't it answer itself? It's similar to asking questions such as "What gender is my niece?" and "How do you spell M-A-T-H-E-M-A-T-I-C-S?". In the text's next section (7.6), the author calls expressions of this type "Grant's tomb properties" (as in "Who's buried in Grant's tomb?"). For more about this, take a look at textbook p. 475.]*

15. For additional details, refer to this manual's solution to Problem 13 (see above) and to your text's explanation for Example 2 (p. 468). Both are very similar to this problem.

a) The logarithmic expression $\log_5 5$ implicitly asks us for the exponent to which we raise 5 to get 5 (i.e., $5^{?} = 5$). Since $5^1 = 5$, $\log_5 5 = \boxed{1}$.

b) $\log_5 25$ can be translated to: $5^{?} = 25$. Since $5^2 = 25$, $\log_5 25 = \boxed{2}$.

c) $\log_5 5^{-3}$ is a "Grant's tomb" expression (as explained above in our solution to Problem 13c). That is to say, this expression asks the following self-answering question: "To what exponent do we raise 5 to get 5^{-3}?" Clearly, this exponent is $\boxed{-3}$.

17. For additional details, refer to textbook Example 2 (p. 468) and this manual's solutions to Problems 13 and 15 (see above).

a) $\log_b b = \boxed{1}$ b) $\log_b b^3 = \boxed{3}$ c) $\log_b b^{-6} = \boxed{-6}$

Please note that these logarithmic expressions are what we (and your text's author) refer to as "Grant's tomb" expressions (i.e., self-answering questions). For details, please refer to this manual's solution to Problem 13c.

19. Problems 19-22 are similar to textbook Example 4 (p. 468). Note that the unknown in all of these equations is the exponent. Since (by definition) logarithms equal exponents, we solve these equations simply by writing them in logarithmic form.

a) In the equation $4^x = 5$, 4 is the base and x is the exponent. So, $x = \boxed{\log_4 5}$.

b) $5^x = 8 \quad \Rightarrow \quad x = \boxed{\log_5 8}$

c) $6^x = 4.5 \quad \Rightarrow \quad x = \boxed{\log_6 4.5}$

21. Please recall that we call a logarithm with a base of 10 a ***common logarithm*** and have agreed to write it without explicit mention of the base (i.e., $\log x = \log_{10} x$).

a) In the equation $10^x = 15$, 10 is the base and x is the exponent. So, by applying the definition of logarithm, we conclude that $x = \log_{10} 15 = \boxed{\log 15}$.

b) $10^x = 2.5 \quad \Rightarrow \quad x = \boxed{\log 2.5}$

c) $10^x = 45 \quad \Rightarrow \quad x = \boxed{\log 45}$

23. a) As discussed below (in this manual's solutions to Problems 31, 33, 35, etc.), we strongly recommend you "mentally estimate" answers **before** using your calculator. Because $\log_{10} 1 = 0$, $\log_{10} 10 = 1$, and $1 < 4.27 < 10$, we expect the common logarithm of 4.27 to be between 0 and 1. The answer obtained with a calculator is consistent with this estimate:

$$\log 4.27 \approx 0.630427875 \approx \boxed{0.63}.$$

b) If you read our previous solutions (to Problem 13c, 15c, and 17), you should immediately recognize $\log_b b^2$ as a <u>self-answering</u> ("Grant's tomb" type) question. Specifically, the logarithmic expression $\log_b b^2$ implicitly asks, "To what exponent do we raise b to get b^2?" So, $\log_b b^2 = \boxed{2.00}$.

c) $\log_t t^3 = \boxed{3.00}$

25. a) As discussed below (in this manual's solutions to Problems 31, 33, 35, etc.), we recommend you always "mentally estimate" answers **before** using your calculator. We expect the common logarithm of 71,600 to be between 4 and 5 because 10^4 = 10,000, 10^5 = 100,000 and 10,000 < 71,600 < 100,000. With this estimate in mind, we then use our calculator to compute the answer:

$$\log 71{,}600 \approx 4.854913022 \approx \boxed{\mathbf{4.85}}$$

b) $\log_3 9 = \boxed{\mathbf{2.00}}$. You should not use a calculator for this question. Instead, use the definition of logarithm.

c) $\log_{19} 1 = \boxed{\mathbf{0.00}}$ *[Once again, a calculator should not be necessary here.]*

27. As extensively discussed below (most notably, in our solutions to Problems 31, 33, and 35), we recommend you always "mentally estimate" answers **before** using your calculator. For example, in the case of this problem, since each argument is less than 1 and since $\log 1 = 0$, we certainly should expect these three logarithms to be less than zero.

a) $\log 0.042 \approx -1.37675071 \approx \boxed{\mathbf{-1.38}}$

b) $\log 0.321 \approx -0.493494968 \approx \boxed{\mathbf{-0.49}}$

c) $\log 0.0532 \approx -1.274088368 \approx \boxed{\mathbf{-1.27}}$

29. As discussed at length in various solutions preceding and succeeding this one, we suggest you attempt to "mentally estimate" the values of these natural logarithms before using your calculator. Although this is harder to do with natural logarithms, by doing so you (i) decrease the likelihood of giving erroneous answers and, more importantly, (ii) significantly enhance your understanding of natural logarithms. We demonstrate this estimation process in part (a) and leave parts (b) and (c) to you and your study partners.

a) Recall that, by definition, $\ln 10 = \log_e 10$ and $e \approx 2.718$. Since e is between 2 and 3, one would expect $\ln 10$ to be between $\log_3 10$ and $\log_2 10$. We know $\log_3 10$ is slightly more than 2 (because $\log_3 9 = 2$) and $\log_2 10$ is more than 3 (because $\log_2 8 = 3$ and $\log_2 16 = 4$). So, we should expect the value of $\ln 10$ to be roughly between 2ish and 3ish. Having made an estimate, we now proceed to compute the actual answer:

$$\ln 10 \approx 2.302585093 \approx \boxed{\mathbf{2.30}}$$

b) $\ln 100 \approx 4.605170186 \approx \boxed{\mathbf{4.61}}$

c) $\ln 1{,}000 \approx 6.907755279 \approx \boxed{\mathbf{6.91}}$

31. **Before** using your calculator, you should always attempt to estimate the answer. By doing so, you will detect calculator keystroke errors and significantly decrease your likelihood of writing erroneous (often completely unreasonable) answers.

To estimate the value of $\log_3 45$, we begin by reminding ourselves of the question it implies: "To what exponent do we raise 3 to obtain 45?" Since 45 is a number between 27 (i.e., 3^3) and 81 (3^4), we expect $\log_3 45$ to be between 3 and 4.

Finally, with our estimate in mind, we use the change-of-base theorem (textbook p. 470) and our calculator to obtain the answer:

$$\log_3 45 = \frac{\log 45}{\log 3} \approx 3.46497352072 \approx \boxed{\mathbf{3.4650}}$$

33. We recommend — as pointed out in our solution to Problem 31 above — that you begin by estimating the answer <u>before</u> using your calculator.

The logarithmic expression $\log_6 10$ asks the question: "To what exponent do we raise 6 to obtain 10?" Since 10 is between 6^1 and 6^2, we expect the value of $\log_6 10$ to be between 1 and 2. Specifically, since 10 is closer to 6 than it is to 36, we estimate that $\log_6 10$ is closer to 1 than it is to 2. With our estimate in mind, we proceed to compute the answer:

$$\log_6 10 = \frac{\log 10}{\log 6} = \frac{1}{\log 6} \approx 1.285097209 \approx \boxed{\mathbf{1.2851}}$$

Notice that one can save calculator keystrokes by realizing that log 10 = 1; so, $\log_6 10$ equals the reciprocal of log 6. More generally, $\log_b 10$ equals the reciprocal of log b.

35. Once again, as demonstrated in our previous two solutions (see Problems 33 and 35 above, among others), we recommend you estimate the solution <u>before</u> using your calculator. However, this time permit us to share a modified form of the estimation procedure shown previously. Since $\log_4 1 = 0$ and $\log_4 4 = 1$, we expect $\log_4 3.05$ to be between 0 and 1, fairly close to 1 (since 3.05 is closer to 4). With the aid of a calculator, the following confirms the accuracy of this estimate:

$$\log_4 3.05 = \frac{\log 3.05}{\log 4} \approx 0.804404621 \approx \boxed{\mathbf{0.8044}}$$

37. In our previous three solutions (see Problems 31, 33, and 35 above), we demonstrated the process of — <u>and</u> conveyed the importance of — making a "mental estimate" of the answer **before** using a calculator. This time, we leave the details to you, but here is a "small" hint: *Since $\log_2 1 = 0$ and since $0.0056 < 1$, what can you conclude about the value of $\log_2 0.0056$?*

$$\log_2 0.0056 = \frac{\ln 0.0056}{\ln 2} \approx -7.480357457 \approx \boxed{\mathbf{-7.4804}}$$

39. Once again, we encourage you to make a quick "mental estimate" of the answer **before** using your calculator. For additional details, refer to this manual's solutions for Problems 31, 33, and 35 (among others).

$$\log_8 10 = \frac{\ln 10}{\ln 8} \approx 1.107309365 \approx \boxed{\mathbf{1.1073}}$$

41. As has been explained in many of our preceding solutions, we begin by mentally estimating the answer and then afterwards compute the actual answer. In this case, since $e \approx 2.718$ (which is a number between 1 and 8), $\log_8 e$ should be between

$\log_8 1 = 0$ and $\log_8 8 = 1$. Indeed, as the following confirms, $\log_8 e$ is between 0 and 1:

$$\log_8 e = \frac{\ln e}{\ln 8} = \frac{1}{\ln 8} \approx 0.480898347 \approx \boxed{\mathbf{0.4809}}$$

Note that one can save calculator keystrokes by knowing $\ln e = 1$; hence, $\log_8 e$ equals the reciprocal of $\ln 8$. More generally, $\log_b e$ equals the reciprocal of $\ln b$.

43. a) $2^x = 128$ *Given exponential equation*

$x = \log_2 128$ *Apply the definition of logarithm.*

$\boxed{x = 7}$ *Simplified exact answer (obtained either via calculator or by hand).*

b) $3^x = 343$ *Given exponential equation*

$x = \log_3 343$ *Apply the definition of logarithm.*

$\boxed{x = 5}$ *Simplified exact answer (obtained either via calculator or by hand).*

45. a) $4^x = \frac{1}{16}$ *Given exponential equation*

$x = \log_4 \frac{1}{16}$ *Apply the definition of logarithm.*

$\boxed{x = -2}$ *Simplified exact answer (obtained either via calculator or by hand).*

b) $27^x = \frac{1}{81}$ *Given exponential equation*

$x = \log_{27} \frac{1}{81}$ *Apply the definition of logarithm.*

$\boxed{x = -\frac{4}{3}}$ *Simplified exact answer (obtained either via calculator or by hand).*

47. a) $e^x = 4$ *Given exponential equation*

$\boxed{x = \ln 4}$ *Definition of logarithm [remember: ln4 = log$_e$4]; ln 4 is the exact answer.*

$\boxed{x \approx 1.386294361}$ *Approximate answer (obtained via calculator).*

b) $e^x = 25$ *Given exponential equation*

$\boxed{x = \ln 25}$ *Apply the definition of logarithm; this is the exact answer.*

$\boxed{x \approx 3.218875825}$ *Approximate answer (obtained via calculator).*

49. a) $10^{5x} = 5$ *Given exponential equation*

$5x = \log 5$ *Apply the definition of logarithm [remember: log 5 = log$_{10}$5].*

$\boxed{x = \frac{\log 5}{5}}$ *Divide both sides by 5. This is the exact answer.*

$\boxed{x \approx 0.1397940009}$ *Approximate answer (obtained via calculator).*

b) $10^{3x} = 0.45$ — *Given exponential equation*

$3x = \log 0.45$ — *Apply the definition of logarithm.*

$\boxed{x = \dfrac{\log 0.45}{3}}$ — *Divide both sides by 3. This is the exact answer.*

$\boxed{x \approx -0.1155958287}$ — *Approximate answer (obtained via calculator).*

51. a) $e^{1-2x} = 3$ — *Given exponential equation*

$1 - 2x = \ln 3$ — *Apply the definition of logarithm.*

$-2x = -1 + \ln 3$ — *Subtract 1 from both sides.*

$x = -\frac{1}{2}(-1 + \ln 3)$ — *Either multiply both sides by -½ or divide by –2.*

$\boxed{x = \frac{1}{2}(1 - \ln 3) \text{ or } \dfrac{1 - \ln 3}{2}}$ — *Simplify (i.e., distribute negative). This is the exact answer.*

$\boxed{x \approx -0.493061443}$ — *Approximate answer (obtained via calculator).*

b) $e^{1-5x} = 25$ — *Given exponential equation*

$1 - 5x = \ln 25$ — *Apply the definition of logarithm.*

$-5x = -1 + \ln 25$ — *Subtract 1 from both sides.*

$x = -\frac{1}{5}(-1 + \ln 25)$ — *Either multiply both sides by –1/5 or divide by –5.*

$\boxed{x = \frac{1}{5}(1 - \ln 25) \text{ or } \dfrac{1 - \ln 25}{5}}$ — *Simplify (i.e., distribute negative). This is the exact answer.*

$\boxed{x \approx -0.443775165}$ — *Approximate answer (obtained via calculator).*

53. $2 \cdot 3^x + 7 = 61$ — *Given exponential equation*

$2 \cdot 3^x = 54$ — *Subtract 7 from both sides.*

$3^x = 27$ — *Divide both sides by 2.*

$\boxed{x = 3}$ — *Solve [since we know $3^3 = 27$, logarithms are unnecessary].*

55. This equation is representative of the type encountered in Chapter 8.

$\left(1 + \dfrac{0.08}{360}\right)^{360x} = 2$ — *Given exponential equation*

$360x = \log_b 2$ where $b = 1 + \frac{0.08}{360}$ — *Apply the definition of logarithm.*

To obtain the exact solution, we divide both sides by 360 and apply the change-of-base theorem:

$$x = \frac{\log_b 2}{360} = \tfrac{1}{360}\log_b 2 = \boxed{\frac{1}{360}\left[\frac{\log 2}{\log\left(1+\frac{0.08}{360}\right)}\right]}$$

When using your calculator to obtain an approximated solution, it is important that you only round once — namely, at the very end. Be sure to <u>avoid</u> rounding in the middle of calculator computations because doing so will result in inaccurate approximations. For this problem, we have:

$$\boxed{x \approx \mathbf{8.665302427}}.$$

57. Note that this problem will help prepare you for Section 7.7. As directed, we solve for the variable t.

$P = P_0 e^{rt}$	*Given exponential equation*
$\dfrac{P}{P_0} = e^{rt}$	*Divide both sides by P_0.*
$rt = \ln\left(\dfrac{P}{P_0}\right)$	*Apply the definition of logarithm.*
$\boxed{t = \dfrac{1}{r}\ln\left(\dfrac{P}{P_0}\right)}$	*Multiply both sides by the reciprocal of r.*

59. Having to "fill up the earth with water" suggests the need to obtain the earth's volume, which, in turn, we use to determine the earth's capacity (in gallons). Given the earth's radius, we know (from our previous work in Section 7.4) that its volume is:

$$V = \tfrac{4}{3}\pi r^3 = \tfrac{4}{3}\pi\,(3{,}963\text{ mi})^3 \approx 2.60711883\times 10^{11}\text{mi}^3$$

From textbook Section 7.3 (specifically, p. 450), we know 1 gal $= 231$ in.^{3} and $1\text{ft}^3 \approx 7.48$ gal. So, in order to convert the earth's volume (presently expressed in mi^3) to liquid capacity (in gal), we first need to convert cubic miles to either cubic feet or cubic inches. *[<u>Disclaimer</u>: We have chosen to use cubic feet, but we recommend you re-work this problem later with cubic inches, which has an exact relationship (unlike ft^3) to gallons.]* Recalling that 1 mi $= 5{,}280$ ft, it follows that

$$1\text{ mi}^3 = (1\text{ mi})^3 = (5{,}280\text{ ft})^3 = (5{,}280^3)\text{ ft}^3.$$

So, we have

$$V = \left(2.60711883\times 10^{11}\text{mi}^3\right)\times\left(\frac{5{,}280^3\text{ft}^3}{1\text{ mi}^3}\right) \approx 3.837625524\times 10^{22}\text{ ft}^3.$$

Then, since $1\text{ft}^3 \approx 7.48$ gal, the earth's (liquid) capacity is

$$\left(3.837625524\times 10^{22}\text{ft}^3\right)\times\left(\frac{7.48\text{ gal}}{1\text{ ft}^3}\right) \approx 2.870543892\times 10^{23}\text{ gal}.$$

The question posed in this problem asks us to determine the moment/instant/time at which this hypothetical machine will pour out a single quantity of water equal to the earth's capacity. This amount of time elapsed before this moment occurs is essentially equal to the number of times the machine goes through its so-called "capacity doubling" process; that is, 1 gal in the 1st minute, 2 gal in the 2nd minute, 4 gal in the 3rd min, 8 gal in the 4th minute, ..., 2^x gal in the $(x+1)^{\text{th}}$ minute.

Hence, we need to solve the following exponential equation:

$$2^x = 2.870543892 \times 10^{23}.$$

By the definition of logarithm,

$$x = \log_2 2.870543892 \times 10^{23} = \frac{\log\left(2.870543892 \times 10^{23}\right)}{\log 2} \approx 77.9256703 \approx 78$$

So, since $x+1 \approx 79$, it takes approximately **79 minutes (i.e., 1 hr 19 min)** for this hypothetical machine to pour out a single quantity of water equivalent to the capacity of the earth's interior.

Problem Set 7.6

1. **Wording varies**. A ***logarithmic equation*** is an equation for which there is a logarithm on one or both sides. See textbook page 476 for more information.

3. **Answers vary**. As necessary, first use algebraic fundamentals and the laws of logarithms to solve for (i.e., isolate) the logarithm so that one of two forms is obtained: (i) $\log_b M = \log_b N$, which implies $M = N$; or (ii) $\log_b M = N$ which implies $b^N = M$. For additional details, refer to textbook Example 2 (pp. 476-477) and Example 5 (pp. 479-480).

5. **False**; a common logarithm (defined on textbook p. 468) is a logarithm in which the base is 10.

7. **False**; $\log_b N$ represents the *exponent*, whereas N is known as the *argument*. Remember: a logarithm is equal to an exponent.

9. **True**; by the third law of logarithms (textbook p. 477), $2\log_3 81 = \log_3 81^2$. Given $2\log_3 81 = 8$ and since $81^2 = 6{,}561$, we know $\log_3 6{,}561 = 8$.

11. **False**; x is the exponent on 1.5, not vice versa. That is to say, via the definition of logarithm, $\log_{1.5} 8 = x$ implies $1.5^x = 8$.

13. **False**; no such law exists. To better understand why, create a counterexample; for example, try $b = 10$, $A = 1{,}000$, and $B = 100$. With these numbers, you will find: $\log_b(A+B) \approx 3.0414$, while $\log_b A + \log_b B = 5$. Refer to textbook pp. 477-478 for the true laws of logarithms.

15. **False**; no such law exists. However, some students — primarily those who blindly memorize — tend to believe this falsity due to having confused the change-of-base theorem (on textbook p. 470) with the second law of logarithms (p. 477). We hope you will not be among this misguided contingent. Examining a counterexample will help; either pick your own numbers or feel free to use those suggested in our solution to Problem 13 (see above).

17. **False**; once again, no such law exists. This erroneous statement exemplifies a common belief among students who do not truly "understand" the second law of logarithms — and, for that matter, might not understand logarithmic notation. As suggested in our solutions to Problems 13 and 15, consider creating a counterexample; the numbers used above (for Problem 13's explanation) will suffice.

19. **False**; $\log_b N$ is undefined when N is negative (or, for that matter, when N equals zero). From the text's definition of logarithm (p. 467), recall that (i) b must be positive (and $b \neq 1$); and (ii) $\log_b N$ implicitly asks us to find the exponent to which we raise the base b to obtain the argument N. Clearly, there is no exponent to which one can raise a positive base to obtain a negative argument.

21. This problem is very similar to textbook Example 1 (p. 475). It relies upon your having acquired an understanding of the fundamental properties of logarithms. These properties are also described in your text as the "Grant's tomb properties," the basis for which is explained in this manual's solution for Problem 13c in Section 7.5.
Instead of repeating your textbook author's eloquent words, permit us to do this: (i) share another perspective in part (a), (ii) rephrase your text's approach in part (b), and (iii) leave part (c) to you.

a) If we choose to temporarily let u represent $\ln 23$, then being asked to evaluate $e^{\ln 23}$ becomes equivalent to evaluating e^u. [Doesn't it?] So, to reiterate, if we can determine the value of e^u, we will have succeeded in evaluating $e^{\ln 23}$.
Well, since $u = \ln 23$, we can apply the definition of logarithm and write the following exponential equation: $e^u = 23$. [Wow! Look at that!]
Because e^u is equivalent to $e^{\ln 23}$ and $e^u = 23$, we must have $e^{\ln 23} = \boxed{23}$. In other words, by making a simple variable substitution and applying the definition of logarithm, the logarithmic expression $e^{\ln 23}$ essentially "unraveled" itself.

b) Take a moment to contemplate the meaning of $\log 3.4$; that is to say, ask yourself what its value represents. The value of $\log 3.4$ represents the particular exponent to which 10 must be raised in order to obtain 3.4. Since the expression $10^{\log 3.4}$ asks us to raise 10 to that very same exponent, the result must be $\boxed{\mathbf{3.4}}$.

c) $4^{\log_4 x} = \boxed{x}$

23. This problem is intended to assess whether you have become sufficiently familiar with the so-called fundamental ("Grant's tomb") properties of logarithms. Permit us to repeat your textbook author's caution (from the margin of p. 475): "Spend some time making sure you **understand** these properties." Please do <u>not</u> blindly accept them. *[<u>Note</u>: We hope your level of understanding is enhanced by this manual's explanations for Problem 21 (see immediately above) and Problem 13c (see solutions for Section 7.5).]*

a) $\log_b b^x = \boxed{x}$ b) $b^{\log_b x} = \boxed{x}$

25. a) Since $3^4 = 81$, $\log_3 81 = 4$. So, $\boxed{x = 4}$.

b) Since $4^3 = 64$, $\log_4 64 = 3$. So, $\boxed{x = 3}$.

27. a) To what exponent does one raise 10 to obtain 1,000? $\boxed{\mathbf{3}}$

b) To what exponent does one raise 10 to obtain the reciprocal of 1,000? $\boxed{\mathbf{-3}}$

29. As explained in textbook Example 2c (p. 477), we apply the definition of logarithm.

a) $\ln x = 2$ implies $\boxed{x = e^2}$ b) $\ln x = 3$ implies $\boxed{x = e^3}$

31. Please take note of the log-of-both-sides theorem (shown on textbook p. 476).

a) $\boxed{x = 9.3}$ b) $\boxed{x = 109}$

33. For additional details, please review textbook pp. 477-479.

a)
$$\begin{aligned} 3\ln 4 - 5\ln 2 + \ln 3 &= \ln 4^3 - \ln 2^5 + \ln 3 && \textit{Third law of logarithms} \\ &= \ln\left(\frac{4^3}{2^5}\right) + \ln 3 && \textit{Second law} \\ &= \ln 2 + \ln 3 && \textit{Simplify } (4^3 = 64;\ 2^5 = 32;\ 64 \div 32 = 2). \\ &= \ln(2 \cdot 3) && \textit{First law} \\ &= \boxed{\ln 6} && \textit{Simplify.} \end{aligned}$$

b) Although the laws of logarithms may be used to do this problem, they are not necessary. That is,

$$\begin{aligned} 3\ln 4 - 2\ln(2+2) &= 3\ln 4 - 2\ln 4 && \textit{Simplify } 2 + 2. \\ &= 1\ln 4 && \textit{Combine like terms [i.e., } 3 - 2 = 1\textit{].} \\ &= \boxed{\ln 4} \end{aligned}$$

Remember the algebraic basis for combining like terms; it is a consequence of the distributive property. Specifically, we can factor $\ln 4$ as follows:

$$3\ln 4 - 2\ln 4 = (3-2)\ln 4 = 1\ln 4 = \ln 4$$

c) $$\begin{aligned} 3\ln 4 - 5(\ln 2 + \ln 3) &= 3\ln 4 - 5\ln 6 && \textit{First law [i.e., } \ln 2 + \ln 3 = \ln(2\cdot 3)\textit{].} \\ &= \ln 4^3 - \ln 6^5 && \textit{Third law} \\ &= \ln\left(\frac{4^3}{6^5}\right) && \textit{Second law} \\ &= \boxed{\ln\left(\frac{2}{243}\right)} && \textit{Simplify [i.e., reduce } \tfrac{64}{7{,}776}\textit{].} \end{aligned}$$

35. a) $$\begin{aligned} \log(x^2-9) - \log(x+3) &= \log\left[\frac{x^2-9}{x+3}\right] && \textit{Second law} \\ &= \log\left[\frac{(x+3)(x-3)}{x+3}\right] && \textit{Factor difference of squares.} \\ &= \boxed{\log(x-3)} && \textit{Reduce.} \end{aligned}$$

b) In a manner comparable to part (a), we find:

$$\begin{aligned} \log(x^2-x-6) - \log(x+2) &= \log\left[\frac{x^2-x-6}{x+2}\right] \\ &= \log\left[\frac{(x-3)(x+2)}{x+2}\right] \\ &= \boxed{\log(x-3)} \end{aligned}$$

c) This problem is virtually identical to part (a):

$$\ln(x^2-4) - \ln(x+2) = \ln\left[\frac{x^2-4}{x+2}\right] = \ln\left[\frac{(x-2)(x+2)}{x+2}\right] = \boxed{\ln(x-2)}$$

37. The author of your text has deliberately chosen to pair these two equations. Can you see why? Do x and $\log x$ play comparable roles?

a) As you continue to increase your level of proficiency, you should become able to "mentally solve" (i.e., solve without written manipulations) simple linear equations like this one. Nevertheless, here are the steps:

$3+2x=11$	*Given linear equation*
$2x=8$	*Subtract 3 from both sides.*
$\boxed{x=4}$	*Divide both sides by 2.*

b)

$\ln e^3+2\log x=11$	*Given logarithmic equation*
$3+2\log x=11$	*Apply fundamental properties of logarithms (see text p. 475).*
$2\log x=8$	*Subtract 3 from both sides.*
$\log x=4$	*Divide both sides by 2.*
$\boxed{x=10^4 \text{ or } 10{,}000}$	*Apply the definition of logarithm.*

39. As was done in Problem 37, your text's author has a good reason for pairing these two equations. Is their relationship apparent to you?

a) $\boxed{x=4}$. Needless to say, $x-2=2$ epitomizes the type of equation to which our previous comment applies [see solution to part (a) of Problem 37 above]. Solving this equation should be — or, at least, it needs to become — as automatic for you as elementary school arithmetic.

b)

$\log 10^x-2=\log 100$	*Given logarithmic equation*
$x-2=2$	*Fundamental props. (log $10^x = x$) and defn. of log (log 100 = 2)*
$\boxed{x=4}$	*"Mentally solve."*

41. Once again (like Problems 37 and 39), please take note of the strong connection between these two equations.

a) In line with the spirit of this manual's comments for part (a) of Problems 37and 39, can you solve this equation mentally? If you're skeptical, take a moment to consider this question: what number should be subtracted from $\frac{3}{2}$ (i.e., $1\frac{1}{2}$ or 1.5) to get 1? For those who prefer the (unduly laborious) algebraic step-by-step approach, here it is:

$1=\frac{3}{2}-x$	*Given linear equation*
$-\frac{1}{2}=-x$	*Subtract $\frac{3}{2}$ from both sides.*
$\boxed{\frac{1}{2}=x}$	*Multiply (or divide) both sides by –1.*

b) Before solving the given equation, permit us to share two (among an assortment of) ways to simplify $\log\sqrt{1{,}000}$:

- $\log\sqrt{1{,}000}=\log 1{,}000^{\frac{3}{2}}=\frac{3}{2}$ *[via fundamental properties of logarithms]*
- $\log\sqrt{1{,}000}=\log 1{,}000^{\frac{1}{2}}=\frac{1}{2}\log 1{,}000=\frac{1}{2}(3)=\frac{3}{2}$ *[via third law, etc.]*

We proceed to solve the equation:

$\log 10 = \log\sqrt{1{,}000} - \log x$	*Given logarithmic equation*
$1 = \frac{3}{2} - \log x$	*See simplification above; log 10 = 1.*
$\log x = \frac{1}{2}$	*"Mentally solve" [see comments regarding part (a)].*
$\boxed{x = 10^{\frac{1}{2}} \text{ or } \sqrt{10}}$	*Apply the definition of logarithm.*

43. a) $0 + x = 2$

$\boxed{x = 2}$

b) $\ln 1 + \ln e^x = 2$

$0 + x = 2$ *[Note: ln 1 = 0; ln $e^x = x$]*

$\boxed{x = 2}$

45.

$\ln[\log(\ln x)] = 0$	*Given logarithmic equation*
$\log(\ln x) = e^0$	*Definition of logarithm*
$\log(\ln x) = 1$	*Simplify e^0.*
$\ln x = 10^1$	*Definition of logarithm*
$\boxed{x = e^{10}}$	*Definition of logarithm*

47. The first step in solving this equation is a delicate one, and as such, we believe its algebraic basis deserves special mention. In specific, the reason why we are permitted to divide both sides of the given equation by 3^x is because 3^x is a <u>nonzero</u> quantity; in other words, no matter the value of x, 3^x will never equal zero. If it were possible for 3^x to equal zero, we could not perform this division step (because division by zero is not defined).

$x^2 3^x = 9(3^x)$	*Given equation*
$x^2 = 9$	*Divide both sides by 3^x (as justified above).*
$x^2 - 9 = 0$	*Subtract 9 from both sides (since the equation is quadratic).*
$(x-3)(x+3) = 0$	*Factor difference of squares.*
$\boxed{x = 3,\ x = -3}$	*Mentally apply the zero-product rule (textbook p. 267).*

49.

$\ln x = 1.8 - \ln 4.8$	*Given logarithmic equation*
$\ln x + \ln 4.8 = 1.8$	*Add ln4.8 to both sides.*
$\ln 4.8x = 1.8$	*Apply the first law of logarithms (textbook p. 477).*
$4.8x = e^{1.8}$	*Apply the definition of logarithm.*
$\boxed{x = \dfrac{e^{1.8}}{4.8}}$	*Divide both sides by 4.8.*

51. $\log x + \log 8 = 12$ *Given logarithmic equation*

$\log 8x = 12$ *Apply the first law of logarithms (textbook p. 477).*

$8x = 10^{12}$ *Apply the definition of logarithm.*

$\boxed{x = \dfrac{10^{12}}{8}}$ *Divide both sides by 8.*

53. This problem is very similar to textbook Example 4 (p. 479).

$\log_8 5 + \frac{1}{2}\log_8 9 = \log_8 x$ *Given logarithmic equation*

$\log_8 5 + \log_8 9^{\frac{1}{2}} = \log_8 x$ *Apply third law of logarithms.*

$\log_8 5 + \log_8 3 = \log_8 x$ *Simplify $9^{1/2}$.*

$\log_8 (5 \cdot 3) = \log_8 x$ *Apply the first law of logarithms.*

$\boxed{x = 15}$ *Apply log-of-both-sides theorem (textbook p. 476).*

55. a) Given that $a = \$1{,}000$, $N = 1{,}500 + 300\ln(1{,}000) \approx 3{,}572.33$. Therefore, if $\$1{,}000$ is spent on advertising, we project — based upon the given formula/model — that approximately **3,572** items will be sold.

b) Before performing any computations, take a moment to estimate a "reasonable" answer to this question. We know from part (a) that $\$1{,}000$ spent on advertising corresponds to a projected sales figure of 3,572 units. Based upon this experience, we should anticipate (i.e., "estimate") that $\$50{,}000$ in advertising would yield a sales level significantly above $3{,}572$ units.

Given $a = \$50{,}000$, $N = 1{,}500 + 300\ln(50{,}000) \approx 4{,}745.93$.

Thus, according to this logarithmic model/formula, one can expect an advertising expenditure of $\$50{,}000$ to result in sales of approximately **4,746** items, which is consistent with our aforementioned estimate.

c) Once again, before attempting to compute the answer, we recommend you use the answers from parts (a) and (b) to estimate the answer. That is to say, since $4{,}000$ units is a quantity between $3{,}572$ and $4{,}746$ (which were the answers to parts (a) and (b), respectively), shouldn't we expect the answer to this question to be no less than $\$1{,}000$ and no more than $\$50{,}000$?

Given $N = 4{,}000$ units, we need to solve for a.

$4{,}000 = 1{,}500 + 300\ln a$ *Given formula (with N = 4,000)*

$2{,}500 = 300\ln a$ *Subtract 1,500 from both sides.*

$\frac{25}{3} = \ln a$ *Divide both sides by 300; reduce.*

$a = e^{25/3}$ *Apply the definition of logarithm.*

$a \approx 4{,}160.26$ *Use calculator.*

So, to the nearest hundred dollars, one needs to spend approximately **$4,200** in advertising to achieve a sales level of $4{,}000$ units. Is this answer consistent with our aforementioned estimate?

57. a) Given $E = 15^{15}$ ergs, $M = \dfrac{\log(15^{15}) - 11.8}{1.5} \approx 3.894245924 \approx$ **3.89** magnitude (on the Richter scale).

b) In the same manner as explained in our solution to Problem 55, please take a moment to "mentally estimate" the answer **before** computing it. To do so, compare this question with part (a). Since 10^{25} is larger than 15^{15} (feel free to confirm this with your calculator), should one expect the magnitude of an earthquake that releases 10^{25} ergs of energy to have a larger or smaller magnitude than 3.89? Your answer should be "larger."

$$M = \frac{\log(10^{25}) - 11.8}{1.5} = \frac{25 - 11.8}{1.5} = \boxed{8.8} \text{ Richter-scale magnitude}$$

c) If you are attempting to mentally estimate the answer (as we hope you are), please note that part (b) of this problem serves as a helpful guidepost. That is, since 8.0 is a little less than 8.8, how should the answer to this question compare to 10^{25}?

Given $M = 8.0$, we need to solve for E.

$$8.0 = \frac{\log E - 11.8}{1.5} \qquad \textit{Given formula (with M = 8.0)}$$

$$12 = \log E - 11.8 \qquad \textit{Multiply both sides by 1.5.}$$

$$23.8 = \log E \qquad \textit{Add 11.8 to both sides.}$$

$$E = 10^{23.8} \qquad \textit{Apply the definition of logarithm.}$$

Therefore, in an 8.0 magnitude earthquake, about $\boxed{10^{23.8} \text{ ergs}}$ of energy is released.

d) Essentially, we perform the same steps as those just shown in part (c). Doing so provides us with an alternative form of Gutenberg and Richter's formula, one that makes it easier to find the energy (E) when given the magnitude (M). This alternative form will prove useful in solving some of the problems in Section 7.7 (e.g., Example 9 on p. 488 and Problems 31 & 39 on p. 491).

$$M = \frac{\log E - 11.8}{1.5} \qquad \textit{Given formula}$$

$$1.5M = \log E - 11.8 \qquad \textit{Multiply both sides by 1.5.}$$

$$1.5M + 11.8 = \log E \qquad \textit{Add 11.8 to both sides.}$$

$$\boxed{E = 10^{1.5M + 11.8}} \qquad \textit{Apply the definition of logarithm.}$$

59. a) By definition, recall that a logarithm's argument must be positive (i.e., $\ln x$ exists when $x > 0$ but is undefined when $x \le 0$). So, in the case of the given formula, it follows that

$$1-\frac{N}{80}>0.$$

By solving for N, you will find that in order to use the formula, N must be less than 80 words per minute (wpm); however, a person with a typing goal of $79\frac{1}{2}$ wpm (or even 79.99 wpm) could still use the formula. Therefore, the maximum number of words per minute is **80**.

b)

$t=-62.5\ln\left(1-\frac{N}{80}\right)$	*Given formula*
$-\frac{t}{62.5}=\ln\left(1-\frac{N}{80}\right)$	*Multiply both sides by – 62.5.*
$e^{-t/62.5}=1-\frac{N}{80}$	*Apply the definition of logarithm.*
$\frac{N}{80}=1-e^{-t/62.5}$	*Isolate $\frac{N}{80}$.*
$\boxed{N=80(1-e^{-t/62.5})}$	*Multiply both sides by 80.*

Problem Set 7.7

1. The ***growth/decay formula*** can be found on textbook page 483.

3. **Wording varies**. As stated on p. 486 in your text, a ***logarithmic scale*** is a scale in which logarithms are used to make data more manageable by expanding small variations and compressing large ones. Problem 58 (p. 492) provides a helpful illustration of this.

5. **Projected population answers vary** (depending on the date).

As to how we determine the annual growth rate based upon the population figures given for July 1998 and July 1999, the process is the same as that used to solve textbook Examples 2 and 3 (pp. 483-484). We use the growth/decay formula:

$$A=A_0e^{rt}$$

$$5{,}996{,}215{,}340=5{,}918{,}624{,}368e^{r(1)}$$

Given: A_0 = 5,918,624,368 in July '98 and A = 5,996,215,340 in July '99, so t = 1.

$$\frac{5{,}996{,}215{,}340}{5{,}918{,}624{,}368} = e^r$$ *Divide both sides by 5,918,624,368*

$$\ln \tfrac{5{,}996{,}215{,}340}{5{,}918{,}624{,}368} = r$$ *Apply the definition of logarithm.*

$$r \approx 0.0130244419$$ *Use calculator to evaluate.*

So, the annual growth rate is approximately 1.3%.

Regarding Problems 7-14

Since these eight problems are identical in terms of the information requested, it makes sense to begin by creating (deriving) formulas that will expedite our efforts to determine this information.

For Part (a): *1980 to 1990 is a ten-year period, so t = 10. Using the growth/decay formula, we derive a formula that will yield the growth rate r:*

$$A = A_0 e^{r(10)}$$

$$\frac{A}{A_0} = e^{10r}$$

$$\ln\left(\frac{A}{A_0}\right) = 10r$$

$$r = \tfrac{1}{10}\ln\left(\frac{A}{A_0}\right) = \boxed{0.1\ln\left(\frac{A}{A_0}\right)}$$

So, to answer part (a), we will determine the growth rate by using the "framed" formula above, with A representing the 1990 population and A_0 the 1980 population.

For Part (b): *To predict the population P for 1994, we use the growth/decay formula $P = Ae^{4r}$, where r and A are as they were in part (a); namely, r is the exact value (not the rounded value) from part (a) and A represents the 1990 population. Obviously, t = 4 because 1990-1994 is a four-year period.*

Alternatively, if preferred, the 1994 population P can be predicted by using the formula $P = A_0 e^{14r}$ where A_0 represents the 1980 population (as it did in part (a)) and, once again, r is the "unrounded" growth rate.

For Part (c): *Comparisons and commentaries will vary.*

For Part (d): *This is similar to part (b). To predict the 2005 population, we can use the growth/decay formula $P = Ae^{15r}$ (i.e., 1990-2005 implies t = 15); or, alternatively, we can use $P = A_0e^{25r}$ with A_0 representing the 1980 population (so, 1980-2005 indicates t = 25).*

7. For Problems 7-14, please **read the information above** (beginning on the previous page). Using this information (and notation), we find:

a) Jacksonville's annual growth rate from 1980 to 1990 was approximately:

$$r = 0.1\ln\left(\frac{A}{A_0}\right) = 0.1\ln\left(\frac{636{,}070}{540{,}920}\right) \approx 0.016203723 \approx \boxed{1.6\%}.$$

b) We use the exact (not the rounded) growth rate from part (a) to predict Jacksonville's 1994 population. There are two (equivalent) ways to do this:

$$\begin{aligned} P &= Ae^{4r} & P &= A_0e^{14r} \\ &= 636{,}070e^{4r} & &= 540{,}920e^{14r} \\ &\approx 678{,}662.2023 & &\approx 678{,}662.2023 \end{aligned}$$

Thus, our prediction for Jacksonville's 1994 population is **678,662**.

c) **Comparisons/comments vary**. Since Jacksonville's actual 1994 population of 665,070 was slightly lower than the predicted number, we can infer that Jacksonville's rate of growth between 1990 and 1994 has slightly decreased compared to what it was in the previous decade.

d) Just like we did in part (b), there are two equivalent ways to predict Jacksonville's population for 2005:

$$\begin{aligned} P &= Ae^{15r} & P &= A_0e^{25r} \\ &= 636{,}070e^{15r} & &= 540{,}920e^{25r} \\ &\approx 811{,}078.1890 & &\approx 811{,}078.189 \end{aligned}$$

Thus, our prediction for Jacksonville's 2005 population is **811,078**.

9. For complete details, refer to this manual's solution to Problem 7 (and, just as importantly, the explanation that preceded it).

a) $r = 0.1\ln\left(\frac{A}{A_0}\right) = 0.1\ln\left(\frac{488{,}374}{455{,}651}\right) \approx 0.006935434 \approx \boxed{0.7\%}$.

b)
$$\begin{aligned} P &= Ae^{4r} & P &= A_0e^{14r} \\ &= 488{,}374e^{4r} & &= 455{,}651e^{14r} \\ &\approx 502{,}112.0198 & &\approx 502{,}112.0198 \end{aligned}$$

Thus, our prediction for Nashville's 1994 population is **502,112**.

c) **Comparisons/comments vary**.

d) $P = Ae^{15r}$ $\qquad$ $P = A_0e^{25r}$

$= 488{,}374e^{15r}$ $\qquad$ $= 455{,}651e^{25r}$

$\approx 541{,}917.0882$ $\qquad$ $\approx 541{,}917.0882$

Thus, our prediction for Nashville's 2005 population is **541,917**.

11. For complete details, refer to this manual's solution to Problem 7 (<u>and</u>, just as importantly, the explanation that preceded it).

Regarding part (c), please note that the data itself — as well as the answer to part (b) — makes it clear that Honolulu's actual growth in the early 1990's was more rapid than in the 80's.

a) **0.006%** $\qquad$ b) **365,362** $\qquad$ d) **365,608**

13. For complete details, refer to this manual's solution to Problem 7 and, just as importantly, the explanation that preceded it.

a) **0.07%** $\qquad$ b) **273,041** $\qquad$ d) **275,270**

15. January 1992 to January 2001 is a 9-year period, so $t = 9$. Based upon the information provided in textbook Example 3 (p. 484), we would predict the cumulative number of AIDS-related deaths at the end of January 2001 to be:

$$A = 209{,}693e^{0.1614549977(9)} \approx \boxed{896{,}716}.$$

17. Although this problem is very similar to text Example 5 (p. 485), we think several concepts are worthy of elaboration. First, remember that a radioactive substance's ***half-life*** refers to the time it takes for a quantity of the substance to decay to half of its original amount. So, with regard to the growth/decay formula's role in half-life problems, please note that it does not matter what values are assigned to A and A_0 <u>provided</u>, of course, that A equals $\frac{1}{2}A_0$ (or $0.5A_0$). For instance, in Example 5, your text's author chose to let $A = 50$ mg and $A_0 = 100$ mg, but he just as easily could have decided to have $A = 3$ lb and $A_0 = 6$ lb — or, more generally, we could leave A_0 as it is and simply replace A with $0.5A_0$. If we choose this latter approach, we would have:

$A = A_0e^{rt}$	*Growth/decay formula*
$0.5A_0 = A_0e^{rt}$	*Substitute appropriate half-life values.*
$0.5 = e^{rt}$	*Divide both sides by A_0.*
$rt = \ln 0.5$	*Apply the definition of logarithm.*

$$\boxed{r = \tfrac{1}{t}\ln 0.5 \text{ or } \frac{\ln 0.5}{t}}$$ *Divide both sides by t.*

So, to solve this particular problem, we use the above (framed) formula. Given that cesium-17 has a 30-year half-life, it follows that its decay constant is:

$$r = \tfrac{1}{t}\ln 0.5 = \tfrac{1}{30}\ln 0.5 \approx \boxed{\mathbf{-0.023104906}}$$

19. Finding the half-life means we need to determine t, and the explanation provided above for Problem 17 can help us do that. In particular, we derived the fact that $rt = \ln 0.5$, from which it follows that

$$t = \left(\frac{1}{r}\right)\ln 0.5 \text{ or } \frac{\ln 0.5}{r}.$$

So, given that its decay constant is -0.0246, we can conclude that strontium-90's half-life is approximately

$$t = \frac{\ln 0.5}{-0.0246} \approx 28.17671466 \approx \boxed{\textbf{28 years}}.$$

21. This problem is very similar to text Example 7 (p. 487). For additional details, please refer to this example <u>and</u> the information preceding it.

$$\begin{aligned} M &= 4.0 + \log A \\ &= 4.0 + \log(5.1 \times 10^2) \\ &\approx \boxed{\mathbf{6.7}} \text{ Richter-scale magnitude} \end{aligned}$$

23. This problem is very similar to text Example 8 (pp. 487-488). "Comparing" the strengths of two earthquakes suggests the need to determine the "difference" between their respective magnitudes (i.e., $M_1 - M_2$). For this particular problem, $M_1 = 6$, $M_2 = 4$, and we proceed in a manner comparable to Example 8:

$$\begin{aligned} 6 - 4 &= \log\frac{A_1}{A_2} \\ 2 &= \log\frac{A_1}{A_2} \\ 10^2 &= \frac{A_1}{A_2} \\ A_1 &= 100A_2 \end{aligned}$$

Therefore, an earthquake with a magnitude of 6 is **100 times stronger** than an earthquake whose magnitude is 4.

25. Problems 25-30 are virtually identical to textbook Example 10 (p. 489). We use the formula

$$D = 10\log\left(\frac{I}{I_0}\right),$$

where D is the number of decibels for a sound of intensity I and I_0 is the intensity of a sound barely audible for a person with normal hearing.

So, given a whisper with an intensity of $I = 10^{-13}$ watts/cm^2, we use $I_0 = 10^{-16}$ watts/cm^2 and obtain a decibel rating of:

$$\begin{aligned} D &= 10\log\frac{10^{-13}}{10^{-16}} \\ &= 10\log 10^3 && \textit{Subtraction law of exponents (see text p. 36)} \\ &= 10(3) && \textit{Fundamental "Grant's tomb" properties (p. 475)} \\ &= \boxed{\textbf{30 dB}} \end{aligned}$$

For information pertaining to the decibel rating scale and for assistance in putting this 30 dB rating in context relative to other sounds, we recommend you examine textbook Table 7.5 (p. 489) and read the author's comments located between Examples 10 and 11.

27. For additional details, refer to textbook Example 10 (p. 489) and this manual's solution to Problem 25 above.

The decibel rating for a rock concert with sound intensity 5.23 × 10^{-6} watts/cm^2 is:

$$D = 10\log\frac{5.23\times10^{-6}}{10^{-16}} = 10\log(5.23\times10^{10}) \approx 107.1850169 \approx \boxed{\textbf{107 dB}}.$$

It is our hope that you anticipated (i.e., "mentally estimated") a significantly higher decibel rating for a rock concert than for Problem 25's whisper. Did you?

29. Before computing the answer, please take a moment to make a "mental estimate." For example, how do you think a rocket engine's decibel rating should compare to a rock concert's rating (as in Problem 27)?

After estimating the answer, we, once again, apply the formula previously explained both above and on textbook page 489:

$$D = 10\log\frac{2.53\times10^{-5}}{10^{-16}} = 10\log(2.53\times10^{11}) \approx 114.0312052 \approx \boxed{\textbf{114 dB}}.$$

31. This problem is similar to parts (c) and (d) of the preceding section's Problem 57 (see Section 7.6, p. 482), a problem for which this manual contains a detailed solution. Since $\log E = 11.8 + 1.5M$, it follows (by the definition of logarithm) that

$$E = 10^{11.8+1.5M}.$$

Thus, given $M = 8.3$, the 1906 San Francisco earthquake released

$$E = 10^{11.8+1.5(8.3)} = \boxed{10^{24.25}\ \textbf{ergs}} \text{ of energy.}$$

33. This problem is similar to various other problems, including text Example 2 (pp. 483-484) and Problems 7-14 (see solutions above).

a) First, we need to compute the annual growth rate r (from 1980 to 1990). Solving the growth/decay formula r, we find:

$$A = A_0 e^{rt}$$

$$\frac{A}{A_0} = e^{rt}$$

$$\ln\left(\frac{A}{A_0}\right) = rt$$

$$r = \left(\frac{1}{t}\right)\ln\frac{A}{A_0}$$

So, for the 1980-1990 period, we have $t = 10$, $A = 3{,}485{,}557$, and $A_0 = 2{,}968{,}528$. Hence, the growth rate is:

$$r = \frac{1}{10}\ln\left(\frac{3{,}485{,}557}{2{,}968{,}528}\right) \approx \underline{\underline{0.01605616522}} \approx 1.6\%$$

Then, to predict Los Angeles' 2004 population, use the growth rate shown on your calculator display (i.e., the double-underlined number above); do not use the rounded value. As for which year's population (1994, 1990, or 1980) should be assigned to A_0, the most recent population data (in this case, 1994) should be used. Since 1994-2004 is a ten-year span, we let $t = 10$ and find:

$$P = 3{,}448{,}613e^{10r}$$
$$\approx 4{,}049{,}258.482$$

Thus, our prediction for Los Angeles' 2004 population is **4,364,109**.

b) By using the formula derived in part (a), we obtain the growth rate for the 1990-1994 period:

$$r = \frac{1}{4}\ln\left(\frac{3{,}448{,}613}{3{,}485{,}557}\right) \approx \underline{\underline{-0.002663935}} \approx -0.27\%$$

So, we use this value of r (in its "unrounded" form) and the growth/decay formula to predict the population of Los Angeles in 2004. There are two (equivalent) ways to do this:

$$P = 3{,}485{,}557e^{14r} \approx 3{,}357{,}957.076$$

$$P = 3{,}448{,}613e^{10r} \approx 3{,}357{,}957.076$$

Thus, our prediction for Los Angeles' 2004 population is **3,357,957**.

c) Once again, by using the formula derived in part (a), we obtain the growth rate for the 1980-1994 period:

$$r = \frac{1}{14}\ln\left(\frac{3{,}448{,}613}{2{,}968{,}528}\right) \approx \underline{\underline{0.010707565298}} \approx 1.07\%$$

Then, we use this "unrounded" value of r and the growth/decay formula to predict the population of Los Angeles in 2004. Here are two (equivalent) ways to do so:

$$P = 3{,}448{,}613e^{10r} \approx 3{,}838{,}369.911 \qquad P = 2{,}968{,}528e^{24r} \approx 3{,}838{,}369.911$$

Thus, our prediction for Los Angeles' 2004 population is **3,838,369**.

35. This problem is comparable to text Example 6 (p. 486). In terms of the growth/decay equation, having 28% of the original ^{14}C remaining means the ratio of A to A_0 equals 0.28.

In the event the reasoning behind this inference is unclear to you, please pause to reflect upon the half-life problems solved previously in this section (e.g., Problem 19) and compare them to the situation presented in this problem. In the half-life problems, the process of finding the time it takes for half the original substance to remain lead us to the realization that the ratio of A to A_0 equaled 0.5. We hope you see the parallel.

So, the probable age of the artifact can be determined as follows:

$$A = A_0e^{rt}$$
$$\frac{A}{A_0} = e^{rt}$$
$$0.28 = e^{rt}$$
$$rt = \ln 0.28$$
$$t = \frac{\ln 0.28}{r}$$

With the decay rate for carbon-14 being $r = -1.209680943 \times 10^{-4}$ (as was previously determined in text Example 5), it follows that:

$$t = \frac{\ln 0.28}{-1.209680943 \times 10^{-4}} \approx 10{,}523.15227$$

Hence, the artifact is about **10,500 years old**.

37. This problem is related to text Examples 5-6 (pp. 485-486) and assorted other problems within this problem set. We first need to determine the decay rate for sodium-22, and we rely upon this manual's explanation for Problem 17 to accomplish this. Given that sodium-22 has a half-life of 2.6 years, its decay constant is:

$$r = \tfrac{1}{t}\ln 0.5 = \tfrac{1}{2.6}\ln 0.5 \approx -0.2665950694$$

The fact that 15.5 g of an original 100-g sample remains implies that the ratio of A to A_0 is 0.155 (i.e., $15.5 \div 100$). So, to determine the elapsed time t, we proceed in a manner comparable to what was shown for Problem 35 (see above solution):

$$t = \frac{\ln 0.155}{-0.2665950694} \approx 6.993115688$$

Accordingly, the elapsed time is about **7 years**.

39. This problem is very similar to the text's Example 9 (p. 488).

$$\frac{E_1}{E_2} = \frac{10^{1.5(8.3)+11.8}}{10^{1.5(9.5)+11.8}} = \frac{10^{24.25}}{10^{26.05}} = 10^{-1.8},$$

so $10^{1.8}E_1 = E_2$.

Therefore, the 1960 Chile quake (which had a magnitude of 9.5) released $10^{1.8} \approx$ **63 times more energy** than the 1906 San Francisco quake (with magnitude 8.3).

41. For additional details, please refer to text Example 6 (p. 486) and this manual's solution to Problem 35.

As was previously determined in Example 5 (p. 485), the decay constant for ^{14}C is $r = -1.209680943 \times 10^{-4}$. So, given that 85% of this artifact's original ^{14}C remains,

$$t = \frac{\ln 0.85}{-1.209680943 \times 10^{-4}} \approx 1{,}343.485904.$$

Hence, rounded to the nearest 100 years, the artifact is about **1,300 years old**.

43. This problem is comparable to text Example 9 (p. 488) and Problems 38-39.

$$\frac{E_1}{E_2} = \frac{10^{1.5(7.0)+11.8}}{10^{1.5(7.1)+11.8}} = 10^{-0.15},$$

so $10^{0.15}E_1 = E_2$.

Therefore, this 0.1 revision in magnitude equated to an increase of $10^{0.15} \approx$ **1.41 times more energy** having been released during the 1989 World Series than was initially believed.

45. Once again, we perform the same procedure as was previously demonstrated in text Example 9 (p. 488) and in this manual's solutions to Problems 39 and 43.

$$\frac{E_1}{E_2}=\frac{10^{1.5(8)+11.8}}{10^{1.5(7)+11.8}}=10^{1.5},$$

so $E_1=10^{1.5}E_2$.

Therefore, a quake having a magnitude of 8 releases $10^{1.5}\approx$ **31.6 times more energy** than a quake with a magnitude of 7.

47. This problem is comparable in some respects to text Examples 5-6 (pp. 485-486) and assorted other problems within this problem set.
We first need to determine the hourly decay rate for gallium-67 (^{67}Ga), and we rely upon this manual's explanation for Problem 17 to accomplish this. Given that ^{67}Ga has a half-life of 46.5 hours, it follows that its decay constant is

$$r=\frac{\ln 0.5}{t}=\frac{\ln 0.5}{46.5}\approx -0.014906391.$$

Next, given an initial quantity of 100 mg, we use the growth/decay formula to compute the amount of gallium-67 remaining after 30 hours and 35 hours:

Amount present after 30 hrs: $A=100e^{30r}\approx 63.94213007$ mg

Amount present after 35 hrs: $A=100e^{35r}\approx 59.34966649$ mg

So, 4.59246358 mg of ^{67}Ga was lost during these 5 hours. Thus, in terms of a percentage, the loss is approximately **7.2%** (i.e., 4.59246358 ÷ 63.94213007).

In response to the problem's follow-up question, **yes**, the percentage lost (7.2%) will remain the same regardless of the 5-hour period under consideration. Here is why:

Amount A_1 present after x hrs: $A_1=100e^{rx}$

Amount A_2 present after $(x+5)$ hrs: $A_2=100e^{r(x+5)}=100e^{rx+5r}=100e^{rx}e^{5r}$

Amount of ^{67}Ga lost:

$$A_1-A_2=100e^{rx}-100e^{rx}e^{5r}=100e^{rx}\left(1-e^{5r}\right)$$

Percentage (proportion) lost:

$$\frac{A_1-A_2}{A_1}=\frac{100e^{rx}\left(1-e^{5r}\right)}{100e^{rx}}=1-e^{5r}$$

Notice that the percentage lost ($1-e^{5r}$) is a quantity that is independent of x (the hours elapsed); instead, it is solely dependent upon r. Therefore, we know the percentage lost remains the same (specifically, 7.2% for ^{67}Ga) regardless of which 5-hour period is considered (or, for that matter, regardless of the initial quantity of gallium-67).

You may wish to begin by confirming that $1-e^{5r}$ equals (in the case of ^{67}Ga) 7.2%. We encourage you to investigate this further, especially since it is applicable to other growth-type applications — including those related to personal finance in Chapter 8.

49. Via Newton's Law of Cooling, we have:

$$T = A + (B - A)10^{-kt} = 74 + (250 - 74)10^{-0.075(10)} \approx 105.2977176$$

So, the temperature of the cookies is approximately **105°F**.

51. First, we determine the constant k by applying the given information to Newton's Law of Cooling (as shown on textbook p. 491):

$$k = \frac{1}{t}\log\frac{B-A}{T-A} = \tfrac{1}{30}\log\frac{100-22}{45-22} = \tfrac{1}{30}\log\tfrac{78}{23} \approx 0.0176788922.$$

So, the object's approximate temperature after 40 minutes of cooling will be

$$T = A + (B - A)10^{-kt} = 22 + (100 - 22)10^{-0.0176788922(40)} \approx 37.308721 \approx \mathbf{37.3°C}.$$

53. Given $P = 11.9$ psi, we solve for a (in miles):

$$P = 14.7e^{-0.21a}$$ *Given formula*

$$11.9 = 14.7e^{-0.21a}$$ *Substitute accordingly.*

$$\frac{11.9}{14.7} = e^{-0.21a}$$ *Divide both sides by 14.7.*

$$\ln\left(\frac{11.9}{14.7}\right) = -0.21a$$ *Apply the definition of logarithm.*

$$a = \frac{\ln\left(11.9/14.7\right)}{-0.21}$$ *Divide both sides by –0.21.*

$$a \approx 1.006233779 \text{ mi}$$ *Via calculator*

So, knowing that 1 mi = 5,280 ft, we conclude that Denver's altitude is approximately **5,313 ft** above sea level.

55. As directed, we begin by solving the given formula for t:

$$P = 50e^{-t/250}$$

$$\frac{P}{50} = e^{-t/250}$$

$$\ln\left(\frac{P}{50}\right) = -\frac{t}{250}$$

$$t = -250\ln\left(\frac{P}{50}\right)$$

If $P = 30$ watts, then $t = -250\ln\left(\frac{30}{50}\right) = -250\ln(0.6) \approx 127.7064$. So, the satellite will operate for about **128 days**.

57. This problem is very similar to text Example 6 (p. 486), as well as Problems 35, 40, and 41.

As was previously determined in Example 5, the decay constant for carbon-14 (denoted ^{14}C) is $r = -1.209680943 \times 10^{-4}$. So, given that 92.3% of the Shroud's original ^{14}C remains,

$$t = \frac{\ln 0.923}{-1.209680943 \times 10^{-4}} \approx 662.3733716.$$

Hence, as of 1988 (the year in which the sample was taken), the Shroud of Turin was about **662 years old**, which would suggest this artifact was made in the 14th century A.D. (i.e., 1988 – 662 = 1326). Assuming that this scientific analysis was performed correctly, the implication would mean that the shroud is not the burial robe of Christ and, in fact, was virtually new when Pierre d'Arcis wrote his memo to the Pope.

59. a) Substitute the given value $x = 75°$ F into the formula, and then use your calculator to compute y, the expected number of eroded/leaky O-rings. Be sure to avoid rounding in the middle of the computation, for doing so can adversely affect the accuracy of your answer.

$$y = \frac{6e^{5.085-0.1156(75)}}{1+e^{5.085-0.1156(75)}} = \frac{6e^{-3.585}}{1+e^{-3.585}} \approx 0.161928645$$

So, since $0.16 \approx 0$, **we expect no O-ring failures**.

b) Similarly, we let $x = 32°$ F (i.e., water's freezing point) and find:

$$y = \frac{6e^{5.085-0.1156(32)}}{1+e^{5.085-0.1156(32)}} = \frac{6e^{1.3858}}{1+e^{1.3858}} \approx 4.799525343.$$

So, at this colder temperature, **we expect 5 O-ring failures**.

Chapter 7 Review Questions

1. a) *[accuracy varies]* ______________________________

b) To the nearest tenth, the given segment's length is **1.1 cm**.

c) **Answers vary** from person to person. The approximate width of my index finger is ¾ in., or roughly 1.9 cm.

2. These concepts are covered in Sections 7.1 and 7.2.

a) Begin by finding the distance around the semicircular (curved) part of the figure. Recall that the circumference C of a circle with diameter d is $C = \pi d$. Since the semicircle's diameter is 8 in. long, the distance around it is

$$\tfrac{1}{2}C = \tfrac{1}{2}\pi d = \tfrac{1}{2}\pi(8) = 4\pi.$$

With that, one can determine the distance around the entire figure:

$$10 + 7 + 4\pi + 7 + 6 = (30 + 4\pi) \text{ in.} \approx 42.5664 \text{ in.},$$

which (to the nearest inch) equals **43 in.**.

b) The given figure may be viewed in two ways: either as a figure consisting of a trapezoid and a semicircle, or as one made up of a triangle, rectangle, and semicircle. Since the first characterization involves only two regions, it is the more efficient approach.

$$A_{\text{trapezoid}} = \tfrac{1}{2}h(b+B) = \tfrac{1}{2}(8)(7+13) = 80$$
$$A_{\text{semicircle}} = \tfrac{1}{2}\pi r^2 = \tfrac{1}{2}\pi(4)^2 = 8\pi$$
$$A_{\text{total}} = A_{\text{trapezoid}} + A_{\text{semicircle}} = 80 + 8\pi \approx \boxed{105 \text{ in.}^2}$$

3. This problem is representative of the type(s) discussed in Section 7.3.

a) $V = \ell wh = (3)(2)(5) = \boxed{30 \text{ dm}^3}$

b) Since 1 L = 1 dm^3 [review, as needed, textbook pp. 450-451], 30 dm^3 = **30 L**.

c) Notice that this rectangular box (also known as a *parallelepiped*) does not have a top. The front and back faces are rectangles, each with dimensions 3 dm by 5dm. So, each has an area of 15 dm^2 (i.e., 3 × 5). In a similar fashion, we find that the area of the bottom face is 6 dm^2 and the area of each of the two side faces is 10 dm^2. Therefore, the surface area of this "topless" box is:

$$\underbrace{(\text{Area of front \& back})}_{(2 \times 15 \text{ dm}^2)} + \underbrace{(\text{Area of 2 sides})}_{(2 \times 10 \text{ dm}^2)} + \underbrace{(\text{Area of bottom})}_{(6 \text{ dm}^2)}$$

which equals $\boxed{56 \text{ dm}^2}$.

4. This problem is very similar to textbook Example 5 (see p. 458 in Section 7.4).

a) $S = 4\pi r^2 = 4\pi(1)^2 = \boxed{4\pi \text{ ft}^2}$

b) $S = 4\pi \approx 12.56637061 \approx \boxed{12.6 \text{ ft}^2}$

c) $V = \tfrac{4}{3}\pi r^3 = \tfrac{4}{3}\pi(1)^3 = \boxed{\tfrac{4}{3}\pi \text{ ft}^3}$

d) $V = \tfrac{4}{3}\pi \approx 4.188790205 \approx \boxed{4.2 \text{ ft}^3}$

5. a) **Feet and meters** would be reasonable units of measurement to use when measuring the height of a building.

b) **Answers vary**; 104°F = 40°C would be such a temperature. To review the standard units of temperature, refer to textbook pp. 460-461.

c) The prefix milli- means **one-thousandth (0.001 or 1/1000)**.

6. a) 8.6 milliliters is a measurement of **capacity**.

b) For additional details, review textbook Section 7.4 (in particular pp. 460-463). If you are sufficiently familiar with the metric system (and we hope you are), then it is understood that a conversion from km to cm requires moving the decimal point 5 places to the right. So, 10 km = **1,000,000 cm**.

7. a) The area of the "top of the pool" corresponds to the area consumed by the surface of the water. So, in light of its rectangular dimensions, the pool's surface area equals $80 \text{ ft} \times 30 \text{ ft} =$ **2,400 ft²**.

b) $V = \ell wh = (80)(30)(3) =$ **7,200 ft³**

c) From textbook p. 451, recall that 1 cu ft ≈ 7.48 gal. So, it follows that

$$7{,}200 \text{ ft}^3 \approx 7{,}200 \text{ ft}^3 \times 7.48 \text{ gal/ft}^3 = \textbf{53,856 gal}.$$

8. Here are two approaches:

I. Convert each of the carpet's dimensions from inches to yards, and then determine its area. That is, since 1 yd = 3 ft, it follows that

$$11 \text{ ft} = \tfrac{11}{3} \text{ yd} = 3\tfrac{2}{3} \text{ yd} \quad \text{and} \quad 16 \text{ ft} = \tfrac{16}{3} \text{ yd} = 3\tfrac{2}{3} \text{ yd}$$

Hence, the carpet's area is $A = \ell w = (\tfrac{11}{3} \text{ yd})(\tfrac{16}{3} \text{ yd}) = \tfrac{176}{9} \text{ yd}^2 = 19.\overline{5} \text{ yd}^2$.

However, since fractions of a square yard cannot be purchased, an 11-ft by 16-ft room requires **20 yd²** of carpeting.

II. Alternatively, first determine the carpet's area in square feet, and then convert square feet to square yards. That is,

$$A = \ell w = (11 \text{ ft})(16\text{ft}) = 176 \text{ ft}^2.$$

We know $1 \text{ yd}^2 = 9 \text{ ft}^2$ [because $1 \text{ yd}^2 = (1 \text{ yd})^2 = (3 \text{ ft})^2 = 9 \text{ ft}^2$]. So, the amount of carpet (in square yards) needed for an 11-ft by 16-ft room is:

$$176 \text{ ft}^2 \div 9 \text{ ft}^2/\text{yd}^2 = 176 \text{ ft}^2 \times \frac{1 \text{ yd}^2}{9 \text{ ft}^2} = 19.\overline{5} \text{ yd}^2 \approx \underline{20 \text{ yd}^2}.$$

9. a) Consider constructing a table.

Size	*Diameter*	*Radius*	*Area*	*Price per square inch*
Mini	6 in.	3 in.	$\pi(3)^2 = 9\pi$ in.²	\$3.71 ÷ 9π in.² ≈ \$0.13/in.²
Small	10 in.	5 in.	$\pi(5)^2 = 25\pi$ in.²	\$7.95 ÷ 25π in.² ≈ \$0.10/in.²
Medium	12 in.	6 in.	$\pi(5)^2 = 36\pi$ in.²	\$11.18 ÷ 36π in.² ≈ \$0.10/in.²
Large	14 in.	7 in.	$\pi(7)^2 = 49\pi$ in.²	\$13.88 ÷ 49π in.² ≈ \$0.09/in.²

Because it has the lowest price (per unit of area), the **large size** pizza is the best value.

b) **Answers vary**. A 16-in. diameter ("extra large") pizza has an approximate area of 201 in.². Since the price for a large is about \$0.09/in.², one might decide to charge \$17.95 for the extra large [note: 201 × 0.09 = 18.09]. Any price between \$16.95 and \$18.25 would be considered appropriate.

10. a) Here is such an illustration:

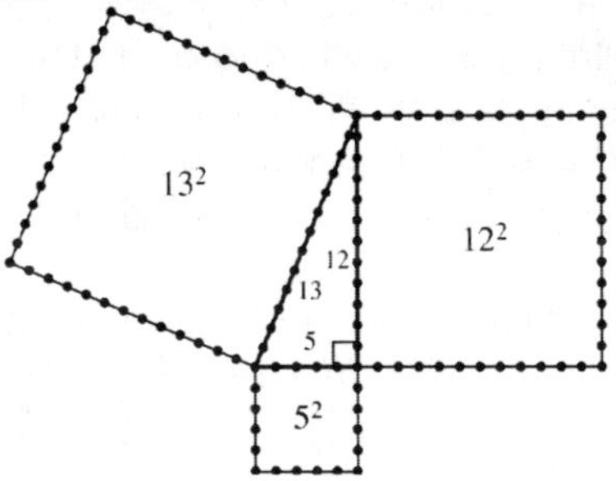

b) **Yes**; since $25^2 + 312^2 = 313^2$, it is possible to form such a right triangle.

11. Begin by creating a diagram.

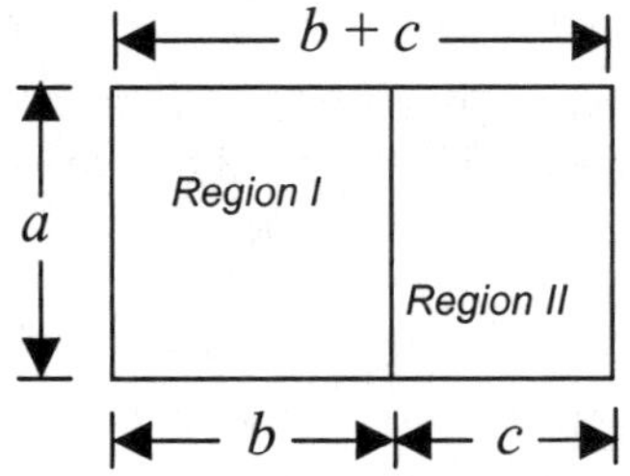

We know the area of the large rectangle is the same as the sum of the areas of regions I and II. That is,

$$\underbrace{\text{Area of Region I}}_{ab} + \underbrace{\text{Area of Region II}}_{ac} = \underbrace{\text{Area of Large Rectangle}}_{a(b+c)}$$

This provides the desired geometric justification for the distributive property.

12. This problem requires a general understanding of logarithms and, more specifically, a comprehension of the fundamental ("Grant's tomb) properties and the laws of logarithms. These properties/laws are introduced on textbook pp. 475 and 477.

a) By the definition of logarithm, recall that $\log 100$ implicitly asks us to determine the exponent to which we raise a base of 10 to obtain 100. Hence, $\log 100 = 2$. Via the Grant's tomb properties (which were explained at length in this manual's solutions for Section 7.6), we know

$$\log\sqrt{10} = \log 10^{\frac{1}{2}} = \tfrac{1}{2} \text{ and } 10^{\log 0.5} = 0.5.$$

Therefore, $\log 100 + \log\sqrt{10} + 10^{\log 0.5} = 2 + \frac{1}{2} + 0.5 = \boxed{3}$.

b) This is similar to part (a). Recall that $\log 1{,}000 = 3$.

$$\left(\ln e + \ln 1 + \ln e^{542}\right) + \ln e^{\log 1{,}000} = (1 + 0 + 542) + \ln e^3 = 543 + 3 = \boxed{546}$$

c) Via the first (additive) law of logarithms, it follows that

$$(\log_8 4+\log_8 16)=\log_8(4\cdot 16)=\log_8 64.$$

Thus, $(\log_8 4+\log_8 16)+\log_8 8^{2.3}=(\log_8 64)+2.3=2+2.3=\boxed{4.3}$.

13. **Before** using your calculator, you should always attempt to "mentally estimate" the answer. By doing so, you will be able to detect calculator keystroke errors and, as a consequence, decrease your likelihood of writing erroneous (often completely unreasonable) answers. More importantly, you will significantly enhance your understanding of logarithms. To review such estimation techniques, please refer to many of this manual's solutions for Section 7.5.

a) $\log 8.43\approx 0.92582757\approx\boxed{0.93}$ b) $\log 9{,}760\approx 3.98944982\approx\boxed{3.99}$

14. Please: **Before** using your calculator, make sure you always attempt to "mentally estimate" the answer. By doing so, you will be able to detect calculator keystroke errors and, as a consequence, decrease your likelihood of writing erroneous (often completely unreasonable) answers. Most importantly, though, you will significantly enhance your understanding of logarithms. To review such estimation techniques, please revisit many of this manual's solutions for Section 7.5.

a) $\ln 2\approx 0.69314718\approx\boxed{0.69}$ b) $\ln 0.125\approx -2.07944154\approx\boxed{-2.08}$

c) $\log_\pi\left(\frac{1}{\pi}\right)=\log_\pi \pi^{-1}=\boxed{-1}$ *[No calculator use necessary.]*

15. This problem is comparable to many of those encountered in Sections 7.5–7.7 (beginning with Examples 9-10 on textbook pp. 471-472).

a) $10^x=85$ *Given exponential equation*

$x=\log 85$ *Apply the definition of logarithm.*

$\boxed{x\approx 1.929418926}$ *Calculator computation*

b) $e^{3x+1}=45$ *Given exponential equation*

$3x+1=\ln 45$ *Apply the definition of logarithm.*

$x=\dfrac{-1+\ln 45}{3}$ *From both sides, subtract 1 and then divide by 3.*

$\boxed{x\approx 0.9355541633}$ *Calculator computation*

c) $435^x=890$ *Given exponential equation*

$x=\log_{435} 890$ *Apply the definition of logarithm.*

$x=\dfrac{\log 890}{\log 435}$ or $\dfrac{\ln 890}{\ln 435}$ *Apply the change-of-base theorem.*

$\boxed{x\approx 1.117832865}$ *Calculator computation*

16. a) By the definition of logarithm, $\log_6 x = 4$ implies $\boxed{\mathbf{x = 6^4 = 1,296}}$.

b) $2^{3x-1} = 6$ *Given exponential equation*

$3x - 1 = \log_2 6$ *Apply the definition of logarithm.*

$\boxed{\mathbf{x = \dfrac{1+\log_2 6}{3}}}$ *From both sides, add 1 and then divide by 3. This is the <u>exact</u> answer.*

17. a) $\log(x+1) = 2 + \log(x-1)$ *Given logarithmic equation*

$\log(x+1) - \log(x-1) = 2$ *Subtract log (x - 1) from both sides.*

$\log\left(\dfrac{x+1}{x-1}\right) = 2$ *Apply the second law of logarithms.*

$\dfrac{x+1}{x-1} = 10^2$ *Apply the definition of logarithm.*

$x + 1 = 100x - 100$ *Multiply both sides by (x - 1); distribute.*

$101 = 99x$ *Solve linear equation accordingly.*

$\boxed{\mathbf{x = \tfrac{101}{99}}}$

b) Here are two ways to solve this equation.

<u>Approach I</u>: $3\ln\dfrac{e}{\sqrt[3]{5}} = 3 - \ln x$ *Given logarithmic equation*

$3\left(\ln e - \ln\sqrt[3]{5}\right) = 3 - \ln x$ *Apply the second law of logarithms.*

$3\left(\ln e - \ln 5^{\frac{1}{3}}\right) = 3 - \ln x$ *Recall that* $\sqrt[3]{5} = 5^{\frac{1}{3}}$.

$3\left(1 - \tfrac{1}{3}\ln 5\right) = 3 - \ln x$ *Via the third law of logarithms.*

$3 - \ln 5 = 3 - \ln x$ *Apply the distributive property.*

$\ln 5 = \ln x$ *Subtract 3 from both sides, etc.*

$\boxed{\mathbf{x = 5}}$ *Via log-of-both-sides theorem (text p. 476)*

<u>Approach II</u>: $3\ln\dfrac{e}{\sqrt[3]{5}} = 3 - \ln x$ *Given logarithmic equation*

$\ln\left(\dfrac{e}{\sqrt[3]{5}}\right)^3 = 3 - \ln x$ *Third law of logarithms*

$\ln\left(\dfrac{e^3}{5}\right) + \ln x = 3$ *Simplify exponents; add ln x to both sides.*

$$\ln\left(\frac{e^3 x}{5}\right) = 3 \qquad \textit{First law of logarithms}$$

$$\frac{e^3 x}{5} = e^3 \qquad \textit{Definition of logarithm}$$

$$x = e^3\left(\frac{5}{e^3}\right) \qquad \textit{Solve for } x.$$

$$x = 5 \qquad \textit{Simplify}$$

18. $A = P(1+i)^x$

$$\frac{A}{P} = (1+i)^x \qquad \textit{Divide both sides by } P.$$

$$\boxed{x = \log_{(1+i)}\left(A/P\right)} \qquad \textit{Apply the definition of logarithm.}$$

19. When half the wound has healed, the area of the unhealed skin (A) equals half the area of the original wound (A_0). In other words, the ratio A/A_0 equals ½.

To find the number of days it takes for half the wound to heal (or, equivalently, half the wound to remain), solve the given model for t and substitute/evaluate accordingly.

$$A = A_0 e^{-0.1t} \qquad \textit{Given}$$

$$\frac{A}{A_0} = e^{-0.1t} \qquad \textit{Divide both sides by } A_0.$$

$$-0.1t = \ln\frac{A}{A_0} \qquad \textit{Definition of logarithm}$$

$$t = -10\ln\frac{A}{A_0} \qquad \textit{Isolate } t.$$

$$t = -10\ln\tfrac{1}{2} \qquad \textit{Substitute accordingly (as explained above).}$$

$$t \approx 6.93147181 \qquad \textit{Calculator computation}$$

Hence, it takes approximately **7 days** for half the wound to heal.

20. Since the AIDS epidemic affects all our lives, we think you may find it informative to consider this problem while you simultaneously re-examine Example 3 (on textbook p. 484). Each addresses an important aspect of this complex societal issue.

The spread of the AIDS virus can be modeled with the growth/decay formula $A = A_0 e^{rt}$ (review Section 7.7 as necessary). Since we desire to create an equation/model that will be used to make projections for the next decade, it makes sense to let t represent the elapsed time in years. As the following shows, the given 1992 data enables us to determine the virus' rate of growth among teenagers:

$2A_0 = A_0 e^{r\left(\frac{7}{6}\right)}$ *Notes: (i) 14 months is equivalent to $\frac{14}{12}$ yr = $\frac{7}{6}$ yr.*

(ii) "Doubling" in 14 months implies $A = 2A_0$.

$2 = e^{\frac{7}{6}r}$ *Divide both sides by A_0.*

$\frac{7}{6}r = \ln 2$ *Apply the definition of logarithm.*

$r = \frac{6}{7}\ln 2$ *Multiply both sides by the reciprocal of $\frac{7}{6}$.*

$r \approx 0.594126155$ *Approximate (annual) rate of infection*

Thus, a suitable equation/model is $\boxed{A = A_0 e^{0.59t}}$ where A_0 is the number of teens infected at a given time and A is the number infected t years later.

CHAPTER 8: The Nature of Sequences, Series, and Financial Management

Problem Set 8.1

1. **Answers vary**. ***Interest*** is a rental fee paid for the use of another's money.

3. **Wording varies**. Interest that is calculated only on the principal (i.e., the amount of the deposit or loan) is known as ***simple interest***. Interest that is calculated both on the principal and on any accrued interest is known as ***compound interest***.

5. **Wording varies**. Example 14 pertains to investing — on behalf of a newborn child — a lump sum of money in an interest-bearing account, with the intention of letting the money grow unimpeded until the child reaches normal retirement age. In dramatic fashion, the example reveals how much money can be accrued if one plans ahead. In today's investment climate, 12% is not an unrealistic rate of return, and Example 14 shows that an investment at such a rate will turn a modest one-time deposit of $410.27 into a retirement nest egg of $1,000,000.
— Yes, with a little foresight, self-discipline, and mathematical knowledge, you too can become a millionaire.

7. **Choice C**. The ***future value*** is the amount of money you have after the interest has been added to the principal, which for this problem is $100. Hence, since the future value must be more than $100, choices A and B are immediately disregarded. Needless to say, we eliminate choice D because (with our knowledge of percentage rates) it is relatively easy to estimate an investment's future value.

9. **Choice B**. Choice A is equivalent to an annual income of $3,600, a sum that is much too little to live on in the U.S. As for choice C, a monthly income of $500,000 equates to $6 million per year, which only the ultra rich might consider a "reasonable" retirement income.

11. **Choice A**. Since $P \approx \$50{,}000$ and $r \approx 10\%$, we can "mentally estimate" the interest for one year as approximately $5,000. Therefore, the interest for two years is estimated to be $10,000.

13. **Choice D**. It is important to remember in all of these investment/loan formulas that t is defined as the time <u>*in years*</u>. So, the decimal portion of 3.52895 (namely, 0.52895) represents half a year. From textbook page 502, please recall that we are using 360 for the number of days in a year (unless otherwise stated). Clearly, choice D is the best answer because it is closest to ½ of 360.

15. **Choice B**. Like several of its predecessors, this problem is intended to heighten your awareness of the fact that t is defined as the time *in years* (not in days, months, etc.). Using 360 for the number of days in a year (as was previously stipulated in your text), a loan period of 180 days implies $t = ½$.

17. To review simple interest, refer to textbook page 501. Prior to performing the actual computations, be sure to express the interest rate r in decimal notation (not as a percentage). $I = Prt = 5{,}000(0.10)(3) = \boxed{\$1{,}500}$

19. $I = Prt = 1{,}000(0.14)(30) = \boxed{\$4{,}200}$

21. For further details, review textbook pp. 504-507.

$$A = P\left(1+\frac{r}{n}\right)^{nt} = 835\left(1+\frac{0.035}{2}\right)^{2\cdot 6} \approx 1{,}028.251828 \approx \boxed{\$1{,}028.25}$$

23. $$A = P\left(1+\frac{r}{n}\right)^{nt} = 9{,}730.50\left(1+\frac{0.076}{12}\right)^{12\cdot 7} \approx 16{,}536.79164 \approx \boxed{\$16{,}536.79}$$

25. The concept of continuous compounding is introduced on textbook pp. 507-508.

$$A = Pe^{rt} = 119{,}400e^{0.075\cdot 30} \approx 1{,}132{,}835.659 \approx \boxed{\$1{,}132{,}835.66}$$

27. Using the present value formula provided on textbook p. 510, we obtain

$$P = A\left(1+\frac{r}{n}\right)^{-nt} = 2{,}500\left(1+\frac{0.082}{12}\right)^{-(12\cdot 5)} \approx 1{,}661.440864 \approx \boxed{\$1{,}661.44}.$$

29. From $A = Pe^{rt}$, it follows that $P = A \div e^{rt} = Ae^{-rt}$. Thus,

$$P = 1{,}000{,}000e^{-(0.06\cdot 30)} \approx 165{,}298.8882 \approx \boxed{\$165{,}298.89}$$

31. Before computing these eight answers, are you able to predict their order from least to most? If not, you do not yet sufficiently understand the main concepts. You may find textbook pp. 507-508 helpful in remedying this deficiency.

Given $P = 12{,}000$, $r = 0.045$, and $t = 20$, we find:

a) $n = 1 \quad \Rightarrow \quad A = P\left(1+\frac{r}{n}\right)^{nt} = 12{,}000(1+0.045)^{20} \approx \boxed{\$28{,}940.57}$

b) $n = 2 \quad \Rightarrow \quad A = P\left(1+\frac{r}{n}\right)^{nt} = 12{,}000\left(1+\frac{0.045}{2}\right)^{2\cdot 20} \approx \boxed{\$29{,}222.27}$

c) $n = 4 \quad \Rightarrow \quad A = P\left(1+\frac{r}{n}\right)^{nt} = 12{,}000\left(1+\frac{0.045}{4}\right)^{4\cdot 20} \approx \boxed{\$29{,}367.30}$

d) $n = 12 \quad \Rightarrow \quad A = P\left(1+\frac{r}{n}\right)^{nt} = 12{,}000\left(1+\frac{0.045}{12}\right)^{12\cdot 20} \approx$ **$29,465.60**

e) $n = 360 \quad \Rightarrow \quad A = P\left(1+\frac{r}{n}\right)^{nt} = 12{,}000\left(1+\frac{0.045}{360}\right)^{360\cdot 20} \approx$ **$29,513.58**

f) $A = P\left(1+\frac{r}{n}\right)^{nt} = 12{,}000\left(1+\frac{0.045}{525{,}600}\right)^{525{,}600\cdot 20} \approx$ **$29,515.22**

g) $A = Pe^{rt} = 12{,}000e^{0.045\cdot 20} = 12{,}000e^{0.9} \approx$ **$29,515.24**

h) $A = P(1+rt) = 12{,}000(1+0.045\times 20) = 12{,}000(1.9) =$ **$22,800**

33. This problem is very similar to text Example 4 (p. 503). Moreover, between Example 4 and Example 5, please take special note of the author's derivation and the resulting future value formula for simple interest.

We begin by identifying P, r, and t. Remember that t is always to be expressed in terms of years. Hence, we are given $P = 1{,}500$, $r = 0.21$, and $t = \frac{55}{360}$. By using the future value formula (for simple interest), we find the amount that must be repaid:

$A = P(1+rt) = 1{,}500\left(1+0.21\cdot\frac{55}{360}\right) = 1{,}548.125 \approx$ **$1,548.13**.

35. The concept of inflation is discussed on textbook page 510, and Example 11 illustrates its relevance to our daily lives. To quote the text's author, "For our purposes in this book, we will assume continuous compounding when working inflation problems."

So, in order to find the future (inflated) costs of these eight individual items, we use the formula $A = Pe^{rt}$ where P is the present cost of the item, $r = 0.09$ (given the assumed 9% rate of inflation), and $t = 5$ (because we are determining the costs 5 years from now). Expressed more simply, each of the inflated costs below is obtained by using the formula $A = Pe^{0.09\cdot 5} = Pe^{0.45}$ and, of course, a calculator.

a) $A = 0.75e^{0.45} \approx 1.176234139 \approx$ **$1.18** for a cup of coffee.

b) $A = 1.25e^{0.45} \approx 1.960390232 \approx$ **$1.96** for a Sunday paper.

c) $A = 1.95e^{0.45} \approx 3.058208762 \approx$ **$3.06** for a Big Mac.

d) $A = 1.55e^{0.45} \approx 2.430883888 \approx$ **$2.43** for a gallon of gas.

e) $A = 600e^{0.45} \approx 940.9873113 \approx$ **$940.99** for a television.

f) $A = 14{,}000e^{0.45} \approx 21{,}956.3706 \approx$ **$21.956.37** for a small car.

g) $A = 28{,}000e^{0.45} \approx 43{,}912.74119 \approx$ **$43.912.74** for a car.

h) $A = 16{,}000e^{0.45} \approx 25{,}092.99497 \approx$ **$25,092.99** for tuition.

37. We are given $A = 1{,}000$, $P = 750$, and $r = 0.08$.

To determine the time necessary to achieve this investment goal, we need to solve the future value (simple interest) formula for t. There are essentially two (equivalent) ways to go about this: (i) first solve the general formula $A = P(1+rt)$ for t, then substitute the given values, and finally evaluate the resulting expression; or (ii) first substitute the given values and then solve the resulting equation for its one remaining unknown (namely, t). Permit us to demonstrate both strategies:

(i) $A = P(1+rt) \Rightarrow$

$$A = P + Prt \qquad \textit{Distribute } P.$$

$$A - P = Prt \qquad \textit{Subtract } P \textit{ from both sides.}$$

$$t = \frac{A-P}{Pr} \qquad \textit{Divide both sides by } Pr.$$

Hence, $t = \dfrac{1{,}000 - 750}{750(0.08)} = 4.1\overline{6}$.

(ii) $A = P(1+rt) \Rightarrow$

$$1{,}000 = 750(1 + 0.08t) \qquad \textit{Substitute the values given.}$$

$$\tfrac{4}{3} = 1 + 0.08t \qquad \textit{Divide both sides by 750; reduce.}$$

$$\tfrac{1}{3} = 0.08t \qquad \textit{Subtract 1 from both sides.}$$

$$t = \tfrac{25}{6} = 4\tfrac{1}{6} = 4.1\overline{6} \qquad \textit{Isolate } t.$$

The directions for Problems 37-40 remind us to consider 365 to be the number of days in a year. Note that $4.1\overline{6}$ years is equivalent to 4 years + $0.1\overline{6}$ year. So, to determine the answer to the nearest day, we multiply $0.1\overline{6} = \frac{1}{6}$ by 365, which gives us $60.8\overline{3}$ days. Thus, to the nearest day, **4 years and 61 days** is the time necessary to achieve this investment goal.

39. Please note that this problem is very similar to textbook Example 10a (p. 509). We are given $A = 5{,}000$, $P = 3{,}500$, $r = 0.05$, and $n = 365$. To determine the time necessary to achieve this investment goal, we need to solve the future value (compound interest) formula for t. Doing so requires a working knowledge of the material in Chapter 7 (in particular, Sections 7.5-7.7). Just as we demonstrated in our solution to Problem 37, one may choose either to first substitute the given values and then solve the resulting equation for t, or vice versa. Since p. 509 of your text demonstrates the former approach, permit us to share the latter:

$$A = P\left(1+\frac{r}{n}\right)^{nt} \Rightarrow \frac{A}{P} = \left(1+\frac{r}{n}\right)^{nt} \qquad \textit{Divide both sides by } P.$$

$$nt = \log_{\left(1+r/n\right)}\left(\frac{A}{P}\right) \qquad \textit{Apply the definition of logarithm.}$$

$$t = \frac{\log_{\left(1+r/n\right)}\left(\frac{A}{P}\right)}{n}$$ *Divide both sides by n.*

We then substitute the given values and with a calculator apply the previous chapter's change-of-base theorem (as introduced on textbook p. 470).

$$t = \frac{\log_{\left(1+0.05/365\right)}\left(\frac{5{,}000}{3{,}500}\right)}{365} \approx 7.133987462$$

To find the number of days, we subtract 7 from this value; multiply the remaining decimal (namely, 0.133987462) by 365; and round accordingly (as explained in Example 10, p. 509). Upon doing so, we conclude **7 years and 49 days** is the time necessary to achieve this particular investment goal.

41. $A = P\left(1+\frac{r}{n}\right)^{nt} = 1{,}000\left(1+\frac{0.04}{4}\right)^{4\cdot 5} \approx 1{,}220.19004 \approx$ **$1,220.19**

43. $A = Pe^{rt} = 1{,}000e^{0.16\cdot 15} \approx 11{,}023.17638 \approx$ **$11,023.18**

45. This problem is done as shown in textbook Example 2 (p. 502). Monthly installments of $410.83 paid over a 4-year period result in a total loan repayment of $19,719.84 (i.e., $410.83/month × 12 months/year × 4 years). Hence, given a sale price of $14,500, the amount of interest paid on this loan equals **$5,219.84**, which is simply the difference between $19,719.84 (the future value A) and $14,500 (the present value P).

47. This problem is akin to textbook Example 2 (p. 502) and its three predecessors, Problems 44-46. First, we compute the loan's future value (i.e., the total amount repaid): A = $1,247.40/mo × 12 mo/yr × 30 yrs = $449,064. Then, given the present value P = $170,000, it follows that the total amount of mortgage interest is

$$A - P = \$449{,}064 - \$170{,}000 = \boxed{\$279{,}064}.$$

49. For problems pertaining to inflation, remember that we are assuming continuous compounding [review, as necessary, textbook page 510 and this manual's solution for Problem 35]. Thus, thirty years from now, under the given assumptions, the monthly rent for this two-bedroom apartment will be

$$A = Pe^{rt} = 650e^{0.10\cdot 30} = 650e^{3} \approx 13{,}055.599 \approx \boxed{\$13{,}055.60}.$$

51. With respect to the simple interest formula $I = Prt$, P represents the amount of the deposit (i.e., the present value). So, via basic algebra, it follows that $P = I \div rt$. Therefore, the amount your friend initially deposited in this 5-year CD was:

$$P = \frac{I}{rt} = \frac{5{,}075}{(0.05)(1)} = \boxed{\$101{,}500}.$$

53. a) $\dfrac{\$5.7 \text{ trillion}}{300 \text{ million people}} = \dfrac{5.7 \times 10^{12}}{3 \times 10^{8}} = 1.9 \times 10^{4} = \boxed{\$19{,}000 \text{ per person}}$

b) We use the simple interest formula $I = Prt$. Since t represents the elapsed time in years, we need to express one second in terms of years. To do so, we first compute the number of seconds in a 365-day year:

$$365 \text{ days} \times (24 \text{ hrs/day}) \times (60 \text{ min/hr}) \times (60 \text{ sec/min}) = 31{,}536{,}000 \text{ sec.}$$

So, 1 sec = $\frac{1}{31{,}536{,}000}$ yr, and it follows that

$$I = Prt = (5.7 \times 10^{12})(0.06)(\tfrac{1}{31{,}536{,}000}) \approx 10{,}844.74886 \approx 10{,}844.75.$$

Thus, the national debt is accumulating interest at an approximate rate of $\boxed{\$10{,}844.75 \text{ per second}}$.

55. It is important to understand that (i) Melissa desires to be retired while earning a monthly income of $50,000/month, and (ii) this monthly income is to be derived solely from the interest generated by a retirement investment. Consequently, because each month the interest is to be removed from the account, we will assume the account does not accrue/compound any interest. As such, we treat Melissa's monthly retirement income as simple interest. Thus, upon noting that 1 month = $\frac{1}{12}$ year, we find that for Melissa to attain her financial goal she will need a principal of

$$I = Prt \quad \Rightarrow \quad P = \frac{I}{rt} = \frac{\$50{,}000}{(0.12)(\frac{1}{12})} = \boxed{\$5{,}000{,}000}.$$

57. a) To review two different approaches for this type of problem, please refer to textbook Example 10 (p. 509) and this manual's solution to Problem 39.

$$t = \frac{\log_{(1+r/n)}\left(\frac{A}{P}\right)}{n} = \frac{\log_{(1+0.07/12)}\left(\frac{1{,}250}{1{,}000}\right)}{12} \approx 3.197053654$$

To find the number of days, we subtract 3 from this value and multiply the remaining decimal (namely, 0.197053654) by 12. The result (≈ 2.36) means that after 3 years and 2 months we have not quite accrued $1,250. Therefore, the necessary time is **3 years, 3 months**.

b) "Doubling" one's money implies $A = 2P = 2(\$1{,}000) = \$2{,}000$. Hence,

$$\begin{aligned} t &= \frac{\log_{\left(1+r/n\right)}\left(\frac{A}{P}\right)}{n} \\ &= \frac{\log_{\left(1+0.07/12\right)}\left(\frac{2{,}000}{1{,}000}\right)}{12} \\ &= \frac{\log_{\left(1+0.07/12\right)} 2}{12} \\ &\approx 9.930955715 \end{aligned}$$

Once again, to find the number of days, we subtract 9 from this value and multiply the remaining decimal (namely, 0.930955715) by 12. The result (≈ 11.17) means that after 9 years and 11 months our initial investment has not quite doubled yet. Therefore, we round up and report the necessary time to achieve this goal as **10 years**.

c) We first substitute the given values and then solve the resulting equation for r.

$A = P\left(1+\frac{r}{n}\right)^{nt}$	*Compound interest formula.*
$2{,}000 = 1{,}000\left(1+\frac{r}{12}\right)^{12\cdot 5}$	*Substitute given values.*
$2 = \left(1+\frac{r}{12}\right)^{60}$	*Divide both sides by 1,000; simplify 12 × 5.*
$(2)^{1/60} = \left[\left(1+\frac{r}{12}\right)^{60}\right]^{1/60}$	*Raise both sides to the 1/60 power.*
$2^{1/60} = 1+\frac{r}{12}$	*Simplify.*
$12\left(2^{1/60}-1\right) = r$	*From both sides, subtract 1 and then multiply by 12.*
$r \approx 0.1394332836 \approx 13.9\%$	*Use calculator.*

Therefore, to the nearest percent, the interest rate is about **14%**.

59. In the spirit of textbook Section 5.6 (p. 277), we begin by translating the problem's directions into a "verbal" equation, and then (as directed) we proceed to solve for Y:

$$\underbrace{\text{SIMPLE-INTEREST FUTURE VALUE}}_{P(1+Y)} = \underbrace{\text{COMPOUND-INTEREST FUTURE VALUE}}_{P\left(1+\frac{r}{n}\right)^{nt}}$$

Given a one-year time period (i.e., $t = 1$), we obtain the desired formula:

$$P(1+Y) = P\left(1+\frac{r}{n}\right)^{n}$$

$$1+Y = \left(1+\frac{r}{n}\right)^{n} \qquad \textit{Divide both sides by } P.$$

$$\boxed{Y = \left(1+\frac{r}{n}\right)^{n} - 1} \qquad \textit{Subtract 1 from both sides.}$$

Problem Set 8.2

1. **Wording varies**. ***Add-on interest*** is the most common method for calculating interest on installment loans. The way it works is: (i) the amount of interest over the life of the loan is first calculated by using the simple interest formula; and (ii) the total debt (i.e., the amount to be repaid) is arrived at by "adding on" this amount of interest to the principal (i.e., the amount borrowed).

3. **Wording varies**. ***Open-ended credit*** (or revolving credit) is the type of consumer credit in which a limit (or "line of credit") is set initially and the consumer may make purchases or cash advances as long as the line of credit is not exceeded. Credit cards are the most common form of open-ended credit. ***Closed-ended credit*** (sometimes called a *fixed installment loan*) is an agreement to borrow a set amount of money and then repay it in accordance with the agreement, usually by making equal payments at regular intervals for some pre-specified period of time. Purchases such as furniture, vehicles, appliances, and homes are often made with closed-ended credit.

5. **Answers vary**. Be forewarned: an inability to answer this question may imperil your (financial) well-being.

7. **Choice B**. Remember: you will owe interest in addition to the $2,400 principal. Consequently, your total debt will exceed $2,400. So, because you will be making 24 equal payments (monthly), your monthly payment will be more than $100.

9. **Choice B**. The author of your text eloquently addresses these issues on pages 519-521. Obviously, interest is the (often exorbitant) cost incurred by those who do not pay off their credit card debt within the grace period, and the ***annual percentage rate*** (APR) is the most important factor in determining this cost. In contrast, if a credit card has an ***annual fee***, this fee will be charged regardless of whether or not the balance of the debt is repaid within the month. You should know that many issuers of credit cards do not charge annual fees; so, it pays to shop around.

11. **Choice B**. Example 6 (on textbook pp. 520-521) clearly illustrates why the **adjusted balance method** best serves the consumer. Additionally, take a look at what your text's author has to say — in the margin adjacent to Example 6 — about an accounting study, cited in *Consumer Reports* magazine, that compared these three interest calculation methods.

13. **Choice B**. Since $P \approx \$1,300$ and $r \approx 10\%$, we can "mentally estimate" the add-on interest (recall $I = Prt$) for one year as approximately $\$130$. Therefore, the interest for 4 years is estimated to be more than $\$500$.

15. **Choice B**. Don't forget that the lender's profit relies largely upon the amount of interest you, the borrower, repays. No lender can afford to lend money at an APR of 1%; so, we immediately disregard choice A. As for choice C, an APR of 32% is awfully high. Often, those who pay outlandish interest rates have poor credit repayment histories to begin with, and many of the lenders who charge such fees prey upon these debt-ridden consumers. Thankfully, an increasing number of state legislatures are writing laws to protect the borrower against these so-called "predatory lending practices."

17. **Choice B**. Needless to say, you always want to offer less than the dealer's sticker price.

19. **Choice C**. The month of August consists of 31 days. The average daily balance method defines t as "the number of days in the billing period divided by 365." For additional details, refer to textbook pp. 520-521.

21. **Choice A**. What does the 'A' in APR stand for? So, if the "annual" percentage rate is 12%, the monthly interest rate must be 1% (i.e., 12% ÷ 12 months/yr). Because credit card interest is charged on a monthly basis, the interest charged equals 1% of the balance — which, in this particular case, amounts to $6.50 on the given $650 balance.

23. **Choice B**. Remember that the adjusted balance method calculates the interest on the previous month's balance *less* any credits and payments. So, if you make a \$50 payment on a previous balance of \$952, the interest will be calculated on the remaining \$902. Then, as explained in our solution to Problem 21 above, an APR of 12% means the monthly interest rate is 1%. Therefore, in accordance with the adjusted balance method, the finance charge equals \$9.02 (i.e., 1% of \$902).

25. Problems 25-30 are comparable to textbook Example 5 (p. 519).
Since there are 12 months in a year, we obtain the APR ("annual" percentage rate) by multiplying the monthly rate by 12: $1\frac{1}{4}\% \times 12 = \boxed{15\%}$ APR.

27. Once again, APR = $1\frac{1}{2}\%$ per month × 12 mo/yr = $\boxed{18\,\%}$.

29. Since there are 365 days in a year, the APR equals:

$$0.02192\% \text{ per day} \times 365 \text{ days/yr} = 8.008\% \approx \boxed{8\%}.$$

31. Problems 31-34 are very similar to text Example 6 (pp. 520-521). Remember that for all three methods we use the simple interest formula $I = Prt$.

a) The ***previous balance method*** treats each month equally (as $\frac{1}{12}$ of a year) — no matter whether its duration is 28 days or 31 days. This method calculates the interest based upon the previous month's balance (which, in this case, is \$300). Hence,

$$I = Prt = (\$300)(0.18)\left(\tfrac{1}{12}\right) = \boxed{\$4.50}.$$

b) The ***adjusted balance method*** calculates the interest on the previous month's balance <u>after</u> any credits and payments have been subtracted. So, if one makes a \$50 payment on a previous balance of \$300, the interest is calculated only on the remaining \$250. Hence,

$$I = Prt = (\$250)(0.18)\left(\tfrac{1}{12}\right) = \boxed{\$3.75}.$$

c) The ***average daily balance method*** requires that we first compute the "average daily balance." We must sum the outstanding balances for each day in the billing period and then divide by the number of days (in this case, 30). Since it takes 10 days for the \$50 payment to be recorded, this account has a balance of \$300 for each of the first 10 days of the 30-day billing period, and then (after the \$50 is recorded) the balance is \$250 for each of the remaining 20 days. So, the average daily balance is

$$\frac{10 \times \$300 + 20 \times \$250}{30} \approx \$266.67.$$

Recall that the average daily balance method defines t differently than the previous two methods: t equals the number of days in the billing cycle divided by 365. Hence,

$$I = Prt = (\$266.67)(0.18)\left(\tfrac{30}{365}\right) \approx \$3.945254795 \approx \boxed{\$3.95}.$$

33. For additional details, please refer to textbook Example 6 (pp. 520-521) and this manual's solution to Problem 31 (see above).

a) $I = Prt = (\$3{,}000)(0.15)(\tfrac{1}{12}) = \boxed{\$37.50}$

b) $I = Prt = (\$2{,}950)(0.15)(\tfrac{1}{12}) = \$36.875 \approx \boxed{\$36.88}$

c) The account balance is \$3000 for the first 10 days of the 30-day billing period and \$2950 for the 20 days thereafter. So, the average daily balance is

$$\frac{10 \times \$3{,}000 + 20 \times \$2{,}950}{30} \approx \$2{,}966.67.$$

Thus, $I = Prt = (\$2{,}966.67)(0.15)(\tfrac{30}{365}) = \$36.57538356 \approx \boxed{\$36.58}$.

35. As explained on textbook pp. 516-517 (see Example 2), an "x% offer" means to make an offer that is x% over the dealer's cost (i.e., the price the dealer paid for the car you want to buy). Notice that such an offer is based solely on the dealer's cost — not on the sticker price, the dealer's preparation charges, etc. So, if you make a 6% offer on a Chevrolet Corsica that costs the dealer \$13,378, you are offering to pay 106% of the dealer's cost, namely

$$13{,}378(1+0.06) = 13{,}378(1.06) = 14{,}180.68 \approx \boxed{\$14{,}181}.$$

37. For additional details, review text Example 2 (pp. 516-517) and this manual's solution to Problem 35 above.

$$17{,}250(1+0.10) = 17{,}250(1.1) = \boxed{\$18{,}975}.$$

39. Once again, refer (if necessary) to text Example 2 and our above solution to Problem 35.

$$18{,}172(1+0.07) = 18{,}172(1.07) = 19{,}444.04 \approx \boxed{\$19{,}444}.$$

41. Problems 41-46 are analogous to textbook Example 3 (p. 518). To find the APR we use the formula

$$APR = \frac{2Nr}{N+1}$$

where N is the number of payments and r is the add-on interest rate (expressed as a decimal). Notice that the purchase price has no affect upon APR.

So, for a \$3,600 living room set purchased at 12% add-on interest for 3 years, we have $N =$ (12 payments/yr) × (3 yrs) = 36 payments, and, hence, the APR is

$$\frac{2Nr}{N+1} = \frac{2(36)(0.12)}{36+1} = 0.23\overline{351} \approx 0.234 = \boxed{23.4\%}.$$

43. $APR = \dfrac{2Nr}{N+1} = \dfrac{2(24)(0.11)}{24+1} = 0.2112 \approx 0.211 = \boxed{21.1\%}$

45. $APR = \frac{2Nr}{N+1} = \frac{2(60)(0.029)}{60+1} = 0.05704918 \approx 0.057 = \boxed{\textbf{5.7\%}}$

47. In terms of the variables used in the text, the \$9,000 advertised price represents P, which is the amount to be financed (i.e., present value); N, the number of payments, equals 36; and m, the amount of the monthly payment, is \$317.50.

a) The amount to be paid is: $A = mN = (\$317.50)(36) = \boxed{\textbf{\$11,430}}$.

b) The amount of interest is: $I = A - P = \$11,430 - \$9,000 = \boxed{\textbf{\$2,430}}$.

c) Given that N = 36 monthly payments, t equals $\frac{36}{12}$ = 3 years. Since $I = Prt$, $r = \frac{I}{Pt}$. So, the interest rate is: $r = \frac{I}{Pt} = \frac{\$2,430}{(\$9,000)(3)} = \boxed{\textbf{0.09}} = 9\ \%$.

d) The annual percentage rate (APR) is:

$$APR = \frac{2Nr}{N+1} = \frac{2(36)(0.09)}{36+1} = 0.175\overline{135} \approx 0.175 = \boxed{\textbf{17.5\%}}.$$

49. Based upon $P = \$14,350$, $N = 48$, and $m = \$488.40$, we have:

a) $A = mN = (\$488.40)(48) = \boxed{\textbf{\$23,443.20}}$

b) $I = A - P = \$23,443.20 - \$14,350 = \boxed{\textbf{\$9,093.20}}$

c) $r = \frac{I}{Pt} = \frac{\$9,093.20}{(\$14,350)(4)} \boxed{\approx \textbf{0.01584181185}} \approx 0.158 = 15.8\%$

d) $APR = \frac{2Nr}{N+1} \approx \frac{2(48)(0.1584181185)}{48+1} = 0.310370191 \approx 0.310 = \boxed{\textbf{31.0\%}}$

51. If one makes an initial (down) payment of \$798 on a \$6,798 car, then the remaining \$6,000 represents the amount to be financed, P. Given that N = 48 and $m = \$168.51$, we find:

a) The total amount paid for this used car equals the sum of the monthly payments (A) added to the amount of the down payment. That is, the total amount paid for this vehicle is:

$$A + \$798 = mN + \$798 = (\$168.51)(48) + \$798 = \boxed{\textbf{\$8,886.48}}$$

b) $I = A - P = \$8,088.48 - \$6,000 = \boxed{\textbf{\$2,088.48}}$

c) $r = \frac{I}{Pt} = \frac{\$2,088.48}{(\$6,000)(4)} = \boxed{\textbf{0.08702}} \approx 0.087 = 8.7\ \%$

d) $APR = \dfrac{2Nr}{N+1} = \dfrac{2(48)(0.08702)}{48+1} \approx 0.170488163 \approx 0.170 = \boxed{\mathbf{17.0\%}}$

53. To make use of the annual percentage rate (APR) formula, all we need to know is the add-on interest rate r and the total number of payments N. Fortunately, both are given: $r = 0.08$ and $N = 48$. Thus, to the nearest tenth of a percent, the APR is

$$\frac{2Nr}{N+1} = \frac{2(48)(0.08)}{48+1} \approx 0.156734693 \approx 0.157 = \boxed{\mathbf{15.7\%}}.$$

55. As in Problems 53-54 (see above solution), we are given the add-on interest rate $r = 11\% = 0.11$ and the number of payments $N = 48$. Hence,

$$APR = \frac{2Nr}{N+1} = \frac{2(48)(0.11)}{48+1} \approx 0.215510204 \approx \boxed{\mathbf{21.6\%}}.$$

57. This problem is comparable to text Example 6 (pp. 520-521) and Problems 31-34.

Previous balance method: Remember that this method calculates the interest based upon the previous month's balance (in this case, \$3,000), and it treats each month equally (as $\frac{1}{12}$ of a year) —regardless of its duration. So, if the credit card Marsha decided to use calculates interest in this way, her bill will include an interest charge of

$$I = Prt = (\$3{,}000)(0.18)\left(\tfrac{1}{12}\right) = \boxed{\mathbf{\$45.00}}.$$

Adjusted balance method: Recall that this method calculates the interest on the previous month's balance *after* any credits and payments have been subtracted. So, given that Marsha made a \$300 payment on her prior balance of \$3,000, the interest is calculated only on the remaining \$2,700. So, if her credit card issuer uses this interest calculation method, then Marsha's interest charges amount to

$$I = Prt = (\$2{,}700)(0.18)\left(\tfrac{1}{12}\right) = \boxed{\mathbf{\$40.50}}.$$

Average daily balance method: To apply this method, we first must compute Marsha's "average daily balance." Since it takes 14 days for her \$300 payment to be recorded, we view her account as having a balance of \$3,000 for 14 days, and then (after the \$300 is recorded) the balance is \$2,700 for the remaining 17 days of this 31-day billing cycle. So, her average daily balance is

$$\frac{14 \times \$3{,}000 + 17 \times \$2{,}700}{31} \approx \$2{,}835.48.$$

Remember that the average daily balance method defines t differently than the previous two methods; namely, t equals the number of days in the billing cycle divided by 365. Hence, if the average daily balance method is used, Marsha's bill reflects finance charges of

$$I = Prt = (\$2{,}835.48)(0.18)\left(\tfrac{31}{365}\right) \approx \$43.34788603 \approx \boxed{\mathbf{\$43.35}}.$$

59. This problem is similar to Example 1 on textbook pp. 515-516.
The total cost of this refrigerator (including sales tax) is \$1,598 × 1.05 = \$1,677.90. Since "their savings account is depleted," Karen and Wayne will need to finance this entire amount; in other words, $P = \$1{,}677.90$. Given that their 3-year loan carries an add-on annual interest rate of 15%, the accrued interest amounts to

$$I = Prt = (\$1{,}677.90)(0.15)(3) = \$755.055 \approx \underline{\$755.06}.$$

To repay this loan in 36 equal installments, Karen and Wayne will have to make monthly payments of

$$m = \frac{A}{N} = \frac{P+I}{N} = \frac{\$1{,}677.90 + \$755.055}{36} = \$67.58208\overline{3} \approx \boxed{\$67.58}.$$

Problem Set 8.3

1. **Wording varies**. A ***sequence*** is a list of numbers having a first number, a second number, a third number, and so on. These numbers are called the *terms* of the sequence. There are many types of sequence, but we shall limit our attention in this section to only three: arithmetic, geometric, and Fibonacci-type.

3. **Wording varies**. An ***arithmetic sequence*** is a sequence whose consecutive terms differ by the same real number, called the *common difference*.

5. **Wording varies**. A ***Fibonacci-type sequence*** is one in which the first and second terms are given and then successive terms are found by adding their two immediate predecessors. The ***Fibonacci sequence*** is 1, 1, 2, 3, 5, 8, 13, 21, ….

7. Given $s_1 = 2$, $s_2 = 4$, $s_3 = 8$, and $s_4 = 16$, it follows that:

a) As shown in part (b), this sequence is **geometric** because all of its successive ratios (i.e., quotients of consecutive terms) are equal.

b) $$\left.\begin{aligned} r_1 &= s_2/s_1 = 4/2 = 2 \\ r_2 &= s_3/s_2 = 8/4 = 2 \\ r_3 &= s_4/s_3 = 16/8 = 2 \end{aligned}\right\}$$ This sequence's common ratio is $\boxed{r = 2}$.

c) There are several ways to determine the next term, s_5.

- We can use its immediate predecessor s_4. That is, since geometric sequences are such that $s_n = s_{n-1}r$, we know $s_5 = s_4 r = 16(2) = \boxed{32}$.
- Alternatively, we can determine s_5 by writing the formula for this sequence's general term [review, as necessary, textbook page 528]. Specifically, since the general term of a geometric sequence can be

expressed as $s_n = s_1 r^{n-1}$, we have $s_n = 2(2)^{n-1} = 2^1 \cdot 2^{n-1} = 2^n$. Hence, the fifth term is $s_5 = 2^5 = 32$.

9. Given $s_1 = 5$, $s_2 = 15$, and $s_3 = 25$, it follows that:

a) As shown in part (b), this sequence is **arithmetic** because its successive differences are equal (i.e., consecutive terms differ by the same real number).

b) $\left.\begin{array}{l} d_1 = s_2 - s_1 = 15 - 5 = 10 \\ d_2 = s_3 - s_2 = 25 - 15 = 10 \end{array}\right\}$ The common difference is $\boxed{d = 10}$.

c) There are several ways to determine the next term, s_4.

- We can use its immediate predecessor s_3; namely, since arithmetic sequences are such that $s_n = s_{n-1} + d$, we know
$$s_4 = s_3 + d = 25 + 10 = \boxed{35}.$$
- Alternatively, we can determine s_4 by writing the formula for this sequence's general term [review, as necessary, textbook page 527]. Since the general term of an arithmetic sequence can be expressed as $s_n = s_1 + (n-1)d$, we have
$$s_n = 5 + (n-1)(10) = 5 + 10n - 10 = 10n - 5.$$
Hence, the fourth term is $s_4 = 10(4) - 5 = \boxed{35}$.

11. Given $s_1 = 5$, $s_2 = 15$, and $s_3 = 20$, it follows that:

a) Unlike Problems 7 and 9 (see above solutions), this sequence does not have a common difference nor does it have a common ratio. Hence, it is not arithmetic and not geometric. However, notice that the third term s_3 is equal to $s_2 + s_1$. So, by definition, we classify this sequence as **Fibonacci-type**.

b) $\boxed{s_1 = 5,\ s_2 = 15}$

c) $s_4 = s_3 + s_2 = 20 + 15 = \boxed{35}$

13. Given $s_1 = 25$, $s_2 = 5$, and $s_3 = 1$, it follows that:

a) Since $s_3 - s_2 \neq s_2 - s_1$, we immediately know the sequence is not arithmetic. Similarly, because $s_3 \neq s_1 + s_2$, we conclude it is not Fibonacci-type. However, as shown in part (b), its successive ratios are equal, so the sequence must be **geometric**.

b) $\left.\begin{array}{l} r_1 = s_2/s_1 = \frac{5}{25} = \frac{1}{5} \\ r_2 = s_3/s_2 = \frac{1}{5} \end{array}\right\}$ This sequence's common ratio is $\boxed{r = \frac{1}{5}}$.

c) As demonstrated in previous solutions (see above), there are several ways to determine the next term, s_4.

- We can use its immediate predecessor s_3; namely, since geometric sequences are such that $s_n = s_{n-1}r$, we know $s_4 = s_3 r = 1\left(\frac{1}{5}\right) = \boxed{\frac{1}{5}}$.
- Alternatively, we can use the formula for this sequence's general term. Specifically, since the general term of a geometric sequence can be expressed as $s_n = s_1 r^{n-1}$, we have $s_n = 25\left(\frac{1}{5}\right)^{n-1}$. Hence, the fourth term is $s_4 = 25\left(\frac{1}{5}\right)^{4-1} = 25\left(\frac{1}{5}\right)^3 = \frac{25}{125} = \frac{1}{5}$.

15. Once again, we employ the same approach as used in the above solutions for Problems 7, 9, 11, and 13. Given $s_1 = 1$, $s_2 = 3$, and $s_3 = 9$, it follows that:

a) This sequence is **geometric**.

b) $r_1 = s_3/s_2 = \frac{3}{1} = 3$
$r_2 = s_3/s_2 = \frac{9}{3} = 3$ } This sequence's common ratio is **r = 3**.

c) There are several ways to determine the next term, s_4. By using its immediate predecessor, we find $s_4 = s_3 r = 9(3) = \boxed{27}$. Alternatively, if preferred, one can use the formula for this sequence's general term. In particular, once we know $s_n = s_1 r^{n-1} = 1(3)^{n-1} = 3^{n-1}$, it becomes easy to find any term we desire (e.g., $s_4 = 3^{4-1} = 3^3 = 27$).

17. a) **None**. This sequence has no common difference; so, it is not arithmetic. This sequence has no common ratio; hence, it cannot possibly be geometric. This sequence's successive terms are not equal to the sum of their two immediate predecessors; hence, it is not Fibonacci-type.

b) There is a discernible pattern ("subtract 2, add 1, subtract 2, add 1, ..."). That is, **the differences between consecutive terms are –2, 1, –2, 1, –2, ...**.

c) $s_7 = \boxed{5}$

19. Once again, by employing the same reasoning as explained in our previous solutions (Problems 7, 9, 11, etc.), we find:

a) This sequence is **geometric**.

b) Its common ratio is **r = 2**.

c) Its sixth term is $s_6 = s_5 r = 48(2) = \boxed{96}$. Alternatively, $s_n = s_1 r^{n-1} = 3(2)^{n-1}$, from which it follows that $s_6 = 3(2)^{6-1} = 3(2)^5 = 3 \cdot 32 = 96$.

21. Given $s_1 = 10$, $s_2 = 10$, and $s_3 = 10$, we continue to adhere to the approach repeatedly demonstrated above (beginning with our solution to Problem 7).

a) Is there a common difference between successive terms? Is there a common ratio? As shown in part (b), your answer to both these questions should be a resounding "yes." Thus, this sequence is **arithmetic and geometric**. Clearly, the sequence is not Fibonacci-type because $s_3 = 10$ is not equal to $s_2 + s_1 = 20$.

b) $d_1 = s_2 - s_1 = 10 - 10 = 0$ and $d_2 = s_3 - s_2 = 10 - 10 = 0$. So, $\boxed{\mathbf{d = 0}}$.
$r_1 = s_2/s_1 = 10/10 = 1$ and $r_2 = s_3/s_2 = 10/10 = 1$. So, $\boxed{\mathbf{r = 1}}$.

c) $s_4 = \boxed{\mathbf{10}}$

23. Given $s_1 = 3$, $s_2 = 6$, $s_3 = 9$, and $s_4 = 15$,

a) At first, it might seem as though this sequence has a common difference because $s_2 - s_1 = s_3 - s_2 = 3$, but notice that $s_4 - s_3 = 6$. Hence, the sequence is not arithmetic. Similarly, because there is no common ratio, it is not geometric. Observe, though, that $s_3 = s_2 + s_1$ and $s_4 = s_3 + s_2$. So, we classify this sequence as **Fibonacci-type**.

b) $\boxed{\mathbf{s_1 = 3,\ s_2 = 6}}$

c) $s_5 = s_4 + s_3 = 15 + 9 = \boxed{\mathbf{24}}$

25. Once again, we apply the same reasoning as explained in our previous solutions (Problems 7, 9, 11, etc.). Given $s_1 = 8$, $s_2 = 12$, $s_3 = 18$, and $s_4 = 27$, we find

a) This sequence is **geometric**.

b) $$\left.\begin{aligned} r_1 &= s_2/s_1 = 12/8 = \tfrac{3}{2} \\ r_2 &= s_3/s_2 = 18/12 = \tfrac{3}{2} \\ r_3 &= s_4/s_3 = 27/18 = \tfrac{3}{2} \end{aligned}\right\}$$ This sequence's common ratio is $\boxed{\mathbf{r = \tfrac{3}{2}}}$.

c) Its fifth term is $s_5 = s_4 r = 27\left(\tfrac{3}{2}\right) = \boxed{\mathbf{\tfrac{81}{2}}}$. Alternatively, $s_n = s_1 r^{n-1} = 8\left(\tfrac{3}{2}\right)^{n-1}$, from which it follows that $s_5 = 8\left(\tfrac{3}{2}\right)^{5-1} = 8\left(\tfrac{3}{2}\right)^4 = \tfrac{81}{2}$.

27. Given $s_1 = 4^5$, $s_2 = 4^4$, $s_3 = 4^3$, and $s_4 = 4^2$,

a) this sequence is **geometric**.

b) $$\left.\begin{aligned} r_1 &= s_2/s_1 = 4^4/4^5 = \tfrac{1}{4} = 4^{-1} \\ r_2 &= s_3/s_2 = 4^3/4^4 = \tfrac{1}{4} = 4^{-1} \\ r_3 &= s_4/s_3 = 4^2/4^3 = \tfrac{1}{4} = 4^{-1} \end{aligned}\right\}$$ This sequence's common ratio is $\boxed{\mathbf{r = \tfrac{1}{4} \text{ or } 4^{-1}}}$.

c) Its fifth term is $s_5 = s_4 r = 4^2\left(4^{-1}\right) = \boxed{4}$. Alternatively, we can determine this sequence's general term: $s_n = s_1 r^{n-1} = 4^5\left(4^{-1}\right)^{n-1} = 4^5\left(4^{-n+1}\right) = 4^{6-n}$. From this, it follows that $s_5 = 4^{6-(5)} = 4^1 = 4$.

29. We continue to adhere to the approach repeatedly demonstrated above (beginning with our solution to Problem 7).
We are given $s_1 = \frac{1}{10}$, $s_2 = \frac{1}{5}$, $s_3 = \frac{3}{10}$, $s_4 = \frac{2}{5}$, and $s_5 = \frac{1}{2}$.

a) Consider writing this sequence so that its first 5 terms share a common denominator (i.e., l.c.d. = 10): $\frac{1}{10}$; $\frac{1}{5} = \frac{2}{10}$; $\frac{3}{10}$; $\frac{2}{5} = \frac{4}{10}$; $\frac{1}{2} = \frac{5}{10}$. By expressing the terms in this manner, a common difference becomes apparent. Hence, we classify this sequence as **arithmetic**.

b) $\boxed{\mathbf{d = \frac{1}{10}}}$

c) Its sixth term is $s_6 = s_5 + d = \frac{1}{2} + \frac{1}{10} = \frac{6}{10} = \boxed{\frac{3}{5}}$. Alternatively, we can determine the formula for this sequence's general term: $s_n = s_1 + (n-1)d = \frac{1}{10} + (n-1)\left(\frac{1}{10}\right) = \frac{1}{10} + \frac{n}{10} - \frac{1}{10} = \frac{n}{10}$. So, since $s_n = \frac{n}{10}$, it follows that $s_6 = \frac{6}{10} = \frac{3}{5}$.

31. Given $s_1 = \frac{7}{12}$, $s_2 = \frac{2}{3}$, $s_3 = \frac{3}{4}$, and $s_4 = \frac{5}{6}$,

a) we begin — like we did in our solution above to Problem 29 — by writing the given terms with a common denominator (i.e., l.c.d. = 12): $\frac{7}{12}$, $\frac{8}{12}$, $\frac{9}{12}$, $\frac{10}{12}$. This representation makes it evident that the sequence has a common difference (namely, $\frac{1}{12}$), so we classify the sequence as **arithmetic**.

b) $\boxed{\mathbf{d = \frac{1}{12}}}$ [Note: If you would not have thought to use a common denominator like we did in part (a), you nonetheless should have discovered the existence of a common difference by subtracting successive pairs of terms (i.e., $s_2 - s_1 = s_3 - s_2 = s_4 - s_3$). On the other hand, since successive quotients are found to be unequal, this sequence does not have a common ratio and, for this reason, is not geometric.]

c) The sequence's fifth term is $s_5 = s_4 + d = \frac{5}{6} + \frac{1}{12} = \boxed{\frac{11}{12}}$. Alternatively, if preferred, one can determine the formula for this sequence's general term: $s_n = s_1 + (n-1)d = \frac{7}{12} + (n-1)\left(\frac{1}{12}\right) = \frac{7}{12} + \frac{n}{12} - \frac{1}{12} = \frac{6+n}{12}$. From this, it follows that $s_5 = \frac{6+(5)}{12} = \frac{11}{12}$.

33. Problems 32-47 are comparable to textbook Example 8. Additionally, please note that part (b) for each of these problems requires the same approach as was used in Problems 6-31 (see detailed solutions above). Given $s_n = -3+3n$,

a) the first three terms of this sequence are:

$s_1 = -3+3(1) = \boxed{0}$, $s_2 = -3+3(2) = \boxed{3}$, and $s_3 = -3+3(3) = \boxed{6}$.

b) The sequence is **arithmetic** with common difference **d = 3**.

35. Given $s_n = 2-n$,

a) the first three terms of this sequence are:

$s_1 = 2-(1) = \boxed{1}$; $s_2 = 2-(2) = \boxed{0}$; and $s_3 = 2-(3) = \boxed{-1}$.

b) The sequence is **arithmetic** with common difference **d = –1**.

37. Given $s_n = 10-10n$,

a) the first three terms of this sequence are:

$s_1 = 10-10(1) = \boxed{0}$; $s_2 = 10-10(2) = \boxed{-10}$; and $s_3 = 10-10(3) = \boxed{-20}$.

b) The sequence is **arithmetic** with common difference **d = –10**.

39. Given $s_n = 1-\dfrac{1}{n}$,

a) the first three terms of this sequence are:

$s_1 = 1-\frac{1}{(1)} = \boxed{0}$; $s_2 = 1-\frac{1}{(2)} = \boxed{\frac{1}{2}}$; and $s_3 = 1-\frac{1}{(3)} = \boxed{\frac{2}{3}}$.

b) Since it has no common difference and no common ratio, this sequence is **neither** arithmetic nor geometric.

41. Given $s_n = \frac{1}{2}n(n+1)$,

a) the first three terms of this sequence are:

$s_1 = \frac{1}{2}(1)[(1)+1] = \boxed{1}$; $s_2 = \frac{1}{2}(2)[(2)+1] = \boxed{3}$; and $s_1 = \frac{1}{2}(3)[(3)+1] = \boxed{6}$.

b) Since it has no common difference and no common ratio, this sequence is **neither** arithmetic nor geometric.

43. Given $s_n = (-1)^n$,

a) the first three terms of this sequence are:

$s_1 = (-1)^1 = \boxed{-1}$; $s_2 = (-1)^2 = \boxed{1}$; and $s_3 = (-1)^3 = \boxed{-1}$.

b) This "oscillating" sequence is **geometric** with common ratio **r = –1**.

45. This problem is akin to Problem 21. Given $s_n = \frac{2}{3}$,

a) this sequence's first three terms are identical to all its other terms; namely,

$$s_1 = \boxed{\frac{2}{3}};\ s_2 = \boxed{\frac{2}{3}};\ \text{and } s_3 = \boxed{\frac{2}{3}}.$$

b) This "constant" sequence is both **arithmetic and geometric** with common difference **d = 0** and common ratio **r = 1**.

47. Given $s_n = (-1)^n(n+1)$,

a) the first three terms of this sequence are:

$s_1 = (-1)^{(1)}[(1)+1] = \boxed{-2}$; $s_2 = (-1)^{(2)}[(2)+1] = \boxed{3}$; and $s_3 = (-1)^{(3)}[(3)+1] = \boxed{-4}$

b) Since it has no common difference and no common ratio, this sequence is **neither** arithmetic nor geometric.

49. The 69th term of the given sequence is $s_{69} = 7 - 3(69) = \boxed{-200}$.

51. The 3rd term of the given sequence is $s_3 = (-1)^{(3)+1}(5)^{(3)+1} = (-1)^4(5)^4 = \boxed{625}$.

53. Given $s_1 = 3$ and $s_n = \frac{1}{3}s_{n-1}$ (for $n \geq 2$), we find:

$$\begin{aligned}
n = 1 &\Rightarrow s_1 = 3 \quad \textit{Given} \\
n = 2 &\Rightarrow s_2 = \tfrac{1}{3}s_1 = \tfrac{1}{3}(3) = 1 \\
n = 3 &\Rightarrow s_3 = \tfrac{1}{3}s_2 = \tfrac{1}{3}(1) = \tfrac{1}{3} \\
n = 4 &\Rightarrow s_4 = \tfrac{1}{3}s_3 = \tfrac{1}{3}\left(\tfrac{1}{3}\right) = \tfrac{1}{9} \\
n = 5 &\Rightarrow s_5 = \tfrac{1}{3}s_4 = \tfrac{1}{3}\left(\tfrac{1}{9}\right) = \tfrac{1}{27}
\end{aligned}$$

So, the first five terms of this (geometric) sequence are respectively: $\boxed{3, 1, \frac{1}{3}, \frac{1}{9}, \frac{1}{27}}$.

55. From the way its general term is defined, you immediately should recognize this as a Fibonacci-type sequence. Given $s_1 = 1$, $s_2 = 2$, and $s_n = s_{n-1} + s_{n-2}$ (for $n \geq 3$), we find:

$$\begin{aligned}
n = 1 &\Rightarrow s_1 = 1 \quad \textit{Given} \\
n = 2 &\Rightarrow s_2 = 2 \quad \textit{Given} \\
n = 3 &\Rightarrow s_3 = s_2 + s_1 = 2 + 1 = 3 \\
n = 4 &\Rightarrow s_4 = s_3 + s_2 = 3 + 2 = 5 \\
n = 5 &\Rightarrow s_5 = s_4 + s_3 = 5 + 3 = 8
\end{aligned}$$

So, respectively, the first five terms of this (geometric) sequence are: **1, 2, 3, 5, 8**.

57. **Yes**, it is Fibonacci.

$$n=1 \quad \Rightarrow \quad \frac{1}{\sqrt{5}}\left[\frac{1+\sqrt{5}}{2}\right]^1 \approx 0.7236 \approx \underline{1=a_1}$$

$$n=2 \quad \Rightarrow \quad \frac{1}{\sqrt{5}}\left[\frac{1+\sqrt{5}}{2}\right]^2 \approx 1.1708 \approx \underline{1=a_2}$$

$$n=3 \quad \Rightarrow \quad \frac{1}{\sqrt{5}}\left[\frac{1+\sqrt{5}}{2}\right]^3 \approx 1.8944 \approx \underline{2=a_3}$$

$$n=4 \quad \Rightarrow \quad \frac{1}{\sqrt{5}}\left[\frac{1+\sqrt{5}}{2}\right]^4 \approx 3.0652 \approx \underline{3=a_4}$$

$$n=5 \quad \Rightarrow \quad \frac{1}{\sqrt{5}}\left[\frac{1+\sqrt{5}}{2}\right]^5 \approx 4.9597 \approx \underline{5=a_5}$$

And so on….
In the event you wish to explore this further, you should know that this pattern follows a form known as the Binet form (or Binet's formula), which is

$$a_n = \frac{1}{\sqrt{5}}\left[\left(\frac{1+\sqrt{5}}{2}\right)^n - \left(\frac{1-\sqrt{5}}{2}\right)^n\right].$$

Both the pattern shown above and the Binet form reveal, once again, the fascinating connections that exist between the golden ratio (discussed in Section 6.6) and the Fibonacci sequence.

59. Suppose x and y represent the two numbers selected by the audience. Then the ten numbers become:

1st number = x
2nd number = y
3rd number = $x+y$
4th number = $x+2y$
5th number = $2x+3y$
6th number = $3x+5y$
7th number = $5x+8y$
8th number = $8x+13y$
9th number = $13x+21y$
10th number = $21x+34y$

SUM = $55x+88y$

Notice that the sum is $55x+88y=11(5x+8y)$ and $5x+8y$ is the 7th number. This proves that — regardless of the two numbers chosen by the audience — the sum of the ten numbers in the sequence will always equal 11 times the 7th number. Since the coefficients of x and y increase in accordance with the Fibonacci sequence, doesn't it make perfect sense to call this "Fibonacci's magic trick"?

Problem Set 8.4

1. **Wording varies**. A ***sequence*** is a list of numbers having a first term, a second term and so on. A ***series*** is the sum of the terms of a given sequence.

3. **Wording varies**. As introduced on textbook page 543, given a sequence with terms g_1, g_2, g_3, ..., the sequence's ***partial sums*** are defined by

1st partial sum = $G_1 = g_1$ 2nd partial sum = $G_2 = g_1 + g_2$

3rd partial sum = $G_3 = g_1 + g_2 + g_3$ kth partial sum = $G_k = g_1 + g_2 + \cdots + g_k$

5. From textbook p. 537, we know S_5 denotes $s_1 + s_2 + s_3 + s_4 + s_5$. While it is true that one could choose to find each of these five terms and then obtain their sum S_5 (as is illustrated by text Example 1, p. 537), such an approach would not be as viable/pleasant if asked to find S_{15}, S_{50}, or S_{500}. This shortcoming motivates us to share with you another (more efficient) approach that will remain viable (if not pleasant) regardless of the number of terms being summed. Moreover, this method will serve to enhance your knowledge of sequences and series.

We begin with an observation: notice that $s_n = 15 - 3n$ forms an arithmetic sequence with common difference $d = -3$, first term $s_1 = 12$, and fifth term $s_5 = 0$. Consequently, we can determine S_5 by using (either form of) the arithmetic series formulas, which were introduced/derived on textbook pp. 538-540 and summarized in Table 8.1, p. 545. That is,

$$\begin{aligned} A_n &= \tfrac{n}{2}(a_1 + a_n) \\ &\Downarrow \\ A_5 &= S_5 \\ &= \tfrac{5}{2}(s_1 + s_5) \\ &= \tfrac{5}{2}(12 + 0) \\ &= 30 \end{aligned}$$

$$\begin{aligned} A_n &= \tfrac{n}{2}[2a_1 + (n-1)d] \\ &\Downarrow \\ A_5 &= S_5 \\ &= \tfrac{5}{2}[2s_1 + (5-1)d] \\ &= \tfrac{5}{2}[2(12) + 4(-3)] \\ &= 30 \end{aligned}$$

Therefore, $S_5 = \boxed{30}$.

7. For the same reasons as those conveyed in our solution to Problem 5 (see details above), we choose not to bother determining s_1, s_2, s_3, and s_4. Instead, we notice that $s_n = 5 \cdot 2^n$ defines a <u>geometric</u> sequence with first term $s_1 = 10$ and common ratio $r = 2$. Consequently, we can determine S_4 by using the finite geometric series formula, which was introduced/derived on textbook pp. 540-541 and summarized in Table 8.1, p. 545. Specifically,

$$G_n = \frac{g_1\left(1-r^n\right)}{1-r} \quad \Longrightarrow \quad \begin{aligned} G_4 &= S_4 \\ &= \frac{s_1\left(1-r^4\right)}{1-r} \\ &= \frac{10\left(1-2^4\right)}{1-2} \\ &= \boxed{\mathbf{150}} \end{aligned}$$

9. Like Problem 7 (explained above), we recognize that $s_n = (-1)^n$ defines a <u>geometric</u> sequence. Its first term is $s_1 = -1$, and its common ratio $r = -1$. Thus,

$$G_7 = S_7 = \frac{s_1(1-r^7)}{1-r} = \frac{(-1)[1-(-1)^4]}{1-(-1)} = \boxed{\mathbf{-1}}.$$

11. Please note that Problems 11-18 make use of summation notation, which is introduced on textbook pp. 537-538 (see Example 2).

$$\sum_{k=3}^{5} k = 3+4+5 = \boxed{\mathbf{12}}$$

13. $\sum_{k=2}^{6} k^2 = 2^2 + 3^2 + 4^2 + 5^2 + 6^2 = 4+9+16+25+36 = \boxed{\mathbf{90}}$

15. In the interest of conserving (your mental) energy, take a moment to consider two questions: What is the value of $\left[1^k + (-1)^k\right]$ when k is an odd integer? What is its value when k is even?

You should find that the given quantity equals zero whenever k is odd and it equals 2 whenever k is even. So, given that we need to find the sum for $k = 1$ to $k = 10$, we do not need to concern ourselves with the five odd values of k (i.e., 1, 3, 5, 7, 9) since 5 times zero is self-evident. All we need to consider are the even values of k. Thus, since there are five of them (2, 4, 6, 8, 10) and each yields a value of 2, we conclude that $\sum_{k=1}^{10}\left[1^k+(-1)^k\right]=5(2)=\boxed{10}$.

17. Here are two approaches:

I. $$\sum_{k=0}^{4} 3(-2)^k = 3(-2)^0+3(-2)^1+3(-2)^2+3(-2)^3+3(-2)^4$$
$$= 3+(-6)+12+(-24)+48$$
$$= 63-30$$
$$= \boxed{33}$$

II. Like Problems 7-10, this problem presents us with a geometric series, one consisting of five terms (i.e., $n=5$) with first term $g_1=3(-2)^0=3$ and common ratio $r=-2$. Hence, we apply the formula for a finite geometric series:

$$\sum_{k=0}^{4} 3(-2)^k = G_5 = \frac{g_1(1-r^5)}{1-r} = \frac{3[1-(-2)^5]}{1-(-2)} = 33.$$

19. To review infinite geometric series, refer to textbook page 543 (and/or Table 8.1 on p. 545). The terms of this infinite series form a geometric sequence with common ratio $r=\frac{1}{2}$ and first term $g_1=1$. With $|r|<1$, the conditions of the infinite geometric series formula are satisfied. Thus, we find

$$1+\tfrac{1}{2}+\tfrac{1}{4}+\cdots=\frac{g_1}{1-r}=\frac{1}{1-\frac{1}{2}}=\boxed{2}.$$

21. With $r=\frac{1}{3}$ and $g_1=1$, we again have a geometric series in which $|r|<1$. Therefore, we determine this sum by applying the infinite geometric series formula, namely

$$1+\tfrac{1}{3}+\tfrac{1}{9}+\cdots=\frac{g_1}{1-r}=\frac{1}{1-\frac{1}{3}}=\boxed{\tfrac{3}{2}}.$$

23. Using the knowledge we acquired in Section 8.3, we first determine the common ratio r by computing <u>either</u> of the two (equivalent) successive ratios. Since $r_1=g_2/g_1=10/-20=-\frac{1}{2}$ (or, if preferred, via r_2), we know $r=-\frac{1}{2}$. Then, upon recognizing that $g_1=-20$ and the condition $|r|<1$ is satisfied, we apply the infinite geometric series formula:

$$-20+10-5+\cdots=\frac{g_1}{1-r}=\frac{-20}{1-\left(-\frac{1}{2}\right)}=\boxed{-\tfrac{40}{3}}.$$

25. The first 5 odd positive integers (1, 3, ...) form an arithmetic sequence with first term $a_1 = 1$ and common difference $d = 2$. Since there are 5 numbers, the sum of these numbers is denoted by A_5:

$$A_n = \tfrac{n}{2}\left[2a_1 + (n-1)d\right] \quad \Rightarrow \quad A_5 = \tfrac{5}{2}\left[2(1) + (5-1)(2)\right] = \boxed{\textbf{25}}.$$

27. Similar to Problem 25 above, the first 5 positive integers form an arithmetic sequence with first term $a_1 = 1$ and common difference $d = 1$. Their sum is denoted by A_5, and we find:

$$A_5 = \tfrac{5}{2}\left[2(1) + (5-1)(1)\right] = \boxed{\textbf{15}}.$$

29. Like Problems 25 and 27 shown above, the first 10 even positive integers (2, 4, ...) form an arithmetic sequence. With first term $a_1 = 2$ and common difference $d = 2$, the sum of this sequence is:

$$A_5 = \tfrac{10}{2}\left[2a_1 + (10-1)d\right] = 5\left[2(2) + 9(2)\right] = \boxed{\textbf{110}}.$$

31. This problem is an extension of its predecessors, most particularly Problem 25 (see above solution) and Problem 28.

The first 100 odd positive integers (1, 3, ...) form an arithmetic sequence with first term $a_1 = 1$ and common difference $d = 2$. Since there are 100 numbers, the sum of these numbers is denoted by A_{100}:

$$A_n = \tfrac{n}{2}\left[2a_1 + (n-1)d\right] \quad \Rightarrow \quad A_{100} = \tfrac{100}{2}\left[2(1) + (100-1)(2)\right] = \boxed{\textbf{10,000}}.$$

33. The first 100 positive integers form a sequence consisting of 1, 2, 3, ..., 99, 100. Specifically, it is an arithmetic sequence with first term $a_1 = 1$, last term $a_{100} = 100$, and common difference $d = 1$. Therefore, the sum of the first 100 positive integers (denoted by A_{100}) can be obtained in either of two ways:

I. $$A_n = \frac{n(a_1 + a_n)}{2} \quad \Rightarrow \quad A_{100} = \frac{100(1+100)}{2} = \boxed{\textbf{5,050}}$$

II. $$A_n = \tfrac{n}{2}\left[2a_1 + (n-1)d\right] \quad \Rightarrow \quad A_{100} = \tfrac{100}{2}\left[2(1) + (100-1)(1)\right] = 5{,}050$$

35. This problem's purpose is to "generalize" Problems 26, 29, and 32. That is to say, we are being asked to determine an expression/formula that will always yield the sum of the first n even positive integers (for any positive integer n). Once determined, we can use this formula to solve problems like Problems 26, 29 and 32 in a more expeditious manner.

We know the first n even positive integers (2, 4, ...) form an arithmetic sequence. With first term $a_1 = 2$ and common difference $d = 2$, the sum of this sequence is:

$$\begin{aligned} A_n &= \tfrac{n}{2}[2a_1 + (n-1)d] \\ &= \tfrac{n}{2}[2(2) + (n-1)(2)] \\ &= \tfrac{n}{2}[4 + 2n - 2] \\ &= \tfrac{n}{2}[2 + 2n] \\ &= \tfrac{n}{2} \cdot 2(1+n) \\ &= n(n+1) \end{aligned}$$

Therefore, the sum of the first n even positive integers is equal to $\boxed{n(n+1)}$.
We now encourage you to experiment with this formula. Begin by revisiting Problem 26 (where $n = 5$), Problem 29 ($n = 10$), and Problem 32 ($n = 100$). Do you obtain the same answers as you did before? Does this newfound formula make the computations significantly easier?

37. First, we "translate" the problem's words into the shorthand notation we have devised for sequences/series: given $n = 20$, $a_1 = 100$, and $d = 50$, we are asked to find A_{20}. With the aid of the appropriate arithmetic series formula, the problem becomes simple:

$$A_n = \tfrac{n}{2}[2a_1 + (n-1)d] \quad \Rightarrow \quad A_{20} = \tfrac{20}{2}[2(100) + (20-1)(50)] = \boxed{11{,}500}$$

39. The even integers between 41 and 99 form the arithmetic sequence 42, 44,..., 98. So, we identify the sequence's first term ($a_1 = 42$), last term ($a_n = 98$), and common difference ($d = 2$). To obtain the sum of this sequence, we need to use the arithmetic series formula. However, note that both forms of the formula require the value of n, the number of terms. We can use our knowledge of sequences (from Section 8.3) to find n; in particular, recall that $a_n = a_1 + (n-1)d$ [as shown in textbook Table 8.1 on p. 545]. After substituting the three known values, we solve the resulting equation for n:

$$a_n = a_1 + (n-1)d \quad \Rightarrow \quad \begin{aligned} 98 &= 42 + (n-1)(2) \\ 56 &= 2n - 2 \\ 58 &= 2n \\ n &= 29 \end{aligned}$$

Now, to find the sum of the even integers between 41 and 99, we may use either of the two arithmetic series formulas:

I. $$A_n = \frac{n(a_1 + a_n)}{2} \quad \Rightarrow \quad A_{29} = \frac{29(42+98)}{2} = \boxed{2{,}030}$$

II. $$A_n = \tfrac{n}{2}[2a_1 + (n-1)d] \quad \Rightarrow \quad A_{29} = \tfrac{29}{2}[2(42) + (29-1)(2)] = 2{,}030$$

41. The total points received by a play who "runs the table" is equal to the sum of the numbers on the balls; that is, $1+2+3+\cdots+15$. Here are two ways to compute this sum:

I. Given $a_1 = 1$, $a_{15} = 15$, and $d = 1$, we use the arithmetic series formula

$$A_n = \frac{n(a_1 + a_n)}{2} \quad \Rightarrow \quad A_{15} = \frac{15(1+15)}{2} = \boxed{120}.$$

II. Equivalently, we know (from Problem 36) that the sum of the first n positive integers is equal to $\frac{n}{2}(n+1)$. Hence, the sum of the first 15 positive integers is $\frac{15}{2}(16) = 120$.

43. The game of pool uses 7 even-numbered balls and 8 odd-numbered balls.
Missy sunk the 7 even-numbered balls, so her point total equals $2+4+\cdots+14$. To compute this sum, we can either: (i) use the arithmetic series formula, or (ii) apply the answer to Problem 35. That is,

(i) With $n = 7$, $a_1 = 2$, and $a_7 = 14$, it follows that

$$A_n = \tfrac{n}{2}(a_1 + a_n) \quad \Rightarrow \quad 2+4+\cdots+14 = \tfrac{7}{2}(2+14) = 56.$$

(ii) In Problem 35 (see solution above), we found that the sum of the first n even positive integers equals $n(n+1)$. Hence, the sum of the first 7 even positive integers equals 7(8) = 56.

So, **Missy's score is 56**.
Shannon sunk the 8 odd-numbered balls, so her point total equals $1+3+\cdots+15$. To compute this sum, we can either: (i) use the arithmetic series formula, or (ii) apply the answer to Problem 34. That is,

(i) With $n = 8$, $a_1 = 1$, and $a_8 = 15$, we obtain

$$A_n = \tfrac{n}{2}(a_1 + a_n) \quad \Rightarrow \quad 1+3+\cdots+15 = \tfrac{8}{2}(1+15) = 64.$$

(ii) In Problem 34, we discovered that the sum of the first n odd positive integers equals n^2. Thus, the sum of the first 8 odd positive integers is $(8)^2 = 64$.

So, **Shannon's score is 64**.

45. For additional background pertaining to chain-letter mailings (and the *Peanuts* cartoon that accompanies it), refer to textbook pages 540-541.

Notice that the number of letters sent in the fifth mailing — just like the prior mailings — forms a finite geometric series with $r = 6$. That is, the total number of letters (denoted by G_5) equals the sum $6+36+216+\cdots = 6^1 + 6^2 + 6^3 + \cdots + 6^5$.

So, with $g_1 = 6$, $n = 5$, and $r = 6$, we find

$$G_5 = \frac{g_1(1-r^5)}{1-r} = \frac{6[1-(6)^5]}{1-(6)} = 9{,}330.$$

Thus, in 5 mailings of this particular chain letter, a total of **9,330 letters** were sent.

47. Figure 8.7 shows that the number of blocks in a given row is one more than the number of blocks in the row immediately above it. So, an 87-row stack of this type requires a total of $(1+2+3+\cdots+87)$ blocks. Here, once again, are two (virtually identical) ways to compute this sum:

I. Given $a_1 = 1$, $a_{87} = 87$, and $d = 1$, we use the arithmetic series formula

$$A_n = \frac{n(a_1 + a_n)}{2} \quad \Rightarrow \quad A_{87} = \frac{87(1+87)}{2} = 3{,}828.$$

II. We know (from Problem 36) that the sum of the first n positive integers is equal to $\frac{n}{2}(n+1)$. Hence, the total number of blocks required for this 87-row stack is $\frac{87}{2}(88) =$ **3,828**.

49. Having a culture of 1 million bacteria that "increases by 100% every 24 hours" means the culture doubles in size every day. In particular, after 1 day, the original culture of 1 million bacteria has grown to 2 million bacteria; after 2 days, we have 4 million bacteria; after 3 days, 8 million bacteria; and so on. This growth pattern is that of a geometric sequence whose common ratio is $r = 2$.

Given that the original culture contains 1 million bacteria (i.e., $a_0 = 1$ million), we let a_1 denote the number of bacteria after 1 day, a_2 the number of bacteria after 2 days, and so on. So, with a first term of $a_1 = 2{,}000{,}000$, we let the number of bacteria present after 10 days be a_{10}, the 10th term of this geometric sequence.

$$g_n = g_1 r^{n-1} \Longrightarrow a_n = a_1 r^{n-1} \Longrightarrow a_{10} = 2{,}000{,}000(2)^{10-1} = 1{,}024{,}000{,}000$$

Thus, we conclude that **1,024,000,000 bacteria** are present after 10 days.

51. This problem is very similar to text Example 5 (pp. 542-543). Here are two approaches:

I. Take note of what the author of your text has to say in the "Look Back" step of Example 5. He writes, "[In] general, if there are 2^n teams in a [two-team elimination] tournament, there will be $2^n - 1$ games." So, a tournament with 32 teams (i.e., $32 = 2^5$) will require $2^5 - 1 =$ **31 games** in all.

II. If the results of Example 5 were not used (or known), we would begin by analyzing the structure of the tournament. In a two-team elimination tournament with 32 teams involved, there are 16 games in the first round (i.e., 2 teams per game); 8 games in the next round; 4 games in the quarter-final round; 2 games in the semi-final round; and 1 championship game. Hence, the total number of games required equals the sum $1+2+4+8+16$. You should recognize this as a geometric series (in which G_5 denotes the sum), with $g_1 = 1$, $r = 2$, and $n = 5$:

$$G_5 = \frac{g_1(1-r^5)}{1-r} = \frac{1(1-2^5)}{1-2} = 31.$$

53. This problem is similar to text Example 5 (pp. 542-543), Problem 51 (see solution above), and, most notably, Problem 52 (note Figure 8.8 on p. 546).

In a three-team elimination tournament (e.g., "Wheel of Fortune" and "Jeopardy"), each game consists of 3 players/teams, the two who lose (and, thus, are eliminated from the competition) and the one who wins (and advances to the next round). If there are 729 teams in all, the first round must consist of 243 games (i.e., 729 ÷ 3); the second round consists of 81 games (243 ÷ 3); the third round consists of 27 games (81 ÷ 3); and so on. Ultimately, of course, the final (championship) round consists of 1 game. Thus, working from the championship round backward, the required number of games in all equals the sum

$$1+3+9+\cdots+243=1+3^1+3^2+\cdots+3^5$$

We should recognize this as a geometric series in which $g_1=1$, $r=3$, and $n=6$:

$$G_6=\frac{g_1\left(1-r^6\right)}{1-r}=\frac{1\left(1-3^6\right)}{1-3}=364$$

Thus, this particular tournament will require **364 games**.

55. This problem closely resembles text Example 6 (p. 544). Given that the first (initial) swing of the pendulum is 25 cm, the second swing is 75% as long as the first swing, namely (25 × 0.75) cm. The third swing is 75% as long as the second swing, that is, [(25 cm × 0.75) × 0.75] cm. The fourth swing is 75% as long as the third swing, and so on. So, the total distance traveled by the tip of the pendulum is an infinite sum of distances. In particular, it is the following infinite geometric series:

$$\textit{TOTAL DISTANCE} = 25+25(0.75)+25(0.75)^2+25(0.75)^3+\cdots$$

Upon identifying $g_1=25$ and $r=0.75$, we apply the formula for an infinite geometric series:

$$\textit{TOTAL DISTANCE}=\frac{g_1}{1-r}=\frac{25}{1-0.75}=\boxed{\textbf{100 cm}}.$$

57. This problem is analogous to Example 6 (p. 544) and Problems 54-56. The given information tells us that the flywheel makes 500 revolutions during the first minute. During the second minute, it makes two-thirds as many revolutions as during the first minute (i.e., $\frac{2}{3}\times 500$ revolutions). Then, during the third minute, the flywheel makes two-thirds as many revolutions as during the second minute (i.e., $\frac{2}{3}\times\left(\frac{2}{3}\times 500\right)$ revolutions). And so on. So, the total number of revolutions made by the flywheel before it comes to rest is an infinite sum of each minute's revolutions. In particular, it is the following infinite geometric series:

$$\textit{TOTAL \# of REVOLUTIONS} = 500+500\left(\tfrac{2}{3}\right)+500\left(\tfrac{2}{3}\right)^2+500\left(\tfrac{2}{3}\right)^3+\cdots$$

Thus, with $g_1=500$ and $r=\frac{2}{3}$ and the infinite geometric series formula, we find:

$$\textit{TOTAL \# of REVOLUTIONS} = \frac{g_1}{1-r}=\frac{500}{1-\frac{2}{3}}=\boxed{\textbf{1500 revolutions}}.$$

59. As you well know, sketching a diagram is frequently a great way to acquire a better understanding of a problem:

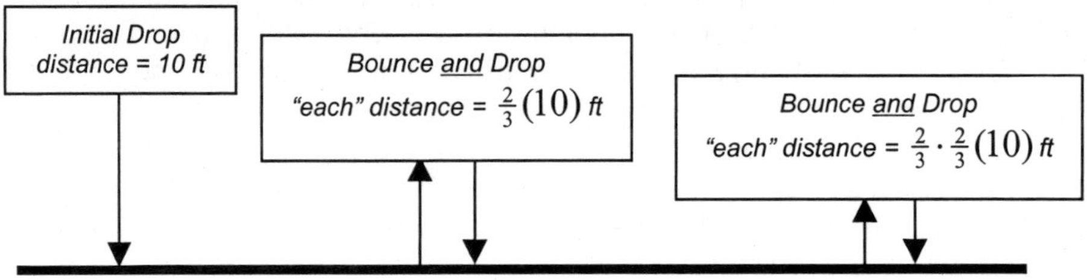

After the initial drop of 10 ft, notice that the ball twice travels a distance of $\frac{2}{3}(10)$ ft, then it twice covers a distance of $\left(\frac{2}{3}\right)^2(10)$, and so on. This pattern of distances forms the following infinite geometric series:

$$\begin{aligned}\textit{Total distance traveled} &= 10 + 2\left(\tfrac{2}{3}\right)(10) + 2\left(\tfrac{2}{3}\right)^2(10) + 2\left(\tfrac{2}{3}\right)^3(10) + \cdots \\ &= 10 + \left[2\left(\tfrac{2}{3}\right)(10) + 2\left(\tfrac{2}{3}\right)^2(10) + 2\left(\tfrac{2}{3}\right)^3(10) + \cdots\right] \\ &= 10 + \left[2\left(\tfrac{2}{3}\right)(10)(1) + 2\left(\tfrac{2}{3}\right)(10)\left(\tfrac{2}{3}\right) + 2\left(\tfrac{2}{3}\right)(10)\left(\tfrac{2}{3}\right)^2 + \cdots\right] \\ &= 10 + 2\left(\tfrac{2}{3}\right)(10)\left[1 + \left(\tfrac{2}{3}\right) + \left(\tfrac{2}{3}\right)^2 + \cdots\right] \\ &= 10 + \tfrac{40}{3}\left[1 + \left(\tfrac{2}{3}\right) + \left(\tfrac{2}{3}\right)^2 + \cdots\right]\end{aligned}$$

Using the formula $G = g_1/(1-r)$ with $r = \frac{2}{3}$ and $g_1 = 1$, we obtain

$$\begin{aligned}\textit{Total distance traveled} &= 10 + \tfrac{40}{3}\left[1 + \left(\tfrac{2}{3}\right) + \left(\tfrac{2}{3}\right)^2 + \cdots\right] \\ &= 10 + \tfrac{40}{3}\left(\frac{1}{1-\frac{2}{3}}\right) \\ &= 10 + \tfrac{40}{3}(3) \\ &= \boxed{\textbf{50 ft}}.\end{aligned}$$

Problem Set 8.5

1. **Wording varies**. A ***lump-sum problem*** is a type of financial problem that deals with a single amount of money. In Section 8.1, we encountered this type of problem; in particular, we solved problems in which there was either (i) a single "lump sum" loan, or (ii) a single "lump sum" deposit into an interest-bearing account (with no other deposits or withdrawals made for a fixed time thereafter).

3. **Wording varies**. In contrast to a lump-sum situation, an ***annuity*** is the future value of <u>periodic</u> payments to an interest bearing account. Since an annuity is an accumulation of payments and interest, it is a periodic payment problem. If the payments are made into an interest-bearing account at the end of each time period, <u>and</u> if the frequency of payments is the same as the frequency of compounding, the annuity is called an ***ordinary annuity***.

5. **Wording varies**. An ***annuity*** seeks the future value, given a periodic (often monthly) payment. In contrast, a ***sinking fund*** seeks the periodic payment, given the future value.

7. Problems 7-22 are solved in exactly the same manner as Example 1 (on p. 549).

$$A = m\left[\frac{\left(1+\frac{r}{n}\right)^{nt}-1}{\frac{r}{n}}\right] = 50\left[\frac{\left(1+\frac{0.05}{12}\right)^{12\cdot 3}-1}{\frac{0.05}{12}}\right] \approx 1{,}937.66678$$

So, if over a 3-year period you deposit $50 per month into an account paying 5% compounded monthly, you will have accumulated (a future value of) **$1,937.67**.

<u>Note</u>: Beginning with Problem 23 (see below), this manual demonstrates a slightly simplified (condensed) form of the **ordinary annuity** formula.

9. **$2,026.28** **11.** **$41,612.93** **13.** **$74,517.97**

15. **$15,528.23** **17.** **$18,294.60** **19.** **$170,413.86**

21. **$540,968.11**

23. Problems 23-34 differ from Problems 7-22 in only one respect: whereas Problems 7-22 assume monthly payments (i.e., n = 12), Problems 23-34 vary the value of n (the number of compounding periods each year) from problem to problem.

First, identify the given variables: $m = \$500$, $n = 1$, $r = 0.08$, $t = 30$. With $N = nt = 1(30) = 30$ and $i = \frac{r}{n} = \frac{0.06}{1} = 0.06$, it follows that the value of this ordinary annuity at the end of the 30-year period is:

$$A = m\left[\frac{(1+i)^N - 1}{i}\right] = 500\left[\frac{(1+0.08)^{30}-1}{0.08}\right] \approx 56{,}641.60556 \approx \$56{,}641.61$$

> **<u>Warning</u>**: *In contrast to our solution to Problem 7 above, notice that in solving this problem we introduced N and I in order to decrease the overall number of variables in the formula. In the event you decide to proceed in a similar manner, please be forewarned: it is critically important in any calculation that you only round <u>once</u> — specifically, at the very end. Always avoid rounding in the middle of these (calculator) computations, because doing so frequently will result in an inaccurate final answer.*

25. First, identify the variables: $m = \$250$, $n = 2$, $r = 0.08$, $t = 30$. So, with $N = nt = 2(30) = 60$, $i = \frac{r}{n} = \frac{0.08}{2} = 0.04$, the ordinary annuity formula yields a future value of

$$A = m\left[\frac{(1+i)^N - 1}{i}\right] = 250\left[\frac{(1.04)^{60} - 1}{0.04}\right] \approx 59{,}497.67130 \approx 59{,}497.67.$$

Therefore, a twenty-year-old student, for example, who deposits \$250 every 6 months (over a 30-year period) into an account that bears 8% interest compounded semiannually will have accumulated **\$59,497.67** by the time (s)he turns fifty.

27. Quarterly compounding implies $n = 4$ (i.e., compounding every three months amounts to four times per year). Given $m = \$300$, $r = 0.06$, and $t = 30$, it follows that $N = nt = 4(30) = 120$ and $i = \frac{r}{n} = \frac{0.06}{4} = 0.015$. By using the ordinary annuity formula either in the manner shown in your text (see Example 1, p. 549) or that of this manual above (see solutions above for Problems 23 and 25), we obtain a future value A of **\$99,386.46**.

29. Given $m = \$200$, $n = 4$, $r = 0.08$, and $t = 20$, it follows that $N = 80$ and $i = 0.02$. Once again, by using the ordinary annuity formula either in the manner shown in your text (see Example 1) or that of this manual (see solutions above for Problems 23 and 25), we obtain the future value: A = **\$38,754.39**.

31. This problem requires a little extra caution, for it is one in which you may be tempted to round prematurely. In the event you have not already done so, please take a moment to read the "warning" stated previously (in our solution for Problem 23).

Given $m = \$30$, $n = 12$, $r = 0.08$, $t = 5$, and $N = 60$, we forego the use of the variable i and instead revert to a more basic form of the annuity formula (comparable to our solution to Problem 7):

$$A = m\left[\frac{\left(1+\frac{r}{n}\right)^N - 1}{\frac{r}{n}}\right] = 30\left[\frac{\left(1+\frac{0.08}{12}\right)^{60} - 1}{\frac{0.08}{12}}\right] \approx 2{,}204.30569 \approx \boxed{\$2{,}204.31}$$

33. Given $m = \$2{,}500$, $n = 2$, $r = 0.085$, and $t = 20$, it follows that $N = 40$ and $i = 0.0425$. Once again, by using the ordinary annuity formula either in the manner shown in your text (see Example 1, p. 549) or that of this manual (see solutions for Problems 23 and 25), we obtain a future value A of **\$252,057.07**.

35. Problems 35-46 are very similar to text Example 4 (p. 552). Each requires the use of the ***sinking fund*** formula and, of course, a suitable calculator.

First, identify the known values: $A = \$7{,}000$, $n = 1$, $r = 0.08$, and $t = 5$. Now, apply the appropriate formula:

$$m = \frac{A\left(\frac{r}{n}\right)}{\left(1+\frac{r}{n}\right)^{nt} - 1} = \frac{7{,}000(0.08)}{(1.08)^5 - 1} \approx 1{,}193.195182.$$

So, in order to accumulate (over a 5-year period) a sum of \$7,000 in an account paying 8% interest compounded annually, one must make annual payments of **\$1,193.20**.

37. In contrast to our solution for Problem 35, let us apply the same (alternative) approach that was previously demonstrated (beginning with our solution to Problem 23). First, identify the variables: $A = \$25{,}000$, $n = 2$, $r = 0.12$, and $t = 5$. So, with $N = nt = 2(5) = 10$, $i = \frac{r}{n} = \frac{0.12}{2} = 0.06$, the sinking fund formula yields a semiannual payment amount of

$$m = \frac{A\left(\frac{r}{n}\right)}{\left(1+\frac{r}{n}\right)^{nt} - 1} = \frac{Ai}{(1+i)^N - 1} = \frac{25{,}000(0.06)}{(1.06)^{10} - 1} \approx 1{,}896.698956 \approx \boxed{\$1{,}896.70}.$$

What does all this mean? A person who makes semiannual payments of \$1,896.70 will have made (over a 5-year span) a total of 10 payments amounting to \$18,967. So, the total amount of interest is \$6,033 (i.e., \$25,000 – \$18,967).

39. Given $A = \$165{,}000$, $n = 2$, $r = 0.02$, and $t = 10$, it follows that $N = nt = 20$ and $i = \frac{r}{n} = 0.01$. Use the sinking fund formula in either of two ways: (i) the manner shown in text Example 4 (and this manual's solution to Problem 35 above) or (ii) the approach shown for Problem 37 (see above). Both approaches yield a required quarterly payment amount m of **\$7,493.53**.

41. Quarterly compounding implies $n = 4$ (i.e., compounding every three months amounts to four times per year). Given $A = \$500{,}000$, $r = 0.08$, and $t = 10$, it follows that $N = nt = 40$ and $i = \frac{r}{n} = 0.02$. Again, use the sinking fund formula in either of two ways: (i) the manner shown in text Example 4 (and this manual's solution to Problem 35 above) or (ii) the approach shown for Problem 37 (see above). Both approaches yield a required quarterly payment amount m of **\$8,277.87**.

43. Given $A = \$100{,}000$, $n = 4$, $r = 0.08$, and $t = 8$, it follows that $N = 32$ and $i = 0.02$. Again, use the sinking fund formula in either of two ways: (i) the manner shown in text Example 4 (and this manual's solution to Problem 35 above) or (ii) the approach shown for Problem 37 (see above). Via either approach, we obtain the required quarterly payment amount: $m =$ **$2,261.06**.

45. Like Problem 31 (see solution), this problem is one that is more easily subject to "premature rounding error." In the event you have not already done so, please take a moment to read the "warning" stated in our solution for Problem 23.

Given $A = \$45{,}000$, $n = 12$, $r = 0.08$, $t = 6$, and $N = 72$, we forego the use of the variable i and instead revert to a more basic form of the sinking fund formula (comparable to our solution to Problem 35):

$$m = \frac{A\left(\frac{r}{n}\right)}{\left(1+\frac{r}{n}\right)^N - 1} = \frac{45{,}000\left(\frac{0.08}{12}\right)}{\left(1+\frac{0.08}{12}\right)^{72} - 1} \approx 488.995828 \approx \boxed{\$489.00}$$

47. Because this problem asks us to determine the future value (at the end of 20 years) based upon annual payments of \$20,000 into a Keogh account, it is, by definition, an ***ordinary annuity*** problem. You may use the ordinary annuity formula either in the manner shown in your text (see Example 1, p. 549) or that of this manual above (see solutions above for Problems 23 and 25, including our "warning"). First, identify the known values: $m = \$20{,}000$, $n = 1$, $r = 0.08$, and $t = 20$. So, with $N = nt = 20$, $i = \frac{r}{n} = 0.08$, the ordinary annuity formula yields a future value of

$$A = m\left[\frac{(1+i)^N - 1}{i}\right] = 20{,}000\left[\frac{(1.08)^{20} - 1}{0.08}\right] \approx 915{,}239.285962 \approx 915{,}239.29.$$

If your employment status makes you eligible to contribute to a Keogh account and if you decide to contribute \$20,000 at the end of each year, you will have accumulated **$915,239.29** at the end of 20 years (assuming an interest rate of 8% compounded annually).

49. Just like Problem 47 (solution above), this is an ordinary annuity problem because it asks us to determine a future value based upon an interest rate that compounds with the same frequency as the frequency of the payments. You may use the ordinary annuity formula either in the manner shown in your text (see Example 1, p. 549) or that of this manual above (see solutions above for Problems 23 and 25, including our "warning").

Once again, we begin by identifying the values of the known variables: $m = \$1{,}000$, $r = 0.08$, $n = 4$ [quarterly payments], and $t = 5.5$ [5yrs, 6mos equals 5 ½ yrs]. It then follows that $N = nt = 22$ and $i = \frac{r}{n} = 0.02$. So, after 5 years and 6 months of quarterly \$1,000 deposits, the owner of Oak Hill Squirrel Farm has an account valued at

$$A = m\left[\frac{(1+i)^N - 1}{i}\right] = 1{,}000\left[\frac{(1.02)^{22} - 1}{0.02}\right] \approx 27{,}298.983539 \approx \boxed{\$27{,}298.98}.$$

51. Problems, like this one, in which we are given a future value and asked to determine the periodic payment are, by definition, ***sinking fund*** problems.

We are given a future value of $A = \$70{,}000$; a 5-year time period, so $t = 5$; an interest rate of 8%, so $r = 0.08$; and a quarterly compounding (and payment) frequency, so $n = 4$. Hence, with $N = 20$, $i = 0.02$, the sinking fund formula reveals that the owners of this particular business must make quarterly payments of

$$m = \frac{A\left(\frac{r}{n}\right)}{\left(1+\frac{r}{n}\right)^{nt} - 1} = \frac{Ai}{(1+i)^N - 1} = \frac{70{,}000(0.02)}{(1.02)^{20} - 1} \approx 2{,}880.970269 \approx \boxed{\$2{,}880.97}.$$

53. The process of finding the future value of the 29 equal installments defines an ordinary annuity problem; in particular, one in which we are given $m = \$20{,}000$, $n = 1$, $r = 0.05$, and $t = 29$. You may use the ordinary annuity formula either in the manner shown in your text (see Example 1) or that of this manual above (see solutions above for Problems 23 and 25, including our "warning"). With $N = nt = 29$, $i = \frac{r}{n} = 0.05$, the ordinary annuity formula yields a future value (after 29 years of \$20,000 annual payments) of

$$A = m\left[\frac{(1+i)^N - 1}{i}\right] = 20{,}000\left[\frac{(1.05)^{29} - 1}{0.05}\right] \approx 1{,}246{,}454.23815 \approx 1{,}246{,}454.24.$$

Thus, after the 30th final payment of \$420,000 is made (and added to the previous \$1,246,454.24), the total value of this annuity becomes $\boxed{\$1{,}666{,}454.24}$.

55. Think of a bond as a type of loan. The bondholders are the lender of the \$200,000, and the owners of Bardoza Greeting Cards are the borrower. In exchange for this loan, Bardoza promises to do two things: (i) pay the bondholders 6% (simple) interest on a semi-annual basis; and (ii) ten years from now reimburse the bondholders their original \$200,000 investment.

We begin by considering the first part of Bardoza's promise. Recall that 6 months equals ½ year. Since $I = Prt = \$200{,}000(0.06)(\frac{1}{2}) = \$6{,}000$, they must pay the bondholders $\underline{\$6{,}000 \text{ in interest every six months}}$.

Next, we turn our attention to the second part of their promise. In order to accumulate the sum necessary to pay (10 years from now) the \$200,000 "face value" of the bonds, the owners of Bardoza set up a sinking fund. That is, in addition to the

$6,000 in interest they pay the bondholders, they will make semiannual payments to an interest-bearing account (with a rate of 8% compounded semiannually), with their goal being to attain a future value of $200,000 after 10 years. We are given $A = \$200{,}000$, $n = 2$, $r = 0.08$, $t = 10$, from which it follows that $N = 20$ and $i = 0.04$. So, every six months Bardoza must make a payment to the sinking fund in the amount of

$$m = \frac{Ai}{(1+i)^N - 1} = \frac{200{,}000(0.04)}{(1.04)^{20} - 1} \approx 6{,}716.350066 \approx \underline{\$6{,}716.35}.$$

Therefore, to fulfill their two-part promise to their bondholders, the owners of Bardoza Greeting Cards will need a sum of money equal to $\$6{,}000 + \$6{,}716.35$ = **$12,716.35**.

57. Similar to text Example 2 (pp. 549-550), this problem presents you with an often-faced financial planning question, one of enormous importance to your so-called "golden" (retirement) years.

Just like Problems 47-49 (among others), this is an *ordinary* annuity problem because you are asked to determine a future value (40 years from now) based upon an interest rate that compounds with the same frequency as the frequency of the payments. Note that a 25-year-old person can afford to take more risk (because retirement is so many decades away). So, the 13% rate given in the problem may be viewed as representing an interest-rate projection (or estimate) for a more aggressive-type investment (e.g., a stock mutual fund).

Begin by identifying the values of the known variables: $m = \$1{,}200$, $r = 0.13$, $n = 1$ [annual payments and compounding], and $t = 40$ [from age 25 to 65]. It then follows that $N = 40$ and $i = 0.13$. So, assuming you are now 25, after forty years of annual $1,000 deposits and a return of 13% compounded annually, you will have accrued at age 65 a sum of

$$A = m\left[\frac{(1+i)^N - 1}{i}\right] = 1{,}200\left[\frac{(1.13)^{40} - 1}{0.13}\right] \approx 1{,}216{,}445.092 \approx \boxed{\$1{,}216{,}445.09}.$$

59. **Answers vary**. Based upon your present age, adjust t (and N) accordingly.

Problem Set 8.6

1. **Wording varies**. ***Amortization*** is the process of paying off a debt by systematically making partial payments until the debt and the accrued interest are repaid. Home mortgages and car loans are two of the most common examples of amortized debt.

3. $\boldsymbol{m}$ is the amount of a periodic (e.g., monthly, quarterly, …) payment; $\boldsymbol{n}$ is the number of payments made each year; $\boldsymbol{t}$ is the number of years payments will be made; $\boldsymbol{r}$ is the annual interest rate (in decimal form); $\boldsymbol{A}$ is the future value; and $\boldsymbol{P}$ is the present value.

5. Problems 4-11 are solved in exactly the same manner as Example 1 (on p. 556).

$$P = m\left[\frac{1-\left(1+\frac{r}{n}\right)^{-nt}}{\frac{r}{n}}\right] = \$50\left[\frac{1-\left(1+\frac{0.06}{12}\right)^{-(12)(5)}}{\frac{0.06}{12}}\right] \approx 2{,}586.278038$$

This means that (given monthly payments of \$50 on a five-year 6% loan) a person can afford to finance **\$2,586.28**.

<u>**Note**</u>: Beginning with Problem 21 (see below), this manual gradually demonstrates simplified (condensed) forms of the **present value of an annuity** formula. Please take special note of the "warning" (regarding "premature rounding error") stated in our solution to Problem 21.

7. **\$27,942.24** **9.** **\$20,442.52** **11.** **\$175,131.20**

13. Problems 12-19 are solved in exactly the same manner as Example 4 (on p. 558).

$$m = \frac{P\left(\frac{r}{n}\right)}{1-\left(1+\frac{r}{n}\right)^{-nt}} = \frac{14{,}000\left(\frac{0.10}{12}\right)}{1-\left(1+\frac{0.10}{12}\right)^{-(12)(5)}} \approx 297.458626$$

What does this answer mean? It means, for example, that in order to pay off a \$14,000 five-year 10% car loan, one must make sixty monthly payments of **\$297.46** each.

— These payments include, of course, interest and principal, but you will need to budget additional money to cover fuel, maintenance, and insurance costs.

<u>**Note**</u>: **Problems 15-19** will enhance your understanding of the costs involved in financing the purchase of a home.

15. **\$1,100.65** **17.** **\$1,316.36** **19.** **\$2,092.02**

21. Problems 20-31 differ from Problems 4-11 in only one respect: the value of n varies from problem to problem. Recall that n represents the number of payments each year (and the frequency of compounding).

First, identify the given variables: $m = \$500$, $n = 1$, $r = 0.06$, $t = 30$. So, with $N = nt = 1(30) = 30$ and $i = \frac{r}{n} = \frac{0.06}{1} = 0.06$, it follows that the present value of this ordinary annuity is:

$$P = m\left[\frac{1-(1+i)^{-N}}{i}\right] = 500\left[\frac{1-(1+0.06)^{-30}}{0.06}\right] \approx 6{,}882.415576 \approx \boxed{\$6{,}882.42}.$$

> **Warning**: *In contrast to our solution to Problem 5 above, note that in solving this problem we introduced* N *and* I *in order to decrease the overall number of variables in the formula. In the event you decide to proceed in a similar manner, please be forewarned: it is critically important in any calculation that you only round once — namely, at the very end. Always avoid rounding in the middle of these (calculator) computations, because doing so often yields less-than-accurate final answers.*

23. First, identify the variables: $m = \$600$, $n = 2$, $r = 0.02$, and $t = 10$. So, with $N = nt = 2(10) = 20$, $i = \frac{r}{n} = \frac{0.02}{2} = 0.01$, we again apply the ***present value of an annuity*** formula:

$$P = m\left[\frac{1-(1+i)^{-N}}{i}\right] = 600\left[\frac{1-(1.01)^{-20}}{0.01}\right] \approx 10{,}827.33178 \approx \boxed{\$10{,}827.33}.$$

25. This problem requires a little extra caution, for it is one in which you may be tempted to round prematurely. In the event you have not already done so, please take a moment to read the "warning" portion of our solution for Problem 21 above.

Given $m = \$100$, $n = 12$, $r = 0.04$, $t = 5$, and $N = 60$, we forego the use of the variable i and instead revert to a more basic form of the present value of an annuity formula (comparable to our solution to Problem 5):

$$P = m\left[\frac{1-\left(1+\frac{r}{n}\right)^{-N}}{\frac{r}{n}}\right] = \$100\left[\frac{1-\left(1+\frac{0.04}{12}\right)^{-60}}{\frac{0.04}{12}}\right] \approx 5{,}429.906890 \approx \boxed{\$5{,}429.91}$$

27. First, identify the variables: $m = \$400$, $n = 4$, $r = 0.11$, and $t = 20$. So, with $N = nt = 80$, $i = \frac{r}{n} = 0.0275$, we again apply the ***present value of an annuity*** formula:

$$P = m\left[\frac{1-(1+i)^{-N}}{i}\right] = 400\left[\frac{1-(1.0275)^{-80}}{0.0275}\right] \approx 12{,}885.1763908 \approx \boxed{\$12{,}885.18}.$$

29. Like Problem 25, because the variable i becomes a repeating decimal, this problem is one that is more readily subject to rounding error (as discussed in the "warning" section of our solution to Problem 21).

Given $m = \$75$, $n = 12$, $r = 0.04$, $t = 10$, and $N = 120$, we obtain:

$$P = m\left[\frac{1-\left(1+\frac{r}{n}\right)^{-N}}{\frac{r}{n}}\right] = \$75\left[\frac{1-\left(1+\frac{0.04}{12}\right)^{-120}}{\frac{0.04}{12}}\right] \approx 7{,}407.763114 \approx \boxed{\$7{,}407.76}$$

31. Given $m = \$100$, $n = 4$, $r = 0.03$, $t = 40$, $N = 160$, and $i = 0.0075$, we find:

$$P = m\left[\frac{1-(1+i)^{-N}}{i}\right] = 100\left[\frac{1-(1.0075)^{-160}}{0.0075}\right] \approx 9{,}299.388390 \approx \boxed{\$9{,}299.39}.$$

Problems 32-43: *Because each of these problems asks for the amount of a given loan's monthly installment, we use the* ***amortization*** *formula.*

33. Given that the amount of the 18% loan is \$100, it follows that $P = \$100$ and $r = 0.18$. Paying on a monthly basis implies $n = 12$; and doing so for 18 months (i.e., 1 ½ years) implies $t = 1.5$ years.

Hence, with $N = nt = 12(1.5) = 18$ and $i = \frac{r}{n} = \frac{0.18}{12} = 0.015$, it follows that the monthly payment for this \$100 loan is

$$m = \frac{Pi}{1-(1+i)^{-N}} = \frac{100(0.015)}{1-(1.015)^{-18}} \approx 6.380578 \approx \boxed{\$6.38}.$$

35. *[See the comment preceding Problem 33's solution.]* Given $P = \$3{,}520$, $r = 0.19$, $n = 12$, $t = \frac{30}{12} = 2.5$, and $N = 30$, we find

$$m = \frac{P\left(\frac{r}{n}\right)}{1-\left(1+\frac{r}{n}\right)^{-N}} = \frac{3{,}520\left(\frac{0.19}{12}\right)}{1-\left(1+\frac{0.19}{12}\right)^{-30}} \approx 148.307225 \approx \boxed{\$148.31}$$

37. *[See the comment preceding Problem 33's solution.]* Given $P = \$12{,}450$, $r = 0.029$, $n = 12$, $t = \frac{30}{12} = 2.5$, and $N = 30$, we find $m \approx 430.726550 \approx$ **$430.73**.

39. *[See the comment preceding Problem 33's solution.]* Given $P = \$985$, $r = 0.17$, $n = 12$, $t = \frac{15}{12} = 1.25$, and $N = 15$, we find $m \approx 73.352988 \approx$ **$73.35**.

41. *[See the comment preceding Problem 33's solution.]* As explained on textbook p. 557 (Example 2), a homebuyer who makes a 30% down payment must finance the remaining 70% of the purchase price. Hence, $P = \$108{,}000(0.70) = \$75{,}600$. Given $r = 0.1205$, $n = 12$, $t = 30$, and $N = nt = 360$, we find $m \approx 780.542027 \approx$ **$780.54**.

43. *[See the comment preceding Problem 33's solution.]* Given $P = \$859{,}000$, $r = 0.132$, $n = 12$, $t = 20$, and $N = nt = 240$, we find $m \approx 10{,}186.473681 \approx$ **$10,186.47**.

45. The answer to Problem 41 implies that the condo buyer will pay — over the lifetime of the 30-year loan — a total of $280,994.40; that is, $780.54/mo × 12 mo/yr × 30 yrs. Based upon the loan amount of $P = \$75{,}600$, it follows that the total interest paid over 30 years is $I = A - P = \$280{,}994.40 - \$75{,}600 = \underline{\$205{,}394.40}$.

Before we can compare this figure to the interest on a 15-year mortgage, we first need to calculate the monthly payment when $t = 15$ and $N = 180$ (with n, r, and P remaining the same as they were in Problem 41). Via the amortization formula, we find $m \approx 909.760356 \approx \$909.76$. So, 15 years worth of $909.76 monthly payments amounts to a total expense of $163,756.80 (i.e., $909.76/mo × 12 mo/yr × 15 yrs); and, in turn, the total interest paid over the life of this 15-year loan is $I = A - P = \$163{,}756.80 - \$75{,}600 = \underline{\$88{,}156.80}$.

Therefore, to the nearest dollar, the amount of interest saved by financing this condominium for 15 (instead of thirty) years is

$$\$205{,}394.40 - \$88{,}156.80 = \$117{,}237.60 \approx \boxed{\$117{,}238}.$$

47. First, we need to identify/classify the problem: it requires that we find the <u>present</u> value of a sequence of payments (i.e., an annuity). We then identify the variables: $m = \$50{,}000$, $n = 1$ [annual payments], $r = 0.1225$, $t = 20$, $N = 20$, and $i = 0.1225$. So, the present value of this lottery prize is:

$$P = m\left[\frac{1-(1+i)^{-N}}{i}\right] = 50{,}000\left[\frac{1-(1.1225)^{-20}}{0.1225}\right] \approx \boxed{\$367{,}695.71}.$$

49. We first need to calculate the present value for the annuity option (i.e., receiving $750 per month for 20 years). Assuming a current rate of return of 9% compounded monthly, we have: $m = \$750$, $n = 12$, $r = 0.09$, and $t = 20$. The present value of an annuity formula yields $P \approx \$83{,}358.72$. Since this is greater than the \$50,000 lump-sum payment, **the annuity** is the better choice.

51. The statement of the problem asks for the "fair price to be paid today" for a \$20 million prize to be "paid in 20 [equal \$1 million] annual installments." This means we need to calculate the present value of an annuity in which the known variables are $m = \$1{,}000{,}000$, $n = 1$, $r = 0.05$, and $t = 20$. So, by using the present value of an annuity formula, we find $P \approx$ **\$12,462,210.34**. This would be a fair price to be paid for the \$20,000,000 lottery prize.

53. This problem is comparable to text Example 2 (pp. 556-557).

This problem states that the bank will allow "you" to pay up to "36% of your gross monthly income (less current monthly payments)." So, in light of your current monthly payments of \$625 and gross monthly income of \$5,500, you can afford to pay no more than 0.36(\$5,500 – \$625) = 0.36(\$4,875) = \$1,755 per month.

We now proceed to find the maximum loan you can obtain, which amounts to finding the present value of an annuity. We know $m = \$1{,}755$, $n = 12$, $r = 0.0965$, and $t = 30$. By using the present value of an annuity formula, we find $P \approx$ **\$206,029.43**, which is the maximum mortgage you can afford (based on your current income level and revolving debt).

55. This problem is similar to Problems 40-41 and Problems 44-45.

If a 25% down payment is made on a \$225,000 home, the homebuyer must finance the remaining 75%; namely, $P = \$225{,}000(0.75) = \$168{,}750$. Given $r = 0.1024$, $n = 12$, and $t = 30$, we find (via the amortization formula) that the monthly payment (on this 30-yr loan) is $m \approx 1{,}510.916625 \approx$ **\$1,510.92**.

Over the 30-year lifetime of the loan, the amount paid (for principal and interest) totals (\$1,510.92 per payment) × (360 payments) = **\$543,931.20**.

Instead, if the home is financed for 15 years ($t = 15$), the monthly payment increases to $m \approx \$1{,}838.25$, but the total spent for principal and interest decreases to \$330,885.00 (i.e., \$1,838.25 × 180 payments). So, financing for 15 (instead of 30) years saves the homeowner a total of \$543,931.20 – \$330,885.00 = **\$213,046.20**.

57. This problem is comparable to text Example 2 (pp. 556-557).

Let's begin by determining the **lower** end of the McBertys' price range. It would be based on a monthly payment amount of \$1,500. To determine the amount of this 20-year loan, we use the present value of an annuity formula. Given $m = \$1{,}500$, $r = 0.11$, $t = 20$, and $n = 12$, we find $P \approx \$145{,}322.31$. So, if they finance this amount and put an additional \$30,000 down, the McBertys can afford a \$175,322.31 home (i.e., financed amount + down payment).

As for the **upper** end of their price range, it would be based on a monthly payment amount of \$1,800. Given $m = \$1{,}800$ (with all other variables unchanged), we find

$P \approx \$174{,}386.77$. So, if they finance this amount and put an additional \$30,000 down, the McBertys can afford a \$204,386.77 home.

Therefore, based on a 20-year loan, they should consider houses in the range of **\$175,322.31 to \$204,386.77**.

59. If you find it difficult to work with these financial concepts in such an abstract setting, consider narrowing the focus of the question by considering a specific scenario; any of those encountered in this section will suffice.

a) Since the future value is (by definition) the amount after the interest is added to the principal, an increase in the interest rate induces an **increase** in the future value.

b) Remember that the present value of an annuity is different from the present value of an investment (like those in Section 8.1). Think, for example, about a mortgage. What do we mean when we refer to its "present value"? The present value represents the (maximum) loan amount you can afford. So, if the interest rate increases, your present value (i.e., the loan amount you qualify for) will **decrease**.

c) Recall (from textbook p. 547) that an annuity is a sequence of payments into or out of an interest-bearing account. If the interest rate increases, the amount of these (annuity) payments must **increase**.

Problem Set 8.7

1. The formula for the ***present value of an annuity*** is

$$P = m\left[\frac{1-\left(1+\frac{r}{n}\right)^{-nt}}{\frac{r}{n}}\right]$$ where P is the unknown.

Note: As shown in this manual's solutions for Sections 8.6's problems, you may find *some* problems conducive to using simplified (condensed) forms of this formula. These forms include

$$P = m\left[\frac{1-(1+i)^{-N}}{i}\right]$$, in which $i = \frac{r}{n}$ and $N = nt$.

For additional details, please return to our explanations (and "warnings") in the previous section.

3. **Answers vary**. A 20% to 30% down payment is typical, although many special programs available to first-time homebuyers offer 5% down (on occasion even 3%). Buyers who put less than 20% down often incur an additional monthly expense known as PMI (private mortgage insurance).

5. **Answers vary**. A good procedure is to ask a series of questions:
   - ***Is it a lump sum problem?***
     If it is, then what is its unknown?
     - *If the FUTURE VALUE is the UNKNOWN, then it is a **future value** problem.*
     - *If the PRESENT VALUE is the UNKNOWN, then it is a **present value** problem.*
   - ***Is it a periodic payment problem?***
     If it is, then is the periodic payment known?
     - *If the PERIODIC PAYMENT is KNOWN and you want to find the future value, then it is an **ordinary annuity** problem.*
     - *If the PERIODIC PAYMENT is KNOWN and you want to find the present value, then it is **a present value of an annuity** problem.*
     - *If the PERIOD PAYMENT is UNKNOWN and you know the future value, then it is a **sinking fund** problem.*
     - *If the PERIOD PAYMENT is UNKNOWN and you know the present value, then it is an **amortization** problem.*

7. If the future value A in a lump-sum problem is known and the present value P is unknown, then it is a **present value problem**.

9. **Ordinary annuity problem**.
[FYI: "Ordinary" annuities are the only type of annuity discussed in this text.]

11. **Amortization problem**.

13. This is a **future value problem**. Given $P = \$1{,}000$, $n = 1$, $t = 3$, and $r = 0.12$, the certificate's (future) value in 3 years is $A = \$1{,}404.928 \approx$ **$1,404.93**.

15. This is a **present value problem**. Given $A = \$10{,}000$, $n = 1$, $t = 5$, and $r = 0.12$, the insurance policy's value today is $P \approx \$5{,}674.268557 \approx$ **$5,674.27**.

17. a) **future value problem**
    b) Given $P = \$1{,}000$, $n = 12$, $t = 2.5$, and $r = 0.12$, the certificate's (future) value in 2½ years is $A = \$1{,}347.848915 \approx$ **$1,347.85**.

19. a) **present value problem**
    b) Given $A = \$10{,}000$, $n = 4$, $t = 5$, and $r = 0.12$, a lump-sum deposit today of $P \approx \$5{,}536.757542 \approx$ **$5,536.76** will yield $10,000 in 5 years.

21. a) **sinking fund problem**

b) Given $A = \$10{,}000$, $n = 1$, $t = 5$, and $r = 0.09$, annual deposits of $m \approx \$1{,}670.924570 \approx$ **\$1,670.92** will accrue to \$10,000 in 5 years.

23. a) **future value problem**

b) Given $P = \$20{,}000$, $n = 1$, $t = 5$, and $r = 0.0625$, we find that the future value $A \approx \$27{,}081.623077$. So, at a growth rate of 6.25% per year, one would expect a \$20,000 automobile to cost **\$27,081.62** five years from now.

25. a) **present value of an annuity problem**

b) Given $m = \$500$, $n = 1$, $t = 33$, and $r = 0.08$, we find that a deposit today of $P \approx \$5{,}756.944184 \approx$ **\$5,756.94** is equal to 33 annual deposits of \$500.

27. a) **amortization problem**

b) Given $P = 0.8(\$125{,}000) = \$100{,}000$, $n = 12$, $t = 30$, and $r = 0.12$, the monthly payment on a \$100,000 mortgage is $m \approx \$1{,}028.612597$ $\approx$ **\$1,028.61**.

29. a) **ordinary annuity problem**

b) It is important to read carefully: Rita faithfully puts her \$2 per day into a jar, and then at the end of **each** year (for a total of 15 years), she deposits \$730 (i.e., \$2/day × 365 days/yr) into an interest-bearing account.
So, given $m = \$730$, $n = 1$, $t = 15$, and $r = 0.10$, Rita will have accumulated after 15 years a total of $A \approx \$23{,}193.911637 \approx$ **\$23,193.91**.

31. This problem is similar to text Example 14 in Section 8.1 (pp. 511-512).

a) **present value problem**

b) Given $A = \$1{,}000{,}000$, $n = 12$, $t = 65 - 10 = 55$, and $r = 0.09$, a lump-sum deposit today (made on behalf of a 10-year-old child) of $P \approx \$7{,}215.456675 \approx$ **\$7,215.46** eventually will grow (over a 55-year span) into a \$1 million retirement gift.

33. a) **ordinary annuity problem**

b) Given $m = \$900$, $n = 4$, $t = 5$, and $r = 0.08$, Thompson Cleaners' account balance (after 5 years) will equal $A \approx \$21{,}867.632819 \approx$ **\$21,867.63**.

35. a) **present value problem**

b) Given $A = \$300{,}000$, $n = 12$, $t = 5$, and $r = 0.12$, a lump-sum deposit today of $P \approx \$165{,}134.884781 \approx$ **\$165,134.88** will cover the future \$300,000 cost of remodeling Thomas' Grocery Store five years from now.

37. a) **amortization problem**

b) Given $P = \$45{,}000$, $n = 12$, $t = 3.5$, and $r = 0.12$, Teal and Associates will have to make monthly payments of $m \approx \$1{,}317.403172 \approx$ **\$1,317.40** on this \$45,000 loan.

39. a) **ordinary annuity problem**

b) Given $m = \$1{,}000$, $n = 12$, $t = \frac{25}{3}$, and $r = 0.18$, the value of Major Magic Corporation's account (after $8\frac{1}{3}$ years) will be $A \approx \$228{,}803.043302 \approx$ **\$228,803.04**.

41. a) **future value problem**

b) Given $P = \$800$, $n = 360$ [see textbook p. 502], $t = 1$, and $r = 0.10$, the future value of \$800 invested for one year (at a rate of 10% compounded daily) is $A = \$884.124457 \approx$ **\$884.12**.

43. a) **future value problem**

b) Given $P = \$5{,}000$, $n = 1$, $t = 12$, and $r = 0.055$, the future value of \$5,000 invested for 12 years (at a rate of 5.5% compounded annually) is $A = \$9{,}506.037429 \approx$ **\$9,506.04**.

45. a) **present value problem**

b) Given $A = \$5{,}000$, $n = 1$, $t = 3$, and $r = 0.110$, a lump-sum payment today of $P \approx \$3{,}655.956907 \approx$ **\$3,655.96** is equivalent to the \$5,000 payment in 3 years.

47. a) **present value problem**

b) Given $A = \$200{,}000$, $n = 12$, $t = 3$, and $r = 0.10$, a lump-sum deposit today (rounded to the nearest dollar) equaling $P \approx \$148{,}347.940691 \approx$ **\$148,348** will cover the future \$200,000 cost of remodeling the Fair View Market.

49. a) **present value problem**

b) Given $A = \$560{,}000$, $n = 360$ [per textbook p. 502], $t = 1.5$ [18 mo = 1.5 yr], and $r = 0.075$, a lump-sum deposit today (rounded to the nearest dollar) equaling $P \approx \$500{,}420.377842 \approx$ **\$500,420** will cover the future \$560,000 cost (18 months from now) of the grocery store's computerized checkout system.

51. No matter whether we choose to compare the present values or the future values, the better choice remains the same: **take the two-payment option**. Here are the details:

Present: Given the future value $A = \$45{,}000$ (with $n = 12$, $t = 1$, and $r = 0.10$), we find (via the present value compound interest formula) that the \$45,000 to be paid one year from now has a present value P of \$40,734.56. Added to the present payment of \$10,000, the two-payment choice has a total (present) value of \$50,734.56, which is better than the \$50,000 one-payment option.

Future: The one-payment option has a present value of $P = \$50{,}000$. Via the future value compound interest formula (with $n = 12$, $t = 1$, and $r = 0.10$), we find that this \$50,000 single payment will have a future value A (one year from now) of \$55,235.65. As for the two-payment option, the \$10,000 received now grows to a (future) value of \$11,047.13 one year from now. Added to the additional \$45,000 to be received at that time, the two-payment option has a total (future) value of \$56,047.13, which is still better than the one-payment option.

53. In Problem 52, we calculated the monthly mortgage payment as \$1,328.94. So, over the 30-year lifetime of the loan, the amount paid for principal and interest combined is \$478,418.40 (i.e., \$1,328.94 per payment × 360 payments). Thus, given that the principal was \$185,500, the total amount of interest paid is

$$I = A - P = \$478{,}418.40 - \$185{,}500 = \boxed{\$292{,}918.40}.$$

55. We first need to calculate the monthly payment on the 20-year loan. Via the amortization formula with $t = 20$ and $N = 240$ (and n, r, and P remaining the same as they were in Problem 52), we find $m \approx \$1{,}522.859587 \approx \$1{,}522.86$. So, 20 years worth of \$1,522.86 in monthly payments amounts to a total expense of \$365,486.40 (i.e., \$1,522.86 per payment × 240 payments); and, in turn, the total interest paid over the life of this 20-year loan is $I = A - P = \$365{,}486.40 - \$185{,}500 = \boxed{\$179{,}986.40}$.

Therefore, the amount of interest saved by financing for 20 years (instead of thirty) is **\$112,932** (i.e., \$292,918.40 – \$179,986.40).

57. This problem is akin to Problem 53 (solved above).

In Problem 56, we obtained the monthly mortgage payment: \$3,180.90. So, over the 30-year lifetime of the loan, the amount paid for principal and interest combined is \$1,145,124 (i.e., \$3,180.90 per payment × 360 payments). Thus, given a loan amount of \$418,500, the total amount of interest paid is

$$I = A - P = \$1{,}145{,}124 - \$418{,}500 = \boxed{\$726{,}624}.$$

59. This problem is analogous to Problem 55 (solved above).

We first calculate the monthly payment on the 22-year loan. Via the amortization formula with $t = 22$ (and n, r, and P the same as they were in Problem 56), we find $m \approx \$3{,}474.795695 \approx \$3{,}474.80$. So, 22 years worth of \$3,474.80 in monthly payments amounts to a total expense of \$917,347.20; and, in turn, the total interest paid over the life of this 22-year loan is $I = A - P = \$917{,}347.20 - \$418{,}500 =$ **\$498,847.20**.

Therefore, by financing for 22 years (instead of thirty), the homeowner saves $\$726{,}624 - \$498{,}847.20 = \$227{,}776.80 \approx$ **\$227,777** in interest.

Chapter 8 Review Questions

1. **Wording varies**. A ***sequence*** is a list of numbers having a first term, a second term and so on. A ***series*** is the sum of the terms of a given sequence. An ***arithmetic sequence*** is a sequence whose consecutive terms differ by the same real number, called the common difference; the *general term* of an arithmetic sequence is given by the formula $a_n = a_1 + (n-1)d$. A ***geometric sequence*** is a sequence whose consecutive terms have the same quotient, called the common ratio; the *general term* of a geometric sequence is $g_n = g_1 r^{n-1}$. A ***Fibonacci-type sequence*** is one in which the first and second terms are given and then successive terms are found by adding their two immediate predecessors; the *general term* of a Fibonacci-type sequence is $s_n = s_{n-1} + s_{n-2}$. The formula for the sum of an arithmetic sequence is

$$A_n = \tfrac{n}{2}(a_1 + a_n) \text{ or } A_n = \tfrac{n}{2}\left[2a_1 + (n-1)d\right],$$

whereas the formula for the sum of a (finite) geometric sequence is $G_n = \dfrac{g_1(1-r^n)}{1-r}$.

2. **Answers vary**. A good procedure is to ask a series of questions:
   - ***Is it a lump sum problem?***
     If it is, then what is its unknown?
     - *If the FUTURE VALUE is the UNKNOWN, then it is a **future value** problem.*
     - *If the PRESENT VALUE is the UNKNOWN, then it is a **present value** problem.*
   - ***Is it a periodic payment problem?***
     If it is, then is the periodic payment known?
     - *If the PERIODIC PAYMENT is KNOWN and you want to find the future value, then it is an **ordinary annuity** problem.*
     - *If the PERIODIC PAYMENT is KNOWN and you want to find the present value, then it is **a present value of an annuity** problem.*
     - *If the PERIOD PAYMENT is UNKNOWN and you know the future value, then it is a **sinking fund** problem.*
     - *If the PERIOD PAYMENT is UNKNOWN and you know the present value, then it is an **amortization** problem.*

3. This problem is like many that we previously encountered in Section 8.3.

a) This sequence is **arithmetic** because its consecutive terms differ by the same real number; specifically, this common difference is $d = 5$. So, given its first term $a_1 = 5$, this arithmetic sequence's general term is:

$$a_n = a_1 + (n-1)d \quad \Rightarrow \quad a_n = 5 + (n-1)(5)$$
$$a_n = 5 + 5n - 5$$
$$\boxed{a_n = 5n}.$$

b) This sequence is **geometric** because all of its successive ratios (i.e., quotients of consecutive terms) are equal; specifically, this common ratio is $r = 2$. So, given its first term $g_1 = 5$, this geometric sequence's general term is:

$$g_n = g_1 r^{n-1} \quad \Rightarrow \quad \boxed{g_n = 5 \cdot 2^{n-1}}.$$

c) Denote $s_1 = 5$, $s_2 = 10$, $s_3 = 15$, and $s_4 = 25$. At first, it might seem as though this sequence has a common difference because $s_2 - s_1 = s_3 - s_2 = 5$, but notice that $s_4 - s_3 = 10$. Hence, the sequence is not arithmetic. Similarly, because there is no common ratio, it is not geometric. Observe, though, that $s_3 = s_2 + s_1$ and $s_4 = s_3 + s_2$. So, we classify this sequence as **Fibonacci-type**; its general term is:

$$\boxed{s_1 = 5,\ s_2 = 10, \text{ and } s_n = s_{n-1} + s_{n-2} \text{ for } n \ge 3}.$$

d) **None of these**. Given $s_1 = 5$, $s_2 = 10$, $s_3 = 20$, and $s_4 = 35$. Take note of this sequence's successive differences: $s_2 - s_1 = 5$, $s_3 - s_2 = 10$, and $s_4 - s_3 = 15$. These successive differences form an arithmetic sequence (namely, 5, 10, 15, 20, 25, ...). So, with fifth term s_5 equal to $s_4 + 20$ and sixth term s_6 equaling $s_5 + 25$, we have: 5, 10, 20, 35, **55, 80**, ...

e) Notice that each term of this sequence is ten times larger than its immediate predecessor. Hence, the sequence is **geometric**, and its general term is

$$\boxed{g_n = 5 \cdot 10^{n-1}}.$$

f) **None of these**. This sequence's terms alternate. So, its fifth and sixth terms are respectively: **5, 50**.

4. To review summation notation, refer to textbook Section 8.4 (pp. 537-538).

a) $$\sum_{k=1}^{3}\left(k^2 - 2k + 1\right) = \left[(1)^2 - 2(1) + 1\right] + \left[(2)^2 - 2(2) + 1\right] + \left[(3)^2 - 2(3) + 1\right]$$
$$= 0 + 1 + 4$$
$$= \boxed{5}$$

b) $$\sum_{k=1}^{4}\frac{k-1}{k+1}=\left(\frac{1-1}{1+1}\right)+\left(\frac{2-1}{2+1}\right)+\left(\frac{3-1}{3+1}\right)+\left(\frac{4-1}{4+1}\right)$$
$$=0+\frac{1}{3}+\frac{1}{2}+\frac{3}{5}$$
$$=\boxed{\frac{43}{30}\text{ or }1.4\overline{3}}$$

5. Before reading this (or, for that matter, any) solution in its entirety, please be sure you have spent an adequate amount of time analyzing the problem. For instance, did you try to sketch a family tree? Have you thought about the sequential pattern formed by the number of parents, grandparents, great-grandparents, etc.?
Obviously, the lady in question has 2 parents. Assuming no intermarriage, each of her parents had 2 parents, so she herself had 4 grandparents. In turn, each of the grandparents had 2 parents, so she had 8 great-grandparents. And so on. In terms of her family's generations, we have the following geometric sequence:

1 generation back	⇔	2 parents
2 generations back	⇔	4 grandparents *[note: 4 = 2^2]*
3 generations back	⇔	8 great-grandparents *[note: 8 = 2^3]*
⋮	⋮	⋮
10 generations back	⇔	2^{10} ancestors
Total # of family members	=	??

So, to find the minimum number of people on her family tree, we need to find the sum of this geometric sequence. With $g_1 = 2$, $r = 2$, and $n = 10$, we use the (finite) geometric series formula:

$$G_n = \frac{g_1\left(1-r^n\right)}{1-r} \quad \Longrightarrow \quad G_{10} = \frac{2\left(1-2^{10}\right)}{1-2} = 2{,}046$$

Thus, she has traced a minimum of **2,046** members of her family tree.

6. At a rate of one division every 20 minutes (i.e., three per hour), 72 divisions of this bacteria culture occur in a 24-hour period. Each division, of course, doubles the size of the culture. Consequently, we have a geometric sequence, one in which the present number of bacteria can be denoted by $g_1 = 1{,}024 = 2^{10}$ and the common ratio $r = 2$ characterizes this culture's division (or doubling) process. Therefore, assuming that no bacteria die during this period, there will be (24 hours from now) a total of $g_{73} = g_1 r^{73-1} = 2^{10}\cdot 2^{72} =$ **2^{82} bacteria** in the culture.

7. This problem is just like Problems 35-40 in Section 8.2 (p. 522). To review the associated terminology and concepts, refer to text Example 2 on pp. 516-517.
An "x% offer" means to make an offer that is x% over the dealer's cost (i.e., the price the dealer paid for the car you want to buy). Notice that such an offer is based solely

on the dealer's cost — not on the sticker price, the dealer's preparation charges, etc. So, if you make a 5% offer on a car that costs the dealer \$18,579, you are offering to pay 105% of the dealer's cost, namely

$$18{,}579(1+0.05)=18{,}579(1.05)=19{,}507.95\,.$$

Rounded to the nearest hundred dollars, you should offer **\$19,500** for this car.

8. For additional details, review textbook pages 515-519 (Section 8.2).

a) In terms of the variables used in the text, we denote the amount to be financed (i.e., the present value) by P = \$13,500; t = 2 years of monthly installments; N, the number of payments, equals 24; and the value of r, the interest rate, is 0.029. It follows that the add-on ***interest*** is $I = Prt$ = (\$13,500)(0.029)(2) = **\$783**. The total amount (i.e., principal and interest) to be paid is known as the future value, which is $A=P+I=\$13{,}500+\$783=\$14{,}283$. So, the ***monthly installment*** m equals $A \div N$ = \$14,283 ÷ 24 = \$595.125 ≈ **\$595.13**.

b) The annual percentage rate (APR) is:

$$APR=\frac{2Nr}{N+1}=\frac{2(24)(0.029)}{24+1}=0.05568=\boxed{5.568\,\%}.$$

9. Like Problem 8 above, we refer you — in the event you need to further review these concepts — to textbook Section 8.2, pp. 515-519.

The \$11,450 offer represents the present value P (since there is no down payment); N = 48 monthly installments; $t=\frac{48}{12}$ = 4 years of monthly installments; and m, the amount of the monthly payment, is \$353.04. Hence,

- the total amount to be paid is $A=mN=(\$353.04)(48)=\$16{,}945.92$;
- the amount of interest is $I=A-P=\$16{,}945.92-\$11{,}450=\$5{,}495.92$;
- and since $I = Prt$, it follows that the interest rate is

$$r=\frac{I}{Pt}=\frac{\$5{,}495.92}{(\$11{,}450)(4)}\approx 0.119998253\,.$$

Therefore, the annual percentage rate (APR) is:

$$APR=\frac{2Nr}{N+1}=\frac{2(48)r}{48+1}=0.235098619\approx 0.235=\boxed{23.5\,\%}.$$

10. The **adjusted balance method** is most advantageous to the consumer.

From textbook pp. 520-521 in Section 8.2, you will recall that all three methods make use of the simple interest formula $I = Prt$.

- The ***previous balance method*** treats each month equally (as $\frac{1}{12}$ of a year) —no matter whether its duration is 28 days, 30 days, or 31 days. This method calculates the interest based upon the previous month's balance (which, in this case, is \$525). Hence,
$$I = Prt = (\$525)(0.09)\left(\tfrac{1}{12}\right) = \$3.9375 \approx \boxed{\$3.94}.$$
- The ***adjusted balance method*** calculates the interest on the previous month's balance <u>after</u> any credits and payments have been subtracted. So, if one makes a \$100 payment on a previous balance of \$525, the interest is calculated only on the remaining \$425. Hence,
$$I = Prt = (\$425)(0.09)\left(\tfrac{1}{12}\right) = \$3.1875 \approx \boxed{\$3.19}.$$
- The ***average daily balance method*** requires that we first compute the "average daily balance." We must sum the outstanding balances for each day in the billing period and then divide by the number of days (in this case, 31). Since it takes 7 days for the \$100 payment to be recorded, this account has a balance of \$525 for each of the first 7 days of the 31-day billing period, and then (after the \$100 is recorded) the balance is \$425 for each of the remaining 24 days. So, the average daily balance is
$$\frac{7\times\$525 + 24\times\$425}{31} \approx \$447.58.$$
Recall that the average daily balance method defines t differently than the previous two methods: t equals the number of days in the billing cycle divided by 365. Hence, $I = Prt = (\$447.58)(0.09)\left(\tfrac{31}{365}\right) \approx \$3.421227945 \approx \boxed{\$3.42}$.

11. This is a ***present value problem***. It is very similar to Example 14 in Section 8.1 (see textbook pp. 511-512) and to Problem 31 in Section 8.7 (refer to this manual's solution).

We are given $A = \$1{,}000{,}000$ (the future value); $n = 12$ (the number of compounding periods per year); $t = 50$ (the time in years); and $r = 0.09$ (the interest rate expressed as a decimal).

So, with $i = \frac{r}{n} = \frac{0.09}{12} = 0.0075$ and $N = nt = 12(50) = 600$, we obtain a present value of

$$P = A\left(1+\frac{r}{n}\right)^{-nt} = A(1+i)^{-N} = 1{,}000{,}000(1.0075)^{-600} \approx 11{,}297.103617.$$

So, it will take a lump-sum deposit today of **\$11,297.10** to achieve this \$1 million goal fifty years from now.

12. a) This ***amortization problem*** is like many we solved previously (in Sections 8.6 and 8.7, beginning on textbook p. 558).
Given $P = \$154{,}000$, $r = 0.08$, $n = 12$, $t = 20$, and $N = nt = 240$, we obtain (via the amortization formula) a monthly payment in the amount of

$$m = \frac{P\left(\frac{r}{n}\right)}{1-\left(1+\frac{r}{n}\right)^{-N}} = \frac{154{,}000\left(\frac{0.08}{12}\right)}{1-\left(1+\frac{0.08}{12}\right)^{-240}} \approx 1{,}288.117706 \approx \boxed{\$1{,}288.12}.$$

b) If one puts 20% down, then the remaining 80% of the home price must be borrowed. So, given that the amount of the loan is \$154,000, it follows that

0.80 × PRICE OF HOME = \$154,000

PRICE OF HOME = \$154,000 ÷ 0.80 = **\$192,500**.

c) For problems pertaining to ***inflation***, remember that we are assuming continuous compounding [review, as necessary, textbook page 510]. Thus, under the given assumptions, the value in 30 years of this \$192,500 home is expected to be

$$A = Pe^{rt} = 192{,}500e^{0.04\cdot 30} = 192{,}500e^{1.2} \approx \boxed{\$639{,}122.51}.$$

13. This is a typical ***present value problem***, like so many we originally encountered in Section 8.1 (see pp. 510-512) and later reviewed in Section 8.7. Given A = \$100,000, n = 12, $t = 3\frac{4}{12} = \frac{40}{12}$, r = 0.064, and $N = nt = 12\left(\frac{40}{12}\right) = 40$, this \$100,00 inheritance has a present value of

$$P = A\left(1+\frac{r}{n}\right)^{-nt} = A\left(1+\frac{r}{n}\right)^{-N} = 100{,}000\left(1+\frac{0.064}{12}\right)^{-40} \approx \boxed{\$80{,}834.49}.$$

14. Here, yet again, is another ***present value problem***. Given A = \$420,000, n = 1 [assume annual compounding], $t = N = 30$, $r = i = 0.05$, the \$420,000 future payment has a present value of

$$P = A\left(1+\frac{r}{n}\right)^{-nt} = A(1+i)^{-N} = 420{,}000(1.05)^{-30} \approx \boxed{\$97{,}178.53}.$$

15. This is a typical ***ordinary annuity problem***, like those found in Section 8.5 (beginning on textbook p. 549) and reviewed in Section 8.7.

First, identify the variables: $m = \$20{,}000$, $n = 1$, $r = 0.05$, $t = 29$. So, with $N = nt = 29$, $i = \frac{r}{n} = 0.05$, the ordinary annuity formula yields:

$$A = m\left[\frac{(1+i)^N - 1}{i}\right] = 20{,}000\left[\frac{(1.05)^{29}-1}{0.05}\right] \approx 1{,}246{,}454.23815 \approx 1{,}246{,}454.24.$$

Therefore, the actual (future) value of the prize's annuity portion is **\$1,246,454.24**.

16. The preferred (less risky) approach is to use the ***present value of an annuity*** formula. Given $m = \$20{,}000$, $n = 1$, $r = 0.05$, $t = 29$, $N = nt = 29$, and $i = \frac{r}{n} = 0.05$, we find that a lump-sum payment today of

$$P = m\left[\frac{1-(1+i)^{-N}}{i}\right] = 20{,}000\left[\frac{1-(1.05)^{-29}}{0.05}\right] \approx \boxed{\$302{,}821.47}$$

is equivalent to 29 annual payments of $20,000 each.

Note: Although the above approach is recommended, there is an alternative method. The same result could be obtained by treating the problem as a ***present value problem*** with the future value A being the answer calculated for Problem 15. However, there are inherent risks in using such a strategy; namely, the strategy relies upon a previously rounded (and possibly erroneous) computation. Nonetheless, for informational purposes only, this method yields the following:

$$P = A\left(1+\frac{r}{n}\right)^{-nt} = A(1+i)^{-N} = 1{,}246{,}454.24(1.05)^{-29} \approx \$302{,}821.47 .$$

17. Just like Problems 13-14 above, this is another ***present value problem***. Given $A = \$420{,}000$, $n = 1$, $t = 30$, and $r = 0.05$, we find that a lump-sum deposit today of $P \approx \$97{,}178.528436 \approx \boxed{\$97{,}178.53}$ eventually will grow (over a 30-year span) into $420,000.

18. $I = Prt = (\$200{,}000)(0.05)(1) = \boxed{\$10{,}000}$

19. To practice with other problems of this type, refer to Problem 51 in Section 8.7 and Problems 48-49 in Section 8.6.

No matter whether we choose to compare the present values or the future values, the better choice remains the same: **take the installment option**. Here are the details:

Present: The one-time (lump-sum) payment option has a (present) cash value of $300,000. As for the installment option, its annuity portion has a present value of $302,821.47 (see Problem 16), and its one-time payment of $420,000 in the 30th year has a present value of $97,178.53 (see Problem 17). So, the installment option has a present value totaling $400,000 (i.e., $302,821.47 + $97,178.53), which is $100,000 more than the one-time payment.

Future: The one-time $300,000 payment option has a future value (30 years from now) of $A = \$300{,}000(1.05)^{30} \approx \$1{,}296{,}582.71$. As for the installment option, its annuity portion has a future value of $1,246,454.24 (see Problem 15). Added to the additional $420,000 to be received in the 30th year, the installment option has a total (future) value of $1,666,454.24, which is nearly $370,000 more than the one-payment option.

20. Let us compare the present values of the two options. To calculate the installment option's present value, we simply repeat Problems 14 and 16, using $r = 0.10$ instead of $r = 0.05$. We find: (i) the present value of the annuity to be $\$187,392.12$, and (ii) the present value of the future $\$420,000$ payment to be $\$24,069.59$. So, the combined present value of the installment option is $\underline{\underline{\$211,461.71}}$.

Since the one-time payment of $\$300,000$ is more than the installment option's present value of $\$211,461.71$, one should **take the one-time payment**.

CHAPTER 9: The Nature of Probability

Problem Set 9.1

1. **Wording varies**. An event's ***empirical probability*** is obtained by way of experimentation/observation (i.e., "empirical" evidence) and represents the event's relative frequency (percent) of occurrence. In contrast, a ***theoretical probability*** is obtained by logical reasoning according to stated definitions.

3. **Arguments vary**. *[Note: A prerequisite for presenting your own thoughtful argument — let alone comprehending the first of the two stated below — is a familiarity with the essential terminology (and notation) developed in this section. Before reading any further, please make sure you are sufficiently familiar with the terminology on textbook p. 575 and the basic concept of probability and its associated notation (pp. 576-578).]*

Below are two arguments. The first makes use of the textbook's notation and necessary formalism and is suitable for mathematics students like you; the other is appropriate for a layperson (unlike yourself) with little mathematical knowledge.

- Recall that: (i) n denotes the number of possible outcomes (i.e., the size of the sample space S); (ii) event E is, by definition, a subset of S; and (iii) s denotes the number of outcomes favorable to E. Since E is a subset of S, the number s must be between 0 and n. Therefore, since $0 \le s \le n$ and $P(E) = \frac{s}{n}$, $0 \le P(E) \le 1$. If the occurrence of E is impossible, then E is the empty set [and, necessarily, $s = 0$ and $P(E) = \frac{0}{n} = 0$]. If the occurrence of event E is an absolute certainty, then E is the sample space [so, $s = n$ and $P(E) = \frac{n}{n} = 1$].
- If the occurrence of an event is *impossible* (has absolutely no chance of occurring), we say it has a probability of 0 (or 0%). If an event's occurrence is *certain* (absolutely will happen), we say it has a probability of 1 (or 100%). Since any event has a likelihood of occurrence between "impossible" and "certain," all probabilities are between 0 and 1.

5. Take a careful look at this spinner. Notice that region F consumes exactly half of the spinner's area, while regions D and E individually comprise one-fourth of its area. Thus, by applying the basic (theoretical) definition of probability (given on p. 578), we obtain the following probabilities:

a) $P(D) = \boxed{\frac{1}{4}}$ b) $P(E) = \boxed{\frac{1}{4}}$ c) $P(F) = \boxed{\frac{1}{2}}$

7. This problem is similar to textbook Example 3 (p. 578). The spinner in this case is divided into eight <u>equal</u> areas, of which three are white, three are gray, and two are black. By applying the theoretical definition of probability, we find:

a) $P(\text{white}) = \boxed{\frac{3}{8}}$ b) $P(\text{black}) = \frac{2}{8} = \boxed{\frac{1}{4}}$ c) $P(\text{gray}) = \boxed{\frac{3}{8}}$

9. Problems 8-11 are comparable to text Example 5 (p. 579); all five of these problems are examples of empirical probabilities.

$$\text{P(grade of A)} = \frac{\text{\# of A's awarded}}{\text{Total \# of students}} = \frac{13}{285} \approx 0.045614035 \boxed{\approx \mathbf{0.05}}$$

11. $\text{P(winning ticket)} = \frac{\text{\# of winning tickets}}{\text{Total \# of tickets sold}} = \frac{285}{1{,}500} = \boxed{\mathbf{0.19}}$

13. From Table 9.1 and the directions for Problems 13-22, we know there are 2,598,960 five-card poker hands, of which 4 are royal flushes.

Thus, $\text{P(royal flush)} = \frac{4}{2{,}598{,}960} \approx \boxed{\mathbf{0.000001539077169}}$

15. $\text{P(four of a kind)} = \frac{624}{2{,}598{,}960} \approx \boxed{\mathbf{0.0002400960384}}$

17. $\text{P(flush)} = \frac{5{,}108}{2{,}598{,}960} \approx \boxed{\mathbf{0.00196540155}}$

19. $\text{P(three of a kind)} = \frac{54{,}912}{2{,}598{,}960} \approx \boxed{\mathbf{0.02112845138}}$

21. $\text{P(one pair)} = \frac{1{,}098{,}240}{2{,}598{,}960} \approx \boxed{\mathbf{0.42256902761}}$

23. This problem is very similar to text Example 7.

a) Only one of the 52 cards in the deck is the five of hearts.

Hence, $\text{P(five of } \heartsuit) = \boxed{\frac{1}{52}}$.

b) Since there is a '5' in each of the four suits (hearts, diamonds, spades, and clubs), there are four possible ways to select a five.

Hence, $\text{P(five)} = \frac{4}{52} = \boxed{\frac{1}{13}}$.

c) Here are two perspectives:

- View the deck in terms of its suits. Since an ordinary deck has 4 suits (hearts, diamonds, spades, and clubs) each of equal size, the probability of selecting a heart is $1/4$.
- View the deck in terms of its cards. Since an ordinary 52-card deck consists of 4 equivalent suits, each suit must have 13 cards (i.e., $52 \div 4$). Hence, there are 13 ways to select a heart when drawing a single card from a standard 52-card deck.

Both perspectives yield the same result: $\text{P}(\heartsuit) = \frac{13}{52} = \boxed{\frac{1}{4}}$.

25. This problem is akin to text Example 10 (pp. 583-584).
In all probability problems, it is critically important to remember the specific way the experiment (act or process) is performed. In this particular problem, the experiment consists of selecting a "**single**" card (not two cards).

a) It is not possible to select a single card that is both (simultaneously) a five and a jack. We refer to events that cannot occur simultaneously as mutually exclusive (see textbook p. 576).
In terms of the notation demonstrated in Example 10, we have:

$$\text{five} = \{\text{five of } \heartsuit, \text{ five of } \spadesuit, \text{ five of } \diamondsuit, \text{ five of } \clubsuit\}$$

$$\text{jack} = \{\text{jack of } \heartsuit, \text{ jack of } \spadesuit, \text{ jack of } \diamondsuit, \text{ jack of } \clubsuit\}$$

$$\text{five} \cap \text{jack} = \varnothing$$

$$\text{P(five and a jack)} = \text{P(five} \cap \text{jack)} = \tfrac{0}{52} = \boxed{0}.$$

b) Since there are four jacks and four 5's, there are 8 possible ways to succeed in obtaining a five **or** a jack when selecting a single card. That is,

$$\text{five} \cup \text{jack} = \{5 \text{ of } \heartsuit, 5 \text{ of } \spadesuit, 5 \text{ of } \diamondsuit, 5 \text{ of } \clubsuit, \text{J of } \heartsuit, \text{J of } \spadesuit, \text{J of } \diamondsuit, \text{J of } \clubsuit\}$$

$$\text{P(five or a jack)} = \text{P(five} \cup \text{jack)} = \tfrac{8}{52} = \boxed{\tfrac{2}{13}}.$$

27. Given $S = \{\text{H1, H2, H3, H4, H5, H6, T1, T2, T3, T4, T5, T6}\}$,

a) Since the only way to obtain a tail and a five is {T5}, P(tails and a five) $= \boxed{\tfrac{1}{12}}$.

b) There are exactly 7 ways for the outcome to be a tail or a five: {T1, T2, T3, T4, T5, T6, H5}. In the event you do not immediately grasp this, it may help you to list the possibilities:

tails = { T1, T2, T3, T4, **T5**, T6}
five = {H5, **T5**}
tails ∪ five = {T1, T2, T3, T4, **T5**, T6, H5}

Therefore, P(tails or a five) = P(tails ∪ five) $= \boxed{\tfrac{7}{12}}$.

c) Like part (a), P(heads and a two) $= \boxed{\tfrac{1}{12}}$.

29. If necessary, take a moment to list the possibilities (as previously demonstrated in the text's solution to Example 10 and this manual's solutions to Problems 25 and 27).

a) There are exactly 3 ways to obtain a head and an odd number: {H1, H3, H5}. Thus, since there are 12 possible outcomes in all, we have

$$\text{P(heads and an odd number)} = \tfrac{3}{12} = \boxed{\tfrac{1}{4}}.$$

b) There are precisely 9 ways to obtain a head or an odd number: {H1, H2, H3, H4, H5, H6, T1, T3, T5}. Thus, P(heads or an odd number) $= \tfrac{9}{12} = \boxed{\tfrac{3}{4}}$.

31. If necessary, take a moment to recreate textbook Figure 9.4 (p. 581). You may find it helpful to replace each ordered (i.e., green/red) outcome with its corresponding sum. That is,

Green die \ Red die	1	2	3	4	5	6
1	2	3	4	5	6	7
2	3	4	5	6	7	8
3	4	5	6	7	8	9
4	5	6	7	8	9	10
5	6	7	8	9	10	11
6	7	8	9	10	11	12

Out of 36 possible outcomes, there are 4 ways to obtain a sum of '5' when rolling a pair of dice: (1,4), (2,3), (3,2), and (4,1). Hence, the probability of rolling a '5' is

$$\text{P(five)} = \frac{\text{\# of ways of rolling a '5'}}{\text{Total \# of possible outcomes}} = \frac{4}{36} = \boxed{\frac{1}{9}}.$$

33. Upon referring either to the table shown above or to the text's Figure 9.4 (p. 581), we realize there are 6 ways to obtain a sum of '7': (1,6), (5,2), (4,3), (3,4), (2,5), (6,1). Thus, P(seven) $= \frac{6}{36} = \boxed{\frac{1}{6}}$.

35. As the table above makes evident, there are exactly 4 ways to obtain a sum of '9'. They are (3,6), (4,5), (5,4), and (6,3). Consequently, P(nine) $= \frac{4}{36} = \boxed{\frac{1}{9}}$.

37. There are 3 ways to obtain a sum of four and 4 ways to obtain a sum of five; respectively, (1,3), (2,2), (3,1) and (1,4), (2,3), (3,2), (4,1). Notice that these two events are mutually exclusive (i.e., it is not possible to simultaneously roll sums of four and five). Hence,

$$\text{P(four } \underline{\text{or}} \text{ five)} = \text{P(four} \cup \text{five)} = \frac{3+4}{36} = \boxed{\frac{7}{36}}.$$

39. Using the same reasoning as demonstrated above for Problem 37, we find:

$$\text{P(eight } \underline{\text{or}} \text{ ten)} = \text{P(eight} \cup \text{ten)} = \frac{5+3}{36} = \frac{8}{36} = \boxed{\frac{2}{9}}.$$

41. A good problem-solving approach is to list the set of possible outcomes (i.e., the sample space) and then assess which spinner/player wins more often.

In this particular problem, the player who picks spinner A can spin a 3 or 5, and the player on C can spin a 2 or 4. So, the sample space consists of four outcomes:

A \ C	2	4
3	*A wins*	*C wins*
5	*A wins*	*A wins*

Since P(A winning) $= \frac{3}{4}$ and P(C winning) $= \frac{1}{4}$, you should pick **spinner A**.

43. The player who picks spinner D can spin a 1, 7, or 8, whereas the player on E can spin a 4, 5, or 6. So, the sample space consists of nine outcomes:

D \ E	4	5	6
1	*E wins*	*E wins*	*E wins*
7	*D wins*	*D wins*	*D wins*
8	*D wins*	*D wins*	*D wins*

Since P(D) $= \frac{2}{3}$ and P(E) $= \frac{1}{3}$, you should pick **spinner D**.

45. The player who picks spinner D can spin a 1, 7, or 8, and the player on F can spin a 2, 3, or 9. So, once again, the sample space consists of nine outcomes:

D \ F	2	3	9
1	*F wins*	*F wins*	*F wins*
7	*D wins*	*D wins*	*F wins*
8	*D wins*	*D wins*	*F wins*

Because P(F) $= \frac{5}{9}$ and P(D) $= \frac{4}{9}$, you should pick **spinner F**.

47. The player who picks spinner F can spin a 2, 3, or 9, whereas the player on C can spin a 2 or 4. If both players spin a 2, they both spin again.

C \ F	2	3	9
2	*Tie*	*F wins*	*F wins*
4	*C wins*	*C wins*	*F wins*

Because P(F) = $\frac{3}{5}$ and P(C) = $\frac{2}{5}$, you should pick **spinner F**.

49. **Answers vary**. Although the problem requests empirical probabilities, here (***for informational purposes only***) are the theoretical probabilities:

a) P(two) $= \frac{1}{36} = 0.02\overline{7} = 2.\overline{7}\%$

b) P(three) $= \frac{1}{18} = 0.0\overline{5} = 5.\overline{5}\%$

c) P(four) $= \frac{1}{12} = 0.08\overline{3} = 8.\overline{3}\%$

d) P(five) $= \frac{1}{9} = 0.\overline{1} = 11.\overline{1}\%$

e) P(six) $= \frac{5}{36} = 0.13\overline{8} = 13.\overline{8}\%$

f) P(seven) $= \frac{1}{6} = 0.1\overline{6} = 16.\overline{6}\%$

g) P(eight) $= \frac{5}{36} = 0.13\overline{8} = 13.\overline{8}\%$

h) P(nine) $= \frac{1}{9} = 0.\overline{1} = 11.\overline{1}\%$

i) P(ten) $= \frac{1}{12} = 0.08\overline{3} = 8.\overline{3}\%$

j) P(eleven) $= \frac{1}{18} = 0.0\overline{5} = 5.\overline{5}\%$

k) P(twelve) $= \frac{1}{36} = 0.02\overline{7} = 2.\overline{7}\%$

51. **Yes**, each outcome has a probability of $\frac{1}{12}$. Remember that empirical probabilities can differ from theoretical probabilities. Of course, in this particular case, it is impossible (after performing this process 100 times) for the empirical probabilities to be equal, since 100 is not divisible by 12.

53. Remembering leap years, we assume that the length of a year is 365.25 days. How many hours are in a year? How many hours per year are you at school? If you can answer these two questions, you can determine the requested probability.

Since 365.25 days equates to 8,766 hrs/yr and since a 5-hour per day 174-day school year equates to 870 hrs/yr in school, the approximate probability (to the nearest percent) of an earthquake occurring while you are at school is:

$$\text{P(earthquake)} = \frac{870}{8{,}766} \approx 0.09924709 \approx \boxed{10\%}.$$

55. This problem is virtually identical to Problems 31-39 (see above solutions).

a) $\text{P(seven } \underline{\text{or}} \text{ eleven)} = \text{P(seven} \cup \text{eleven)} = \frac{6+2}{36} = \frac{8}{36} = \boxed{\frac{2}{9}}.$

b) $\text{P(two } \underline{\text{or}} \text{ three } \underline{\text{or}} \text{ twelve)} = \text{P(two} \cup \text{three} \cup \text{twelve)} = \frac{1+2+1}{36} = \frac{4}{36} = \boxed{\frac{1}{9}}.$

57. This problem is very similar to Problem 31 (see solution above).

a) As previously suggested, you may find it helpful to accompany the sample space (shown on the left below) with a comparable table (like the one on the right) in which each outcome's corresponding sum is displayed.

Sample Space

	1	2	3	4
1	(1,1)	(1,2)	(1,3)	(1,4)
2	(2,1)	(2,2)	(2,3)	(2,4)
3	(3,1)	(3,2)	(3,3)	(3,4)
4	(4,1)	(4,2)	(4,3)	(4,4)

$\Leftrightarrow$

Corresponding Sums

	1	2	3	4
1	2	3	4	5
2	3	4	5	6
3	4	5	6	7
4	5	6	7	8

b) There is only 1 way to obtain a sum of two: (1,1). Hence, P(two) $= \boxed{\frac{1}{16}}$.

c) There are 2 ways to obtain a sum of three: (1,2) and (2,1).
Therefore, P(three) $= \frac{2}{16} = \boxed{\frac{1}{8}}$.

d) P(four) $= \boxed{\frac{3}{16}}$

e) P(five) $= \frac{4}{16} = \boxed{\frac{1}{4}}$

f) P(six) $= \boxed{\frac{3}{16}}$

g) P(seven) $= \frac{2}{16} = \boxed{\frac{1}{8}}$

59. There are three possible cases to consider:

I. If in the jar there were originally two brains (call them B_1 and B_2) and if B_1 was the brain the madman already removed, then B_2 is the one organ remaining in the jar.

II. Similar to Case I, if there were originally two brains in the jar and if the madman had already removed B_2, then B_1 is the one organ remaining in the jar.

III. If the original contents of the jar consisted of a brain and kidney (and we already know the brain was removed), then the one organ remaining in the jar is the kidney.

In two of these three (equally likely) cases, there is a brain in the jar. Therefore, your response to the madman should be: P(brain) $= \boxed{\frac{2}{3}}$.

Problem Set 9.2

1. **False**. As Figure 9.6 (page 592) shows, two of the 38 spots are green. Half of the 36 remaining spots are black. So, P(black) $= \frac{18}{38} = \frac{9}{19}$.

3. **True**. As explained on the top of textbook p. 588, you should want to play games with positive expectations (i.e., "in your favor"); avoid games with negative expectations (not in your favor). A game whose expectation equals zero is said to be a "fair" game.

5. **False**. Remember that there is a cost of playing. This cost must be taken into consideration when calculating expected values. Review, as necessary, textbook pp. 590-591. So, the actual expected value for this lottery is:

$$\begin{aligned} E &= \text{AMOUNT TO WIN(PROBABILITY OF WINNING)} - \text{COST OF PLAYING} \\ &= \$50{,}000\left(\tfrac{1}{1{,}000{,}000}\right) - \$1 \\ &= \underline{\underline{-\$0.95}}. \end{aligned}$$

7. **Choice B**. We know the probability of correctly calling heads or tails on **one** flip of a coin is ½, so its expectation is ½ of $950 (or approximately $500). Since the probability of correctly calling heads or tails **five times in a row** is significantly less than ½, its expectation is substantially less than $500.

9. **Choice B**. An expected value of –$0.05 means that if you were to play roulette a large number of times, you would expect your *average loss per game* to be $0.05. So, if you played 100 times, you should have lost about $5 (i.e., $0.05 per game × 100 games).

11. **Choice B**. Proportionally speaking, there are many more people who have car accidents (on a daily basis, weekly basis,...) than there are winners of state lottery grand prizes. Similarly, one would expect the proportion of *Jeopardy* applicants who go on to become actual contestants to be significantly larger than the proportion of lottery players who go on to win their state's grand prize.

13. For additional details, refer to Examples 1-2 (textbook p. 588).
Each of the 1 million entries has an equally likely chance of being chosen, so the probability of winning is 1/1,000,000. Therefore, with there being no required cost of playing, the expectation is:

$$E = \overbrace{\$80{,}000}^{\text{amount of prize}} \times \underbrace{\left(\tfrac{1}{1{,}000{,}000}\right)}_{P(\text{winning the prize})} = \boxed{\$0.08}.$$

15. Each of the five bills has an equally likely chance of being chosen; hence, each outcome has a probability of $\frac{1}{5}$ (0.2 or 20%). So, the expected value is:

$$\begin{aligned} E &= \left[\$1\left(\tfrac{1}{5}\right) + \$5\left(\tfrac{1}{5}\right) + \$10\left(\tfrac{1}{5}\right) + \$20\left(\tfrac{1}{5}\right) + \$100\left(\tfrac{1}{5}\right)\right] - \overbrace{\$20}^{\text{cost of playing}} \\ &= \tfrac{1}{5}(\$1 + \$5 + 10 + \$20 + \$100) - \$20 \\ &= \tfrac{1}{5}(\$136) - \$20 \\ &= \$27.20 - \$20 \\ &= \boxed{\$7.20} \end{aligned}$$

17. Recall (from your experiences in Section 9.1) that a good first step to solving probability-related problems is to list the sample space. In this problem, the experiment's sample space is {HH, TH, HT, TT}. This list indicates that the probability of winning (i.e., tossing two heads) is $\frac{1}{4}$. So, the expected value is:

$$E = \overbrace{\$0.50}^{\text{amount of prize}} \times \underbrace{\left(\tfrac{1}{4}\right)}_{P(2\text{ heads})} = \$0.125 .$$

This amount represents the _average_ (i.e., over the long run) _winnings per play_ of this game, where each "play" consists of two tosses.

Remember that a so-called ***fair*** game is one whose expected value equals the cost of playing the game. So, a fair amount to pay for the privilege of playing this particular coin game would be **$0.25 for 2 plays**.

19. The expected value is $\$100\left(\frac{1}{100}\right)+\$10\left(\frac{5}{100}\right)=\$1.50$. So, for it to be a _fair_ game, you should anticipate the cost of playing to be **$1.50**.

21. This problem is very similar to text Example 7 (p. 591).

If the home is sold, let us assume that its actual sale price is the same as the original listing price of $85,000. So, if the seller's realtor succeeds in selling the home herself, her commission/payoff would be 6% of $85,000; that is, 0.06($85,000) = _$5,100_. If another realtor sells this home, then the seller's realtor earns a 3% commission; specifically, 0.03($85,000) = _$2,550_. Finally, if the home remains unsold after 6 months, then the realtor's contract with the seller expires and the realtor has a net loss of $800 (i.e., the amount expended trying to sell the home).

Thus, based upon the probabilities given in the statement of the problem, the realtor's profit expectation is:

$$\begin{aligned} E &= \overbrace{0.5(0.06\cdot\$85{,}000)}^{\text{She sells the house.}} + \overbrace{0.3(0.03\cdot\$85{,}000)}^{\text{Another agent sells the house.}} + \overbrace{0.20(\$0)}^{\text{No sale}} - \overbrace{\$800}^{\text{Realtor's expenses}} \\ &= 0.5(\$5{,}100) + 0.3(\$2{,}550) - \$800 \\ &= \boxed{\$2{,}515}. \end{aligned}$$

23. Problems 22-23 are analogous to Problem 21 and text Example 7.

To determine whether the oil-drilling company should sink the well, we determine the expected value:

$$E = \overbrace{\$825{,}000\left(\tfrac{1}{40}\right)}^{\text{Test well hits oil.}} + \overbrace{\$225{,}000\left(\tfrac{1}{20}\right)}^{\text{Test well hits natural gas.}} - \overbrace{\$25{,}000}^{\text{Cost of sinking test well}} = \$6{,}875 .$$

So, their expectation is **$6,875**. Therefore, they should **dig the well** because the expectation is positive.

25. Once again, we make this decision based upon the mathematical expectation. Recall that a standard 52-card deck includes 4 aces (one from each suit), 12 face cards (4 jacks, 4 queens, 4 kings), and, of course, only 1 two of spades. Hence,

$$E = \overbrace{\$0.50\left(\tfrac{4}{52}\right)}^{\text{An ace is drawn.}} + \overbrace{\$0.25\left(\tfrac{12}{52}\right)}^{\text{A face card is drawn.}} + \overbrace{\$1.00\left(\tfrac{1}{52}\right)}^{\text{The two of spades is drawn}} - \overbrace{\$0.10}^{\text{Cost of playing}}$$

$$= \$0.0\overline{153846}$$

$$\approx \$0.02\,.$$

Yes, since the expectation is positive (i.e., the game is in your favor), **you should play** the game.

27. To make this decision, the contestant should compare the expected value of playing against the $1,000 sure prize for not playing. Recall that the four prizes are hidden on a 20-space game board. So, the expected value of playing is

$$E = \$10{,}000\left(\tfrac{1}{20}\right) + \$5{,}000\left(\tfrac{2}{20}\right) + \$1{,}000\left(\tfrac{1}{20}\right) = \$1{,}050\,.$$

Since this is more than the $1,000 (guaranteed for not playing), we conclude that the **contestant should play the game**.

29. $E = \$25\left(\tfrac{1}{1{,}635}\right) + \$3\left(\tfrac{1}{163}\right) + \$1\left(\tfrac{1}{68}\right) \approx 0.0484013102$. So, to the nearest cent, the expectation from playing one card 13 times is **$0.05**.

31. The expected number of tardies per child is

$$E = 0(0.15) + 1(0.25) + 2(0.31) + 3(0.21) + 4(0.08) = \boxed{1.82}.$$

33. Based upon the given table, the merchant's expected gross income is

$$E = \$1{,}000(0.12) + \$800(0.38) + \$500(0.45) + \$200(0.05) = \$659\,.$$

If the merchant's purchasing cost is $500, then the expected profit is only $159, which is less than their desired $200 minimum. So, our answer to the question is: **No**, **the merchant should not purchase the collectables**. If the merchant can purchase the collectables for $459 or less, then this transaction is worthwhile.

35. Problems 34-43 are akin to text Example 8 (p. 592).

The roulette wheel has 38 numbered slots, 18 of which are odd (so, the remaining 20 are even). As shown in Figure 9.6, betting the odd numbers (or, for that matter, evens or blacks or reds or …) is classified as an "even money bet." The payoff is listed as "1 to 1." In the case of this particular problem, "1 to 1" means that, when the ball lands in an odd-numbered slot, you receive $1 for each $1 bet and you get to keep your original $1 wager. If the ball lands in an even-numbered slot, you lose your $1. Hence, the expectation for a $1 even-money type bet is:

$$E = \$1\left(\tfrac{18}{38}\right) + (-\$1)\left(\tfrac{20}{38}\right) \approx -\$0.0526315789 \approx \boxed{-\$0.05}.$$

37. Double-number bets [see "Two Numbers or Split" in Figure 9.6] cover two of the 38 numbered slots. The payoff is listed as 17 to 1. This means, when the ball lands in one of the two slots you selected, you receive \$17 (for every \$1 bet) and you keep your original wager. Of course, if the ball lands in any of the other 36 remaining slots, you lose the amount you wagered. Accordingly, the expectation for a \$1 double-number bet is:

$$E = \$17\left(\tfrac{2}{38}\right) + (-\$1)\left(\tfrac{36}{38}\right) \approx -\$0.0526315789 \approx \boxed{-\$0.05}.$$

39. From Figure 9.6, we learn that a four-number bet covers four joining numbers and has a payoff of 8 to 1. So, if the ball lands in one of the four slots you wagered on, then you receive \$8 (per \$1 bet) and you keep your original wager. Of course, if the ball lands on any of the remaining 34 slots, you lose. Thus, the expectation for a \$1 four-number bet is:

$$E = \$8\left(\tfrac{4}{38}\right) + (-\$1)\left(\tfrac{34}{38}\right) \approx -\$0.0526315789 \approx \boxed{-\$0.05}.$$

41. In the same manner as the previous problems, we find the expected value of a \$1 six-number bet to be (based upon a payoff of 5 to 1):

$$E = \$5\left(\tfrac{6}{38}\right) + (-\$1)\left(\tfrac{32}{38}\right) \approx -\$0.0526315789 \approx \boxed{-\$0.05}$$

43. As explained in textbook Figure 9.6, a column bet covers a column of 12 numbers and pays 2 to 1. Hence, a \$1 column bet's expectation is:

$$E = \$2\left(\tfrac{12}{38}\right) + (-\$1)\left(\tfrac{26}{38}\right) \approx -\$0.0526315789 \approx \boxed{-\$0.05}$$

45. Notice that the \$5 region consumes exactly half of the spinner's area, while the \$0 and \$10 regions individually comprise one-fourth of its area. Given that the cost to spin the wheel is \$5.00, this game's mathematical expectation is:

$$E = \left[\$5\left(\tfrac{1}{2}\right) + \$0\left(\tfrac{1}{4}\right) + \$10\left(\tfrac{1}{4}\right)\right] - \$5 = \$0$$

Since the expectation is zero (i.e., the expected amount won equals the cost of playing), this is a $\boxed{\textbf{fair game}}$.

47. Upon observing the proportion of the spinner's area consumed by each region, we obtain an expectation of $\left[\$2\left(\tfrac{1}{2}\right) + \$4\left(\tfrac{1}{4}\right) + \$8\left(\tfrac{1}{8}\right) + \$14\left(\tfrac{1}{8}\right)\right] - \$5 = \underline{\underline{-\$0.25}}$.

Because its expectation is negative, this is $\boxed{\textbf{not a fair game}}$. As such, it should not be played.

49. Similar to Problems 44-47, the each outcome's probability equals the proportion of the area it consumes. Hence, this dart game has an expectation of

$$E = \tfrac{1}{2}(-\$5) + \tfrac{1}{4}(\$16) + \tfrac{1}{8}(\$8) + \tfrac{1}{16}(\$4) + \tfrac{1}{32}(\$2) + \tfrac{1}{32}(\$1) = \$2.84375 \approx \$2.84.$$

Remember: you should be unwilling to play games that are not in your favor. Thus, to play this game, you should be willing to pay (no more than) $\boxed{\textbf{\$2.84}}$.

51. So far, the contestant has picked twice and has won a total of \$350,000. This is the amount at risk if he/she picks again.

We make the decision of whether (or not) to pick a third time based upon the mathematical expectation of this third pick. With two hidden windows remaining, the probability of picking the window containing \$1,000,000 is the same as picking the window containing the "stopper"; each has a probability of ½. Hence, the expected value of the third pick is:

$$E = \$1{,}000{,}000\left(\tfrac{1}{2}\right) + (\$0)\left(\tfrac{1}{2}\right) = \$500{,}000.$$

Since this is more than the amount at risk (i.e., the game is in the player's favor), the **contestant should continue**.

53. In order to win the \$1 prize, your (first) coin toss must come up heads; of course, the probability of heads is P(H) = $\frac{1}{2}$. If it came up heads, you receive \$1 and the game stops.

To win \$2, your first toss must be tails and the second toss must be heads. Upon recalling the sample space for two coin tosses S = {HH, HT, TH, TT}, it becomes clear that the probability of this outcome is P(TH) = $\frac{1}{4}$. If this outcome occurs, the game stops.

To win \$4, your three tosses must be in order tails, tails, and finally heads. This is one of the eight (equally) possible outcomes for three tosses; that is, P(TTH) = $\frac{1}{8}$.

Thus, the mathematical expectation for this game is:

$$E = \overbrace{\$1\left(\tfrac{1}{2}\right)}^{\text{Outcome } H} + \overbrace{\$2\left(\tfrac{1}{4}\right)}^{\text{Outcome } TH} + \overbrace{\$4\left(\tfrac{1}{8}\right)}^{\text{Outcome } TTH} = \$0.50 + \$0.50 + \$0.50 = \boxed{\$1.50}.$$

55. This problem is just a continuation of the pattern experienced in Problems 53-54.

$$\begin{aligned} E &= \overbrace{\$1\left(\tfrac{1}{2}\right)}^{\text{Outcome } H} + \overbrace{\$2\left(\tfrac{1}{4}\right)}^{\text{Outcome } TH} + \overbrace{\$4\left(\tfrac{1}{8}\right)}^{\text{Outcome } TTH} + \overbrace{\$8\left(\tfrac{1}{16}\right)}^{\text{Outcome } TTTH} + \cdots + \overbrace{\$2^{999}\left(\frac{1}{2^{1,000}}\right)}^{\text{Outcome 999 } T\text{'s followed by 1 } H} \\ &= \underbrace{\$0.50 + \$0.50 + \$0.50 + \$0.50 + \cdots + \$0.50}_{\text{1000 times}} \\ &= 1000(\$0.50) \\ &= \boxed{\$500} \end{aligned}$$

57. **Answers vary**.

59. **Answers vary**.

Problem Set 9.3

1. **Wording varies**. The property of complements states $P(E) = 1 - P(\overline{E})$; or equivalently, it also says $P(\overline{E}) = 1 - P(E)$ and $P(E) + P(\overline{E}) = 1$. To quote the textbook's author (p. 597), "Don't underestimate the importance of this simple property; it has far-reaching applications."

3. **Wording varies**. A ***conditional probability*** of an event E is the probability of E occurring, given/assuming that event F has already occurred. We denote this by $P(E|F)$, which is read as "probability of E given F." To determine a conditional probability, we adjust/modify the experiment's sample space based upon how it has been altered by the (past) assumed occurrence of event F. We then reevaluate E (and its probability) by viewing it in terms of this new (adjusted) sample space.

5. Doesn't the game of tick-tack-toe involve a multi-step procedure (task)? Consequently, to determine the number of possible different sequences for its first five moves, we use the fundamental counting principle.

There are 9 possible squares in which to place the first move; followed by 8 possibilities for the second move; then 7 for the third move; and so on. Therefore, the total number of possibilities for the first five moves is $9 \cdot 8 \cdot 7 \cdot 6 \cdot 5 =$ **15,120**.

7. **Choice C**. The odds of winning the lottery are approximately 1 to 5,000,000; being struck by lightning about 1 to 600,000; and of appearing on the *Tonight Show* about 1 to 490,000.

9. **Choice B**. What do the words "**at least** 2 times" mean?

11. **Choice B**. The odds of choice A occurring are about 1 to 10,000,000, whereas the odds of choice B are approximately 1 to 2,000,000. For additional perspective regarding choice B, refer to text Example 4 (pp. 599-600).

13. Via the property of complements, $P(\overline{B}) = 1 - P(B) = 1 - \frac{4}{5} = \boxed{\frac{1}{5}}$. Make sure you can express (in clear language) what such an answer means. It tells us that if the probability of an event's occurrence (call the event B) is $\frac{4}{5}$ (or 80%), then the probability of its nonoccurrence is $\frac{1}{5}$ (i.e., 20%).

15. $P(D) = 1 - P(\overline{D}) = 1 - 0.005 =$ **0.995**

17. *[**Important Strategy Tip**: Whenever a probability problem seems difficult to solve directly (perhaps because there are too many possibilities to consider), try to solve it indirectly by using the property of complements.]*

There are 20 multiples of 5 between 1 and 100 (namely, 5, 10, 15, …, 95, 100). So,

$$\begin{aligned}\text{P(not a multiple of 5)} &= 1 - \text{P(multiple of 5)}\\ &= 1 - \tfrac{20}{100}\\ &= 1 - \tfrac{1}{5}\\ &= \boxed{\tfrac{4}{5}}\end{aligned}$$

19. This problem is very similar to text Example 2 (p. 598).

First, make sure you have read the question carefully and understand its full implication. What is meant by the words "**at least one head** in four flips"? The question asks us to find the probability of obtaining one *or more* heads in four flips. Since this involves the consideration of a large number of possibilities (namely, four time-consuming cases to assess), we adhere to our above-stated strategy tip (see Problem 17's solution); namely, instead of trying to solve the problem directly, we solve it indirectly (by using the property of complements).

Via the fundamental counting principle, we know for four flips of a coin that there are 16 possible outcomes (i.e., $2 \cdot 2 \cdot 2 \cdot 2 = 2^4$). Note that there is only one way to obtain all tails (TTTT). Finally, with this (first) solution, permit us to share many more steps ("rephrasings") than are necessary:

$$\begin{aligned}\text{P(at least one head)} &= 1 - \text{P}\overline{\text{(at least one head)}}\\ &= 1 - \text{P(\textbf{\textit{not}} at least one head)}\\ &= 1 - \text{P(no heads)}\\ &= 1 - \text{P(all tails)}\\ &= 1 - \text{P(TTTT)}\\ &= 1 - \tfrac{1}{16}\\ &= \boxed{\tfrac{15}{16}}.\end{aligned}$$

21. In terms of probability, the genders of four children are no different than the numbers of heads/tails obtained when tossing four coins. The fundamental counting principle tells us that, since there are 2 ways of having a child (**B**oy or **G**irl), there are a total of $2 \cdot 2 \cdot 2 \cdot 2 = 2^4 = 16$ ways for 4 children. Of these 16 ways, there is only 1 way to have 4 boys (BBBB) — so there are 15 ways not to have 4 boys.

To determine the odds against having four boys, we apply the concepts and notation introduced on textbook p. 600. Here are **two** approaches:

I. With $n = 16$, $s = 15$, and $f = 1$, the odds against (f to s) having four boys are $\boxed{\textbf{15 to 1}}$.

II. Since P(four boys) $= \frac{1}{16}$ and P($\overline{\text{four boys}}$) $= \frac{15}{16}$, the odds against having four boys are

$$\frac{P(\overline{\text{four boys}})}{P(\text{four boys})} = \frac{\frac{15}{16}}{\frac{1}{16}} = \frac{15}{16} \cdot \frac{16}{1} = \frac{15}{1} \quad \Rightarrow \quad \underline{\textbf{15 to 1}}.$$

23. This problem is virtually identical to text Example 5 (p. 600). Given that $P(E) = 0.82$, it follows that $P(\overline{E}) = 1 - 0.82 = 0.18$. Hence, the **odds in favor** of this event are

$$\frac{P(E)}{P(\overline{E})} = \frac{0.82}{0.18} = \frac{41}{9} \approx \frac{5}{1} \quad \Rightarrow \quad \boxed{\textbf{about 5 to 1}}.$$

25. From textbook p. 600, recall that the **odds against** an event E are $\boldsymbol{f}$ **to** $\boldsymbol{s}$ (failures to successes), whereas the **probability of E occurring** is

$$P(E) = \frac{s}{n} = \frac{s}{s+f} \text{ (successes divided by total number of possible outcomes).}$$

Given each horse's "odds against" winning, we determine their respective **probabilities of winning** as follows:

Horse #1: $f = 2,\ s = 1 \quad \Rightarrow \quad P(\#1 \text{ winning}) = \frac{1}{1+2} = \boxed{\frac{1}{3}}$

Horse #2: $f = 15,\ s = 1 \quad \Rightarrow \quad P(\#2 \text{ winning}) = \frac{1}{1+15} = \boxed{\frac{1}{16}}$

Horse #3: $f = 3,\ s = 2 \quad \Rightarrow \quad P(\#3 \text{ winning}) = \frac{2}{2+3} = \boxed{\frac{2}{5}}$

Horse #4: $f = 7,\ s = 5 \quad \Rightarrow \quad P(\#4 \text{ winning}) = \frac{5}{5+7} = \boxed{\frac{5}{12}}$

Horse #5: $f = 1,\ s = 1 \quad \Rightarrow \quad P(\#5 \text{ winning}) = \frac{1}{1+1} = \boxed{\frac{1}{2}}$

27. Given $f = 33$ and $s = 1$, $\frac{f}{s+f} = \frac{33}{1+33} = \boxed{\frac{33}{34}}$.

29. In some respects, this problem is similar to text Example 11 (pp. 602-603). There are **26** letters (A to Z) to choose from, **10** numerals (0 to 9), **1** blank space, and the **4** available symbols. Hence, there are **41** possibilities ("choices") for each of the seven spaces on a California license plate. Via the fundamental counting principle, it follows that the total number of possible license plates is:

$$41 \times 41 \times 41 \times 41 \times 41 \times 41 \times 41 = \boxed{\mathbf{41^7}}.$$

31. This problem is an extension of text Example 12 (p. 604).

This is a ***conditional probability*** problem. As such, we must pay special attention to its precondition: "given that at least two [of the four flips] are heads." Remember that the words "at least two" mean "two or more" (i.e., 2 heads, 3 head, or 4 heads). This condition reduces the size of the sample space from 16 possible outcomes to 11; that is,

Altered Sample Space: HHHH, HHHT, HHTH, HHTT, HTHH, HTHT, HTTH, ~~HTTT~~, THHH, THHT, THTH, ~~THTT~~, TTHH, ~~TTHT~~, ~~TTTH~~, ~~TTTT~~

Of these remaining 11possible outcomes, four of them consist of "exactly three heads" (namely, HHHT, HHTH, HTHH, THHH). Thus,

$$\text{P(exactly 3H} \mid \text{at least 2H)} = \boxed{\frac{4}{11}}.$$

33. *[Suggestion: After completing this problem, make sure you do Problem 34, which will enhance your understanding of conditional probabilities.]*

Recall that the notation P(face card | jack) asks us to determine the probability of the single card drawn being a face card given that (i.e., "already knowing that") this drawn card is a jack.

Since we already know the single card drawn was a jack and all jacks are (by definition) face cards, it is a certainty that the card (jack) drawn was a face card. Hence, P(face card | jack) = $\boxed{1}$.

35. Given that the single card drawn was not a spade, the original 52-card (four-suit) sample space is reduced/altered to 39 cards (i.e., 13 ♥'s, 13 ♦'s, 13 ♣'s). Thus,

$$\text{P(heart} \mid \text{not a spade)} = \frac{13}{39} = \boxed{\frac{1}{3}}.$$

37. Given that the single card drawn was a jack, the sample space is reduced from 52 cards to 4 cards, namely {jack of ♥, jack of ♠, jack of ♦, jack of ♣}. Of course, spades and clubs are black. Consequently,

$$\text{P(black} \mid \text{jack)} = \frac{2}{4} = \boxed{\frac{1}{2}}.$$

39. Problems 39-44 are analogous to parts (b) - (c) of text Example 13 (pp. 604-605).

Given that the first card drawn was a two, this card's removal reduces the size of the sample space (for the second draw) to 51 cards, 4 of which are aces. Thus,

$$\text{P(ace} \mid \text{two)} = \boxed{\frac{4}{51}}.$$

41. Given that the first card drawn was one of the 13 hearts, the removal of this heart reduces the size of the sample space (for the second draw) to 51 cards, 12 of which are the remaining hearts. Accordingly,

$$\text{P(heart} \mid \text{heart)} = \frac{12}{51} = \boxed{\frac{4}{17}}.$$

43. Given that the first card selected was one of the 26 red cards, this red card's removal reduces the size of the sample space (for the second draw) to 51 cards, 26 of which are black and 25 are red. Consequently,

$$\text{P(black} \mid \text{red)} = \boxed{\frac{26}{51}}.$$

45. This problem is virtually identical to text Example 11 (pp. 602-603).

[<u>As stated in previous solutions</u>: Whenever a probability problem appears difficult to solve directly (because there are too many possibilities to consider), try to solve it indirectly by using the property of complements.]

Let **R** denote the event of getting a license plate that has a repeated letter or digit. The problem asks us to determine P(R). The complement of R is denoted by $\overline{\mathbf{R}}$, which represents the event of getting a license plate with <u>no</u> repeated letters or digits. From the property of complements, we know $P(R) = 1 - P(\overline{R})$. So, if we can find $P(\overline{R})$, it will be easy to compute P(R). By applying the fundamental counting principle, we find:

$$P(\overline{R}) = \frac{\textit{\# of possible license plates without repeated letters/digits}}{\textit{Total \# of possible license plates}}$$

The number of (two-letter, four-numeral) license plates without repeated letters/digits is: $26 \times 25 \times 10 \times 9 \times 8 \times 7 = 3{,}276{,}000$. The total number of possible license plates is: $26 \times 26 \times 10 \times 10 \times 10 \times 10 = 26^2 \times 10^4 = 6{,}760{,}000$.

Therefore, the probability of getting a (two-letter, four-digit) license plate that has a repeated letter or digit is:

$$P(R) = 1 - P(\overline{R}) = 1 - \frac{3{,}276{,}000}{6{,}760{,}000} = 1 - \frac{63}{130} = \frac{67}{130} \approx 0.5153 \boxed{\approx \mathbf{51.5\%}}.$$

47. This problem is similar to text Example 14 (pp. 605-606).

a) From the table, we know 25 of the 80 executives surveyed are from NBC. Thus, $P(\text{NBC}) = \frac{25}{80} = \frac{5}{16} = \boxed{\mathbf{0.3125}}$ or 31.25%.

b) Of the 80 executives surveyed, a total of 54 said they're satisfied. So, $P(\text{satisfied}) = \frac{54}{80} = \frac{27}{40} = \boxed{\mathbf{0.675}}$ or 67.5%.

c) Pay special attention to the phrase "if we know…." It implies that a ***conditional*** probability needs to be determined. Given that the person selected is among those satisfied with current programming, the sample space is reduced to only the 54 satisfied executives. Among these 54, exactly 21 are from CBS. Hence, $P(\text{CBS} \mid \text{satisfied}) = \frac{21}{54} = \frac{7}{18} = 0.3\overline{8} \boxed{\approx \mathbf{0.389}}$ or about 38.9%.

d) Knowing that the person selected is from CBS, the sample space is reduced to only the 30 CBS executives. Among these 30, exactly 21 are satisfied. Therefore, $P(\text{satisfied} \mid \text{CBS}) = \frac{21}{30} = \frac{7}{10} = \boxed{\mathbf{0.70}}$ or 70.0%.

49. a) The phrase "given that the rolled number was a 6" implies that the sample space is reduced/altered. Specifically, the sample space becomes $S = \{6\}$. So, it is impossible for the outcome of this roll to be odd; that is,

$$\text{P(odd} \mid 6) = \boxed{0}.$$

b) The phrase "given that the rolled number was a 5" implies that the reduced sample space becomes $S = \{5\}$. So, it is an absolute certainty that the outcome of this roll is an odd number; that is,

$$\text{P(odd} \mid 5) = \boxed{1}.$$

51. If necessary, refer to textbook Figure 9.4 (p. 581) while solving this problem.

Each part of this problem includes the precondition "given that exactly one die came up 2." So, all are ***conditional*** probability problems. This precondition reduces the size of the original sample space (that originally consisted of 36 ordered pairs) to a set consisting of the following ten outcomes:

$$S = \{(2,1), (2,3), (2,4), (2,5), (2,6), (1,2), (3,2), (4,2), (5,2), (6,2)\}$$

a) Among these ten outcomes, only two yield a sum of 5; namely, (2,3) and (3,2). Thus,

$$\text{P(sum of 5} \mid \text{exactly one die came up 2)} = \tfrac{2}{10} = \boxed{\tfrac{1}{5}}.$$

b) Among these ten outcomes, only two yield a sum of 3; namely, (2,1) and (1,2). Thus,

$$\text{P(sum of 3} \mid \text{exactly one die came up 2)} = \tfrac{2}{10} = \boxed{\tfrac{1}{5}}.$$

c) Among these ten outcomes, none yields a sum of 2. Remember that (1,1) is not included in the altered sample space. Thus, given the precondition, it is impossible to obtain such a sum. Accordingly, we write:

$$\text{P(sum of 2} \mid \text{exactly one die came up 2)} = \boxed{0}.$$

53. $$\frac{P(\overline{E})}{P(E)} = \frac{\left(\frac{f}{n}\right)}{\left(\frac{s}{n}\right)} = \frac{f}{n} \div \frac{s}{n} = \frac{f}{n} \cdot \frac{n}{s} = \frac{f}{s} = \text{Odds against.}$$

55. This problem is analogous to text Example 15 (pp. 606-607).

Let L_1 = {1st member selected is a pledge} and L_2 = {2nd member selected is a pledge}.

a) Since 10 of the 35 members are pledges, $P(L_1) = \frac{10}{35} = \boxed{\frac{2}{7}}$.

b) To determine this probability, one can either use the given information (i.e., 25 of the 35 sorority sisters are full members) or apply the property of complements to part (a)'s answer. In either case, $P(\overline{L_1}) = \frac{25}{35} = \boxed{\frac{5}{7}}$.

c) Note the phrase "if the first person selected was also a pledge." It is a precondition that tells us this problem involves a ***conditional*** probability. After the selection of the first person (who was one of the 10 pledges), we are left with 34 remaining members to choose from, 9 of whom are pledges. Hence,

$$P(L_2 \mid L_1) = \boxed{\frac{9}{34}}.$$

d) If the first person selected was a pledge, then for the second selection there are 34 remaining members (including all of the 25 full members). So,

$$P(\overline{L}_2 | L_1) = \boxed{\frac{25}{34}}.$$

e) If the first person selected was a full member, then for the second selection there are 34 remaining members (including all 10 of the original pledges). So,

$$P(L_2 | \overline{L}_1) = \tfrac{10}{34} = \boxed{\frac{5}{17}}.$$

f) If the first person selected was a full member, then for the second selection there are 34 remaining members (including 24 of the original 25 full members).

$$P(\overline{L}_2 | \overline{L}_1) = \tfrac{24}{34} = \boxed{\frac{12}{17}}.$$

g) This problem does not reveal the outcome of the first selection. If we do not know what happened on the first selection, then this is not a conditional probability and we consider the probability to be the same as $P(L_1)$, which we know from part (a) is $\boxed{\frac{2}{7}}$. Alternatively, as will be explained more clearly in Section 9.5 (ask your instructor), we may calculate this probability by considering the following possible scenarios:

- If the first selection was also a pledge, then the probability of selecting two pledges consecutively is:

$$P(L_1 \text{ and } L_2) = P(L_1 \cap L_2) = P(L_1) \cdot P(L_2 | L_1) = \tfrac{10}{35} \cdot \tfrac{9}{34} = \tfrac{9}{119}.$$

- If the first selection was full member, then the probability of selecting a full member first and a pledge second is:

$$P(\overline{L}_1 \text{ and } L_2) = P(\overline{L}_1 \cap L_2) = P(\overline{L}_1) \cdot P(L_2 | \overline{L}_1) = \tfrac{25}{35} \cdot \tfrac{10}{34} = \tfrac{25}{119}.$$

The probability of the second selection being a pledge is the sum of these two mutually exclusive events. That is,

$$P(L_2) = P(L_1 \text{ and } L_2) + P(\overline{L}_1 \text{ and } L_2) = \tfrac{9}{119} + \tfrac{25}{119} = \tfrac{34}{119} = \tfrac{2}{7}.$$

h) The procedure for constructing (and using) these tree diagrams is explained on textbook p. 607.

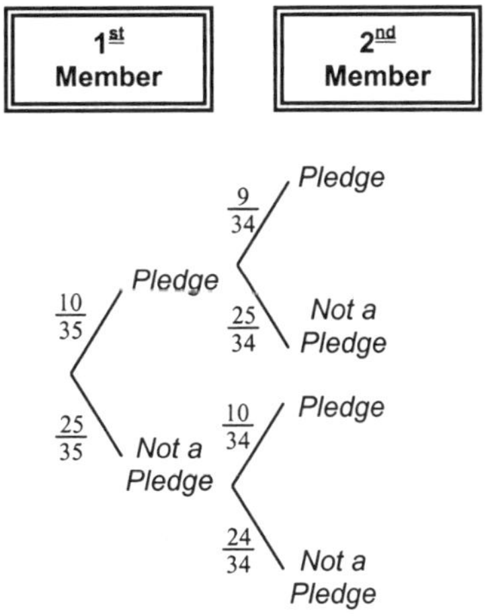

57. **Answers vary**. A typical classroom is 20 ft × 30 ft × 10 ft = 10,368,000 in.^{3}. Assume that a ping-pong ball has a volume of 1 in.^{3}, and imagine filling the entire classroom with ping-pong balls — where only one of these millions of balls is painted red. The odds against winning the lottery are about the same as the odds of reaching into this ball-filled classroom and drawing out the one red ball.

59. Regarding the fallacy with this betting system, **answers vary**. One concern is the limitations of the bettor's resources; another is the betting limit imposed on the game.

Turning our attention to the system itself, we recall (from textbook p. 592) that when the bettor wins, he/she keeps the amount of the original wager. Said differently, the amount you bet is collected only when you lose.

If a bettor lost on the first spin (with a \$1 wager) but won on the second spin (a \$2 wager), the bettor's net winnings equal: (–\$1) + (\$2) = **\$1**.

If a bettor lost on the first spin (\$1 wager) and the second spin (\$2 wager) but won on the third spin (with a \$4 wager), the bettor's net winnings equal:

(–\$1) + (–\$2) + (\$4) = **\$1**.

If a bettor lost \$1 on the first spin and \$2 on the second spin and \$4 on the third spin but won \$8 on the fourth spin, the bettor's net winnings equal:

(–\$1) + (–\$2) + (–\$4) + (\$8)= **\$1**.

And so on....

To show that in every case (with this roulette system) the net winnings amount to \$1, we apply our knowledge of geometric series (from Section 8.4). If a bettor loses every bet until winning on the kth bet, then his/her net winnings equal:

$$\left[\underbrace{\overbrace{(-\$1)}^{\text{1st bet}}+\overbrace{(-\$2)}^{\text{2nd bet}}+\overbrace{(-\$2^2)}^{\text{3rd bet}}+\overbrace{(-\$2^3)}^{\text{4th bet}}+\cdots+\overbrace{(-\$2^{k-2})}^{(k-1)\text{th bet}}}_{\text{consecutive losses}}\right]+\overbrace{(\$2^{k-1})}^{k\text{th bet (finally a winner!)}}$$

This can be rewritten as:

$$\overbrace{-\left[\$1+\$2+\$2^2+\$2^3+\cdots+\$2^{k-2}\right]}^{\text{losses}}+\overbrace{\$2^{k-1}}^{\text{win}}$$

The ***consecutive losses*** form a geometric sequence consisting of $(k-1)$ terms [recall: the kth bet is a win] with the sequence's first term being $g_1 = 1$ and common ratio $r = 2$. From our previous work in Chapter 8, we know the sum of such a sequence is

$$G_n = \frac{g_1(1-r^n)}{1-r} \implies G_{k-1} = \frac{(1)(1-2^{k-1})}{1-2} = \frac{1-2^{k-1}}{-1} = \underline{\underline{2^{k-1}-1}}.$$

Hence, the net winnings are:

$$\text{Net winnings} = -\overbrace{\left[1+2+2^2+2^3+\cdots+2^{k-2}\right]}^{\text{losses}} + \overbrace{2^{k-1}}^{\text{win}}$$
$$= -\left[2^{k-1}-1\right]+2^{k-1}$$
$$= -2^{k-1}+1+2^{k-1}$$
$$= \underline{\underline{\$1}}\ \checkmark$$

Problem Set 9.4

1. **Answers vary**. The ***fundamental counting principle*** counts the total number of ways a multi-step task (or process) can be performed. The principle can be applied when the number of choices at any given step is not affected by the choices made on a previous step; in other words, repetition is allowed. In contrast, we use a ***permutation*** to count the total number of possible ways to arrange r distinct objects out of a given set of n items. As the word "distinct" implies, the formula for permutation can be applied only when repetitions are not allowed.

3. **Answers vary**. Table 9.2 (on textbook page 617) summarizes the four types of counting techniques and under what circumstances each applies.

5. **Answers vary**.

Reminder: ***Many calculators include built-in keys for factorials, combinations, and permutations. For details, check your calculator user's manual (or, if applicable, its accompanying "quick reference" card).***

7. For additional information, refer to text Example 4 (p. 613).
   a) $10! = 10\cdot 9\cdot 8\cdot 7\cdot 6\cdot 5\cdot 4\cdot 3\cdot 2\cdot 1 = \boxed{3{,}628{,}800}$
   b) $4! = \boxed{24}$
   c) $10! - 4! = 3{,}628{,}800 - 24 = \boxed{3{,}628{,}776}$
   d) $(10-4)! = 6! = \boxed{720}$

9. a) $\dfrac{10!}{9!} = \dfrac{10\cdot 9!}{9!} = \boxed{10}$
   b) $\dfrac{10!}{8!} = \dfrac{10\cdot 9\cdot 8!}{8!} = 10\cdot 9 = \boxed{90}$
   c) $\dfrac{10!}{7!} = \dfrac{10\cdot 9\cdot 8\cdot 7!}{7!} = 10\cdot 9\cdot 8 = \boxed{720}$
   d) $\dfrac{10!}{6!} = 10\cdot 9\cdot 8\cdot 7 = \boxed{5{,}040}$

11. We note that these four problems are intended to prepare you for (the computation of) ***combinations***. For instance, part (a), in essence, asks for the number of different possible 5-card poker hands, given a standard 52-card deck. The answer may look familiar to you; it was required for Problems 13-22 in Section 9.1 (see textbook p. 585).

a) *[Note: This is equivalent to ${}_{52}C_5$.]*

$$\frac{52!}{5!47!} = \frac{(52 \cdot 51 \cdot 50 \cdot 49 \cdot 48) \cdot 47!}{(5 \cdot 4 \cdot 3 \cdot 2 \cdot 1) \cdot 47!}$$

$$= 52 \cdot 51 \cdot \left(\frac{50}{5}\right) \cdot 49 \cdot \left(\frac{48}{4 \cdot 3 \cdot 2 \cdot 1}\right)$$

$$= 52 \cdot 51 \cdot 10 \cdot 49 \cdot 2$$

$$= \boxed{\mathbf{2{,}598{,}960}}$$

b) *[This is equivalent to ${}_{52}C_3$.]*

$$\frac{52!}{3!(52-3)!} = \frac{52!}{3!49!} = \frac{(52 \cdot 51 \cdot 50) \cdot 49!}{(3 \cdot 2 \cdot 1) \cdot 49!} = 52 \cdot 51 \cdot 50 = \boxed{\mathbf{22{,}100}}$$

c) *[This is equivalent to ${}_{12}C_3$.]*

$$\frac{12!}{3!(12-3)!} = \frac{12!}{3!9!} = \frac{(12 \cdot 11 \cdot 10) \cdot 9!}{(3 \cdot 2 \cdot 1) \cdot 9!} = 4 \cdot 11 \cdot 5 = \boxed{\mathbf{220}}$$

d) After enough practice, you should begin to increase the number of "mentally computed" steps. Let us demonstrate:

$$\frac{156!}{3!(156-3)!} = \frac{156 \cdot 155 \cdot 154}{3 \cdot 2 \cdot 1} = \boxed{\mathbf{620{,}620}}$$

Note: This is equivalent to ${}_{156}C_3$.

13. For additional details, refer to text Examples 3, 6, and 8 (pp. 612, 614-615), in which three different approaches for computing permutations ("arrangements") are demonstrated.
Please remember three things: (i) permutations are a special case of the fundamental counting principle (specifically, one without repetitions); (ii) many calculators include a permutation key; and (iii) some problems [e.g., parts (c), (e), and (f) below] can be computed more readily (or at least differently) than others.

a) What does ${}_6P_4$ mean? It asks you to find the number of different ways of arranging 4 objects selected from a set of 6 (distinct) objects. One approach is to view this as a four-step task (i.e., position the first object, then the second, then …) and, consequently, apply the fundamental counting principle. That is, there are 6 choices for which object will be the first to be placed, then 5 choices for the position of the second object, and so on. Via the fundamental counting principle, this leads to ${}_6P_4 = 6 \cdot 5 \cdot 4 \cdot 3 = \boxed{\mathbf{360}}$. Alternatively,

$${}_6P_4 = 4!\cdot {}_6C_4 = 360 \ \underline{\text{or}}\ {}_6P_4 = \frac{6!}{(6-4)!} = \frac{6!}{2!} = 360.$$

b) ${}_{52}P_4 =$ **6,497,400**

c) ${}_6P_2 =$ **30** *[Think "fundamental counting principle": 6 × 5 = 30 arrangements.]*

d) ${}_9P_4 =$ **3,024**

e) ${}_{50}P_1 =$ **50**

Before proceeding to other problems, please be sure you **fully** understand this answer. Does it make sense? Ask yourself this: how many ways are there to select "groups of one" from a set of 50 objects? In how many ways can each of these single-object "groups" be arranged?

f) ${}_{40}P_{40} =$ **40!**

15. For additional information, refer to text Examples 3, 5, and 8 (pp. 612, 614-615), in which several approaches for computing combinations ("groups", "subsets") are demonstrated. Make sure you understand the difference between combinations and permutations: combinations disregard differences in order/arrangement.

a) ${}_5C_4 =$ **5**. Does it make sense to you that ${}_5C_4 = {}_5C_1$? If not, consider this: each time we form a subset of 4 elements from a set of five elements, 1 of the five elements gets left out. How many ways are there to leave 1 of the five elements out? There are 5 ways.

b) ${}_5C_3 =$ **10**

c) ${}_7C_2 =$ **21**

d) ${}_4C_4 =$ **1**. You should not need to perform any computations here. Make sure this answer make (immediate) intuitive sense to you. Ask yourself: how many ways are there to choose a group (subset) of 4 elements from a set of 4 elements? Once again, remember that the arrangement of the elements in a subset is unimportant.

e) ${}_{100}C_1 =$ **100**. Once again, you should not need to perform any arithmetic here. It should make intuitive sense that there are 100 ways to select so-called "groups of one" from a set of 100 objects.

f) ${}_{1000}C_2 =$ **499,500**

17. a) ${}_8P_4 =$ **1,680**

b) ${}_{25}C_1 =$ **25**. This answer (like that of Problem 15e above) should not require any computation. Doesn't it make sense that there are 25 ways to form one-element subsets from a set of 25 elements?

c) ${}_nC_0 = \boxed{1}$. How many ways are there for you to choose nothing? There is one way to choose nothing: simply, choose nothing! Said differently, there is only one subset consisting of zero elements (i.e., the empty set $\varnothing$) that can be taken from a set of n elements.

d) ${}_nP_0 = \boxed{1}$. Similar to part (c), there is only one way to "arrange" nothing.

e) ${}_gC_h = \boxed{\dfrac{g!}{h!(g-h)!}}$

f) ${}_sP_t = \boxed{\dfrac{s!}{(s-t)!}}$

19. Problems 18-21 are equivalent to text Example 9 (p. 616). Additionally, if you desire to learn more about the traveling salesperson problem (TSP), review textbook Section 6.7 (beginning on p. 383).

From part (b) of Example 9, please recall that we found there to be

$$\frac{(n-1)!}{2} \text{ possible routes for } n \text{ cities.}$$

a) For 6 cities (i.e., $n = 6$), there are $\dfrac{(6-1)!}{2} = \dfrac{5!}{2} = \boxed{60}$ possible routes.

b) For 10 cities ($n = 10$), there are $\dfrac{(10-1)!}{2} = \dfrac{9!}{2} = \boxed{181{,}440}$ possible routes.

21. a) For 8 cities, there are $\dfrac{(8-1)!}{2} = \dfrac{7!}{2} = \boxed{2{,}520}$ possible routes.

b) For q cities, there are $\boxed{\dfrac{(q-1)!}{2}}$ possible routes.

23. $$\frac{\frac{n!}{(n-1)!}}{r!} = \frac{n!}{(n-1)!} \div r! = \frac{n!}{(n-1)!} \cdot \frac{1}{r!} = \frac{n \cdot (n-1)!}{(n-1)!} \cdot \frac{1}{r!} = \boxed{\frac{n}{r!}}$$

25. Via the multiplication property of factorials (textbook p. 613), we find

$$\frac{(n+1)!}{n!} = \frac{(n+1) \cdot n!}{n!} = \boxed{n+1}.$$

27. $\dfrac{1}{n!}+\dfrac{1}{(n-1)!}=\dfrac{(n-1)!+n!}{n!(n-1)!}$ *Add the fractions; LCD = n! (n – 1)! .*

$=\dfrac{(n-1)!+n(n-1)!}{n!(n-1)!}$ *Apply multiplication property of factorials.*

$=\dfrac{(n-1)![1+n]}{n!(n-1)!}$ *Factor the numerator; GCF = (n – 1)! .*

$=\boxed{\dfrac{\mathbf{1+n}}{\mathbf{n!}}}$ or $\boxed{\dfrac{\mathbf{n+1}}{\mathbf{n!}}}$ *Reduce the fraction.*

29. a) Remember that ***subsets*** consist of distinct elements (i.e., "no repetition allowed") and the order of the elements does not matter. Therefore, to determine the number of subsets of size 3 that can be formed from a set of five elements, we use the ***combination*** formula: ${}_5C_3=\boxed{\mathbf{10}}$.

b) Note the words "arrangement of size 3." The word "arrangement" tells us that order matters, and the words "of size 3" imply that no repetition is allowed (i.e., a group of size 3 consists of 3 distinct elements). As such, this problem is in complete accord with the definition of a ***permutation***. So, we find there to be ${}_5P_3=\boxed{\mathbf{60}}$ arrangements of size 3 one can form from a set of five elements.

[General reminder: The number of permutations can also be found by using the fundamental counting principle, if preferred.]

31. This problem is concerned about the "order of his visits." Obviously, these 15 offices are different (i.e., distinct), so repetition does not occur. Therefore, this is a ***permutation*** problem, and we find:

${}_{15}P_{15}=\boxed{\mathbf{15!}}$ [or 1,307,674,368,000] ways the watchman can visit these 15 offices.

33. Think of the formation of a social security number as a nine-step task. Since repetition of digits is allowed, we use the ***fundamental counting principle*** for these three problems.

a) There are ten ways to select each digit (0 through 9). Thus, there is a total of

$$10\cdot10\cdot10\cdot10\cdot10\cdot10\cdot10\cdot10\cdot10=\boxed{\mathbf{10^9 \text{ or } 1 \text{ billion}}}$$

possible (9-digit) social security numbers.

b) There are 9 ways to select the first digit (1 through 9) and ten ways to select each of the remaining eight digits (0 through 9). Hence, we have:

$$9\cdot10\cdot10\cdot10\cdot10\cdot10\cdot10\cdot10\cdot10=\boxed{\mathbf{9\times10^8 \text{ or } 900 \text{ million}}}$$ possibilities.

c) $9\cdot9\cdot10^7=\boxed{\mathbf{810 \text{ million}}}$ possible social security numbers.

35. Needless to say, Peppermint Patty must answer 10 different (distinct) questions (i.e., repetition is not allowed). This problem asks us to determine the number of possible groups (subsets) of ten questions she can choose. The order in which she chooses to answer the 10 questions she selects is unimportant. Consequently, we use the combination formula to solve the problem and discover: ${}_{12}C_{10} = \boxed{66}$ ways to select the questions.

37. Without any regard to their genders, the question asks us to find the number of ways to arrange groups of 4 chosen from among these 6 (distinct) people. This conforms to the definition of a permutation. By using the permutation formula (or, if preferred, the fundamental counting principle), one finds that there are ${}_6P_4 = 6 \cdot 5 \cdot 4 \cdot 3 = \boxed{360}$ ways these partygoers can be seated in a row of four.

39. This TSP problem is comparable to Problem 18-21. Once again, recall (from part (b) of Example 9) that we found there to be

$$\frac{(n-1)!}{2} \text{ possible routes for } n \text{ cities.}$$

So, Dave and Brad can visit the given 8 cities in $\dfrac{(8-1)!}{2} = \dfrac{7!}{2} = \boxed{2{,}520}$ ways.

41. a) P(lemon drop) = $\dfrac{\text{\# of ways to select a lemon drop candy}}{\text{Total \# of possible selections}} = \boxed{\dfrac{5}{12}}$

b) Envision yourself reaching (one time) into this bag and pulling out two pieces of candy (or, for that matter, any other type of item). The order of the items is indistinguishable. In other words, it is not possible to classify one piece as "first" and the other as the "second." Therefore, since two (distinct) pieces of candy are being chosen without regard to order, the number of ways to make this selection equals the combination ${}_{12}C_2 = \boxed{66}$.

c) By way of the same reasoning as applied to part (b), we find ${}_5C_2 = \boxed{10}$.

d) From parts (b) and (c), we know there are 66 ways to select two pieces of candy from this bag, and 10 of these 66 ways consist exclusively of lemon drop candy. Thus,

$$\text{P(2 lemon drops)} = \frac{{}_5C_2}{{}_{12}C_2} = \frac{10}{66} = \boxed{\frac{5}{33} \approx 0.15}.$$

43. Like the selection of candy in Problem 41 (explained above), the order in which cards are received is unimportant, so finding the number of ways of receiving 4 cards is a combination problem.

a) There are ${}_{52}C_4 = \boxed{270{,}725}$ ways to draw four cards from a standard 52-card deck.

b) Since there are four aces in the deck, there is only $\boxed{1}$ way to draw all four of them (i.e., ${}_4C_4$).

c) Using the answers from parts (a) and (b), it follows that

$$\text{P(4 aces)} = \frac{\text{\# of ways to draw 4 aces}}{\text{Total \# of possible 4-card hands}} = \boxed{\frac{1}{270{,}725}}.$$

45. Assuming the given information, it follows that:

$$(50 \text{ jumps/min}) \times (60 \text{ mins/hr}) = 3{,}000 \text{ jumps per hour;}$$

$$(3{,}000 \text{ jumps/hr}) \times (8 \text{ hr/day}) = 24{,}000 \text{ jumps per day;}$$

$$(24{,}000 \text{ jumps/day}) \times (5 \text{ days/wk}) = 120{,}000 \text{ jumps per week;}$$

$$(120{,}000 \text{ jumps/wk}) \times (50 \text{ wk/yr}) = 6{,}000{,}000 \text{ (or } 6 \times 10^6\text{) jumps per year;}$$

$$(6 \times 10^6 \text{ jumps/yr}) \times (10^2 \text{ yr/cent}) = 6 \times 10^8 \text{ jumps per century.}$$

From the cartoon (shown on textbook page 611), we know thus far she has jumped 5,480,523,297,162 times. So, at a rate of 6×10^8 jumps per century, it would take a person **about 9,134.2 centuries** to have achieved this.

47. The marginal note to which this problem refers is on textbook page 614. Since there are 12 links per foot and each link takes 1 minute to construct, each foot required 12 minutes of labor. Since the paper chain is 6,077 ft long, it took 72,924 minutes (i.e., 6077 ft × 12 min/ft) to construct it. This equates to 1,215.4 hour. Assuming a 40-hr workweek, this amounts to **30.385 weeks**.

If we instead assume a 24-hour workday (which would be highly unlikely), then it would take approximately 50.64 days or 7.235 weeks.

49. Since this problem requires the arrangement of distinct objects (in this case, books), we apply the permutation formula (or, alternatively, the fundamental counting principle).

a) ${}_7P_7 = 7! =$ **5,040 ways** b) ${}_3P_3 = 3! =$ **6 ways** c) ${}_4P_4 = 4! =$ **24 ways**

d) Here are two perspectives:

I. Since the math books must remain together, we view the four individual science books and the "group" of math books as five objects to be arranged on the shelf. We find ${}_5P_5 = 5! = 120$.

Now, within the "group" of math books, we know (from part b) that these 3 math books can be arranged in 6 different ways. Therefore, there is a total of $6 \cdot {}_5P_5 = 6 \cdot 120 =$ **720 ways** to arrange the seven books so that all the math books will be together.

II. First think of this arrangement task as a two-step process: imagine arranging the 3 math books (MMM) on the left side of the shelf and the four science books (SSSS) to the right. There are ${}_3P_3 \cdot {}_4P_4 = 3!4! = 144$ ways to obtain an "MMMSSSS" arrangement.

This is not, however, the only way to keep the math books together. There are actually a total of five ways: MMMSSSS, SMMMSSS, SSMMMSS, SSSMMMS, and SSSSMMM. Therefore, since each of these 5 approaches has 144 possible arrangements, there is a total of $5 \cdot 144 =$ **720 ways** to arrange the seven books so that all the math books will remain together.

51. **Yes**, this claim is correct. This equalizer has 5 dials ("zones"), and each dial has 13 ("control") settings. So, by the fundamental counting principle, we find the SEA graphic equalizer indeed can create a total of

$13 \cdot 13 \cdot 13 \cdot 13 \cdot 13 =$ **13^5 or 371,293** different possible sounds.

53. Before calculating the number of possible two-pizza orders, we first must understand the number of possibilities for one pizza. A given pizza can have "up to" 5 (out of 11 available) toppings. Of course, the order (arrangement) of the toppings on the pizza is unimportant; for example, a pizza with "mushrooms and anchovies" is no different than a pizza with "anchovies and mushrooms." Additionally, we assume that no toppings can be repeated; that is to say, a three-topping pizza, for example, has three distinct (different) toppings. Thus, it follows that this is a combination problem, and the number of possibilities for one pizza is:

$$\overbrace{{}_{11}C_5 + {}_{11}C_4 + {}_{11}C_3 + {}_{11}C_2 + {}_{11}C_1 + {}_{11}C_0}^{\text{Respectively, the \# of 5-topping, 4-topping, 3-topping, 2-topping, 1-topping, and plain pizzas possible}} = 1{,}024.$$

Now, regarding the given advertisement (pertaining to two-pizza orders), one can order either two *identical* pizzas or two *different* pizzas. Since there are 1,024 possible pizzas one can order, there are 1,024 possible ways to order two identical pizzas. As for those who want two different pizzas, they can place their order in ${}_{1,024}C_2 =$ 523,776 possible ways. Thus, the total number of possible ways to order two pizzas is 523,776 + 1,024 = **524,800**.

55. Once again (like we did for Problem 53), we begin by considering the number of possibilities for one pizza. This time, however, we address these possibilities on a case-by-case basis:

0-topping pizza: Of course, there is only **1** such pizza (${}_{11}C_0 = 1$).

1-topping pizza: Needless to say, such a pizza cannot have a double topping. Given that there are 11 toppings to choose from, it follows that there are **11** (i.e., ${}_{11}C_1 = 11$) ways to order a 1-topping pizza.

2-topping pizza: A two-topping pizza can either have (i) two different toppings or (ii) the same topping "doubled" (i.e., one double topping). When two different toppings are chosen, there are ${}_{11}C_2 = 55$ possibilities. When one double topping is chosen, there are as many possible pizzas as there are possible toppings; so, there are ${}_{11}C_1 = 11$ possibilities.

Hence, there are **66** different possibilities for a 2-topping pizza.

3-topping pizza: A three-topping pizza can either have (i) 3 different toppings or (ii) one double topping with one single topping. In the case of a pizza with 3 different toppings, there are ${}_{11}C_3 = 165$ possibilities.

As for a single pizza with one double topping and one single topping, this is a more delicate computation. There are several ways to view the possibilities. First, because it makes a difference which of the toppings is doubled and which is not, it become clear that the order (arrangement) matters. Hence, we compute a permutation (namely, ${}_{11}P_2$). Alternatively, one can view this pizza selection as a two-step task (i.e., first select the double topping, then the single topping, or vice-versa). This perspective lends itself to the fundamental counting principle, namely ${}_{11}C_1 \cdot {}_{10}C_1$. Both approaches yield $\underline{\underline{110}}$ possibilities.

Thus, there are $\underline{\underline{\mathbf{275}}}$ $(110 + 165)$ possible 3-topping pizzas.

4-topping pizza: There are 3 different types of four-topping pizza: (i) a pizza with four different toppings; (ii) a pizza with two double-toppings (i.e., ${}_{11}C_2 = 55$ possibilities); or (iii) a pizza with 1 double topping and 2 other (different) single toppings.

First, as for a pizza with four different toppings, there are ${}_{11}C_4 = \underline{\underline{330}}$ possibilities.

Second, the selection of a pizza with two double-toppings is (in terms of the selection process) equivalent to the selection of two single toppings. Hence, there are ${}_{11}C_2 = \underline{\underline{55}}$ possibilities.

Third, to compute the number of possibilities for a pizza with 1 double topping and 2 other (different) single toppings, we view this as a two-step task and, as such, apply the fundamental counting principle. Assume that the first task is the selection of the double topping. It can be done in ${}_{11}C_1 = 11$ ways. Then, the second task becomes the selection of two other single toppings, which can be performed in ${}_{10}C_2 = 45$ ways. Therefore, via the fundamental counting principle, there are ${}_{11}C_1 \cdot {}_{10}C_2 = \underline{\underline{495}}$ ways to order such a pizza.

All together, there are $\underline{\underline{\mathbf{880}}}$ $(330 + 55 + 495)$ possibilities for a 4-topping pizza.

5-topping pizza: A five-topping pizza can consist of either (i) five different single toppings (i.e., ${}_{11}C_5 = 462$); (ii) one double topping and three different single toppings (i.e., ${}_{11}C_1 \cdot {}_{10}C_3 = 1{,}320$); or (iii) two double toppings and one single topping (i.e., ${}_{11}C_2 \cdot {}_{9}C_1 = 495$). So, there are $\underline{\underline{\mathbf{2{,}277}}}$ total ways to order a 5-topping pizza.

Summing each of the previous subtotals, we find there are $\underline{\underline{\mathbf{3{,}510}}}$ $(1 + 11 + 66 + 275 + 880 + 2{,}277)$ possible ways to order a pizza.

Finally, regarding the given advertisement (pertaining to two-pizza orders), one can order either two <u>identical</u> pizzas or two <u>different</u> pizzas. Since there are 3,510 possible pizzas one can order, there are $\underline{\underline{3{,}510}}$ possible ways to order two identical pizzas. As for those who want two different pizzas, they can place their order in

${}_{3,510}C_2 = \underline{6,158,295}$ possible ways. Thus, the total number of possible ways to order two pizzas is 6,158,295 + 3,510 = **6,161,805**.

57. As shown below, a BINGO card consists of 5 columns.

B *1-15*	I *16-30*	N *31-45*	G *46-60*	O *61-75*
		FREE		

With the exception of the middle "N" column (which has a "free" space), each column consists of 5 distinctly numbered squares (i.e., no repetition allowed). In each column, its five numbers are chosen from among a set of 15 possible numbers; for example, the rightmost "O" column consists of 5 distinct numbers ranging from 61 to 75. Furthermore, because a different arrangement would result in a "different'" card, the order of the numbers within a column matters. Thus, since repetition is not allowed and order matters, the number of possibilities (arrangements) for each column is a ***permutation*** problem, namely ${}_{15}P_5$ (except for the middle column, which is ${}_{15}P_4$).

Finally, to determine the total number of possible BINGO cards, we view the selection of each column's numbers as an independent "task," and, hence, the making of an entire BINGO card is a five-step task. Therefore, via the fundamental counting principle, we find that there are:

$$\underbrace{\overbrace{{}_{15}P_5}^{\text{\# of possibilities for the B column}} \cdot \overbrace{{}_{15}P_5}^{\text{\# of possibilities for the I column}} \cdot \overbrace{{}_{14}P_5}^{\text{\# of possibilities for the N column}} \cdot \overbrace{{}_{15}P_5}^{\text{\# of possibilities for the G column}} \cdot \overbrace{{}_{15}P_5}^{\text{\# of possibilities for the O column}}}$$

$\approx$ **5.524464741 × 10^{26}** possible BINGO cards.

59. From Problem 58, we know ${}_nP_r = n(n-1)(n-2)\cdots(n-r+1)$. Hence,

$$\begin{aligned}
{}_nP_r &= n(n-1)(n-2)\cdots(n-r+1) \\
&= \frac{n(n-1)(n-2)\cdots(n-r+1)}{1}\cdot\left[\frac{(n-r)(n-r-1)\cdots 3\cdot 2\cdot 1}{(n-r)(n-r-1)\cdots 3\cdot 2\cdot 1}\right] \\
&= \frac{n(n-1)(n-2)\cdots(n-r+1)(n-r)(n-r-1)\cdots 3\cdot 2\cdot 1}{(n-r)(n-r-1)\cdots 3\cdot 2\cdot 1} \\
&= \frac{n!}{(n-r)!} \quad \checkmark
\end{aligned}$$

Problem Set 9.5

1. **Wording varies**. Events E and F are said to be ***independent*** if the occurrence of one of these in no way influences the probability of the other. In terms of *conditional* probabilities, events E and F are independent if $P(E|F) = P(E)$. For additional information, refer to textbook page 625-627.

3. **Labels vary**. As derived on textbook page 629, we know (for any events A and B) the formula for the probability of their union is:

$$P(A \cup B) = P(A) + P(B) - P(A \cap B).$$

5. **Answers vary**.

7. $P(\overline{A}) = 1 - P(A) = 1 - \frac{1}{2} = \boxed{\frac{1}{2}}.$

Review, if necessary, the concept of ***complementary probabilities*** and the property of complements (discussed in Section 9.3, see p. 597).

9. $P(\overline{C}) = 1 - P(C) = 1 - \frac{1}{6} = \boxed{\frac{5}{6}}.$

11. Via the multiplication property of probability (textbook p. 627), it follows that

$$P(A \cap C) = P(A) \cdot P(C) = \tfrac{1}{2} \cdot \tfrac{1}{6} = \boxed{\tfrac{1}{12}}.$$

13.
$$\begin{aligned} P(A \cup B) &= P(A) + P(B) - P(A \cap B) && \textit{Addition property of probability (text p. 629)} \\ &= P(A) + P(B) - P(A) \cdot P(B) && \textit{Multiplication property of probability} \\ &= \tfrac{1}{2} + \tfrac{1}{3} - \tfrac{1}{2} \cdot \tfrac{1}{3} \\ &= \boxed{\tfrac{2}{3}} \end{aligned}$$

15.
$$\begin{aligned} P(B \cup C) &= P(B) + P(C) - P(B \cap C) \\ &= P(B) + P(C) - P(B) \cdot P(C) \\ &= \tfrac{1}{3} + \tfrac{1}{6} - \tfrac{1}{3} \cdot \tfrac{1}{6} \\ &= \boxed{\tfrac{4}{9}} \end{aligned}$$

17. Via the property of complements and the answer to Problem 11 (see solution above), we obtain:

$$P(\overline{A \cap C}) = 1 - P(A \cap C) = 1 - \tfrac{1}{12} = \boxed{\tfrac{11}{12}}.$$

19. Using the answer to Problem 12 (see solution), we obtain:

$$P(\overline{A \cup B}) = 1 - P(A \cup B) = 1 - \tfrac{2}{3} = \boxed{\tfrac{1}{3}}.$$

21. Using the answer to Problem 15 (see solution), we obtain:

$$P(\overline{B \cup C}) = 1 - P(B \cup C) = 1 - \tfrac{4}{9} = \boxed{\tfrac{5}{9}}.$$

23. Recalling the answer to Problem 22 (namely, $\frac{1}{36}$), we obtain:

$$P(\overline{A \cap B \cap C}) = 1 - P(A \cap B \cap C) = 1 - \tfrac{1}{36} = \boxed{\tfrac{35}{36}}$$

25. Since three of the six numbered faces of a die are prime (namely, 2, 3, and 5), it follows that $P(A) = \frac{3}{6} = \frac{1}{2}$. By the definition of probability, we know $P(B) = \frac{1}{6}$.

Remember that two events are said to be ***independent*** if the occurrence of one of these in no way influences the probability of the other. Consequently, a good way to assess "influence" is to use our experience (from Section 9.3) with conditional probabilities. That is to say, events A and B will be independent if $P(A \mid B) = P(A)$ and $P(B \mid A) = P(B)$. In this particular case, we find:

$$P(A \mid B) = P(\text{prime} \mid \text{three}) = \tfrac{1}{1} = 1 \neq P(A)$$
$$P(B \mid A) = P(\text{three} \mid \text{prime}) = \tfrac{1}{3} \neq P(B)$$

Therefore, the answer to the question is: **NO**, events A and B are not independent.

27. **Yes**, events A and D are independent because the outcome of the first roll in no way influences the second roll.

29. **Yes**; once again, because the outcome of the first roll in no way influences the second roll, events B and D are independent.

31. Since events A and B are not independent, we cannot find this probability by multiplying their respective probabilities. Instead, we find this probability by looking at the sample space {1, 2, 3, 4, 5, 6}. How many of these outcomes are both a prime and a 3? We hope the answer is obvious: there is only one such number (specifically, 3).

Hence, $P(A \cap B) = P(A \text{ and } B) = P(\text{first toss is prime and three}) = \boxed{\tfrac{1}{6}}$.

33. Here are two perspectives:

I. Using the addition property of probability (and the answer to Problem 31), we obtain: $P(A \cup B) = P(A) + P(B) - P(A \cap B) = \frac{3}{6} + \frac{1}{6} - \frac{1}{6} = \frac{3}{6} = \boxed{\tfrac{1}{2}}$.

II. Alternatively, since event $B = \{3\}$ is a subset of event $A = \{2, 3, 5\}$, their union is the same as A; that is, $A \cup B = \{2, 3, 5\}$. Therefore, because $A \cup B$ can occur in three ways (out of six total possibilities), we know $P(A \cup B) = \frac{3}{6} = \frac{1}{2}$.

35. Again, here are two perspectives:

I. Since events B and D are independent, we can use the multiplication property of probability to find $P(B \cap D) = P(B) \cdot P(D) = \frac{1}{6} \cdot \frac{1}{6} = \frac{1}{36}$. Then, by using the addition property of probability, we obtain:

$$P(B \cup D) = P(B) + P(D) - P(B \cap D) = \frac{1}{6} + \frac{1}{6} - \frac{1}{36} = \boxed{\frac{11}{36}}.$$

II. We know (from previous experience beginning with Section 9.1, p. 581) that there are 6 · 6 = 36 possible outcomes when rolling a die twice. Of these 36 outcomes, how many include a 3 on the first toss **or** a 3 on the second toss? There are eleven: (3,1), (3,2), **(3,3)**, (3,4), (3,5), (3,6), (1,3), (2,3), (4,3), (5,3), (6,3). Therefore, because $B \cup D$ can occur in 11 ways (out of 36 total possibilities), $P(B \cup D) = \frac{11}{36}$.

37. Since the three wheels are independent, we can use, when appropriate, the multiplication property of probability.

a) P(3 bars) = P(bar) · P(bar) · P(bar) $= \frac{1}{13} \cdot \frac{1}{13} \cdot \frac{1}{13} = \left(\frac{1}{13}\right)^3 = \frac{1}{2{,}197} \approx \boxed{0.000455}$

b) P(3 oranges) $= \left(\frac{3}{13}\right)^3 = \frac{27}{2{,}197} \approx \boxed{0.01229}$

c) P(3 plums) $= \left(\frac{3}{13}\right)^3 = \frac{27}{2{,}197} \approx \boxed{0.01229}$

d) P(1st cherry) $= \frac{2}{13} \approx \boxed{0.1538}$

e) P(cherries on first two wheels) $= \left(\frac{2}{13}\right)^2 = \frac{4}{169} \approx \boxed{0.0237}$

39. To review the concept of mathematical expectation, please refer to textbook Section 9.2. Based upon Figure 9.9 (shown in Problem 38), we obtain:

$$E = \overbrace{2\left(\tfrac{2}{20}\right)}^{\text{First one cherry}} + \overbrace{5\left(\tfrac{2}{20}\cdot\tfrac{5}{20}\right)}^{\text{First two cherries}} + \overbrace{10\left(\tfrac{2}{20}\cdot\tfrac{5}{20}\cdot\tfrac{2}{20}\right)}^{\text{First two cherries, third a bar}} + \overbrace{10\left(\tfrac{2}{20}\cdot\tfrac{5}{20}\cdot\tfrac{8}{20}\right)}^{\text{Three cherries}} + \overbrace{14\left(\tfrac{6}{20}\cdot\tfrac{2}{20}\cdot\tfrac{3}{20}\right)}^{\text{Three oranges}} +$$

$$\overbrace{14\left(\tfrac{3}{20}\cdot\tfrac{2}{20}\cdot\tfrac{6}{20}\right)}^{\text{Three plums}} + \overbrace{20\left(\tfrac{1}{20}\cdot\tfrac{6}{20}\cdot\tfrac{1}{20}\right)}^{\text{Three bells}} + \overbrace{50\left(\tfrac{2}{20}\cdot\tfrac{4}{20}\cdot\tfrac{2}{20}\right)}^{\text{Three bars}} - \overbrace{1 \text{ coin}}^{\text{Cost of playing}} = \underline{\underline{-\ 0.309}} \text{ coins}$$

For playing the game with quarter dollars, the mathematical expectation is

$$\$0.25(-0.309) = -\$0.07725 \approx \boxed{-\ \$0.08}.$$

41. This problem assumes familiarity with the lotto game ***Keno***, which is explained in text Examples 1-2 (pp. 622-624).

We first compute the probability of winning. In this case of special "catch all" Keno, a player wins only if all six of his/her chosen numbers are included among the 20 removed from the pot (consisting of 80 numbers). Since the order of the numbers is unimportant and repetition is not allowed, we use a combination to determine the probability of winning:

$$\text{P(winning)} = \frac{\textit{\# of ways to win}}{\textit{Total \# of possible outcomes}} = \frac{\textit{\# of ways to select 6 numbers out of 20}}{\textit{\# of ways to select 6 numbers out of 80}} = \frac{{}_{20}C_6}{{}_{80}C_6}.$$

Via the property of complements, it follows that:

$$\text{P(losing)} = 1 - \frac{{}_{20}C_6}{{}_{80}C_6}.$$

Finally, recall that in Keno, the cost of playing the game is collected in advance (i.e., not returned regardless of the outcome). Thus, the mathematical expectation for this game of special "catch all" Keno is:

$$E = \overbrace{(\$27{,}777 - \$5)\left(\frac{{}_{20}C_6}{{}_{80}C_6}\right)}^{\textit{win (if you pick all six)}} + \overbrace{(-\$5)\left(1 - \frac{{}_{20}C_6}{{}_{80}C_6}\right)}^{\textit{lose (otherwise)}}$$

$$\approx \$27{,}772(1.289849391 \times 10^{-4}) - \$5(0.9998710151)$$

$$\approx \boxed{-\$1.42}.$$

Note: The same result could have been obtained by computing the expectation as follows:

$$E = \overbrace{(\$27{,}777)\left(\frac{{}_{20}C_6}{{}_{80}C_6}\right)}^{\textit{amount to win(prob. of winning)}} - \overbrace{\$5}^{\textit{cost of playing}}$$

43. This problem is very similar to text Example 2 (p. 624).
From Table 9.3 (p. 623), we first obtain the Keno payoffs for playing a three-spot $3 Keno ticket:

Pick 0 numbers, win $0.
Pick 1 number, win $0.
Pick 2 numbers, win $3 (i.e., break even).
Pick 3 numbers, win $120.

Then, by using the same approach (and probabilities) as was explained in text Example 2, we find:

$$E = (120 - 3) \cdot P(\text{pick 3}) + (3 - 3) \cdot P(\text{pick 2}) + (0 - 3) \cdot P(\text{pick 1}) + (0 - 3) \cdot P(\text{pick 0})$$

$$\approx 117(0.0138753651) + 0 - 3(0.4308666018) - 3(0.4165043817)$$

$$\approx -0.9187$$

So, the expectation is approximately $\boxed{-\$0.92}$.

45. *[Correction: In an early printing of this edition of the textbook, there was an omission in the statement of this problem. The author's intent was to ask for the probability that at least 2 of 5 unrelated persons share the same birthday.]*
This problem is patterned after the so-called ***birthday problem*** (explained in text Example 4, p. 628, and again encountered in Problem 46). It asks for the probability

that **at least two** (i.e., 2 or more) of these five persons share the same birthday. Applying the same plan as developed in text Example 4, we find:

$$\begin{aligned}\text{P(match)} &= 1 - \text{P(no match)}\\ &= 1 - \frac{365}{365}\times\frac{364}{365}\times\frac{363}{365}\times\frac{362}{365}\times\frac{361}{365}\\ &= 1 - \frac{{}_{365}P_5}{365^5}\\ &\approx 0.0271355737\end{aligned}$$

So, the probability of at least 2 of 5 unrelated people sharing the same birthday is approximately $\boxed{2.7\,\%}$.

47. Since each toss is an independent event, we can use the multiplication property of probability:

$$\text{P(five tails)} = \text{P(tails)}\cdot\text{P(tails)}\cdot\text{P(tails)}\cdot\text{P(tails)}\cdot\text{P(tails)} = \left(\tfrac{1}{2}\right)^5 = \boxed{\frac{1}{32}}.$$

49. This problem is analogous to text Example 7 (pp. 630-631).

Recall that drawing the marbles ***with replacement*** means the events (i.e., the marbles) are independent (and we may use the multiplication property of probability), whereas ***without replacement*** means the events are dependent (and we will use combinations when the order does not matter and permutations otherwise). Notice in part (d) that the order of the selections is important

	With Replacement	***Without Replacement***
a) **P(R ∩ R)**	$P(R)\cdot P(R) = \frac{5}{8}\cdot\frac{5}{8} = \boxed{\frac{25}{64}}$	$\frac{{}_5C_2}{{}_8C_2} = \frac{10}{28} = \boxed{\frac{5}{14}}$
b) **P(B ∩ B)**	$P(B)\cdot P(B) = \frac{3}{8}\cdot\frac{3}{8} = \boxed{\frac{9}{64}}$	$\frac{{}_3C_2}{{}_8C_2} = \boxed{\frac{3}{28}}$
c) **P(R ∩ B)**	Two approaches ↓ 1 − P(2 of the same color) $1 - \left(\frac{25}{64} + \frac{9}{64}\right) = \frac{30}{64} = \boxed{\frac{15}{32}}$ **or** P(R) · P(B) + P(B) · P(R) $\frac{5}{8}\cdot\frac{3}{8} + \frac{3}{8}\cdot\frac{5}{8} = \frac{30}{64} = \frac{15}{32}$	$\frac{{}_5C_1\cdot{}_3C_1}{{}_8C_2} = \boxed{\frac{15}{28}}$ **or** $1 - \left(\frac{5}{14} + \frac{3}{28}\right) = \frac{15}{28}$
d) **P(R₁ ∩ B₂)**	$P(R_1)\cdot P(B_2) = \frac{5}{8}\cdot\frac{3}{8} = \boxed{\frac{15}{64}}$	$\frac{{}_5P_1\cdot{}_3P_1}{{}_8P_2} = \frac{5\cdot 3}{8\cdot 7} = \boxed{\frac{15}{56}}$

51. This problem is akin to text Example 8 (p. 631).
This game has four possible outcomes: (i) if a heart turns up on the first cut, you win and the game stops; (ii) if a heart doesn't turn up on the first cut but it does on the second cut, you win and the game stops; (iii) if a heart doesn't turn up on the first two cuts but it does on the third (i.e., the last) cut, you win; (iv) you lose because a heart failed to come up on any of the three cuts.
Recall that a deck of cards consists of four suits (hearts, diamonds, clubs, and spades). So, the probability of a heart turning up on a given cut is $\frac{1}{4}$; the probability of it not turning up is $1-\frac{1}{4}=\frac{3}{4}$.

a) To make this decision, we compute the mathematical expectation:

$$\begin{aligned} E &= \overbrace{\$1\left(\tfrac{1}{4}\right)}^{\text{Heart on the 1st cut}} + \overbrace{\$1\left(\tfrac{3}{4}\cdot\tfrac{1}{4}\right)}^{\text{Heart on the 2nd cut}} + \overbrace{\$1\left(\tfrac{3}{4}\cdot\tfrac{3}{4}\cdot\tfrac{1}{4}\right)}^{\text{Heart on the 3rd cut}} + \overbrace{(-\$1)\left(\tfrac{3}{4}\cdot\tfrac{3}{4}\cdot\tfrac{3}{4}\right)}^{\text{No hearts on 3 cuts}} \\ &= \$0.15625 \end{aligned}$$

Since the expectation is positive, the answer is: **yes**, you should play.

b) Because the "non-hearts" are not put back into the deck after each cut, the likelihood of obtaining a heart on the subsequent cut is increased. Therefore, since the expectation with replacement (in part a) was positive, we should anticipate the expectation without replacement to be even more positive. Therefore, **yes**, you should play.
To determine the expectation of this game, first consider the implications of "without replacement": (i) on the second cut, the deck will consist of 51 cards, of which 13 cards will be hearts and 38 will not (i.e., a "non-heart" was removed in the first cut); and (ii) on the third cut, the deck will consist of 50 cards (13 hearts and 37 others). Hence,

$$\begin{aligned} E &= \overbrace{\$1\left(\tfrac{1}{4}\right)}^{\text{Heart on the 1st cut}} + \overbrace{\$1\left(\tfrac{3}{4}\cdot\tfrac{13}{51}\right)}^{\text{Heart on the 2nd cut}} + \overbrace{\$1\left(\tfrac{3}{4}\cdot\tfrac{38}{51}\cdot\tfrac{13}{50}\right)}^{\text{Heart on the 3rd cut}} + \overbrace{(-\$1)\left(\tfrac{3}{4}\cdot\tfrac{38}{51}\cdot\tfrac{37}{50}\right)}^{\text{No hearts on 3 cuts}} \\ &\approx \$0.17294117647. \end{aligned}$$

53. Just like rolling a pair of dice once is equivalent to rolling a single die twice, two roulette balls spinning at once is equivalent to two independent rolls of a usual game of roulette. Thus, the probability is $\boxed{\frac{1}{38}}$. Look at it this way: the first ball specifies the slot, so the second one has a 1 out of 38 chance of dropping into that slot.

55. In order for the <u>sucker</u> to win, he must roll out three black olives. The probability of this occurring is:

$$\text{P(3 black)} = \frac{\textit{\# of favorable outcomes}}{\textit{Total \# of possible outcomes}}$$

$$= \frac{\textit{\# of ways to obtain 3 black out of 5 black olives}}{\textit{\# of ways to obtain 3 olives out of 7 in all}}$$

$$= \frac{{}_5C_3}{{}_7C_3}$$

$$= \frac{10}{35}$$

$$= \frac{2}{7}$$

You win if the sucker loses. Therefore, the probability **<u>you</u>** win is:

$$\text{P(win)} = 1 - \tfrac{2}{7} = \boxed{\tfrac{5}{7} \approx \mathbf{0.71 \text{ or } 71\%}}.$$

57. Problems 56-60 are very similar to text Examples 1-2 (see p. 622-624) and Problems 40-44 (see solutions).

As shown at the top of the ticket, \$1 is the cost of playing. Based upon the indicated payoffs for the four-spot game, the expectation is:

$$E = \overbrace{(\$50 - \$1)\left(\frac{{}_{20}C_4}{{}_{80}C_4}\right)}^{\textit{match 4 numbers}} + \overbrace{(\$4 - \$1)\left(\frac{{}_{20}C_3 \cdot {}_{60}C_1}{{}_{80}C_4}\right)}^{\textit{match 3 numbers}} +$$

$$\overbrace{\$0}^{\textit{match 2 numbers}} + \overbrace{(-\$1)\left(\frac{{}_{20}C_1 \cdot {}_{60}C_3}{{}_{80}C_4}\right)}^{\textit{match 1 number}} + \overbrace{(-\$1)\left(\frac{{}_{60}C_4}{{}_{80}C_4}\right)}^{\textit{match 0 numbers}}$$

$$\approx -\$0.4612033536$$

$$\approx \boxed{-\mathbf{\$0.46}}.$$

59. Once again, based upon the indicated payoffs for the six-spot game and the \$1 cost of playing, the expectation is:

$$E = \overbrace{(\$1{,}000 - \$1)\left(\frac{{}_{20}C_6}{{}_{80}C_6}\right)}^{\textit{match 6 numbers}} + \overbrace{(\$25 - \$1)\left(\frac{{}_{20}C_5 \cdot {}_{60}C_1}{{}_{80}C_6}\right)}^{\textit{match 5 numbers}} + \overbrace{(\$4 - \$1)\left(\frac{{}_{20}C_4 \cdot {}_{60}C_2}{{}_{80}C_6}\right)}^{\textit{match 4 numbers}} +$$

$$\overbrace{\$0}^{\textit{match 3 numbers}} + \overbrace{(-\$1)\left(\frac{{}_{20}C_2 \cdot {}_{60}C_4}{{}_{80}C_6}\right)}^{\textit{match 2 numbers}} + \overbrace{(-\$1)\left(\frac{{}_{20}C_1 \cdot {}_{60}C_5}{{}_{80}C_6}\right)}^{\textit{match 1 number}} + \overbrace{(-\$1)\left(\frac{{}_{60}C_6}{{}_{80}C_6}\right)}^{\textit{match 0 numbers}}$$

$$\approx -\$0.5496528788$$

$$\approx \boxed{-\mathbf{\$0.55}}.$$

Problem Set 9.6

1. **Estimates vary**.

$(331{,}776 \text{ sec}) \div (60 \text{ sec/min}) = 5{,}529.6$ minutes *[0.6 min = 36 sec]*

$(5{,}529 \text{ min}) \div (60 \text{ min/hr}) = 92.15$ hours *[0.15 hr = 9 min]*

$92 \text{ hr} \div (24 \text{ hr/day}) = 3.8\overline{3}$ days *[$0.8\overline{3}$ day = 20 hr]*

So, at a rate of one different move per second, it would take you precisely

3 days, 20 hours, 9 minutes, and 36 seconds

to go through all the possible arrangements.

3. a)

b)

c)

d)

5. Recall, for example, that F^2 denotes a 180° clockwise rotation (i.e., two consecutive 90° clockwise rotations) of the front face.

a)

b)

c)

d)

7. RL means first rotate the right face 90° clockwise and then rotate the left face 90° clockwise. The result is:

9. FB^{-1} means first rotate the front face 90° clockwise and then rotate the back face 90° counterclockwise. These two steps yield:

11.

13.

15.

17. There are two ways to reverse a 180° clockwise rotation of the top face: (i) rotate the top face an additional 180° clockwise because doing so means the top face will have rotated a total of 360° (i.e., a complete revolution); or (ii) rotate the top face 180° <u>counter</u>clockwise. We denote these two moves as $\boxed{\boldsymbol{T^2} \text{ or } \boldsymbol{(T^{-1})^2}}$.

19. To undo the two-step move TU, we first need to reverse the second step (U) <u>*and then*</u> reverse the first step (T). Hence, the reversal of TU is $\boxed{\boldsymbol{U^{-1}T^{-1}}}$.

21. To reverse L^{-1} (a 90° counterclockwise rotation of the left face), we perform a 90° clockwise rotation of the left face; that is, $\boxed{\boldsymbol{L}}$.

23. As explained in our solution above to Problem 19, we undo multi-step moves by performing <u>*in reverse order*</u> the opposite of each of the original steps. So, in the case of $F^{-1}B^{-1}$, its reversal is $\boxed{\boldsymbol{BF}}$.

25. **No**, the resulting arrangements (in general) will be different.

27. Before finding the number of different possibilities, two stipulations are in order:

I. Do <u>not</u> count changes in the <u>*ordering*</u> of the cubes (because once we have found a solution, it obviously does not matter how we switch the four cubes around — as long as we don't rotate any of them).

II. Do <u>not</u> count as a different solution the rotation of the final $1 \times 1 \times 4$ block around its long axis; that is to say, treat the final $1 \times 1 \times 4$ block as <u>1 solution</u> (not 4 solutions).

Now, with these two disclaimers in mind, we see that the first cube can be placed in any one of ***three*** different ways (of course, only one of these ways is correct, but nonetheless there are three possibilities). The key here is to get the "proper faces" along the axis that is never shown. That is, if this first cube looks like the one shown below, it will have an R-R that "doesn't come up."

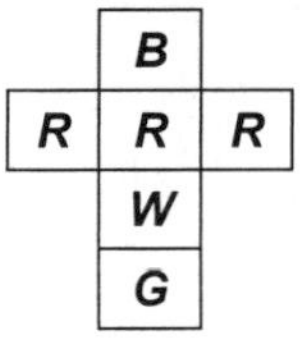

As shown below, the other two possibilities have W-B and R-G along the axis:

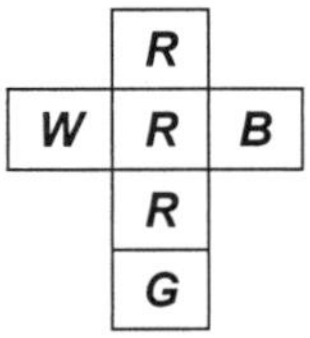

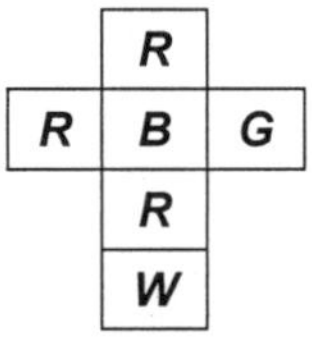

So, the first cube can be arranged in three ways. Now, since the first cube has been "fixed" in position, the other cubes do not have this much freedom of choice. That is, each of the remaining cubes cannot be rotated independently; they must be rotated in conjunction with the first cube. Hence, each of the remaining cubes has 24 (i.e., 4!) possibilities. This "arrangement process" is a multi-step (specifically, four-step) procedure, so we use the fundamental counting principle to compute the result:

There are $3 \cdot 24 \cdot 24 \cdot 24 = \boxed{\textbf{41,472}}$ possible arrangements, of which only one is the correct solution.

29.

Chapter 9 Review Questions

1. To review additional problems of this type, refer to Section 9.1.

$$\text{P(defective)} = \frac{\textit{\# of ways to obtain a defective item}}{\textit{Total \# of possible outcomes}} = \frac{4}{1{,}000} = \boxed{\mathbf{0.004}}$$

2. The experiment of rolling a single die has a sample space of $S = \{1, 2, 3, 4, 5, 6\}$. Three of these six possibilities are prime numbers (namely, 2, 3, and 5). Hence,

$$\text{P(prime)} = \frac{3}{6} = \boxed{\mathbf{\frac{1}{2}}}.$$

3. Refer to Figure 9.4 (textbook p. 581) and/or this manual's solutions for Problem 31-39 in Section 9.1. Of the 36 possible outcomes for the experiment of rolling a pair of dice, 5 yield a sum of 36 [namely, (2,6), (3,5), (4,4), (5,3), and (6.2)]. Therefore,

$$\text{P(eight)} = \boxed{\frac{5}{36}}.$$

4. To review the sample space for a deck of cards, see textbook Figure 9.3 (p. 580).

a) P(jack or better) $= \frac{16}{52} = \boxed{\frac{4}{13}}$

b) Pay special attention to the phrase "if you know...." It implies that a ***conditional*** probability needs to be determined. Given that one ace has already been removed from the 52-card deck, the sample space is reduced to 51 cards, of which only 3 are aces. Hence,

$$P(A_2 \mid A_1) = \frac{3}{51} = \boxed{\frac{1}{17}}.$$

5. Once again, refer, if necessary, to Figure 9.4 (textbook p. 581) and/or this manual's solutions for Problem 31-39 in Section 9.1.

a) Out of the 36 possible outcomes, 11 have a 5 on <u>at least one</u> of the dice. So,

$$\text{P(5 on at least one of the dice)} = \boxed{\frac{11}{36}}.$$

b) P(5 on one die **<u>or</u>** 4 on the other) $= \frac{20}{36} = \boxed{\frac{5}{9}}$.

c) P(5 on one die **<u>and</u>** 4 on the other) $= \frac{2}{36} = \boxed{\frac{1}{18}}$.

6. Apply the property of complements (introduced on textbook p. 597):

P(not dropping) = 1 – P(dropping) = 1 – 0.01 = $\boxed{0.99}$.

7. To review the concept of **odds**, refer to textbook pp. 600-601 and this manual's solutions for assorted problems in Section 9.3.

Given $P(E) = 0.9$, it follows that $P(\overline{E}) = 1 - 0.9 = 0.1$. Therefore, the odds <u>in favor</u> of this event are:

$$\frac{P(E)}{P(\overline{E})} = \frac{0.9}{0.1} = \frac{9}{1} \quad \Rightarrow \quad \boxed{\textbf{9 to 1}}.$$

8. From textbook pp. 600-601, recall that the **odds against** an event E are ***f* to *s*** (failures to successes), whereas the **probability of *E* occurring** is

$$P(E) = \frac{s}{n} = \frac{s}{s+f} \quad \text{(successes divided by total number of possible outcomes).}$$

Hence, with E representing the event and given $f = 1{,}000$ and $s = 1$, it follows that:

$$P(E) = \frac{1}{1 + 1{,}000} = \boxed{\frac{1}{1{,}001}}$$

9. Given that one ace has already been removed from the 52-card deck, the sample space is reduced to 51 cards, of which only 3 are aces. So,

$$P(A_2 \mid A_1) = \frac{3}{51} = \boxed{\frac{1}{17}}.$$

10. For additional details, refer to Examples 1-2 (textbook p. 588, Section 9.2).

Each of the 6 possible faces of a single die has an equally likely chance of being the outcome, so the probability of a one being rolled is 1/6. Therefore, with there being no required cost of playing, the expectation is:

$$E = \overbrace{\$12}^{\textit{amount of prize}} \times \underbrace{\left(\tfrac{1}{6}\right)}_{P(\textit{rolling a '1'})} = \boxed{\$2}.$$

11. To broaden your understanding of combinations and permutations, we encourage you to: (i) review textbook Section 9.4 (pp. 611-615) **and** (ii) revisit this manual's solutions for that section (most notably, Problems 13, 15, and 17). Each resource shares a variety of approaches (e.g., Pascal's triangle, factorial formulas, fundamental counting principle, calculator usage).

a) ${}_5C_3 = \boxed{10}$

b) ${}_8P_3 = 8 \cdot 7 \cdot 6 = \boxed{336}$

c) ${}_{12}P_0 = \boxed{1}$ *[Ask yourself this: How many ways are there to arrange no objects? One way?]*

d) ${}_{14}C_4 = \boxed{1{,}001}$

e) ${}_{100}P_3 = 100 \cdot 99 \cdot 98 = \boxed{970{,}200}$

12. For additional information, revisit Section 9.5 (e.g., Example 6, p. 629).

$$\begin{aligned} P(\overline{A \cup B}) &= 1 - P(A \cup B) && \textit{Complementary probabilities} \\ &= 1 - [P(A) + P(B) - P(A \cap B)] && \textit{Addition property of probability} \\ &= 1 - P(A) - P(B) + P(A \cap B) && \textit{Distributive property} \\ &= 1 - P(A) - P(B) + P(A) \cdot P(B) && \textit{Multiplication property; A \& B \underline{independent}} \end{aligned}$$

Now, given $P(A) = \frac{2}{3}$ and $P(B) = \frac{3}{5}$, we evaluate:

$$P(\overline{A \cup B}) = 1 - P(A) - P(B) + P(A) \cdot P(B) = 1 - \frac{2}{3} - \frac{3}{5} + \frac{2}{3} \cdot \frac{3}{5} = \boxed{\frac{2}{15}}.$$

13. Since the order of selection for a committee does not matter and we do not select the same person twice (i.e., no repetition allowed), this is a ***combination*** problem. Therefore, there are ${}_{12}C_3 = \boxed{220}$ ways to choose a three-member committee from a group of 12 people.

14. Lining people up entails "arranging" them, so this is a ***permutation*** problem (or, alternatively, it may be viewed as a fundamental counting principle problem). Therefore, it follows that there are ${}_5P_5 = 5! = \boxed{\mathbf{120}}$ ways for five people to line up at a bank teller's window.

15. This problem should be viewed as a multi-step decision-making process; that is: does the customer want catsup (yes/no)? How about onion? Mustard? Pickles? Lettuce? Tomato? Mayo? Relish?
We use the ***fundamental counting principle*** to determine the total number of possibilities. Each item can be included or excluded (i.e., there are 2 possibilities per item). Since there are 8 items in all, there are indeed $2^8 = 256$ ways to prepare a Wendy's hamburger. So, **yes**, Wendy's claim is correct.

16. To review the distinctions between "with replacement" and "without replacement," refer to Section 9.5 (beginning with p. 630).

a) After the first orange draw is replaced, the box's contents are restored to its original state. So, the second draw is independent of the first. Thus,

$$P(O_2 \mid O_1) = P(O_2) = \boxed{\frac{3}{5}}.$$

b) In this case, the second draw is dependent on the first. The removal (and non-replacement) of the first ball leaves the box with 4 balls (both of the original purple balls and two remaining orange balls). Therefore,

$$P(O_2 \mid O_1) = \tfrac{2}{4} = \boxed{\frac{1}{2}}.$$

17. Consider the sample space for two births: S = {MM, MF, FM, FF}. In this case, the vet has ruled out FF, so the sample space is reduced to {MM, MF, FM}. Therefore, there is one way (out of 3 possibilities) for the other kitten to be a male; that is, P(other a male) = $\boxed{\frac{1}{3}}$.

18. After writing out the possible sample spaces for each pairing (A versus D, B versus, and C versus D), we find that the best choice is **die C**.
As shown below, there are 64 possible outcomes, and C defeats D in 40 of these outcomes. In other words, P(C wins) $= \frac{40}{64} = \frac{5}{8} = 0.625$ or 62.5%.

Die C \ Die D	5	5	5	5	1	1	1	1
2	D	D	D	D	C	C	C	C
2	D	D	D	D	C	C	C	C
2	D	D	D	D	C	C	C	C
2	D	D	D	D	C	C	C	C
2	D	D	D	D	C	C	C	C
2	D	D	D	D	C	C	C	C
6	C	C	C	C	C	C	C	C
6	C	C	C	C	C	C	C	C

You will then find that: (i) the probability of B defeating D is 50%, and (ii) the probability of A defeating D is 37.5%. We leave the details to you.
[Note: There are several (quick) ways to compute these probabilities without listing the sample spaces. Can you find one? Give it some thought.]

19. After you pick a key, one of five cars will fit that key. Therefore, P(winning) = $\boxed{\frac{1}{5}}$.
Caution: If you viewed this as a two-step process (i.e., pick a key and then try an ignition), then you may have been tempted to give an answer of 1 out of 25 (because $5 \cdot 5 = 25$, via the fundamental counting principle). However, with this perspective, you must remember that there are 5 ways to be a winner (not just one way) and, accordingly, $5 \cdot \frac{1}{25} = \frac{1}{5}$.

20. Here are two ways to view this problem:

I. As explained in our solution to Problem 19, there are 25 possible ways the contestant can perform this two-step task (of picking a key and trying an ignition). If you are in the audience and you selected a car, there is only 1 way for the contestant to win that particular car (namely, pick its key and then choose to try the key in that car's ignition). Thus, by the basic definition of probability, it follows that the probability is $\boxed{\frac{1}{25}}$.

II. Alternatively, since these two steps are independent, one can solve this problem by using the multiplication property of probability, namely

P(picking a particular key **and** key fitting in the selected car) $= \frac{1}{5} \cdot \frac{1}{5} = \frac{1}{25}$.

CHAPTER 10: The Nature of Statistics

Problem Set 10.1

1. **Wording varies**. A ***frequency distribution*** tabulates the number of outcomes in each class, whereas a ***stem-and-leaf plot*** is a systematic listing of each outcome.

3. **Answers vary**. Without additional information, the advertisement has very little meaning. Any astute person should be extremely skeptical of Whiskas' claim and should have an almost immediate (mental) list of questions for the advertiser. This list might include: Upon what statistical methodology is the claim based? Who are these so-called owners? How many are there? How were they chosen? Were they compensated (financially or otherwise)? Are they representative of the overall population of cat owners? And so on....

We encourage you, as a savvy consumer of data, to think of (at the very least) three more questions pertaining to this advertisement that you would want answered.

5. a)

Salary	*Tally*	*Frequency*
$60,000	|	*1*
$50,000	|	*1*
$35,000	|	*1*
$30,000	||	*2*
$25,000	|||	*3*
$20,000	|	*1*
$18,000	||	*2*
$16,000	|	*1*
$14,000	|	*1*

b) The stems below represent the ten-thousands place (e.g., $<u>1</u>4,000), and the leaves denote the thousands place (e.g., $1**<u>4</u>**,000). After preparing a frequency distribution (as done in part a), it then becomes easy to draw a stem-and-leaf plot.

Stems	*Leaves*
1	*4 6 8 8*
2	*0 5 5 5*
3	*0 0 5*
4	
5	*0*
6	*0*

c)

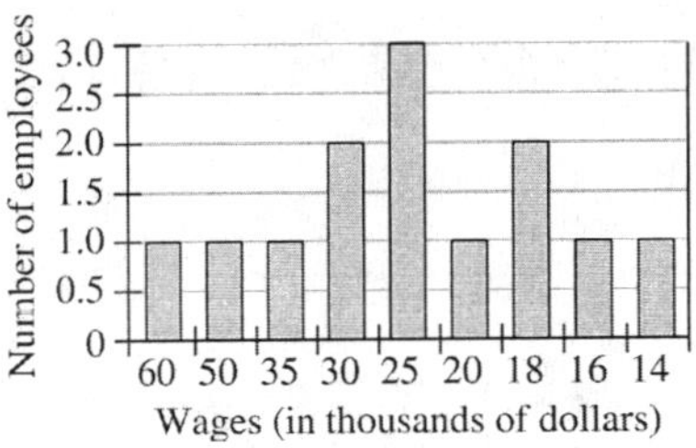
Number of employees
3.0
2.5
2.0
1.5
1.0
0.5
0
60 50 35 30 25 20 18 16 14
Wages (in thousands of dollars)

d)

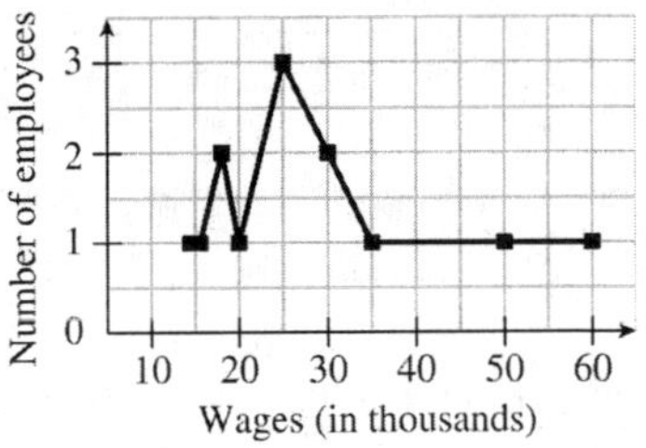
Number of employees
3
2
1
0
10 20 30 40 50 60
Wages (in thousands)

7.

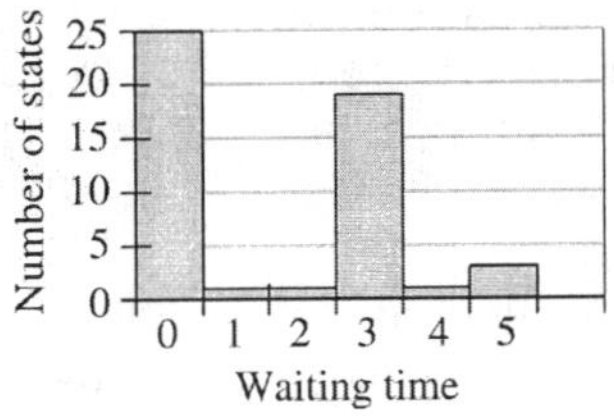
Number of states
25
20
15
10
5
0
0 1 2 3 4 5
Waiting time

9.

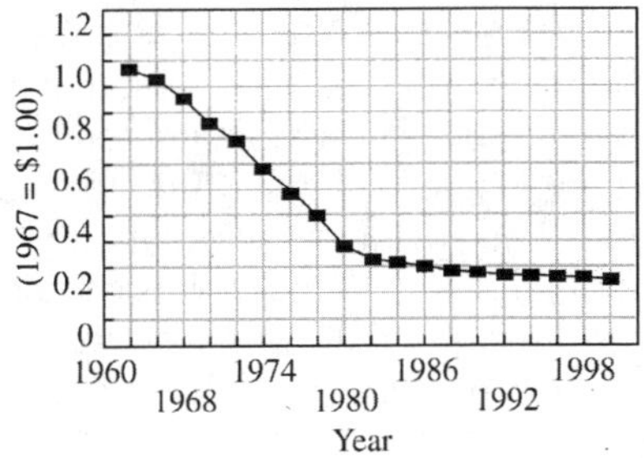
(1967 = $1.00)
1.2
1.0
0.8
0.6
0.4
0.2
0
1960 1968 1974 1980 1986 1992 1998
Year

11.

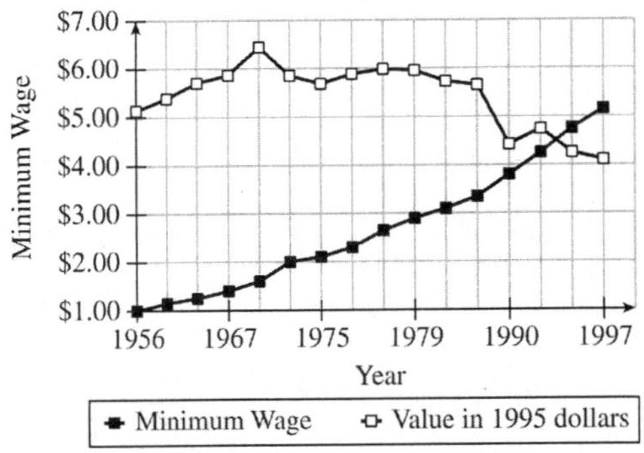
Minimum Wage
$7.00
$6.00
$5.00
$4.00
$3.00
$2.00
$1.00
1956 1967 1975 1979 1990 1997
Year
Minimum Wage
Value in 1995 dollars

13.

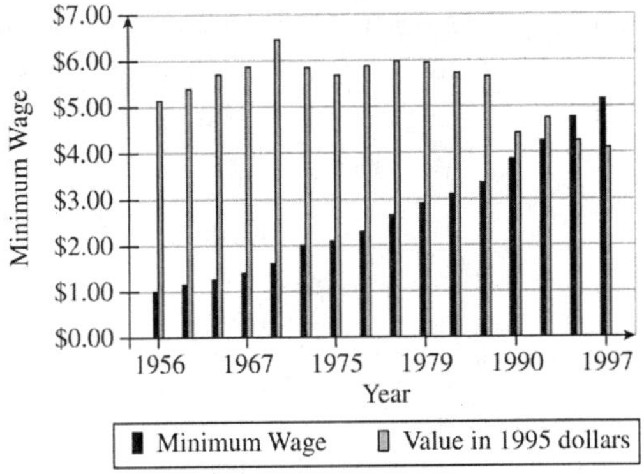
Minimum Wage
$7.00
$6.00
$5.00
$4.00
$3.00
$2.00
$1.00
$0.00
1956 1967 1975 1979 1990 1997
Year
Minimum Wage
Value in 1995 dollars

15.

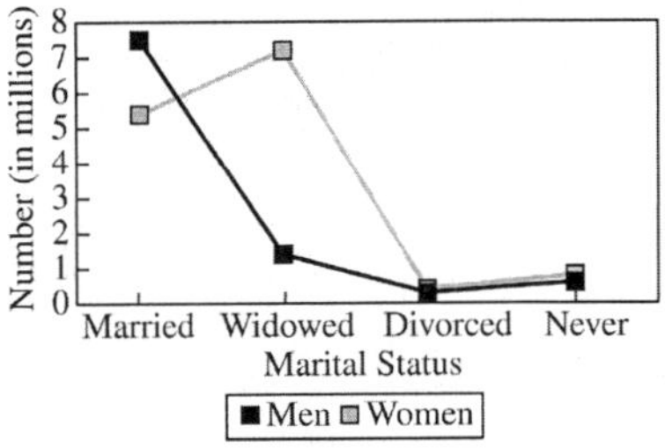
Number (in millions)
8
7
6
5
4
3
2
1
0
Married Widowed Divorced Never
Marital Status
Men Women

17. a) **Gonorrhea cases by race and gender (from 1981 to 1988)**.

The four categories are **black males, black females, white males, and white females**.

b) Black females and white males are almost the same in 1981 (acceptable answer), but the only point on the graph at which the number of cases are precisely the same is **white males and white females in 1986**.

c) Black female cases were increasing **between 1984 and 1986**.

d) About **100,000 cases**

19. **Choice B** is closest to the actual number (4 states). Remember that $1,000 million is the same as $1 billion.

21. **1965 and 1990**

23. **NBC** had a **30%** share in the early 1950's.

25. **Answers vary**; the ratings for all three networks has generally declined from 1955 to 1997.

27. **More imports**

29. **$90 billion**

31. **More imports**; despite the fact that in 1991 the United States had more exports to Mexico than imports, the U.S. had an ***overall*** trade <u>deficit</u> (during this four-year period) with Mexico of $6.7 billion [i.e., (–3.1) + (–2.6) + (–2.4) + 1.4].

33. **$25 billion**

35. If you weigh 115 lbs, you fall into the "110-129" category. Within this weight category, if you have had four drinks in 3 hours, your BAC (blood alcohol concentration) is in the <u>red shaded</u> area, which denotes that your BAC is <u>at least</u> 0.10%. Hence, you are **25 times** more likely to have an accident than if you had refrained from drinking.

37. You would be a **likely DUI**. Your BAC falls into the gray shaded area within your weight class (of 190-209 lbs), which means your BAC is probably in the 0.05% – 0.09% range.

39. **No**; six drinks within a 3-hour period puts you in the red shaded area, which indicates that your BAC would be <u>at least</u> 0.10% ("definitely illegal').

41.

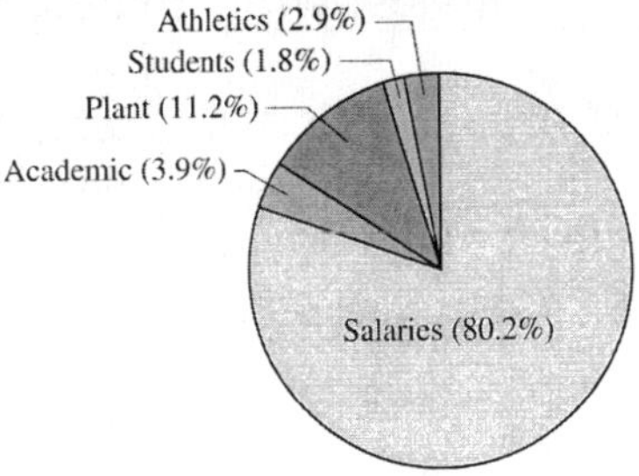

43.

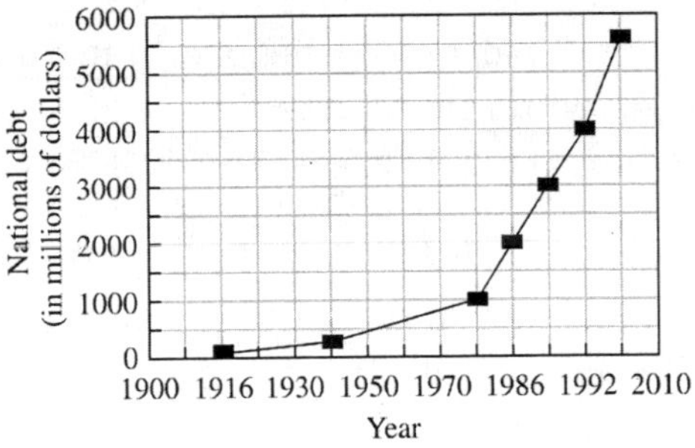

45. **Answers vary**; without a scale, the graph is meaningless.

47. The **graph is accurate and appropriate**. It is refreshing to see a graph in which conflicting points of view are presented honestly and without prejudice.

49. The SAAB99E's **width is misstated**. The advertisement claims that the car is 57 in. wide on the outside, but a full 5 ft (60 in.) on the inside.

51. **3.4 persons**; the lower-right quadrant of Figure 10.24 provides this information.

53. Quoting the lower-left quadrant of Figure 10.24, "… the [social security] trust fund will run out of money by the year **2036**."

55. The line graph in the upper-right quadrant begins at 50 rather than 0.

57.

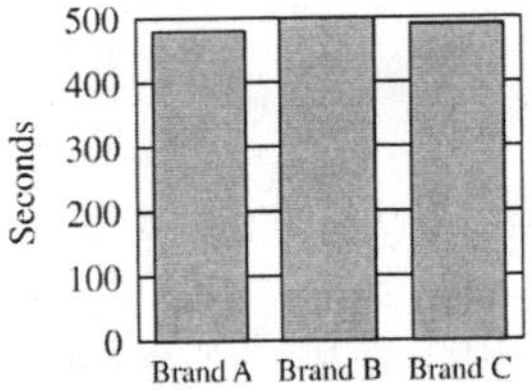

59. The **graph in Figure 10.25b** (with a 10-sec scale) would give the product a more distinct advantage than Figure 10.25a (which has a 100-sec scale). By choosing a smaller scale, the advertiser provides a graph that inflates the relatively small difference of 10-20 seconds between Brand A and its competitors — consequently giving consumers a false impression of the difference's true magnitude.

Problem Set 10.2

1. **Wording varies**. In the study of statistics, the word ***average*** is used in a broad sense. "Average" connotes a measure of the center (or middle) of a set of data values. In this section, three types of averages (i.e., measures of central tendency) are discussed: mean, median, and mode.

3. **Wording varies**. A ***measure of dispersion*** is a single number that quantifies how spread or clustered a data set is (i.e., how much overall variability/deviation there is). In this section, three measures of dispersion are considered: the range, the standard deviation, and the variance. The standard deviation and the variance are measures that indicate how closely the data is related to the mean.

5. **No**, it does not. Although Andy was better than Dave in both years, their respective combined batting averages tell a different story. Andy's combined (1989-1990) batting average was 0.261 (based on 253 hits in 969 at bats). Meanwhile, Dave's combined batting average was 0.278 (136 hits in 490 at bats). This scenario illustrates what is known as *Simpson's paradox* (see Problem 59).

7. **Wording varies**. A variance of zero means the data set has no dispersion (no spread, no variability). In order for this to occur, the data values must all be the same.

9. **Examples vary**. The first student's scores could have been 80, 80, 80, 80, 80, and 80 (so, $\Sigma x = 480$, $\overline{x} = 80$ and $s = 0$), and the second student's scores could have been 100, 100, 100, 60, 50, 70 ($\Sigma x = 480$, $\overline{x} = 80$ and $s \approx 22.80$).

11. **Comments vary**. The person scored better in English than 78% of the people taking the test and received the median score in math.

Note: Problems 12-21 are patterned after text Example 1 (p. 666).

13. For your information: You should not need to perform any computations for this problem. Simply by looking at the data values, the mean should become readily apparent.

$\overline{x}$ = 19; median = 19; no mode

15. Like Problem 13, you should not need to do any arithmetic for this problem.

$\overline{x}$ = 767; median = 767; no mode

17. $\overline{x} = \dfrac{\Sigma x}{n} = \dfrac{55}{5} = 11$

$\overline{x}$ = 11; median = 9; no mode

19. Mean: $\bar{x} = \frac{\Sigma x}{n} = \frac{410}{5} = 82$

Median: Remember to arrange the data values from smallest to largest.

70, 79, **81**, 85, 95

In summary, **$\bar{x}$ = 82; median = 81; no mode**

21. Mean: $\bar{x} = \frac{\Sigma x}{n} = \frac{48}{8} = 6$

Median: Since the median is the middle data value, and since there is an even number of values (namely, 8), the median is the mean of the middle two data values (in this case, the 4th and 5th). That is,

$$0, 1, 1, \mathbf{2, 3}, 4, 16, 21 \Longrightarrow \text{Median} = \frac{2+3}{2} = 2.5$$

Mode: The mode is, by definition, the most frequently occurring data value. Since the value 1 occurs twice, 1 is this data set's mode.

In summary, **$\bar{x}$ = 6; median = 2.5; mode = 1**

Reminder: ***Many calculators include built-in statistical functions. After inputting the data, one can instantaneously obtain (with the pressing of a button or two) the standard deviation, variance, mean, and assorted other statistical information. Check your calculator user's manual (or, if applicable, its accompanying "quick reference" card) for specific details.***

23. For additional details, see textbook pages 669-672 (including Examples 4 and 6). The ***range*** is the difference between the largest and the smallest data values:

$$\text{Range} = \text{Hi} - \text{Lo} = 21 - 17 = \mathbf{4}.$$

As for the variance and the standard deviation, **if** you do not have a calculator with built-in statistical capabilities, use the text's six-step procedure (explained on p. 671).

Step 1: First, determine the mean: $\bar{x} = 19$ (previously obtained in Problem 13)

Step 2: Subtract the mean from each data value, which results in a list of deviations from the mean (of the form $x - \bar{x}$).

Step 3: Square each of these differences; the result is a list of squared deviations of the form $(x - \bar{x})^2$.

For Steps 2 and 3, you may find it helpful to construct a table:

Data values x	***Squared deviations*** $(x-\bar{x})^2$
17	$(17-19)^2=(-2)^2=4$
18	$(18-19)^2=(-1)^2=1$
19	$(19-19)^2=(0)^2=0$
20	$(20-19)^2=(1)^2=1$
21	$(21-19)^2=(2)^2=4$

Step 4: Find the sum of these squared differences.

$$\Sigma(x-\bar{x})^2=4+1+0+1+4=10$$

Step 5: Divide this sum by one less than the number of data values; the result represents the ***variance of the sample*** (also known as the *sample variance*).

$$\text{Sample Variance} = s^2 = \frac{\Sigma(x-\bar{x})^2}{n-1}=\frac{10}{5-1}=\frac{10}{4}=2.5$$

Step 6: Obtain the ***standard deviation of the sample*** (also known as the *sample standard deviation*) by finding the squaring root of the sample variance (from Step 5).

$$\text{Sample Standard Deviation} = s = \sqrt{\frac{\Sigma(x-\bar{x})^2}{n-1}}=\sqrt{2.5}\approx 1.58$$

In summary, **range = 4; s^2 = 2.5; s ≈ 1.58**

25. Compare this problem's data set to Problems 22-24. How are they similar? Notice that the data sets in all four problems share the same amount of dispersion (spread, variation, scattering). So, it should come as no surprise that they share the same measures of dispersion (range, variance, and standard deviation). Thus, based upon our work for Problem 23 (shown above), it follows that the data set {765, 766, 767, 768, 769} has:

range = 4; s^2 = 2.5; s ≈ 1.58

27. *[Reminder: Whenever possible, try to make a "mental estimate"* ***before*** *performing any computations. In this problem, since its data set is more "spread out" than those in Problems 22-25, you should expect the values for its measures of dispersion to be larger than the values obtained above.]*

The ***range*** is: Hi – Lo = $25-1=$ **24**.

Once again, **if** you do not have a calculator with built-in statistical capabilities, use the following six-step procedure to obtain the variance and standard deviation.

Step 1: The mean is $\bar{x}=11$ (previously obtained in Problem 17).

Step 2: Compute the deviations $x-\overline{x}$; that is, subtract the mean from each data value.

Step 3: Compute the squared deviations $(x-\overline{x})^2$; in other words, square each of the differences obtained in Step 2.

Here are Steps 2 and 3:

Data values x	*Squared deviations* $(x-\overline{x})^2$
1	$(1-11)^2=(-10)^2=100$
4	$(4-11)^2=(-7)^2=49$
9	$(9-11)^2=(-2)^2=4$
16	$(16-11)^2=(5)^2=25$
25	$(25-11)^2=(14)^2=196$

Step 4: Find the sum of these squared differences.

$$\Sigma(x-\overline{x})^2=100+49+4+25+196=374$$

Step 5: Obtain the sample variance s^2 by dividing this sum by $n-1$ (i.e., 1 less than the total number of data values).

$$\text{Sample Variance} = s^2 = \frac{\Sigma(x-\overline{x})^2}{n-1}=\frac{374}{5-1}=\frac{374}{4}=93.5$$

Step 6: Obtain the sample standard deviation s by computing the squaring root of the sample variance.

$$\text{Sample Standard Deviation} = s = \sqrt{\frac{\Sigma(x-\overline{x})^2}{n-1}}=\sqrt{93.5}\approx 9.67$$

In summary, **range = 24; s^2 = 93.5; s ≈ 9.67**

29. The ***range*** is: Hi – Lo = $95-70=\underline{\underline{25}}$.

Once again (for those who do not have a calculator with built-in statistical functions), here is the procedure for finding the variance and the standard deviation. For additional details/explanation, see textbook pages 669-672 and this manual's solutions for Problems 23 and 27.

Step 1: The mean is $\overline{x}=82$ (previously computed in Problem 19).

Step 2: Compute the deviations $x-\overline{x}$.

Step 3: Square each of these deviations $(x-\overline{x})^2$.

Here are Steps 2 and 3:

Data values x	***Squared deviations*** $(x-\bar{x})^2$
70	$(70-82)^2 = (-12)^2 = 144$
81	$(81-82)^2 = (-1)^2 = 1$
95	$(95-82)^2 = (13)^2 = 169$
79	$(79-82)^2 = (-3)^2 = 9$
85	$(85-82)^2 = (3)^2 = 9$

***Step 4**:* Find the sum of these squared differences.

$$\Sigma(x-\bar{x})^2 = 144+1+169+9+9 = 332$$

***Step 5**:* Obtain the sample variance s^2 by dividing this sum by $n-1$.

$$\text{Sample Variance} = s^2 = \frac{\Sigma(x-\bar{x})^2}{n-1} = \frac{332}{5-1} = \frac{332}{4} = 83$$

***Step 6**:* Obtain the sample standard deviation s by computing the squaring root of the sample variance.

$$\text{Sample Standard Deviation} = s = \sqrt{\frac{\Sigma(x-\bar{x})^2}{n-1}} = \sqrt{83} \approx 9.11$$

In summary, **range = 25; s^2 = 83; s ≈ 9.11**

31. The ***range*** is equal to Hi – Lo $= 21 - 0 = \underline{\underline{21}}$.
Again (for those who do not have a calculator with built-in statistical capabilities), here is the procedure for finding the variance and the standard deviation. For a more detailed explanation, please refer to this manual's solutions for Problems 23 and 27, as well as textbook pp. 670-672.

***Step 1**:* The mean is $\bar{x} = 6$ (previously computed in Problem 21).

***Step 2**:* Compute the deviations $x-\bar{x}$.

***Step 3**:* Square each of these deviations $(x-\bar{x})^2$.

Here are Steps 2 and 3:

Data values x	*Squared deviations* $(x-\bar{x})^2$
0	$(0-6)^2=(-6)^2=36$
1	$(1-6)^2=(-5)^2=25$
1	$(1-6)^2=(-5)^2=25$
2	$(2-6)^2=(-4)^2=16$
3	$(3-6)^2=(-3)^2=9$
4	$(4-6)^2=(-2)^2=4$
16	$(16-6)^2=(10)^2=100$
21	$(21-6)^2=(15)^2=225$

<u>Step 4</u>: Find the sum of these squared differences.

$$\Sigma(x-\bar{x})^2=36+2(25)+16+9+4+100+225=440$$

<u>Step 5</u>: Obtain the <u>sample variance</u> s^2 by dividing this sum by $n-1$.

$$\text{Sample Variance}=s^2=\frac{\Sigma(x-\bar{x})^2}{n-1}=\frac{440}{8-1}=\frac{440}{7}\approx 62.86$$

<u>Step 6</u>: Obtain the <u>sample standard deviation</u> s by computing the squaring root of the sample variance.
<u>Caution</u>: Use the exact value from Step 5, <u>not</u> the rounded 62.86.

$$\text{Sample Standard Deviation}=s=\sqrt{\frac{\Sigma(x-\bar{x})^2}{n-1}}=\sqrt{\tfrac{440}{7}}\approx 7.93$$

In summary, **range = 21; s^2 = 62.86; s ≈ 7.93**

33. This problem is akin to text Example 5 (pp. 669-670), which explains the procedure for constructing a ***box plot*** (also known as a <u>*box-and-whisker plot*</u>).
First, it is important to note that the data set is already arranged in order (from least to most). Consequently, we can begin by finding the median (also known as the second quartile Q_2). Since there is an even number of data values (i.e., 10 monthly costs), the median is the mean of the two middle numbers. In this case, the 5th cost is \$431.52 and the 6th is \$433.50, so the median (or Q_2) is

$$\frac{431.52+433.50}{2}=\boxed{\$432.51}.$$

Next, we find the first quartile (Q_1), which is defined (on textbook p. 670) as the median of all items below Q_2. That is, there are 5 costs below \$432.51 (namely, \$415.24, \$424.77, \$427.48, \$428.05, and \$431.52), so their median is the 3rd item among these five. Hence, Q_1 = **\$427.48**.

Similarly, the third quartile (Q_3) is the median of the data values above Q_2. Beginning with \$433.50 and ending with \$457.94, there are five of these, and their median is Q_3 = **\$442.28**.

The minimum and maximum costs are respectively \$415.24 and \$457.94, so, putting all this information together, the box plot is:

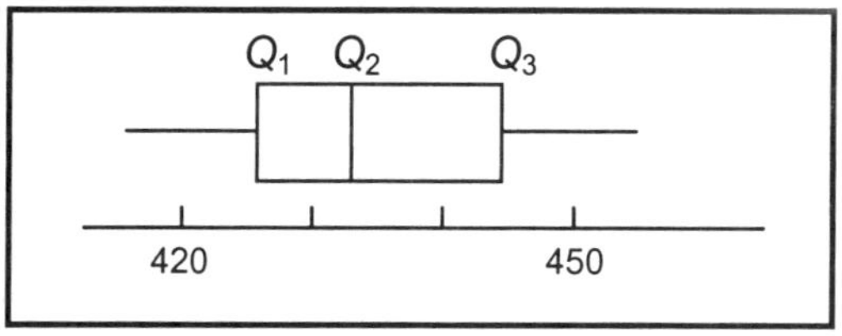

35. It is essential that you understand the meaning (significance) of your answers. For this problem, the mean, median, and mode give us three different ways to view the "average" (or typical) number of clients per day contacted by G. Thumb for the week of October 23-28.

Mean: $\bar{x} = \dfrac{\Sigma x}{n} = \dfrac{12+9+2(10)+16+21}{6} = \dfrac{78}{6} = \boxed{13}$

Median: Since the median is the middle data value, and since there is an even number of values (namely, 6), the median is the mean of the middle two data values (in this case, the 3rd and 4th). That is,

$$9,\ 10,\ \underline{\mathbf{10,\ 12}},\ 16,\ 21 \implies \text{Median} = \frac{10+12}{2} = \boxed{11}$$

Mode: The mode is, by definition, the most frequently occurring data value. Since the value 10 occurs twice, **10** is this data set's mode.

37. Problems 36-38 are very similar to text Example 3 (pp. 667-668).
In order to answer problems like these correctly, you must understand the meaning of "frequency." In this problem, the given frequency distribution tells us that the test scores are:

50, 50, 60, 60, 60, 60, 60, 70, 70, 70, 70, 70, 70, 70, 70, 70, 70, 80, 80, 80, 90.

Recall that we determine a ***weighted mean*** as follows:

$$\bar{x} = \frac{\Sigma(w \cdot x)}{\Sigma w} = \frac{2(50)+5(60)+10(70)+3(80)+1(90)}{21} = \frac{1{,}430}{21} \approx 68.0952 \approx \boxed{68.1}$$

As for the ***median***, there are 21 test scores in all, so the median is the 11th score (when they are arranged in order). Hence, the median score is **70**.

Finally, the most frequently occurring score is 70 (which has a frequency of 10), so the ***mode*** is **70**.

39. Mean: $\bar{x} = \dfrac{\Sigma x}{n} = \dfrac{2{,}198}{6} = 366\tfrac{1}{3}$. Since the original data is given "in thousands," we report the county fair's ***mean*** annual attendance over the past six-year period as being (approximately) **366,000**.

In order to obtain the median, the data must be arranged in order:

346, 358, **364, 366**, 373, 391 ⟹ ***median*** = **365,000**

Since no attendance figure is repeated, there is **no mode**.

41. **Choice A**. Because it has the tallest bar (i.e. the highest frequency), a score of 10 is the mode. Due to the fact that all the remaining scores are lower than 10, the mean is less than 10. Finally, judging by the relative heights of the bars, it is clear that the middle score (i.e., the median) is less than 10.

43. **Choice C**. Based upon the given information, if you were to create a bar graph representing the number of students in this class it **might** look something like one of these:

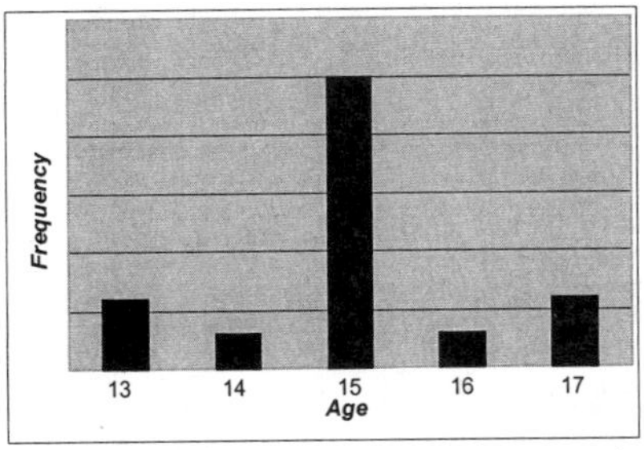

or

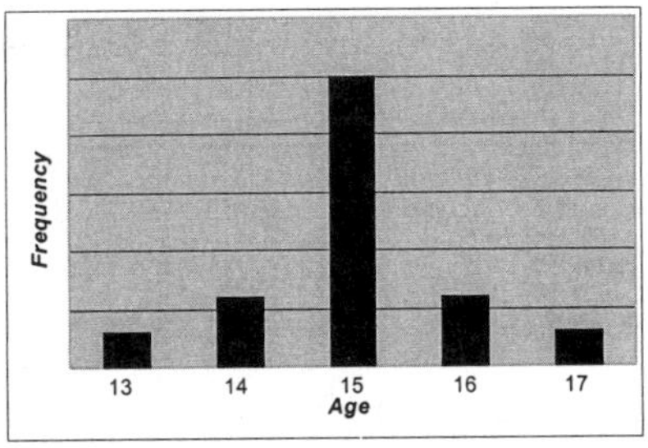

[Disclaimer: Of course, a scale should always be marked along the y-axis.]

Since the problem tells us "most of the students are 15 years old," we immediately know the mode is 15. Since there are as many ballet students below 15 years of age as there are students above 15 years of age, the median is 15. Finally, the symmetry of this distribution of ages causes the mean to "balance out" to 15.

45. Just like the preceding problems, the answers for this problem (and Problem 46) can either be calculated directly or by using a calculator that has built-in statistical capabilities.

a) The mean calories from fat is:

$$\bar{x} = \frac{35 + 3(70) + 2(90) + 100 + 6(110) + 3(120) + 2(130) + 140 + 160}{20}$$

$$= \frac{2{,}105}{20}$$

$$= \boxed{105.25 \text{ cal}}.$$

b) First, arrange the data in order (from least to most). Since there are 20 data values in all, the median is the mean of the 10th and 11th values, which are both equal to 12 grams of total fat. Hence, the median is **12 grams**.

c) The mode is **42 grams**, which occurs 3 times.

47. Once again (for those who do not have a calculator with built-in statistical capabilities), here is the procedure for finding the standard deviation. For a more detailed explanation, please refer to this manual's solutions for Problems 23 and 27, as well as textbook pp. 670-672.

Recall (from Problem 35) that $\overline{x} = 13$.

Data values x	***Squared deviations*** $(x-\overline{x})^2$
12	$(12-13)^2 = (-1)^2 = 1$
9	$(9-13)^2 = (-4)^2 = 16$
10	$(10-13)^2 = (-3)^2 = 9$
16	$(16-13)^2 = (3)^2 = 9$
10	$(10-13)^2 = (-3)^2 = 9$
21	$(21-13)^2 = (8)^2 = 64$

$$s = \sqrt{\frac{\Sigma(x-\overline{x})^2}{n-1}} = \sqrt{\frac{1+3(9)+16+64}{6-1}} = \sqrt{\frac{108}{5}} = \sqrt{21.6} \approx 4.6476$$

Therefore, rounded to the nearest unit, the sample standard deviation is **5**.

49. In determining the standard deviation for Problem 37, we must make sure not to overlook the given frequencies.

First, note that there are 25 data values in all — not 6. Second, recall that we previously found $\overline{x} = 68$.

Once again (for those who do not have a suitable calculator), here is the procedure for finding the standard deviation. In the following computations, please pay special attention to the inclusion of each data value's frequency:

Frequency	Data values x	Squared deviations $(x-\bar{x})^2$
1	90	$(90-68)^2=(22)^2=484$
6	80	$(80-68)^2=(12)^2=144$
10	70	$(70-68)^2=(2)^2=4$
4	60	$(60-68)^2=(-8)^2=64$
3	50	$(50-68)^2=(-18)^2=324$
1	40	$(40-68)^2=(-28)^2=784$

$$\begin{aligned} s &= \sqrt{\frac{\Sigma(x-\bar{x})^2}{n-1}} \\ &= \sqrt{\frac{1(484)+6(144)+10(4)+4(64)+3(324)+1(784)}{25-1}} \\ &= \sqrt{\frac{3{,}400}{24}} \\ &\approx 11.9024 \end{aligned}$$

Therefore, rounded to the nearest unit, the sample standard deviation is **12**.

51. Upon performing the same procedure as demonstrated in previous problems (or by using a calculator), we find the sample standard deviation is $s \approx 15.1085 \approx$ **15**.

53. One approach is to view the data in terms of thousands of miles; that is, perform the computations with the data set {17, 19, 19, 20, 21} and then rewrite the final answers in terms of miles (i.e., by "re-attaching" 000).

Mean: $\bar{x} = [17 + 2(19) + 20 + 21] \div 5 = 19.2 \Rightarrow$ **19,200 miles**

Range: Hi – Lo $= 21 - 17 = 4 \Rightarrow$ **4,000 miles**

Standard Deviation:
$$\begin{aligned} s &= \sqrt{\frac{\Sigma(x-\bar{x})^2}{n-1}} \\ &= \sqrt{\frac{(17-19.2)^2+2(19-19.2)^2+(20-19.2)^2+(21-19.2)^2}{5-1}} \\ &= \sqrt{\frac{8.8}{4}} \\ &= \sqrt{2.2} \\ &\approx 1.483239697 \Rightarrow \quad 1000\times 1.483239697 = 1{,}483.239697 \text{ mi} \end{aligned}$$

Therefore, rounded to the nearest unit, the standard deviation is **1,483 miles**.

55. **Answers vary**. It is, of course, a good idea to make <u>in advance</u> some predictions about (i.e., estimates of) the mean, median, mode, and range. Ask yourself: what would be a reasonable expectation?

57. a) $\bar{x} = \frac{\Sigma x}{n} = \frac{2(2)+2(5)+7+2(8)+2(9)+10}{10} = \frac{65}{10} = \boxed{6.5}$

$$\text{H.M.} = \frac{n}{\Sigma\frac{1}{x}} = \frac{10}{\frac{1}{2}+\frac{1}{2}+\frac{1}{5}+\frac{1}{5}+\frac{1}{7}+\frac{1}{8}+\frac{1}{8}+\frac{1}{9}+\frac{1}{9}+\frac{1}{10}} = \frac{10}{\frac{533}{252}} = \frac{2{,}520}{533} \approx \boxed{4.7}$$

b) $\bar{x} = \frac{52+61}{2} = \boxed{56.5}$ $\quad\text{H.M.} = \frac{n}{\Sigma\frac{1}{x}} = \frac{2}{\frac{1}{52}+\frac{1}{61}} = \frac{2}{\frac{113}{3{,}172}} = \frac{6{,}344}{133} \approx \boxed{56.1}$

59. a) Population A is Player A; population B is Player B; $r_1 = 0.223$; $r_2 = 0.284$; $r = 0.257$; $R_1 = 0.232$; $R_2 = 0.296$; and $R = 0.251$. Finally, $C_1 =$ against right-handed pitchers and $C_2 =$ against left-handed pitchers.

b) **Examples vary**.

Problem Set 10.3

1. Recall (from textbook page 676) that a ***dictatorship*** is a unilateral decision; that is, a decision made without the involvement/input of others.

a) **Yes**, a dictatorship; dad made this purchase without consulting anyone in the family.

b) **Yes**, a dictatorship; the cellar master is the sole decision-maker.

c) **No**, not a dictatorship; the chairperson is obtaining the committee members' opinions.

d) **Yes**, a dictatorship; there is no indication that the department sought anyone's input.

3. **Answers vary**. Your response should include mention of the apportionment plans of Adams, Jefferson, Webster, Hamilton, and Huntington-Hill (HH). These are discussed on textbook pp. 684-686.

5. As explained on textbook pp. 683-684, a ***Borda count*** is a commonly used vote-counting method that is applied when no candidate has a majority of the votes. The voter ranks the candidates in order of preference. When there are n candidates, the voters assign n points to their respective first (favorite) choice candidate, then $n-1$ to their second (next favorite) choice candidate, $n - 2$ for the third choice, and so on. The voter's last choice (least favorite) candidate is awarded only 1 point.

a) When there are 5 candidates, the voters individually award 5 points to their favorite candidate, 4 points to their next favorite candidate, and so on. Consequently, each voter awards a total of $5 + 4 + 3 + 2 + 1 = \underline{15 \text{ points}}$. Thus, given that there are ten voters in all, a grand total of

$$10(5+4+3+2+1) = \boxed{\textbf{150 points}} \text{ is awarded.}$$

b) Given 20 voters and 4 candidates, each voter awards a total of 10 points (i.e., 4 + 3 + 2 + 1). Thus, the total amount of points awarded to the candidates is $20(4+3+2+1) = \boxed{\textbf{200 points}}$.

c) $200(3+2+1) = 200(6) = \boxed{\textbf{1,200 points}}$.

7. In this problem, we generalize the pattern experienced in Problem 5.
If there were m candidates, then each voter would award a total of

$$m + (m-1) + (m-2) + \cdots + 3 + 2 + 1 = \frac{m(m+1)}{2} \text{ points.}$$

Thus, given that there are n voters, the total points awarded by the voters is

$$n[m + (m-1) + (m-2) + \cdots + 3 + 2 + 1] = \boxed{n\left[\frac{m(m+1)}{2}\right] \text{ or } \tfrac{1}{2}nm(m+1)}.$$

We encourage you to revisit Problem 5; use this formula to obtain its answers.
[Note: If you previously studied arithmetic series (which is a topic covered in Section 8.4 of your text), the formula $\frac{m(m+1)}{2}$ should seem familiar to you.]

9. This problem is almost identical to text Example 2 (p. 677).
By the plurality rule, the winner of an election is deemed to be the candidate with the highest number of votes. In this case, we have the following outcome:

Candidate A [via (ABC) and (ACB)]:	2 + 4 = 6 votes
Candidate B [via (BAC) and (BCA)]:	2 + 1 = 3 votes
Candidate C [via (CAB) and (CBA)]:	2 + 1 = 3 votes

Hence, **A is the winner**.

11. This problem is comparable to text Example 4 (pp. 678-679).
First, we attempt to eliminate a candidate by dropping the one with the least first-place votes, but this cannot be done because (as shown in the solution to Problem 9) B and C are tied with 3 votes each.
Since that did not work, we try another approach: drop the candidate with the most last place votes. We find A has 2 last place votes, B has 6, and C has 4. So, we drop B from the race.

Now, the election is to be decided between A and C. Removing B from the original vote tabulations, we have:

(AC)	(AC)	(AC)	(CA)	(CA)	(CA)
2	4	2	1	2	1

We see that A gets 8 votes, while C gets 4. Thus, **A is the winner**.

13. The majority rule (see textbook p. 676) requires that the winner must receive more than 50% of the votes. In this case, there are 17 voters. So, in order to obtain a winner, a candidate must receive 9 or more votes. The outcome of this election was:

Candidate A [via (ABC) and (ACB)]:	1 + 3 = 4 votes
Candidate B [via (BAC) and (BCA)]:	4 + 3 = 7 votes
Candidate C [via (CAB) and (CBA)]:	5 + 1 = 6 votes

Since no candidate received at least 9 votes, there is **no winner** (according to the majority rule)

15. Introduced on textbook p. 680, the ***binary voting*** method pairs two competitors at a time and then successively pairs up the winner of each round with the next competitor, until a final winner ("king-of-the-hill") emerges.

Let us consider all the possible pairings of the competitors:

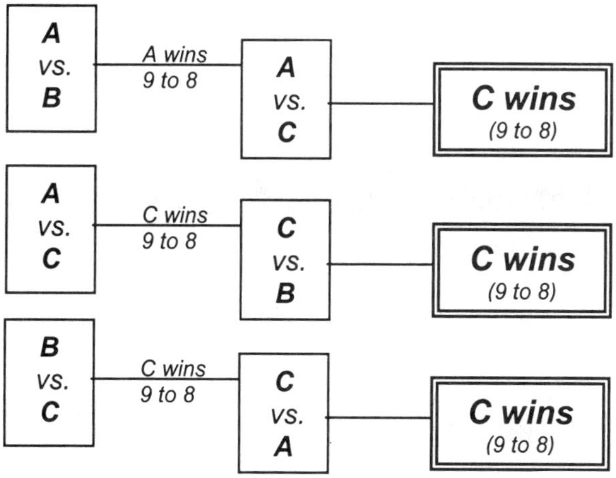

Therefore, the binary voting result yields: **C is the winner**.

17. Using a Borda count with three candidates means the voters award their individual favorite 3 points, their next favorite 2 points, and their least favorite 1 point. An easy way to tabulate the results is to construct a table:

Outcome	A	B	C		A	B	C
(ABC), 1 vote	1(3	2	1)	⇒	3	2	1
(ACB), 3 votes	3(3	1	2)	⇒	9	3	6
(BAC), 4 votes	4(2	3	1)	⇒	8	12	4
(BCA), 3 votes	3(1	3	2)	⇒	3	9	6
(CAB), 5 votes	5(2	1	3)	⇒	10	5	15
(CBA), 1 vote	1(1	2	3)	⇒	1	2	3
				Totals	**34**	**33**	**35***

*So, with a total of 35 points, **C is the winner**.

19. a) With 1,100 convention delegates voting, a majority requires a minimum of 551 votes (i.e., more than 50% of 1,100). Since the leading candidate (namely, candidate A) only has 360 votes, the answer to the question is: **No**, there is no majority winner.

b) **Yes, candidate A** is the plurality winner (with 360 votes).

21. Using a Borda count with five candidates means the convention delegates award their individual favorite 5 points, their next favorite 4 points, and so on (ending with their least favorite candidate who is awarded 1 point). Once again, we construct a table to tally the results:

Outcome	A	B	C	D	E		A	B	C	D	E
(ADECB), 360	360(5	1	2	4	3)	⇒	1,800	360	720	1,440	1,080
(BEDCA), 240	240(1	5	2	3	4)	⇒	240	1,200	480	720	960
(CBEDA), 200	200(1	4	5	2	3)	⇒	200	800	1,000	400	600
(DCEBA), 180	180(1	2	4	5	3)	⇒	180	360	720	900	540
(EBDCA), 80	80(1	4	2	3	5)	⇒	80	320	160	240	400
(ECDBA), 40	40(1	2	4	3	5)	⇒	40	80	160	120	200
						Totals	**2,540**	**3,120**	**3,240**	**3,820***	**3,780**

*So, with 3,820 total points, **candidate D** wins the Borda count.

23. For additional details, refer to text Example 10 (pp. 684-687).

The total population is 62,000, and number of representatives is 26. So, the ***standard divisor*** is $D = 62{,}000 \div 26 \approx \underline{2{,}384.62}$.

We begin with the three simplest representation plans, those of Adams, Jefferson, and Webster:

Region (Pop.)	Std. Quota [Pop. ÷ D]	**Adams** [round up]	**Jefferson** [round down]	**Webster** [round to nearest]
North (18,200)	7.632	**8**	**7**	**8**
South (12,900)	5.410	**6**	**5**	**5**
East (17,600)	7.381	**8**	**7**	**7**
West (13,300)	5.577	**6**	**5**	**6**
Total = 62,000	Total # of Reps. ⇒	**28**	**24**	**26**

Next, we turn our attention to Hamilton's plan, which achieves the desired total of representatives (in this case, 26) by allocating seats based upon the decreasing order of the decimal portions of the standard quota. Notice that Jefferson's plan resulted in a total representation of 24 seats. So, in order for Hamilton's plan to achieve 26 seats, two regions will be awarded an additional seat.

Region	Std. Quota	Decimal Portion	Decimal Rank	Rounding Decision	**Hamilton's Apportionment**
N	7.632	0.632	1st (highest)	up	**8**
S	5.410	0.410	3rd	down	**5**
E	7.381	0.381	4th (lowest)	down	**7**
W	5.577	0.577	2nd	up	**6**
				Total # of Reps. ⇒	**26**

Finally, we use the Huntington-Hill (HH) plan (explained in detail on textbook p. 685 and demonstrated on p. 687):

Region	Std. Quota	a	b	Geometric Mean	Rounding Decision	**HH's Apportionment**
N	7.632	7	8	$\sqrt{7 \cdot 8} \approx 7.48$	Up (because 7.632 > 7.48)	**8**
S	5.410	5	6	$\sqrt{5 \cdot 6} \approx 5.48$	Down (because 5.410 < 5.48)	**5**
E	7.381	7	8	$\sqrt{7 \cdot 8} \approx 7.48$	Down (because 7.381 < 7.48)	**7**
W	5.577	5	6	$\sqrt{7 \cdot 8} \approx 5.48$	Up (because 5.577 > 5.48)	**6**
					Total # of Reps. ⇒	**26**

25. Once again, for additional details, we refer you to text Example 10 (pp. 684-687). The total population is 16,630,000, and number of representatives is 475. So, the ***standard divisor*** is $D = 16{,}630{,}000 \div 475 \approx \underline{35{,}010.53}$.

Region (Pop.)	Std. Quota	**Adams** [round up]	**Jefferson** [round down]	**Webster** [round to nearest]
North (1,820,000)	51.98	**52**	**51**	**52**
Northeast (2,950,000)	84.26	**85**	**84**	**84**
East (1,760,000)	50.27	**51**	**50**	**50**
Southeast (1,980,000)	56.55	**57**	**56**	**57**
South (1,200,000)	34.28	**35**	**34**	**34**
Southwest (2,480,000)	70.84	**71**	**70**	**71**
West (3,300,000)	94.26	**95**	**94**	**94**
Northwest (1,140,000)	32.56	**33**	**32**	**33**
Total = 16,630,000	Total # of Reps. ⇒	**479**	**471**	**475**

Turning our attention to Hamilton's apportionment method, we first notice that Jefferson's plan resulted in a total representation of 471 seats. Consequently, in order for Hamilton's plan to achieve 475 seats, four of the eight regions will need to be awarded an additional seat.

Region	Std. Quota	Decimal Portion	Decimal Rank	Rounding Decision	**Hamilton's Apportionment**
N	51.98	0.98	1st (highest)	up	**52**
NE	84.260	0.260	7th	down	**84**
E	50.27	0.27	6th	down	**50**
SE	56.55	0.55	4th	up	**57**
S	34.28	0.28	5th	down	**34**
SW	70.84	0.84	2nd	up	**71**
W	94.257	0.257	8th (lowest)	down	**94**
NW	32.56	0.56	3rd	up	**33**
				Total # of Reps. ⇒	**475**

Finally, we use the Huntington-Hill (HH) apportionment method:

Region	Std. Quota	a	b	Geometric Mean	Rounding Decision	HH's Apportionment
N	51.98	51	52	$\sqrt{51\cdot 52}\approx 51.5$	Up (51.59 > 51.5)	**52**
NE	84.26	84	85	$\sqrt{84\cdot 85}\approx 84.5$	Down (84.26 < 84.5)	**84**
E	50.27	50	51	$\sqrt{50\cdot 51}\approx 50.5$	Down (50.27 < 50.5)	**50**
SE	56.55	56	57	$\sqrt{56\cdot 57}\approx 56.5$	Up (56.55 > 56.5)	**57**
S	34.28	34	35	$\sqrt{34\cdot 35}\approx 34.5$	Down (34.28 < 34.5)	**34**
SW	70.84	70	71	$\sqrt{70\cdot 71}\approx 70.5$	Up (70.84 > 70.5)	**71**
W	94.26	94	95	$\sqrt{94\cdot 95}\approx 94.5$	Down (94.26 < 94.5)	**94**
NW	32.56	32	33	$\sqrt{32\cdot 33}\approx 32.5$	Up (32.56 > 32.5)	**33**
					Total # of Reps. ⇒	**475**

27. **Answers vary**.

[__Hint__: Before you read this solution, please read the first paragraph of Problem 28 and then re-attempt the problem.]

At first, since 17 is not divisible by 2, 3, nor 9, dividing this stable of horses among the rancher's three heirs appears to be a difficult task. However, by noting that the least common multiple of 2, 3, and 9 is 18, a resolution to this dilemma arises. Suppose a wise person gives an old nag horse to the estate. By doing so, the rancher's estate will now consist of 18 horses. We then find $\frac{1}{2}(18)=9$ horses, $\frac{1}{3}(18)=6$ horses, and $\frac{1}{9}(18)=2$ horses. Since the eldest picked 9 horses, and the middle child 6 horses, and the youngest 2 horses, they picked a total of 17 horses. Certainly, none picked the old nag horse, which was then returned to the wise person.

29. **Apportionments vary**. The average of 1, 2, 3, …, N is $\dfrac{N+1}{2}$, so for N = 17, the average ranking is 9. The youngest child must receive 2 horses whose ranks add up to 18 (i.e., $9\times 2=18$); the middle one must receive 6 horses whose ranks add up to 54 ($9\times 6=54$); and the eldest must receive 9 horses whose ranks add up to 81 ($9\times 9=81$). Based upon this ranking agreement, here is __one__ possible solution:

Eldest receives:	**2, 3, 4, 7, 11, 12, 13, 14, 15**	Total rank = 81; Avg rank = 9
Middle receives:	**5, 6, 8, 9, 10, 16**	Total rank = 54; Avg rank = 9
Youngest receives:	**1, 17**	Total rank = 18; Avg rank = 9

Problem Set 10.4

1. **Wording varies**. A ***cumulative frequency*** is the total number of cases having any score or a score that is lower. Said differently, cumulative frequencies tell us, for each data value, how many items have that value or a lower value.

3. **Wording varies**. A ***normal curve*** (sometimes called a *bell curve*) is a graphical representation of a normal frequency distribution. Its shape resembles the outline of a flared bell, with its high point occurring at the mean and its shape being symmetric with respect to this mean. As shown in textbook Figure 10.34 (page 691), about 68.1% (i.e., 34.1× 2) of all data values of a normal curve lie within 1 standard deviation (on either side) of the mean. About 95.4% lie within 2 standard deviations (on either side) of the mean, and virtually all (about 99.8%) lie within 3 standard deviations (on either side) of the mean.

5. Problems 4-9 are very similar to text Example 1 (p. 690).
An easy way to obtain the cumulative distribution is to begin with the lowest interval (in this case, 0.00-0.99) and subtotal one interval at a time:

GPA	*Interval Mean*	*Percent of Class*	***Cumulative Percent***
0.00 – 0.99	*0.50*	*11%*	***11%***
1.00 – 1.74	*1.37*	*6%*	***17%***
1.75 – 2.49	*2.12*	*19%*	***36%***
2.50 – 3.24	*2.87*	*41%*	***77%***
3.25 – 4.00	*3.625*	*23%*	***100%***

First, to determine the mean GPA for these graduating seniors, we compute a weighted mean by using the mean (or midpoint) of each interval as a representative value of that interval and the respective percent as the interval's weight. Since the sum of the percents is 1, we divide by 1. Hence, the **mean** GPA is

$$\bar{x} = \frac{0.11(0.50) + 0.06(1.37) + 0.19(2.12) + 0.41(2.87) + 0.23(3.625)}{1}$$

$$= 2.55045$$

$$\approx \boxed{2.55}.$$

Next, in order to better understand how to determine the median when percentages are used and why it is done in the manner it is, we make two suggestions: (i) remember that the median is equivalent to the 50th percentile (or the middle quartile Q_2); or (ii) if you need to, momentarily think of the percentages as "numbers" of students.

Let us illustrate this second suggestion. Pretend there is a total of 100 students, with 11 having GPAs between 0.00 and 0.99, 6 between 1.00 and 1.74, and so on. That is to say, envision the data as follows:

0.00 …0.99, 1.00…1.74, 1.75………2.49, 2.50………………..3.24,3.25…………..4.00

11 values *6 values* *19 values* *41 values* *23 values*

100 values in all

Ask yourself: where are the 50th and 51st values? Your answer should be "somewhere in the 2.50 – 3.24 interval." Notice that this is the interval in which the cumulative percent first exceeds 50%; so, from this point forward, we will use the cumulative percent column to determine the median. Since we do not know the GPAs of each individual student, we report the median as the midpoint of the 2.50 – 3.24 interval; thus, the **median** GPA is **2.87**.

Finally, the mode is the most frequently occurring interval. With 41% of the students having GPAs between 2.50 and 3.24, we report the **mode** as the midpoint of that interval, namely **2.87**.

7. Of course, proportions are treated no differently than percentages, so we obtain the following cumulative distribution:

# of Motor Vehicles per Household	*Proportion of Households*	***Cumulative Proportion***
0	*0.18*	***0.18***
1	*0.24*	***0.42***
2	*0.32*	***0.74***
3	*0.16*	***0.90***
4	*0.09*	***0.99***
5	*0.01*	***1.00***

Using the same approach as demonstrated in our solution to Problem 5 (see above) and in text Example 1 (see p. 690), we find:

Mean: $$\bar{x} = \frac{0.18(0)+0.24(1)+0.32(2)+0.16(3)+0.09(4)+0.01(5)}{1}$$

$= \boxed{1.77}$ vehicles/household

Median: Remember that the median is the same as the 50th percentile (i.e., the data value at which 50% of the data set lies below). Since 2 is the number of motor vehicles for which the cumulative proportion first exceeds 0.50, we conclude that the median number of vehicles per household is **2**.

Mode: Since two-vehicle households comprise the largest proportion (0.32) of households, the mode is **2**.

9.

Years of Schooling	*Proportion of Employees*	***Cumulative Proportion***
11 or fewer	*0.02*	***0.02***
12	*0.25*	***0.27***
13	*0.11*	***0.38***
14	*0.08*	***0.46***
15	*0.06*	***0.52***
16	*0.32*	***0.84***
17	*0.02*	***0.86***
18	*0.12*	***0.98***
19 or more	*0.02*	***1.00***

Once again, we use the same approach as explained in our solution to Problem 5 (see above) and in text Example 1 (see p. 690), we find:

Mean: If we assume 11 years of education is a representative value for those with "11 or fewer" and assume 19 years as representative of those with "19 or more," we obtain the following:

$$\bar{x} = \frac{0.02(11) + 0.25(12) + 0.11(13) + 0.08(14) + 0.06(15) + 0.32(16) + 0.02(17) + 0.12(18) + 0.02(19)}{1}$$

$= \boxed{14.67}$ years of formal education.

Median: Remember that the median is the same as the 50th percentile (i.e., the data value at which 50% of the data set lies below). Since 15 is the number of years of education for which the cumulative proportion first exceeds 0.50, we conclude that the median number of years of education is $\boxed{15}$.

Mode: Since employees with 16 years of schooling comprise the largest proportion (0.32) of employees, the mode is $\boxed{16}$.

11. If we view percentiles as cumulative (percentage) frequencies, the given GPA data can be expressed as follows:

Percentile rank	*GPA*	***Percent of Students***
99	*4.00*	***11%***
88	*3.50*	***20%***
68	*3.00*	***18%***
50	*2.50*	***24%***
26	*2.00*	***10%***
16	*1.50*	***4%***
12	*1.00*	***7%***
5	*0.50*	***5%***
	Total:	***99%***

Clarification: Remember, for instance, that the 99th percentile means 99% of the students have GPAs below 4.00. That is, the 11% noted in the above table signifies the percentage of students with GPAs of at least 3.50 but less than 4.00 (i.e., between 3.50 and 3.99). So, if we were to view each group as an interval, we would obtain the interval means (like was done in Problem 5). However, since we are (somewhat casually) estimating the mean in part (a) below, we will use each of the GPAs indicated in the table. Furthermore, for purposes of this problem, you should assume that the values cited in the table are the only GPA's possible for a student to achieve.

a) Using the 99% total as noted in the above table, we estimate that the mean GPA is

$$\bar{x} = \frac{0.11(4.00) + 0.20(3.50) + 0.18(3.00) + 0.24(2.50) + 0.10(2.00) + 0.04(1.50) + 0.07(1.00) + 0.05(0.50)}{0.99}$$

$$= 2.66\overline{16} \approx \boxed{2.66}.$$

b) As for the mode, the highest percentage of students in any one category is the 24% listed next to a GPA of **2.50**. Please note that (in a more formal statistics course) the 24% would be treated as the percentage of this college's students having a GPA of at least 2.00 but less than 2.50.

c) From the given data and the definition of a "percentile" (which, we remind you, is very different from a "percentage"), we know that **26%** of GPAs are less than 2.00.

13. As explained on textbook pp. 693-695, the percent of the total population between the mean and a given z-score is found by using the table in textbook Appendix F (p. 901). In preparation for the more difficult problems in this section, it is a good idea to make a sketch of the normal curve showing the area you are trying to find. Let us demonstrate:

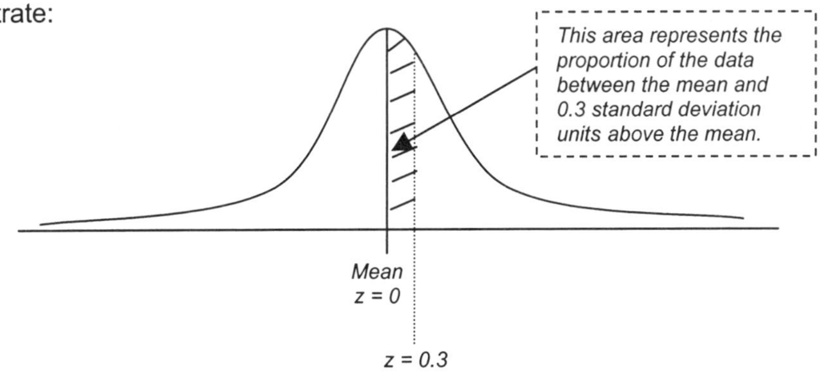

To find z = 0.3 (i.e., 0.30), look at the row labeled 0.3 and the column headed 0.00 to find the entry 0.1179. This means that **11.79%** of the values in a normal distribution are between the mean and 0.3 standard deviations above the mean.

15. If you have not already done so, please read this manual's solution to Problem 13 (immediately above). To find z = 1.86, look at the row labeled 1.8 and the column headed 0.06 to find the entry 0.4686. This means that **46.86%** of the values in a normal distribution are between the mean and 1.86 standard deviations above the mean.

17. Due to the symmetry of the normal curve/distribution, the percent of the data between the mean and z = – 0.6 is the same as the percent of the data between the mean and z = + 0.6. As recommended in our solution to Problem 13 above, it is a good idea (in preparation for the more difficult problems in this section) to make a sketch of the normal curve showing this area. That is,

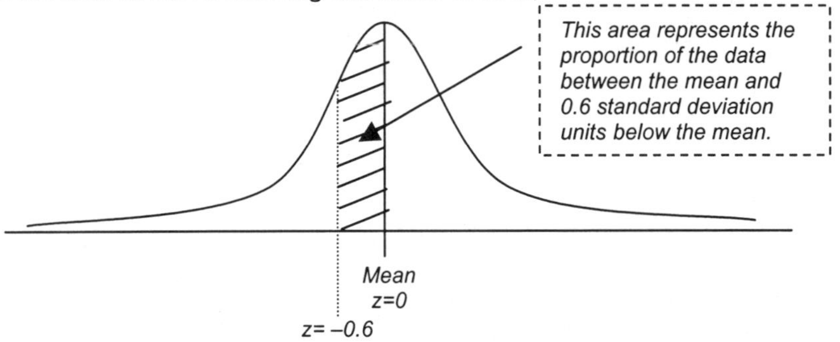

To find z = – 0.6, look at the row labeled 0.6 and the column headed 0.00 to find the entry 0.2257. This means that **22.57%** of the values in a normal distribution are between the mean and 0.6 standard deviations below the mean.

19. **19.15%**. Use the same approach as explained in the solutions for Problems 13, 15, and 17.

21. **38.30%**. Use the same approach as explained in the solutions for Problems 13, 15, and 17.

23. **49.99%**. Take note of the words below the table given in Appendix F (p. 901): "For value of z above 3.09, use 0.4999."

25. *[__Suggestion__: Drawing the normal curve with labeling/shading appropriate to the problem you are solving can help alleviate confusion. To review how this is done, refer to textbook pp. 694-695 (see Example 4) and this manual's solutions to assorted problems, including Problems 13, 17, and part (b) of this problem.]*

a) For Problems 24-28, we are given that the mean height is 170 cm; in other words, $z = 0$ for $x = 170$ cm. For $x = 180$ cm, the z-score is

$$z = \frac{x - \mu}{\sigma} = \frac{180 - 170}{5} = 2 .$$

Based upon either textbook Figure 10.34 (p. 691) or Appendix F (p. 901), we know that approximately 47.7% of the area under a normal curve is between the mean ($z = 0$) and two standard deviations above the mean ($z = 2$). Therefore, out of the 50 people in this sample, we would expect approximately **24** of them [note: 0.477 × 50 = 23.85] to be between 170 and 180 cm tall.

b) We first compute the z-score for 176 cm:

$$z = \frac{x-\mu}{\sigma} = \frac{176-170}{5} = 1.2\,.$$

To determine the proportion of the population that is taller than 176 cm, we begin with a sketch of the normal curve and then use Appendix F accordingly:

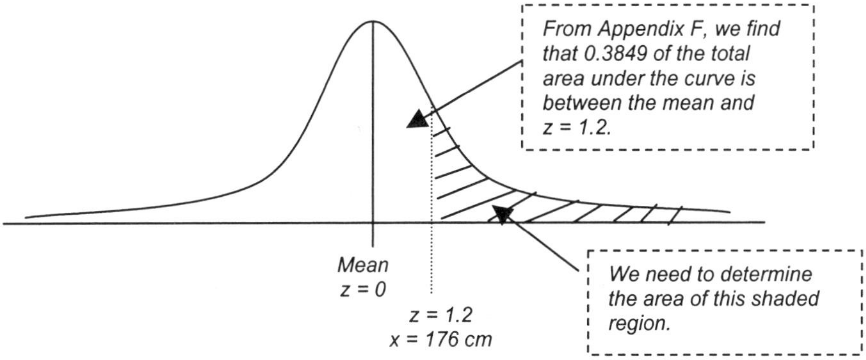

Due to the symmetry of the normal curve, we know 50% of the area under the curve is to the right of the mean. So, since 38.49% is between the mean and $z = 1.2$, we find the indicated shaded area by computing the difference $0.5000 - 0.3849 = 0.1151$. Thus, upon noting that $0.1151 \times 50 = 5.755$, we should expect approximately **6** of the 50 people in this sample to be taller than 176 cm.

27. From Section 10.2, recall that the ***variance*** is the square of the standard deviation. So, given that $\sigma = 5$ in Problems 24-28, we know the variance is $\sigma^2 =$ **25**.

29. Problems 29-34 are related to text Example 3 (pp. 692-693).
When the author of your text makes mention in these problems of "grading on a curve," he is referring to the grade distribution depicted in Figure 10.36 (p. 692). According to that figure, the students who receive a C on the exam are those whose scores are within 1 standard deviation (on either side) of the mean, which encompasses approximately 68.2% (i.e., 34.1× 2) of the class. Hence, since there are 45 students in the class and we know 0.682 × 45 = 30.69, approximately **31** students should receive a C.

31. As shown in Figure 10.36 (p. 692), a student who receives an A must have earned a test score that is at least 2 standard deviations above the mean; that is, x must be greater than or equal to $\mu + 2\sigma$. We are given in Problems 29-34 that $\mu = 50$ and $\sigma = 5$, so it follows that $\mu + 2\sigma = 50 + 2(5) = 60$. Therefore, a score of **60 or above** is necessary to obtain an A.

33. Once again (as previously noted in our solution to Problem 27), you should know that the variance is the square of the standard deviation. So, given that $\sigma = 5$ in Problems 29-34, we know the variance is $\sigma^2 = \boxed{25}$.

35. We begin by sketching a normal curve that depicts the top 6% of the students who receive the grade of A:

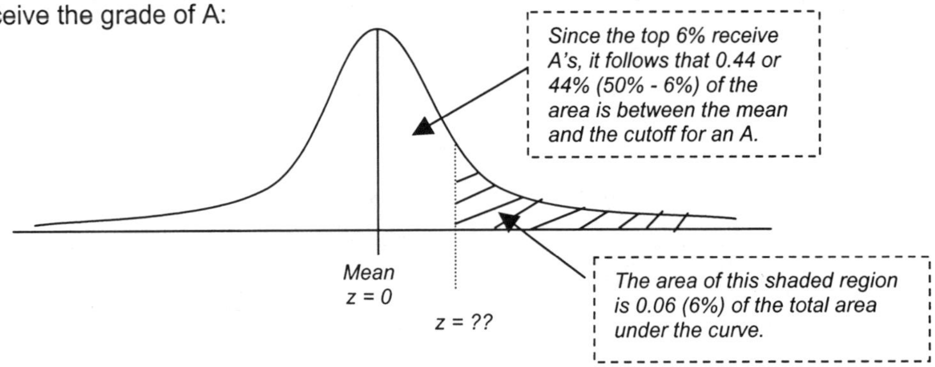

In Appendix F (p. 901), we find that the area closest to 0.44 is 0.4406, which corresponds to a z-score of 1.56. We then apply our algebra skills to solve $z = \dfrac{x-\mu}{\sigma}$ for x; the result is $x = \mu + z \cdot \sigma$ [for the details, see this manual's solution to Problem 43]. Therefore, given that the scores are normally distributed with a mean of 75 and a standard deviation of 8 and given that one must score at least 1.56 standard deviations above the mean to receive an A, the bottom cutoff for an A is $x = \mu + z \cdot \sigma = 75 + (1.56)(8) = 87.48 \approx \boxed{\textbf{87 points}}$.

37. Here are two approaches:

I. We begin by sketching a normal curve that depicts the bottom cutoff for a C:

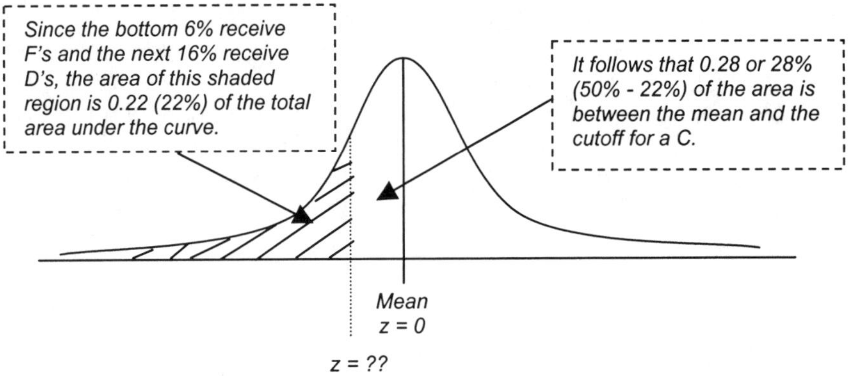

In Appendix F (p. 901), we find that the area closest to 0.28 is 0.2794, which corresponds to a z-score of –0.67. We then apply our algebra skills to solve

$z = \dfrac{x - \mu}{\sigma}$ for x; the result is $x = \mu + z \cdot \sigma$ [for the details, see this manual's solution to Problem 43]. Therefore, given that the scores are normally distributed with a mean of 75 and a standard deviation of 8 and given that one must score no less than 0.67 standard deviations below the mean to receive a C, the bottom cutoff for a C is

$$x = \mu + z \cdot \sigma = 75 + (-0.67)(8) = 69.64 \approx \boxed{\textbf{70 points}}.$$

II. If you solved Problem 36, you know the bottom cutoff for a grade of B is 80 points; of course, this cutoff is the threshold between a high C and a low B. The difference between 80 and the mean is 80 – 75 = 5 points. Therefore, based upon the symmetric nature of this grade distribution, the bottom cutoff for a C must be 5 points below the mean, which is 75 – 5 = 70 points. In other words, C's are awarded to those with test scores ranging from 70 to 79.

39.

Letter Grade	*Scores*	*Percent of Students*	***Cumulative Percent***
A	*87 and ↑*	*6%*	***6%***
B	*80-86*	*16%*	***22%***
C	*70-79*	*56%*	***78%***
D	*65-69*	*16%*	***94%***
F	*64 and ↓*	*6%*	***100%***

41. **Explanations vary**. As shown in Figure 10.40 (p. 696), the **mode** has the greatest value in a distribution that is ***skewed to the left*** (sometimes called *negatively skewed*). Remember that the mean is affected by extreme scores (data values). So, if a distribution includes scores "in the left tail" (i.e., scores that are extremely small in comparison to the rest of the values), the mean will be pulled toward those low scores. However, in such a distribution, the majority of the scores will be above the mean, so the median will be to the right of the mean. Finally, the mode indicates the highest frequency, and in a skewed-to-the-left distribution, the mode is positioned at the higher (right) end of the distribution.

43. We begin by solving the formula $z = \dfrac{x - \mu}{\sigma}$ for x:

$$z \cdot \sigma = x - \mu \quad \text{Multiply both sides by } \sigma.$$

$$z \cdot \sigma + \mu = x \quad \text{Add } \mu \text{ to both sides.}$$

$$x = \mu + z \cdot \sigma \quad \text{Exchange sides, etc.}$$

Given that $\mu = 85.7$ and $\sigma = 4.85$, the data value corresponding to $z = 2.55$ is

$$\begin{aligned} x &= \mu + z \cdot \sigma \\ &= 85.7 + (2.55)(4.85) \\ &= \boxed{\textbf{98.0675}}. \end{aligned}$$

45. Via the same approach as used in Problem 43, we obtain the data value corresponding to $z = -3.46$:

$$\begin{aligned} x &= \mu + z \cdot \sigma \\ &= 85.7 + (-3.46)(4.85) \\ &= \boxed{\mathbf{68.919}}. \end{aligned}$$

47. This problem is analogous to text Example 4c (pp. 694-695).

*[**General Reminder**: Drawing the normal curve with labeling/shading appropriate to the problem you are solving can help alleviate confusion. To review how this is done, refer to textbook pp. 694-695 (see Example 4) and this manual's solutions to Problems 13, 17, 25b, 35, 37, and 51.]*

We are given $\mu = 0.9$ inch and $\sigma = 0.01$ inch. For $x = 0.91$ in., the z-score is

$$z = \frac{x - \mu}{\sigma} = \frac{0.91 - 0.90}{0.01} = 1.$$

The probability of a randomly selected cable having a diameter that exceeds 0.91 is equal to the proportion of the area under the normal curve to the right of $z = 1$. There are two ways to obtain this area:

I. From our familiarity with the normal curve (namely, the information provided in textbook Figure 10.34, p. 691), we know that about 0.159 = 15.9% (13.6% + 2.2% + 0.1%) of the area under the curve is to the right of $z = 1$.

II. In Appendix F (p. 901), we find that 0.3413 is the proportion of the area between the mean and $z = 1$. Due to the symmetry of the normal curve, we know 50% of the area under the curve is to the right of the mean. So, since 34.13% is between the mean and $z = 1$, the area to the right of $z = 1$ is the difference $0.5000 - 0.3413$, which equals 0.1587.

Thus, using the more accurate of our two answers, we report the probability of a randomly selected electric cable having a diameter greater than 0.91 in. as $\boxed{\mathbf{0.1587}}$ or 15.87%.

49. *[**General Reminder**: Drawing the normal curve with labeling/shading appropriate to the problem you are solving can help alleviate confusion. To review how this is done, refer to textbook pp. 694-695 (see Example 4) and this manual's solutions to Problems 13, 17, 25b, 35, 37, and 51.]*

We are given (in Problem 48) $\mu = 35.5$ inches and $\sigma = 2.5$ inches.

For $x = 36$ in., the z-score is

$$z = \frac{x - \mu}{\sigma} = \frac{36 - 35.5}{2.5} = 0.2.$$

In Appendix F (p. 901), we find that 0.0793 is the proportion of the area between the mean and $z = 0.2$. Due to the symmetry of the normal curve, we know 50% of the area under the curve is to the right of the mean. So, since 7.93% is between the mean and $z = 0.2$, the area to the right of $z = 0.2$ is the difference $0.5000 - 0.0793$, which equals 0.4207.

Therefore, based upon the assumption that annual rainfall is normally distributed with a mean 35.5 in. and standard deviation of 2.5 in., the probability that Ferndale will receive more than 36 inches of rain this year is **0.4207** or 42.07%.

51. It is a good idea to begin by drawing the normal curve and shading the appropriate region:

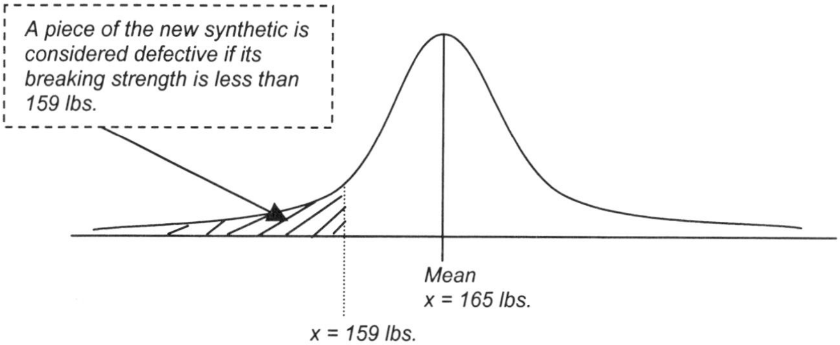

We are given that $\mu = 165$ lbs. Remember that the standard deviation is the square root of the variance. Hence, since the variance is 9 lbs, $\sigma = 3$ lbs.
For $x = 159$ lbs, the z-score is

$$z = \frac{x - \mu}{\sigma} = \frac{159 - 165}{3} = -2.$$

In Appendix F (p. 901), we find that 0.4772 is the proportion of the area between the mean and $z = -2$. Due to the symmetry of the normal curve, we know 50% of the area under the curve is to the left of the mean. So, since 47.72% is between the mean and $z = -2$, the area to the left of $z = -2$ is 0.0228 (i.e., $0.5000 - 0.4772$).

Therefore, based upon the assumption that the breaking strength of this synthetic material is normally distributed with the above-mentioned mean and standard deviation, we conclude that the probability of a single, randomly selected piece of this material being defective is **0.0228** or 2.28%.

53. *[__General Reminder__: Drawing the normal curve with labeling/shading appropriate to the problem you are solving can help alleviate confusion. To review how this is done, refer to textbook pp. 694-695 (see Example 4) and this manual's solutions to Problems 13, 17, 25b, 35, 37, and 51.]*

We are given that $\mu = 0.4$ inches. Remember that the standard deviation is the square root of the variance. Hence, since the variance is 0.0004 in., $\sigma = 0.02$ in.

For $x = 0.41$ in., the z-score is

$$z = \frac{x-\mu}{\sigma} = \frac{0.41-0.4}{0.02} = 0.5.$$

In Appendix F (p. 901), we find that 0.1915 is the proportion of the area between the mean and $z = 0.5$. Due to the symmetry of the normal curve, we know 50% of the area under the curve is to the right of the mean. So, since 19.15% is between the mean and $z = 0.5$, the area to the right of $z = 0.5$ is the difference $0.5000 - 0.1915$, which equals 0.3085.

Thus, based upon the assumption that the diameter of a pipe is normally distributed with the above-mentioned mean and standard deviation, we conclude that the probability of a randomly selected pipe having a diameter greater than 0.41 in. is **0.3085** or 30.85%.

55. Because the variance is 0.04 oz and the standard deviation is (by definition) the square root of the variance, it follows that $\sigma = 0.2$ oz.

a) From our familiarity with the characteristics of the normal curve (in particular, its symmetry), we know 50% of the area under the curve is to the left of the mean. So, given that the mean weight of packages of Toys Galore Cereal is 12 oz and given that the weights are normally distributed, we expect **50%** of the packages to weigh less than 12 oz.

b) Review, if necessary, textbook Figure 10.34 (p. 691). We recognize 2.3% as the percent of the data (in a normal distribution) that is greater than 2 standard deviations above the mean (i.e., z = 2). From this manual's solution to Problem 43, we know $z = \dfrac{x-\mu}{\sigma}$ implies $x = \mu + z \cdot \sigma$. Therefore, given that $\mu = 12$ oz, $\sigma = 0.2$ oz, and $\text{z} = 2$, it follows that the weight exceeded by 2.3% of the cereal packages is

$$x = \mu + z \cdot \sigma = 12 + 2(0.2) = \boxed{\textbf{12.4 oz}}.$$

57. With the aid of a calculator, you will obtain the following approximations for y:

x	−4	−3	−2	−1	0
y	0.00013	0.00443	0.05399	0.24197	0.39894

x	1	2	3	4
y	0.24197	0.05399	0.00443	0.00013

Three observations are worthy of mention:

(i) the largest y-value occurs when $x = 0$;

(ii) the y-values become infinitesimally small as the x-values get farther from zero; and

(iii) the y-values are equal for $x = \pm 1$, and again equal for $x = \pm\, 2$, and so on.

In light of your knowledge of the characteristics of the normal curve, all three of these observations should make (intuitive) sense to you. Make sure they do.

59. With $\mu = 50$ and $\sigma = 10$, the equation for the normal curve becomes

$$y = \frac{e^{-(x-50)^2/200}}{10\sqrt{2\pi}}.$$

With the aid of a calculator, you will obtain the following approximations for y:

x	*20*	*30*	*40*	*50*	*60*	*70*	*80*
y	*0.0004*	*0.0054*	*0.0242*	*0.0399*	*0.0242*	*0.0054*	*0.0004*

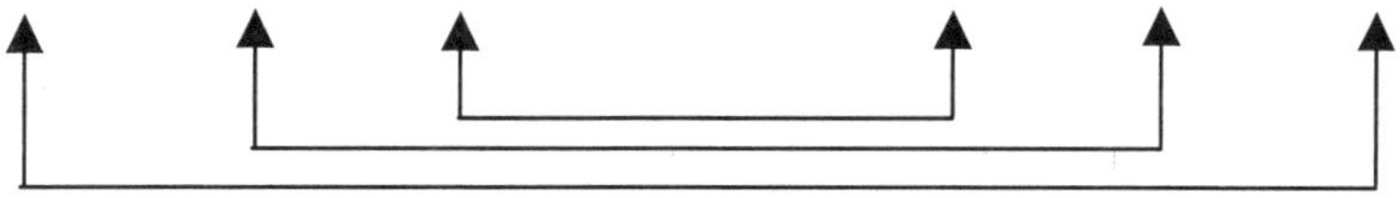

As was encouraged in our solution to Problem 57, please make sure you understand why these y-values are what they are. That is,

(i) Why is the largest y-value associated with $x = 50$?

(ii) Why do the y-values become infinitesimally small as the x-values get farther from 50?

(iii) Why are the y-values equal for $x = 40$ and $x = 60$, and again equal for $x = 30$ and $x = 70$, and also equal for $x = 20$ and $x = 80$?

While contemplating these questions, you may wish to convert these x-values to z-scores.

Problem Set 10.5

1. **Answers vary**. ***Correlation*** is a number which measures (quantifies) the relationship between two variables. In this text, we consider only linear relationships, which means the points, when graphed, approximate a straight line. In the case of linear relationships, correlation is measured by r, the *linear correlation coefficient* (also known as the *Pearson product moment correlation coefficient*), defined on textbook pages 700-702.
As your text's author eloquently points out (on the bottom of p. 703), a "significant" linear correlation means only that the variables are related — not that there is a cause-and-effect relationship. Even if there is a cause-and-effect relationship, correlation analysis cannot identify which variable is the cause and which is the effect.

3. **Wording varies**. To determine whether there is a statistically significant linear correlation between two variables, we first obtain the value of the linear correlation coefficient, r, and then compare the absolute value of r to Table 10.9 (textbook p. 702). If $|r|$ is greater than the given table value, then you may assume a correlation exists (at the given significance level) between the variables.

5. **Wording varies**. There appears to be a **strong positive correlation** between fat intake and life expectancy.

7. **No**. From Table 10.9 (textbook p. 702), we see the critical value is 0.765 (for $\alpha = 0.01$ and $n = 10$). Since $r = 0.7 < 0.765$, there is no significant linear correlation at the 1% significance level.

9. **Yes**. From Table 10.9 (textbook p. 702), we see the critical value is 0.361 (for $\alpha = 0.05$ and $n = 30$). Since $r = 0.4 > 0.361$, there is a significant linear correlation at the 5% significance level.

11. **No**. From Table 10.9 (textbook p. 702), we see the critical value is 0.430 (for $\alpha = 0.01$ and $n = 35$). Since $|r| = 0.413 < 0.430$, there is no significant linear correlation at the 1% significance level.

13. **Yes**. From Table 10.9 (textbook p. 702), we see the critical value is 0.505 (for $\alpha = 0.01$ and $n = 25$). Since $r = 0.521 > 0.505$, there is a significant linear correlation at the 1% significance level.

15. **Yes**. From Table 10.9 (textbook p. 702), we see the critical value is 0.312 (for $\alpha = 0.05$ and $n = 40$). Since $r = 0.416 > 0.312$, there is a significant linear correlation at the 5% significance level.

17. **No**. From Table 10.9 (textbook p. 702), we see the critical value is 0.765 (for $\alpha = 0.01$ and $n = 10$). Since $|r| = 0.56 < 0.765$, there is no significant linear correlation at the 1% significance level.

Reminder: *Many calculators include two-variable (bivariate) statistical functions. After inputting the paired data, one can instantaneously obtain — without having to endure any of the tedious computations demonstrated below — the linear correlation coefficient r, the slope (m) and y-intercept (b) of the least squares (regression) line, and assorted other statistical information. Check your calculator user's manual (or, if applicable, its accompanying "quick reference" card) for specific details.*

19. We begin with a scatter diagram (i.e., plot the four given points):

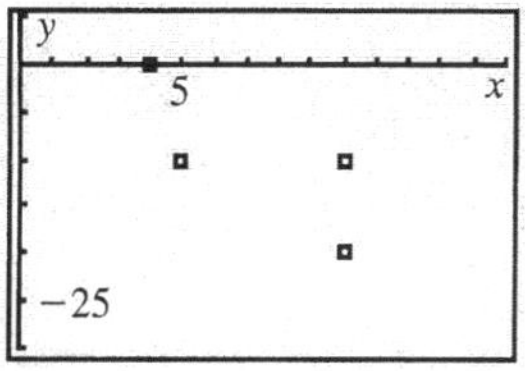

Based upon the positioning of these points, ask yourself: do you expect a negative or positive value for the correlation coefficient? If you are unsure, please take a moment to review textbook Figure 10.43 (p. 701).

Next, if you do not have a calculator that automatically does all these computations for you, then here is how you must go about determining r:

$n = 4$ $\quad \Sigma y = -40$ $\quad \Sigma x^2 = 241$

$\Sigma x = 29$ $\quad \Sigma xy = -350$ $\quad \Sigma y^2 = 600$

$$r = \frac{n\Sigma xy - (\Sigma x)(\Sigma y)}{\sqrt{n(\Sigma x^2) - (\Sigma x)^2}\sqrt{n(\Sigma y^2) - (\Sigma y)^2}}$$

$$= \frac{4(-350) - (29)(-40)}{\sqrt{4(241) - (29)^2}\sqrt{4(600) - (-40)^2}}$$

$$\approx \boxed{-0.765}$$

From Table 10.9 (on textbook p. 702), we see that the critical value (for $n = 4$) at the 5% level is 0.950. Since $|r| = 0.765$ is not greater than 0.950, we report that there is **no significant linear correlation** between the values of x and y.

21. Notice that the scatter diagram's points *appear* to depict a negative linear trend.

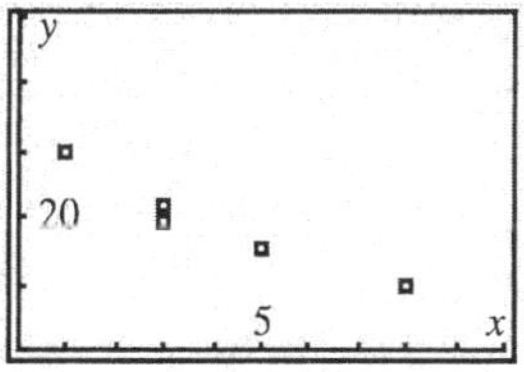

$n = 5$ $\quad \Sigma y = 96$ $\quad \Sigma x^2 = 108$

$\Sigma x = 20$ $\quad \Sigma xy = 308$ $\quad \Sigma y^2 = 2{,}070$

$$r = \frac{n\Sigma xy - (\Sigma x)(\Sigma y)}{\sqrt{n(\Sigma x^2) - (\Sigma x)^2}\sqrt{n(\Sigma y^2) - (\Sigma y)^2}}$$

$$= \frac{5(308) - (20)(96)}{\sqrt{5(108) - (20)^2}\sqrt{5(2{,}070) - (96)^2}}$$

$$\approx \boxed{-0.954}$$

From Table 10.9 (on textbook p. 702), we see that the critical value (for $n = 5$) at the 5% level is 0.878. Since $|r| = 0.954$ is greater than 0.878, we conclude there is a **significant linear correlation at the 5% level**. However, since $|r| = 0.954$ is not greater than 0.959 (i.e., the critical value at the 1% level), there is not a statistically significant correlation at the 1% level.

23. Once again, the scatter diagram's points _appear_ to depict a negative linear trend.

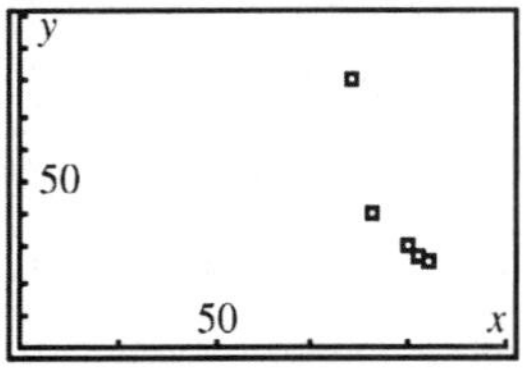

$n = 5$ $\quad \Sigma y = 203$ $\quad \Sigma x^2 = 46{,}754$

$\Sigma x = 482$ $\quad \Sigma xy = 18{,}881$ $\quad \Sigma y^2 = 10{,}309$

$$r = \frac{n\Sigma xy - (\Sigma x)(\Sigma y)}{\sqrt{n(\Sigma x^2) - (\Sigma x)^2}\sqrt{n(\Sigma y^2) - (\Sigma y)^2}}$$

$$= \frac{5(18{,}881) - (482)(203)}{\sqrt{5(46{,}754) - (482)^2}\sqrt{5(10{,}309) - (203)^2}}$$

$$\approx \boxed{-0.890}$$

From Table 10.9 (on textbook p. 702), we see that the critical value (for $n = 5$) at the 5% level is 0.878. Since $|r| = 0.890$ is greater than 0.878, we conclude there is a **significant linear correlation at the 5% level**. However, since $|r| = 0.890$ is not greater than 0.959 (which is the critical value at the 1% level), there is not a statistically significant correlation at the 1% level.

25. Problems 25-30 are done either by using the formulas given on textbook p. 704 or by using a calculator with built-in linear regression capability.
From our previous work in Problem 19, we know: $n = 4$, $\Sigma x = 29$, $\Sigma y = -40$, $\Sigma x^2 = 241$, and $\Sigma xy = -350$.
So, it follows that the slope of the regression line is:

$$m = \frac{n\Sigma xy - (\Sigma x)(\Sigma y)}{n(\Sigma x^2) - (\Sigma x)^2} = \frac{4(-350) - (29)(-40)}{4(241) - (29)^2} = -\tfrac{80}{41} \approx -1.95121951 \approx -1.95.$$

Using the ("unrounded") value for m, we obtain the y-intercept of the regression line:

$$b = \frac{\Sigma y - m(\Sigma x)}{n} = \frac{(-40) - \left(-\frac{80}{41}\right)(29)}{4} = \tfrac{170}{41} \approx 4.14634146 \approx 4.146.$$

So, the equation of the regression line is $\boxed{\mathbf{y' = -1.95x + 4.146}}$.

27. From our previous work in Problem 21, we know: $n = 5$, $\Sigma x = 20$, $\Sigma y = 96$, $\Sigma x^2 = 108$, and $\Sigma xy = 308$. So,

$$m = \frac{n\Sigma xy - (\Sigma x)(\Sigma y)}{n(\Sigma x^2) - (\Sigma x)^2} = \frac{5(308) - (20)(96)}{5(108) - (20)^2} = -\tfrac{19}{7} = -2.7\overline{142857} \approx -2.7143$$

$$b = \frac{\Sigma y - m(\Sigma x)}{n} = \frac{(96) - \left(-\frac{19}{7}\right)(20)}{5} = \tfrac{1052}{35} \approx 30.0571$$

Thus, the equation of the regression line is $\boxed{\mathbf{y' = -2.7143x + 30.0571}}$.

29. From our previous work in Problem 23, we know: $n = 5$, $\Sigma x = 482$, $\Sigma y = 203$, $\Sigma x^2 = 46{,}754$, and $\Sigma xy = 18{,}881$. So,

$$m = \frac{n\Sigma xy - (\Sigma x)(\Sigma y)}{n(\Sigma x^2) - (\Sigma x)^2} = \frac{5(18{,}881) - (482)(203)}{5(46{,}754) - (482)^2} = -\tfrac{1{,}147}{482} \approx -2.380$$

$$b = \frac{\Sigma y - m(\Sigma x)}{n} = \frac{(203) - \left(-\frac{1{,}147}{482}\right)(482)}{5} = \tfrac{1{,}350}{5} = 270$$

Thus, the equation of the regression line is $\boxed{\mathbf{y' = -2.380x + 270}}$.

31. Observe that graphs A, D, and E illustrate a <u>positive</u> correlation. Accordingly, they must be paired (in some order) with linear equations in which $m > 0$; namely, those given in Problems 31-33. Out of the equations in Problems 31-33, notice that $r = 0.9$ is the strongest positive correlation. Out of graphs A, D, and E, the strongest positive

correlation is exhibited in Graph D. Therefore, the given equation/correlation ($y' = 0.6x + 2$ and $r = 0.9$) matches **graph D**.
Additionally, please note that out of Problems 31-33, the equation in this problem has the largest slope ($m = 0.6$). Consequently, its regression line must be the steepest line among A, D, and E, which happens to be graph D.
[Reminder: If you need to review the concepts associated with graphing lines, see Section 11.1.]

33. Employ the same reasoning as explained in our solution to Problem 31.
Among Problems 31-33, notice that (i) $r = 0.5$ is the weakest positive correlation, and (ii) $m = 0.4$ is the smallest positive slope. The only graphs that exhibit positive correlations are graphs A, D, and E. Of these, the points in graph A exhibit the most scattering (i.e., weakest correlation) and its regression line is the least steep (i.e., smallest slope). Therefore, the given equation/correlation ($y' = 0.4x + 2$ and $r = 0.5$) matches **Graph A**.
[Reminder: If you need to review the concepts associated with graphing lines, see Section 11.1.]

35. Employ the same reasoning as explained in our solutions to Problems 31 and 33.
Among the correlations given in Problems 34-36, notice that (i) $r = -0.6$ is between Problem 36's correlation of $r = -0.8$ (which is the strongest negative correlation among the three) and Problem 34's correlation of $r = -0.4$ (the weakest negative correlation among the three). Also, among the equations for these three problems, observe that the given slope of $m = -0.5$ is steeper than the one in Problem 34 ($m = -0.4$) but not as steep as the slope in Problem 36 ($m = -0.7$). The only graphs that exhibit negative correlations are graphs B, C, and F. Therefore, based upon this comparative analysis of these correlations and slopes, we match the given equation/correlation ($y' = -0.5x + 2$ and $r = -0.6$) with **graph C**.
[Reminder: If you need to review the concepts associated with graphing lines, see Section 11.1.]

37. We first determine the slope m and the y-intercept b:

$$m = \frac{n\Sigma xy - (\Sigma x)(\Sigma y)}{n(\Sigma x^2) - (\Sigma x)^2} = \frac{8(240) - (40)(48)}{8(210) - (40)^2} = \frac{0}{80} = 0$$

$$b = \frac{\Sigma y - m(\Sigma x)}{n} = \frac{(48) - (0)(40)}{8} = 6$$

So, since the slope is $m = 0$ and the y-intercept is $b = 6$, the equation for the least squares line is $\boxed{y' = 6}$.
As for the value of the linear correlation coefficient, we obtain:

$$r = \frac{n\Sigma xy - (\Sigma x)(\Sigma y)}{\sqrt{n(\Sigma x^2) - (\Sigma x)^2}\sqrt{n(\Sigma y^2) - (\Sigma y)^2}} = \frac{8(240) - (40)(48)}{\sqrt{8(200) - (40)^2}\sqrt{8(288) - (48)^2}} = \boxed{0}.$$

This value tells us that there is no linear relationship whatsoever between x and y. Consequently, the best (safest) predicted value of y is the sample mean $\bar{y}$. We encourage you to take a moment to compute $\bar{y}$ (from the given information) and then compare it to the equation of the least squares line determined above.

39. We first determine the slope m and the y-intercept b:

$$m = \frac{n\Sigma xy - (\Sigma x)(\Sigma y)}{n(\Sigma x^2) - (\Sigma x)^2} = \frac{6(322) - (36)(48)}{6(250) - (36)^2} = \frac{204}{204} = 1$$

$$b = \frac{\Sigma y - m(\Sigma x)}{n} = \frac{(48) - m(36)}{6} = \frac{12}{6} = 2$$

So, since the slope is $m = 1$ and the y-intercept is $b = 2$, the equation for the least squares line is $\boxed{y' = x + 2}$.

As for the value of the linear correlation coefficient, we obtain:

$$r = \frac{n\Sigma xy - (\Sigma x)(\Sigma y)}{\sqrt{n(\Sigma x^2) - (\Sigma x)^2}\sqrt{n(\Sigma y^2) - (\Sigma y)^2}} = \frac{6(322) - (36)(48)}{\sqrt{6(250) - (36)^2}\sqrt{6(418) - (48)^2}} = \boxed{1}.$$

A correlation coefficient value of 1 tells us that there is a perfect positive linear relationship between x and y. As such, the least squares line contains every one of the data points.

41. Like Problems 19-24, this problem may be done either by using the formula for the linear correlation coefficient or by using a calculator with built-in linear correlation/regression capability.

In this problem, the order of the variables does <u>not</u> matter. If we choose to let x represent beer consumption (from Example 5, p. 703) and y represent wine consumption, we obtain:

Year	***1988***	***1989***	***1990***	***1991***	***1993***	***1995***
x	24.7	25.4	24	23.1	22.8	22.6
y	2.24	2.09	2.05	1.85	1.74	1.80

$n = 6$ $\quad \Sigma y = 11.77$ $\quad \Sigma x^2 = 3395.46$

$\Sigma x = 142.6$ $\quad \Sigma xy = 280.701$ $\quad \Sigma y^2 = 23.2783$

$$r = \frac{n\Sigma xy - (\Sigma x)(\Sigma y)}{\sqrt{n(\Sigma x^2) - (\Sigma x)^2}\sqrt{n(\Sigma y^2) - (\Sigma y)^2}}$$

$$= \frac{6(280.701) - (142.6)(11.77)}{\sqrt{6(3{,}395.46) - (142.6)^2}\sqrt{6(23.2783) - (11.77)^2}}$$

≈ 0.883027995

$\boxed{r \approx \mathbf{0.8830}}$

From Table 10.9 (on textbook p. 702), we see that the critical value (for $n = 6$) at the 5% level is 0.811 and the critical value at the 1% level is 0.917. Since $|r| = 0.8830$ is greater than 0.811 but not greater than 0.917, we conclude there is a **significant linear correlation at the 5% level** between beer drinking and wine drinking.

43. Assume x is the difficulty level and y is the time. If you do not have a calculator that automatically does the computation for you, then here is how you must go about determining r:

$n = 8$ $\qquad \Sigma y = 115$ $\qquad \Sigma x^2 = 109$

$\Sigma x = 27$ $\qquad \Sigma xy = 432$ $\qquad \Sigma y^2 = 1{,}805$

$$r = \frac{n\Sigma xy - (\Sigma x)(\Sigma y)}{\sqrt{n(\Sigma x^2) - (\Sigma x)^2}\sqrt{n(\Sigma y^2) - (\Sigma y)^2}}$$

$$= \frac{8(432) - (27)(115)}{\sqrt{8(109) - (27)^2}\sqrt{8(1{,}805) - (115)^2}}$$

≈ 0.8420753583

$\boxed{r \approx \mathbf{0.8421}}$

From Table 10.9 (on textbook p. 702), we see that the critical value (for $n = 8$) at the 1% level is 0.834. Since $|r| = 0.8421$ is greater than 0.834, we conclude there is a **significant linear correlation at the 1% level**.

45. There are two approaches, either use a calculator with built-in linear regression capability or apply the formulas given on textbook p. 704.
First, with the use of a graphing calculator, we obtain the following screen displays:

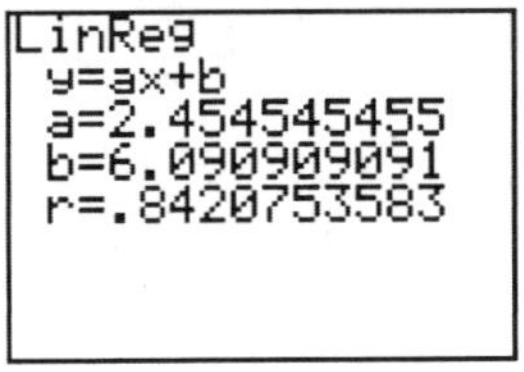

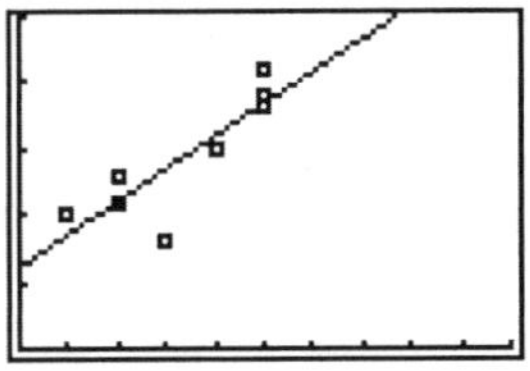

Alternatively, from our previous work in Problem 43, we know: $n = 8$, $\Sigma x = 27$, $\Sigma y = 115$, $\Sigma x^2 = 109$, and $\Sigma xy = 432$. So,

$$m = \frac{n\Sigma xy - (\Sigma x)(\Sigma y)}{n(\Sigma x^2) - (\Sigma x)^2} = \frac{8(432) - (27)(115)}{8(109) - (27)^2} = \tfrac{27}{11} = 2.\overline{45} \approx 2.4545$$

$$b = \frac{\Sigma y - m(\Sigma x)}{n} = \frac{(115) - \left(\frac{27}{11}\right)(27)}{8} = \tfrac{67}{11} = 6.\overline{09} \approx 6.0909$$

Via either approach, we obtain the following regression line equation:

$$\boxed{y' = \mathbf{2.4545}x + \mathbf{6.0909}}.$$

47. Once again, there are two approaches: either use a calculator with built-in linear correlation/regression capability or apply the formula given on textbook p. 701.
First, after inputting the data and pressing the appropriate key(s), we automatically obtain the following output:

```
LinReg
 y=ax+b
 a=.46761113
 b=19.23530709
 r=.3584296566
```

Alternatively, if you do not have a suitably equipped calculator, then here is how you must go about determining r:

$n = 15$	$\Sigma y = 405.9$	$\Sigma x^2 = 4{,}593$
$\Sigma x = 251$	$\Sigma xy = 6{,}975.8$	$\Sigma y^2 = 11{,}652.43$

$$r = \frac{15(6{,}975.8) - (251)(405.9)}{\sqrt{15(4{,}593) - (251)^2}\sqrt{15(11{,}652.43) - (405.9)^2}}$$

$$\approx 0.358429657$$

$$\boxed{r \approx \mathbf{0.358}}$$

From Table 10.9 (on textbook p. 702), we see that the critical value (for $n = 15$) at the 5% level is 0.514. Since $|r| = 0.358$ is not greater than 0.514, we conclude there is **no significant linear correlation** (based upon this small sample of 15 persons) between formal full-time education and annual salary.

49. Once again, there are two approaches: either use a calculator with built-in linear correlation/regression capability or apply the formula given on textbook p. 701.
First, after inputting the data and pressing the appropriate key(s), we automatically obtain the following output:

```
LinReg
 y=ax+b
 a=⁻6.186388358
 b=87.17895046
 r=⁻.9356261053
```

Alternatively, if you do not have a suitably equipped calculator, then here is how you go about determining r:

$n = 16$ $\quad \Sigma y = 251$ $\quad \Sigma x^2 = 2{,}158.01$

$\Sigma x = 184.9$ $\quad \Sigma xy = 2{,}769.1$ $\quad \Sigma y^2 = 4{,}867$

$$r = \frac{16(2{,}769.1) - (184.9)(251)}{\sqrt{16(2{,}158.01) - (184.9)^2}\sqrt{16(4{,}867) - (251)^2}}$$

$$\approx -0.935626105$$

$\boxed{r \approx -0.936}$

From Table 10.9 (on textbook p. 702), we see that the critical value (for $n = 16$) at the 1% level is 0.623. Since $|r| = 0.936$ is greater than 0.623, we conclude there is a **significant linear correlation at the 1% level** between the current interest rate and the number of mortgage applicants.

Based on your intuition alone, you should not be surprised by this strong negative correlation. As interest rates fall, the number of applicants rises; and conversely, as interest rates increase, the number of applicants decreases.

51. Once again, there are two approaches: either use a calculator with built-in linear regression capability or apply the formulas given on textbook p. 704.

First, with the use of a graphing calculator, we obtain the following screen displays:

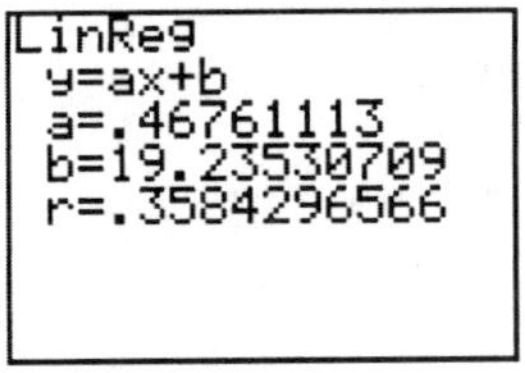

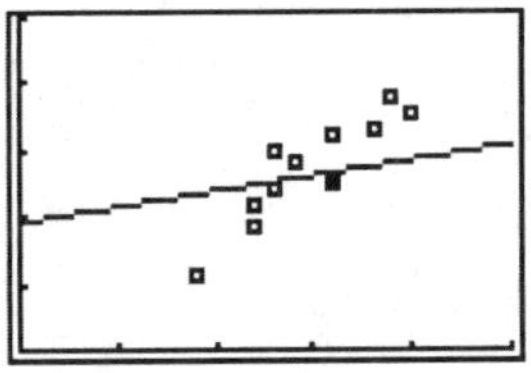

Alternatively, from our previous work in Problem 47, we know: $n = 15$, $\Sigma x = 251$, $\Sigma y = 405.9$, $\Sigma x^2 = 4{,}593$, and $\Sigma xy = 6{,}975.8$. So,

$$m = \frac{n\Sigma xy - (\Sigma x)(\Sigma y)}{n(\Sigma x^2) - (\Sigma x)^2} = \frac{15(6{,}975.8) - (251)(405.9)}{15(4{,}593) - (251)^2} \approx 0.46761113$$

$$b = \frac{\Sigma y - m(\Sigma x)}{n} = \frac{(405.9) - m(251)}{15} \approx 19.23530709$$

Via either approach, we obtain the following regression line equation:

$\boxed{y' = 0.468x + 19.235}$.

53. Once again, there are two approaches: either use a calculator with built-in linear regression capability or apply the formulas given on textbook p. 704.

First, with the use of a graphing calculator, we obtain the following screen displays:

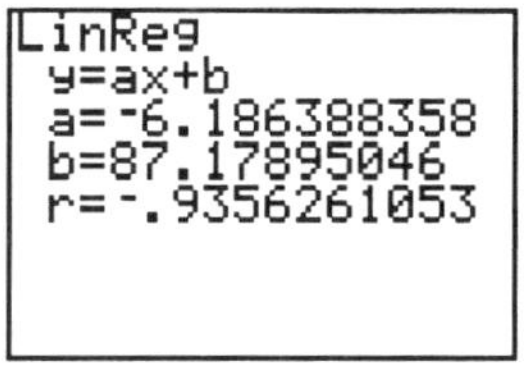

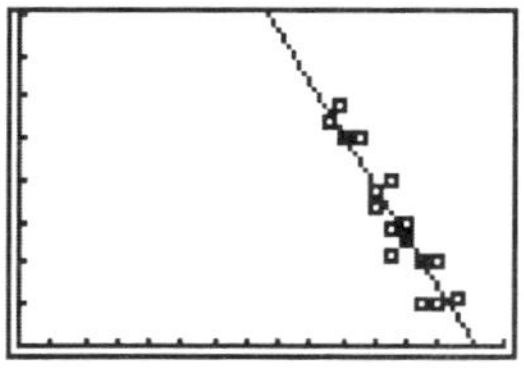

Alternatively, from our previous work in Problem 49, we know: $n = 16$, $\Sigma x = 184.9$, $\Sigma y = 251$, $\Sigma x^2 = 2{,}158.01$, and $\Sigma xy = 2{,}769.1$. So,

$$m = \frac{n\Sigma xy - (\Sigma x)(\Sigma y)}{n(\Sigma x^2) - (\Sigma x)^2} = \frac{16(2{,}769.1) - (184.9)(251)}{16(2{,}158.01) - (184.9)^2} \approx -6.186388358$$

$$b = \frac{\Sigma y - m(\Sigma x)}{n} = \frac{(251) - m(184.9)}{16} \approx 87.178950463$$

Via either approach, we obtain the following regression line equation:

$$\boxed{y' = -6.186x + 87.179}.$$

55. Assume x is the time in (decimal) hours and y is the distance in miles. As the footnote shown on the bottom of textbook p. 707 explains, we should first convert the given times from hours/minutes to decimal hours. Upon doing so (and rounding to two decimal places), we have the following data for Scott Hirst:

Aid Station	***#1***	***#2***	***#3***	***#4***	***#5***	***#6***
x *(in hours)*	*2.95*	*5.05*	*7.37*	*9.43*	*10.17*	*11.27*
y *(in miles)*	*12.2*	*19.0*	*28.0*	*33.4*	*36.6*	*43.8*

Like Problem 54, to predict Scott Hirst's arrival time at the finish of the race, we need the equation of the line of best fit (i.e., the regression line). We obtain this equation in the same manner as demonstrated previously (see assorted solutions above and/or text Examples 6-7). Leaving the details for you, we obtain the equation $y' = 3.6x + 1.04$ with a strong positive correlation of $r = 0.993$.

Now, in terms of this equation, the race's distance of 100.1 miles represents a y-value. To estimate (predict) Hirst's arrival time, we substitute 100.1 for y and solve for x:

$$100.1 = 3.6x + 1.04$$

$$99.06 = 3.6x$$

$$x = 27.51\overline{6}$$

This is equivalent to 27 hours, 31 minutes [Note: $0.51\overline{6}$ hr × 60 min/hr = 31 min]. Therefore, **the predicted time of his arrival is 27:31, which is 4 hours and 29 minutes less than his actual time**. Clearly, his pace slowed during the latter stages of the race, most likely due to exhaustion.

57. This problem is similar to its three predecessors (Problems 54, 55, and 56). Assume x is the time in (decimal) hours and y is the distance in miles. As the footnote shown on the bottom of textbook p. 707 explains, we should first convert the given times from hours/minutes to decimal hours. Upon doing so (and rounding to two decimal places), we have the following data for John DeWalt:

Aid Station	***#1***	***#2***	***#3***	***#4***	***#5***	***#6***
x *(in hours)*	*3.88*	*6.93*	*10.6*	*14.2*	*15.43*	*17.83*
y *(in miles)*	*12.2*	*19.0*	*28.0*	*33.4*	*36.6*	*43.8*

To predict DeWalt's arrival time at the finish of the race, we need the equation of the line of best fit (i.e., the regression line). Using the same procedure as demonstrated previously (see assorted solutions above and/or text Examples 6-7), we obtain the equation $y' = 2.17x + 3.87$ with a strong positive correlation of $r = 0.996$.

Now, in terms of this equation, the race's distance of 100.1 miles represents a y-value. We predict DeWalt's arrival time by substituting 100.1 for y and solving for x:

$$100.1 = 2.17x + 3.87$$
$$96.23 = 2.17x$$
$$x \approx 44.345622$$

This is equivalent to 44 hours, 21 minutes [Note: 0.345622 hr × 60 min/hr ≈ 20.7 min]. Therefore, **the predicted time of his arrival is 44:21, which is 3 hours and 29 minutes less than his actual time**. Like the first-place runner Scott Hirst, the last-place finisher, John DeWalt, ran more slowly during the latter stages of the race, most likely due to fatigue.

59. Once again, we begin by heeding the text's recommendation (see the footnote on p. 708 that accompanies this problem); namely, we first convert the given times from hours/minutes to decimal hours.

Since our sole purpose is to determine the correlation, the order of the variables does not matter. If we choose to let x represent age (in years) and y represent finishing time (in decimal hours, rounded to two decimal places), we obtain:

$$n = 36 \qquad \Sigma y = 1{,}499.91 \qquad \Sigma x^2 = 65{,}847$$
$$\Sigma x = 1{,}513 \qquad \Sigma xy = 63{,}579.39 \qquad \Sigma y^2 = 63{,}391.5471$$

$$r = \frac{36(63{,}579.39) - (1{,}513)(1{,}499.91)}{\sqrt{36(65{,}847) - (1{,}513)^2}\sqrt{36(63{,}391.5471) - (1{,}499.91)^2}}$$
$$\approx 0.379976$$

$$\boxed{r \approx \mathbf{0.380}}$$

[Note: To avoid performing this laborious computation by hand, we recommend you either use a calculator (with correlation/regression capability) or computer software.]

From Table 10.9 (on textbook p. 702), we see (for $n = 35$) that the critical value at the 5% level is 0.335 and the critical value at the 1% level is 0.430. Additionally, we notice that the critical value (for $n = 40$) at the 1% level is 0.402. We know the critical values for $n = 36$ must be between those for $n = 35$ and $n = 40$. Therefore, since $|r| = 0.380$ is greater than 0.335 (for $\alpha = 0.05$ and $n = 35$) but not greater than 0.402 ($\alpha = 0.01$ and $n = 40$), we conclude there is a **significant linear correlation at the 5% level (but not at the 1% level)** between age and running time.

Problem Set 10.6

1. **Answers vary**. Consider these three definitions of the word ***statistics*** (the first two are from *Webster's New World College Dictionary*, the third is from the *Random House College Dictionary*): (1) facts or data of a numerical kind, assembled, classified, and tabulated so as to present significant information about a given subject; (2) the calculation, description, manipulation, and interpretation of the mathematical attributes of sets or populations too numerous or extensive for exhaustive measurements; (3) the science that deals with the collection, classification, analysis, and interpretation of numerical facts or data, and that, by use of mathematical theories of probability, imposes order and regularity on aggregates of more or less disparate elements.

3. **Answers vary**. The ***fallacy of the exception*** occurs when one forms a conclusion based upon a particular case, which may (or may not) be "typical." Text Example 3 (pp. 712-713) provides an illustration of this fallacy.

5. **Choice B**. Sprint wants to survey Pacific Bell's customers, so the target population is Pacific Bell's customers and choice B is the only procedure that targets this population.

7. **Choice C**. The target population is the local party's members.

9. **Choice D**. Since the target population is the union members, we eliminate choices A and C. In the case of choice B, although the members of the union's negotiating committee are members, they are not a representative (unbiased) random sample of union members.

11. **Choice A**. The target population is the students of a certain college. We refer you to textbook page 709 (above Example 1). Of the four choices, option A is the only procedure that is sufficiently random and takes into account the intended target population.

13. **Choice C**. Since the target population is the customers of a particular supermarket, choice A is eliminated. Also, as noted on textbook page 709, asking people to voluntary mail or phone in their preferences is not a random (unbiased) sampling procedure, so we disregard choices B and D.

15. Survey AT&T's customers by randomly choosing from a list of AT&T's customers.

17. Survey a selection of people whose names are randomly chosen from a local list of registered voters who are party members.

19. Survey a selection of people whose names are randomly chosen from a list of all union members.

21. Survey a random selection of students chosen from a list of all students.

23. Survey a selection of people whose names are randomly chosen from a list of all customers.

25. For additional details, revisit textbook pp. 710-712 (including Example 2).
I would **not accept** the conclusion. The four possibilities are:

- You accept the claim that 72 is the mean, and the actual mean is 72.
- You accept the claim that 72 is the mean, and the actual mean is not 72. This is known as ***Type II error*** (accepting a hypothesis when it is actually false).
- You reject the claim that 72 is the mean, and the actual mean is not 72.
- You reject the claim that 72 is the mean, and the actual mean is 72. This is known as ***Type I error*** (rejecting a hypothesis when it is actually true).

27. **Survey results vary**.

29. **Surveys and their results vary**.

Chapter 10 Review Questions

1. a)

Outcome	Tally	Frequency
Heads	𝍸 𝍸 𝍸 |||	*18*
Tails	𝍸 𝍸 𝍸 𝍸 ||	22

b)

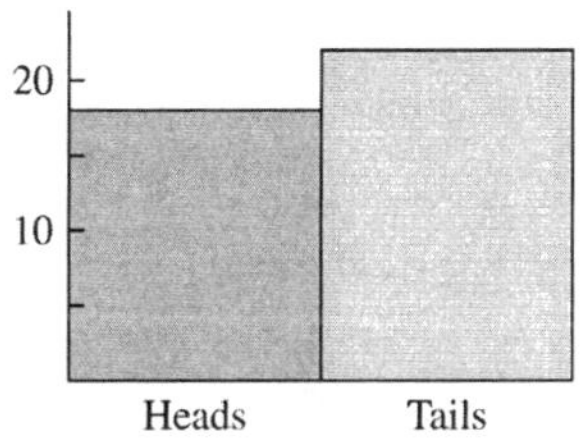

2. a)

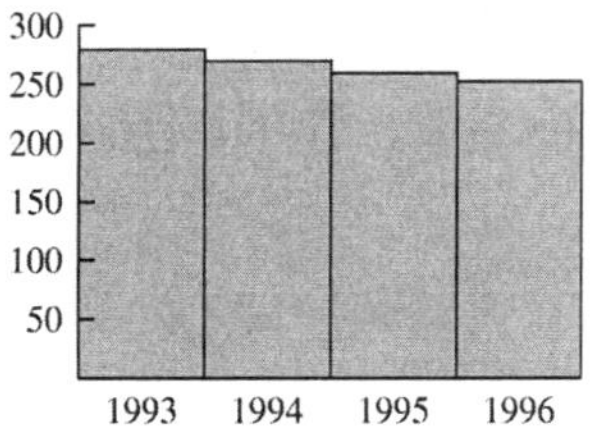

b) If you want to diminish the size (and relative importance) of the spending decrease, you should choose a large scale for the y-axis.

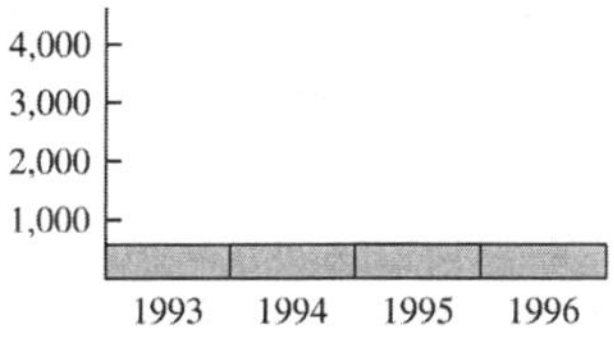

c) If you want to exaggerate the size (and relative importance) of the spending decrease, you should choose a small scale for the y-axis.

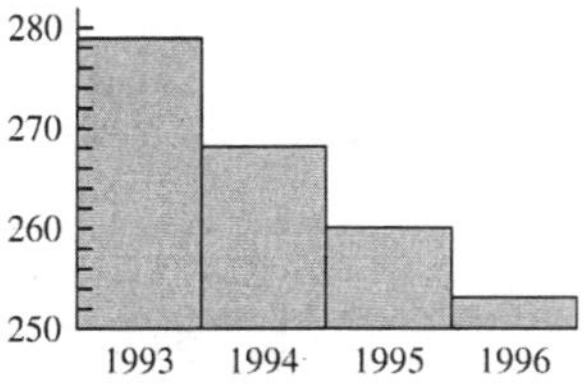

d) **Answers vary**. As demonstrated in parts (b) and (c), impressions can be greatly influenced by using a misleading scale — or, even worse, no scale at all.

3. A store manager is concerned with stocking the most popular (best selling) items. Consequently, in making such a decision, the **mode** is the most useful measure of central tendency.

The values of the mean, median, and mode are:

Mean: $\bar{x} = \dfrac{\Sigma(w \cdot x)}{\Sigma w} = \dfrac{5(6)+10(10)+35(12)}{50} = \dfrac{550}{50} = \boxed{11 \text{ oz}}$

Median: Always make sure the data are in order (from low to high) before determining the median. That is,

$$\underbrace{6, 6, 6, 6, 6,}_{\text{5 values}}\ \underbrace{10, 10, 10, \ldots\ldots, 10,}_{\text{10 values}}\ \underbrace{12, 12, 12, \ldots\ldots\ldots\ldots\ldots, 12}_{\text{35 values}}$$

The median is defined as the middlemost data value. Since this data set consists of an even number of values (namely, 50), the median is the mean of the middle two values (in this case, the 25th and 26th). Both of these values are 12 oz, so the median is **12 oz**.

Mode: The mode is, by definition, the most frequently occurring data value. Since 35 cases of the 12 oz size were sold, **12 oz** is the mode.

4. The values of the mean, median, and mode are:

Mean: $\bar{x} = \dfrac{\Sigma x}{n} = \dfrac{395}{5} = \boxed{79}$

Median: The five data values are already arranged in order (from smallest to largest). So, the median is the middle value, namely

72, 73, **74**, 85, 91.

Mode: The mode is, by definition, the most frequently occurring data value. Since no value occurs more than once, there is **no mode**.

The **mean** is most representative of the student's scores. It is the only measure of the three that takes into consideration all of the student's scores. A student's overall (average) performance should be positively affected by high scores and negatively affected by low scores.

5. The values of the mean, median, and mode are:

Mean: $\bar{x} = \dfrac{\Sigma x}{n} = \dfrac{2{,}460{,}172}{38} \approx \boxed{\$64{,}741.37}$

Median: Reading left to right and top to bottom, notice that the 38 data values are already arranged in order (from largest to smallest). The median is defined as the middlemost data value. Since this data set consists of an even number of values (namely, 38), the median is the mean of the middle two values (in this case, the 19th and 20th):

$$\frac{\$67{,}691 + \$69{,}000}{2} = \boxed{\$68{,}345.50}.$$

Mode: The mode is, by definition, the most frequently occurring data value. Since no salary occurs more than once, there is **no mode**.

The **median** is most representative of the superintendents' salaries. Unlike the mean, the median's value is not affected (skewed) by unusually high or low salaries. Additionally, remember that the median is equivalent to the 50th percentile, which is the point at which 19 of the 38 salaries are below and the other 19 are above.

6. a) **Yes**; as the graph below clearly illustrates, the survival rates were related to the class of passenger. Approximately 75% of third-class passengers (528 of 706) died in the *Titanic* disaster, whereas the approximate fatality rates for first-class and second-class passengers were respectively 38% (122 of 325) and 59% (167 of 285).

b) **Yes**; the survival rates were related to the gender/age. In terms of relative survival rates, while the majority of child-age (52%) and adult female (74%) passengers survived, only approximately 18% of the adult male passengers survived this tragedy. In terms of the raw number of deaths, 659 of the 817 fatalities were adult males, whereas 106 were adult females and 52 were children.

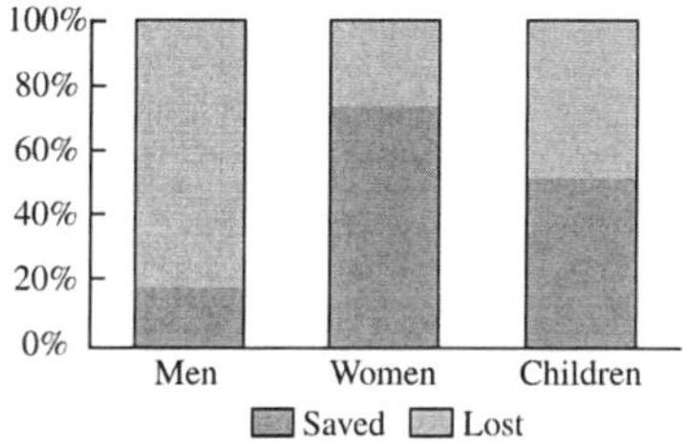

7. a) The median family income (in 1990 dollars) was the least in **1982** (approximately $31,600).

b) The median family income was the greatest in **1994**, the most recent year included in the line graph.

c) The median family income in 1994 was approximate **$41,500**.

> ***Reminder:*** *Many calculators include two-variable (bivariate) statistical functions. After inputting the paired data, one can instantaneously obtain — without having to endure any of the tedious computations demonstrated below — the linear correlation coefficient r, the slope (m) and y-intercept (b) of the least squares (regression) line, and assorted other statistical information. Check your calculator user's manual (or, if applicable, its accompanying "quick reference" card) for specific details.*

8. a) There are two approaches one may use to determine the equation of the regression line: either use a calculator with built-in linear regression capability (as noted above) or apply the formulas given on textbook p. 704 (as we demonstrate below).

Letting x represent the age (in years) and y represent the blood pressure, we obtain:

$n = 8$ $\qquad \Sigma y = 725$ $\qquad \Sigma x^2 = 14{,}899$

$\Sigma x = 323$ $\qquad \Sigma xy = 29{,}677$ $\qquad \Sigma y^2 = 65{,}907$

$$m = \frac{n\Sigma xy - (\Sigma x)(\Sigma y)}{n(\Sigma x^2) - (\Sigma x)^2} = \frac{8(29{,}677) - (323)(725)}{8(14{,}899) - (323)^2} \approx 0.218058265$$

$$b = \frac{\Sigma y - m(\Sigma x)}{n} = \frac{(725) - m(323)}{8} \approx 81.820897531$$

Thus, the equation of the best-fitting line is $\boxed{y' = \mathbf{0.22}x + \mathbf{81.8}}$. Here is the scatter diagram and sketch of this regression line:

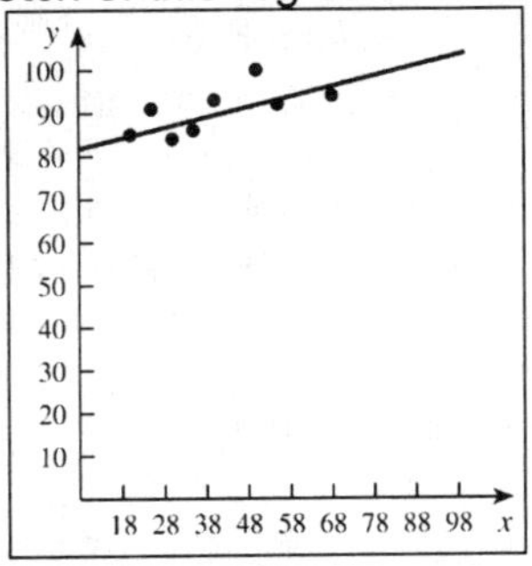

b) If you do not have a calculator that automatically does the computation for you, then here is how you must go about determining the linear correlation coefficient:

$$r = \frac{n\Sigma xy - (\Sigma x)(\Sigma y)}{\sqrt{n(\Sigma x^2) - (\Sigma x)^2}\sqrt{n(\Sigma y^2) - (\Sigma y)^2}}$$

$$= \frac{8(29{,}677) - (323)(725)}{\sqrt{8(14{,}899) - (323)^2}\sqrt{8(65{,}907) - (725)^2}}$$

$$\approx 0.6582620404$$

$\boxed{r \approx \mathbf{0.6583}}$

From Table 10.9 (on textbook p. 702), we see that the critical value (for $n = 8$) at the 5% level is 0.707. Since $|r| = 0.6583$ is not greater than 0.707, we conclude (based upon this small sample of data) that there is **no statistically significant linear correlation** (at the 1% level or 5% level) between age and blood pressure.

9. Given that $\mu = 266$ days and $\sigma = 16$ days. For $x = 314$ days, the z-score is

$$z = \frac{x - \mu}{\sigma} = \frac{314 - 266}{16} = \frac{48}{16} = 3.$$

This tells us that a 314-day pregnancy is 3 standard deviations above the mean gestation period. From our knowledge of the standard normal curve [if necessary, review textbook Figure 10.34, p. 691] we know that approximately 0.1% (0.001) of normally distributed data lies 3 or more standard deviations above the mean. In other words, **approximately 1 out of 1,000** pregnancies will have a 314-day duration.

10. For additional details, refer to text Example 10 (pp. 684-687).
The total population is 24,600,000, and number of representatives is 100. So, the ***standard divisor*** is $D = 24{,}600{,}000 \div 100 = \underline{246{,}000}$.
We begin with the three simplest representation plans, those of Adams, Jefferson, and Webster:

Region (Pop.)	Std. Quota	**Adams** [round up]	**Jefferson** [round down]	**Webster** [round to nearest]
North (2,400,000)	9.76	**10**	**9**	**10**
Northeast (4,000,000)	16.26	**17**	**16**	**16**
East (2,700,000)	10.98	**11**	**10**	**11**
Southeast (2,500,000)	10.16	**11**	**10**	**10**
South (2,800,000)	11.38	**12**	**11**	**11**
Southwest (3,500,000)	14.23	**15**	**14**	**14**
West (4,100,000)	16.67	**17**	**16**	**17**
Northwest (2,600,000)	10.57	**11**	**10**	**11**
Total = 24,600,000	Total # of Reps. ⇒	**104**	**96**	**100**

Next, we turn our attention to Hamilton's plan, which achieves the desired total of representatives (in this case, 100) by allocating seats based upon the decreasing order of the decimal portions of the standard quota. Notice that Jefferson's plan resulted in a total representation of 96 seats. So, in order for Hamilton's plan to achieve 100 seats, four regions will be awarded an additional seat.

Region	Std. Quota	Decimal Portion	Decimal Rank	Rounding Decision	Hamilton's Apportionment
N	9.76	0.76	2nd	up	**10**
NE	16.26	0.26	6th	down	**16**
E	10.98	0.98	1st (highest)	up	**11**
SE	10.16	0.16	8th (lowest)	down	**10**
S	11.38	0.38	5th	down	**11**
SW	14.23	0.23	7th	down	**14**
W	16.67	0.67	3rd	up	**17**
NW	10.57	0.57	4th	up	**11**
				Total # of Reps. ⇒	**100**

Finally, we use the Huntington-Hill (HH) plan (explained in detail on textbook p. 685 and demonstrated on p. 687):

Region	Std. Quota	a	b	Geometric Mean	Rounding Decision	HH's Apportionment
N	9.76	9	10	$\sqrt{9\cdot 10}\approx 9.49$	Up (9.76 > 9.49)	**10**
NE	16.26	16	17	$\sqrt{16\cdot 17}\approx 16.49$	Down (16.26 < 16.49)	**16**
E	10.98	10	11	$\sqrt{10\cdot 11}\approx 10.49$	Up (10.98 > 10.49)	**11**
SE	10.16	10	11	$\sqrt{10\cdot 11}\approx 10.49$	Down (10.16 < 10.49)	**10**
S	11.38	11	12	$\sqrt{11\cdot 12}\approx 11.49$	Down (11.38 < 11.49)	**11**
SW	14.23	14	15	$\sqrt{14\cdot 15}\approx 14.49$	Down (14.23 < 14.49)	**14**
W	16.67	16	17	$\sqrt{16\cdot 17}\approx 16.49$	Up (16.67 > 16.49)	**17**
NW	10.57	10	11	$\sqrt{10\cdot 11}\approx 10.49$	Up (10.57 > 10.49)	**11**
					Total # of Reps. ⇒	**100**

CHAPTER 11: The Nature of Graphs and Functions

Problem Set 11.1

1. a) Aphrodite b) Maxwell Montes
c) Atalanta Planitia d) Rhea Mons e) Lavina Planitia

3. Wording varies. The ***y*-intercept** of a line is the point where the line crosses the y-axis; it has coordinates $(0, b)$. ***Slope*** is a measure of a line's steepness (slant, tilt, grade, incline, ...); it is the ratio of the change in the y-coordinate ("rise") to the change in the x-coordinate ("run"). For additional details, refer to textbook page 727.

5. ***Horizontal lines*** have zero slope ($m = 0$), whereas ***vertical lines*** have no slope (m does not exist).
*[**Caution**: Please do not blindly accept these facts. To avoid confusion, make sure you have developed a true understanding of at least one of the several reasons why the respective slopes of horizontal/vertical lines are what they are. We encourage you to spend some time exploring this concept on your own — after which, if necessary, seek the assistance of your instructor.]*

Problems 7-18: Ordered pairs may vary.

7. Points include (0, 5), (1, 6), and (2, 7). The slope is $m = 1$, so $\frac{\text{rise}}{\text{run}} = \frac{+1}{+1}$ or $\frac{-1}{-1}$.

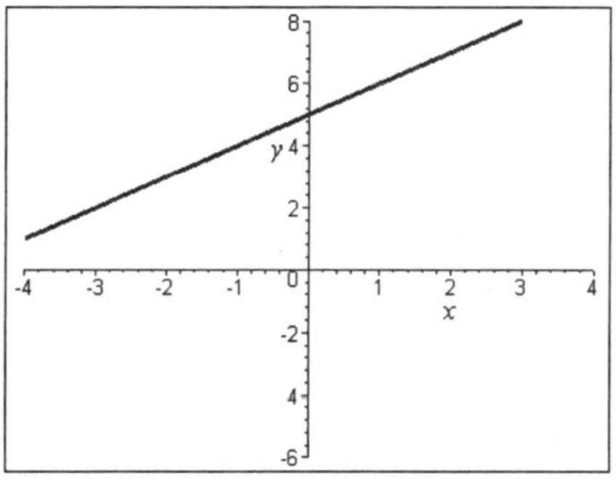

9. Points include (0, 5), (1, 7), and (–1, 3). The slope is $m = 2$, so $\frac{\text{rise}}{\text{run}} = \frac{+2}{+1}$ or $\frac{-2}{-1}$.

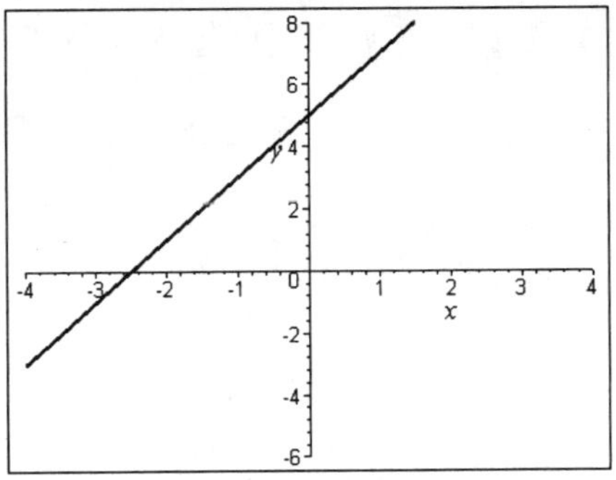

11. Points include (0, –1), (1, 0), and (2, 1). The slope is $m = 1$, so $\frac{\text{rise}}{\text{run}} = \frac{+1}{+1}$ or $\frac{-1}{-1}$.

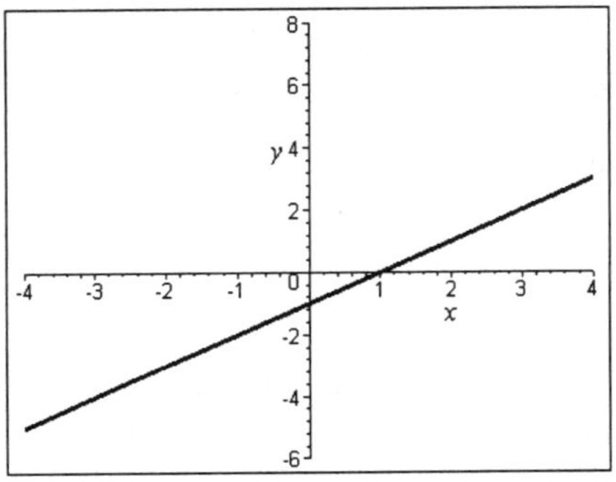

13. Points include (0, 1), (–1, 3), and (1, –1). The slope is $m = -2$, so $\frac{\text{rise}}{\text{run}} = \frac{+2}{-1}$ or $\frac{-2}{+1}$.

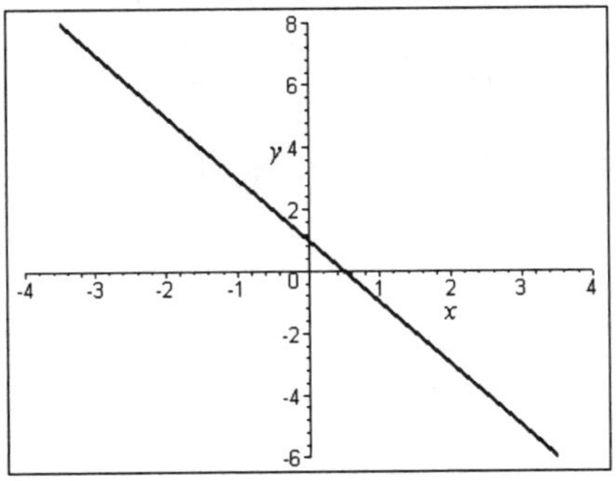

15. Points include (0, 1), (1, 3), and (2, 5). The slope is $m = 2$, so $\frac{\text{rise}}{\text{run}} = \frac{+2}{+1}$ or $\frac{-2}{-1}$.

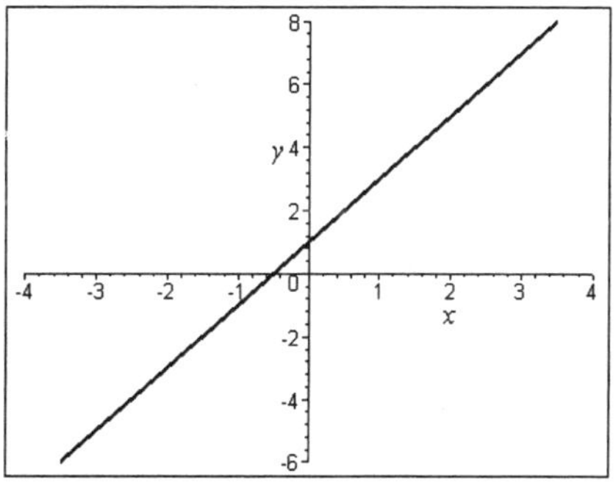

17. As demonstrated in textbook Example 5-7, first solve the equation for y, which puts it in slope-intercept form. The result is $y = -\frac{3}{4}x + 2$. This reveals that the y-intercept is (0, 2) and the slope is $m = \frac{\text{rise}}{\text{run}} = \frac{+3}{-4}$ or $\frac{-3}{+4}$. Other points include (4, –1), (2, $\frac{1}{2}$), and (–4, 5).

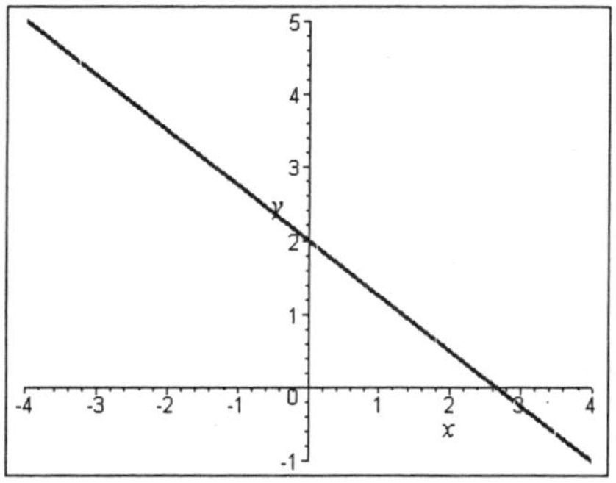

19. Problems 19-30 are very similar to text Example 4 (p. 728). It is very important that you understand (and can apply) the general procedure for graphing (see Figure 11.8, p. 728).

First, plot the point (2, 5). Then, with $m = \frac{\text{rise}}{\text{run}} = \frac{+1}{+3}$ or $\frac{-1}{-3}$, plot additional points by counting out the rise and the run.

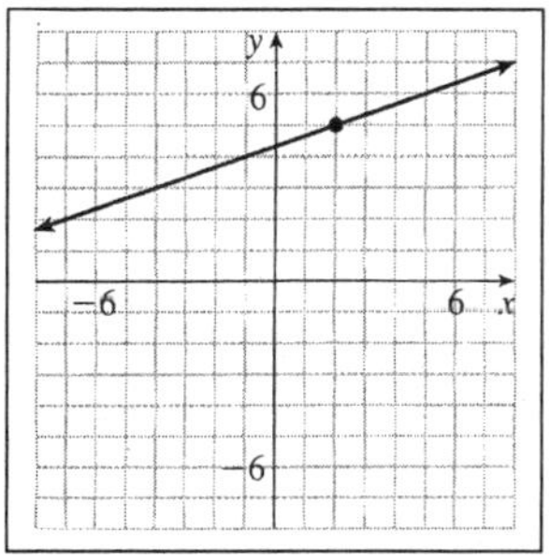

21. First, plot the point (–4, –3). Then, with $m = \frac{\text{rise}}{\text{run}} = \frac{+3}{+4}$ or $\frac{-3}{-4}$, plot additional points by counting out the rise and the run.

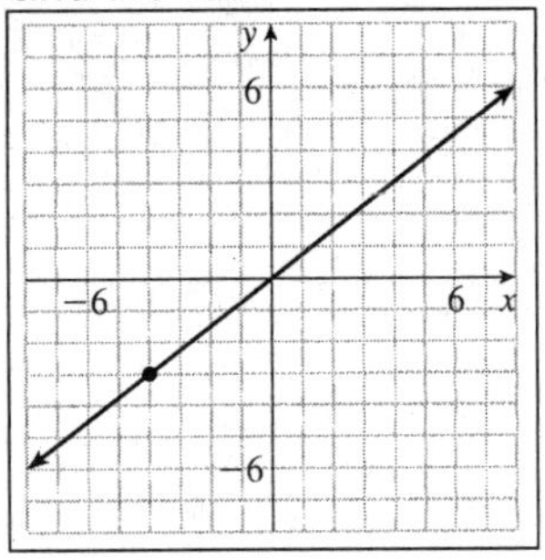

23. First, plot the point (1, –1). Then, with $m = \frac{\text{rise}}{\text{run}} = \frac{-1}{+7}$ or $\frac{+1}{-7}$, plot additional points by counting out the rise and the run.

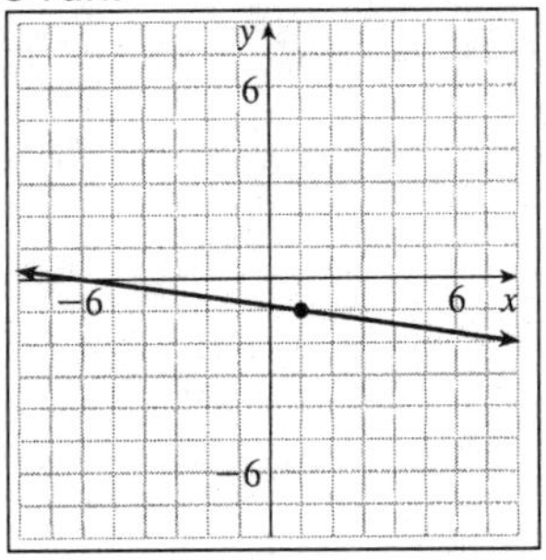

25. First, plot the point (–1, 3). Then, with $m = \frac{\text{rise}}{\text{run}} = \frac{+1}{+1}$ or $\frac{-1}{-1}$, plot additional points by counting out the rise and the run.

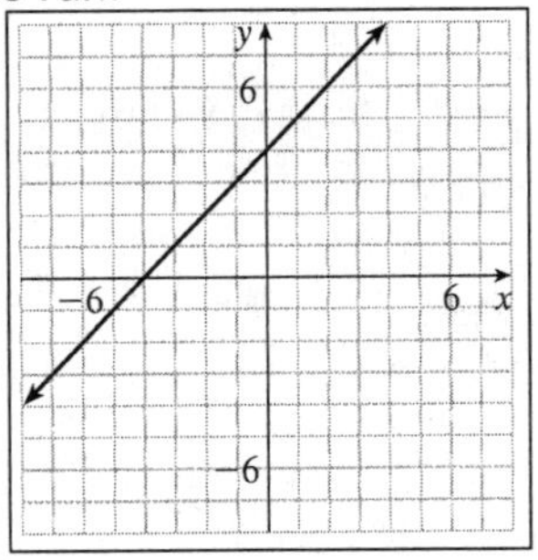

27. First, plot the point (1, 3). Then, with $m = \frac{\text{rise}}{\text{run}} = \frac{+3}{+1}$ or $\frac{-3}{-1}$, plot additional points by counting out the rise and the run.

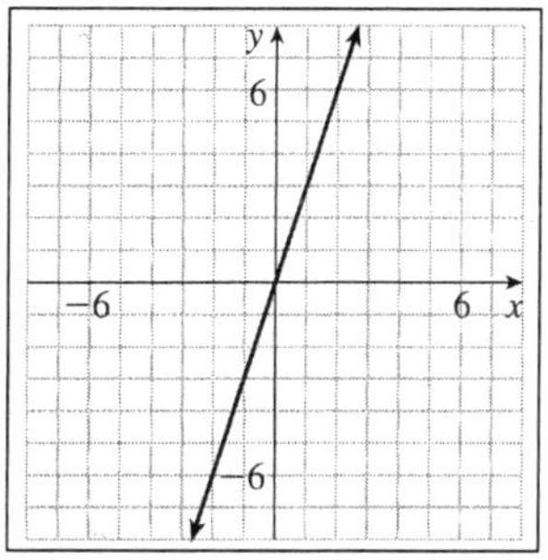

29. Recall that a line with "no slope" is a vertical line. So, we graph a vertical line that runs through the point (2, 3).

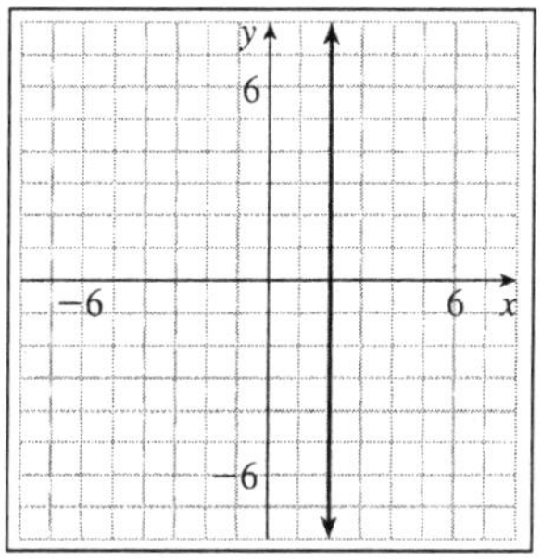

31. The slope is $m = \frac{\text{rise}}{\text{run}} = \frac{+2}{+1}$ or $\frac{-2}{-1}$, and the y-intercept is $b = 3$.

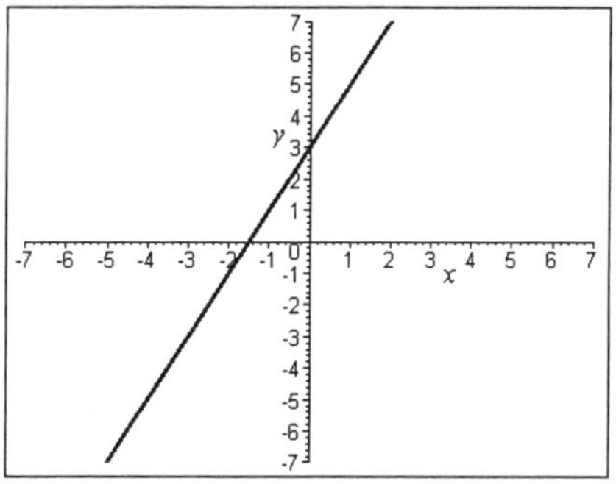

33. The slope-intercept form equation is $y = -x - 4$.

So, the slope is $m = \frac{\text{rise}}{\text{run}} = \frac{-1}{+1}$ or $\frac{+1}{-1}$, and the y-intercept is $b = -4$.

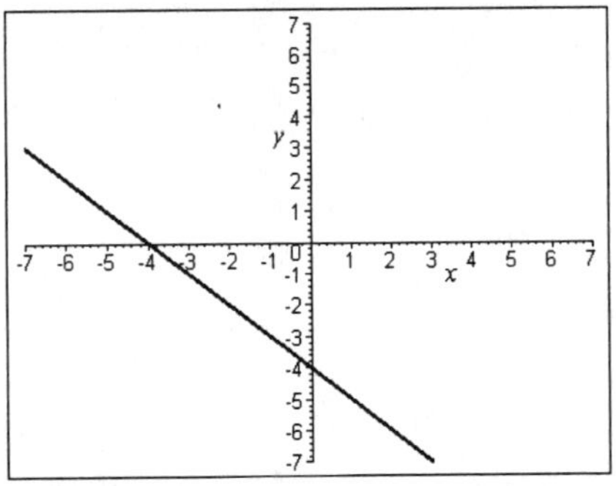

35. The slope-intercept form equation is $y = -\frac{2}{3}x - 2$.

So, the slope is $m = \frac{\text{rise}}{\text{run}} = \frac{-2}{+3}$ or $\frac{+2}{-3}$, and the y-intercept is $b = -2$.

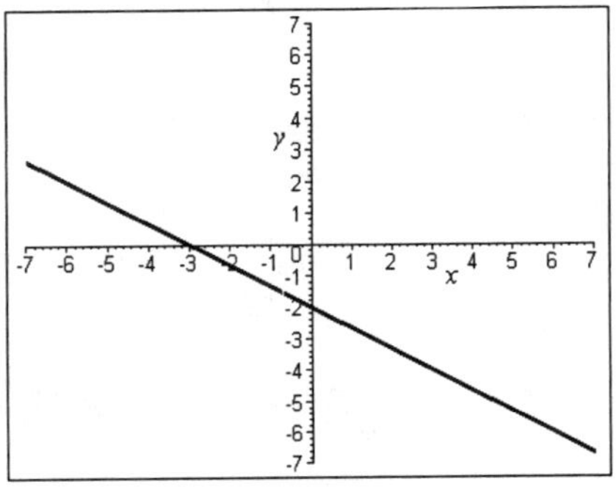

37. The slope-intercept form equation is $y = -\frac{2}{3}x + \frac{5}{2}$.

So, the slope is $m = \frac{\text{rise}}{\text{run}} = \frac{-2}{+3}$ or $\frac{+2}{-3}$, and the y-intercept is $b = \frac{5}{2}$.

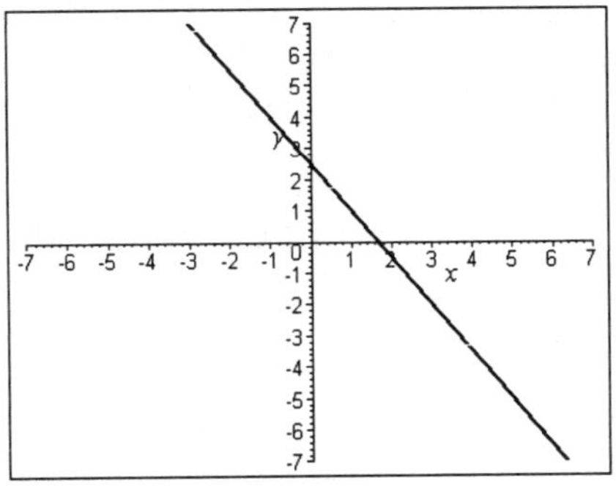

39. Since the only requirement is that x (the first component) equal 5, we see that there is no restriction on the choice for y. Said differently, for <u>any</u> real number y, the point $(5,\ y)$ satisfies the condition that $x=5$. Upon plotting and connecting such points, you will see that the result is a vertical line.

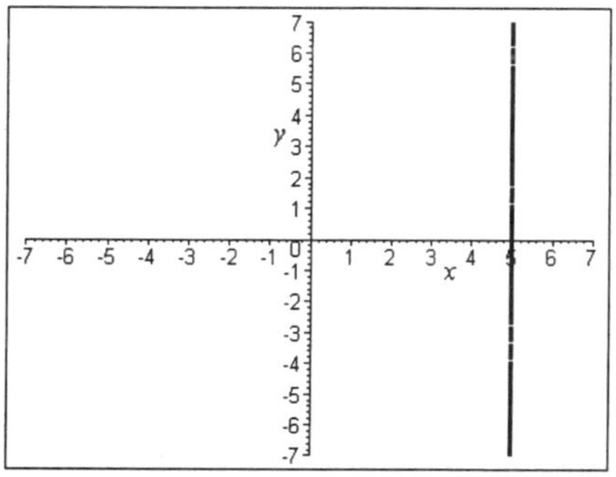

41. The slope-intercept form equation is $y=-x+100$.
So, the slope is $m=\frac{\text{rise}}{\text{run}}=\frac{+1}{-1}$ or $\frac{-1}{+1}$, and the y-intercept is $b=100$.

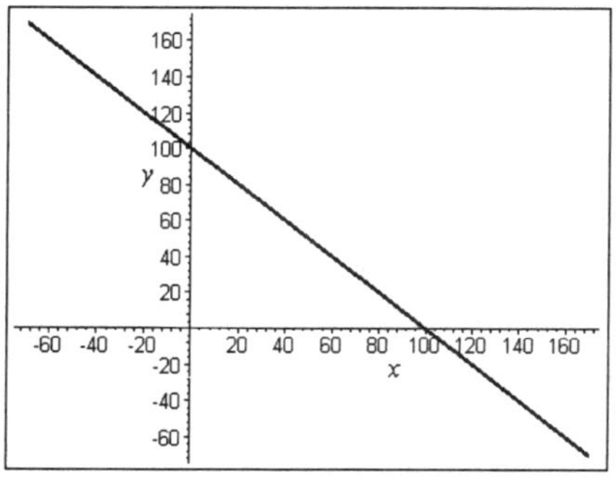

43. The slope is $m=\frac{\text{rise}}{\text{run}}=\frac{+100}{-1}$ or $\frac{-100}{+1}$, and the y-intercept is $b=0$ (i.e., the line contains the origin).

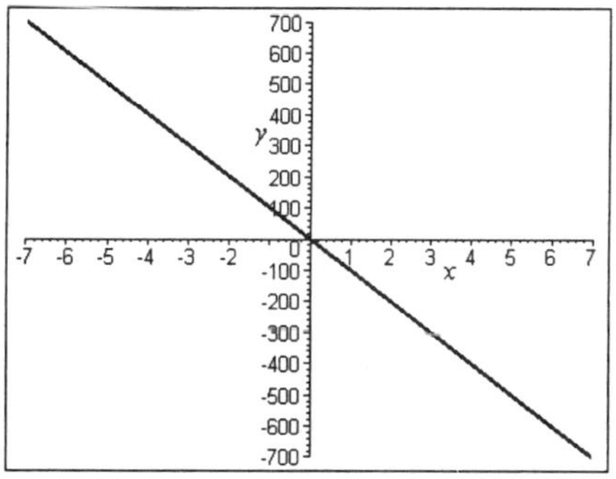

45. Since the line's y-intercept appears to be -2, we immediately narrow our choices to C and F. We then attempt to estimate how much we "rise and run" to get from the y-intercept to other points; for example, it appears that the line's x-intercept is slightly less than 3, and the line also appears to contain the point $(10, 5)$. Based upon this analysis, it appears that the slope is $\frac{7}{10}$ (rather than $\frac{9}{10}$). Hence, we choose **equation C**, $y = 0.7x - 2$.

47. Because the line's y-intercept appears to be -4, we immediately narrow our choices to A and H. We then attempt to estimate how much we "rise and run" to get from the y-intercept to other points; for example, it appears that the line's x-intercept is between 7 and 8. So, the slope is closer to $\frac{5}{10}$ than it is to $\frac{7}{10}$. Therefore, the best choice is **equation A**, $y = 0.5x - 4$.

49. The line's y-intercept appears to be 2, so we immediately narrow our choices to B and E. We then attempt to estimate how much we "rise and run" to get from the y-intercept to other points; for example, the line's x-intercept is close to (if not slightly greater than) 2, and the point $(10, 7)$ is either on the line or close to it. Consequently, we know the slope is certainly closer to $-\frac{9}{10}$ than it is to $-\frac{5}{10}$. Thus, the best choice is **equation E**, $y = -0.9x + 2$.

51. a) **Yes**, the fertility rate in Brazil is declining at a rate that is nearly (but not exactly) linear. It is important to remember that sketching a graph does not provide sufficient enough evidence to support such a claim. Instead, you should use slopes to prove that it is not linear.

b) **Answers vary**. Slopes range from a low of $-\frac{1}{6} = -0.1\overline{6}$ (with data points from 1991 and 1994) to a high of $-\frac{7}{50} = -0.14$ (with data points from 1970 and 1980).

Suppose we choose 1970 as the base year; that is, let $x = 0$ represent 1970. If we use the data points $(0, 5.8)$ and $(10, 4.4)$, we obtain a slope of

$$m = \frac{y_2 - y_1}{x_2 - x_1} = \frac{4.4 - 5.8}{10 - 0} = -0.14.$$

Thus, with y-intercept $b = 5.8$, we obtain the equation $y = -0.14x + 5.8$.

c) **Predictions vary**.

Yes, depending upon the data points selected in part b, the linear equation will vary and, hence, the predicted (estimated) fertility rate in 2005 will vary. Using our equation from part b (namely, $y = -0.14x + 5.8$), the year 2005 is represented by $x = 35$ (i.e., 35 years after the base year 1970). Therefore, we predict that the Brazilian fertility rate in 2005 will be

$$y = -0.14(35) + 5.8 = \underline{\underline{0.9}}.$$

53. a) Given the future value financial formula for simple interest $A = P(1+rt)$. With $P = \$10{,}000$ and $r = 0.08$, we obtain the linear equation $A = 10{,}000(1+0.08t)$. By applying the distributive and commutative properties, we obtain the slope-intercept form (i.e., $A = mt + b$) of the equation: $A = 800t + 10{,}000$. Its graph is:

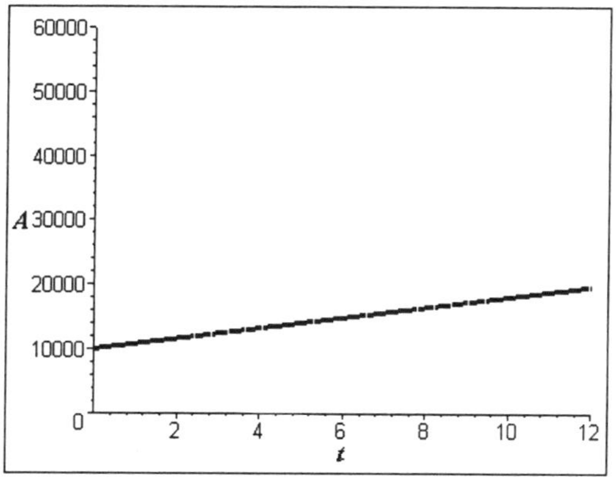

b) From the slope-intercept form of the equation written in our solution to part (a), it follows that the slope is **800**. If you are familiar with the concept of simple interest [if not, see textbook Section 8.1], this slope value should make sense to you. A person who invests \$10,000 in an account bearing 8% simple interest will earn \$800 per year in interest, and this amount of interest will remain unchanged year after year (assuming no additional deposits or withdrawals).

c) The A-intercept is **10,000**.
There are four ways to explain this:
(i) the graph in part (a) includes the point $(0,\ 10{,}000)$;
(ii) we obtain $A = 10{,}000$ when $t = 0$ is substituted in the equation;
(iii) the slope-intercept form $A = 800t + 10{,}000$ [written in part (a)] reveals that $b = 10{,}000$; and
(iv) the initial investment (principal) is \$10,000.

55. a) The graph of the profit equation $P = 1.25x - 850$ is:

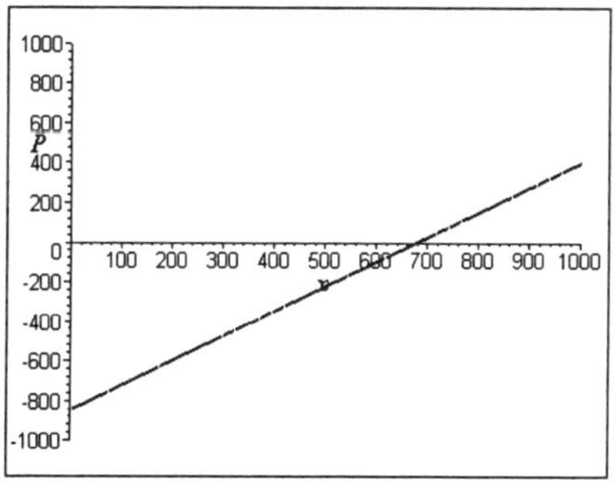

b) When the number of units sold $x = 0$, the profit (in dollars) $P = -850$. This means **if no items are sold, there is a loss of \$850**.

c) Notice that the given equation is written in slope-intercept form. Hence, the marginal profit is **1.25**; it is the profit increase corresponding to each unit increase in number of items sold. In other words, the marginal profit (for a linear profit model) is the amount of profit per item sold.

57. Given the data points (0, 17) and (0, 19.8), it follows that the slope of the line containing these points is:

$$m = \frac{19.8 - 17.0}{8 - 0} = 0.35.$$

With a y-intercept of 17, the resulting linear equation is $\boxed{y = 0.35x + 17}$.

For the year 2005, we have $x = 15$ (i.e., 15 years past the base year 1990), so the equation yields $y = 0.35(15) + 17 = 22.25$. Hence, the projected population of Texas in the year 2005 is about **22.3 million**.

59. Given $y - k = m(x - h)$. Since the line passes through (x_1, y_1) and (x_2, y_2), the slope is $m = \dfrac{y_2 - y_1}{x_2 - x_1}$. Let $(h, k) = (x_1, y_1)$ be the known point. Then, by substitution, we obtain:

$$y - y_1 = \left(\frac{y_2 - y_1}{x_2 - x_1}\right)(x - x_1).$$

Problem Set 11.2

1. **Descriptions vary**. A ***linear inequality in two variables***, say x and y, is an inequality of the form $Ax + By + C \le 0$, $Ax + By + C < 0$, $Ax + By + C \ge 0$, or $Ax + By + C > 0$.

3. **False**. All linear inequalities have a boundary line. In this particular problem, the boundary is $2x + 5y = 2$, which is excluded (so, a dashed line is used).

5. **False**. (0, 0) does not work because it is on the boundary line; a test point must be a point that is not on the boundary.

7. **True**; $3(0) - 2(0) = 0$ and it is true that $0 \ge -1$.

9. **True**; $4 > 2(-2) - 1$ simplifies to $4 > -5$, which is a true statement.

11. **True**; $4(-2) < 3(4)$ simplifies to $-8 < 12$, which is a true statement.

13.

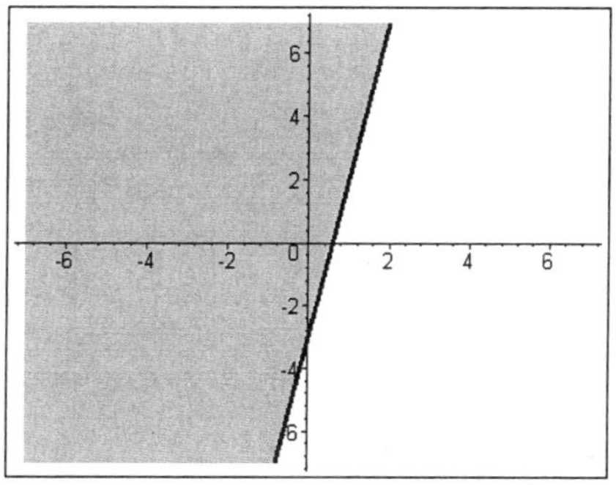

15.

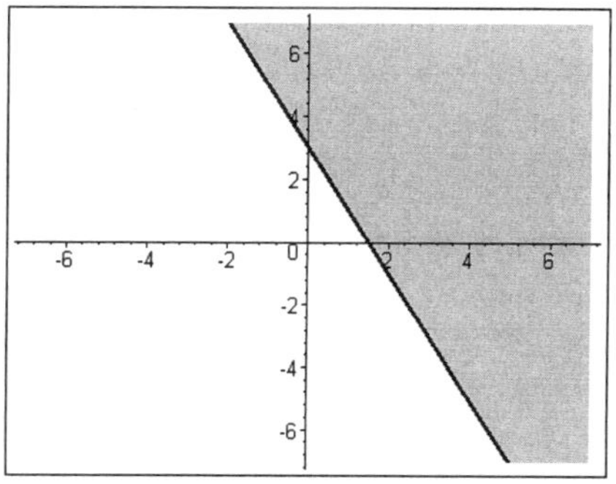

17.

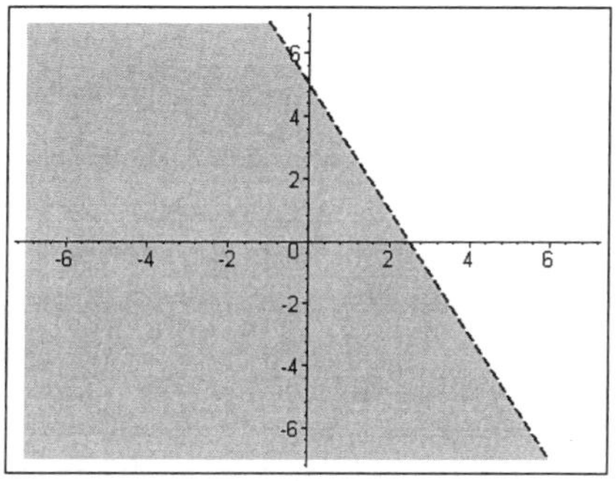

19.

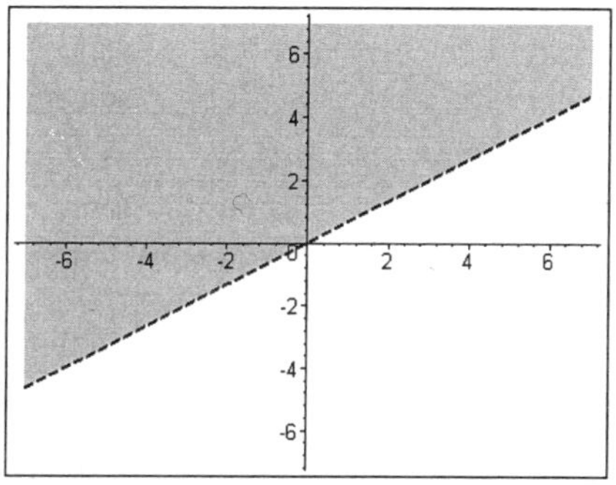

21.

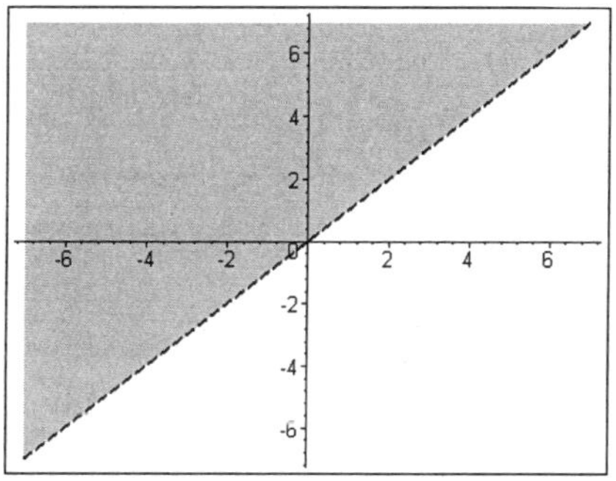

23.

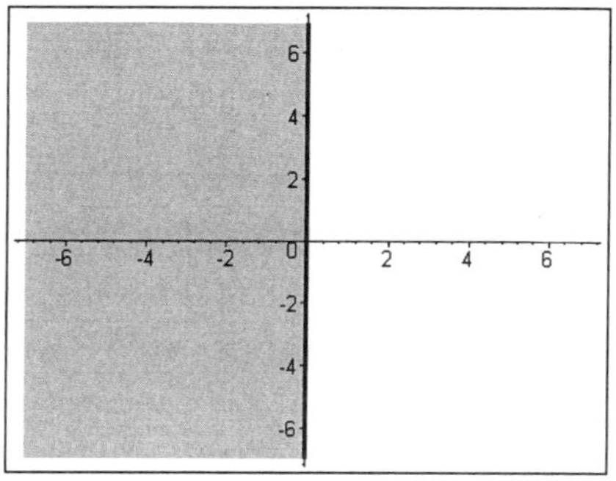

25.

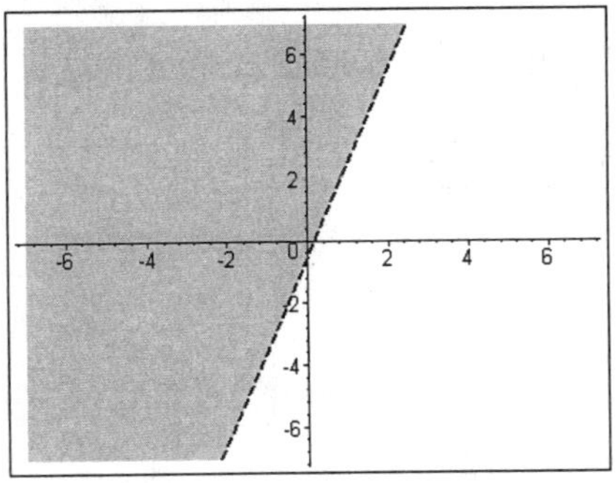

27.

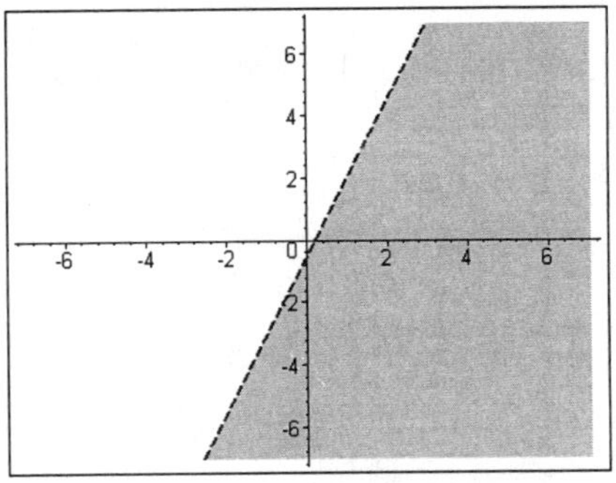

29.

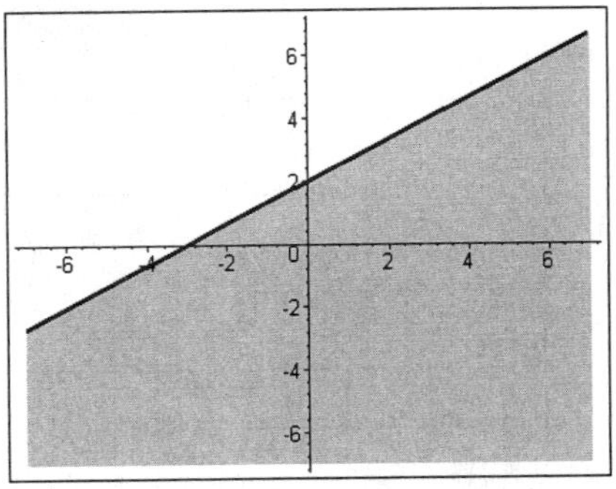

Problem Set 11.3

1. **Wording varies**. As shown in textbook Figure 11.16 (page 736), a ***parabola*** is a curve that describes the path of a projectile. Please note that we will encounter in Section 11.4 (see p. 743) a geometric definition of a parabola.

3. **Answers vary**. One approach is to use the following population growth/decay formula (from Section 7.7, pp. 483-484): $A = A_0e^{rt}$. Text Example 5 (pp. 739-740) provides a superb demonstration of population forecasting applicable to everyday life.

5.

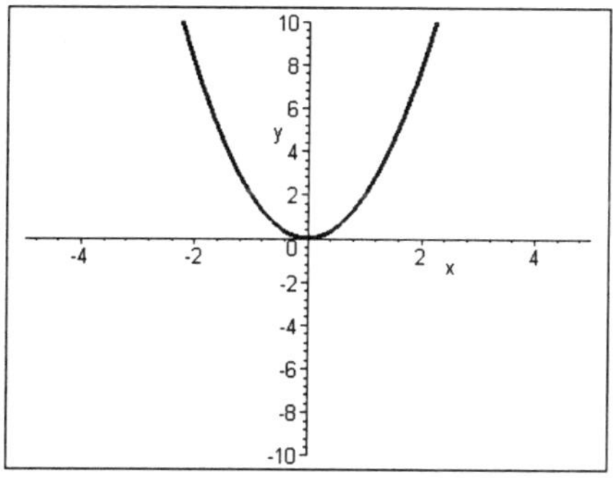

10
8
6
y 4
2
0
-2
-4
-6
-8
-10
-4
-2
2
x
4

7.

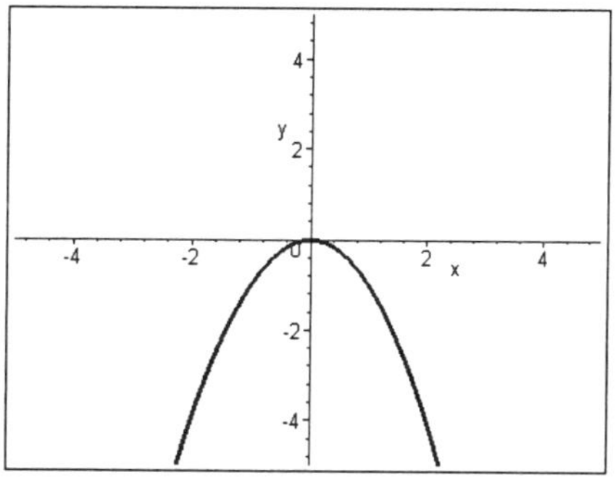

4
y 2
0
-2
-4
-4
-2
2
x
4

9.

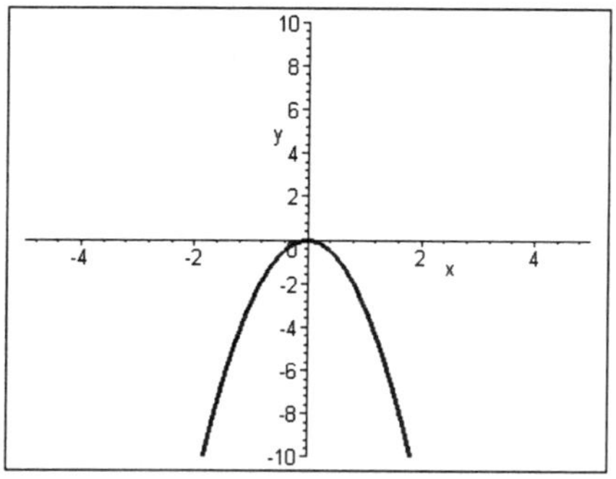

10
8
6
y 4
2
0
-2
-4
-6
-8
-10
-4
-2
2
x
4

11.

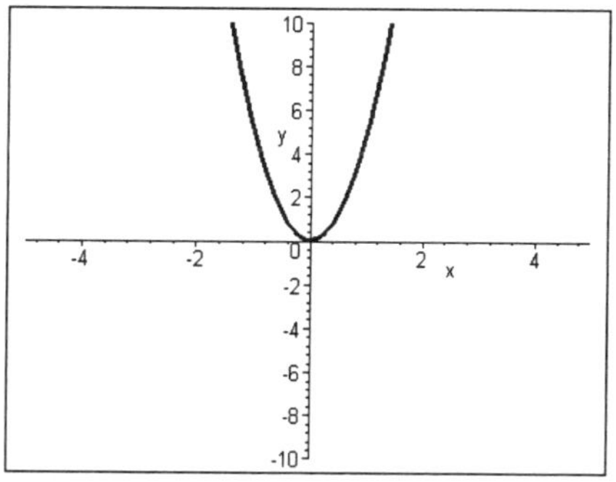

10
8
6
y 4
2
0
-2
-4
-6
-8
-10
-4
-2
2
x
4

13.

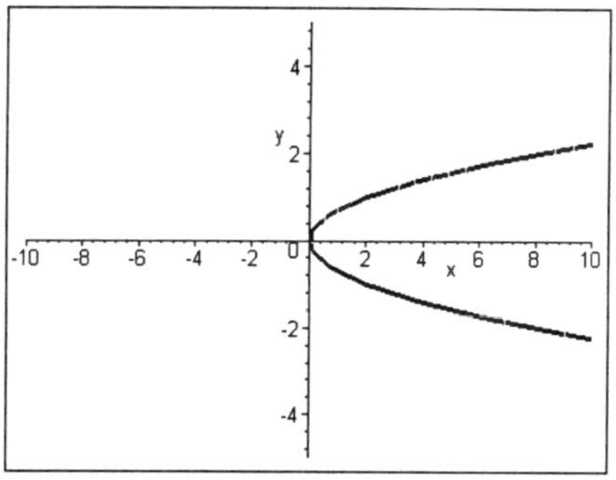

4
y 2
0
-2
-4
-10
-8
-6
-4
-2
2
4
x
6
8
10

15.

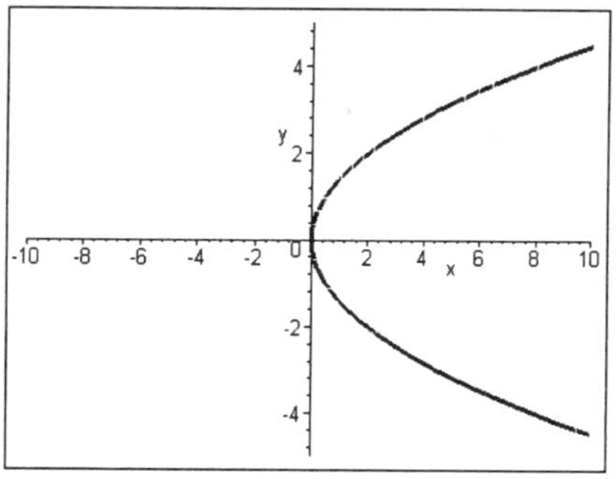

4
y 2
0
-2
-4
-10
-8
-6
-4
-2
2
4
x
6
8
10

17.

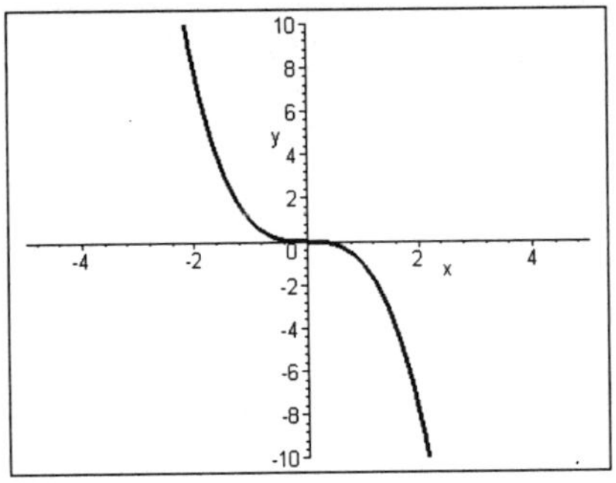

19.

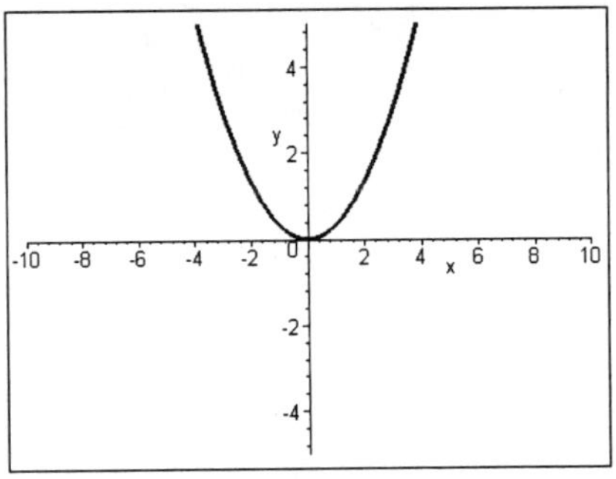

21.

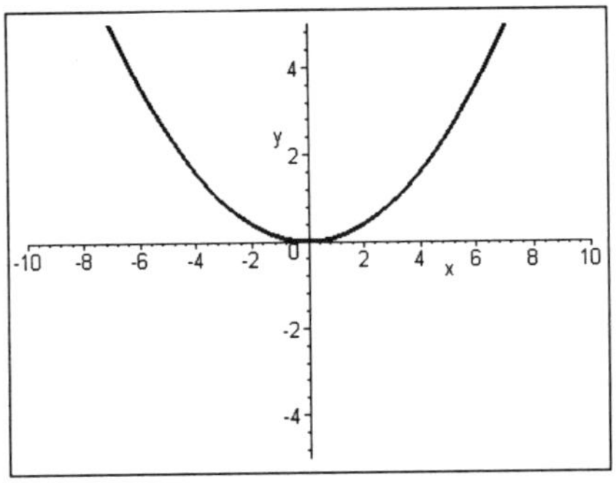

23.

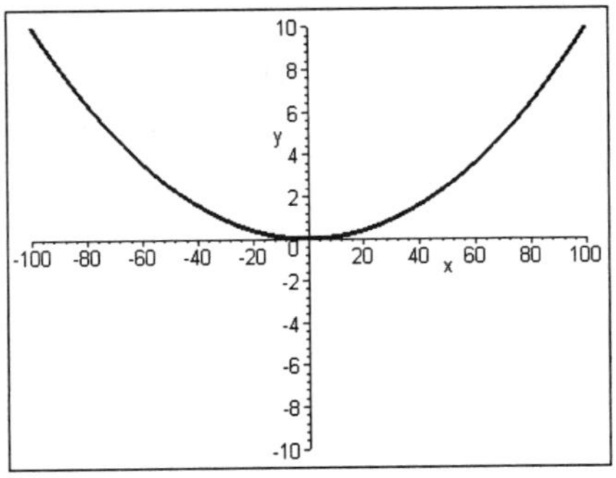

25.

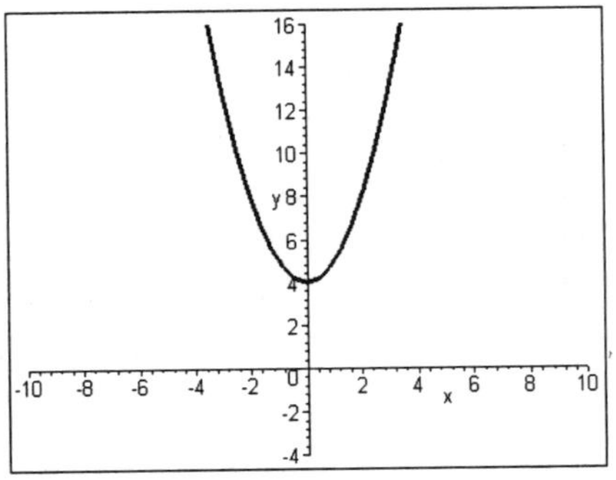

27.

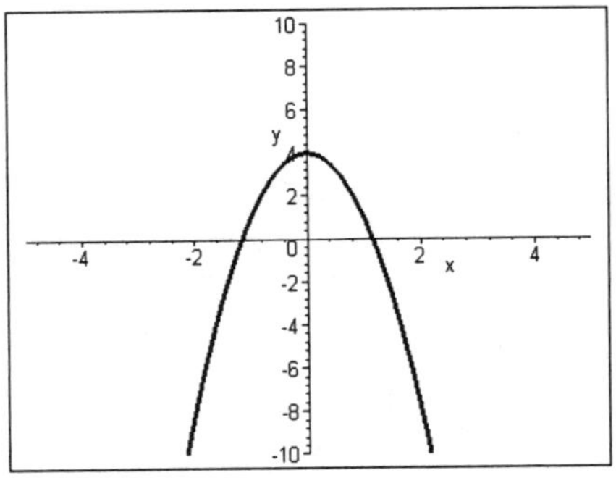

29.

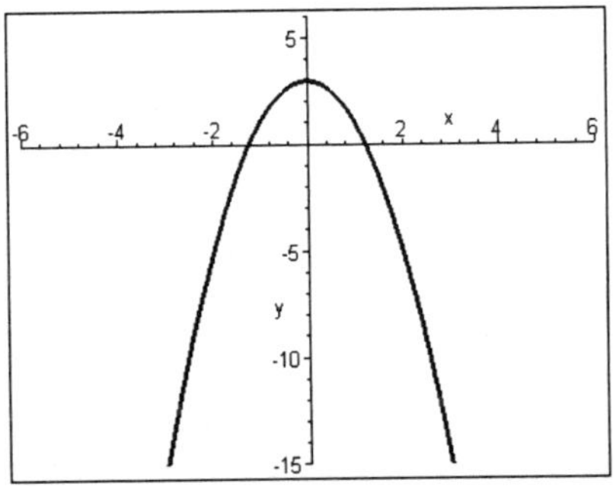

31.

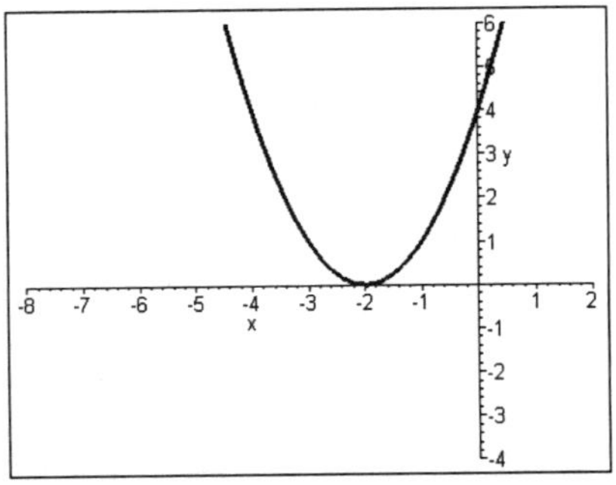

33.

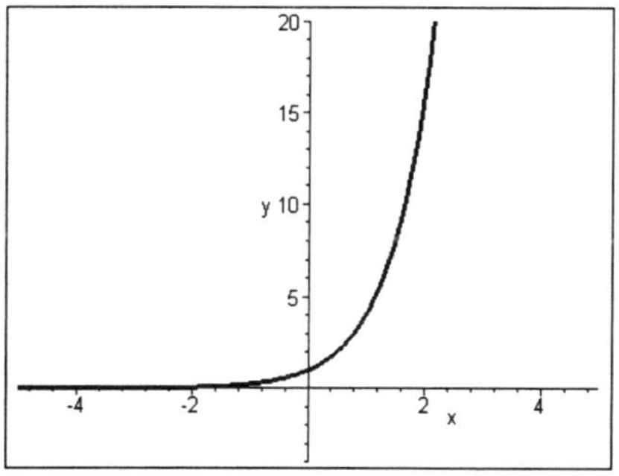

35.

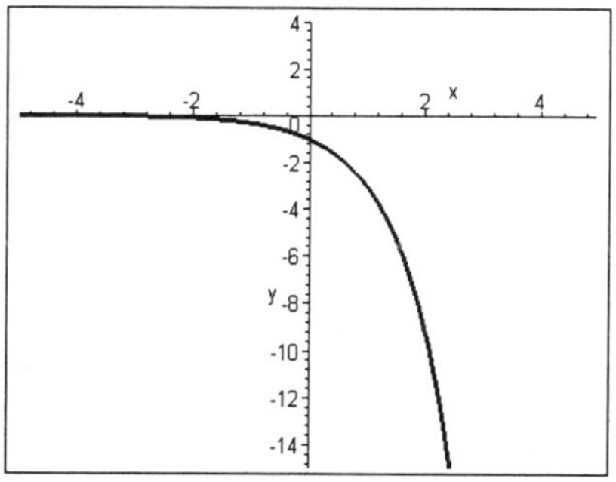

37.

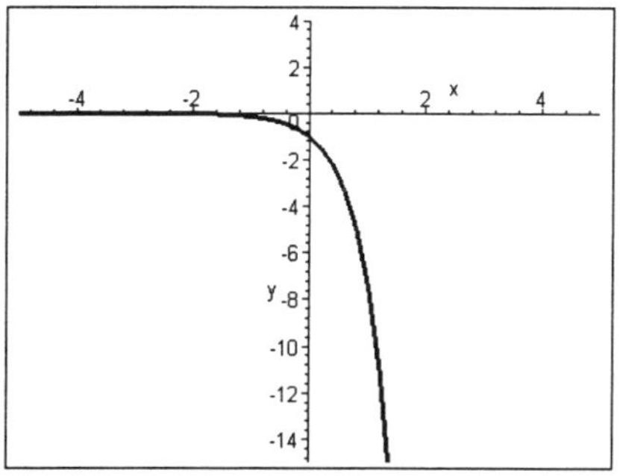

39.

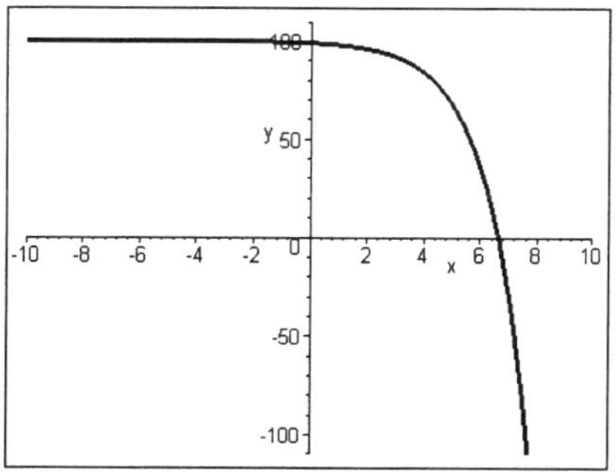

41.

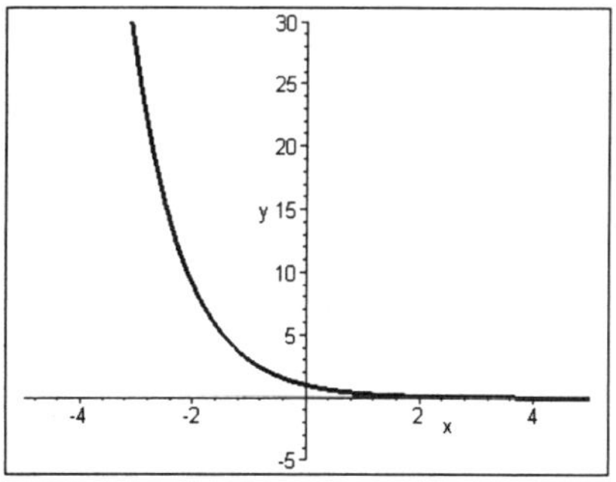

43.

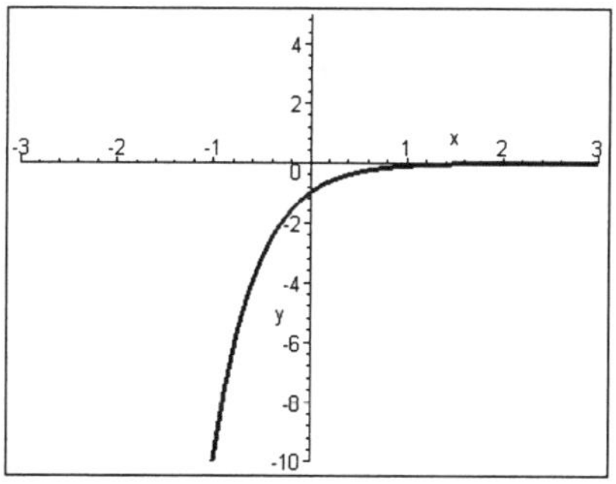

45. a)

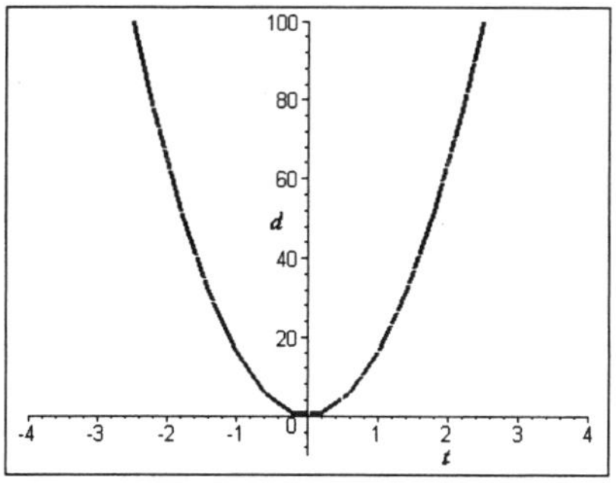

b) The parabola opens **upward**. Note that the number a in $d = at^2$ is positive.

c)

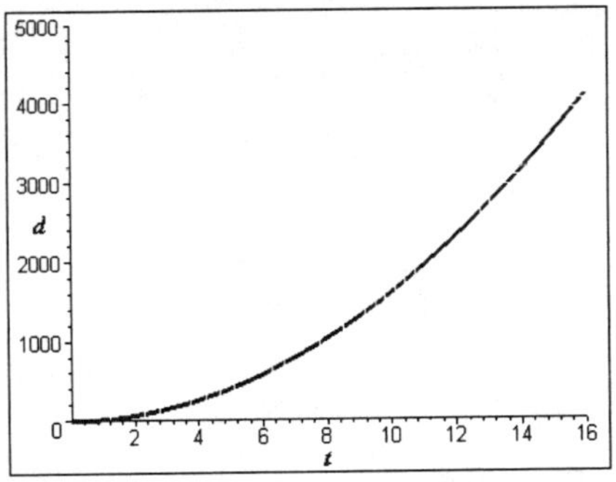

47. The point (1, 0) is the vertex.

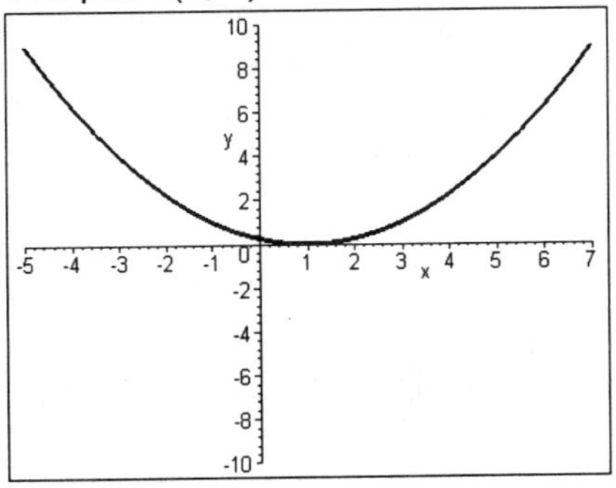

49. (1,2) is the vertex.

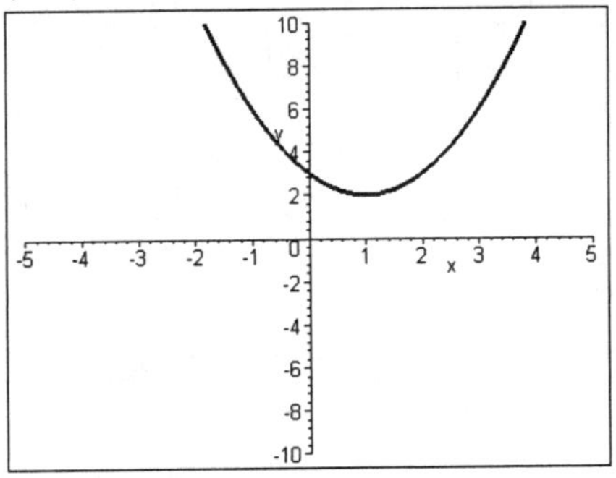

51. (0, 2) is the vertex.

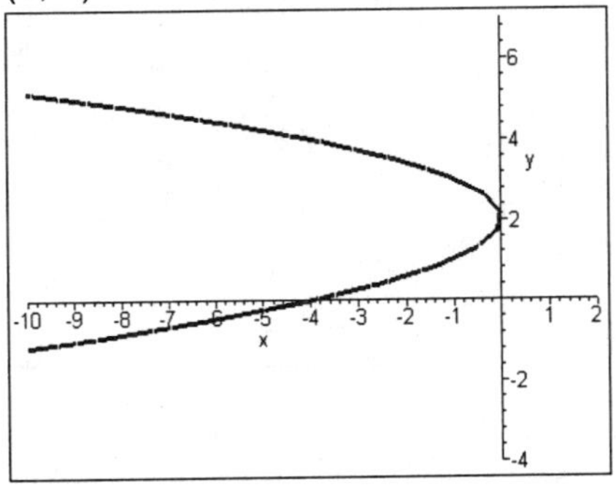

53. Intuitively, it should make sense that a slower growth rate results in a less rapidly increasing population. With $r = 0.015$, the graph of the exponential equation $A = 2{,}500e^{0.015t}$ has a y-intercept of (0, 2500) and approximately contains the points (10, 2905) and (20, 3375). Hence, its graph is as follows:

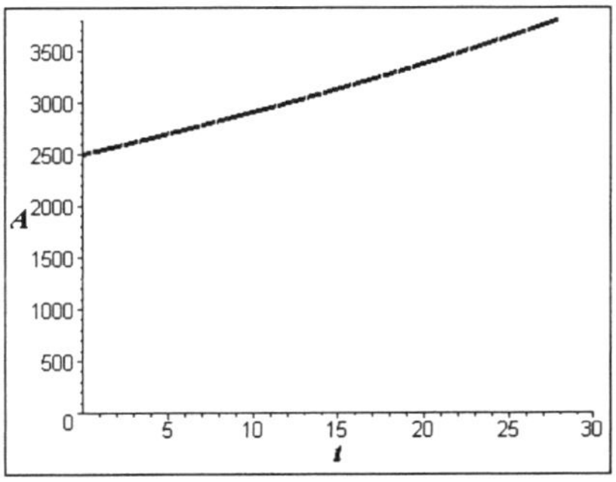

55. With $r = 0.05$, the simple interest formula in Example 6 (p. 740) becomes $A = 100(1 + 0.05t)$, which is a linear equation. As for the compound interest formula, it becomes $A = 100(1 + 0.05)^t$ or $A = 100(1.05)^t$, which is an exponential equation. Graphing them on the same axes, we obtain:

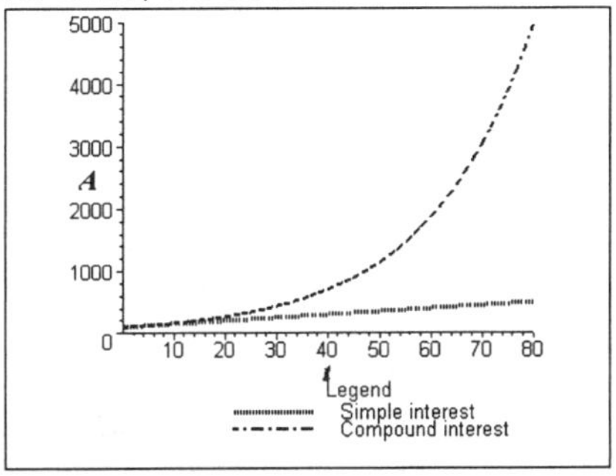

As the years elapse, the amount accrued in an account bearing compound interest far exceeds that of a simple-interest account.

57. With $A_0 = 50$ cm², the healing law for skin wounds gives us the exponential equation $A = \dfrac{A_0}{e^{0.1t}}$. Observe that this equation can be rewritten as $A = A_0 e^{-0.1t}$. Since the area A of the unhealed skin continues to shrink as time t increases, it makes sense that the growth/decay rate of a skin wound is negative (i.e., $r = -0.1$) and that the curve (when viewed from a left-to-right perspective) is decreasing.

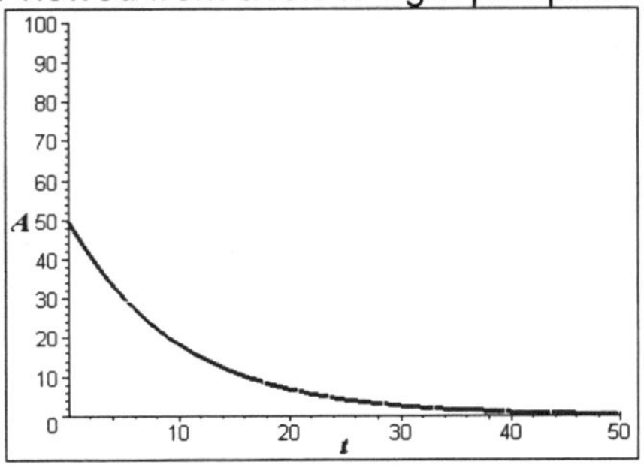

59. Based upon our previous experiences with the normal curve in Section 10.4, the graph below should come as no surprise.
For additional details pertaining to plotting points for this problem, we encourage you to revisit Problem 57 in Section 10.4 (textbook p. 699), a problem for which this manual contains a detailed solution.

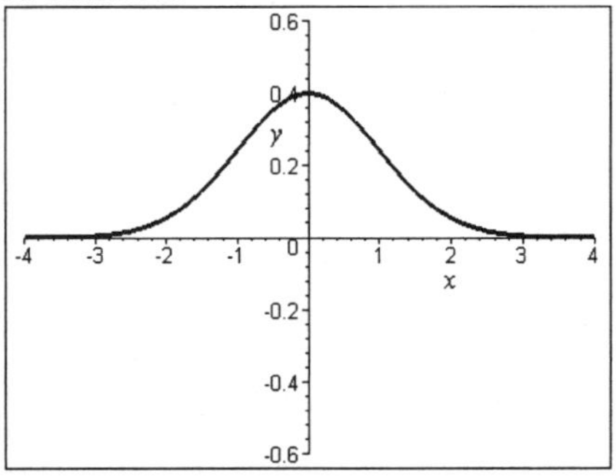

Problem Set 11.4

1. **Wording varies**. A ***conic section*** is the set of points that results from the intersection of a cone and a plane. The conic sections include lines, parabolas, ellipses, circles, and hyperbolas. On pages 742-743, your textbook provides two sets of illustrations showing how the conic sections are generated. Figure 11.24 is a traditional-type depiction, whereas the flashlight-basketball illustrations make for a fun mathematical activity appropriate even for grade-school students.

3. **Wording varies**. An ***ellipse*** is the set of all points in a plane such that, for each point on the ellipse, the sum of its distances from two fixed points (called the *foci*) is a constant. To acquire a richer understanding of this definition, study text Example 3 (pp. 744-745).

5. **Answers vary**. As explained on textbook p. 748 and demonstrated in Examples 6 and 7, the procedure for sketching an ellipse is as follows:

   I. Write the equation in standard form, so that there is a 1 on the right side of the equation and the coefficients of the squared terms are also 1.

   II. Remember: we are restricting our attention to ellipses with center at (0, 0).

   III. Plot the x-intercepts and the y-intercepts.

   For the x-intercepts, plot $\pm$ the square root of the number under the x^2.

   For the y-intercepts, plot $\pm$ the square root of the number under the y^2.

   IV. Finally, using these intercepts as guiding points, draw the ellipse.

7. **Wording varies**. For a specific example, review text Example 5 (p. 747). Remember that we are not considering rotated conic sections in this course; consequently, there is no xy term (i.e., $B = 0$).

a) The equation of a line has no squared (second-degree) terms. In other words, a linear equation is a first-degree equation.
b) The equation of a parabola has exactly one second-degree variable. Its other variable is first-degree.
c) Both variables in the equation of an ellipse are second-degree (squared). When the equation is written in general form, the coefficients of the squared terms have the same sign.
[Reminder: We treat the circle as a special case of the ellipse.]
d) Like an ellipse, both variables in the equation of a hyperbola are second-degree. However, when the equation is written in general form, the coefficients of the squared terms have opposite signs.

9. The general-form equation of a hyperbola is such that **A and C have opposite signs**. Review, as necessary, the box on textbook p. 747, which summarizes the general form equation for each conic section.

11. There are **two** possibilities for the general-form equation of a parabola:

(i) $\mathbf{A = 0}$ **and** $\mathbf{C \neq 0}$ (in which case, the parabola opens rightward if $C > 0$ or opens leftward if $C < 0$);

(ii) $\mathbf{A \neq 0}$ **and** $\mathbf{C = 0}$ (in which case, the parabola opens upward if $A > 0$ or opens downward if $A < 0$).

Once again, we suggest you review, as necessary, the box on textbook p. 747, which summarizes the general form equation for each conic section.

13. A linear equation is, by definition, a first-degree equation. So, $\mathbf{A = C = 0}$. Review, as necessary, the box on textbook p. 747, which summarizes the general form equation for each conic section.

15. As was pointed out in text Example 1 (p. 743), one can use a compass to help draw a circle.

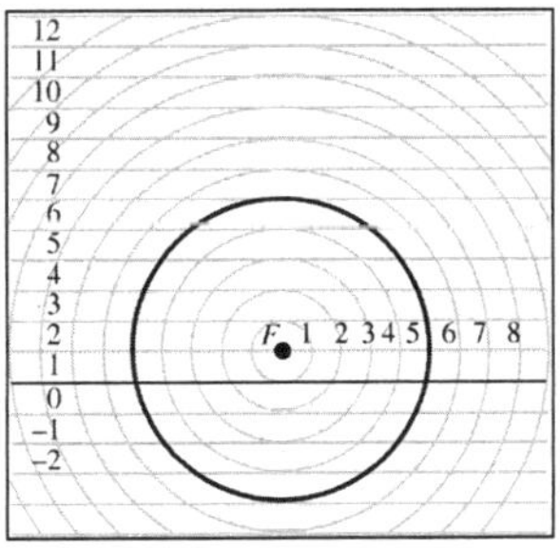

17. This problem is very similar to text Example 2 (see p. 744).

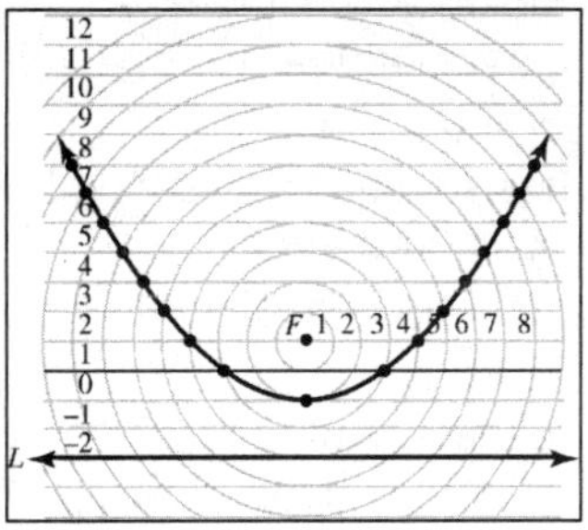

19. Once again, you may find it helpful to take note of text Example 2 (p. 744).

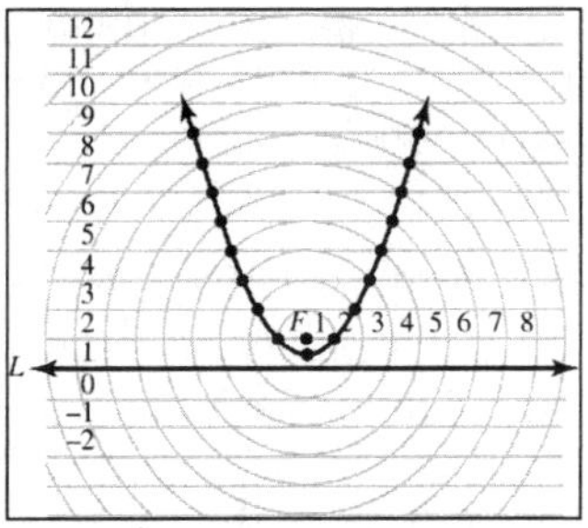

21. This problem is very similar to text Example 3 (pp. 744-745), which explains the graphing process in great detail.

For this particular ellipse, we begin by positioning the foci 10 units apart, and we then plot all the points so that the sum of their distances from the foci is 14, which is equal to the length of the ellipse's major axis.

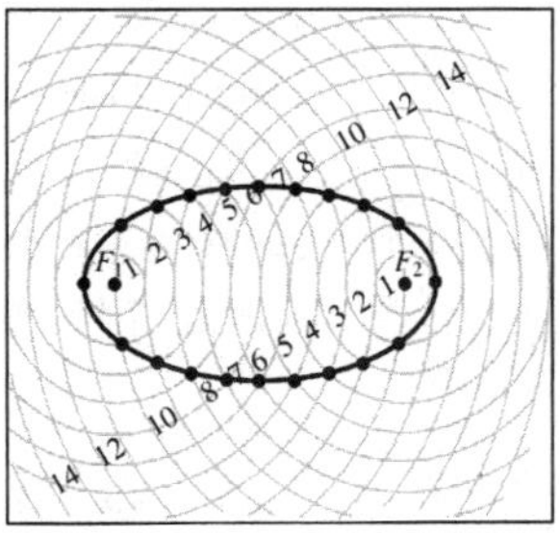

23. This problem is comparable to text Example 4 (p. 746). Notice that the length of the transverse axis (i.e., the distance between the vertices) is equal to 6, which is the difference of the distances from the foci.

To better understand the points shown on the sketch below, let us consider an example. Suppose you plot a point 3 units from F_1. Then the point must be 9 units from F_2, because the difference of these two distances (i.e., 9 – 3) must equal the given constant difference of 6. Note that there will be two possible points that are 3 units from F_1 and 9 units from F_2; both are on the left branch of the hyperbola.

Conversely, there are two points on the right branch that are 9 units from F_1 and 3 units from F_2. As you can see below, the concentric circles on this special graph paper make it easy to continue this point-plotting process.

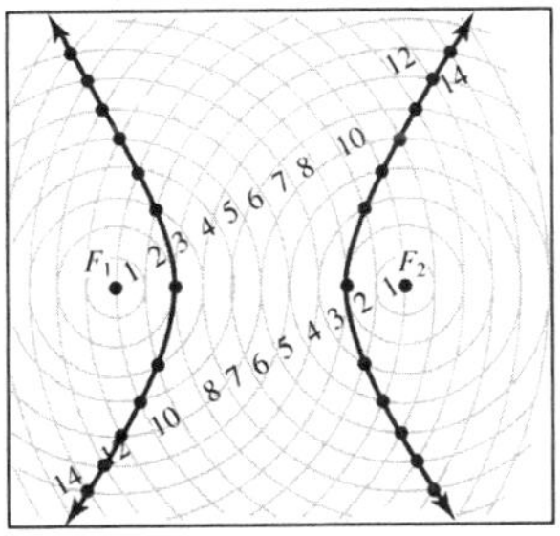

25. Problems 24-27 are akin to text Example 5 (p. 747).

a) Notice that this is a first-degree equation. Thus, the graph is a **line**.

b) Although this equation is not written in general form, it is clear that it is second-degree in both x and y, and the coefficients of the squared terms have the same sign (namely, $A = 1$ and $C = 8$). So, the graph is an **ellipse**.

c) Since there is no x^2 term ($A = 0$), the graph is a **parabola**.

27. Notice that all three of these equations are written in general form (review textbook p. 747, as necessary).

a) Because the coefficient of the y^2 term is nonzero (i.e., $C \neq 0$) and there is no x^2 term ($A = 0$), the graph is a **parabola**.

b) This equation is second-degree in both x and y, and the coefficients of the squared terms have opposite signs ($A = 9$ and $C = -6$). Therefore, the graph is a **hyperbola**.

c) This equation is second-degree in both x and y, and the coefficients of the squared terms have the same sign ($A = 9$ and $C = 6$). Therefore, the graph is an **ellipse**.

29. We see that $A = C = 1$ and $r^2 = 64$; hence, the graph is a circle with its center at (0, 0) and radius 8 units long.

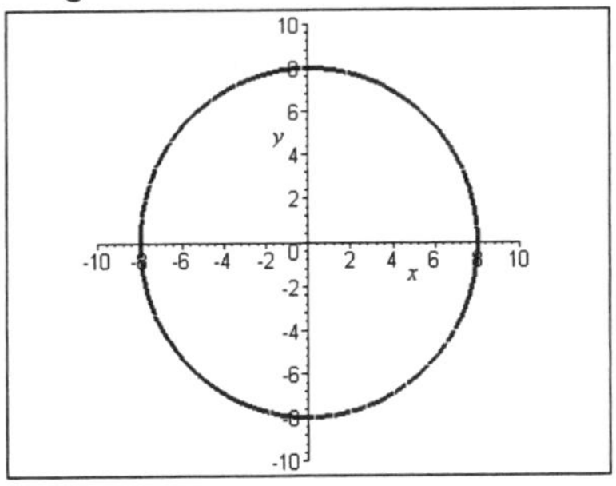

31. Since $A = C = 1$, the graph is a circle. Given that $r^2 = 250$, it follows that the length of the radius is $r = \sqrt{250} \approx 15.8$.

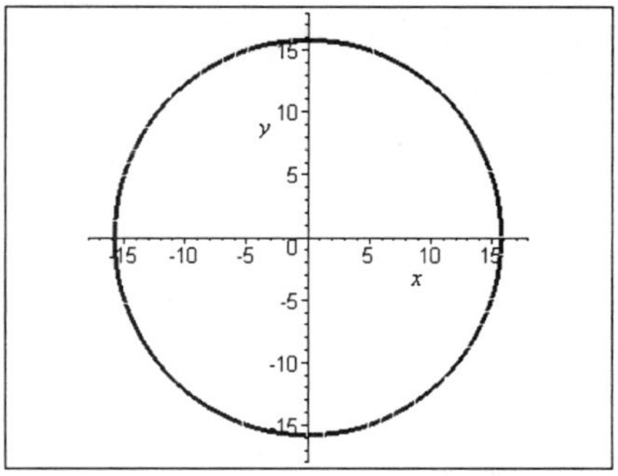

33. This problem is very similar to text Example 6 (p. 748).
We recognize this problem's equation as the standard-form equation of an ellipse. Because the larger number is under the y^2-term, we know the major axis is the y-axis; that is to say, this is a vertical ellipse. With $a^2 = 36$ and $b^2 = 25$, it follows that the y-intercepts (which, because they are the endpoints of the major axis, are called the ***vertices***) are ± 6 and the x-intercepts are ± 5. We use the four intercepts to sketch the ellipse.

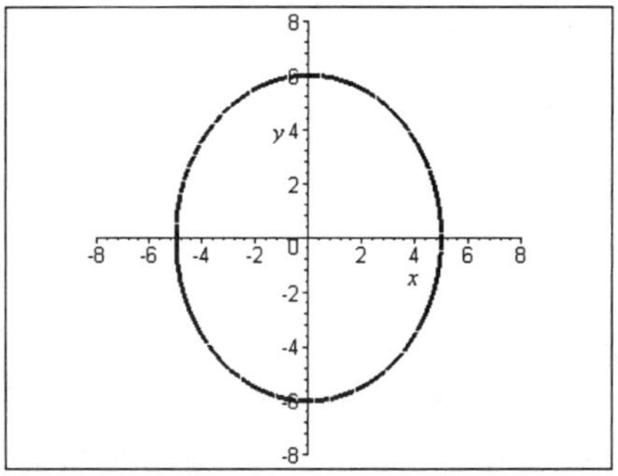

35. This problem is comparable to text Example 7 (pp. 748-749).
By dividing both sides of the equation by 36, we obtain $\frac{4x^2}{36} + \frac{9y^2}{36} = \frac{36}{36}$, which simplifies to the standard-form equation $\frac{x^2}{9} + \frac{y^2}{4} = 1$. Because the larger number is under the x^2-term, the major axis is the x-axis (i.e., this is a horizontal ellipse). With $a^2 = 9$ and $b^2 = 4$, it follows that the x-intercepts (which, because they are the endpoints of the major axis, are called the ***vertices***) are ± 3 and the y-intercepts are ± 2. Using these four intercepts, we obtain the following sketch:

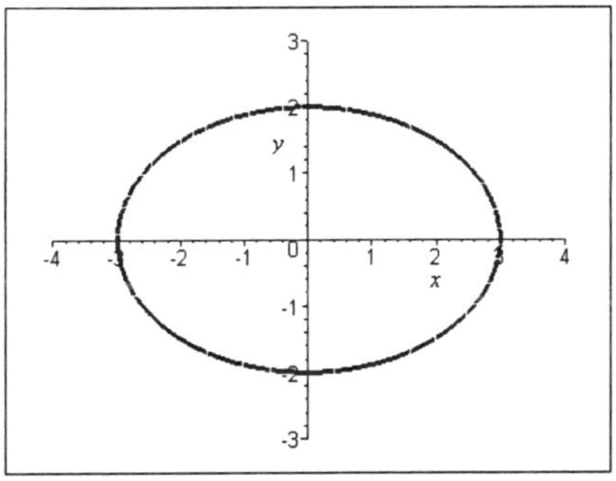

37. Once again, we begin by converting the equation into standard form. By dividing both sides by 900 and simplifying accordingly, we obtain the standard-form equation $\frac{x^2}{25}+\frac{y^2}{36}=1$. The larger number is under the y^2-term, so this is a vertical ellipse (i.e., the major axis is the y-axis). With $a^2=36$ and $b^2=25$, it follows that the y-intercepts (the vertices) are ± 6 and the x-intercepts are ± 5. After plotting these four intercepts, we obtain the following sketch of the ellipse:

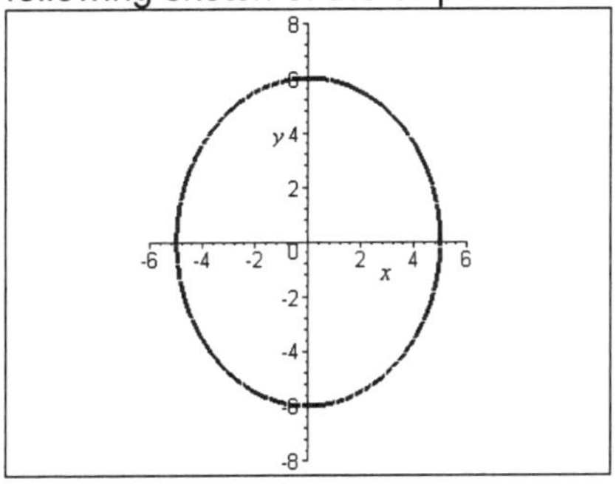

39. *[Note: In the event you have not yet become familiar with the terminology associated with hyperbolas, we have underlined the relevant vocabulary in this solution — in the hope that you will "look them up" in your text.]*

Notice that the coefficients of x^2 and y^2 are opposite in sign; hence, the graph is a hyperbola. Dividing both sides by 4, we obtain the standard-form equation of a hyperbola [review, as necessary, textbook pp. 751-752], specifically $\frac{y^2}{4}-\frac{x^2}{4}=1$. Its vertices are its y-intercepts $(0,\pm 2)$. The hyperbola is a vertical hyperbola. Said differently, the transverse axis is along the x-axis, and the conjugate axis is along the y-axis. Although the graph has no x-intercepts, the points $(\pm 2,0)$ are known as its pseudovertices, which are used (along with the vertices) to form the hyperbola's so-called central rectangle. As demonstrated in text Example 9 (p. 752), we then draw the slant asymptotes (i.e., the diagonal lines passing through the corners of the central rectangle). Finally, using the slant asymptotes (not shown in the graph below) and the vertices, we sketch the hyperbola.

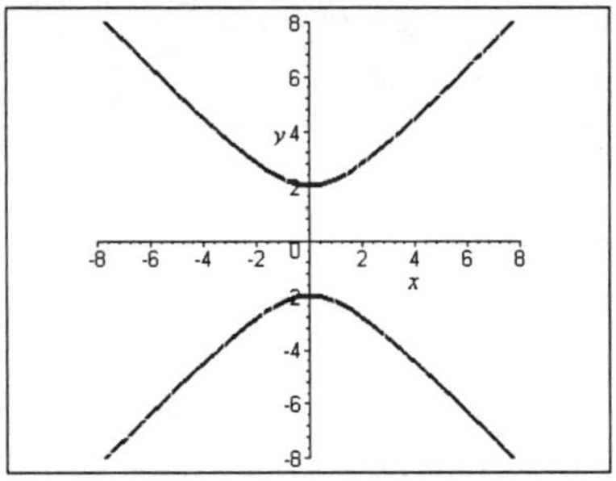

41. For additional details, revisit text Example 9 (p. 752).
Since $a^2 = 4$, $b^2 = 9$, and the x^2-term is positive, we know that this is the equation of a horizontal hyperbola with vertices at $(\pm 2,0)$ and pseudovertices at $(0,\pm 3)$. So, the transverse axis is along the x-axis, and the conjugate axis is along the y-axis. By using the central rectangle and slant asymptotes (neither of which is shown in the graph below) along with the vertices, we obtain the following graph:

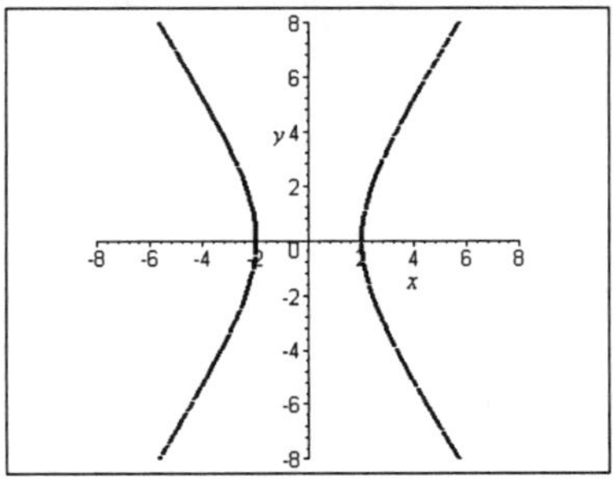

43. For additional details, see text Example 9 (p. 752).
Given that $a^2 = 16$, $b^2 = 36$, and the y^2-term is positive, it follows that this is the equation of a vertical hyperbola with vertices at $(0,\pm 4)$ and pseudovertices at $(\pm 6,0)$. So, the transverse axis is along the y-axis, and the conjugate axis is along the x-axis. By using the central rectangle and slant asymptotes (neither of which is shown in the graph below) along with the vertices, we obtain the following graph:

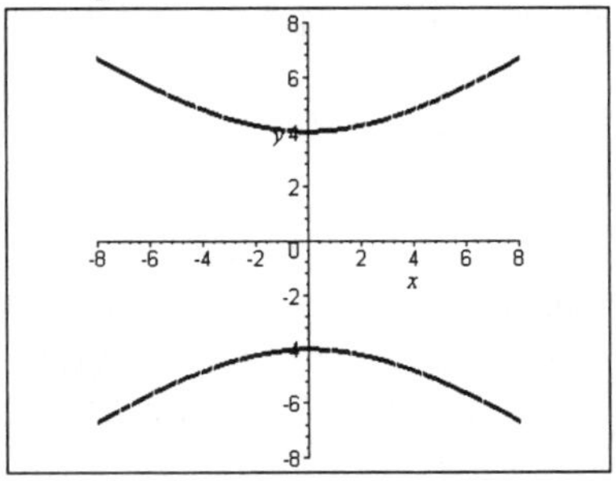

45. We begin by writing the equation in standard form:

$$3y^2 = 4x^2 + 12 \qquad \textit{Given equation}$$

$$3y^2 - 4x^2 = 12 \qquad \textit{Subtract } 4x^2 \textit{ from both sides.}$$

$$\frac{3y^2}{12} - \frac{4x^2}{12} = 1 \qquad \textit{Divide both sides by 12.}$$

$$\frac{y^2}{4} - \frac{x^2}{3} = 1 \qquad \textit{Simplify.}$$

Since $a^2 = 4$, $b^2 = 3$, and the y^2-term is positive, it follows that this is the equation of a vertical hyperbola with vertices at $(0,\pm 2)$ and pseudovertices at $(\pm\sqrt{3},0)$. So, the transverse axis is along the y-axis, and the conjugate axis is along the x-axis. Once again, by using the central rectangle and slant asymptotes (neither of which is shown in the graph below) along with the vertices, we obtain the following graph:

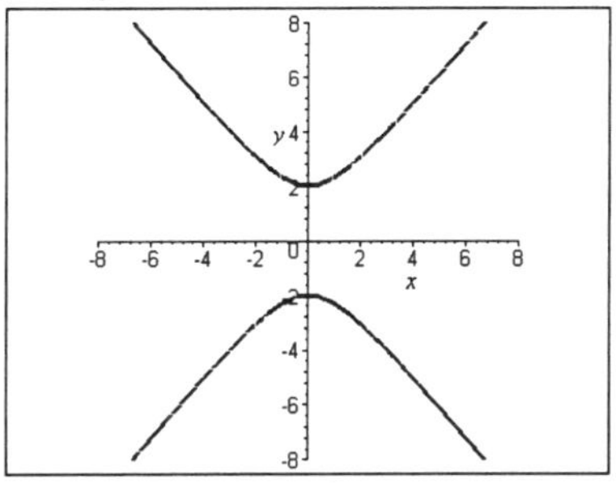

47. By performing the same algebraic steps as shown above in the solution to Problem 45, we obtain the standard-form equation for this hyperbola:

$$\frac{3y^2}{5} - \frac{4x^2}{5} = 1, \text{ which is equivalent to } \frac{y^2}{\frac{5}{3}} - \frac{x^2}{\frac{5}{4}} = 1.$$

Since $a^2 = \frac{5}{3}$, $b^2 = \frac{5}{4}$, and the y^2-term is positive, it follows that this is the equation of a vertical hyperbola with vertices at $(0,\pm\sqrt{\frac{5}{3}})$ [note: $\sqrt{\frac{5}{3}} \approx 1.29$] and pseudovertices at $(\pm\sqrt{\frac{5}{4}},0)$ [note: $\sqrt{\frac{5}{4}} \approx 1.12$]. So, the transverse axis is along the y-axis, and the conjugate axis is along the x-axis. Once again, by using the central rectangle and slant asymptotes (neither of which is shown in the graph below) along with the vertices, we obtain the following graph:

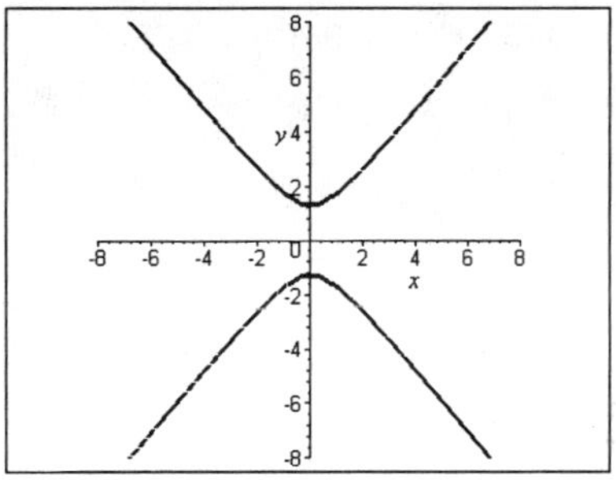

49. The only difference between this problem and Problem 47 (see detailed solution above) is the reversal of the axes.

After writing its standard-form equation $\frac{x^2}{\frac{5}{3}} - \frac{y^2}{\frac{5}{4}} = 1$, it becomes evident that this is the equation of a horizontal hyperbola with vertices at $(\pm\sqrt{\frac{5}{3}},0)$ [note: $\sqrt{\frac{5}{3}} \approx 1.29$] and pseudovertices at $(0,\pm\sqrt{\frac{5}{4}})$ [note: $\sqrt{\frac{5}{4}} \approx 1.12$]. So, the transverse axis is along the x-axis, and the conjugate axis is along the y-axis.

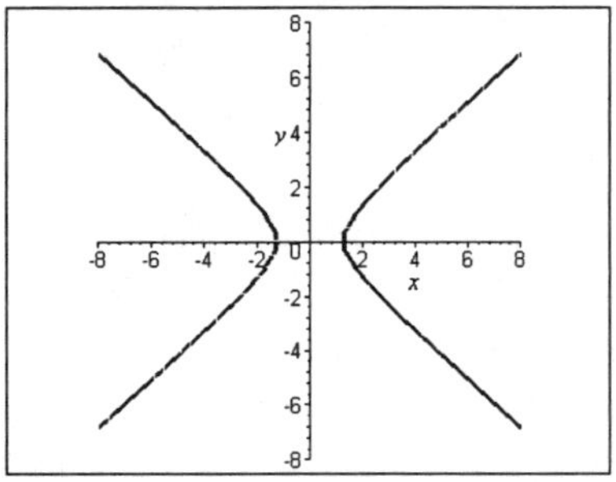

51. The standard-form equation of this hyperbola is $\frac{x^2}{9} - \frac{y^2}{9} = 1$. Since the x^2-term is positive and $a^2 = b^2 = 9$, this is the equation of a horizontal hyperbola with vertices at $(\pm 3,0)$ and pseudovertices at $(0,\pm 3)$. So, the transverse axis is along the x-axis, and the conjugate axis is along the y-axis. Once again, by using the central rectangle and slant asymptotes (neither of which is shown in the graph below) along with the vertices, we obtain the following graph:

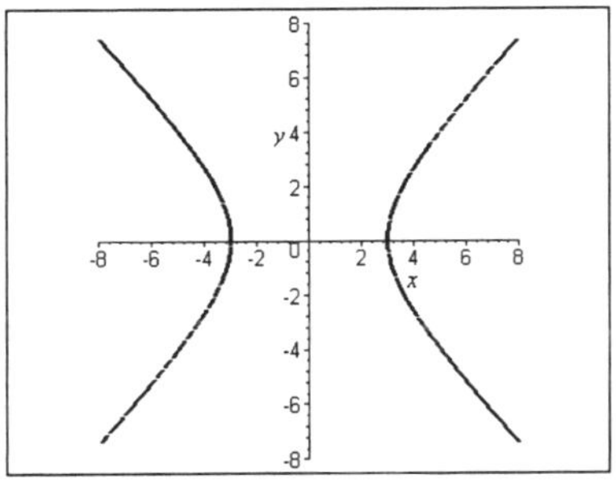

53. This problem is analogous to text Example 8 (pp. 749-750).
The orbit of mars — or, for that matter, any planet — can be described by an ellipse with the sun positioned at one focus. In terms of the notation used in the textbook, the length of the semimajor axis is $a = 1.4 \times 10^8$ mi and the eccentricity is $\in = 0.093$. Recall that eccentricity, a measure of the roundness of an ellipse, is defined as $\in = \dfrac{c}{a}$, where c is the distance from the center to the foci. Substituting the given values and solving accordingly, we obtain:

$$0.093 = \frac{c}{1.4 \times 10^8} \quad \Longrightarrow \quad c = 0.093(1.4 \times 10^8) = 1.302 \times 10^7 \text{ mi}$$

A sketch like the following helps to clarify the elliptical path of mars about the sun, as well as the desired distances:

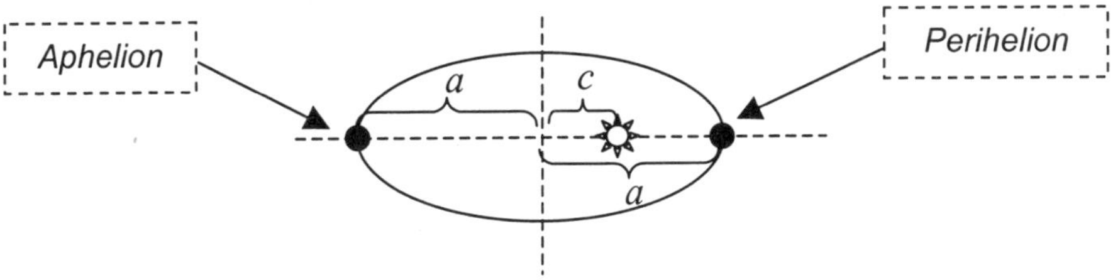

$$\begin{aligned}\text{Greatest distance (from sun to aphelion)} &= a + c \\ &= (1.4 \times 10^8) + (1.302 \times 10^7) \\ &= 1.5302 \times 10^8 \\ &\approx \boxed{\mathbf{1.5 \times 10^8 \text{ mi}}}\end{aligned}$$

$$\begin{aligned}\text{Least distance (from sun to perihelion)} &= a - c \\ &= (1.4 \times 10^8) - (1.302 \times 10^7) \\ &= 1.2698 \times 10^8 \\ &\approx \boxed{\mathbf{1.3 \times 10^8 \text{ mi}}}\end{aligned}$$

55. Your understanding of this solution may be enhanced by first reading the solution to Problem 53 above (including its diagram). Instead of mars orbiting the sun, this problem has to do with the moon orbiting the earth.
We begin by expressing the given information in terms of the variables used in the text:

Greatest distance (from earth to apogee) $= a + c = 199{,}000$ mi

Length of major axis $= 2a = 378{,}000$ mi

It then follows that $a = 189{,}000$. Substituting this value in the first equation gives us the equation $189{,}000 + c = 199{,}000$, which we then solve to obtain $c = 10{,}000$. Finally, with $a = 189{,}000$ and $c = 10{,}000$, we calculate the eccentricity:

$$\epsilon = \frac{c}{a} = \frac{10{,}000}{189{,}000} = \frac{10}{189} \approx \boxed{\mathbf{0.053}}.$$

57. Let the vertex be (0, 0). Since the parabola opens rightward (as depicted in Figure 11.38, p. 755), the equation $y^2 = 4cx$ describes its parabolic-shaped cross section along its axis, with c representing the distance from the focus to the vertex.

Since the parabola passes through a point that is 4 meters to the right and 6 meters above the vertex, the parabola passes through the point (4, 6). From textbook p. 744, we remind you that the points (4, 6) and (4, –6) are the endpoints of what is known as the focal chord of this parabola.

We substitute $x = 4$ and $y = 6$ into the equation $y^2 = 4cx$ and then solve for c:

$$(6)^2 = 4c(4)$$
$$36 = 16c$$
$$c = \tfrac{9}{4} \text{ or } 2.25$$

So, **the focus is 2.25 m from the vertex on the axis of the parabola**.

59. Let the vertex be (0, 0). Since the parabola opens upward, we use the equation $x^2 = 4cy$, where c represents the distance from the focus to the vertex. As Figure 11.39 (textbook p. 755) makes evident, the parabola passes through the point (4, 6) — and also, for that matter, (–4, 6). We substitute $x = 4$ and $y = 6$ into the equation $x^2 = 4cy$ and then solve for c:

$$(4)^2 = 4c(6)$$
$$16 = 24c$$
$$c = \tfrac{2}{3}$$

Since $\frac{2}{3}$ ft equals 8 in., we conclude:

the focus is 8 in. from the vertex on the axis of the parabola.

Problem Set 11.5

1. **Wording varies**. A ***function*** is a set of ordered pairs in which each value of the first component of the ordered pairs is associated with exactly one value of the second component.

You may find it necessary to re-read the above definition several times. In the interest of retention, we suggest you consider adopting less formal language; for example, you might like the "fickle picker" notion explained by the author of your text in Example 1 (see pages 757-758). The idea of a function is essentially this: in term of x's and y's, for a set of ordered pairs to be a function, every value of x must be paired with no more than one value of y. That is to say, with a function, each x-value must be "monogamous" (i.e., it can "date" no more than one y-value). The y-values, however, have no such limitation on the number of "love interests" they may have.

3. Problems 3-14 are patterned after text Example 1 (pp. 757-758). Prior to addressing these problems, you must acquire an understanding of the definition of a function. To review this definition, refer both to your text and this manual's solution to Problem 1 above.
In regards to {(1, 4), (2, 5), (4, 7), (9, 12)}, our answer is: yes, it is a **function**. Each first component of the ordered pairs is associated with exactly one second component.

5. Yes, it is a **function**; each x-value is paired with exactly one y-value. Remember that the definition of a function imposes no pairing limitations on the y-values; only the x-values must be "monogamous."

7. No, it is **not a function** because the first component value 4 picks two values to partner with, namely 3 and 19. In the words of your text's author (Example 1, p. 757), "a function is a set of ordered pairs for which there are no fickle pickers."

9. Yes, it is a **function**; each first component of the ordered pairs is associated with exactly one second component. The fact that 19 and 3 both pick 4 as their second component partner is irrelevant.

11. Yes, it is a **function**. The definition of a function begins with the words "a set of ordered pairs," and a set can consist of a single element (member).

13. No, it is **not a function**. A function must be "a set of order pairs…." However, the set {1, 2, 3, 4, 5} does not consist of ordered pairs.

15. To review the basic concepts associated with function notation and the function machine, refer to textbook pp. 758-759.

a) $f(4)=(4)+8=\boxed{\mathbf{12}}$ b) $f(6)=(6)+8=\boxed{\mathbf{14}}$

c) $f(-8)=(-8)+8=\boxed{\mathbf{0}}$ d) $f(\frac{1}{2})=(\frac{1}{2})+8=\boxed{\mathbf{8\frac{1}{2}}}$

e) $f(x)=\boxed{\mathbf{x+8}}$

17. For additional details pertaining to function notation and the function machine, please revisit textbook pp. 758-759.

a) $M(4)=(4)^2+1=\boxed{\mathbf{17}}$ b) $M(6)=(6)^2+1=\boxed{\mathbf{37}}$

c) $M(-8)=(-8)^2+1=\boxed{\mathbf{65}}$ d) $M(\frac{1}{2})=(\frac{1}{2})^2+1=\frac{1}{4}+1=\boxed{\mathbf{1\frac{1}{4}}}$

e) $M(x)=(x)^2+1=\boxed{\mathbf{x^2+1}}$

19. For additional details pertaining to function notation and the function machine, please revisit textbook pp. 758-759.

a) $g(4)=2(4)-1=\boxed{\mathbf{7}}$ b) $g(6)=2(6)-1=\boxed{\mathbf{11}}$

c) $g(-8)=2(-8)-1=\boxed{\mathbf{-17}}$ d) $g(\frac{1}{2})=2(\frac{1}{2})-1=\boxed{\mathbf{0}}$

e) $g(x)=2(x)-1=\boxed{\mathbf{2x-1}}$

21. a) $f(15)=(15)-7=\boxed{\mathbf{8}}$ b) $f(-9)=(-9)-7=\boxed{\mathbf{-16}}$

c) $f(p)=\boxed{\mathbf{p-7}}$

23. a) $h(0)=3(0)-1=\boxed{\mathbf{-1}}$ b) $h(-10)=3(-10)-1=\boxed{\mathbf{-31}}$

c) $h(a)=\boxed{\mathbf{3a-1}}$

25. a) $g(10)=\dfrac{(10)}{2}=\boxed{\mathbf{5}}$ b) $g(-4)=\dfrac{(-4)}{2}=\boxed{\mathbf{-2}}$

c) $g(3)=\dfrac{(3)}{2}=\boxed{\mathbf{\frac{3}{2}}}$

27. The ***vertical line test*** is introduced on p. 759, and Problems 27-32 are related to text Example 5 (pp. 760-761). The vertical line test is a notion that you should not blindly accept. Instead, based upon the definition of a function, it is a concept/tool you should be able to make perfect sense of. We encourage you to spend the time necessary to do so.

The graph given in this problem does not pass the vertical line test; therefore, it is **not a function**.

To determine the domain and range, we use the information provided below the graph. The graph was created with a graphing calculator, and the information below the graph indicates the dimensions and scale of the so-called "viewing rectangle." The dimensions and scale are values the calculator user can choose to preset. Xmin and Ymin are the respective minimum values for the x-axis and y-axis, while Xmax and Ymax are the respective maximum values for the x-axis and y-axis. Xscl and Yscl are each axis' respective scale (or unit size); in other words, they indicate how far apart each axis' "tick marks" are to be made.

The viewing rectangle in this problem consists of x-axis values ranging from –2 to 2 with a scale of 1. We can see that this particular graph's minimum and maximum x-values are its x-intercepts. Since these intercepts appear to have coordinates (–1, 0) and (1, 0), the ***domain*** is seemingly $\boxed{-1 \le x \le 1}$. Similarly, with the viewing rectangle consisting of y-values ranging from –5 to 5 (with Yscl = 1), the graph itself suggests a ***range*** of $\boxed{-3 \le y \le 3}$.

29. For details regarding the information provided below the graph, please read this manual's solution to Problem 27 immediately above.

The graph does not pass the vertical line test, so it is $\boxed{\textbf{not a function}}$.

Although we cannot be certain (due to the limitations of the given information), this graph appears to be a rightward opening parabola. Its vertex is its x-intercept, which appears to be –3. Because –3 appears to be the graph's minimum x-value and the graph appears to extend without bound to the right, we conclude that the ***domain*** is $\boxed{x \ge -3}$.

As for the y-values, notice the graph extends up and down without bound. Said differently, every possible value (without exception!) on the y-axis corresponds to a point on the graph. Therefore, the ***range*** is the set of all real numbers, $\boxed{\mathbb{R}}$.

31. For details regarding the information provided below the graph, please read this manual's solution to Problem 27 above.

The graph passes the vertical line test, so it is a $\boxed{\textbf{function}}$.

Once again, by using the information provided below the graph and studying the graph itself (e.g., its "high" and "low" values on each axis), we conclude:

Domain: $\boxed{-2 \le x \le 3}$ ***Range***: $\boxed{-9 \le y \le 4}$

33. Note that the forms of some important functions are summarized on textbook p.762. The function $f(x) = 2x^2$ is a $\boxed{\textbf{quadratic}}$ function.

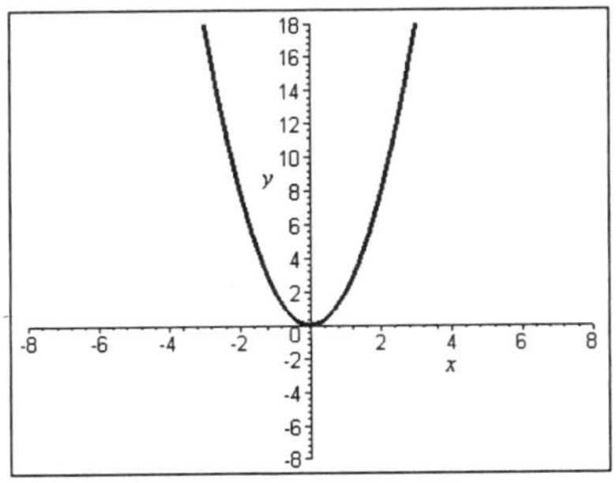

35. This is a **logarithmic** function. To review logarithms, refer to textbook Sections 7.5 – 7.6.

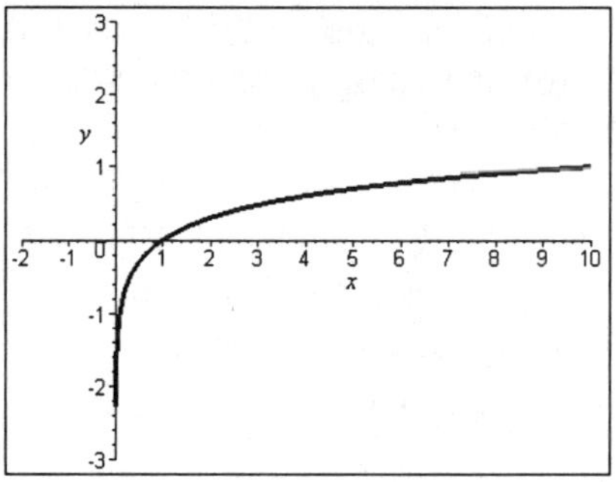

37. We previously encountered this function in Section 10. 4 (Problem 57, p. 699) and in Section 11.3 (Problem 59, p. 741). Based upon the list of function types summarized on p. 762, we classify this function as a **probability** function.

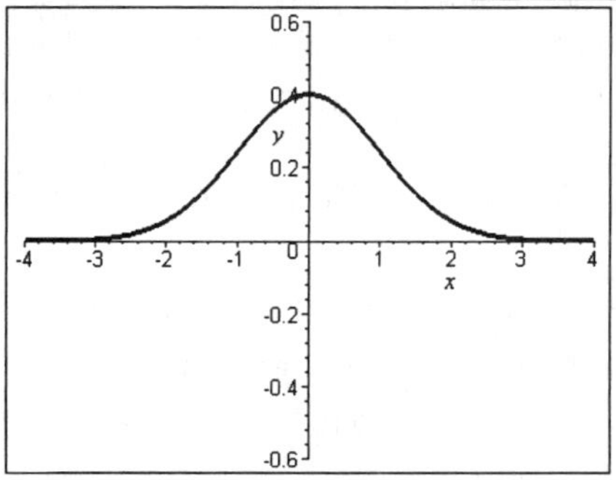

39. The ***difference quotient*** is introduced on textbook p. 759 and demonstrated in Example 4. This concept will be explored in greater depth in Section 11.8.

For $f(x) = 3x - 5$, the difference quotient is:

$$\frac{f(x+h)-f(x)}{h} = \frac{[3(x+h)-5]-[3x-5]}{h}$$

$$= \frac{3x+3h-5-3x+5}{h}$$

$$= \frac{3h}{h}$$

$$= \boxed{3}.$$

41. Given $f(x)=x^3$, the difference quotient is:

$$\frac{f(x+h)-f(x)}{h}=\frac{[(x+h)^3]-[x^3]}{h}$$

$$=\frac{[x^3+3x^2h+3xh^2+h^3]-[x^3]}{h}$$

$$=\frac{3x^2h+3xh^2+h^3}{h}$$

$$=\frac{h(3x^2+3xh+h^2)}{h}$$

$$=\boxed{\mathbf{3x^2+3xh+h^2}}.$$

43. Given $f(x)=\frac{1}{x}$, the difference quotient is:

$$\frac{f(x+h)-f(x)}{h}=\frac{\left[\frac{1}{(x+h)}\right]-\left[\frac{1}{x}\right]}{h}$$

Multiplying both numerator and denominator by the least common multiple $x(x+h)$ yields:

$$=\frac{\frac{1}{(x+h)}-\frac{1}{x}}{h}\cdot\frac{x(x+h)}{x(x+h)}$$

$$=\frac{\frac{x(x+h)}{(x+h)}-\frac{x(x+h)}{x}}{hx(x+h)}$$

$$=\frac{x-(x+h)}{hx(x+h)}$$

$$=\frac{x-x-h}{hx(x+h)}$$

$$=\frac{-h}{hx(x+h)}$$

$$=\boxed{\frac{\mathbf{-1}}{\mathbf{x(x+h)}}}$$

45. Based upon the given formula $v = 32t$, we find:

a) Two seconds ($t = 2$) after having been dropped, the velocity of the rock is $v = 32(2) = \boxed{\textbf{64}}$ ft/sec.

b) Three seconds ($t = 3$) after having been dropped, the velocity of the rock is $v = 32(3) = \boxed{\textbf{96}}$ ft/sec.

c) Four seconds ($t = 4$) after having been dropped, the velocity of the rock is $v = 32(4) = \boxed{\textbf{128}}$ ft/sec.

d) Eight seconds ($t = 8$) after having been dropped, the velocity of the rock is $v = 32(8) = \boxed{\textbf{256}}$ ft/sec.

e) Sixteen seconds ($t = 16$) after having been dropped, the velocity of the rock is $v = 32(16) = \boxed{\textbf{512}}$ ft/sec.

47. Notice that the given formula $V = 2{,}000 - 190n$ is linear with a slope of –190. This tells us that each year the vending machine is losing \$190 of its remaining value. Based upon this formula, we find:

a) The three-year old ($n = 3$) vending machine has a depreciated value of $V = 2{,}000 - 190(3) = \$\boxed{\textbf{1,430}}$.

b) The five-year old ($n = 5$) vending machine has a depreciated value of $V = 2{,}000 - 190(5) = \$\boxed{\textbf{1,050}}$.

c) The seven-year old ($n = 7$) vending machine has a depreciated value of $V = 2{,}000 - 190(7) = \$\boxed{\textbf{670}}$.

d) The nine-year old ($n = 9$) vending machine has a depreciated value of $V = 2{,}000 - 190(9) = \$\boxed{\textbf{290}}$.

e) Consistent with the statement of the problem, we find that the ten-year old ($n = 10$) vending machine does indeed have a depreciated value of $V = 2{,}000 - 190(10) = \$\boxed{\textbf{100}}$.

49. If you forgot what a prime number is, please refer to either your textbook's glossary (p. 924) or Section 4.2 (beginning on p. 166).

a) There are 4 prime numbers less than 10, namely 2, 3, 5, and 7. Thus, $P(10) = \boxed{\textbf{4}}$.

b) By definition, a prime number is a natural number. Negative numbers are not natural numbers, so there are no prime numbers less than –10. Consequently, $P(-10) = \boxed{\textbf{0}}$.

c) As Table 4.2 (textbook p. 170) shows, there are 25 prime numbers less than 100. Hence, $P(100) = \boxed{\textbf{25}}$.

51. The verbal description translates to: $\boxed{y = 3x - 5}$. Since any real number can be used for x, the domain is $\boxed{\mathbb{R}}$.

53. "The difference of the domain value subtracted from 5" translates to $5 - x$. Therefore, the "square root of [this] difference..." leads us to write $\boxed{y = \sqrt{5 - x}}$.

As for the domain, note that $\sqrt{5 - x}$ is a real number only if $5 - x \geq 0$, or $5 \geq x$. Therefore, the domain is $\boxed{x \leq 5}$.

55. a) Let the letters a, b, c, d, and e respectively denote the cities of Arm, Bone, Cheek, Doublehead, and Elbow Lake. By using Figure 11.43a, we obtain the following set of ordered pairs of cities connected by direct service:

$\boxed{\{(a, e), (e, a), (b, e), (e, b), (b, c), (c, b), (c, e), (e, c), (d, e), (e, d)\}}$.

No, it is not a function because several of its first component cities (specifically, b, c, and e) are paired with more than one second component city.

b) By using Figure 11.43b, we obtain the following set of ordered pairs of cities connected by one-way direct service:

$\boxed{\{(a, e), (e, b), (b, c), (c, e), (d, e)\}}$.

Yes, it is a function because each of its first component cities is paired with exactly one second component city.

57. The problem faced by Tom Sawyer and Huckleberry Finn is similar in many respects to that posed in the B.C. comic strip on textbook p. 756.

Tom and Huck could drop a rock into the well and listen for when it hits the bottom. We know 1 mile equals 5,280 feet, and (from p. 756) we know $d = 16t^2$. So, if we assume the well is a mile deep, the time it should take for the rock to hit the bottom can be found by solving the equation $5{,}280 = 16t^2$.

One way to solve the equation is by "trial and error" (i.e., repeatedly substitute values for t into $16t^2$ until $16t^2$ becomes sufficiently close to 5,280). Alternatively, the equation can be solved using the following analytic approach:

$5{,}280 = 16t^2$	*Substitute the depth/distance (in feet).*
$330 = t^2$	*Divide both sides by 16.*
$\pm\sqrt{330} = \sqrt{t^2}$	*Square root both sides.*
$t \approx 18.17$	*Evaluated via calculator. We disregard negative value because time cannot be negative.*

Regardless of the solving method applied, you should reach the following conclusion: it should take approximately **18 seconds** for the rock to hit the bottom.

59. The concepts of perimeter and area were discussed in Sections 7.1 and 7.2.

For a square, we know $A = s^2$ and $P = 4s$. Solving $P = 4s$ for s gives us $s = \boldsymbol{P/4}$. Finally, to obtain the desired function, we substitute $\boldsymbol{P/4}$ for s in the formula $A = s^2$.

Hence, $\boxed{\boldsymbol{A = \left(\frac{P}{4}\right)^2}}$ or $A = \dfrac{P^2}{16}$.

Problem Set 11.6

1. As introduced on textbook pages 765-769, the three main topics of calculus are **limits, derivatives, and integrals**. These topics will be addressed in greater depth in Sections 11.7, 11.8, and 11.9.

3. **Answers vary**. Theoretically, the woman never reaches the wall. However, in reality, she at some point must hit the wall because she has physical dimensions.

5. **Answers vary**. It is reasonable to believe that the sequence of decimals is approaching $0.\overline{3} = \boxed{\frac{1}{3}}$.

7. **Answers vary**. It is reasonable to believe that the sequence of decimals is approaching $0.\overline{9} = \boxed{1}$. We realize you may be skeptical about why the repeating decimal $0.\overline{9} = 0.999...$ is equal to 1. There are a variety of ways to prove this intriguing fact. Given this limited space, however, permit us to share one approach: Let $x = 0.999....$ By dividing both sides by 3, we obtain

$$\frac{x}{3} = 0.333....$$

Since $0.333... = \frac{1}{3}$, we have $\frac{x}{3} = \frac{1}{3}$. Finally, multiplying both sides by 3 yields

$$3 \cdot \frac{x}{3} = \frac{1}{3} \cdot 3$$
$$x = 1$$

Therefore, $0.999... = 1$.

9. **Answers vary**. It is reasonable to believe that the sequence of decimals is approaching the irrational number $\boxed{\pi}$ (pi).

11.

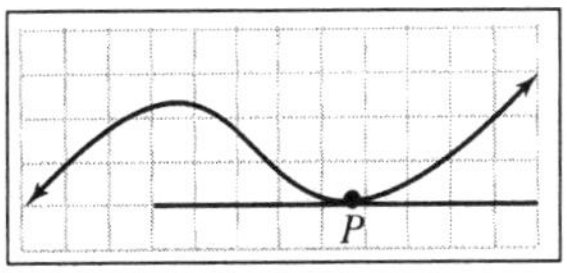

13.

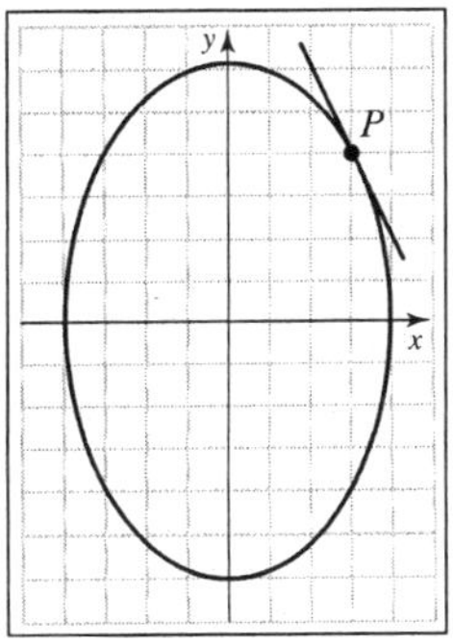

15. No single tangent line exists.

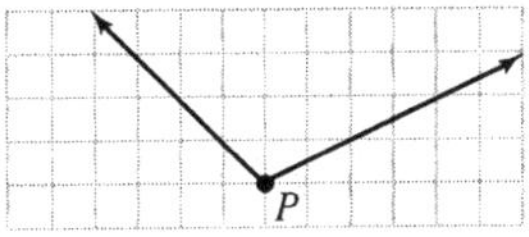

17. Problems 16-21 are related to text Example 1 (p. 767). Please note that in Section 11.7 the topic of limits will be addressed in greater depth.

To make an educated guess about $\lim_{n\to\infty}\frac{2n}{3n+1}$, we consider increasingly larger values for n:

n	100	1,000	10,000	100,000
$\frac{2n}{3n+1}$	$\frac{200}{301}\approx 0.664452$	$\frac{2,000}{3,001}\approx 0.666445$	$\frac{20,000}{30,001}\approx 0.666644$	$\frac{200,000}{300,001}\approx 0.666664$

Based upon this analysis, it is reasonable to guess that the sequence of fractions is approaching $0.666... = 0.\overline{6}$, which is equal to $\boxed{\frac{2}{3}}$.

19. To make an educated guess about $\lim_{n\to\infty} \dfrac{n+1}{2n}$, we consider increasingly larger values for n:

n	100	1,000	10,000	100,000
$\dfrac{n+1}{2n}$	$\frac{101}{200} = 0.505$	$\frac{1,001}{2,000} = 0.5005$	$\frac{10,001}{20,000} = 0.50005$	$\frac{100.001}{200,000} = 0.500005$

Based upon this analysis, it is reasonable to guess that the sequence of fractions is approaching $0.5 = \boxed{\frac{1}{2}}$.

21. To make an educated guess about $\lim_{n\to\infty} \dfrac{3n^2+1}{2n^2-1}$, we consider increasingly larger values for n:

n	10	100	1,000
$\dfrac{3n^2+1}{2n^2-1}$	$\frac{301}{199} \approx 1.512563$	$\frac{30,001}{19,999} \approx 1.500125$	$\frac{3,000,001}{1,999,999} \approx 1.50000125$

So, it is reasonable to guess that the sequence of fractions is approaching $1.5 = \boxed{\frac{3}{2}}$.

23. One approach to estimate this area (among several) is to make use of the graph's underlying grid. As indicated by the scale marked on the axes, the dimensions of each of the grid's underlying "squares" (actually, they are rectangles) are ½ unit wide by 4 units high. Hence, each has an area of 2 square units [i.e., $\ell w = 4(\frac{1}{2}) = 2$].

We then count the number of squares under the curve. There are approximately 10 squares; specifically, there are 6 whole squares and the remaining partial squares account for approximately 4 more.

Thus, since there are approximately 10 squares each with an area of 2 square units, we estimate the total area to be 10 × 2 = $\boxed{\mathbf{20}}$.

[Note: An alternative approach to this estimating problem is to construct rectangles comparable to those illustrated in Figure 11.51 (textbook p. 769) and then calculate the sum of their areas. Also, you may wish to refer to our solution to Problem 27, in which we demonstrate yet another approach.]

25. In the same manner as explained in our solution to Problem 23 above, we can choose to estimate the area by making use of the graph's underlying grid. Alternatively, of course, one can (i) construct rectangles comparable to those illustrated in Figure 11.51 (textbook p. 769), or (ii) apply the strategy demonstrated in our solution to Problem 27 below.

Note that each of the grid's underlying squares has an area of 0.5 square units (i.e., 0.5 units wide by 1 unit high). As for the total number of underlying squares within the shaded region below the curve, one's estimates are likely to range between 8 and 10. Hence, we estimate the total area to be $\boxed{\textbf{4 or 5}}$.

27. Once again, in the same manner as explained in our solutions to Problems 23 and 25 (see solutions), you can estimate this area by counting the squares within the graph's underlying grid. In so doing, it is important to note that each square is ½ unit on a side; hence, the area of each square is is $\frac{1}{2} \times \frac{1}{2} = \frac{1}{4}$ square unit. There are approximately 12 squares in all, so the area of the shaded region is approximately $12 \times \frac{1}{4} = \boxed{3}$.

[Note: This graph has symmetry with respect to the y-axis; that is, the right half is a mirror image (exact replica) of the left half. Can you devise an estimation strategy that makes use of this helpful attribute?]

Finally, please allow us to expose you to an approach that is a bit more sophisticated than those explained previously. We begin by considering two rectangles:

I. Sketch a rectangle with vertical sides containing $x = -1$ and $x = 1$ and horizontal sides $y = 0$ (i.e., the x-axis) and $y = 1$. This rectangle has a width of 2 units (from –1 to 1) and a height (length) of 1 unit (from 0 to 1). Hence, its area is 2 square units [i.e., $\ell w = 1(2) = 2$]. Notice that the given graph, which is a semi-elliptical shape, has an area greater than the area of this rectangle.

II. Now, sketch a rectangle again with vertical sides containing $x = -1$ and $x = 1$ but this time with horizontal sides $y = 0$ (i.e., the x-axis) and $y = 2$. This rectangle has a width of 2 units (from –1 to 1) and a height (length) of 2 unit (from 0 to 2). Hence, its area is 4 square units [i.e., $\ell w = 2(2) = 4$]. Notice that the given semi-elliptical graph is completely contained within this rectangle. Hence, the area under the semi-elliptical curve is less than the area of the rectangle.

Therefore, since the area of the given semi-elliptical shape is greater than the area of Rectangle I (area = 2 sq units) and less than the area of Rectangle II (area = 4 sq units), we estimate the shaded area to be 3.

29. **Answers vary**. You should find text Example 2 (pp. 771-774) to be quite instructive in the process involved in setting up a personal budget. Remember that the purpose behind this example was to demonstrate — in a very "real world" way — the (abstraction) process of mathematical modeling.

Problem Set 11.7

1. **Answers vary**. The ***limit of a sequence*** means that successive terms of the sequence are getting closer and closer to ("converging to") the number called the limit. Two notes are worthy of mention: (i) to review the concept of a sequence, please revisit Section 8.3; and (ii) the formal (mathematical) definition of a limit is beyond the scope of this text.

3. Problems 3-6 are comparable to text Example 1 (p. 777).

We observe that when n is even $1+(-1)^n$ equals 2, whereas when n is odd $1+(-1)^n$ equals 0. Hence, the first five terms of this sequence (i.e., for $n = 1, 2, 3, 4, 5$) are respectively **0, 2, 0, 2, 0**.

5. A table like the following can be a helpful way to organize one's work:

n	1	2	3	4	5
$\dfrac{3n+1}{n+2}$	$\frac{3(1)+1}{(1)+2} = \boxed{\frac{4}{3}}$	$\frac{3(2)+1}{(2)+2} = \boxed{\frac{7}{4}}$	$\frac{3(3)+1}{(3)+2} = \frac{10}{5} = \boxed{2}$	$\frac{3(4)+1}{(4)+2} = \boxed{\frac{13}{6}}$	$\frac{3(5)+1}{(5)+2} = \boxed{\frac{16}{7}}$

7. <u>**Remember**</u>: There are a variety of ways to determine limits. They include:

- Graphing the sequence either in one or two dimensions;
- Constructing a table of values in an effort to detect a pattern;
- Performing algebraic manipulations, as necessary, and then making use of the conclusions stated on the bottom of textbook p. 778.

You would be well advised to become familiar with <u>all</u> three of these approaches — as well as any others your instructor happens to provide. Doing so will greatly enhance your understanding of limits and, in the process, will decrease the likelihood of your experiencing confusion/anxiety on an upcoming exam.
In this first solution, let us share two approaches; afterwards, we encourage you to try another approach.

We begin with the table-of-values method:

n	100	1,000	10,000	100,000	1,000,000
$\dfrac{8{,}000n}{n+2}$	$\approx 7{,}873$	$\approx 7{,}984$	$\approx 7{,}998$	$\approx 7{,}999.8$	$\approx 7{,}999.98$

It, therefore, appears that $\lim\limits_{n\to\infty} \dfrac{8{,}000n}{n+2} = \boxed{8{,}000}$.

You might have noticed that, as n gets larger, the effect of adding 2 in the denominator becomes increasingly negligible (unimportant). In fact, as n approaches infinity, we find the '2' can be ignored entirely; that is to say,

$$\frac{8{,}000n}{n+2} \approx \frac{8{,}000n}{n}.$$

Hence, $\lim\limits_{n\to\infty} \dfrac{8{,}000n}{n+2} = \lim\limits_{n\to\infty} \dfrac{8{,}000\not{n}}{\not{n}} = \lim\limits_{n\to\infty} 8{,}000 = 8{,}000$.

9. If you have not done so already, please take a moment to read this manual's solution to Problem 7, in which we remind you of the variety of approaches available to you.

In the interest of reinforcement, we begin this solution with the way our solution to Problem 7 ended. As n becomes increasingly larger, we find that adding 1 in the numerator and subtracting 4 in the denominator have a very minor effect on the overall fraction obtained; that is, as $n \to \infty$,

$$\frac{2n+1}{3n-4} \approx \frac{2n}{3n}.$$

Based upon this line of reasoning, it follows that $\lim_{n\to\infty} \frac{2n+1}{3n-4} = \lim_{n\to\infty} \frac{2\not{n}}{3\not{n}} = \lim_{n\to\infty} \frac{2}{3} = \boxed{\frac{2}{3}}$.

[Suggestion: We strongly encourage you to re-attempt this problem, using a different approach (e.g., a table of values, a graph, algebraic manipulations).]

11. This problem is very similar to text Example 3b (p. 779) and Example 2d (p. 778). We begin by algebraically rewriting the given expression:

$$\frac{5n+8}{n} = \frac{5n}{n} + \frac{8}{n} = 5 + \frac{8}{n}$$

We then apply the limit to infinity property (see the bottom of p. 778) and assume that the limit of a sum is the sum of the limits:

$$\lim_{n\to\infty} \frac{5n+8}{n} = \lim_{n\to\infty}\left[5 + \frac{8}{n}\right] = \lim_{n\to\infty} 5 + \lim_{n\to\infty} \frac{8}{n} = 5 + 0 = \boxed{5}.$$

13. This problem is comparable to text Examples 3b, 3c, and 4c (pp. 779-780). Viewing the numerator and denominator as two separate polynomials, we notice that the highest degree term is n^2. So, we divide each term of the numerator and denominator by n^2 and proceed to apply the limit to infinity property (stated on the bottom of textbook p. 778):

$$\lim_{n\to\infty} \frac{8n^2 + 800n + 5{,}000}{2n^2 - 1{,}000n + 2} = \lim_{n\to\infty} \frac{\frac{8n^2}{n^2} + \frac{800n}{n^2} + \frac{5{,}000}{n^2}}{\frac{2n^2}{n^2} - \frac{1{,}000n}{n^2} + \frac{2}{n^2}}$$

$$= \lim_{n\to\infty} \frac{8 + \frac{800}{n} + \frac{5{,}000}{n^2}}{2 - \frac{1{,}000}{n} + \frac{2}{n^2}}$$

$$= \frac{8+0+0}{2-0+0}$$

$$= \boxed{4}.$$

Note: Alternatively, we could use the approach previously detailed in this manual's solutions to Problem 7 and 9. In particular, as n becomes very, very large, we find that adding $5{,}000$ in the numerator and adding 2 in the denominator have a very minor effect on the overall fraction obtained. Moreover, in the case of the numerator, $800n$ will lag further and further behind the leading term $8n^2$, rendering it increasingly negligible as $n \to \infty$. Similarly, in the denominator, $-1{,}000n$ becomes miniscule in comparison to $2n^2$ as $n \to \infty$. Thus, only the highest-degree terms are worthy of consideration; in other words, as n approaches infinity we find

$$\frac{8n^2+800n+5{,}000}{2n^2-1{,}000n+2} \approx \frac{8n^2}{2n^2} = 4,$$

which is consistent with the answer indicated above.

15. This problem, too, is comparable to text Examples 3b, 3c, and 4c (pp. 779-780). Viewing the numerator and denominator as two separate polynomials, we notice that the highest degree term is n^3. This observation prompts us to divide each term of the numerator and denominator by n^3 and then apply the limit to infinity property (stated on the bottom of textbook p. 778):

$$\lim_{n\to\infty} \frac{8n^2+6n+4{,}000}{n^3+1} = \lim_{n\to\infty} \frac{\frac{8n^2}{n^3}+\frac{6n}{n^3}+\frac{4{,}000}{n^3}}{\frac{n^3}{n^3}+\frac{1}{n^3}} = \lim_{n\to\infty} \frac{\frac{8}{n}+\frac{6}{n^2}+\frac{4{,}000}{n^3}}{1+\frac{1}{n^3}} = \frac{0+0+0}{1+0}.$$

Therefore, $\lim_{n\to\infty} \frac{8n^2+6n+4{,}000}{n^3+1} = \boxed{0}$.

Alternatively, please note that we could use the approach previously detailed in this manual's solutions to Problem 7, 9, and 13. To be specific, as n becomes very, very large, we find that adding $4{,}000$ in the numerator and adding 1 in the denominator have a very minor effect on the overall fraction obtained. Moreover, in the case of the numerator, $6n$ will lag further and further behind the leading term $8n^2$, rendering it increasingly negligible as $n \to \infty$. Consequently, only the highest-degree terms are worthy of consideration; in other words, as n approaches infinity we find

$$\frac{8n^2+6n+4{,}000}{n^3+1} \approx \frac{8n^2}{n^3} = \frac{8}{n} \text{ and } \lim_{n\to\infty} \frac{8}{n} = 0,$$

which is consistent with the answer indicated above.

17. This problem is similar to text Example 3a (p. 779).

As n becomes larger and larger, $\frac{1}{3n}$ becomes smaller and smaller; that is, $\frac{1}{3n}$ gets closer and closer to zero. In the event this is difficult for you to understand, take a moment to construct either a table of values or a graphical representation of this sequence. Ultimately, you will agree that

$$\lim_{n\to\infty}\frac{1}{3n}=\boxed{\mathbf{0}}.$$

19. It is reasonable to believe that the sequence of decimals is approaching $0.6\overline{9}=\boxed{\mathbf{0.7}}$. We realize you may be skeptical about why the repeating decimal $0.6\overline{9}=0.699...$ is equal to 0.7, yet there are various ways to formally prove it. In this manual's solution to Problem 7 in Section 11.6, we demonstrated one approach (among many) to establish the fact that $0.\overline{9}=0.999...$ equals 1. As for $0.6\overline{9}$ equaling 0.7, let us share a different method.

Let $x=0.69999...=0.6\overline{9}$. It then follows that $10x=6.99999...=6.\overline{9}$. So,

$$\begin{aligned}10x-x&=6.9\overline{9}-0.6\overline{9}\\9x&=6.3\\x&=\frac{6.3}{9}\\x&=0.7\end{aligned}$$

Therefore, $0.69999...=0.7.$

21. Once again, we begin by viewing the numerator and denominator as two separate polynomials. In this particular case, we notice that the highest degree term is n^4. This observation prompts us to divide each term of the numerator and denominator by n^4 and then apply the limit to infinity property:

$$\lim_{n\to\infty}\frac{3n^2-7n+2}{5n^4+9n^2}=\lim_{n\to\infty}\frac{\frac{3n^2}{n^4}-\frac{7n}{n^4}+\frac{2}{n^4}}{\frac{5n^4}{n^4}+\frac{9n^2}{n^4}}$$

$$= \lim_{n\to\infty} \frac{\frac{3}{n^2} - \frac{7}{n^3} + \frac{2}{n^4}}{5 + \frac{9}{n^2}}$$

$$= \frac{0 - 0 + 0}{5 + 0}$$

$$= \boxed{\mathbf{0}}.$$

Alternatively, please note that we could make use of the approach previously detailed in this manual's solutions to Problem 7, 9, 13, and 15. Specifically, as n becomes increasingly large, we find that the lower-degree terms in the numerator (i.e., $-7n$ and 2) have a very minor effect on the overall fraction obtained. Similarly, in the case of the denominator, $9n^2$ will lag further and further behind the leading term $5n^4$, rendering it increasingly negligible as $n \to \infty$. Thus, only the highest-degree terms are worthy of consideration; in other words, as n approaches infinity we find

$$\frac{3n^2 - 7n + 2}{5n^4 + 9n^2} \approx \frac{3n^2}{5n^4} = \frac{3}{5n^2},$$

and $\lim_{n\to\infty} \frac{3}{5n^2} = 0$, which is consistent with the answer indicated above.

23. This problem is very similar to text Example 4c (pp. 780-781).

Dividing each term of the numerator and denominator by n^4 and applying the limit to infinity property yields:

$$\lim_{n\to\infty} \frac{2n^4 + 5n^2 - 6}{3n + 8} = \lim_{n\to\infty} \frac{\frac{2n^4}{n^4} + \frac{5n^2}{n^4} - \frac{6}{n^4}}{\frac{3n}{n^4} + \frac{8}{n^4}}$$

$$= \lim_{n\to\infty} \frac{2 + \frac{5}{n^2} - \frac{6}{n^4}}{\frac{3}{n^3} + \frac{8}{n^4}}$$

As $n \to \infty$, the numerator approaches 2, and the denominator tends toward 0. The consequence of the denominator tending toward 0 is that the fraction (quotient) becomes an ever-increasing positive number. Hence, the fraction fails to converge upon a specific number L, instead it diverges and we declare:

$$\lim_{n\to\infty} \frac{2n^4 + 5n^2 - 6}{3n + 8} = \boxed{\infty}.$$

Alternatively, please note that this conclusion could have been reached by comparing the degrees of the numerator and the denominator. [From previous solutions in this manual, you will recall that the lower-degree terms become insignificant as n approaches infinity.] Since the numerator is a 4th degree polynomial and the denominator is a 1st degree polynomial, we know as $n \to \infty$ that the numerator will grow at a much more rapid rate than the denominator. Consequently, the resulting fraction grows arbitrarily large. Additionally, it should be noted that this particular fraction takes on positive values as $n \to \infty$. Therefore, the limit does not exist, and (since this fraction grows towards positive infinity) we write "∞."

25. This problem is analogous to those encountered previously. For a more detailed solution as well as an explanation of alternative approaches, please refer to this manual's solutions for previous problems within this section; most notably, Problems 7, 9, 11, 13, 15, and 21.

Dividing each term of the numerator and denominator by n^2 and applying the limit to infinity property yields:

$$\lim_{n\to\infty}\frac{15+9n-6n^2}{2n^2-4n+1}=\lim_{n\to\infty}\frac{\frac{15}{n^2}+\frac{9n}{n^2}-\frac{6n^2}{n^2}}{\frac{2n^2}{n^2}-\frac{4n}{n^2}+\frac{1}{n^2}}=\lim_{n\to\infty}\frac{\frac{15}{n^2}+\frac{9}{n}-6}{2-\frac{4}{n}+\frac{1}{n^2}}=\frac{0+0-6}{2-0+0}=\boxed{-3}.$$

27. This problem also is analogous to those encountered previously. For a more detailed solution as well as an explanation of alternative approaches, please refer to this manual's solutions for previous problems within this section; most notably, Problems 7, 9, 11, 13, 15, and 21.

Dividing each term of the numerator and denominator by n^3 and applying the limit to infinity property yields:

$$\lim_{n\to\infty}\frac{-21n^3+52}{-7n^3+n^2+20n-9}=\lim_{n\to\infty}\frac{-21+\frac{52}{n^3}}{-7+\frac{1}{n}+\frac{20}{n^2}-\frac{9}{n^3}}=\frac{-21+0}{-7+0+0-0}=\boxed{3}.$$

29. After 1 hour, 12 mg (i.e., half of 24 mg) of the drug is present within the body. After 2 hours, 6 mg remain (i.e., half of 12 mg); after 3 hours, there are 3 mg; and so on. Thus, the first five terms of this geometric sequence are $a_1 = 12$, $a_2 = 6$, $a_3 = 3$, $a_4 = \frac{3}{2} = 1.5$, and $a_5 = \frac{3}{4} = 0.75$. Hence, after 5 hours, **0.75 mg** of the drug is present. More generally, after n hours, $\boxed{24\left(\frac{1}{2}\right)^n}$ of the drug is present.

Problem Set 11.8

1. Problems 1-5 are related to text Example 1 (p. 783) and Example 3 (pp. 784-785). In this problem, we are calculating what is known as the ***average speed***. In terms of a graph, the average rate of change is equal to the slope of the secant line (as shown in Figure 11.56, p. 783).

a) From the given graph (11.60), we know that after 1 second the projectile is positioned at a height of 40 feet, and after 8 sec, its position is 80 feet. Hence, the average rate of change during that 7 sec span is:

$$\text{Average } = \frac{80-40}{8-1} = 5.\overline{714285} \approx \boxed{\textbf{5.71 ft/sec}}.$$

b) $\text{Average } = \dfrac{100-40}{5-1} \approx \boxed{\textbf{15 ft/sec}}$

c) $\text{Average } = \dfrac{80-40}{8-1} \approx \boxed{\textbf{5.71 ft/sec}}$

d) Unlike parts (a)-(c), this question asks us to find (or at least "estimate") the **instantaneous** rate of change — not the "average" rate of change — at $t=1$. In terms of the graph, this means we need to determine the slope of the **tangent** line — instead of the "secant" line. To determine this slope, you begin by sketching a line tangent to the curve at the point $(1,40)$. It looks like the slope of this tangent line has a rise of 4 squares (i.e., 40 feet) and a run of 1 square (i.e., 1 sec), so the instantaneous rate of change at $t=1$ is $\boxed{\textbf{40 ft/sec}}$.

3. Table 11.1 is shown on textbook p. 783. Since this problem (just like Example 1) pertains to average driving speed, the desired units are miles per hour (mph).

a) From 6:09 AM to 6:36 AM is a span of 27 min (or $\frac{27}{60}$ hr). Hence,

$$\text{Average speed } = \frac{26.7-0}{\frac{27}{60}} = 59.\overline{3} \approx \boxed{\textbf{59 mph}}.$$

b) $\text{Average speed } = \dfrac{35-26.7}{7:03-6:36} = \dfrac{8.3}{\frac{27}{60}} = 18.\overline{4} \approx \boxed{\textbf{18 mph}}$

c) $\text{Average speed } = \dfrac{50-35}{7:28-7:03} = \dfrac{15}{\frac{25}{60}} = \boxed{\textbf{36 mph}}$

d) $\text{Average speed } = \dfrac{50-0}{7:28-6:09} = \dfrac{50}{\frac{79}{60}} \approx \boxed{\textbf{38 mph}}$

5. This problem is similar to its predecessors (Problems 1-4) and the text's Examples 1 and 3. Please note that \$ represents <u>trillions</u> of dollars. Using Table 11.3, we obtain the following yearly rates of change of the Gross National Product (GNP):

a) 1960 to 1996: Average $= \dfrac{7.6360-0.5153}{1996-1960} \approx$ **0.198 \$/yr**

b) 1970 to 1996: Average $= \dfrac{7.6360-1.0155}{1996-1970} \approx$ **0.255 \$/yr**

c) 1980 to 1996: Average $= \dfrac{7.6360-2.7320}{1996-1980} \approx$ **0.307 \$/yr**

d) 1990 to 1996: Average $= \dfrac{7.6360-5.5461}{1996-1990} \approx$ **0.348 \$/yr**

e) 1995 to 1996: Average $= \dfrac{7.6360-7.2654}{1996-1995} \approx$ **0.371 \$/yr**

f) This question asks us to estimate the instantaneous rate of change, which refers to the slope of the tangent line at the indicated point, which in this case happens to be (1996, 7.6360). Of the secant line slopes obtained in parts (a)-(e), the best approximation to the tangent line slope at this point is the slope obtained in part (e) because (1995, 7.2654) is the closest given point to (1996, 7.6360). Therefore, we estimate that in 1996 the GNP is changing at a rate of **\$371 billion per year**.

7. Based upon the sketch of the tangent line shown below (and, in particular, its slope), we estimate that the slope of the curve at point P is $\boxed{\frac{3}{2}}$.

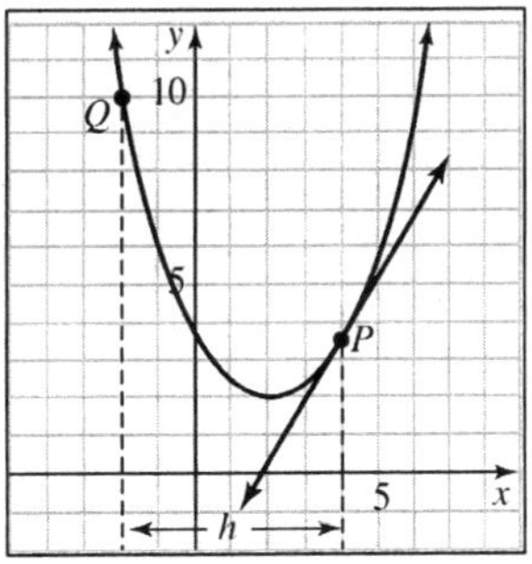

9. The slope is **−2**.

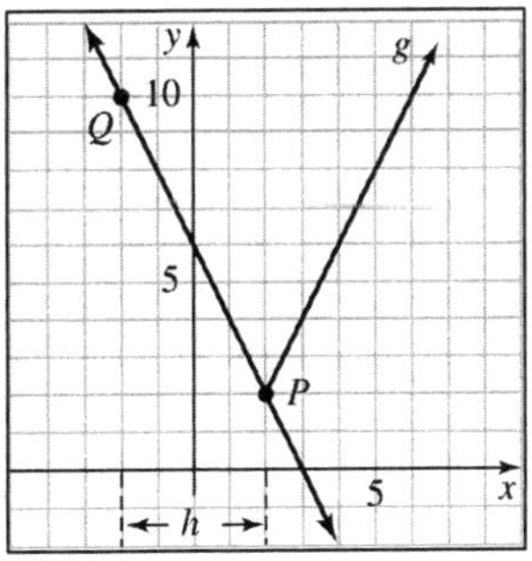

11. The slope is **−2**.

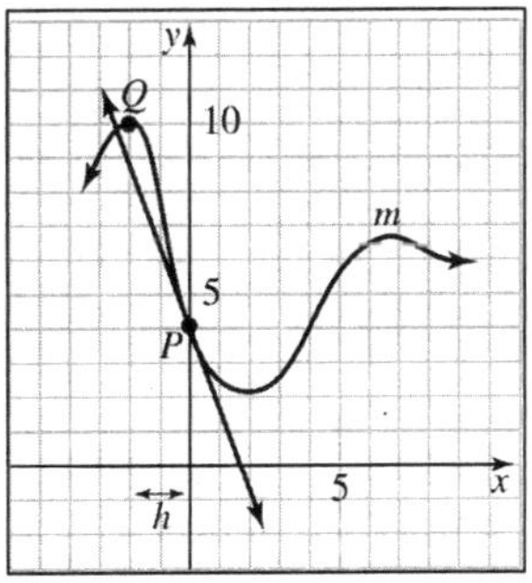

13. If you are familiar with the shape of the graph of $f(x)=5$, we should be mentioned that the algebraic steps shown below become largely unnecessary. That is to say, from our previous work (in particular, see Example 8a in Section 11.1, pp. 730-731), we know the graph of this function is a horizontal line and the slope of such lines equals zero. Thus, we should expect this function's average and instantaneous rates of change to equal 0; you will find that the following computations confirm this expectation.

a) This problem is very similar to text Example 2 (p. 784).
For $x=-3$ to $x=3$, we have $x=-3$ and $h=3-(-3)=6$. So, the average rate of change from $x=-3$ to $x=3$ is

$$\frac{f(x+h)-f(x)}{h}=\frac{f(3)-f(-3)}{6}=\frac{5-5}{6}=\boxed{0}.$$

b) The concept of instantaneous rates of change is introduced on textbook pp. 785-787. For $f(x)=5$, the instantaneous rate of change at $x=-3$ is

$$\begin{aligned}\lim_{h\to 0}\frac{f(x+h)-f(x)}{h}&=\lim_{h\to 0}\frac{f(-3+h)-f(-3)}{h}\\&=\lim_{h\to 0}\frac{5-5}{h}\\&=\lim_{h\to 0}0\\&=\boxed{0}.\end{aligned}$$

15. Given $f(x)=-2x^2+x+4$,

a) The average rate of change from $x=4$ to $x=9$ is

$$\begin{aligned}\frac{f(x+h)-f(x)}{h}&=\frac{f(9)-f(4)}{9-4}\\&=\frac{[-2(9)^2+(9)+4]-[-2(4)^2+(4)+4]}{5}\\&=\frac{-149-(-24)}{5}\\&=\boxed{-25}.\end{aligned}$$

b) The instantaneous rate of change at $x=4$ is

$$\begin{aligned}\lim_{h\to 0}\frac{f(x+h)-f(x)}{h}&=\lim_{h\to 0}\frac{f(4+h)-f(4)}{h}\\&=\lim_{h\to 0}\frac{[-2(4+h)^2+(4+h)+4]-[-2(4)^2+(4)+4]}{h}\end{aligned}$$

$$= \lim_{h\to 0} \frac{[-2(16+8h+h^2)+4+h+4]-[-24]}{h}$$

$$= \lim_{h\to 0} \frac{[-32-16h-2h^2+4+h+4]+24}{h}$$

$$= \lim_{h\to 0} \frac{-15h-2h^2}{h}$$

$$= \lim_{h\to 0} (-15-2h)$$

$$= \boxed{-15}.$$

17. This problem is similar to text Examples 4-5 (pp. 787-788). We perform the five-step process:

(i) $f(x) = \frac{1}{3}x^3$ is given.

(ii) $f(x+h) = \frac{1}{3}(x+h)^3 = \frac{1}{3}(x^3+3x^2h+3xh^2+h^3) = \frac{1}{3}x^3+x^2h+xh^2+\frac{1}{3}h^3$

(iii) $f(x+h)-f(x) = (\frac{1}{3}x^3+x^2h+xh^2+\frac{1}{3}h^3)-(\frac{1}{3}x^3) = x^2h+xh^2+\frac{1}{3}h^3$

(iv) $\dfrac{f(x+h)-f(x)}{h} = \dfrac{x^2h+xh^2+\frac{1}{3}h^3}{h} = \dfrac{\not{h}(x^2+xh+\frac{1}{3}h^2)}{\not{h}} = x^2+xh+\frac{1}{3}h^2$

(v) $\displaystyle\lim_{h\to 0}\frac{f(x+h)-f(x)}{h} = \lim_{h\to 0}(x^2+xh+\tfrac{1}{3}h^2) = x^2$

Therefore, the derivative of $f(x) = \frac{1}{3}x^3$ is $\boxed{\boldsymbol{f'(x) = x^2}}$.

19. Based upon the rule for the derivative of the exponential function $y = e^{ax}$ (stated on textbook p. 788 and demonstrated in Example 6), it follows that the derivative of $y = e^{-6x}$ is $\boxed{\boldsymbol{y' = -6e^{-6x}}}$.

21. We again perform the five-step process (previously shown in text Examples 4-5 and this manual's solution to Problem 19):

(i) $f(x) = 3+2x-3x^2$ is given.

(ii) $f(x+h) = 3+2(x+h)-3(x+h)^2 = 3+2x+2h-3x^2-6xh-3h^2$

(iii) $f(x+h)-f(x) = (3+2x+2h-3x^2-6xh-3h^2)-(3+2x-3x^2)$

$$= 3+2x+2h-3x^2-6xh-3h^2-3-2x+3x^2$$

$$= 2h-6xh-3h^2$$

(iv) $\dfrac{f(x+h)-f(x)}{h}=\dfrac{2h-6xh-3h^2}{h}=2-6x-3h$

(v) $\lim\limits_{h\to 0}\dfrac{f(x+h)-f(x)}{h}=\lim\limits_{h\to 0}(2-6x-3h)=2-6x$

Therefore, the derivative of $f(x)=3+2x-3x^2$ is $\boxed{f'(x)=2-6x}$.

23. This problem is almost identical to text Example 7 (p. 789).
When $x=4$, $y=2(4)^2=32$. So, the point of tangency is $(4,32)$. Remember that the slope of a curve at a point is the value of the derivative at that point. From Example 5, we know the derivative of $y=2x^2$ is $y'=4x$; so, the slope of the line tangent to the curve $y=2x^2$ at any point (x,y) is $4x$. Hence, at the point (4, 32), the slope of the tangent line is $m=4(4)=16$.

Now, to write the equation of the line tangent to $y=2x^2$ at $x=4$, we may use the point-slope form (introduced in Problem 58, p. 733) with $m=16$ and $(h,k)=(4,32)$:

$$y-k=m(x-h)$$
$$y-32=16(x-4)$$
$$y-32=16x-64$$
$$\boxed{16x-y-32=0}$$

Note: Alternatively, one instead could apply the slope-intercept form (from textbook p. 728) to obtain this standard-form linear equation. Specifically, with $(x,y)=(4,32)$ and $m=16$, one can solve for the y-intercept:

$$y=mx+b$$
$$(32)=(16)(4)+b$$
$$32=64+b$$
$$-32=b$$

Substituting $b=-32$ and $m=16$ into the slope-intercept form equation yields:

$$y=16x+(-32)$$
$$0=16x-y-32$$
$$16x-y-32=0.$$

25. We first must find the derivative of $y=3x^2+4x$. By using the definition of the derivative in the same manner as previously demonstrated in textbook Example 5

(p. 788) and this manual's solutions to Problems 17 and 21 (see above), one finds that $y' = 6x + 4$. We leave the details to you.

When $x = 0$, $y = 3(0)^2 + 4(0) = 0$; so, the point of tangency is the origin $(0, 0)$. Remember that the slope of a curve at a point is the value of the derivative at that point; so, the slope of the line tangent to the curve $y = 3x^2 + 4x$ at any point (x, y) is $6x + 4$. Hence, at the point $(0, 0)$, the slope of the tangent line is $m = 6(0) + 4 = 4$.

Now, to write the equation of the line tangent to $y = 3x^2 + 4x$ at $x = 0$, we may use the point-slope form (introduced in Problem 58, p. 733) with $m = 4$ and $(h, k) = (0, 0)$:

$$y - k = m(x - h)$$
$$y - 0 = 4(x - 0)$$
$$y = 4x$$
$$\boxed{\mathbf{4x - y = 0}}$$

[Reminder: We could have chosen to use the slope-intercept form (introduced on textbook p. 728) to obtain this linear equation. If interested, please refer to this manual's solution for Problem 25, in which we demonstrated the details of this approach.]

27. a) Recall that the act of tossing the ball begins with its initial release from the top of the tower. Therefore, the height of the tower is found by evaluating the function $h(t) = -16t^2 + 96t + 176$ at $t = 0$:

$$h(0) = -16(0)^2 + 96(0) + 176 = \boxed{\textbf{176 ft}}.$$

b) Recall that the concept of instantaneous velocity is introduced on textbook pp. 789-791. To express the (instantaneous) velocity as a function of time [call it $v(t)$], we obtain the derivative of $h(t)$. By using the definition of the derivative in the same manner as previously demonstrated in textbook Example 5 (p. 788) and this manual's solutions to Problems 17 and 21 (see above), one finds that $h'(t) = -32t + 96$ — we leave the details to you. Therefore, the velocity function is $\boxed{\mathbf{v(t) = -32t + 96}}$.

29. This problem is similar to text Example 1 and 3, as well as Problems 1-5. Given the cost function $C(x) = 30x^2 - 100x$, we obtain the following average rates of change:

a) Average $= \dfrac{C(200) - C(100)}{200 - 100} = \dfrac{1{,}180{,}000 - 290{,}000}{100} = \boxed{\textbf{\$8,900 per item}}$

b) Average $= \dfrac{C(110) - C(100)}{110 - 100} = \dfrac{352{,}000 - 290{,}000}{10} = \boxed{\textbf{\$6,200 per item}}$

c) Average $= \dfrac{C(101)-C(100)}{101-100} = \dfrac{295{,}930-290{,}000}{1} = \boxed{\textbf{\$5,930 per item}}$

d) Average $= \dfrac{C(100+h)-C(100)}{(100+h)-100}$

$$= \frac{[30(100+h)^2 - 100(100+h)] - 290{,}000}{h}$$

$$= \frac{[30(10{,}000+200h+h^2) - 10{,}000 - 100h] - 290{,}000}{h}$$

$$= \frac{300{,}000 + 6{,}000h + 30h^2 - 10{,}000 - 100h - 290{,}000}{h}$$

$$= \frac{5{,}900h + 30h^2}{h}$$

$$= \boxed{\$(5{,}900 + 30h)} \text{ per item}$$

We encourage you to use this expression to check your answers to parts (a) through (c) above, in which the values of h were 100, 10, and 1, respectively.

Problem Set 11.9

1. This problem is almost identical to text Example 1 on page 794.
$A(8)$ is the area below the horizontal line $y = 5$, above the x-axis, to the right of the y-axis, and to the left of the line $x = 8$. If necessary, take a moment to sketch the region. It is a rectangle, and we know the area of a rectangle is the product of its length and width. Therefore, $A(8) = 5(8) = \boxed{\textbf{40}}$.

3. This problem is almost identical to text Example 2 (p. 795).
$A(4)$ is the area of the triangular region bounded by the line $y = 3x$, the x-axis, and the vertical line $x = 4$. If necessary, take a moment to sketch the region. You will find that the height of this triangle (call it h) is $3(4) = 12$ and the length of its base b is 4. So, upon recalling that the area of a triangle equals $\frac{1}{2}bh$, we obtain:

$$A(4) = \tfrac{1}{2}4(12) = \boxed{\textbf{24}}.$$

5. Take a moment to sketch this region. Like textbook Figure 11.67 on p. 797, you will obtain a trapezoid bounded by the x-axis, the y-axis, the line $y = 3x + 2$, and the vertical line $x = 3$, like the following:

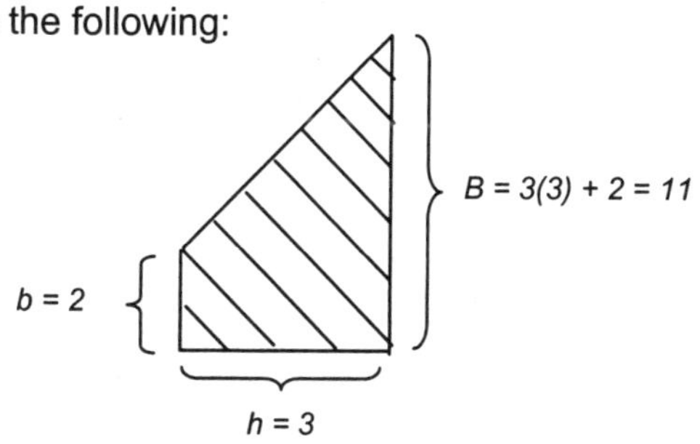

From our previous work in Section 7.2, we know the area formula for a trapezoid is $\frac{1}{2}h(b+B)$. Hence, the value of the area function is $A(3) = \frac{1}{2}(3)(2+11) = \boxed{\frac{39}{2}}$.

7. From text Example 4 (p. 797), you will recall that antiderivatives can be determined by using areas.

To find $\int 6\,dx$, we begin by letting $f(x) = 6$. Consider the region bounded by the x-axis, the y-axis, the horizontal line $y = 6$, and the vertical line x units to the right of the origin, as shown below.

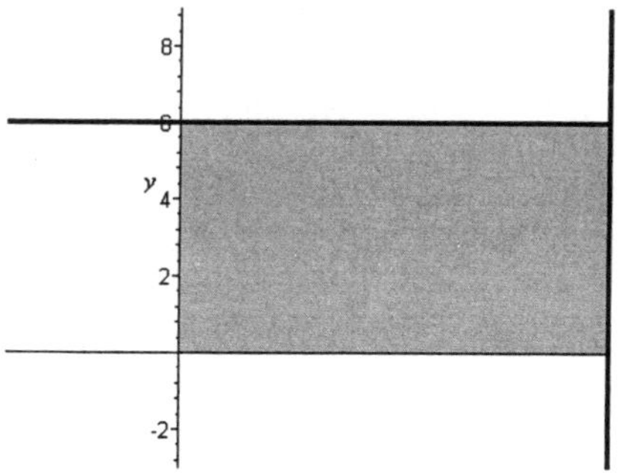

This is a rectangular region x units wide by 6 units high, so its area is $6x$. Therefore, since antiderivatives are not unique, but differ by a constant, we write

$$\int 6\,dx = \boxed{\mathbf{6x + C}}.$$

9. This problem is very similar to text Example 4, p. 797.

To find the antiderivative $\int (x+5)\,dx$, we consider the area of the region bounded by the x-axis, the y-axis, the line $y = x + 5$, and the vertical line x units to the right of the origin, as follows:

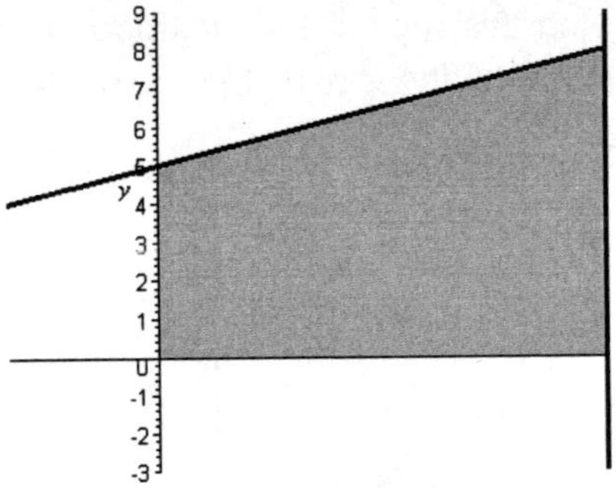

This region is a trapezoid with $h = x$, $b = 5$, and $B = x + 5$. From our previous work in Section 7.2, we know the area formula for a trapezoid is $A = \frac{1}{2}h(b + B)$. Hence, the area of the trapezoid shown above is

$$A = \tfrac{1}{2}h(b + B) = \tfrac{1}{2}(x)[5 + (x + 5)] = \tfrac{1}{2}x(x + 10) = \tfrac{1}{2}x^2 + 5x.$$

Finally, as established on textbook p. 796, we know antiderivatives are not unique, but differ by a constant. Therefore, we write

$$\int (x + 5)\,dx = \boxed{\tfrac{1}{2}x^2 + 5x + C}.$$

11. This problem is very similar to Example 4 (textbook p. 797) and Problem 9 (see solution above).

To find the antiderivative $\int (3x + 4)\,dx$, we compute the area of the region bounded by the x-axis, the y-axis, the line $y = 3x + 4$, and the vertical line x units to the right of the origin. This region, as shown below, is trapezoidal with $h = x$, $b = 4$, and $B = 3x + 4$.

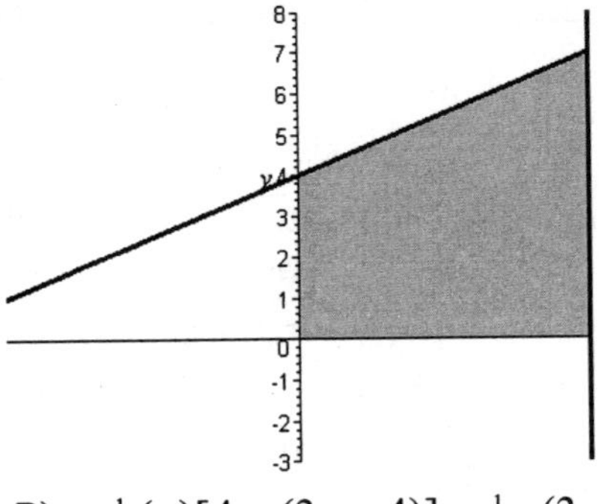

Its area is $A = \frac{1}{2}h(b + B) = \frac{1}{2}(x)[4 + (3x + 4)] = \frac{1}{2}x(3x + 8) = \frac{3}{2}x^2 + 4x$. Since antiderivatives are not unique (differing only by a constant), we write

$$\int (3x + 4)\,dx = \boxed{\tfrac{3}{2}x^2 + 4x + C}.$$

13. **Wording varies**. As stated on textbook p. 787, for a given function *f*, we define the ***derivative of f at x***, denoted by $f'(x)$, to be

$$f'(x)=\lim_{h\to 0}\frac{f(x+h)-f(x)}{h}$$

provided this limit exists. The slope of the line tangent to the graph of *f* at a point is the value of the derivative at that point.

15. **Wording varies**. As depicted in textbook Figure 11.64 (p. 794), the ***area function***, $A(t)$, is the area bounded below by the *x*-axis, above by function $y=f(x)$, on the left by the *y*-axis, and on the right by the vertical line $y=t$.

17. Problems 17-20 are akin to text Example 3 (pp. 795-796).
If $\frac{1}{2}x^2$ is an antiderivative of x, then x must be the derivative of $\frac{1}{2}x^2$. So, one way to confirm that $\frac{1}{2}x^2$ is an antiderivative of x is to show that the derivative of the function $f(x)=\frac{1}{2}x^2$ is $f'(x)=x$. We do this by using the definition of the derivative and the five-step process, both detailed in Section 11.8:

(i) $f(x)=\frac{1}{2}x^2$ is given.

(ii) $f(x+h)=\frac{1}{2}(x+h)^2=\frac{1}{2}(x^2+2xh+h^2)=\frac{1}{2}x^2+xh+\frac{1}{2}h^2$

(iii) $f(x+h)-f(x)=(\frac{1}{2}x^2+xh+\frac{1}{2}h^2)-\frac{1}{2}x^2=xh+\frac{1}{2}h^2$

(iv) $\dfrac{f(x+h)-f(x)}{h}=\dfrac{xh+\frac{1}{2}h^2}{h}=\dfrac{\cancel{h}(x+\frac{1}{2}h)}{\cancel{h}}=x+\frac{1}{2}h$

(v) $\lim\limits_{h\to 0}\dfrac{f(x+h)-f(x)}{h}=\lim\limits_{h\to 0}(x+\frac{1}{2}h)=x$

Therefore, since x is the derivative of $\frac{1}{2}x^2$, we have shown that $\frac{1}{2}x^2$ is an antiderivative of x.

19. Once again, as was demonstrated in text Example 3 (pp. 795-796) and Problem 17 (see solution above), the fact that $2x^3$ is an antiderivative of $6x^2$ can be proven by applying the definition of the derivative and the five-step process, both of which were first introduced in Section 11.8.

So, we let $f(x)=2x^3$ and establish that the derivative of f is $f'(x)=6x^2$ by applying the five-step process:

(i) $f(x)=2x^3$ is given.

(ii) $f(x+h)=2(x+h)^3=2(x^3+3x^2h+3xh^2+h^3)=2x^3+6x^2h+6xh^2+2h^3$

(iii) $f(x+h)-f(x)=(2x^3+6x^2h+6xh^2+2h^3)-2x^3=6x^2h+6xh^2+2h^3$

(iv) $\dfrac{f(x+h)-f(x)}{h}=\dfrac{6x^2h+6xh^2+2h^3}{h}=6x^2+6xh+2h^2$

(v) $\lim\limits_{h\to 0}\dfrac{f(x+h)-f(x)}{h}=\lim\limits_{h\to 0}(6x^2+6xh+2h^2)=6x^2+6x(0)+2(0)^2=6x^2$ ✓

21. This problem is virtually identical to text Example 5 (pp. 799-801).

From previous problems, we know that the antiderivative of x^2 is $\frac{1}{3}x^3$. We can use this knowledge — along with the definite integral (defined on textbook p. 798) — to determine the area of the region bounded by the curve $y=x^2$, the x-axis, and the lines $x=1$ and $x=9$; in specific, we have:

$$\text{Area} = \int_1^9 x^2\,dx=\tfrac{1}{3}x^3\Big|_1^9=\left[\tfrac{1}{3}(9)^3\right]-\left[\tfrac{1}{3}(1)^3\right]=243-\tfrac{1}{3}=\boxed{\mathbf{242\tfrac{2}{3}}}.$$

23. Here are two approaches:

I. We know from Problem 17 that the antiderivative of x is $\frac{1}{2}x^2$. So, by applying the constant multiple property of antiderivatives (see textbook p. 796), one can show that the antiderivative of $3x$ is $\frac{3}{2}x^2$. Hence, the area under the curve $y=3x$ on the interval $[3,8]$ equals the following definite integral:

$$\text{Area} = \int_3^8 3x\,dx=\tfrac{3}{2}x^2\Big|_3^8=\left[\tfrac{3}{2}(8)^2\right]-\left[\tfrac{3}{2}(3)^2\right]=96-\tfrac{27}{2}=\boxed{\mathbf{82.5}}.$$

II. Let $f(x)=3x$. Take a moment to sketch the enclosed region bounded above by the line $y=3x$, below by the x-axis, on the left by the line $x=3$, and on the right by the line $x=8$. Like previous problems (including text Example 4 and Problem 9-12), you will find that this region is trapezoidal; in particular, we have $h=8-3=5$, $b=f(3)=3(3)=9$, and $B=f(8)=3(8)=24$. Thus,

$$\text{Area}=\tfrac{1}{2}h(b+B)=\tfrac{1}{2}(5)(9+24)=\tfrac{165}{2}=82.5.$$

25. Via the formula for the antiderivative of e^{ax} (stated on textbook p. 796), we find

$$\int e^{0.5x}\,dx=\frac{e^{0.5x}}{0.5}+C=\tfrac{1}{0.5}e^{0.5x}+C=2e^{0.5x}+C.$$

Therefore, the given definite integral equals

$$\int_0^2 e^{0.5x}\,dx==2e^{0.5x}\Big|_0^2=\left[2e^{0.5(2)}\right]-\left[2e^{0.5(0)}\right]=2e^1-2e^0=\boxed{\mathbf{2e-2}}.$$

27. Like Problem 23 (see solution above), one can determine the value of $\int_1^5 (1+6x)\,dx$ in either of two ways: (i) by using an antiderivative, or (ii) by using areas (namely, for a trapezoid). Let us demonstrate the former approach and leave the latter for you.

The properties of antiderivatives (on textbook p. 796) tell us that

$$\int (1+6x)\,dx = \int 1\,dx + \int 6x\,dx = \int 1\,dx + 6\int x\,dx.$$

By applying the methods used in previous problems/examples or using the results thereof, we find

$$\int (1+6x)\,dx = \int 1\,dx + 6\int x\,dx = x + 6(\tfrac{1}{2}x^2) + C = x + 3x^2 + C.$$

Thus,

$$\int_1^5 (1+6x)\,dx = (x+3x^2)\Big|_1^5 = \left[(5)+3(5)^2\right] - \left[(1)+3(1)^2\right] = 80 - 4 = \boxed{76}.$$

29. *[__Hint__: This problem is similar in many respects to the oil consumption scenario discussed on textbook pp. 797-798. So, __before__ reading the rest of this solution, please revisit these two pages in your text and see if you can apply the ideas they communicate to bacteria growth.]*

From Problem 27, we know the antiderivative of $e^{0.5t}$ is $2e^{0.5t}$. So, via the constant multiple property of integrals (textbook p. 796), it follows that the antiderivative of $200e^{0.5t}$ is $200(2e^{0.5t}) = 400e^{0.5t}$.

Hence, the area between the graph of $R'(t) = 200e^{0.5t}$ and the horizontal t-axis on the interval $[0, 10]$ equals

$$\int_0^{10} 200e^{0.5t}\,dt = 400e^{0.5t}\Big|_0^{10} = 400e^{0.5(10)} - 400e^{0.5(0)} = 400e^5 - 400 \approx \boxed{58{,}965.3}.$$

As to what you might think this area represents, answers/wording vary. This area represents the **__total increase (net change) in the number of bacteria during this 10-hour span__**. Since $R'(t) = 200e^{0.5t}$ is an instantaneous rate of change that models the growth rate of a bacteria culture, its antiderivative $R(t) = 400e^{0.5t}$ is a growth function (a la Section 7.7, textbook p. 483) that indicates the number of bacteria present after t hours.

Chapter 11 Review Questions

1. Review Section 11.1, as necessary. The slope-intercept form equation is

$y = 5x - 15$. So, the slope is

$m = \frac{\text{rise}}{\text{run}} = \frac{+5}{+1}$ or $\frac{-5}{-1}$, and the y-

intercept is $b = -15$.

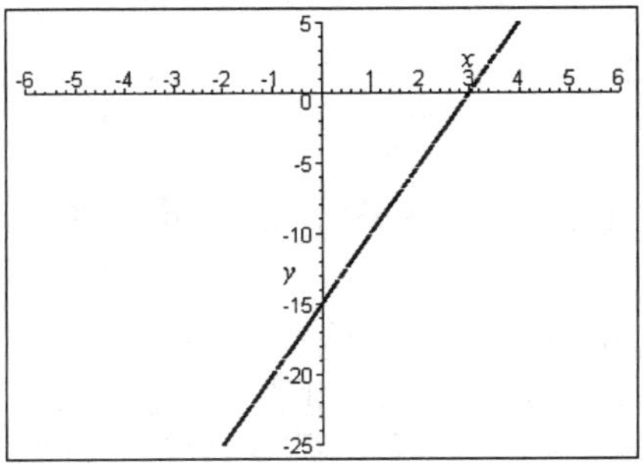

2. Review Section 11.1, as necessary. The slope is $m = \frac{\text{rise}}{\text{run}} = \frac{-4}{+5}$ or $\frac{+4}{-5}$,

and the y-intercept is $b = -3$.

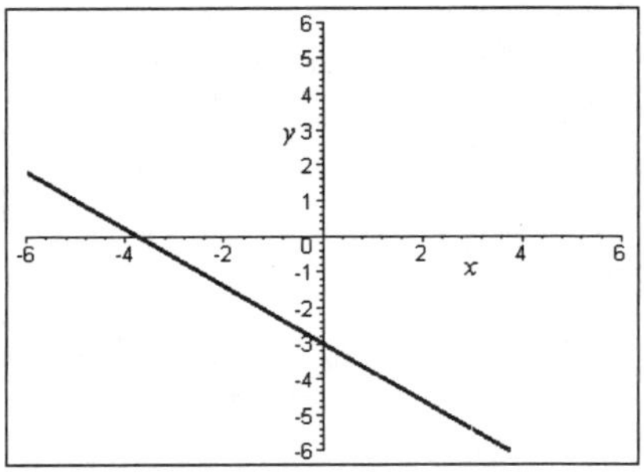

3. Review Section 11.1, as necessary. The slope-intercept form equation is $y = -\frac{2}{3}x + 5$. So, the slope is

$m = \frac{\text{rise}}{\text{run}} = \frac{-2}{+3}$ or $\frac{+2}{-3}$, and the y-

intercept is $b = 5$.

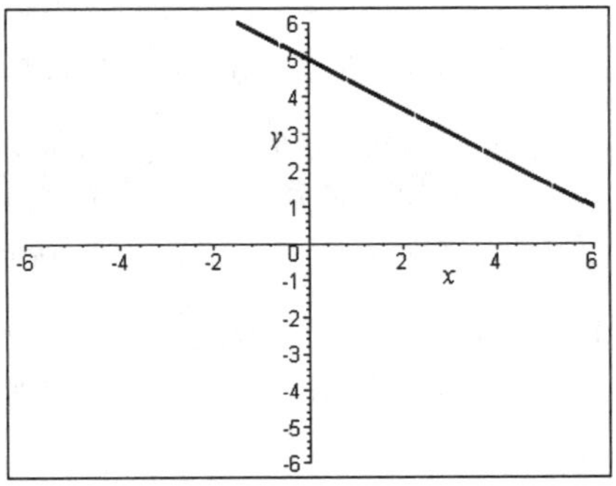

4. Review Section 11.1, as necessary. The slope-intercept form equation is $y = -\frac{3}{2}x + \frac{3}{2}$. So, the slope is

$m = \frac{\text{rise}}{\text{run}} = \frac{-3}{+2}$ or $\frac{+3}{-2}$, and the y-

intercept is $b = \frac{3}{2}$.

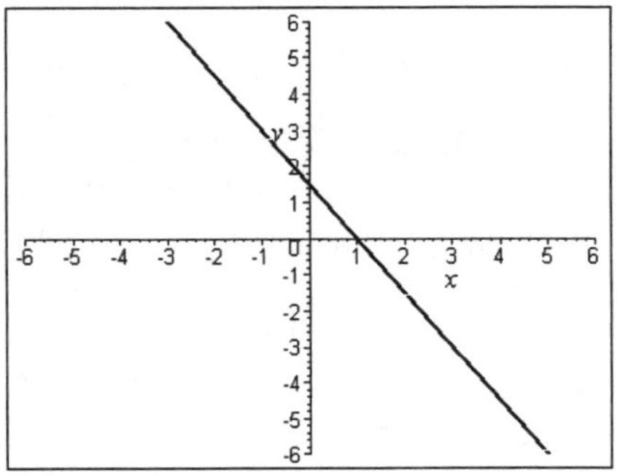

5. Review Section 11.1, as necessary.
Since the only requirement is that x (the first component) equal 150, we see that there is no restriction on the choice for y. Said differently, for <u>any</u> real number y, the point $(150, y)$ satisfies the condition that $x = 150$. Upon plotting and connecting such points, you will see that the result is a vertical line:

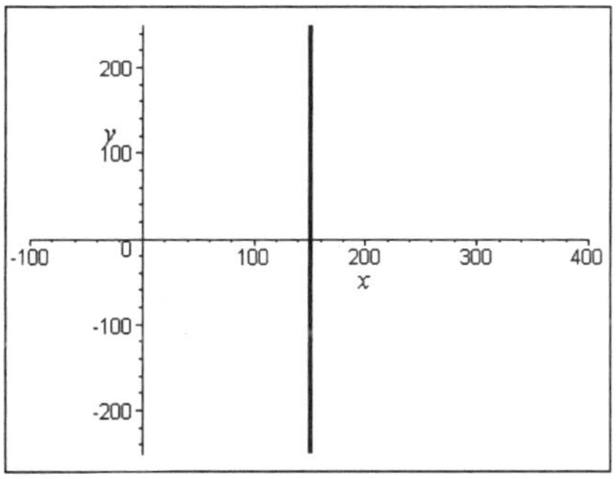

6. Review Section 11.2, as necessary

First, graph the dashed boundary line $x = 3y$, which is equivalent to the slope-intercept form equation $y = \frac{1}{3}x$. Then, pick a test point; note that one cannot choose (0, 0) because it is on the boundary line. Suppose you chose (0, 3). It then would follow that $x < 3y$ becomes $0 < 3(3)$, which is a true statement. Therefore, we shade the half-plane containing this test point.

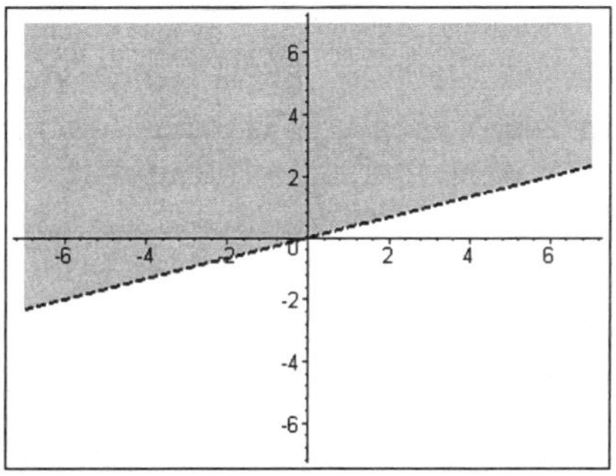

7. Review Sections 11.3 and 11.4, as needed. Simply by looking at its equation, you immediately should be able to identify the curve as a downward opening parabola.

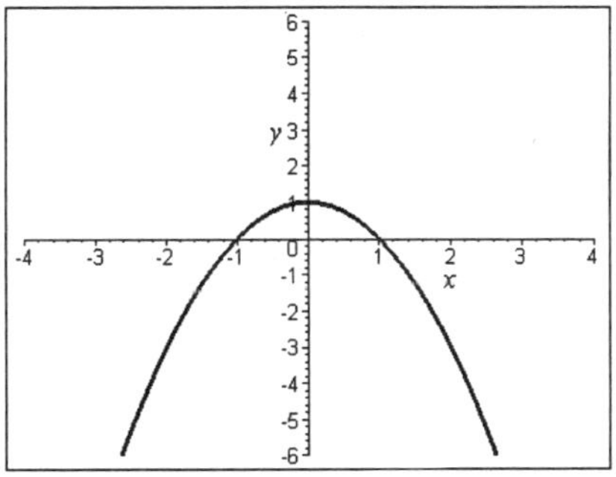

8. Reminder: $y = -2^x = -(2^x)$.
This exponential equation is similar to text Example 4 in Section 11.3 (see p. 739).

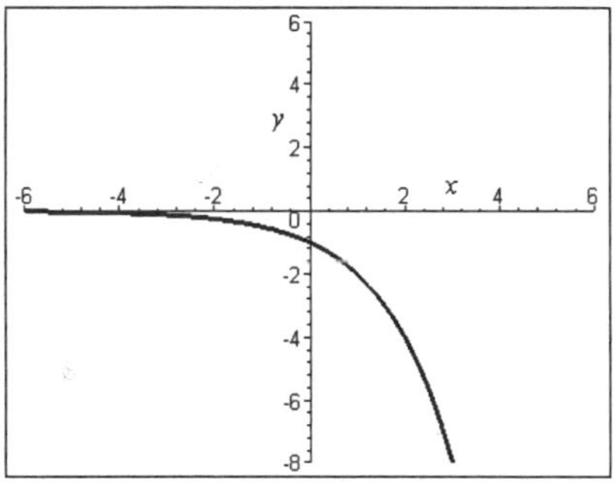

9. This hyperbola is representative of those encountered previously in Section 11.4. For additional reinforcement, we encourage you to revisit text Example 9 (see p. 752) and Section 11.4's Problems 41 and 43 (see this manual's detailed solutions).

Since $a^2 = 16$, $b^2 = 4$, and the x^2-term is positive, we know that this is the equation of a horizontal hyperbola with vertices at $(\pm 4,0)$ and pseudovertices at $(0,\pm 2)$. So, the transverse axis is along the x-axis, and the conjugate axis is along the y-axis. By using the central rectangle and slant asymptotes (neither of which is shown in the graph below) along with the vertices, we obtain the following graph:

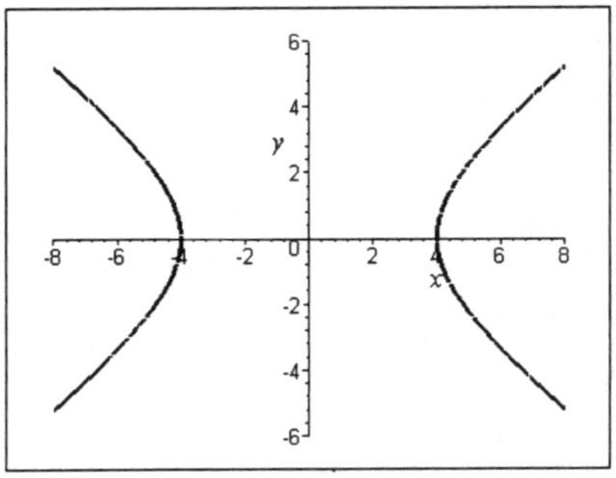

10. This ellipse is representative of those encountered previously in Section 11.4. For additional reinforcement, we encourage you to revisit text Examples 6 and 7 (see pp. 748-749) and Section 11.4's Problems 33, 35, and 43 (see this manual's detailed solutions).

We recognize this problem's equation as the standard-form equation of an ellipse. Because the larger number is under the x^2-term, we know the major axis is the x-axis; that is to say, this is a horizontal ellipse. With $a^2 = 20$ and $b^2 = 10$, it follows that the x-intercepts (which, because they are the endpoints of the major axis, are called the ***vertices***) are $\pm\sqrt{20} \approx 4.5$ and the y-intercepts are $\pm\sqrt{10} \approx 3.2$. We use the four intercepts to sketch the ellipse.

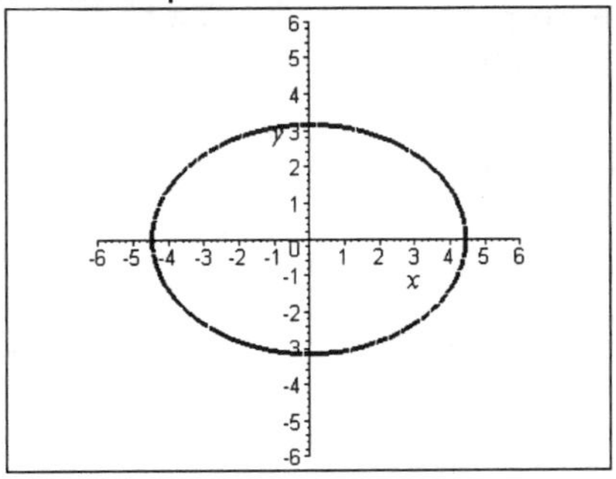

11. From our experience with conic sections in Section 11.4, we recognize $A = C = 1$ and $r^2 = 1$; hence, the graph is a circle with its center at (0, 0) and radius 1 unit in length.

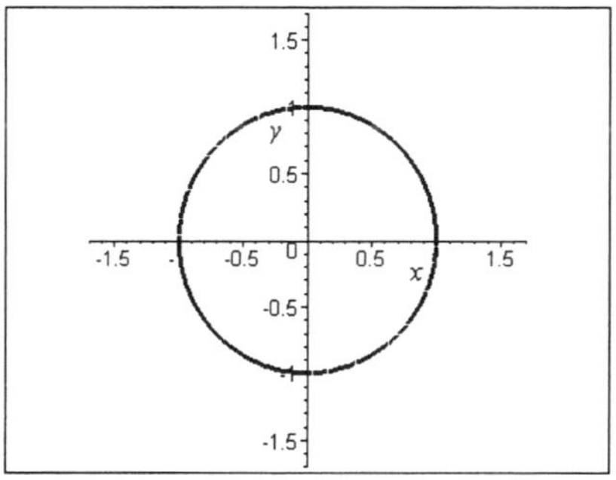

12. Once again, as necessary, revisit Section 11.4. Since the x^2-term is positive, this is the equation of a horizontal hyperbola. Upon writing the standard-form equation as $\frac{x^2}{1} - \frac{y^2}{1} = 1$, we see that $a^2 = b^2 = 1$; so, the vertices are at $(\pm 1, 0)$ and pseudovertices at $(0, \pm 1)$. The transverse axis is along the x-axis, and the conjugate axis is along the y-axis. By using the central rectangle and slant asymptotes (neither of which is shown in the graph below) along with the vertices, we obtain the following graph:

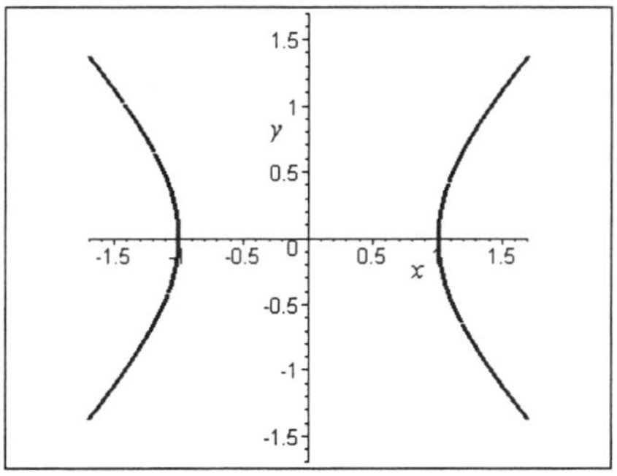

13. This question is comparable to Section 11.5's Example 1 (see pp. 757-758) and the same section's Problems 3-14 (see this manual's solutions). Prior to addressing this type of problem, you must acquire an understanding of the definition of a *function*. If necessary, consult your textbook and/or this manual's solutions for Section 11.5.

As for the question regarding {(4, 3), (5, –2), (6, 3)}, our answer is: **yes**, it is a function. Each first component of the ordered pairs is associated with exactly one second component. The fact that 4 and 6 both pick 3 as their second component partner is irrelevant.

14. The concept of ***functional notation*** was introduced in Section 11.5 (pp. 758-759). For the linear function $f(x) = 3x + 2$, $f(6) = 3(6) + 2 = \boxed{20}$.

15. For the quadratic function $g(x) = x^2 - 3$, $g(0) = (0)^2 - 3 = \boxed{-3}$.

16. For the linear function $F(x) = 5x + 25$, $F(10) = 5(10) + 25 = \boxed{75}$.

17. For any constant function $f(x) = k$, the function value is always k — regardless of the specific value of x chosen. Hence, given the constant function $m(x) = 5$, it follows that $m(10) = \boxed{5}$.

18. The concept of a limit is introduced in Section 11.6 and discussed in greater depth in Section 11.7. As the value of n gets larger and larger (i.e., as $n \to \infty$), the value of $\frac{1}{n}$ gets smaller and smaller, converging to 0. Expressed in terms of limit notation, we write $\lim\limits_{n\to\infty} \frac{1}{n} = \boxed{0}$.

[Reminder: This particular limit and, more generally, $\lim\limits_{n\to\infty} \frac{A}{n^k} = 0$ (for any real number A and positive integer k) were of great importance in this chapter's work with limits. If necessary, please revisit textbook pp. 778-779.]

19. Remember: There are a variety of ways to determine limits. They include:

- Graphing the sequence either in one or two dimensions;
- Constructing a table of values in an effort to detect a pattern;
- Performing algebraic manipulations, as necessary, and then making use of the conclusions stated on the bottom of textbook p. 778.

You would be well advised to become familiar with all three of these approaches — as well as any others your instructor happens to provide. Doing so will greatly enhance your understanding of limits and, in the process, will decrease the likelihood of your experiencing confusion/anxiety on an upcoming exam.

This particular problem is similar to several encountered previously in Section 11.7; they include Example 3b (p. 779), Example 2d (p. 778), and Problem 11 (p. 781). Let us begin by using our algebra skills to rewrite the given expression:

$$\frac{3n^4 + 20}{7n^4} = \frac{3n^4}{7n^4} + \frac{20}{7n^4} = \frac{3}{7} + \frac{20}{7n^4}$$

Applying the limit to infinity property (from textbook p. 778) and assuming that the limit of a sum is the sum of the limits, we obtain:

$$\lim_{n\to\infty} \frac{3n^4 + 20}{7n^4} = \lim_{n\to\infty}\left[\frac{3}{7} + \frac{20}{7n^4}\right] = \left[\lim_{n\to\infty}\frac{3}{7}\right] + \left[\lim_{n\to\infty}\frac{20}{7n^4}\right] = \tfrac{3}{7} + 0 = \boxed{\tfrac{3}{7}}.$$

20. Notice that $a_n = 2n+3$ is a linearly increasing sequence, whose graph has a slope of $m = 2$. So, as n grows increasingly large, $2n+3$ gets larger and larger. Therefore, $\lim_{n\to\infty}(2n+3) = \boxed{\infty}$.

21. To determine $\lim_{n\to\infty}\left(1+\frac{1}{n}\right)^n$, we construct a table of values:

n	1	10	10^3	10^{10}
$\left(1+\frac{1}{n}\right)^n$	2	≈ 2.5937424601	≈ 2.71692393224	≈ 2.71828182832

It looks like the sequence is converging to Euler's number, so we write

$$\lim_{n\to\infty}\left(1+\frac{1}{n}\right)^n = \boxed{e}.$$

In fact, this limit is often given as the definition of e.

22. **Responses vary**. Take note of the title: "*instantaneous* growth rate." This series of pictures illustrates the idea of a tangent line. It exhibits how the slopes of secant lines represent average growth rates and how the slopes of tangent lines depict instantaneous growth rates. The first picture shows four measurements, the second 20 measurements, and the third 40 measurements.

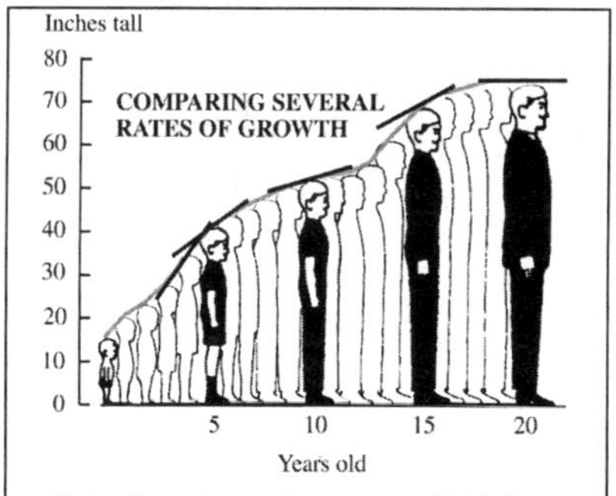

Each point on the curve has a tangent which indicates rate of growth at that point. The steepest tangent shown here occurs at about age three: the boy's growth was most rapid then.

Notice that the last picture measures the growth rate at an instant. The closer together the measurements are, the better the approximation of the growth rate at a particular point/instant.

23. **Descriptions vary**. Introduced on textbook pages 765-769 and discussed in greater depth in Sections 11.7, 11.8, and 11.9, the three main topics of calculus are **limits, derivatives, and integrals**. In brief,

- The ***limit*** $\lim_{n\to\infty} a_n = L$ means the sequence a_n becomes closer and closer ("converges to") the number L as n becomes larger and larger.
- The ***derivative*** illustrates the idea of the slope of a tangent line and has a seemingly endless number of applications. It is the instantaneous rate of change of a function with respect to the variable.
- The ***integral*** enables us to find the area under a curve. Like the derivative, its applications are abundant.

24. The definition of the derivative, which is among the most important notions in all of mathematics, is introduced in Section 11.8 (see textbook p. 787). Recall that previously we employed a five-step process to determine the derivative. In this solution, however, permit us to share a more condensed (stream-lined) version of that process:

$$\begin{aligned}\lim_{h\to 0}\frac{f(x+h)-f(x)}{h} &= \lim_{h\to 0}\frac{[6-4(x+h)^2]-[6-4x^2]}{h}\\ &= \lim_{h\to 0}\frac{6-4(x^2+2xh+h^2)-6+4x^2}{h}\\ &= \lim_{h\to 0}\frac{6-4x^2-8xh-4h^2-6+4x^2}{h}\\ &= \lim_{h\to 0}\frac{-8xh-4h^2}{h}\\ &= \lim_{h\to 0} -8x-4h\\ &= -8x\end{aligned}$$

So, the derivative of $f(x)=6-4x^2$ is $f'(x)=\boxed{-8x}$.

25. To review the concept of the definite integral, revisit Section 11.9.
As established previously (e.g., Problem 17 on p. 801), the antiderivative of x is $\frac{1}{2}x^2$. Similarly, it can be shown that the antiderivative of $4x^3$ is x^4. Therefore,

$$\int_{-1}^{0}\left(x-4x^3\right)dx=\left(\tfrac{1}{2}x^2-x^4\right)\Big|_{-1}^{0}=\left[\tfrac{1}{2}(0)^2-(0)^4\right]-\left[\tfrac{1}{2}(-1)^2-(-1)^4\right]=-(\tfrac{1}{2}-1)=\boxed{\tfrac{1}{2}}$$

26. From textbook p. 788 and p. 796, you will recall that e^x is its own derivative. So, rounding the answer to two decimal places, we find:

$$\int_1^2 e^x\,dx = e^x\Big|_1^2 = e^2 - e^1 \approx 4.67077427047 \approx \boxed{\mathbf{4.67}}.$$

27. Label the horizontal axis P and the vertical axis A. The equation $A = 1.7P$ is linear with a y-intercept of 0 and a slope of 1.7 — that is, $m = \frac{\text{rise}}{\text{run}} = \frac{+1.7}{+1}$ or $\frac{+17}{+10}$.

Hence, the graph is:

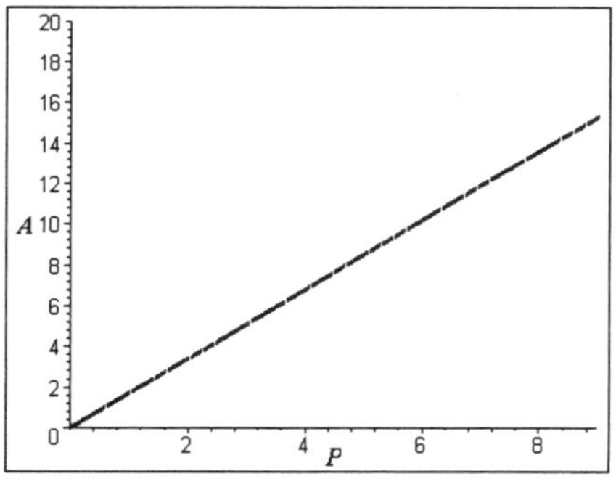

28. The path of a projectile was discussed on several occasions in this chapter. Feel free to revisit textbook p. 736, Problems 44-45 (p. 741), and Problems 26-27 (p. 793).

From your experience with assorted functions and the conic sections, it should come as no surprise that the graph of $y = 128t - 16t^2$ is a downward opening parabola.

We construct a table of values; plot the points (with the horizontal axis labeled t, in seconds, and vertical axis y, in feet); and then connect them.

t	0	1	2	3	4	5	6	7	8
y	0	112	192	240	256	240	192	112	0

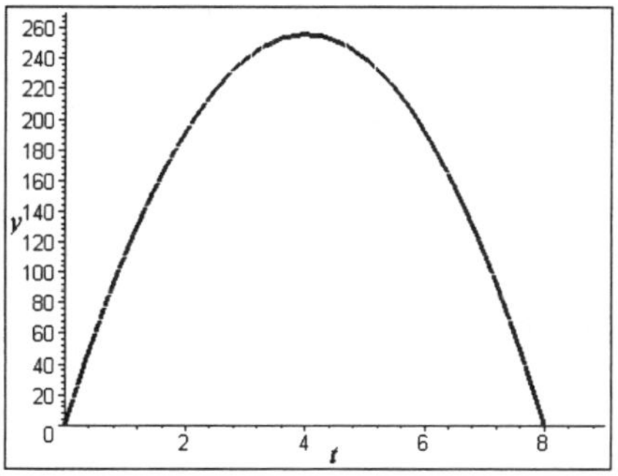

29. This problem is similar to text Example 5 (pp. 739-740 of Section 11.3) and Problems 52-53 (p. 741). Additionally, the formula for population growth was first introduced in Section 7.7 (beginning on p. 483).

Given that the initial population is $P_0 = 53{,}000$ and the rate of growth (expressed as a decimal) is $r = 0.013$, we must graph the curve $P = 53{,}000e^{0.013t}$. We then form a table of values:

t	0	10	20	30	40	50
P	53,000	$\approx 60{,}358$	$\approx 68{,}737$	$\approx 78{,}280$	$\approx 89{,}147$	$\approx 101{,}524$

Finally, we label the horizontal axis t (in years) and the vertical axis P (numbers of people), plot the points calculated above, and connect them with a smooth curve:

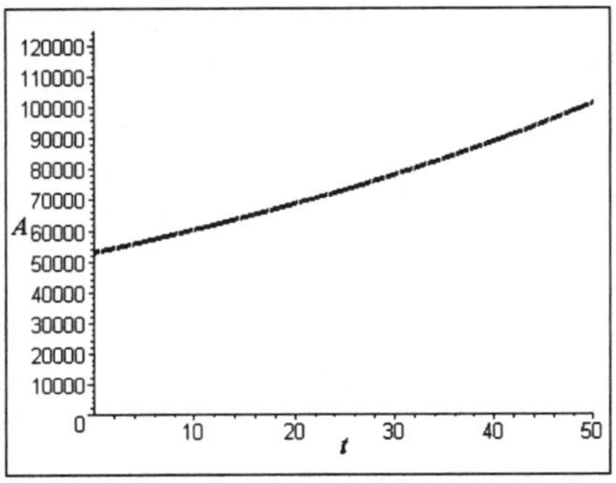

30. *[**Hint**: Before reading this solution, you should revisit textbook pp. 797-798, where this oil consumption scenario is discussed, and then attempt to apply the information it imparts to this problem. Solving this problem will not only require an understanding of the definite integral; you also will need to use your skills with solving exponential/logarithmic equations (from the latter portion of Chapter 7).]*

Since $R(t) = 32.4e^{6t/125}$ is the rate of change of a function (in particular, the total consumption function, which we will call T), the function representing the total amount of oil used (in billions of barrels) since the year 2000 is the antiderivative of R; that is,

$$T(t) = \int_0^t 32.4e^{6x/125}\,dx.$$

Using the antiderivative properties (of p. 796) in the same manner as employed on textbook pp. 797-798, we obtain:

$$\int 32.4e^{6t/125}\,dt = 32.4\int e^{6t/125}\,dt$$ *Apply the constant multiple property.*

$$= 32.4\frac{e^{6t/125}}{\frac{6}{125}} + C$$ *Apply the Formula for antiderivative of e^{ax}.*

$$= 32.4\left(\tfrac{125}{6}\right)e^{6t/125} + C$$

$$= 675e^{6t/125} + C$$

Therefore, the total amount of oil consumption since 2000 is expressed by the function $T(t)$:

$$T(t) = \int_0^t 32.4e^{6x/125}\,dx = = 675e^{6x/125}\Big|_0^t = 675\left(e^{6t/125} - 1\right)$$

To find the length of time (t) it should take to consume 670 billion barrels, we set $T(t)$ equal to 670 and proceed to solve the resulting exponential equation:

$$670 = 675\left(e^{6t/125} - 1\right)$$

$$\tfrac{670}{675} = e^{6t/125} - 1$$ *Divide both sides by 675.*

$$\tfrac{1,345}{675} = e^{6t/125}$$ *Add 1 to both sides; simplify.*

$$\frac{6t}{125} = \ln\left(\frac{1,345}{675}\right)$$ *Apply the definition of logarithm (text p. 467).*

$$t = \frac{125}{6}\ln\left(\frac{1,345}{675}\right)$$ *Multiply both sides by the reciprocal of $\frac{6}{125}$.*

$$t \approx 14.362$$ *Evaluate (via calculator).*

So, assuming that the rate of oil consumption does not change and no new reserves are discovered, we estimate that the oil reserves will be depleted in the year 2014, **approximately 14 years** after the year 2000.

CHAPTER 12: The Nature of Mathematical Systems

Problem Set 12.1

1. Wording varies. A ***system of linear equations*** is two or more first-degree equations considered at the same time.

3. Wording varies. Introduced on textbook pages 808-809 and demonstrated in Example 1, the ***graphing method*** for solving a system of equations is an approach that finds the solution by looking for the points at which the individual graphs intersect. It is important to remember that it is an approximate (not exact) method that is dependent upon the accuracy of the graph drawn.

5. Wording varies. The ***addition method*** (also known as the ***linear combination method***) consists of five steps, each of which is clearly delineated on textbook pp. 810-811 and illustrated in Example 3.

7. To review the techniques for graphing linear equations, we suggest you revisit Section 11.1 (pp. 724-733).
The graphs of $3x-4y=16$ and $-x+2y=-6$ are shown below; the former is depicted by a solid line, the latter as a dashed line. Looking at the point of intersection, we conclude that the solution appears to be **(4, –1)**.

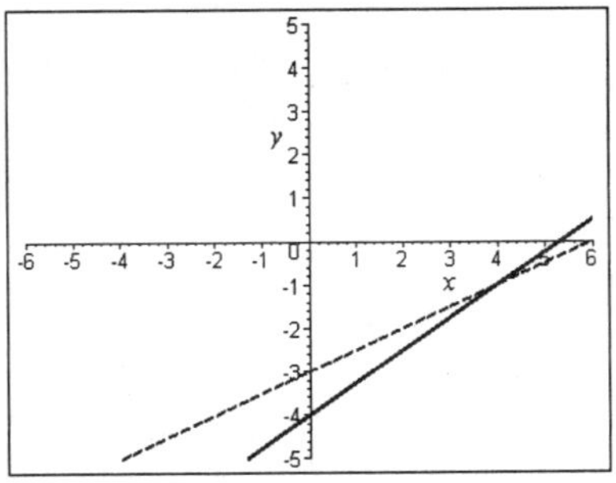

9. The graphs of $2x-3y=12$ and $-4x+6y=18$ are shown below; the former is depicted as a solid line, the latter as a dashed line.

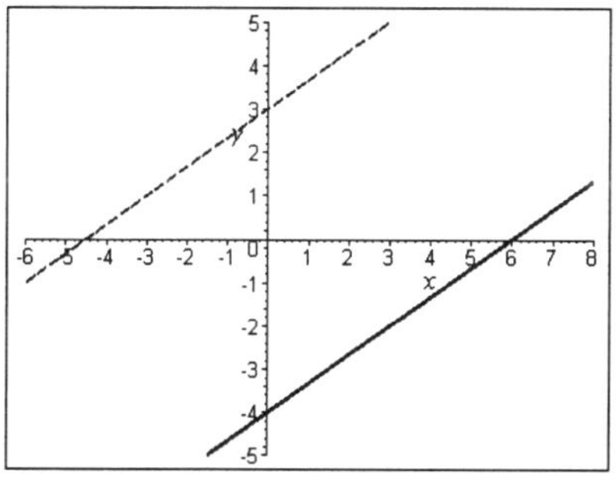

Writing both equations in slope-intercept form confirms the fact that the slopes of the lines are the same (specifically, $m=\frac{2}{3}$). So, since their slopes are equal, these two distinct lines are parallel, and this system of equations is what is called an **inconsistent system**. Accordingly, it has **no solution**; or said differently, its solution is the empty (or null) set, which we denote by $\varnothing$.

11. The graphs of $6x+y=-5$ and $x+3y=2$ are shown below; the former is graphed as a solid line, the latter as a dashed line. Looking at the point of intersection, we conclude that the solution appears to be **(−1, 1)**.

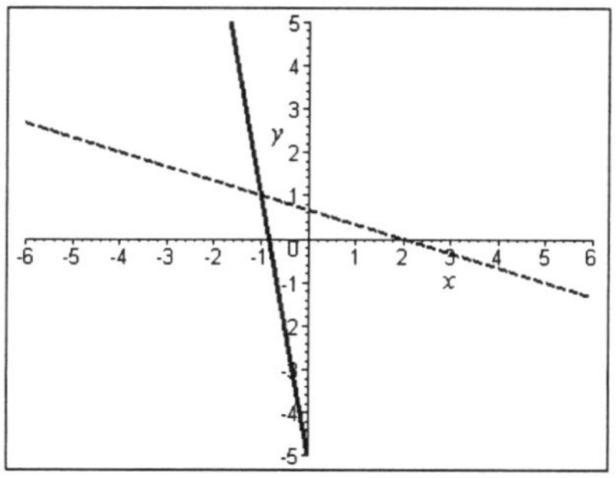

13. The graphs of $3x+2y=5$ and $4x-3y=1$ are shown below; the former is a solid line, the latter as a dashed line. Looking at the point of intersection, we conclude that the solution appears to be **(1, 1)**.

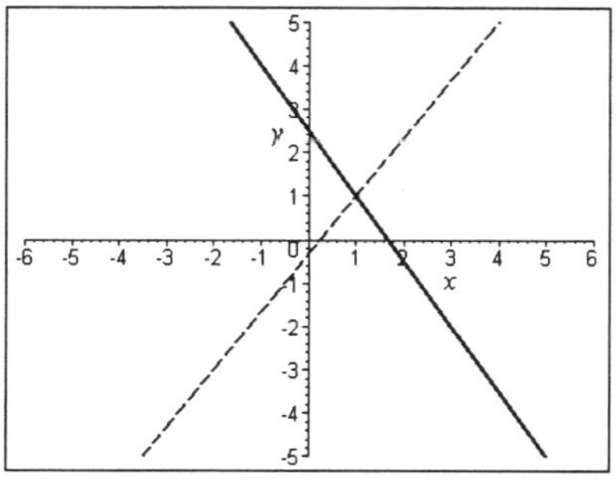

15. The ***substitution method*** is introduced on textbook pp. 809-810.
Given the second equation $x = 3y + 4$, replace x in the first equation with $3y + 4$:

$$\begin{aligned} 5x - 2y &= -19 \\ 5(3y + 4) - 2y &= -19 \\ 15y + 20 - 2y &= -19 \\ 13y &= -39 \\ y &= -3 \end{aligned}$$

Now, to find x, substitute the value of -3 for y in either of the given equations. In this problem, it is more efficient to use the second equation:

$$\begin{aligned} x &= 3y + 4 \\ &= 3(-3) + 4 \\ &= -5 \end{aligned}$$

Therefore, the solution to this linear system is the ordered pair **(–5, –3)**. As always, you should take the time to check the solution, using <u>both</u> equations to do so.

17. Since the second equation is $x = 2y + 3$, we replace x in the first equation with $2y + 3$:

$$\begin{aligned} 3x - y &= -1 \\ 3(2y + 3) - y &= -1 \\ 6y + 9 - y &= -1 \\ 5y &= -10 \\ y &= -2 \end{aligned}$$

Now, to find x, we substitute the value of -2 for y in either of the given equations. Because the second equation has x isolated, we choose to use it:

$$\begin{aligned} x &= 2y + 3 \\ &= 2(-2) + 3 \\ &= -1 \end{aligned}$$

Thus, the solution to this linear system is the ordered pair **(–1, –2)**. Be sure to check this solution in <u>both</u> equations.

19. Noting that the second equation is an equation in <u>one</u> variable, we begin by solving it:

$$\begin{aligned} 0.6y &= 0.5(12) \\ 0.6y &= 6 \\ y &= \frac{6}{0.6} = 10 \end{aligned}$$

Now, to solve for x, substitute the value of 10 for y in the first equation:

$$x + y = 12$$
$$x + 10 = 12$$
$$x = 2$$

Thus, the solution to this linear system is the ordered pair **(2, 10)**. Be sure to check your answer.

21. Let us share one approach (among several). Solving the second equation for x, we find $x = 7 - \frac{y}{2}$. Then, we substitute $7 - \frac{y}{2}$ for x in the first equation:

$$\frac{x}{3} - y = 7$$
$$\frac{\left(7 - \frac{y}{2}\right)}{3} - y = 7$$
$$\frac{7}{3} - \frac{y}{6} - y = 7$$

Multiplying both sides by 6 (i.e., the least common multiple of 3 and 6) yields:

$$14 - y - 6y = 42$$
$$-7y = 28$$
$$y = -4$$

Finally, substituting -4 for y in the equation $x = 7 - \frac{y}{2}$ gives us the value of x:

$$x = 7 - \frac{y}{2} = 7 - \frac{(-4)}{2} = 7 + 2 = 9$$

Hence, the solution set is the ordered pair **(9, –4)**. We encourage you to perform a check of this answer.

23. Since the first equation is $x = -7y - 3$, we replace x in the second equation with $-7y - 3$:

$$2x + 5y = 3$$
$$2(-7y - 3) + 5y = 3$$
$$-14y - 6 + 5y = 3$$

$$9y = 9$$
$$y = -1$$

Now, to find x, we substitute the value of -1 for y in either of the given equations. Because the first equation has x already isolated, we choose to use it:

$$x = -7y - 3 = -7(-1) - 3 = 4$$

Thus, the solution to this linear system is the ordered pair $\boxed{\textbf{(4, –1)}}$. Remember to check this solution in <u>both</u> of the given equations.

25. Since the second equation is $x = 5y + 16$, we replace x in the first equation with $5y + 16$:

$$x + 3y = 0$$
$$(5y + 16) + 3y = 0$$
$$8y = -16$$
$$y = -2$$

Now, to find x, we substitute the value of -2 for y in either of the given equations:

$$x + 3y = 0$$
$$x = -3y$$
$$x = -3(-2)$$
$$x = 6$$

Therefore, the solution to this linear system is the ordered pair $\boxed{\textbf{(6, –2)}}$. Always remember to check your solution in <u>both</u> of the given equations.

27. The ***addition method*** (also known as the ***linear combination method***) is introduced on textbook pp. 810-811.

Notice that the coefficients of y are opposites (i.e., 1 and -1), so there is no need to multiply either equation by a constant. Instead, we immediately add the equations:

$$+\begin{cases} x + y = 560 \\ x - y = 490 \end{cases}$$
$$2x = 1{,}050$$
$$x = 525$$

If $x = 525$, then $x + y = 560$ means $525 + y = 560$, which implies $y = 35$. Thus, the solution set is the ordered pair $\boxed{\textbf{(525, 35)}}$. Once again, we encourage you to check your solution (using <u>both</u> of the given equations).

29. Multiplying both sides of the first equation by 2 forces the coefficients of y to be opposites:

$$2\begin{cases}3x+y=13\\x-2y=9\end{cases}\quad\rightarrow\quad +\begin{cases}6x+2y=26\\x-2y=9\end{cases}$$

$$7x=35$$
$$x=5$$

Now, to obtain the value of y, we substitute the value of 5 for x in either of the given equations and proceed to solve as follows:

$$3x+y=13$$
$$3(5)+y=13$$
$$y=-2$$

So, the solution is **(5, −2)**. Once again, we encourage you to check your solution (using both of the given equations).

31. First, we rewrite the second equation as $6u+4v=-7$. We then multiply both sides of the first equation by -2:

$$-2\begin{cases}3u+2v=5\\6u+4v=-7\end{cases}\quad\rightarrow\quad +\begin{cases}-6u-4v=-10\\6u+4v=-7\end{cases}$$

$$0=-17$$

The statement $0=-17$ is a falsity (a contradiction). Thus, this system is **inconsistent** and accordingly it has **no solution** (i.e., $\varnothing$).
Alternatively, by looking closely at the two equations, you might have noticed — before we attempted to apply the addition method — that their graphs are two distinct lines that share the same slope (i.e., the lines are parallel, so they do not intersect).

33. The coefficients of s_1 (5 and 2) have a least common multiple (l.c.m.) of 10. We multiply the first equation by -2 and the second equation by 5. This procedure turns the coefficients of s_1 into opposites:

$$\begin{matrix}-2\\5\end{matrix}\begin{cases}5s_1+2s_2=23\\2s_1+7s_2=34\end{cases}\quad\rightarrow\quad +\begin{cases}-10s_1-4s_2=-46\\10s_1+35s_2=170\end{cases}$$

$$31s_2=124$$
$$s_2=4$$

To obtain the value of s_1, we then substitute the value of 4 for s_2 in either of the two equations. We have chosen arbitrarily to use the second equation:

$$2s_1 + 7s_2 = 34$$
$$2s_1 + 7(4) = 34$$
$$2s_1 = 6$$
$$s_1 = 3$$

So, the solution is $(s_1, s_2) =$ **(3, 4)**. Always remember to check your solution in <u>both</u> of the original equations.

35. By multiplying the first equation by 2 and the second equation by 3, the coefficients of v become opposites:

$$\begin{matrix} 2 \\ 3 \end{matrix}\begin{cases} 2u - 3v = 16 \\ 5u + 2v = 21 \end{cases} \quad \rightarrow \quad +\begin{cases} 4u - 6v = 32 \\ 15u + 6v = 63 \end{cases}$$
$$19u = 95$$
$$u = 5$$

To obtain the value of v, we substitute the value of 5 for u in either of the two equations. We have chosen arbitrarily to use the first equation:

$$2u - 3v = 16$$
$$2(5) - 3v = 16$$
$$-3v = 6$$
$$v = -2$$

So, the solution is $(u, v) =$ **(5, –2)**. You should then check this answer in <u>both</u> of the original equations.

37. By multiplying the second equation by 2, we force the coefficients of y to be opposites:

$$\begin{matrix} \\ 2 \end{matrix}\begin{cases} 5x + 4y = 5 \\ 15x - 2y = 8 \end{cases} \quad \rightarrow \quad +\begin{cases} 5x + 4y = 5 \\ 30x - 4y = 16 \end{cases}$$
$$35x = 21$$
$$x = \tfrac{21}{35} = \tfrac{3}{5}$$

To obtain the value of y, we substitute the value of $\frac{3}{5}$ for x in either of the two equations. We have chosen arbitrarily to use the first equation:

$$5x + 4y = 5$$
$$5(\tfrac{3}{5}) + 4y = 5$$
$$4y = 2$$
$$y = \tfrac{1}{2}$$

So, the solution is $\boxed{\left(\frac{3}{5}, \frac{1}{2}\right)}$. Once again, using <u>both</u> of the given equations to do so, we encourage you to check the answer.

39. This problem is very similar to Problem 27 (see this manual's solution). The solution is $\boxed{(5, -3)}$. As always, we recommend you check the answer, using <u>both</u> equations to do so.

41. $\boxed{(3, 1)}$. This problem is conducive to both the substitution and addition methods. In the interest of becoming as adept a problem-solver as possible, we suggest you try both methods.

43. $\boxed{(3, 1)}$. As its second equation indicates, this system is especially well suited to the substitution method. For additional reinforcement, refer to this manual's solutions to Problems 15, 17, 23, and 25, all four of which are virtually identical (in form) to this problem.

45. $\boxed{(4, -1)}$. This problem lends itself to either of our two analytic methods. The second equation can be solved readily for x, after which the substitution method may be applied. Alternatively, if you prefer the addition method, you can choose to multiply the second equation either by -3 (to eliminate x) or -2 (to eliminate y). As always, we suggest you take the time to try both methods because by doing so you will enhance your problem-solving skills and, in the process, decrease the likelihood of "going blank" on an upcoming exam.

47. Choosing to use the addition method, we begin by multiplying the second equation by -3. This procedure makes the coefficients of y opposites.

$$\begin{array}{r} \\ -3 \end{array}\begin{cases} 6x+9y=-4 \\ 9x+3y=1 \end{cases} \quad \rightarrow \quad +\begin{cases} 6x+9y=-4 \\ -27x-9y=-3 \end{cases}$$

$$-21x=-7$$

$$x=\tfrac{-7}{-21}=\tfrac{1}{3}$$

To obtain the value of y, we substitute the value of $\frac{1}{3}$ for x in either of the two equations. Arbitrarily, we have chosen to use the second equation, whereby solving the equation $9(\frac{1}{3})+3y=1$ reveals that $y=-\frac{2}{3}$. Hence, the solution is $\boxed{\left(\frac{1}{3}, -\frac{2}{3}\right)}$. As always, you should take the time to check this answer, using <u>both</u> of the given equations to do so.

49. Using the addition method, we begin by multiplying the first equation by 3 and the second equation by -4. Doing so forces the coefficients of y to be opposites.

$$\begin{matrix} 3 \\ -4 \end{matrix}\begin{cases} 5x+4y=9 \\ 9x+3y=12 \end{cases} \rightarrow \quad +\begin{cases} 15x+12y=27 \\ -36x-12y=-48 \end{cases}$$

$$-21x=-21$$

$$x=1$$

To obtain the value of y, we substitute the value of 1 for x in either of the two given equations. If we choose to use the first equation $5x+4y=9$, we find that $5(1)+4y=9$ leads us to conclude $y=1$. Hence, the solution is **(1, 1)**. Of course, before moving on, you should check the answer, using both equations to do so.

51. Based on the form of its first equation, this system appears to be well suited to the substitution method. Substituting $\frac{3}{4}y-2$ for x in the second equation yields:

$$3y-4(\tfrac{3}{4}y-2)=5$$

$$3y-3y+8=5$$

$$8=5$$

The statement $8=5$ is a falsity (a contradiction). Thus, this system is **inconsistent**, and accordingly it has **no solution** (i.e., $\varnothing$).
Alternatively, by looking closely at the two equations (and perhaps even algebraically manipulating one or both), you might have noticed that their graphs are two distinct lines that share the same slope (i.e., the lines are parallel, so they do not intersect). We encourage you to confirm the fact that both lines have a slope of $\frac{4}{3}$.

53. By multiplying both sides by 10, the second equation can be rewritten, if so desired, without the appearance of decimals. That is,

$$\begin{matrix} \\ 10 \end{matrix}\begin{cases} x+y=10 \\ 0.4x+0.9y=0.5(10) \end{cases} \rightarrow \begin{cases} x+y=10 \\ 4x+9y=50 \end{cases}$$

Please note that this step is optional; in other words, solving this system with the decimals present is not an unmanageable task.
Regardless, we find either method (substitution or addition) to be an efficient approach to solving this system of equations. Leaving the details to you (see previous detailed solutions within this manual, if needed), we obtain a solution set of **(8, 2)**. As always, before moving on, you should check the answer, using both equations to do so.

55. Although either approach (substitution or addition) may be applied, let us demonstrate the substitution method. Since both equations have the variable y isolated, we can set $2x-1$ equal to $-3x-9$, which leads to the following:

$$2x-1=-3x-9$$
$$5x=-8$$
$$x=-\tfrac{8}{5}$$

Now, to obtain the value of y, we substitute the value of $-\frac{8}{5}$ for x in either of the given equations and proceed to solve as follows:

$$y=2x-1 \qquad \Rightarrow \qquad y=2(-\tfrac{8}{5})-1=-\tfrac{16}{5}-\tfrac{5}{5}=-\tfrac{21}{5}$$

So, the solution is $\boxed{\left(-\frac{8}{5},-\frac{21}{5}\right)}$. As always, you should take the time to check this answer, using both of the given equations to do so.

57. First, we need to determine Karen's exact age (in terms of human years) on the day her cat, Sydney, was born. The elapsed timespan between Karen's birth date (12/7/70) and Sydney's birth date (4/18/92) consists of:

- Exactly 21 years (that is to say, 365-day years) from 12/7/70 to 12/7/91;
- Exactly 133 days from 12/8/91 to 4/18/92, inclusive [Note: 1992 is a leap year, so remember to count February 29,1992];
- Exactly 5 leap days between 12/7/70 and 12/7/91 (in 1972, 1976, 1980, 1984, and 1988).

Hence, on Sydney's date of birth, Karen was exactly $21+\frac{133}{365}+\frac{5}{365}=21\frac{138}{365}$ years old.

Now, if cats age four times as quickly as humans, then Sydney ages 4 days for every 1 day Karen ages. In terms of 365-day years, this means when x days elapse, Karen ages $\frac{x}{365}$ yr while Sydney ages $4\left(\frac{x}{365}\right)=\frac{4x}{365}$ yr.

To answer the question, we let n represent the number of days from Sydney's birth to the date on which Karen and Sydney will share the exact same age. So, n days after Sydney's birth, Sydney will be $\frac{4n}{365}$ year(s) old, while Karen (who was $21\frac{138}{365}$ years old on Sydney's birth date) will be $\left(21\frac{138}{365}+\frac{n}{365}\right)$ years old. Since n represents the number of days required for them to be the same age, we set Sydney's $\frac{4n}{365}$ equal to Karen's $\left(21\frac{138}{365}+\frac{n}{365}\right)$ and solve for n:

$$\frac{4n}{365}=21\tfrac{138}{365}+\frac{n}{365}$$
$$\frac{3n}{365}=\frac{7{,}803}{365} \qquad \textit{Subtract } \tfrac{n}{365} \textit{ from both sides; write } 21\tfrac{138}{365} \textit{ as an improper fraction.}$$
$$3n=7{,}803$$
$$n=2{,}601$$

Remember that 1996 is a leap year, so 2,601 days equates to 7 years and 45 (not 46) additional days. Thus, on **June 2, 1999**, 2,601 days after Sydney's birth, Sydney the cat and Karen the human were precisely the same age.

59. *[<u>**Note**</u>: If you have forgotten the quadratic formula, we suggest you either use your text's glossary (see p. 925) or refer to textbook Section 5.4 (starting on p. 267.]*

Begin by solving the second equation for one of its variables; we arbitrarily have chosen to solve for x, which gives $x = a - y$. Next, we substitute $a - y$ for x in the first equation:

$$\begin{aligned} xy &= 1 \\ (a-y)y &= 1 \\ ay - y^2 &= 1 \\ y^2 - ay + 1 &= 0 \end{aligned}$$

To solve this quadratic equation, we apply the quadratic formula (using coefficient values — in descending order of the terms' degrees — of $1, -a,$ and 1):

$$\begin{aligned} y &= \frac{-(-a) \pm \sqrt{(-a)^2 - 4(1)(1)}}{2(1)} \\ &= \frac{a \pm \sqrt{a^2 - 4}}{2} \end{aligned}$$

This means there are two values for y, specifically

$$y_1 = \frac{a + \sqrt{a^2 - 4}}{2} \quad \text{and} \quad y_2 = \frac{a - \sqrt{a^2 - 4}}{2}.$$

Next, to obtain their corresponding values of x, we substitute (individually) the above values for y in the equation $x = a - y$:

$$\begin{aligned} y_1 = \frac{a + \sqrt{a^2 - 4}}{2} \Rightarrow \quad x = a - y &= a - \left(\frac{a + \sqrt{a^2 - 4}}{2} \right) \\ &= \frac{2a}{2} - \left(\frac{a + \sqrt{a^2 - 4}}{2} \right) \\ &= \frac{2a - a - \sqrt{a^2 - 4}}{2} \\ &= \frac{a - \sqrt{a^2 - 4}}{2} \end{aligned}$$

Similarly, substituting $y_2 = \dfrac{a - \sqrt{a^2 - 4}}{2}$ leads to an x-value of $\dfrac{a + \sqrt{a^2 - 4}}{2}$.

Therefore, our solution set consists of two ordered pairs:

$$\boxed{\left(\frac{a+\sqrt{a^2-4}}{2},\frac{a-\sqrt{a^2-4}}{2}\right) \text{ and } \left(\frac{a-\sqrt{a^2-4}}{2},\frac{a+\sqrt{a^2-4}}{2}\right)}.$$

As always, we encourage you to check these solutions (in both equations). Doing so will give you a good self-assessment of your basic algebra skills.

Problem Set 12.2

1. This problem is almost identical to text Example 1 (pp. 813-814).
Let $d =$ NUMBER OF DIMES and $n =$ NUMBER OF NICKELS.
In cents, the value of the dimes is $10d$, and the value of the nickels is $5n$. The total value is 715 cents; hence, we have the equation $10d + 5n = 715$. Given that there are 25 more dimes than nickels, it follows that $d = n + 25$. So, we have the following system of two equations in two unknowns:

$$\begin{cases} 10d + 5n = 715 \\ d = n + 25 \end{cases}$$

Because of the form of its second equation, we decide to use the substitution method to solve the system. Substituting $n + 25$ for d in the first equation, we obtain:

$$\begin{aligned} 10(n+25) + 5n &= 715 \\ 10n + 250 + 5n &= 715 \\ 15n &= 465 \\ n &= 31 \end{aligned}$$

If $n = 31$, then $d = n + 25 = 31 + 25 = 56$.

Therefore, **the box contains 31 nickels and 56 dimes**.

Check: There are 25 more dimes than nickels (i.e., $56 - 31 = 25$), and the value of the coinage is $56(\$0.10) + 31(\$0.05) = \$7.15$. ✓

3. Let $q =$ NUMBER OF QUARTERS and $n =$ NUMBER OF NICKELS.
The fact that there are 42 coins in all leads us to write the equation $q + n = 42$. With $25q$ and $5n$ being the value (in cents) of the quarters and nickels respectively, the total value of 690 cents can be expressed by the equation $25q + 5n = 690$. Hence, we have the following system of two equations in two unknowns:

$$\begin{cases} q+n=42 \\ 25q+5n=690 \end{cases}$$

This system lends itself to either of our two analytic methods. The first equation can be solved readily for either variable, after which the substitution method may be applied. Since the question asks only for the number of nickels, one should seek to substitute $42-n$ for q. We leave the details to you.

As for the addition method, one can choose to multiply the first equation either by -25 (to eliminate q) or -5 (to eliminate n). Because the question asks for the number of nickels, multiplying by -25 would be the more efficient approach. Again, we leave the details to you.

Ultimately, regardless of the approach employed, one finds that $n=18$ and $q=24$. Therefore, **there are 18 nickels**.

<u>Check</u>: There are 42 coins in all (i.e., $18 + 24 = 42$), and the value of the coinage is $24(\$0.25)+18(\$0.05)=\$6.90$. ✓

5. This problem is very similar to text Example 3 (p. 815).

Let $p=$ PLANE'S RATE (in still air) and $w=$ WIND'S RATE.

So, when the plane is traveling "with the wind," its speed is $(p+w)$ mph. When the plane is traveling "against the wind," its speed is $(p-w)$ mph. Given that the plane travels 200 mph with the wind and only 150 mph against the wind, we have the following system of equations:

$$\begin{cases} p+w=200 \\ p-w=150 \end{cases}$$

Adding these two equations together gives us the equation $2p=350$, which leads to $p=175$. Thus, **the plane's speed in still air is 175 mph**.

<u>Check</u>: By completing the equation-solving procedure, we find $w=25$. Given a wind speed of 25 mph, a plane whose speed is 175 mph will travel 200 mph with the wind (i.e., $175+25$) and 150 mph against the wind (i.e., $175-25$). ✓

7. For additional background regarding the concepts of supply and demand, read textbook pp. 816-817.

Let $p=$ PRICE, $n=$ NUMBER OF ITEMS, and the ordered pairs be (p, n).

The given information implies that the demand curve (which we will assume is a line) passes through the points (110, 150000) and (20, 300000). As for the supply curve (also assumed to be linear), we have the points (90, 300000) and (10, 200000).

Recall that the equilibrium point is the point at which supply and demand are equal (i.e., where their curves intersect). Like text Example 4, you could find the equilibrium point by using the graphing method (we suggest you give it a try), but let us instead share another approach. Using point-slope form or two-point form (see Problems 58 and 59 in Section 11.2, p. 733), we can write the equations for the demand line and the supply line.

Beginning with the demand line, the points $(110,\ 150000)$ and $(20,\ 300000)$ lead us to a slope of

$$m = \frac{\text{rise}}{\text{run}} = \frac{y_2 - y_1}{x_2 - x_1} = \frac{300{,}000 - 150{,}000}{20 - 110} = \frac{150{,}000}{-90} = -\frac{5{,}000}{3}.$$

Substituting either of the given points, we use the point-slope form equation to write the demand equation:

$$\begin{aligned} y - k &= m(x - h) \\ n - k &= m(p - h) \\ n - 150{,}000 &= -\tfrac{5{,}000}{3}(p - 110) \\ 3n - 450{,}000 &= -5{,}000p + 550{,}000 \\ 5{,}000p + 3n &= 1{,}000{,}000 \end{aligned}$$

Similarly, for the supply line with given points $(90,\ 300000)$ and $(10,\ 200000)$, we obtain a slope of $m = \frac{100{,}000}{80} = 1{,}250$. Then, using point-slope form in the same manner as shown above, we obtain a supply equation of $1{,}250p - n = -187{,}500$.

So, to find the point of equilibrium, we proceed to solve the resulting linear system as follows (via the addition method):

$$\begin{cases} 5{,}000p + 3n = 1{,}000{,}000 \\ 1{,}250p - n = -187{,}500 \end{cases}$$

$$\begin{matrix} \\ 3 \end{matrix}\begin{cases} 5{,}000p + 3n = 1{,}000{,}000 \\ 1{,}250p - n = -187{,}500 \end{cases} \quad \rightarrow \quad + \begin{cases} 5{,}000p + 3n = 1{,}000{,}000 \\ 3{,}750p - 3n = -562{,}500 \end{cases}$$

$$\begin{aligned} 8{,}750p &= 437{,}500 \\ p &= 50 \end{aligned}$$

Substituting 50 for p in either of the equations and then solving for n, we find $n = 250{,}000$.

Therefore, **the equilibrium point for the system is (50, 250,000)**. This means the price charged should be \$50, and one should expect to sell $250{,}000$ items.

9. Once again, as mentioned in our solution to Problem 7 (please see above), one could use the graphing method — despite its shortcomings — to find (or at least estimate) the optimum price (i.e., the p-coordinate of the equilibrium point). Instead, we again use point-slope form to write the supply and demand equations.

Let $p =$ PRICE, $n =$ NUMBER OF ITEMS, and the ordered pairs be (p, n).

Beginning with the demand line, the given information implies that it passes through the points $(1, 900)$ and $(7, 300)$. This leads us to a slope of

$$m = \frac{\text{rise}}{\text{run}} = \frac{y_2 - y_1}{x_2 - x_1} = \frac{300 - 900}{7 - 1} = \frac{-600}{6} = -100.$$

Substituting either of the given points, we use the point-slope form equation to write the demand equation:

$$\begin{aligned} y - k &= m(x - h) \\ n - k &= m(p - h) \\ n - 900 &= -100(p - 1) \\ n - 900 &= -100p + 100 \\ 100p + n &= 1{,}000 \end{aligned}$$

Similarly, for the supply line with given points $(1, 600)$ and $(9, 1000)$, we obtain a slope of $m = \frac{400}{8} = 50$. Then, using point-slope form in the same manner as shown above, we obtain a supply equation of $50p - n = -550$.

So, we proceed to solve the resulting linear system as follows (via the addition method):

$$+\begin{cases} 100p + n = 1{,}000 \\ 50p - n = -550 \end{cases}$$

$$\begin{aligned} 150p &= 450 \\ p &= 3 \end{aligned}$$

Therefore, **the optimum price for the item is \$3**.

Since the problem asks for the optimum "price," it was not necessary to solve for n. In order to check your answer, though, finding n would be a good idea.

11. Let $x =$ Debra Winger's birth year and $y =$ Meryl Streep's birth year.

Since Debra Winger is 6 years younger, we do one of two things: (i) subtract 6 years from her birth year to obtain Streep's birth year (i.e., $x - 6 = y$), or equivalently (ii) add 6 years to Streep's birth year to obtain Winger's birth year (i.e., $y + 6 = x$).

Given that the sum of their birth years is $3{,}904$, we have the following system:

$$\begin{cases} x - 6 = y \\ x + y = 3{,}904 \end{cases}$$

We solve this system by substituting $x - 6$ for y in the second equation:

$$x+(x-6)=3{,}904$$
$$2x-6=3{,}904$$
$$2x=3{,}910$$
$$x=1955$$

So, **Debra Winger was born in 1955**.

13. Mixture **a** weighs 15 lbs, consisting of

- 5 lbs of Ingredient I (80% of which is micoden), and
- 10 lbs of Ingredient II (30% of which is micoden).

So, the amount of micoden in mixture **a** is the sum of 80% of 5 lbs and 30% of 10 lbs: $0.80(5)+0.30(10)=4+3=7$. Therefore, **mixture a has 7 lb of micoden**.

15. From the given information, we know there is no bixon in Ingredient I, while half of Ingredient II is bixon. Since mixture **c** consists of p liters of Ingredient II and 50% of Ingredient II is bixon, **mixture c has 0.5p L of bixon**.

17. First, we determine the pounds of bixon in mixture **a**. From the given information, we know there is no bixon in Ingredient I, while half of Ingredient II is bixon. Since mixture **a** is made up of 10 lb of Ingredient II and 50% of Ingredient II is bixon, mixture **a** has 5 lb (i.e., 50% of 10 lb) of bixon.
To find the proportion of bixon in mixture **a**, we divide the weight of the bixon (5 lb) by the total weight of the mixture (15 lb). So, $\frac{5}{15}=\frac{1}{3}$ of mixture **a** is bixon. Since $\frac{1}{3}=0.\overline{3}=33\frac{1}{3}\%$, we conclude that **mixture a contains $33\frac{1}{3}\%$ bixon**.

19. Let $x=$ AMOUNT (in ounces) OF BASE METAL TO BE ADDED TO 21% ALLOY
$y=$ AMOUNT (in ounces) OF FINAL MIXTURE (i.e., the 15% alloy)

Often, a table summarizing the relevant facts can be helpful:

	Amount	*% Silver*	*Amount of Silver*
21% alloy	100 oz	21%	$0.21(100)=21$ oz
Base metal	x oz	0%	$0.00(x)=0$ oz
Final alloy	y oz	15%	$(0.15y)$ oz

The y oz of the final mixture (alloy) consist of x ounces of a base metal and 100 ounces of the 21% alloy. We represent this idea by writing the equation

$$y=x+100.$$

We then need to write another equation to represent the total amount of silver. In the table above, notice that the final alloy's $(0.15y)$ oz of silver must equal 21 oz because the base metal added to the 21% alloy contains no silver. Hence,

$$0.21(100) + 0(x) = 0.15y$$
$$21 + 0 = 0.15y$$
$$21 = 0.15y$$

We now solve the resulting system of equations:

$$\begin{cases} y = x + 100 \\ 21 = 0.15y \end{cases}$$

We solve the second equation for y:

$$21 = 0.15y \quad \rightarrow \quad y = \frac{21}{0.15} = 140$$

Finally, to find x, we substitute 140 for y in the first equation and solve accordingly:

$$y = x + 100$$
$$140 = x + 100$$
$$40 = x$$

Therefore, to obtain a 15% silver alloy, **40 ounces of the base metal** must be alloyed with 100 ounces of the 21% silver alloy.

21. This problem is almost identical to text Example 10 (p. 821).
Let $m =$ AMOUNT (in gallons) OF MILK
and $c =$ AMOUNT OF CREAM.
The following table summarizes the given information:

	Gallons of Solution	% Butterfat	Gallons of Butterfat
Milk	m	20%	$0.20m$
Cream	c	60%	$0.60c$
Half-and-Half	180	50%	$0.50(180) = 90$

The following system of equations represents (in order) the gallons of solution and the gallons of butterfat:

$$\begin{cases} m + c = 180 \\ 0.20m + 0.60c = 90 \end{cases}$$

Of course, various methods may be used to solve this system. If we choose to divide the second equation by -0.20, we readily can apply the addition method:

$$+\begin{cases} m + c = 180 \\ -m - 3c = -450 \end{cases}$$
$$-2c = -270$$
$$c = 135$$

If $c = 135$, then $m + c = 180$ implies $m = 180 - c = 180 - 135 = 45$.

Therefore, one must **mix 45 gal of milk with 135 gal of cream**.

23. Let $r =$ RENTAL COST and $m =$ MILEAGE.

Under **Option A**, a person who rents a car for four days, pays \$40 per day plus 50¢ per mile. This can be represented by the equation $r = 40(4) + 0.50m$, which simplifies to $r = 160 + 0.50m$.

Under **Option B**, a person who rents a car for four days pays a flat \$60 per day (with no mileage-based fees assessed). This can be represented by the equation $r = 60(4) = 240$, which means the rental cost is \$240.

To find the mileage at which both options cost the same, we solve the following system of equations:

$$\begin{cases} r = 160 + 0.50m \\ r = 240 \end{cases}$$

Substituting 240 for r in the first equation, we find:

$$240 = 160 + 0.50m$$
$$80 = 0.50m$$
$$160 = m$$

Hence, **both rate are the same if you drive 160 miles**. If you plan to drive less than 160 miles over four days, then Option A is the better choice. Otherwise, if your mileage is likely to exceed 160 miles, you should select Option B.

25. From textbook p. 816, we know the equilibrium point is the point for which the supply and demand are equal. We use the substitution method to find this point:

$$\begin{cases} p = 0.0005n - 3 \\ p = 8 - 0.0006n \end{cases} \quad \rightarrow \quad \begin{aligned} 0.0005n - 3 &= 8 - 0.0006n \\ 0.0011n &= 11 \\ n &= 10{,}000 \end{aligned}$$

Substituting $n = 10{,}000$ in either of the given equations, we find:

$$p = 0.0005n - 3 = 0.0005(10{,}000) - 3 = 2.$$

Therefore, **the equilibrium point is (2, 10000)**.

27. This problem is analogous to text Example 1 (pp. 813-814) and Problems 1-3.
Let $q =$ NUMBER OF QUARTERS and $d =$ NUMBER OF DIMES.

In cents, the value of the quarters is $25q$, and the value of the dimes is $10d$. Hence, the total value of the coinage is 840 cents can be represented by the equation

$25q + 10d = 840$. Since the number of quarters is twice the number of dimes, it follows that $q = 2d$. These two linear equations form the following system:

$$\begin{cases} 25q + 10d = 840 \\ q = 2d \end{cases}$$

The form of its second equation facilitates the use of the substitution method. Substituting $2d$ for q in the first equation, we obtain:

$$25(2d) + 10d = 840$$
$$50d + 10d = 840$$
$$60d = 840$$
$$d = 14$$

If $d = 14$, then $q = 2d = 2(14) = 28$.

Hence, **the box contains 14 dimes and 28 quarters**.

<u>Check</u>: There are twice as many quarters as dimes (i.e., $28 = 2 \times 14$), and the value of the coinage is $28(\$0.25) + 14(\$0.10) = \$8.40$. ✓

29. This problem is similar to text Examples 2-3 (p. 815) and Problems 4-6.
Recall that RATE = DISTANCE ÷ TIME. Since the plane makes the 660-mile trip in $2\frac{1}{2}$ hours (when flying with the wind) and 3 hours (against the wind), it follows that:

- RATE WITH THE WIND is $660 \div 2\frac{1}{2} = 660 \div 2.5 = 264$ mph
- RATE AGAINST THE WIND is $660 \div 3 = 220$ mph

Let $p =$ PLANE'S RATE (in still air) and $w =$ WIND'S RATE.

When the plane is traveling "with the wind," its speed is $(p + w)$ mph. When the plane is traveling "against the wind," its speed is $(p - w)$ mph. Since the plane travels 264 mph with the wind and 220 mph against the wind, we have the following system of equations:

$$\begin{cases} p + w = 264 \\ p - w = 220 \end{cases}$$

Adding these two equations together yields the equation $2p = 484$, which leads to $p = 242$. Then substitute 242 for p in either equation and solve accordingly. We arbitrarily have chosen to use the first equation:

$$p + w = 264$$
$$242 + w = 264$$
$$w = 22$$

Thus, **the wind's speed is 22 mph**.

Check: When traveling in a 22 mph wind, a plane whose speed (in still air) is 242 mph will travel $242 + 22 = 264$ mph with the wind and $242 - 22 = 220$ mph against the wind. Recall that TIME = DISTANCE ÷ RATE. So, a 660-mile trip with the wind will require $660 \div 264 = 2.5$ hours of flying time, whereas the same distance when flying against the wind will require $660 \div 220 = 3$ hours. ✓

31. This problem is very similar to text Example 5 (p. 817) and Problems 10-12.
Let $x =$ Kristy McNichol's birth year and $y =$ Tatum O'Neal's birth year.

Since Kristy McNichol is 1 year older, we do either of two things: (i) add 1 year to her birth year to obtain O'Neal's birth year (i.e., $x+1=y$), or equivalently (ii) subtract 1 year from O'Neal's birth year to obtain McNichol's birth year (i.e., $y-1=x$).

Given that the sum of their birth years is $3{,}935$, we have the following system:

$$\begin{cases} x+1=y \\ x+y=3{,}935 \end{cases}$$

To solve this system, substitute $x+1$ for y in the second equation:

$$\begin{aligned} x+(x+1) &= 3{,}935 \\ 2x+1 &= 3{,}935 \\ 2x &= 3{,}934 \\ x &= 1967 \end{aligned}$$

So, **Kristy McNichol was born in 1967**.

33. This problem is very similar to text Example 9 (p. 820).
Let $x =$ AMOUNT OF WATER TO BE ADDED (in gallons)
$y =$ AMOUNT OF FINAL MIXTURE (in gallons)

The following table summarizes the given information:

	Amount of Solution	***% Antifreeze***	***Amount of Antifreeze***
Water	x gal	0%	$0(x) = 0$ gal
80% antifreeze	1 gal	80%	$0.80(1) = 0.8$ gal
Final Mixture	y gal	60%	$0.60y$ gal

The y gal of the final 60% mixture consist of x gal of water and 1 gal of the 80% antifreeze solution. This fact can be represented by the equation

$$x+1=y.$$

We then write another equation to represent the amount of antifreeze. In the table above, recall that water contains no antifreeze and notice that the final mixture consists of $(0.60y)$ gal of antifreeze. Hence, the amount of antifreeze can be represented by the equation

$$0(x)+0.80(1)=0.60y$$
$$0.8=0.6y$$

Now, we solve the resulting system:

$$\begin{cases} x+1=y \\ 0.8=0.6y \end{cases}$$

The second equation $0.8=0.6y$ implies $y=\frac{0.8}{0.6}=\frac{8}{6}=\frac{4}{3}=1\frac{1}{3}$. Substituting this value in the first equation yields:

$$x+1=y$$
$$x+1=1\tfrac{1}{3}$$
$$x=\tfrac{1}{3}$$

Therefore, in order to obtain 60% antifreeze, **add $\frac{1}{3}$ gal of water** to 1 gal of 80% antifreeze.

35. In square miles, let $n=$ AREA OF NEW YORK and $c=$ AREA OF CALIFORNIA. The fact that New York and California's combined area is 204,192 mi^2 can be represented by the equation $n+c=204{,}192$. Since the area of California is 108,530 mi^2 more than that of New York, we write $c=n+108{,}530$.

To find the area of each state, we must solve the system $\begin{cases} n+c=204{,}192 \\ c=n+108{,}530 \end{cases}$.

Substituting $n+108{,}530$ for c in the first equation yields:

$$n+c=204{,}192$$
$$n+(n+108{,}530)=204{,}192$$
$$2n+108{,}530=204{,}192$$
$$2n=95{,}662$$
$$n=47{,}831$$

If $n=47{,}831$, then $c=n+108{,}530=47{,}831+108{,}530=156{,}361$.

Therefore,

New York's area is 47,831 sq mi and California's is 156,361 sq mi.

37. This problem is almost identical to Problem 3.
Let $q =$ NUMBER OF QUARTERS and $n =$ NUMBER OF NICKELS.

The fact that there are 42 coins in all leads us to write the equation $q + n = 42$. With $25q$ and $5n$ being the value (in cents) of the quarters and nickels respectively, the total value of 950 cents can be expressed by the equation $25q + 5n = 950$.
Hence, we have the following system of two equations in two unknowns:

$$\begin{cases} q + n = 42 \\ 25q + 5n = 950 \end{cases}$$

This system lends itself to either of our two analytic methods. The first equation can be solved readily for either variable, after which the substitution method may be applied. Since the question asks only for the number of quarters, one should seek to substitute $42 - q$ for n. We leave the details to you.

As for the addition method, one can choose to multiply the first equation either by -25 (to eliminate q) or -5 (to eliminate n). Because the question asks for the number of quarters, multiplying by -5 would be the more efficient approach. Again, we leave the details to you.

Ultimately, regardless of the method you choose, you will find $q = 37$ and $n = 5$.

Therefore, **there are 37 quarters**.

Check: There are 42 coins in all (i.e., $37 + 5 = 42$), and the value of the coinage is $37(\$0.25) + 5(\$0.05) = \$9.50$. ✓

39. This problem is an extension of text Example 1 (pp. 813-814) and Problems 1-3, 26-27, and 37-38.

Let $n =$ NUMBER OF NICKELS, $d =$ NUMBER OF DIMES, and $q =$ NUMBER OF QUARTERS.

Given that there are 98 coins in all, it follows that $n + d + q = 98$. Since there is an equal number of dimes and quarters, we write $d = q$. There are 8 more nickels than either dimes or quarters, so $n = d + 8$ and $n = q + 8$.

In cents, the respective values of the nickels, dimes, and quarters are $5n$, $10d$, and $25q$. The fact that the total value of the coinage equals $1{,}240$ cents can be expressed by the equation $5n + 10d + 25q = 1{,}240$.

All of the given information may be summarized by the following system:

$$\begin{cases} n + d + q = 98 \\ d = q \\ n = d + 8 \\ n = q + 8 \\ 5n + 10d + 25q = 1{,}240 \end{cases}$$

There are numerous ways to condense this system so that it consists of only two unknowns. For instance, since $d = q$, we can substitute d for q in the first equation, which gives $n + d + (d) = 98$ or $n + 2d = 98$. This equation and the third equation ($n = d + 8$) form an easy-to-solve system:

$$\begin{cases} n + 2d = 98 \\ n = d + 8 \end{cases}$$

Substituting $d + 8$ for n in the equation $n + 2d = 98$, we obtain:

$$\begin{aligned} (d + 8) + 2d &= 98 \\ 3d + 8 &= 98 \\ 3d &= 90 \\ d &= 30 \end{aligned}$$

Hence, **there are 30 dimes**.

Check: By completing the equation-solving process, we find $q = 30$ and $n = 38$. We then confirm that:

- ✓ there are the same number of dimes as quarters (specifically, 30 of each);
- ✓ there are eight more nickels than either dimes or quarters;
- ✓ there are $38 + 30 + 30 = 98$ coins in all; and
- ✓ the value of the coinage is $38(\$0.05) + 30(\$0.10) + 30(\$0.25) = \$12.40$.

41. This problem is similar to its two immediate predecessors, Problems 39 and 40. Let f = NUMBER OF \$5 BILLS, t = NUMBER OF \$10 BILLS, and w = NUMBER OF \$20 bills.

The respective dollar values of these three denominations are $\$(5f)$ in \$5 bills, $\$(10t)$ in \$10 bills, and $\$(20w)$ in \$20 bills. The fact that the bag contains \$1,390 can be expressed by the equation $5f + 10t + 20w = 1{,}390$. Sherlock Holmes' declaration that "there were five times as many \$10 bills as \$5 bills" can be represented by the equation $t = 5f$. Similarly, his deduction that there were "three more than twice as many \$20 bills as \$5 bills" translates to $w = 2f + 3$.

So, to solve the Great Bank Robbery, one must solve the following system:

$$\begin{cases} 5f + 10t + 20w = 1{,}390 \\ t = 5f \\ w = 2f + 3 \end{cases}$$

Use $t = 5f$ and $w = 2f + 3$ to substitute in the first equation:

$$\begin{aligned} 5f + 10t + 20w &= 1{,}390 \\ 5f + 10(5f) + 20(2f + 3) &= 1{,}390 \end{aligned}$$

$$5f + 50f + 40f + 60 = 1{,}390$$
$$95f = 1{,}330$$
$$f = 14$$

If $f = 14$, then $t = 5(14) = 70$ and $w = 2(14) + 3 = 31$.

Therefore, **there are 14 \$5 bills, 70 \$10 bills, and 31 \$20 bills**.

__Check__: The total value of the currency is $14(\$5) + 70(\$10) + 31(\$20) = \$1{,}390$. Since there are fourteen \$5 bills and seventy \$10 bills and we know $14 \times 5 = 70$, it is true that there were five times as many \$10 bills as \$5 bills. Similarly, we note that $14 \times 2 + 3 = 31$, so the thirty-one \$20 bills is consistent with Holmes' conclusion that there were three more than twice as many \$20 bills as \$5 bills. ✓

43. This problem is akin to Problem 29, which is similar to text Examples 2-3 (p. 815) and Problems 4-6. Recall that RATE = DISTANCE ÷ TIME. Since the plane makes the 945-mile trip in 3 hours (when flying with the aid of a tail wind) and $3\frac{1}{2}$ hours (when flying against a head wind), it follows that:

- RATE WITH THE WIND is $945 \div 3 = 315$ mph
- RATE AGAINST THE WIND is $945 \div 3\frac{1}{2} = 945 \div 3.5 = 270$ mph

Let $p =$ PLANE'S RATE (in still air) and $w =$ WIND'S RATE.

When the plane is traveling "with a tail wind," its speed is $(p + w)$ mph. When the plane is traveling "against the wind," its speed is $(p - w)$ mph. Since the plane travels 315 mph with the wind and 270 mph against the wind, we have the following system of equations:

$$\begin{cases} p + w = 315 \\ p - w = 270 \end{cases}$$

Adding these two equations together yields the equation $2p = 585$, which leads to $p = 292.5$. We then substitute 292.5 for p in either equation and solve accordingly:

$$p + w = 315$$
$$292.5 + w = 315$$
$$w = 22.5$$

Thus, **the wind's speed is 22.5 mph**.

__Check__: When traveling in a 22.5 mph wind, a plane whose speed (in still air) is 292.5 mph will travel $292.5 + 22.5 = 315$ mph with the wind and $292.5 - 22.5 = 270$ mph against the wind. Recall that TIME = DISTANCE ÷ RATE. So, a 945-mile trip with the wind will require $945 \div 315 = 3$ hours of flying time, whereas the same distance when flying against the wind will require $945 \div 270 = 3.5$ hours. ✓

45. This problem is similar to text Example 10 (p. 821) and Problem 21.
In gallons, let $m =$ AMOUNT OF MILK and $c =$ AMOUNT OF CREAM.

The following table summarizes the given information:

	Gallons of Solution	***% Butterfat***	***Gallons of Butterfat***
Cream	c	23%	$0.23c$
Milk	m	3%	$0.03m$
Final Mixture *("Richer Milk")*	30	4%	$0.04(30) = 1.2$

The following system of equations represents (in order) the gallons of solution and the gallons of butterfat:

$$\begin{cases} c + m = 30 \\ 0.23c + 0.03m = 1.2 \end{cases}$$

By multiplying both sides by 100, the second equation can be rewritten, if so desired, without the appearance of decimals. That is,

$$100\begin{cases} c + m = 30 \\ 0.23c + 0.03m = 1.2 \end{cases} \quad \rightarrow \quad \begin{cases} c + m = 30 \\ 23c + 3m = 120 \end{cases}$$

Please note that this step is optional; that is to say, solving this system with the decimals present is not an unmanageable task.

Realize that various methods may be used to solve this system. We find substitution and addition to be equally efficient approaches to solving this system of equations. We leave the remaining details to you [refer, if necessary, to the many fully detailed solutions shown previously within this manual]. Regardless of the method employed, one obtains a solution of $(c, m) = (1.5, 28.5)$.

This tells us that one needs to **mix 28.5 gal of milk with 1.5 gal of cream**.

47. Let $x =$ AMOUNT OF 12% ASPIRIN SOLUTION (in mg)
and $y =$ AMOUNT OF 25% ASPIRIN SOLUTION (in mg).

The following table summarizes the given information:

	Amount of Solution	***% aspirin***	***Amount of Aspirin***
12% aspirin	x mg	12%	$0.12x$ mg
25% aspirin	y mg	25%	$0.25y$ mg
Final Mixture *(20% aspirin)*	100 mg	20%	$0.20(100) = 20$ mg

The following system represents respectively the milligrams of solution and the milligrams of pure aspirin:

$$\begin{cases} x + y = 100 \\ 0.12x + 0.25y = 20 \end{cases}$$

By multiplying both sides by 100, the second equation can be rewritten, if so desired, without the appearance of decimals. That is,

$$100\begin{cases} x+y=100 \\ 0.12x+0.25y=20 \end{cases} \rightarrow \begin{cases} x+y=100 \\ 12x+25y=2{,}000 \end{cases}$$

Once again, we remind you that this step is optional; said differently, solving this system <u>with</u> the decimals present is <u>not</u> an unmanageable task.

Assorted methods may be used to solve this system. The substitution and addition are equally efficient approaches. Arbitrarily choosing the addition method, we find:

$$-12\begin{cases} x+y=100 \\ 12x+25y=2{,}000 \end{cases} \rightarrow \begin{cases} -12x+-12y=-1{,}200 \\ \quad 12x+25y=2{,}000 \end{cases}$$

$$13y=800$$

$$y=\tfrac{800}{13}\approx 38.46$$

Finally, we substitute $y=\frac{800}{13}$ into the equation $x+y=100$:

$$x=100-y=100-\tfrac{800}{13}=\tfrac{1{,}300-800}{13}=\tfrac{500}{13}\approx 61.54.$$

We now know the chemist needs to **mix approximately 61.54 mg of 25% aspirin and 38.46 mg of 12% aspirin**.

49. Let $x=$ HEIGHT OF STANDARD OIL BLDG (in ft)
and $y=$ HEIGHT OF SEARS TOWER (in ft).

The fact that the two buildings have a combined height of $2{,}590$ ft can be represented by the equation $x+y=2{,}590$. Since the Sears tower is 318 ft taller than the Standard Oil building, we write $y=x+318$.

To find the height of each building, we solve the system $\begin{cases} x+y=2{,}590 \\ y=x+318 \end{cases}$.

Substituting $x+318$ for y in the first equation implies:

$$x+y=2{,}590$$

$$x+(x+318)=2{,}590$$

$$2x+318=2{,}590$$

$$2x=2{,}272$$

$$x=1{,}136$$

If $x=1{,}136$, then $y=x+318=1{,}136+318=1{,}454$. So, **the height of the Sears Tower is 1,454 ft and the Standard Oil building is 1,136 ft**.

51. Let $v =$ LENGTH OF VERRAZANO-NARROWS BRIDGE (in ft)
and $g =$ LENGTH OF GEORGE WASHINGTON BRIDGE (in ft).

The fact that the combined span of the two bridges is $7,760$ ft suggests the equation $v+g=7,760$. Since the problem asks for the length of the Verrazano-Narrows and since its length is the longer than the George Washington by 760 ft, we write $v-760=g$. So, the following system summarizes the problem:

$$\begin{cases} v+g=7,760 \\ v-760=g \end{cases}.$$

Substituting $v-760$ for g in the first equation yields:

$$\begin{aligned} v+g &= 7,760 \\ v+(v-760) &= 7,760 \\ 2v-760 &= 7,760 \\ 2v &= 8,520 \\ v &= 4,260 \end{aligned}$$

Hence, **the length of the Verrazano-Narrows bridge is 4,260 ft**.

53. For additional information pertaining to the concepts of supply and demand, revisit textbook pp. 816-817.
Let $p =$PRICE, $n =$NUMBER OF ITEMS, and the ordered pairs be (p, n).
The given information implies that the linear demand curve passes through the points $(2,\ 40000)$ and $(1,\ 100000)$. As for the supply curve (which we also assume is linear), the given points are $(0.50,\ 20000)$ and $(5,\ 80000)$.
Recall that the equilibrium point is the point at which supply and demand are equal (i.e., where their curves intersect). Like text Example 4, you could find the equilibrium point by using the graphing method (we suggest you give it a try), but let us instead share another approach. Using point-slope form or two-point form (see Problems 58 and 59 in Section 11.2, p. 733), we can write the equations for the demand line and the supply line.
Beginning with the demand line, the points $(2,\ 40000)$ and $(1,\ 100000)$ lead us to a slope of

$$m=\frac{\text{rise}}{\text{run}}=\frac{y_2-y_1}{x_2-x_1}=\frac{100,000-40,000}{1-2}=\frac{60,000}{-1}=-60,000.$$

Substituting either of the given points, we use the point-slope form equation to write the demand equation:

$$\begin{aligned} y-k &= m(x-h) \\ n-k &= m(p-h) \\ n-100,000 &= -60,000(p-1) \end{aligned}$$

$$n - 100{,}000 = -60{,}000p + 60{,}000$$
$$60{,}000p + n = 160{,}000$$

Similarly, for the supply line with given points $(0.50,\ 20000)$ and $(5,\ 80000)$, we obtain a slope of $m = \frac{60{,}000}{4.5} = \frac{40{,}000}{3}$. Then, using point-slope form in the same manner as shown above, we obtain a supply equation of $40{,}000p - 3n = -40{,}000$.
To find the point of equilibrium, we solve the resulting linear system (via the addition method):

$$\begin{cases} 60{,}000p + n = 160{,}000 \\ 40{,}000p - 3n = -40{,}000 \end{cases}$$

$$3\begin{cases} 60{,}000p + n = 160{,}000 \\ 40{,}000p - 3n = -40{,}000 \end{cases} \rightarrow +\begin{cases} 180{,}000p + 3n = 480{,}000 \\ 40{,}000p - 3n = -40{,}000 \end{cases}$$

$$220{,}000p = 440{,}000$$
$$p = 2$$

Substituting 2 for p in either of the equations and then solving for n, we find $n = 40{,}000$.

Therefore, **the equilibrium point is \$2 for 40,000 items**.

55. Assume the car owner begins with a full radiator, consisting of 17 quarts of 15% antifreeze solution.
A portion of these 17 quarts is to be removed and then replaced with pure (100%) antifreeze. Since the amount to be removed/replaced is unknown, we let

$x =$ AMOUNT TO BE REMOVED/REPLACED (in quarts).

Removing an unknown amount of the 15% antifreeze from this 17-qt radiator leaves the owner with another unknown; namely, the amount of 15% antifreeze remaining in the radiator. Hence, we let

$y =$ AMOUNT OF 15% ANTIFREEZE REMAINING AFTER DRAINING (in qt).

A table is a helpful way to summarize this information:

	Amount of Solution	***% antifreeze***	***Amount of Antifreeze***
Remaining liquid (after draining of 15% antifreeze)	y qt	15%	$0.15y$ qt
Pure Antifreeze (to be added after draining)	x qt	100%	$1.00x = x$ qt
Final Radiator Contents	17 qt	60%	$0.6(17) = 10.2$ qt

The column entitled "Amount of Solution" gives rise to the equation $y + x = 17$, and the column "Amount of Antifreeze" leads to the equation $0.15y + x = 10.2$. So, this problem amounts to solving the following system of equations:

$$\begin{cases} x+y=17 \\ x+0.15y=10.2 \end{cases}$$

Via either the addition or substitution method, we find $x=9$ and $y=8$.

Therefore, **9 qt of the liquid are to be replaced with antifreeze**.

57. The concepts of supply and demand were introduced on textbook pp. 816-817; related problems included Problems 7-9, 24-25, and 52-53.

a) Substitute $p=250$ in both the supply and demand equations:

$$n = 2.5(250)-500=125 \Rightarrow \text{125 items would be supplied}$$

$$n = 200-0.5(250)=75 \Rightarrow \text{75 items would be demanded}$$

b) If no items are supplied, the value of n in the supply equation becomes zero. To find the price at this supply level, substitute $n=0$ and solve for p:

$$\begin{aligned} n &= 2.5p-500 \\ 0 &= 2.5p-500 \\ 500 &= 2.5p \\ 200 &= p \end{aligned}$$

So, at a price of **$200 per item**, no items would be supplied.

c) When no items are demanded, the value of n in the demanded equation becomes zero. To find the price at this demand level, substitute $n=0$ and solve for p:

$$\begin{aligned} n &= 200-0.5p \\ 0 &= 200-0.5p \\ 0.5p &= 200 \\ p &= 400 \end{aligned}$$

So, at a price of **$400 per item**, no items would be demanded.

d) Recall (from textbook p. 816) that the equilibrium point is the point for which supply and demand are equal. We find such (intersection) points by solving systems of equations. Here, substitution is a well-suited approach:

$$\begin{cases} n = 2.5p-500 \\ n = 200-0.5p \end{cases} \rightarrow \quad \begin{aligned} 2.5p-500 &= 200-0.5p \\ 3p &= 700 \\ p &= \tfrac{700}{3} \end{aligned}$$

Thus, $p=\frac{700}{3}\approx$ **$233.33 is the equilibrium price for this product**.

e) Find the value of n, the number of items, when $p = \frac{700}{3}$. To do so, substitute $p = \frac{700}{3}$ in <u>either</u> equation and then solve accordingly:

$$n = 200 - 0.5(\tfrac{700}{3}) = 200 - \tfrac{350}{3} = \tfrac{600-350}{3} = \tfrac{250}{3}.$$

Hence, $\frac{250}{3} \approx$ **83 items will be produced at the equilibrium price**.

59. **The stated answer is incorrect**. The bottle is supposed to cost \$1 more than the cork, but the difference between 10¢ and \$1.00 is only \$0.90. The actual answer is the cork costs 5¢ and the bottle \$1.05.

Problem Set 12.3

1. **Wording varies**. A ***matrix*** is a rectangular array of terms, which are called the matrix's *elements*.

3. **Wording varies**. ***Gauss-Jordan elimination*** is a matrix method for solving a system of equations. It consists of three steps, each of which is explained on textbook page 830 and then demonstrated in Example 4 (on p. 831).

5. **False**. The order of the matrix for this system is 4×5, not 4×4. Since there are four variables <u>and</u> constant terms, the augmented matrix requires 5 columns.

7. **True**. To review the notation for these elementary row operations, see textbook pp. 828-829.

9. **True**. If necessary, please review textbook p. 829, where this notation is first introduced.

11. **False**. Remember: the pivot row is the first row listed, and the target row is the last one listed. So, the notation ∗Row+(3^{-1},[C],4,2) actually means multiply row 4 (<u>not</u> row 2) by $\frac{1}{3}$ and add the corresponding entries to the entries in row 2 (<u>not</u> row 4).

13. **False**. For this matrix [A], the first step should be to interchange (swap) the first and second rows — <u>not</u> the first and third. We denote this by RowSwap([A],1,2). Doing so results in a '1' in the first row, first column entry with zeros below it.

15. For additional information, review textbook pp. 825-826.
[Note: Choice of variable labels may vary.]

a) $\boxed{\begin{cases} 6x+7y+8z=3 \\ x+2y+3z=4 \\ y+3z=4 \end{cases}}$ b) $\boxed{\begin{cases} x_1=3 \\ x_2+2x_3=4 \end{cases}}$ c) $\boxed{\begin{cases} x_1=32 \\ x_2=27 \\ x_3=-5 \\ 0=3 \end{cases}}$

17. The most efficient way to obtain a 1 in the row 1, column 1 position is to interchange the first and second rows. We indicate this row operation by writing **RowSwap([B],1,2)**, the result of which is the following:

$$\boxed{\left[\begin{array}{ccc|c} 1 & 0 & 2 & -8 \\ -2 & 3 & 5 & 9 \\ 0 & 1 & 0 & 5 \end{array}\right]}.$$

19. For this particular matrix, the simplest way to obtain a 1 in the row 1, column 1 position is to multiply the first row by $\frac{1}{5}$ (i.e., the reciprocal of 5). We indicate this row operation by writing ***Row($\frac{1}{5}$,[A],1)**, the result of which is the following:

$$\boxed{\left[\begin{array}{ccc|c} 1 & 4 & 3 & \frac{6}{5} \\ 7 & -5 & 3 & 2 \\ 12 & 0 & 1 & 4 \end{array}\right]}.$$

21. Notice that the number in the row 3, column 1 position is already zero, so our only task is to turn the number in the row 2, column 1 position into a zero. Multiplying the first row by 3 and adding the result to the second row accomplishes this task. We denote this row operation by writing ***Row+(3,[B],1,2)**, and the result is

$$\boxed{\left[\begin{array}{ccc|c} 1 & 3 & -5 & 6 \\ 0 & 13 & -14 & 20 \\ 0 & 5 & 1 & 3 \end{array}\right]}.$$

23. We use the first row as the pivot row. First, to obtain a zero in the row 2, column 1 position, multiply the first row by -2 and add the result to the second row. Next, to obtain a zero in the row 3, column 1 position, multiply the first row by -3 and add the result to the third row. Respectively, we indicate these 2 steps by:

***Row+(–2,[A],1,2)** and ***Row+(–3,[Ans],1,3)**.

*[Remember: The reason why the second operation makes mention of [Ans] ("**ans**wer"), instead of [A], is because this second operation is being performed on the result of (i.e., the "answer" from) the first operation — not on the original matrix [A].]*

Performing these two operations yields the following matrix:

$$\left[\begin{array}{ccc|c} 1 & 5 & 3 & 2 \\ 0 & -7 & -7 & 0 \\ 0 & -13 & -8 & -6 \end{array}\right].$$

25. To obtain a 1 in the row 2, column 2 position, we multiply the second row by $\frac{1}{3}$ (i.e., the reciprocal of 3). We indicate this row operation by writing **∗Row($\frac{1}{3}$,[B],2)**, the result of which is the following:

$$\left[\begin{array}{ccc|c} 1 & 5 & -3 & 5 \\ 0 & 1 & 3 & -5 \\ 0 & 2 & 1 & 5 \end{array}\right]$$

27. While one could choose to multiply the second row by $\frac{1}{4}$ [i.e., ∗Row($\frac{1}{4}$,[A],2)], let us use another approach, one that avoids the appearance of fractions. In specific, if we multiply the third row by -1 and add the result to the second row, we obtain a 1 in the row 2, column 2 position, and, meanwhile, the other numbers in the second row remain integers. We indicate this row operation by **∗Row+(–1,[A],3,2)**. Its outcome is

$$\left[\begin{array}{ccc|c} 1 & 3 & -2 & 0 \\ 0 & 1 & -4 & 8 \\ 0 & 3 & 6 & 1 \end{array}\right].$$

29. We use the row 2, column 2 number as the pivot entry. By multiplying the second row by -3 and adding the result to the first row, we obtain a zero in the first row, second column position. Then, to attain a zero in the row 3, column 2 position, we multiply the second row by 2 and add the result to the first row. Both steps can be indicated by the shorthand notation

∗Row+(–3,[A],2,1) and then **∗Row+(2,[Ans],2,3)**.

The matrix resulting from the execution of these two operations is

$$\left[\begin{array}{ccc|c} 1 & 0 & 12 & 27 \\ 0 & 1 & -2 & -5 \\ 0 & 0 & -2 & -4 \end{array}\right].$$

31. Again, we use the row 2, column 2 number as the pivot entry. By multiplying the second row by -5 and adding the result to the first row, we obtain a zero in the first row, second column position. Then, to attain a zero in the row 3, column 2 position, we multiply the second row by -1 and add the result to the first row. Finally, we multiply the second row by -2 and add the result to the fourth row in order to turn the row 4, column 2 entry into a zero. In succession, these three row operations can be indicated by the following shorthand notation:

***Row+(–5,[D],2,1)** , then ***Row+(–1,[Ans],2,3)**,

and lastly ***Row+(–2,[Ans],2,4)**.

The matrix resulting from the execution of these three steps is

$$\left[\begin{array}{cccc|c} 1 & 0 & -26 & -8 & 8 \\ 0 & 1 & 5 & 2 & 0 \\ 0 & 0 & -1 & -2 & 5 \\ 0 & 0 & -13 & -3 & 7 \end{array}\right].$$

33. To obtain a 1 in the row 3, column 3 position, multiply the third row by $\frac{1}{8}$. We indicate this row operation by writing ***Row($\frac{1}{8}$,[B],3)**, the result of which is

$$\left[\begin{array}{ccc|c} 1 & 0 & 4 & -5 \\ 0 & 1 & 3 & 6 \\ 0 & 0 & 1 & \frac{3}{2} \end{array}\right] \text{ or } \left[\begin{array}{ccc|c} 1 & 0 & 4 & -5 \\ 0 & 1 & 3 & 6 \\ 0 & 0 & 1 & 1.5 \end{array}\right].$$

35. To obtain a 1 in the row 3, column 3 position, multiply the third row by $\frac{1}{2}$ (i.e., the reciprocal of 2). We indicate this row operation by writing ***Row($\frac{1}{2}$,[B],3)**, the result of which is the following:

$$\left[\begin{array}{cccc|c} 1 & 0 & -8 & 2 & 8 \\ 0 & 1 & -1 & 3 & 2 \\ 0 & 0 & 1 & 0 & 5 \\ 0 & 0 & -2 & 1 & 6 \end{array}\right]$$

37. We use the row 3, column 3 number as the pivot entry. To obtain a zero in the row 2, column 3 position, we multiply the third row by -4 and add the result to the second row. Then, by multiplying the third row by 3 and adding the result to the first row, we attain a zero in the first row, third column position. In sequence, both steps are indicated by ***Row+(–4,[B],3,2)** and ***Row+(3,[Ans],3,1)**. The resulting matrix is:

$$\boxed{\left[\begin{array}{ccc|c} 1 & 0 & 0 & 7 \\ 0 & 1 & 0 & -7 \\ 0 & 0 & 1 & 3 \end{array}\right]}.$$

39. Again, we use the row 3, column 3 number as the pivot entry. By multiplying the third row by 8 and adding the result to the first row, we obtain a zero in the first row, third column position. Then, to attain a zero in the row 2, column 3 position, we multiply the third row by -4 and add the result to the second row. Finally, to turn the row 4, column 3 entry into a zero, we can add the third row to the fourth row, placing the sum in the fourth row (the target row). In succession, we indicate these three row operations by the following shorthand notation:

∗Row+(8,[D],3,1), then **∗Row+(–4,[Ans],3,2)**,

and lastly **Row+([Ans],3,4)**.

The matrix resulting from the execution of these three steps is

$$\boxed{\left[\begin{array}{cccc|c} 1 & 0 & 0 & 2 & 24 \\ 0 & 1 & 0 & 2 & -8 \\ 0 & 0 & 1 & 0 & 2 \\ 0 & 0 & 0 & 1 & 9 \end{array}\right]}.$$

41. For additional reinforcement regarding the ***Gauss-Jordan elimination method***, please refer to text Example 3 (p. 829), which is analogous to Problems 40-47.
We write the given system of equations in augmented matrix form, and then, by applying elementary row operations, we seek to transform it into row-reduced form.

$$[A] = \left[\begin{array}{cc|c} 1 & -1 & 8 \\ 1 & 1 & 2 \end{array}\right] \rightarrow \left[\begin{array}{cc|c} 1 & -1 & 8 \\ 0 & 2 & -6 \end{array}\right] \rightarrow \left[\begin{array}{cc|c} 1 & -1 & 8 \\ 0 & 1 & -3 \end{array}\right]$$

∗Row+(–1,[A],1,2) ∗Row($\frac{1}{2}$,[Ans],2)

$$\rightarrow \left[\begin{array}{cc|c} 1 & 0 & 5 \\ 0 & 1 & -3 \end{array}\right]$$

Row+([Ans],2,1)

This final matrix represents the system $\begin{cases} 1x + 0y = 5 \\ 0x + 1y = -3 \end{cases}$.

Hence, the solution to the given system is **(5, –3)**. You should then check this answer, using both of the original equations to do so.

43. $[B]=\left[\begin{array}{cc|c}1 & -6 & -3\\ 2 & 3 & 9\end{array}\right] \to \left[\begin{array}{cc|c}1 & -6 & -3\\ 0 & 15 & 15\end{array}\right] \to \left[\begin{array}{cc|c}1 & -6 & -3\\ 0 & 1 & 1\end{array}\right]$

∗Row+(–2,[B],1,2) ∗Row($\frac{1}{15}$,[Ans],2)

$\to \left[\begin{array}{cc|c}1 & 0 & 3\\ 0 & 1 & 1\end{array}\right] \Rightarrow \begin{cases}1x+0y=3\\ 0x+1y=1\end{cases}$

∗Row+(6,[Ans],1,2)

Thus, the solution to the given system is **(3, 1)**. You should then check this answer, using both of the original equations to do so.

45. $[C]=\left[\begin{array}{cc|c}3 & 7 & 5\\ 4 & 9 & 7\end{array}\right] \to \left[\begin{array}{cc|c}1 & \frac{7}{3} & \frac{5}{3}\\ 4 & 9 & 7\end{array}\right] \to \left[\begin{array}{cc|c}1 & \frac{7}{3} & \frac{5}{3}\\ 0 & -\frac{1}{3} & \frac{1}{3}\end{array}\right]$

∗Row($\frac{1}{3}$,[C],1) ∗Row+(–4,[Ans],1,2)

$\to \left[\begin{array}{cc|c}1 & \frac{7}{3} & \frac{5}{3}\\ 0 & 1 & -1\end{array}\right] \to \left[\begin{array}{cc|c}1 & 0 & 4\\ 0 & 1 & -1\end{array}\right] \Rightarrow \begin{cases}1x+0y=4\\ 0x+1y=-1\end{cases}$

∗Row(–3,[Ans],2) ∗Row+($-\frac{7}{3}$,[Ans],2,1)

So, the solution to the given system is **(4, –1)**. As always, we recommend you check the answer, using both equations to do so.

47. $[D]=\left[\begin{array}{cc|c}3 & -4 & 16\\ -1 & 2 & -6\end{array}\right] \to \left[\begin{array}{cc|c}-1 & 2 & -6\\ 3 & -4 & 16\end{array}\right] \to \left[\begin{array}{cc|c}1 & -2 & 6\\ 3 & -4 & 16\end{array}\right]$

RowSwap([D],1,2) ∗Row(–1,[Ans],1)

$\to \left[\begin{array}{cc|c}1 & -2 & 6\\ 0 & 2 & -2\end{array}\right] \to \left[\begin{array}{cc|c}1 & -2 & 6\\ 0 & 1 & -1\end{array}\right]$

∗Row+(–3,[Ans],1,2) ∗Row($\frac{1}{2}$,[Ans],2)

$\to \left[\begin{array}{cc|c}1 & 0 & 4\\ 0 & 1 & -1\end{array}\right] \Rightarrow \begin{cases}1x+0y=4\\ 0x+1y=-1\end{cases}$

∗Row+(2,[Ans],2,1)

Hence, the solution to the given system is **(4, –1)**. We hope you will take the time to check this answer.

49. For additional reinforcement regarding the application of the ***Gauss-Jordan elimination method*** to systems of three equations with three unknowns, please refer to text Example 4 (p. 831), which is akin to Problems 48-55.
We write the given system of equations in augmented matrix form, and then, by applying elementary row operations, we transform it into row-reduced form:

$$[A]=\left[\begin{array}{ccc|c}1&1&0&2\\1&0&-1&1\\0&-1&1&1\end{array}\right]\to\left[\begin{array}{ccc|c}1&1&0&2\\0&-1&-1&-1\\0&-1&1&1\end{array}\right]\to\left[\begin{array}{ccc|c}1&1&0&2\\0&1&1&1\\0&-1&1&1\end{array}\right]$$

∗Row+(–1,[A],1,2) ∗Row(–1,[Ans],2)

$$\to\left[\begin{array}{ccc|c}1&0&-1&1\\0&1&1&1\\0&0&2&2\end{array}\right]\to\left[\begin{array}{ccc|c}1&0&-1&1\\0&1&1&1\\0&0&1&1\end{array}\right]$$

∗Row+(–1,[Ans],2,1) ∗Row($\frac{1}{2}$,[Ans],2)
Row+([Ans],2,3)

$$\to\left[\begin{array}{ccc|c}1&0&0&2\\0&1&0&0\\0&0&1&1\end{array}\right]\Rightarrow\begin{cases}1x+0y+0z=2\\0x+1y+0z=0\\0x+0y+1z=1\end{cases}$$

Row+([Ans],3,1)
∗Row+(–1,[Ans],3,2)

So, the solution to the given system is **(2, 0, 1)**. Please make sure you check this answer, using all three original equations to do so.

51. $$[B]=\left[\begin{array}{ccc|c}1&0&2&13\\2&1&0&8\\0&-2&9&41\end{array}\right]\to\left[\begin{array}{ccc|c}1&0&2&13\\0&1&-4&-18\\0&-2&9&41\end{array}\right]\to\left[\begin{array}{ccc|c}1&0&2&13\\0&1&-4&-18\\0&0&1&5\end{array}\right]$$

∗Row+(–2,[B],1,2) ∗Row+(2,[Ans],2,3)

$$\to\left[\begin{array}{ccc|c}1&0&0&3\\0&1&0&2\\0&0&1&5\end{array}\right]\Rightarrow\begin{cases}1x+0y+0z=3\\0x+1y+0z=2\\0x+0y+1z=5\end{cases}$$

∗Row+(–2,[Ans],3,1)
∗Row+(4,[Ans],3,2)

Thus, the solution to the given system is **(3, 2, 5)**. As always, you should take the time to check this answer in all three of the original equations.

53. $[C]=\left[\begin{array}{ccc|c}5 & 0 & 1 & 9\\ 1 & 0 & -5 & 7\\ 1 & 1 & -1 & 0\end{array}\right] \to \left[\begin{array}{ccc|c}1 & 0 & -5 & 7\\ 5 & 0 & 1 & 9\\ 1 & 1 & -1 & 0\end{array}\right] \to \left[\begin{array}{ccc|c}1 & 0 & -5 & 7\\ 0 & 0 & 26 & -26\\ 0 & 1 & 4 & -7\end{array}\right]$

RowSwap([C],1,2) ∗Row+(–5,[Ans],1,2)
∗Row+(–1,[Ans],1,3)

$\to \left[\begin{array}{ccc|c}1 & 0 & -5 & 7\\ 0 & 1 & 4 & -7\\ 0 & 0 & 26 & -26\end{array}\right] \to \left[\begin{array}{ccc|c}1 & 0 & -5 & 7\\ 0 & 1 & 4 & -7\\ 0 & 0 & 1 & -1\end{array}\right]$

RowSwap([Ans],2,3) ∗Row($\frac{1}{26}$,[Ans],3)

$\to \left[\begin{array}{ccc|c}1 & 0 & 0 & 2\\ 0 & 1 & 0 & -3\\ 0 & 0 & 1 & -1\end{array}\right] \Rightarrow \begin{cases}1x+0y+0z=2\\ 0x+1y+0z=-3\\ 0x+0y+1z=-1\end{cases}$

∗Row+(5,[Ans],3,1)
∗Row+(–4,[Ans],3,2)

Hence, the solution to the given system is **(2, –3, –1)**. Once again, we encourage you check this answer, using all three original equations to do so.

55. $[D]=\left[\begin{array}{ccc|c}1 & 1 & 0 & -1\\ 0 & 1 & 1 & -1\\ 1 & 1 & 1 & 1\end{array}\right] \to \left[\begin{array}{ccc|c}1 & 1 & 0 & -1\\ 0 & 1 & 1 & -1\\ 0 & 0 & 1 & 2\end{array}\right] \to \left[\begin{array}{ccc|c}1 & 0 & -1 & 0\\ 0 & 1 & 1 & -1\\ 0 & 0 & 1 & 2\end{array}\right]$

∗Row+(–1,[D],1,3) ∗Row+(–1,[Ans],2,1)

$\to \left[\begin{array}{ccc|c}1 & 0 & 0 & 2\\ 0 & 1 & 0 & -3\\ 0 & 0 & 1 & 2\end{array}\right] \Rightarrow \begin{cases}1x+0y+0z=2\\ 0x+1y+0z=-3\\ 0x+0y+1z=2\end{cases}$

Row+([Ans],3,2)
∗Row(–1,[Ans],3,2)

So, the solution to the given system is **(2, –3, 2)**. Please make sure you check this answer, using all three original equations to do so.

57. $[E]=\left[\begin{array}{ccc|c}1 & 0 & 2 & 0\\ 3 & -1 & 2 & 0\\ 4 & 1 & 0 & 6\end{array}\right] \rightarrow \left[\begin{array}{ccc|c}1 & 0 & 2 & 0\\ 0 & -1 & -4 & 0\\ 0 & 1 & -8 & 6\end{array}\right] \rightarrow \left[\begin{array}{ccc|c}1 & 0 & 2 & 0\\ 0 & 1 & 4 & 0\\ 0 & 1 & -8 & 6\end{array}\right]$

∗Row+(–3,[E],1,2)   ∗Row(–1,[Ans],2)
∗Row+(–4,[Ans],1,3)

$$\rightarrow \left[\begin{array}{ccc|c}1 & 0 & 2 & 0\\ 0 & 1 & 4 & 0\\ 0 & 0 & -12 & 6\end{array}\right] \rightarrow \left[\begin{array}{ccc|c}1 & 0 & 2 & 0\\ 0 & 1 & 4 & 0\\ 0 & 0 & 1 & -0.5\end{array}\right]$$

∗Row+(–1,[Ans],2,3)   ∗Row($-\frac{1}{12}$,[Ans],3)

$$\rightarrow \left[\begin{array}{ccc|c}1 & 0 & 0 & 1\\ 0 & 1 & 0 & 2\\ 0 & 0 & 1 & -0.5\end{array}\right] \Rightarrow \begin{cases}1x+0y+0z=1\\ 0x+1y+0z=2\\ 0x+0y+1z=-\frac{1}{2}\end{cases}$$

∗Row+(–2,[Ans],3,1)
∗Row+(–4,[Ans],3,2)

Thus, the solution to the given system is $\boxed{\left(\mathbf{1},\ \mathbf{2},\ -\frac{1}{2}\right)}$. Please make sure you check this answer, using all three original equations to do so.

59. This problem is similar to text Example 5 (pp. 831-832).

Let $x=$ Number of units of candy I
$y=$ Number of units of candy II
$z=$ Number of units of candy III

Since each unit of candy I requires 7 lb of chocolate, it follows that x units of candy I require $7x$ lb of chocolate. Similarly, since each unit of candy II and candy III respectively require 3 lb and 4 lb of chocolate, it follows that y units of candy II require $3y$ lb of chocolate and z units of candy III require $4z$ lb of chocolate. Given that 67 lb of chocolate are available, its allocation can be expressed in terms of the equation $7x+3y+4z=67$.

By applying this same line of reasoning to the milk and mint extract required by each kind of candy, we obtain two more equations: $5x+2y+3z=48$ and $x+2y+3z=32$.

So, we have the following system of equations: $\begin{cases} 7x+3y+4z=67 \\ 5x+2y+3z=48 \\ x+2y+3z=32 \end{cases}$

By using the Gauss-Jordan method (or, for that matter, any other technique with which you are familiar), you will find the solution to this system to be $(x, y, z) = (4, 5, 6)$. This means — with the limited amounts of ingredients available — the candy maker can produce

4 units of candy I, 5 units of candy II, and 6 units of candy III.

Problem Set 12.4

1. **Wording varies**. To paraphrase the definition stated on textbook page 835, two matrices are said to be ***equal*** if they have the same order and if all corresponding entries are equal.

3. **Wording varies**. The procedure for multiplying matrices is stated on the top half of textbook p. 836 and then demonstrated in Examples 1 and 2 (pp. 836-838).

5. **Wording varies**. When we simply speak of the "inverse" of a matrix, we are referring to the inverse for multiplication. Strictly speaking, however, as introduced on textbook pp. 840-841, there are two types of matrix inverses:

- The ***additive inverse*** of a matrix $[M]$ is denoted by $[-M]$ and is defined by $(-1)[M]$. The sum $[M]+[-M]=[-M]+[M]$ equals $[0]$, the zero (or additive identity) matrix.
- The ***multiplicative inverse*** of a matrix $[M]$ is denoted by $[M]^{-1}$ (read "M inverse"), if it exists. The product $[M][M]^{-1}=[M]^{-1}[M]$ equals $[I]$, the multiplicative identity matrix.

  Remember: The -1 in $[M]^{-1}$ is not an exponent; that is to say, $[M]^{-1}$ does not denote the reciprocal of $[M]$. Your text's author makes mention of this in the margins of p. 835 and p. 841.

7. **Wording varies**. The procedure for finding the inverse of a square matrix is stated on textbook p. 844 and subsequently applied to Examples 9-11 (pp. 844-845).

9. **Wording varies**. A ***communication matrix*** is a square matrix in which the entries symbolize the occurrence of some facet or event with a 1 and the nonoccurrence with a 0.

11. Matrix multiplication is explained in greater detail in text Example 2 (pp. 836-838).

a) $$[E][F]=\begin{bmatrix}1&0&2\\3&-1&2\\4&1&0\end{bmatrix}\begin{bmatrix}1&4&0\\3&-1&2\\-2&1&5\end{bmatrix}$$

$$=\begin{bmatrix}1(1)+0(3)+2(-2)&1(4)+0(-1)+2(1)&1(0)+0(2)+2(5)\\3(1)+(-1)(3)+2(-2)&3(4)+(-1)(-1)+2(1)&3(0)+(-1)(2)+2(5)\\4(1)+1(3)+0(-2)&4(4)+1(-1)+0(1)&4(0)+1(2)+0(5)\end{bmatrix}$$

$$=\boxed{\begin{bmatrix}\mathbf{-3}&\mathbf{6}&\mathbf{10}\\\mathbf{-4}&\mathbf{15}&\mathbf{8}\\\mathbf{7}&\mathbf{15}&\mathbf{2}\end{bmatrix}}$$

b) $$[E][G]=\begin{bmatrix}1&0&2\\3&-1&2\\4&1&0\end{bmatrix}\begin{bmatrix}8&1&6\\3&5&7\\4&9&2\end{bmatrix}$$

$$=\begin{bmatrix}1(8)+0(3)+2(4)&1(1)+0(5)+2(9)&1(6)+0(7)+2(2)\\3(8)+(-1)(3)+2(4)&3(1)+(-1)(5)+2(9)&3(6)+(-1)(7)+2(2)\\4(8)+1(3)+0(4)&4(1)+1(5)+0(9)&4(6)+1(7)+0(2)\end{bmatrix}$$

$$=\boxed{\begin{bmatrix}\mathbf{16}&\mathbf{19}&\mathbf{10}\\\mathbf{29}&\mathbf{16}&\mathbf{15}\\\mathbf{35}&\mathbf{9}&\mathbf{31}\end{bmatrix}}$$

13. a) $$[A][B]=\begin{bmatrix}1&2\\4&0\\-1&3\\2&1\end{bmatrix}\begin{bmatrix}4&2\\-1&3\end{bmatrix}=\begin{bmatrix}4-2&2+6\\16+0&8+0\\-4-3&-2+9\\8-1&4+3\end{bmatrix}=\boxed{\begin{bmatrix}\mathbf{2}&\mathbf{8}\\\mathbf{16}&\mathbf{8}\\\mathbf{-7}&\mathbf{7}\\\mathbf{7}&\mathbf{7}\end{bmatrix}}$$

b) $$[B][D]=\begin{bmatrix}4&2\\-1&3\end{bmatrix}\begin{bmatrix}4&1&3&6\\-1&0&-2&3\end{bmatrix}$$

$$=\begin{bmatrix}16-2&4+0&12-4&24+6\\-4-3&-1+0&-3-6&-6+9\end{bmatrix}=\boxed{\begin{bmatrix}\mathbf{14}&\mathbf{4}&\mathbf{8}&\mathbf{30}\\\mathbf{-7}&\mathbf{-1}&\mathbf{-9}&\mathbf{3}\end{bmatrix}}$$

15. a) $[B]^2 = [B][B] = \begin{bmatrix} 4 & 2 \\ -1 & 3 \end{bmatrix}\begin{bmatrix} 4 & 2 \\ -1 & 3 \end{bmatrix} = \begin{bmatrix} 16-2 & 8+6 \\ -4-3 & -2+9 \end{bmatrix} = \boxed{\begin{bmatrix} \mathbf{14} & \mathbf{14} \\ \mathbf{-7} & \mathbf{7} \end{bmatrix}}$

b) Notice that matrix $[C]$ is the 4 × 4 (multiplicative) identity matrix $[I_4]$. The identity for matrix multiplication is just like the number 1, which is the multiplicative identity for real numbers (a la textbook Section 4.6, p. 209). Just as $1^n = 1$, $[I_4]^n = [I_4]$ for any natural number n. So,

$$[C]^3 = \begin{bmatrix} 1 & 0 & 0 & 0 \\ 0 & 1 & 0 & 0 \\ 0 & 0 & 1 & 0 \\ 0 & 0 & 0 & 1 \end{bmatrix}^3 = [I_4]^3 = [I_4] = \boxed{\begin{bmatrix} \mathbf{1} & \mathbf{0} & \mathbf{0} & \mathbf{0} \\ \mathbf{0} & \mathbf{1} & \mathbf{0} & \mathbf{0} \\ \mathbf{0} & \mathbf{0} & \mathbf{1} & \mathbf{0} \\ \mathbf{0} & \mathbf{0} & \mathbf{0} & \mathbf{1} \end{bmatrix}}$$

17. a) In order to add two matrices, they must share the same order. Since matrix $[B]$ is order 2×2 matrix while $[C]$ is 4×4. Since $[B]$ and $[C]$ are **nonconformable**, the matrix expression $([B]+[C])[A]$ cannot be evaluated.

b) Matrix $[B]$ is a $2 \times \mathbf{2}$ matrix, while $[A]$ is $\mathbf{4} \times 2$. Since $[B]$ and $[A]$ are **nonconformable** $(2 \neq 4)$, the product $[B][A]$ cannot be computed. Therefore, the matrix expression $[B][A]+[C][A]$ cannot be evaluated.

19. $[A][X]=[B]$ represents the system of equations $\boxed{\begin{cases} \mathbf{x+2y+4z=13} \\ \mathbf{-3x+2y+z=11} \\ \mathbf{2x+z=0} \end{cases}}$.

21. Problems 21-22 are just like text Example 6 (pp. 841-842).

$$[A][B] = \begin{bmatrix} 2 & 7 \\ 1 & 4 \end{bmatrix}\begin{bmatrix} 4 & -7 \\ -1 & 2 \end{bmatrix} = \begin{bmatrix} 8-7 & -14+14 \\ 4-4 & -7+8 \end{bmatrix} = \begin{bmatrix} 1 & 0 \\ 0 & 1 \end{bmatrix} = [I]$$

and

$$[B][A] = \begin{bmatrix} 4 & -7 \\ -1 & 2 \end{bmatrix}\begin{bmatrix} 2 & 7 \\ 1 & 4 \end{bmatrix} = \begin{bmatrix} 8-7 & 28-28 \\ -2+2 & -7+8 \end{bmatrix} = \begin{bmatrix} 1 & 0 \\ 0 & 1 \end{bmatrix} = [I]$$

Since $[A][B]=[B][A]=[I]$, matrix $[A]$ and matrix $[B]$ are inverses of each other.

23. Let $[A]=\begin{bmatrix}4 & -7\\ -1 & 2\end{bmatrix}$.

Using the same approach as demonstrated in text Example 10 (pp. 844-845), we write the augmented matrix $[A \mid I]$ and proceed to transform, if possible, the left-hand side A into the identity matrix $[I_2]$:

$$[B]=\left[\begin{array}{rr|rr}4 & -7 & 1 & 0\\ -1 & 2 & 0 & 1\end{array}\right] \to \left[\begin{array}{rr|rr}-1 & 2 & 0 & 1\\ 4 & -7 & 1 & 0\end{array}\right]$$

RowSwap([B],1,2)

$$\to \left[\begin{array}{rr|rr}1 & -2 & 0 & -1\\ 4 & -7 & 1 & 0\end{array}\right] \to \left[\begin{array}{rr|rr}1 & -2 & 0 & -1\\ 0 & 1 & 1 & 4\end{array}\right] \to \left[\begin{array}{rr|rr}1 & 0 & 2 & 7\\ 0 & 1 & 1 & 4\end{array}\right]$$

∗Row(–1,[Ans],1)   ∗Row+(–4,[Ans],1,2)   ∗Row+(2,[Ans],2,1)

The right half of this final augmented matrix is the inverse of $[A]$. That is,

$$[A]^{-1}=\boxed{\begin{bmatrix}\mathbf{2} & \mathbf{7}\\ \mathbf{1} & \mathbf{4}\end{bmatrix}}.$$

If you previously worked Problem 21, this answer should not have surprised you.

25. Let $[A]=\begin{bmatrix}1 & 0 & 2\\ 2 & 1 & 0\\ 0 & -2 & 9\end{bmatrix}$.

Using the same approach as demonstrated in text Example 11 (p. 845), we write the augmented matrix $[A \mid I]$ and proceed to transform, if possible, the left-hand side A into the corresponding identity matrix $[I_3]$:

$$[B]=\left[\begin{array}{rrr|rrr}1 & 0 & 2 & 1 & 0 & 0\\ 2 & 1 & 0 & 0 & 1 & 0\\ 0 & -2 & 9 & 0 & 0 & 1\end{array}\right] \to \left[\begin{array}{rrr|rrr}1 & 0 & 2 & 1 & 0 & 0\\ 0 & 1 & -4 & -2 & 1 & 0\\ 0 & -2 & 9 & 0 & 0 & 1\end{array}\right]$$

∗Row+(–2,[B],1,2)

$$\to \left[\begin{array}{rrr|rrr}1 & 0 & 2 & 1 & 0 & 0\\ 0 & 1 & -4 & -2 & 1 & 0\\ 0 & 0 & 1 & -4 & 2 & 1\end{array}\right] \to \left[\begin{array}{rrr|rrr}1 & 0 & 0 & 9 & -4 & -2\\ 0 & 1 & 0 & -18 & 9 & 4\\ 0 & 0 & 1 & -4 & 2 & 1\end{array}\right]$$

∗Row+(2,[Ans],2,3)   ∗Row+(–2,[Ans],3,1)
∗Row+(4,[Ans],3,2)

The right half of this final augmented matrix is the inverse of $[A]$. In particular,

$$[A]^{-1} = \boxed{\begin{bmatrix} \mathbf{9} & \mathbf{-4} & \mathbf{-2} \\ \mathbf{-18} & \mathbf{9} & \mathbf{4} \\ \mathbf{-4} & \mathbf{2} & \mathbf{1} \end{bmatrix}}.$$

We encourage you to check this answer by confirming $[A][A]^{-1} = [A]^{-1}[A] = [I]$.

27. Let $[A] = \begin{bmatrix} 1 & 0 & 0 & 1 \\ 0 & 2 & 0 & 0 \\ 0 & 0 & 0 & 1 \\ 2 & 0 & 1 & 0 \end{bmatrix}$.

Once again, we write the augmented matrix $[A \,\vdots\, I]$ and proceed to transform, if possible, the left-hand side A into the corresponding identity matrix $[I_4]$:

$$[B] = \left[\begin{array}{cccc|cccc} 1 & 0 & 0 & 1 & 1 & 0 & 0 & 0 \\ 0 & 2 & 0 & 0 & 0 & 1 & 0 & 0 \\ 0 & 0 & 0 & 1 & 0 & 0 & 1 & 0 \\ 2 & 0 & 1 & 0 & 0 & 0 & 0 & 1 \end{array}\right] \to \left[\begin{array}{cccc|cccc} 1 & 0 & 0 & 1 & 1 & 0 & 0 & 0 \\ 0 & 2 & 0 & 0 & 0 & 1 & 0 & 0 \\ 0 & 0 & 0 & 1 & 0 & 0 & 1 & 0 \\ 0 & 0 & 1 & -2 & -2 & 0 & 0 & 1 \end{array}\right]$$

∗Row+(–2,[B],1,4)

$$\to \left[\begin{array}{cccc|cccc} 1 & 0 & 0 & 1 & 1 & 0 & 0 & 0 \\ 0 & 1 & 0 & 0 & 0 & \frac{1}{2} & 0 & 0 \\ 0 & 0 & 0 & 1 & 0 & 0 & 1 & 0 \\ 0 & 0 & 1 & -2 & -2 & 0 & 0 & 1 \end{array}\right] \to \left[\begin{array}{cccc|cccc} 1 & 0 & 0 & 1 & 1 & 0 & 0 & 0 \\ 0 & 1 & 0 & 0 & 0 & \frac{1}{2} & 0 & 0 \\ 0 & 0 & 1 & -2 & -2 & 0 & 0 & 1 \\ 0 & 0 & 0 & 1 & 0 & 0 & 1 & 0 \end{array}\right]$$

∗Row ($\frac{1}{2}$,[Ans],2) RowSwap([Ans],3,4)

$$\to \left[\begin{array}{cccc|cccc} 1 & 0 & 0 & 0 & 1 & 0 & -1 & 0 \\ 0 & 1 & 0 & 0 & 0 & \frac{1}{2} & 0 & 0 \\ 0 & 0 & 1 & 0 & -2 & 0 & 2 & 1 \\ 0 & 0 & 0 & 1 & 0 & 0 & 1 & 0 \end{array}\right]$$

∗Row+(–1,[Ans],4,1)
∗Row+(2,[Ans],4,3)

The right half of this final augmented matrix is the inverse of $[A]$. That is,

$$[A]^{-1}=\boxed{\begin{bmatrix} 1 & 0 & -1 & 0 \\ 0 & \frac{1}{2} & 0 & 0 \\ -2 & 0 & 2 & 1 \\ 0 & 0 & 1 & 0 \end{bmatrix}}.$$

We encourage you to check this answer by confirming $[A][A]^{-1} = [A]^{-1}[A]=[I]$.

29. The concept of using the inverse matrix to solve linear systems is explained on textbook pp. 846-847.

First, we write the given system $\begin{cases} 4x-7y=-2 \\ -x+2y=1 \end{cases}$ in matrix form:

$$[A]=\begin{bmatrix} 4 & -7 \\ -1 & 2 \end{bmatrix},\ [X]=\begin{bmatrix} x \\ y \end{bmatrix},\text{ and } [B]=\begin{bmatrix} -2 \\ 1 \end{bmatrix}.$$

The solution is $[X]=[A]^{-1}[B]$.

From Problem 23 (solution provided above), we know $[A]^{-1}=\begin{bmatrix} 2 & 7 \\ 1 & 4 \end{bmatrix}$.

So, $[X]=[A]^{-1}[B]=\begin{bmatrix} 2 & 7 \\ 1 & 4 \end{bmatrix}\begin{bmatrix} -2 \\ 1 \end{bmatrix}=\begin{bmatrix} -4+7 \\ -2+4 \end{bmatrix}=\begin{bmatrix} 3 \\ 2 \end{bmatrix}$

Hence, the solution to the given system of equations is the ordered pair **(3, 2)**. You should always check the answer, using both equations to do so.

31. Like the rest of Problems 29-34 (see solution to Problem 29 above), we have

$$[A]=\begin{bmatrix} 4 & -7 \\ -1 & 2 \end{bmatrix}\text{ and } [X]=\begin{bmatrix} x \\ y \end{bmatrix},\text{ with } [A]^{-1}=\begin{bmatrix} 2 & 7 \\ 1 & 4 \end{bmatrix}.$$

The only difference among these six problems is the constants. In this problem, we have $[B]=\begin{bmatrix} 48 \\ -13 \end{bmatrix}$, which yields the following:

$$[X]=[A]^{-1}[B]=\begin{bmatrix} 2 & 7 \\ 1 & 4 \end{bmatrix}\begin{bmatrix} 48 \\ -13 \end{bmatrix}=\begin{bmatrix} 96-91 \\ 48-52 \end{bmatrix}=\begin{bmatrix} 5 \\ -4 \end{bmatrix}$$

Hence, the solution to the given system of equations is the ordered pair **(5, −4)**. We again encourage you to check this answer, using both equations to do so.

33. Once again, like the rest of Problems 29-34 (see solution to Problem 29 above), we have

$$[A]=\begin{bmatrix}4 & -7\\ -1 & 2\end{bmatrix} \text{ and } [X]=\begin{bmatrix}x\\ y\end{bmatrix}, \text{ with } [A]^{-1}=\begin{bmatrix}2 & 7\\ 1 & 4\end{bmatrix}.$$

The only difference among these six problems is the constants. In this problem, we have $[B]=\begin{bmatrix}5\\ 4\end{bmatrix}$, which yields the following:

$$[X]=[A]^{-1}[B]=\begin{bmatrix}2 & 7\\ 1 & 4\end{bmatrix}\begin{bmatrix}5\\ 4\end{bmatrix}=\begin{bmatrix}10+28\\ 5+16\end{bmatrix}=\begin{bmatrix}38\\ 21\end{bmatrix}$$

Hence, the solution to the given system of equations is the ordered pair **(38, 21)**. It is easy (and important) to check this solution; we encourage you to do so.

35. We remind you that the concept of using the inverse matrix to solve linear systems is explained on textbook pp. 846-847.

First, we write the given system $\begin{cases}8x+6y=12\\ -2x+4y=-14\end{cases}$ in matrix form:

$$[A]=\begin{bmatrix}8 & 6\\ -2 & 4\end{bmatrix}, [X]=\begin{bmatrix}x\\ y\end{bmatrix}, \text{ and } [B]=\begin{bmatrix}12\\ -14\end{bmatrix}.$$

The solution is $[X]=[A]^{-1}[B]$.

From Problem 24, we find $[A]^{-1}=\begin{bmatrix}\frac{1}{11} & -\frac{3}{22}\\ \frac{1}{22} & \frac{2}{11}\end{bmatrix}$ or $\frac{1}{22}\begin{bmatrix}2 & -3\\ 1 & 4\end{bmatrix}$. So,

$$[X]=[A]^{-1}[B]=\tfrac{1}{22}\begin{bmatrix}2 & -3\\ 1 & 4\end{bmatrix}\begin{bmatrix}12\\ -14\end{bmatrix}=\tfrac{1}{22}\begin{bmatrix}24+42\\ 12-56\end{bmatrix}=\tfrac{1}{22}\begin{bmatrix}66\\ -44\end{bmatrix}=\begin{bmatrix}\frac{66}{22}\\ -\frac{44}{22}\end{bmatrix}=\begin{bmatrix}3\\ -2\end{bmatrix}.$$

Thus, the solution to the given system of equations is the ordered pair **(3, −2)**. You should then check the answer, using both equations to do so.

37. Like the rest of Problems 35-40 (see solution to Problem 35 above), we have

$$[A]=\begin{bmatrix}8 & 6\\ -2 & 4\end{bmatrix} \text{ and } [X]=\begin{bmatrix}x\\ y\end{bmatrix}, \text{ with } [A]^{-1}=\tfrac{1}{22}\begin{bmatrix}2 & -3\\ 1 & 4\end{bmatrix}.$$

The only difference among these six problems is the constants. In this problem, we have $[B]=\begin{bmatrix}-6\\ -26\end{bmatrix}$, which leads to the following:

$$[X]=[A]^{-1}[B]=\tfrac{1}{22}\begin{bmatrix}2 & -3\\ 1 & 4\end{bmatrix}\begin{bmatrix}-6\\ -26\end{bmatrix}=\tfrac{1}{22}\begin{bmatrix}-12+78\\ -6-104\end{bmatrix}=\tfrac{1}{22}\begin{bmatrix}66\\ -110\end{bmatrix}=\begin{bmatrix}\frac{66}{22}\\ -\frac{110}{22}\end{bmatrix}=\begin{bmatrix}3\\ -5\end{bmatrix}$$

Hence, the solution to the given system of equations is the ordered pair **(3, –5)**. Your final step should be to check this answer in both of the original equations.

39. Once again, like the rest of Problems 35-40 (see solution to Problem 35 above), we have

$$[A]=\begin{bmatrix}8 & 6\\ -2 & 4\end{bmatrix} \text{ and } [X]=\begin{bmatrix}x\\ y\end{bmatrix}, \text{ with } [A]^{-1}=\tfrac{1}{22}\begin{bmatrix}2 & -3\\ 1 & 4\end{bmatrix}.$$

The only difference among these six problems is the constants. In this problem, we have $[B]=\begin{bmatrix}-26\\ 12\end{bmatrix}$, which leads to the following:

$$[X]=[A]^{-1}[B]=\tfrac{1}{22}\begin{bmatrix}2 & -3\\ 1 & 4\end{bmatrix}\begin{bmatrix}-26\\ 12\end{bmatrix}=\tfrac{1}{22}\begin{bmatrix}-52-36\\ -26+48\end{bmatrix}=\tfrac{1}{22}\begin{bmatrix}-88\\ 22\end{bmatrix}=\begin{bmatrix}-\frac{88}{22}\\ \frac{22}{22}\end{bmatrix}=\begin{bmatrix}-4\\ 1\end{bmatrix}$$

So, the solution to the given system of equations is the ordered pair **(–4, 1)**. Your final step should be to check this answer in both of the original equations.

41. We remind you that the concept of using the inverse matrix to solve linear systems is explained on textbook pp. 846-847.

First, we write the given system $\begin{cases}2x+3y=9\\ x-6y=-3\end{cases}$ in matrix form:

$$[A]=\begin{bmatrix}2 & 3\\ 1 & -6\end{bmatrix},\ [X]=\begin{bmatrix}x\\ y\end{bmatrix}, \text{ and } [B]=\begin{bmatrix}9\\ -3\end{bmatrix}.$$

The solution is $[X]=[A]^{-1}[B]$, so we need to determine $[A]^{-1}$. To do so, we apply the same approach as used previously in text Example 10 (pp. 844-845) and Problems 23-28. That is, we write the augmented matrix $[A \,\vdots\, I]$ and proceed to transform, if possible, the left-hand side A into the identity matrix $[I_2]$:

$$[C]=\left[\begin{array}{cc|cc}2 & 3 & 1 & 0\\ 1 & -6 & 0 & 1\end{array}\right] \rightarrow \left[\begin{array}{cc|cc}1 & -6 & 0 & 1\\ 2 & 3 & 1 & 0\end{array}\right]$$

RowSwap([C],1,2)

$$\rightarrow \left[\begin{array}{cc|cc}1 & -6 & 0 & 1\\ 0 & 15 & 1 & -2\end{array}\right] \rightarrow \left[\begin{array}{cc|cc}1 & -6 & 0 & 1\\ 0 & 1 & \frac{1}{15} & -\frac{2}{15}\end{array}\right]$$

∗Row+(–2,[Ans],1,2) ∗Row($\frac{1}{15}$,[Ans], 2)

$$\rightarrow \left[\begin{array}{cc|cc} 1 & -6 & \frac{6}{15} & \frac{3}{15} \\ 0 & 1 & \frac{1}{15} & -\frac{2}{15} \end{array}\right]$$

*Row+(6,[Ans],2,1)

The right half of this final augmented matrix is the inverse of $[A]$. That is,

$$[A]^{-1} = \begin{bmatrix} \frac{6}{15} & \frac{3}{15} \\ \frac{1}{15} & -\frac{2}{15} \end{bmatrix} = \frac{1}{15}\begin{bmatrix} 6 & 3 \\ 1 & -2 \end{bmatrix}.$$

So, the solution to the system of equations is

$$[X] = [A]^{-1}[B] = \frac{1}{15}\begin{bmatrix} 6 & 3 \\ 1 & -2 \end{bmatrix}\begin{bmatrix} 9 \\ -3 \end{bmatrix} = \frac{1}{15}\begin{bmatrix} 54-9 \\ 9+6 \end{bmatrix} = \frac{1}{15}\begin{bmatrix} 45 \\ 15 \end{bmatrix} = \begin{bmatrix} 3 \\ 1 \end{bmatrix}.$$

Hence, the solution to the given system of equations is the ordered pair **(3, 1)**. As always, you should check the answer, using both equations to do so.

43. Like the rest of Problems 41-46 (see solution to Problem 41 above), we have

$$[A] = \begin{bmatrix} 2 & 3 \\ 1 & -6 \end{bmatrix} \text{ and } [X] = \begin{bmatrix} x \\ y \end{bmatrix}.$$

In solving Problem 41, we established that $[A]^{-1} = \frac{1}{15}\begin{bmatrix} 6 & 3 \\ 1 & -2 \end{bmatrix}$.

The only difference among Problems 41-46 is the constants. In this problem, we have $[B] = \begin{bmatrix} 2 \\ -14 \end{bmatrix}$, which leads to the following:

$$[X] = [A]^{-1}[B] = \frac{1}{15}\begin{bmatrix} 6 & 3 \\ 1 & -2 \end{bmatrix}\begin{bmatrix} 2 \\ -14 \end{bmatrix} = \frac{1}{15}\begin{bmatrix} 12-42 \\ 2+28 \end{bmatrix} = \frac{1}{15}\begin{bmatrix} -30 \\ 30 \end{bmatrix} = \begin{bmatrix} -2 \\ 2 \end{bmatrix}$$

So, the solution to the given system of equations is the ordered pair **(–2, 2)**. The final step, of course, should be to check this answer, which is a task we leave to you.

45. Once again, like the rest of Problems 41-46 (see solution to Problem 41 above), we have

$$[A] = \begin{bmatrix} 2 & 3 \\ 1 & -6 \end{bmatrix} \text{ and } [X] = \begin{bmatrix} x \\ y \end{bmatrix}.$$

In solving Problem 41, we established that $[A]^{-1} = \frac{1}{15}\begin{bmatrix} 6 & 3 \\ 1 & -2 \end{bmatrix}$.

The only difference among Problems 41-46 is the constants. In this problem, we have $[B]=\begin{bmatrix}-22\\49\end{bmatrix}$, which leads to the following:

$$[X]=[A]^{-1}[B]=\tfrac{1}{15}\begin{bmatrix}6 & 3\\1 & -2\end{bmatrix}\begin{bmatrix}-22\\49\end{bmatrix}=\tfrac{1}{15}\begin{bmatrix}-132+147\\-22-98\end{bmatrix}=\tfrac{1}{15}\begin{bmatrix}15\\-120\end{bmatrix}=\begin{bmatrix}1\\-8\end{bmatrix}$$

Hence, the solution to the given system of equations is the ordered pair **(1, –8)**. The final step, of course, should be to check this answer, which is a task we leave to you.

47. We remind you that the concept of using the inverse matrix to solve linear systems is explained on textbook pp. 846-847.

First, we write the given system $\begin{cases}x+2z=7\\2x+y=16\\-2y+9z=-3\end{cases}$ in matrix form:

$$[A]=\begin{bmatrix}1 & 0 & 2\\2 & 1 & 0\\0 & -2 & 9\end{bmatrix},\ [X]=\begin{bmatrix}x\\y\\z\end{bmatrix},\text{ and } [B]=\begin{bmatrix}7\\16\\-3\end{bmatrix}.$$

The solution is $[X]=[A]^{-1}[B]$. From Problem 25 (see this manual's solution), we know

$$[A]^{-1}=\begin{bmatrix}9 & -4 & -2\\-18 & 9 & 4\\-4 & 2 & 1\end{bmatrix}.$$

So,

$$[X]=[A]^{-1}[B]=\begin{bmatrix}9 & -4 & -2\\-18 & 9 & 4\\-4 & 2 & 1\end{bmatrix}\begin{bmatrix}7\\16\\-3\end{bmatrix}=\begin{bmatrix}63-64+6\\-126+144-12\\-28+32-3\end{bmatrix}=\begin{bmatrix}5\\6\\1\end{bmatrix}.$$

Thus, the solution to the given system of equations is $(x,y,z)=$ **(5, 6, 1)**. You should then check the answer, using all three given equations to do so.

49. Like the rest of Problems 47-52 (see solution to Problem 47 above), we have

$$[A]=\begin{bmatrix}1 & 0 & 2\\2 & 1 & 0\\0 & -2 & 9\end{bmatrix}\text{ and } [X]=\begin{bmatrix}x\\y\\z\end{bmatrix},\text{ with } [A]^{-1}=\begin{bmatrix}9 & -4 & -2\\-18 & 9 & 4\\-4 & 2 & 1\end{bmatrix}.$$

The only difference among Problems 47-52 is the constants. In this problem, we have $[B]=\begin{bmatrix}4\\0\\31\end{bmatrix}$, which leads to the following:

$$[X]=[A]^{-1}[B]=\begin{bmatrix}9 & -4 & -2\\-18 & 9 & 4\\-4 & 2 & 1\end{bmatrix}\begin{bmatrix}4\\0\\31\end{bmatrix}=\begin{bmatrix}36+0-62\\-72+0+124\\-16+0+31\end{bmatrix}=\begin{bmatrix}-26\\52\\15\end{bmatrix}.$$

Thus, the solution to the given system of equations is $(x,y,z)=$ **(–26, 52, 15)**. The final step, of course, should be to check this answer, which is a task we leave to you.

51. Once again, like the rest of Problems 47-52 (see solution to Problem 47 above), we have

$$[A]=\begin{bmatrix}1 & 0 & 2\\2 & 1 & 0\\0 & -2 & 9\end{bmatrix} \text{ and } [X]=\begin{bmatrix}x\\y\\z\end{bmatrix}, \text{ with } [A]^{-1}=\begin{bmatrix}9 & -4 & -2\\-18 & 9 & 4\\-4 & 2 & 1\end{bmatrix}.$$

The only difference among Problems 47-52 is the constants. In this problem, we have $[B]=\begin{bmatrix}12\\0\\10\end{bmatrix}$, which leads to the following:

$$[X]=[A]^{-1}[B]=\begin{bmatrix}9 & -4 & -2\\-18 & 9 & 4\\-4 & 2 & 1\end{bmatrix}\begin{bmatrix}12\\0\\10\end{bmatrix}=\begin{bmatrix}108+0-20\\-216+0+40\\-48+0+10\end{bmatrix}=\begin{bmatrix}88\\-176\\-38\end{bmatrix}.$$

So, the solution to the given system of equations is $(x,y,z)=$ **(88, –176, –38)**. The final step should be to check this answer, which is a task we leave to you.

53. We remind you that the concept of using the inverse matrix to solve linear systems is explained on textbook pp. 846-847.

First, we write the given system $\begin{cases}6x+y+20z=27\\ x-y=0\\ y+3z=4\end{cases}$ in matrix form:

$$[A]=\begin{bmatrix}6 & 1 & 20\\ 1 & -1 & 0\\ 0 & 1 & 3\end{bmatrix},\ [X]=\begin{bmatrix}x\\ y\\ z\end{bmatrix},\text{ and }[B]=\begin{bmatrix}27\\ 0\\ 4\end{bmatrix}.$$

The solution is $[X]=[A]^{-1}[B]$.

In Problem 26, we find $[A]^{-1}=\begin{bmatrix}3 & -17 & -20\\ 3 & -18 & -20\\ -1 & 6 & 7\end{bmatrix}$.

So, $[X]=[A]^{-1}[B]=\begin{bmatrix}3 & -17 & -20\\ 3 & -18 & -20\\ -1 & 6 & 7\end{bmatrix}\begin{bmatrix}27\\ 0\\ 4\end{bmatrix}=\begin{bmatrix}81+0-80\\ 81+0-80\\ -27+0+28\end{bmatrix}=\begin{bmatrix}1\\ 1\\ 1\end{bmatrix}$.

Thus, the solution to the given system of equations is $(x,y,z)=$ **(1, 1, 1)**. The final step should be to check this answer, which is a task we leave to you.

55. Like the rest of Problems 53-56 (see solution to Problem 53 above), we have

$$[A]=\begin{bmatrix}6 & 1 & 20\\ 1 & -1 & 0\\ 0 & 1 & 3\end{bmatrix}\text{ and }[X]=\begin{bmatrix}x\\ y\\ z\end{bmatrix},\text{ with }[A]^{-1}=\begin{bmatrix}3 & -17 & -20\\ 3 & -18 & -20\\ -1 & 6 & 7\end{bmatrix}.$$

The only difference among Problems 53-56 is the constants. In this problem, we have

$[B]=\begin{bmatrix}-12\\ 6\\ -7\end{bmatrix}$, which leads to the following:

$$[X]=[A]^{-1}[B]=\begin{bmatrix}3 & -17 & -20\\ 3 & -18 & -20\\ -1 & 6 & 7\end{bmatrix}\begin{bmatrix}-12\\ 6\\ -7\end{bmatrix}=\begin{bmatrix}-36-102+140\\ -36-108+140\\ 12+36-49\end{bmatrix}=\begin{bmatrix}2\\ -4\\ -1\end{bmatrix}.$$

So, the solution to the given system of equations is $(x,y,z)=$ **(2, −4, −1)**. You should then check this answer, using all three given equations to do so.

57. From text Example 5 (pp. 839-840), recall that the matrix $[A]^3$ reveals how many ways the countries can communicate if they use two intermediaries.

$$[A]^3 = \begin{bmatrix} 0 & 1 & 0 & 1 \\ 1 & 0 & 1 & 1 \\ 0 & 1 & 0 & 0 \\ 1 & 1 & 0 & 0 \end{bmatrix}^3 = \begin{bmatrix} 0 & 1 & 0 & 1 \\ 1 & 0 & 1 & 1 \\ 0 & 1 & 0 & 0 \\ 1 & 1 & 0 & 0 \end{bmatrix}^2 \cdot \begin{bmatrix} 0 & 1 & 0 & 1 \\ 1 & 0 & 1 & 1 \\ 0 & 1 & 0 & 0 \\ 1 & 1 & 0 & 0 \end{bmatrix}$$

In Example 5, we learned

$$\begin{bmatrix} 0 & 1 & 0 & 1 \\ 1 & 0 & 1 & 1 \\ 0 & 1 & 0 & 0 \\ 1 & 1 & 0 & 0 \end{bmatrix}^2 = \begin{bmatrix} 2 & 1 & 1 & 1 \\ 1 & 3 & 0 & 1 \\ 1 & 0 & 1 & 1 \\ 1 & 1 & 1 & 2 \end{bmatrix}.$$

So, it follows that

$$[A]^3 = \begin{bmatrix} 0 & 1 & 0 & 1 \\ 1 & 0 & 1 & 1 \\ 0 & 1 & 0 & 0 \\ 1 & 1 & 0 & 0 \end{bmatrix}^2 \begin{bmatrix} 0 & 1 & 0 & 1 \\ 1 & 0 & 1 & 1 \\ 0 & 1 & 0 & 0 \\ 1 & 1 & 0 & 0 \end{bmatrix}$$

$$= \begin{bmatrix} 2 & 1 & 1 & 1 \\ 1 & 3 & 0 & 1 \\ 1 & 0 & 1 & 1 \\ 1 & 1 & 1 & 2 \end{bmatrix} \begin{bmatrix} 0 & 1 & 0 & 1 \\ 1 & 0 & 1 & 1 \\ 0 & 1 & 0 & 0 \\ 1 & 1 & 0 & 0 \end{bmatrix}$$

$$= \begin{bmatrix} 0+1+0+1 & 2+0+1+1 & 0+1+0+0 & 2+1+0+0 \\ 0+3+0+1 & 1+0+0+1 & 0+3+0+0 & 1+3+0+0 \\ 0+0+0+1 & 1+0+1+1 & 0+0+0+0 & 1+0+0+0 \\ 0+1+0+2 & 1+0+1+2 & 0+1+0+0 & 1+1+0+0 \end{bmatrix}$$

$$= \boxed{\begin{bmatrix} \mathbf{2} & \mathbf{4} & \mathbf{1} & \mathbf{3} \\ \mathbf{4} & \mathbf{2} & \mathbf{3} & \mathbf{4} \\ \mathbf{1} & \mathbf{3} & \mathbf{0} & \mathbf{1} \\ \mathbf{3} & \mathbf{4} & \mathbf{1} & \mathbf{2} \end{bmatrix}}$$

This matrix tells us a great deal about diplomatic relations (in particular, the communication channels) among the U.S., Russia, Cuba, and Mexico. For example, look at the entry either in the first row, third column or the third row, first column; it reveals that the U.S. can talk with Cuba through two intermediaries in 1 way, namely U.S. to Mexico to Russia to Cuba.

59. a) $[C][W] = \begin{bmatrix} 1 & 4 & 6 \end{bmatrix} \begin{bmatrix} 2 & 1 & 3 \\ 4 & 3 & 6 \\ 1 & 2 & 4 \end{bmatrix}$

$$= \begin{bmatrix} 2+16+6 & 1+12+12 & 3+24+24 \end{bmatrix}$$
$$= \begin{bmatrix} 24 & 25 & 51 \end{bmatrix}$$

So, the respective costs of producing each of the three types of wine are: **Riesling, 24; Charbono, 25; and Rosé, 51**.

b) $[W][D] = \begin{bmatrix} 2 & 1 & 3 \\ 4 & 3 & 6 \\ 1 & 2 & 4 \end{bmatrix} \begin{bmatrix} 40 \\ 60 \\ 30 \end{bmatrix} = \begin{bmatrix} 80+60+90 \\ 160+180+180 \\ 40+120+120 \end{bmatrix} = \begin{bmatrix} 230 \\ 520 \\ 280 \end{bmatrix}$

Hence, the respective costs of producing these wine types according to the production method used are:

Outside bottling, 230;
produced/bottled at the winery, 520;
and estate bottled, 280.

c) From part (a), we know $[C][W] = \begin{bmatrix} 24 & 25 & 51 \end{bmatrix}$. So,

$$([C][W])[D] = \begin{bmatrix} 24 & 25 & 51 \end{bmatrix} \begin{bmatrix} 40 \\ 60 \\ 30 \end{bmatrix} = \begin{bmatrix} 960 + 1{,}500 + 1{,}530 \end{bmatrix} = \boxed{[3{,}990]}.$$

Similarly, from part (b), we know $[W][D] = \begin{bmatrix} 230 \\ 520 \\ 280 \end{bmatrix}$, so it follows that

$$[C]([W][D]) = \begin{bmatrix} 1 & 4 & 6 \end{bmatrix} \begin{bmatrix} 230 \\ 520 \\ 280 \end{bmatrix} = \begin{bmatrix} 230 + 2{,}080 + 1{,}680 \end{bmatrix} = \boxed{[3{,}990]}.$$

This answer represents the **total cost of production** for all three wines (produced via all three methods).

Problem Set 12.5

1. **Wording varies**. A ***system of linear inequalities*** is two or more first-degree inequalities considered at the same time.

3. a)

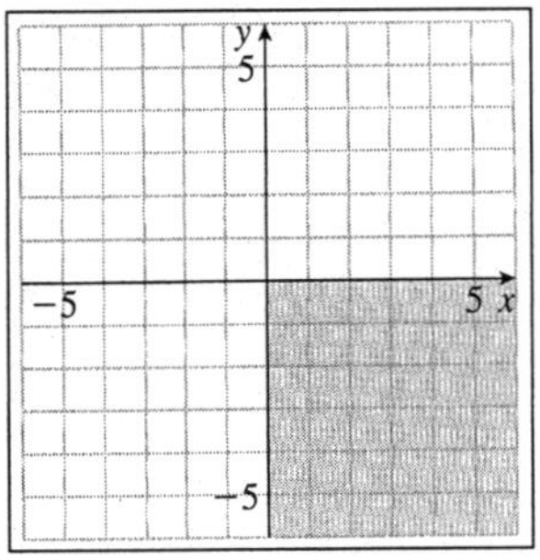

b)

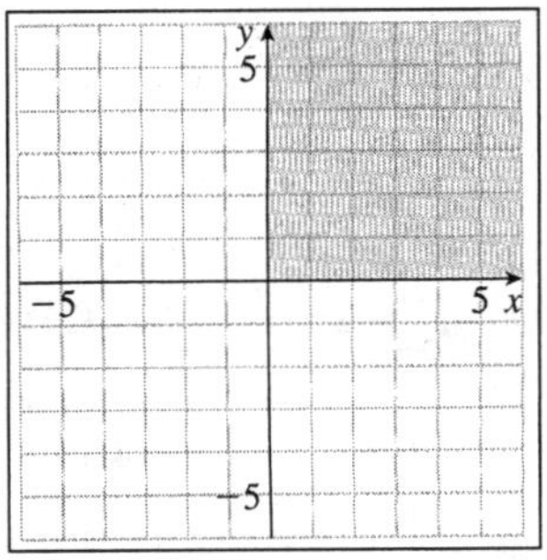

5.

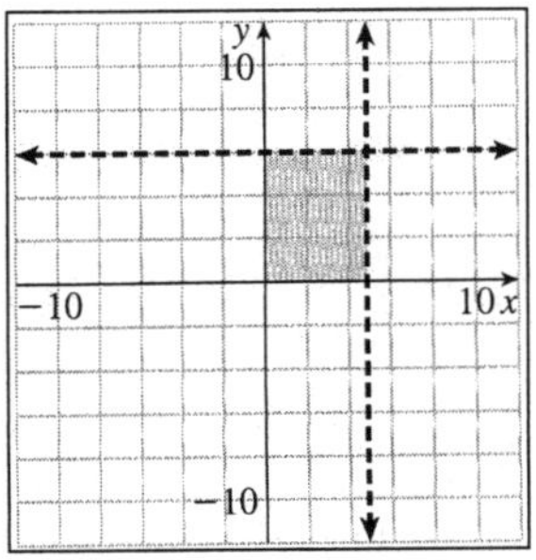

7.

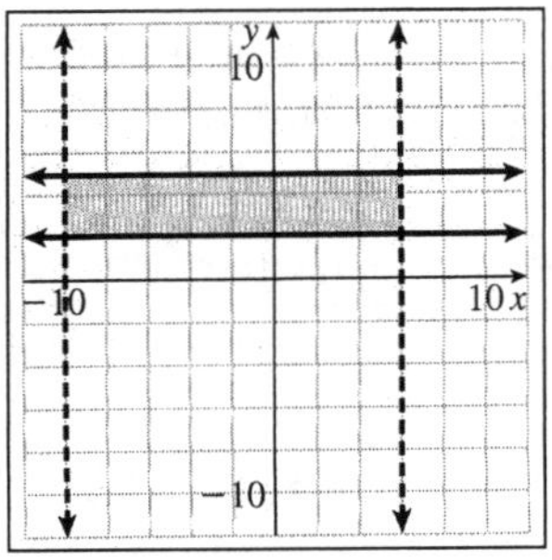

9.

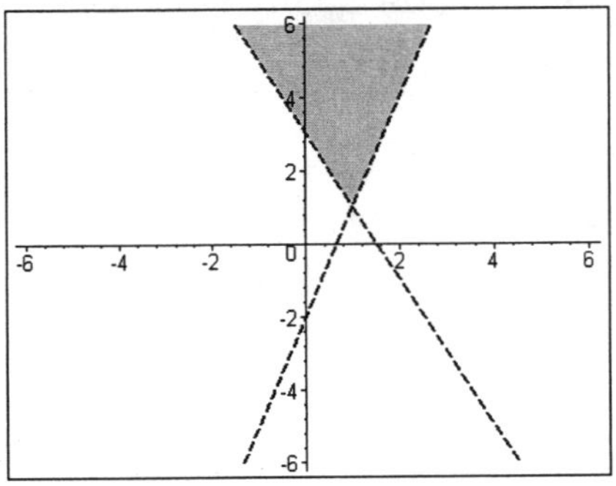

11.

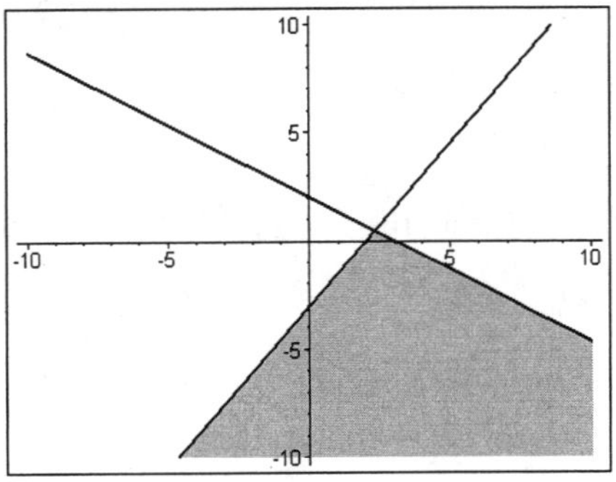

13.

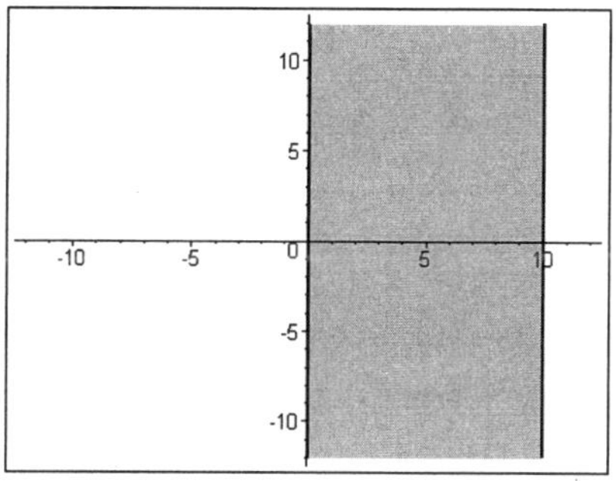

15.

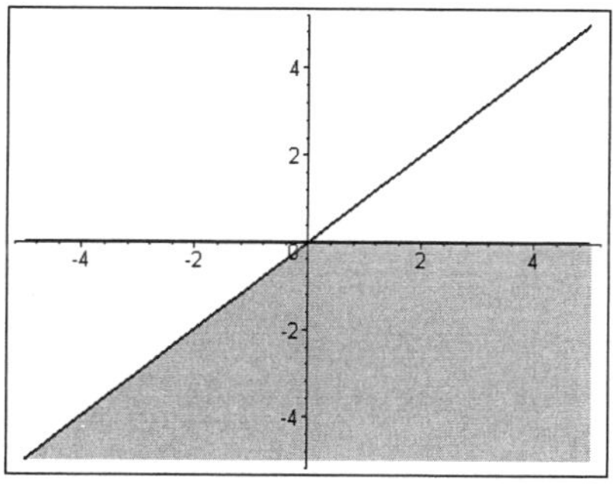

17.

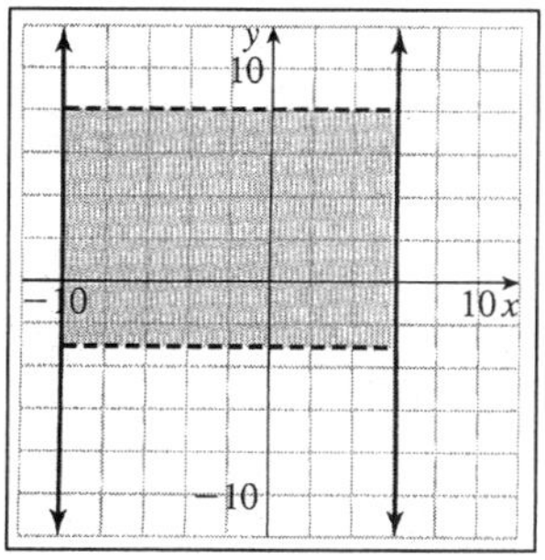

19.

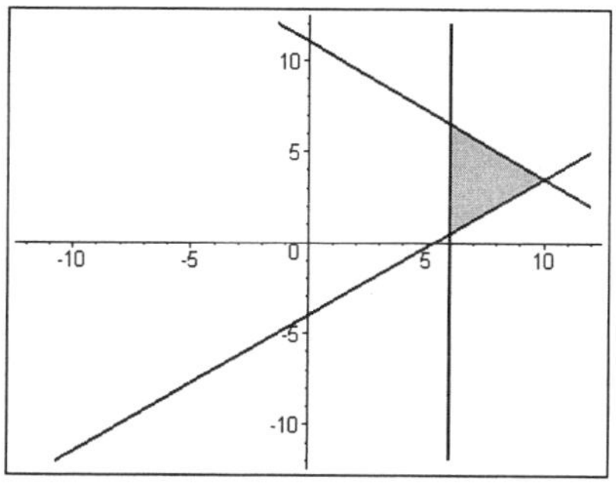

21.

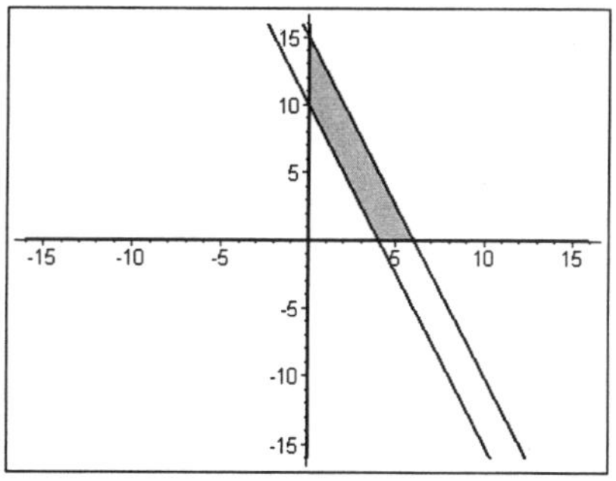

23.

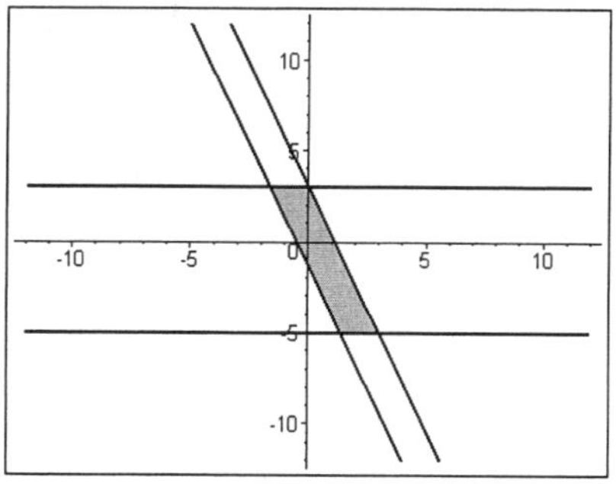

25.

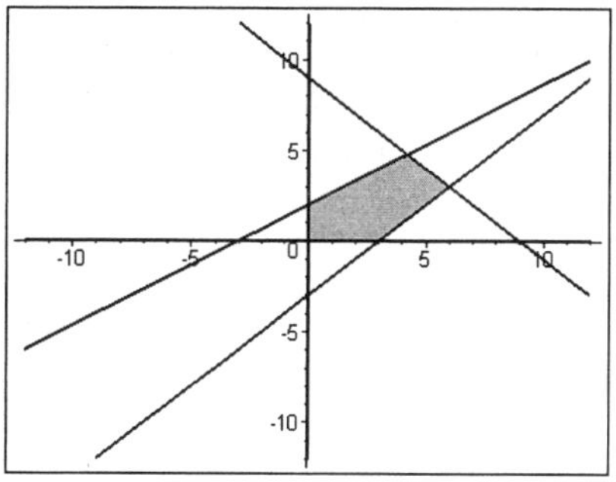

27.

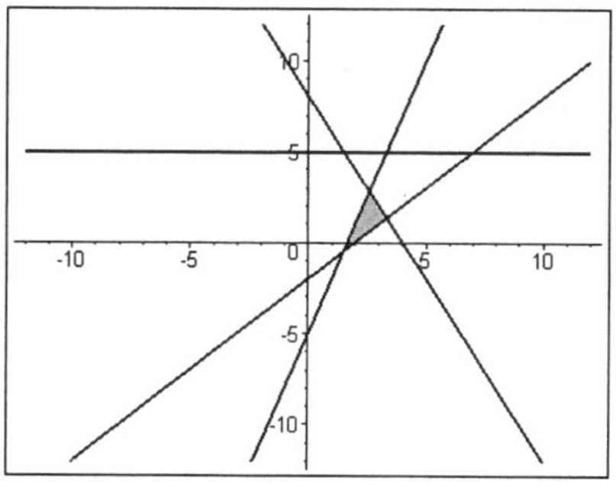

29.

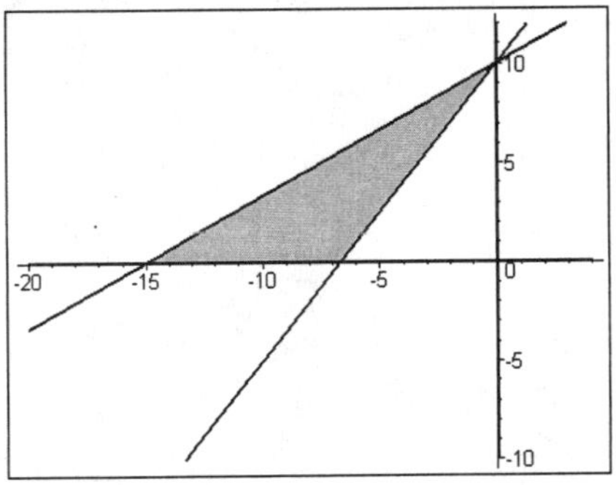

Problem Set 12.6

1. **Wording varies**. A ***linear programming problem*** is a problem which seeks the maximum or minimum value of an expression, called the *objective function*. This solution is subject to a set of *constraints*, which take the form of linear inequalities.

3. There are two ways to decide whether a given point is a feasible solution: (i) graph the set of constraints (see below) and visually assess whether the point is located within the graph's feasible region; or (ii) check if the point satisfies all of the constraints. In this solution, we have chosen to employ the latter approach:

a) **No**, the point $(1,\ 3)$ is not a feasible solution because it fails to satisfy the constraint $2x+4y\le 8$.

b) **Yes**, the point $(2,\ 1)$ is a feasible solution because it satisfies all four of the constraints.

c) **No**, the point $(1,\ 2)$ is not a feasible solution because it fails to satisfy the constraint $2x+4y\le 8$.

d) **No**, the point $(-1,\ 4)$ is not a feasible solution because it fails to satisfy the constraints $x\ge 0$ and $2x+4y\le 8$.

e) **No**, the point $(2,\ 2)$ is not a feasible solution because it fails to satisfy the constraint $2x+4y\le 8$.

f) **No**, the point $(0,\ 4)$ is not a feasible solution because it fails to satisfy the constraint $2x+4y\le 8$.

On the next page, for informational purposes only, is the graph of the given constraints:

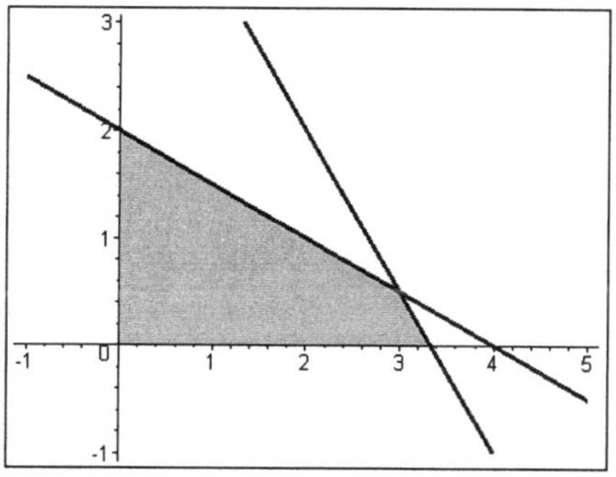

5. First, graph the set of constraints:

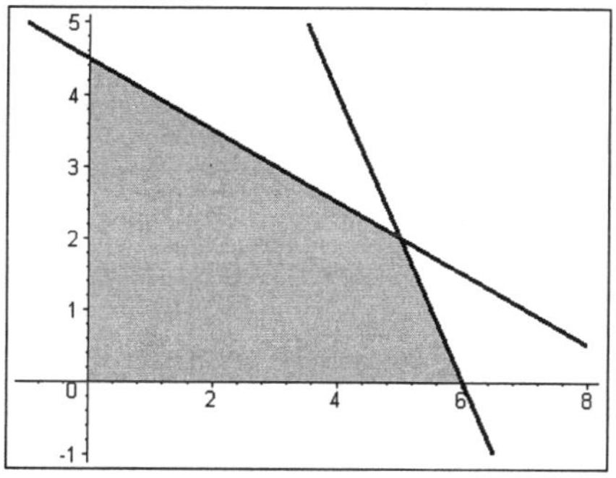

Then, to find the coordinates of this feasible region's four corner points, solve the following systems of equations:

Corner Point	***System to Solve***	***Suggested Method***	***Solution***
Lower left	$\begin{cases} x = 0 \\ y = 0 \end{cases}$	n/a	$(0,0)$
Upper left	$\begin{cases} x = 0 \\ x + 2y = 9 \end{cases}$	Substitution	$(0, \frac{9}{2})$
Lower Right	$\begin{cases} y = 0 \\ 2x + y = 12 \end{cases}$	Substitution	$(6,0)$
Upper Right	$\begin{cases} x + 2y = 9 \\ 2x + y = 12 \end{cases}$	Substitution or Addition	$(5,2)$

To review the suggested solving methods, please revisit textbook Section 12.1 (including this manual's solutions thereof). Additionally, remember that systems of linear equations can be solved via the Gauss-Jordan method (Section 12.3) or the inverse matrix method (Section 12.4).

In summary, the four corner points are $\boxed{\mathbf{(0,0), (0,\frac{9}{2}), (6,0), (5,2)}}$.

7. The graph of the set of constraints is as follows:

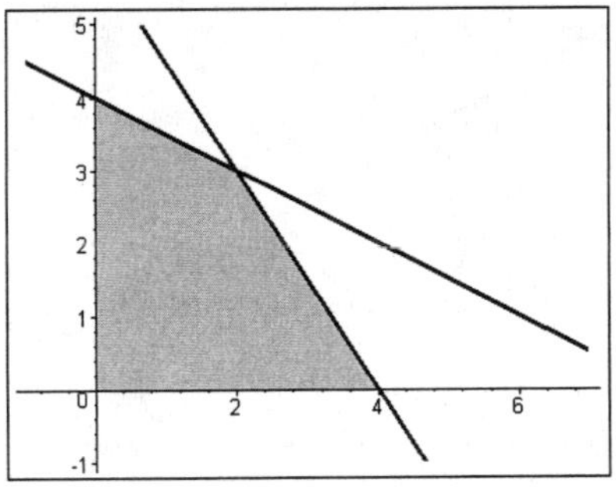

To find the coordinates of this feasible region's four corner points, solve the following systems of equations:

Corner Point	*System to Solve*	*Suggested Method*	*Solution*
Lower left	$\begin{cases} x=0 \\ y=0 \end{cases}$	n/a	$(0,0)$
Upper left	$\begin{cases} x=0 \\ x+2y=8 \end{cases}$	Substitution	$(0,4)$
Lower Right	$\begin{cases} y=0 \\ 3x+2y=12 \end{cases}$	Substitution	$(4,0)$
Upper Right	$\begin{cases} x+2y=8 \\ 3x+2y=12 \end{cases}$	Substitution or Addition	$(2,3)$

To review the suggested solving methods, please revisit textbook Section 12.1 (including this manual's solutions thereof). Additionally, remember that systems of linear equations can be solved via the Gauss-Jordan method (Section 12.3) or the inverse matrix method (Section 12.4).

In summary, the four corner points are **(0,0), (0,4), (4,0), (2,3)**.

9. First, graph the set of constraints:

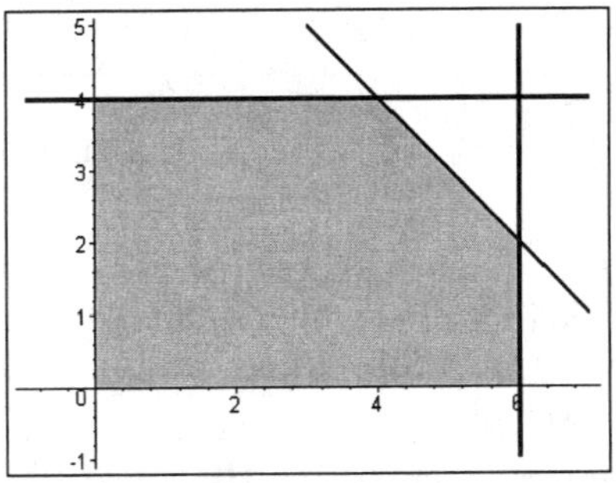

Three of the five corner points are obvious; specifically, the origin $(0,0)$, the y-intercept $(0,4)$, and the x-intercept $(6,0)$. As for the remaining two points, one point is the intersection of $y = 4$ and $x + y = 8$; the other is the intersection of $x = 6$ and $x + y = 8$. Via the substitution method (of Section 12.1), we find that the first of these two points has coordinates $(4,4)$, while the other has coordinates $(6,2)$.

In summary, the five corner points are **(0,0), (0,4), (6,0), (4,4), (6,2)**.

11. Again, begin by graphing the set of constraints:

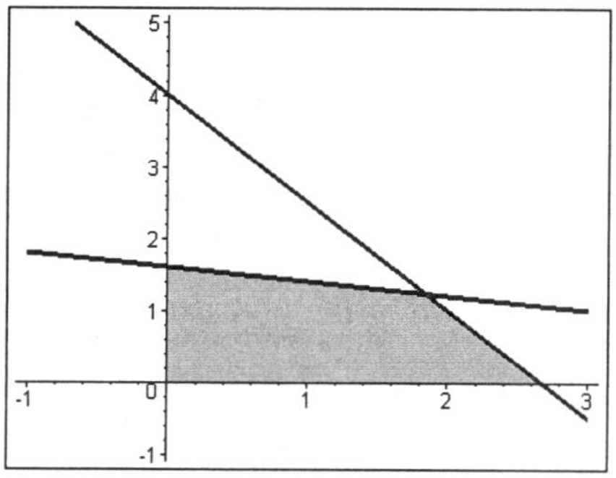

To find the coordinates of this feasible region's four corner points, solve the following systems of equations:

Corner Point	*System to Solve*	*Suggested Method*	*Solution*
Lower left	$\begin{cases} x = 0 \\ y = 0 \end{cases}$	n/a	$(0,0)$
Upper left	$\begin{cases} x = 0 \\ x + 5y = 8 \end{cases}$	Substitution	$(0, \frac{8}{5})$
Lower Right	$\begin{cases} y = 0 \\ 3x + 2y = 8 \end{cases}$	Substitution	$(\frac{8}{3}, 0)$
Upper Right	$\begin{cases} 3x + 2y = 8 \\ x + 5y = 8 \end{cases}$	Substitution or Addition	$(\frac{24}{13}, \frac{16}{13})$

To review the suggested solving methods, please revisit textbook Section 12.1 (including this manual's solutions thereof). Additionally, remember that systems of linear equations can be solved via the Gauss-Jordan method (Section 12.3) or the inverse matrix method (Section 12.4).

In summary, the four corner points are $\mathbf{(0,0), (0,\frac{8}{5}), (\frac{8}{3},0), (\frac{24}{13},\frac{16}{13})}$.

13. Begin by graphing the set of constraints:

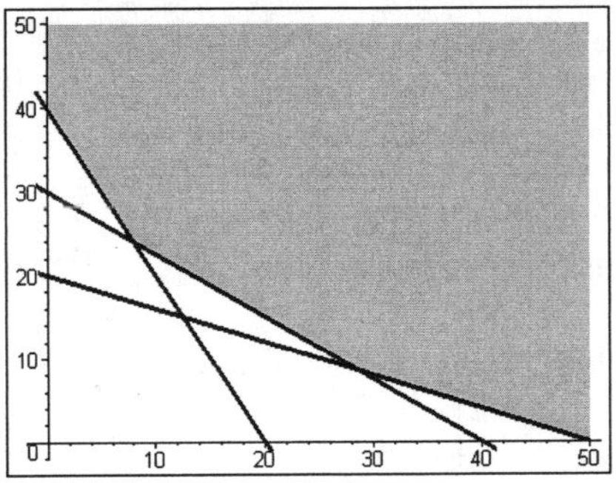

We have in this case an unbounded feasible region. To find the coordinates of its four corner points, we once again solve systems of equations. With reference to the graph (in top-left to bottom-right order), we have:

Corner Point	*System to Solve*	*Suggested Method*	*Solution*
Leftmost	$\begin{cases} x = 0 \\ 10x + 5y = 200 \end{cases}$	Substitution	$(0,40)$
Next to leftmost	$\begin{cases} 10x + 5y = 200 \\ 3x + 4y = 120 \end{cases}$	Addition	$(8,24)$
Next to rightmost	$\begin{cases} 2x + 5y = 100 \\ 3x + 4y = 120 \end{cases}$	Addition	$(\frac{200}{7}, \frac{60}{7})$
Rightmost	$\begin{cases} y = 0 \\ 2x + 5y = 100 \end{cases}$	Substitution or Addition	$(50,0)$

To review the suggested solving methods, please revisit textbook Section 12.1 (including this manual's solutions thereof). Remember that systems of linear equations also can be solved via the Gauss-Jordan method (Section 12.3) or the inverse matrix method (Section 12.4).

In summary, the four corner points are $\boxed{(0,40), (8,24), (\frac{200}{7}, \frac{60}{7}), (50,0)}$.

15. This problem's six constraints lead to the following graph:

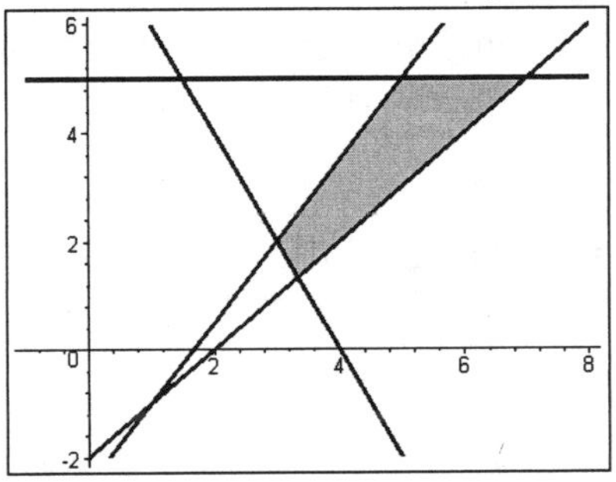

To find the coordinates of this feasible region's four corner points, solve the following systems of equations:

Corner Point	***System to Solve***	***Suggested Method***	***Solution***
Lower left	$\begin{cases} 2x+y=8 \\ 3x-2y=5 \end{cases}$	Substitution or Addition	$(3,2)$
Upper left	$\begin{cases} y=5 \\ 3x-2y=5 \end{cases}$	Substitution	$(5,5)$
Upper Right	$\begin{cases} y=5 \\ x-y=2 \end{cases}$	Substitution	$(7,5)$
Lower Right	$\begin{cases} x-y=2 \\ 2x+y=8 \end{cases}$	Addition	$(\frac{10}{3},\frac{4}{3})$

To review the suggested solving methods, please revisit textbook Section 12.1 (including this manual's solutions thereof). Additionally, remember that systems of linear equations can be solved via the Gauss-Jordan method (Section 12.3) or the inverse matrix method (Section 12.4).

In summary, the four corner points are $\boxed{(0,0),\ (5,5),\ (7,5),\ (\frac{10}{3},\frac{4}{3})}$.

17. From our work in Problem 5 (see solution above), we know the region of feasible solutions has four corner points: $(0,0)$, $(0,\frac{9}{2})$, $(6,0)$, and $(5,2)$. The linear programming theorem (stated on textbook p. 852) tells us that the objective function's optimum value (maximum or minimum) will occur at one of these corner points. So, to find the maximum value of the given objective function $W=30x+20y$, we evaluate this function for the coordinates of each corner point:

Corner Point	$W=30x+20y$
$(0,0)$	$W=30(0)+20(0)=0$
$(0,\frac{9}{2})$	$W=30(0)+20(\frac{9}{2})=90$
$(6,0)$	$W=30(6)+20(0)=180$
$(5,2)$	$W=30(5)+20(2)=190$

Hence, the maximum occurs at the point $(5,2)$ and has a value of $W=\boxed{190}$.

19. From our work in Problem 7 (see solution above), we know the region of feasible solutions has four corner points: $(0,0)$, $(0,4)$, $(4,0)$, and $(2,3)$. The linear programming theorem (textbook p. 852) tells us that the objective function's optimum value (maximum or minimum) will occur at one of these corner points. So, to find the maximum value of the given objective function $P = 100x + 100y$, we evaluate this function for the coordinates of each corner point:

Corner Point	$P = 100x + 100y$
(0,0)	$P = 100(0) + 100(0) = 0$
(0,4)	$P = 100(0) + 100(4) = 400$
(4,0)	$P = 100(4) + 100(0) = 400$
(2,3)	$P = 100(2) + 100(3) = 500$

Hence, the maximum occurs at the point (2,3) and has a value of $P = \boxed{500}$.

21. From our work in Problem 9 (see solution above), we know the region of feasible solutions has five corner points: (0,0), (0,4), (6,0), (4,4), and (6,2). The linear programming theorem (textbook p. 852) tells us that the objective function's optimum value (maximum or minimum) will occur at one of these corner points. So, to find the minimum value of the given objective function $A = 2x - 3y$, we evaluate this function for the coordinates of each corner point:

Corner Point	$A = 2x - 3y$
(0,0)	$A = 2(0) - 3(0) = 0$
(0,4)	$A = 2(0) - 3(4) = -12$
(6,0)	$A = 2(6) - 3(0) = 12$
(4,4)	$A = 2(4) - 3(4) = -4$
(6,2)	$A = 2(6) - 3(2) = 6$

Hence, the maximum occurs at the point (0,4) and has a value of $A = \boxed{-12}$.

23. The convalescent hospital's goal ("objective") is to minimize its patient food costs, while satisfying (the "constraints" of) the patients' minimum dietary requirements. This scenario can be expressed as a linear programming problem. Its objective function represents the food cost for each patient per day, which depends on the number of servings of each food type purchased. The constraints consist of a set of inequalities that express the minimum carbohydrate, protein, and fat requirements (in grams) for each patient per day. So, we let

x = number of servings of food A for each patient per day

and y = number of servings of food B for each patient per day.

Given that food A costs \$0.29 per serving and food B costs \$0.15 per serving, the objective function becomes $C = 0.29x + 0.15y$, where C is the total cost of both foods per patient per day. We now turn our attention to the constraints.

If one serving of food A contains 10g of carbohydrates, then x servings of food A contain $(10x)$g of carbohydrates. Since one serving of food B contains 5g of carbohydrates, y servings of food B contain $(5x)$g of carbohydrates. If a patient eats x servings of food A and y servings of food B, a total of $(10x+5y)$g of carbohydrates are ingested. Given that each patient's daily diet must include no less than 200g of carbohydrates, we express this constraint (in terms of x and y) as $10x+5y \geq 200$.

By way of similar reasoning, we express the given protein and fat requirements in terms of two more inequalities (constraints):

Protein: $2x+5y \geq 100$ Fat: $3x+4y \geq 20$

Finally, it is important to remember two obvious (but often overlooked) constraints:

$x \geq 0$ and $y \geq 0$ (i.e., the number of servings of food cannot be negative).

In summary, we have formulated the following linear programming model:

x = number of servings of food A for each patient per day
y = number of servings of food B for each patient per day
Minimize $C = 0.29x + 0.15y$

Subject to the constraints $\begin{cases} x \geq 0 \\ y \geq 0 \\ 10x+5y \geq 200 \\ 2x+5y \geq 100 \\ 3x+4y \geq 20 \end{cases}$

25. The goal ("objective") is clear: maximize the yearly return on the pension fund's investments, subject to certain restrictions ("constraints").

To achieve this goal, one must decide the allocation of the investments; namely, how much is to be invested in stocks and how much in long-term bonds. Since these amounts are the unknowns, we let

x = amount invested in stocks (in millions of dollars)

and y = amount invested in bonds (in millions of dollars).

The objective function (call it T) represents the total annual return on the investments. So, given that the stocks yield a 12% return and the bonds 8%, we write $T = 0.12x + 0.08y$.

As for the constraints, we have:

- A maximum of \$10 million combined will be invested, so $x+y \leq 10$;
- No more than \$8 million can be invested in stocks, so $x \leq 8$;
- At least \$2 million must be invested in bonds, so $y \geq 2$;

- The stock-to-bond investment ratio cannot be more than 3 to 1, so $\frac{x}{y} \le 3$, which is algebraically equivalent to $x \le 3y$.

Finally, it is important to remember that the amounts invested cannot be negative; in terms of our unknowns, that means $x \ge 0$ and $y \ge 0$. Please note, however, that we already have the constraint that $y \ge 2$. Therefore, we exclude mention of $y \ge 0$.

In summary, we have formulated the following linear programming model:

x = amount invested in stocks (in millions of dollars)
y = amount invested in bonds (in millions of dollars)
Maximize $T = 0.12x + 0.08y$

$$\textbf{Subject to the constraints } \begin{cases} x \ge 0 \\ x \le 8 \\ y \ge 2 \\ x + y \le 10 \\ x \le 3y \end{cases}$$

27. With $x =$ number of acres to be planted in corn and $y =$ number of acres to be planted in wheat, we recall the set of constraints from text Example 2 (see p. 855):

$$\begin{cases} x \ge 0 \\ y \ge 0 \\ x + y \le 100 \\ 120x + 210y \le 15{,}000 \\ 110x + 30y \le 4{,}000 \end{cases}$$

If the farmer decides to contract to have "at least" 4,000 bushels stored at a neighboring farm, the last constraint becomes $110x + 30y \ge 4{,}000$, instead of $110x + 30y \le 4{,}000$. This, in turn, affects the region of feasibility. Instead of the graph shown in Figure 12.9 (textbook p. 855), we now have:

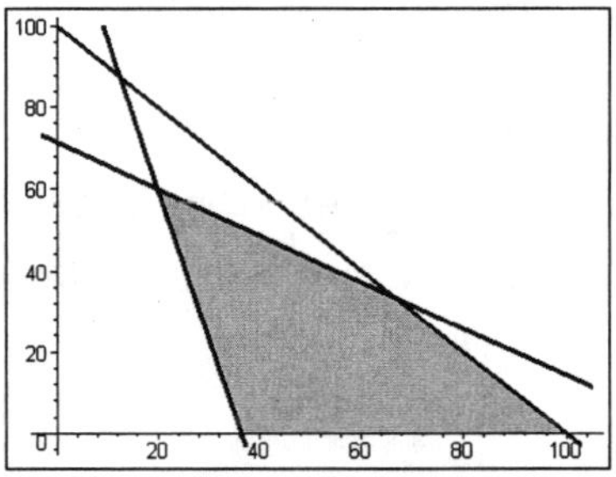

From Example 2, we know the coordinates for two of this feasible region's four corner points; specifically, $(\frac{400}{11},0)$ is the lower left corner and $(20,60)$ is the upper left corner. The remaining two are:

- $(\frac{200}{3},\frac{100}{3})$, which is the upper right corner and is determined by solving (via the addition or substitution method) the system
$$\begin{cases} x+y=100 \\ 120x+210y=15{,}000 \end{cases}$$
- $(100,0)$, which is the lower right corner and is determined by solving (via the substitution method) the system
$$\begin{cases} y=0 \\ x+y=100 \end{cases}$$

By the linear programming theorem (textbook p. 852), we know the objective function's optimum value (maximum or minimum) will occur at one of these corner points. So, to find the acres of corn and of wheat that should be planted to maximize profit, we evaluate the objective (profit) function $P=143x+60y$ at each of these corner points:

Corner Point	$P=143x+60y$
$(\frac{400}{11},0)$	$P=143(\frac{400}{11})+60(0)=5{,}200$
$(20,60)$	$P=143(20)+60(60)=6{,}460$
$(\frac{200}{3},\frac{100}{3})$	$P=143(\frac{200}{3})+60(\frac{100}{3})\approx 11{,}533.33$
$(100,0)$	$P=143(100)+60(0)=14{,}300$

Thus, we conclude: **the maximum profit P=\$14,300 is achieved by planting all 100 acres in corn**.

29. Let $x=$ number of ounces of Kellogg's Corn Flakes
$y=$ number of ounces of Post Honeycombs

The problem's stated objective is to minimize cost. So, since Corn Flakes cost \$0.07 per ounce and Honeycombs \$0.19 per ounce, we write the following objective function: $C=0.07x+0.19y$.

As for the constraints, there are four:

- Given that an ounce of Corn Flakes contains 23g of starch while an ounce of Honeycombs contains 14g of starch, we represent the constraint that at least 322g of starch must be received by the inequality $23x+14y\geq 322$;

- Given that an ounce of Corn Flakes contains 7g of sucrose while an ounce of Honeycombs contains 17g of sucrose, we represent the constraint that at least 119g of sucrose must be received by $7x+17y \geq 119$;
- The number of ounces of Corn Flakes cannot be negative, so $x \geq 0$;
- The number of ounces of Honeycombs cannot be negative, so $y \geq 0$;

These four constraints constitute a system of linear inequalities:

$$\begin{cases} x \geq 0 \\ y \geq 0 \\ 23x+14y \geq 322 \\ 7x+17y \geq 119 \end{cases}$$

The graph of these constraints form the following unbounded region of feasible solutions:

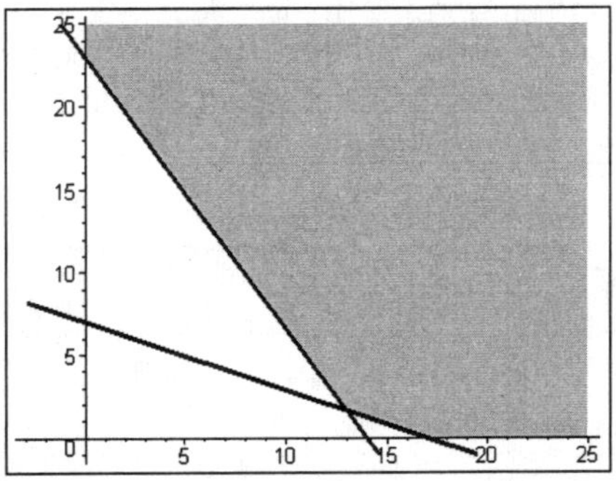

To find the coordinates of this unbounded feasible region's three corner points, we solve the following systems of equations:

Corner Point	***System to Solve***	***Suggested Method***	***Solution***
Upper left	$\begin{cases} x=0 \\ 23x+14y=322 \end{cases}$	Substitution	$(0,23)$
Middle	$\begin{cases} 23x+14y=322 \\ 7x+17y=119 \end{cases}$	Addition	$(\frac{3,808}{293}, \frac{483}{293})$
Lower Right	$\begin{cases} y=0 \\ 7x+17y=119 \end{cases}$	Substitution	$(17,0)$

The linear programming theorem (textbook p. 852) tells us that the objective function's optimum value (maximum or minimum) will occur at one of these corner points. So, to find the minimum value of the given objective function $C=0.07x+0.19y$, we evaluate this function for the coordinates of each corner point:

Corner Point	$C = 0.07x + 0.19y$
$(0,23)$	$C = 0.07(0) + 0.19(23) = 4.37$
$(\frac{3,808}{293}, \frac{483}{293})$	$C = 0.07(\frac{3,808}{293}) + 0.19(\frac{483}{293}) \approx 1.22$
$(17,0)$	$C = 0.07(17) + 0.19(0) = 1.19$

Therefore, **the minimum cost is $1.19 and is obtained with 17 oz of Corn Flakes and no Honeycombs**.

Chapter 12 Review Questions

1. $[C][B] - 3[D] = \begin{bmatrix} 2 & 0 \\ 1 & 2 \\ -1 & 1 \end{bmatrix} \begin{bmatrix} 2 & -1 & 0 \\ 1 & 0 & 1 \end{bmatrix} + (-3)\begin{bmatrix} 0 & 1 & 0 \\ -1 & 0 & 0 \\ 0 & 1 & -1 \end{bmatrix}$

$$= \begin{bmatrix} 4+0 & -2+0 & 0+0 \\ 2+2 & -1+0 & 0+2 \\ -2+1 & 1+0 & 0+1 \end{bmatrix} + \begin{bmatrix} 0 & -3 & 0 \\ 3 & 0 & 0 \\ 0 & -3 & 3 \end{bmatrix}$$

$$= \begin{bmatrix} 4 & -2 & 0 \\ 4 & -1 & 2 \\ -1 & 1 & 1 \end{bmatrix} + \begin{bmatrix} 0 & -3 & 0 \\ 3 & 0 & 0 \\ 0 & -3 & 3 \end{bmatrix}$$

$$= \boxed{\begin{bmatrix} \mathbf{4} & \mathbf{-5} & \mathbf{0} \\ \mathbf{7} & \mathbf{-1} & \mathbf{2} \\ \mathbf{-1} & \mathbf{-2} & \mathbf{4} \end{bmatrix}}$$

2. $[A][B][C] = \begin{bmatrix} 1 & 0 \\ 2 & -1 \end{bmatrix} \begin{bmatrix} 2 & -1 & 0 \\ 1 & 0 & 1 \end{bmatrix} \begin{bmatrix} 2 & 0 \\ 1 & 2 \\ -1 & 1 \end{bmatrix}$

$$= \begin{bmatrix} 2 & -1 & 0 \\ 3 & -2 & -1 \end{bmatrix} \begin{bmatrix} 2 & 0 \\ 1 & 2 \\ -1 & 1 \end{bmatrix} = \begin{bmatrix} 4-1+0 & 0-2+0 \\ 6-2+1 & 0-4-1 \end{bmatrix} = \boxed{\begin{bmatrix} \mathbf{3} & \mathbf{-2} \\ \mathbf{5} & \mathbf{-5} \end{bmatrix}}$$

3. Matrix $[B]$ is a $2 \times \underline{3}$ matrix, while $[A]$ is $\underline{2} \times 2$. Since $[B]$ and $[A]$ are **nonconformable** ($3 \neq 2$), the product $[B][A][D]$ cannot be computed.

4.
$$[C][A][B] = \begin{bmatrix} 2 & 0 \\ 1 & 2 \\ -1 & 1 \end{bmatrix} \begin{bmatrix} 1 & 0 \\ 2 & -1 \end{bmatrix} \begin{bmatrix} 2 & -1 & 0 \\ 1 & 0 & 1 \end{bmatrix}$$
$$= \begin{bmatrix} 2+0 & 0+0 \\ 1+4 & 0-2 \\ -1+2 & 0-1 \end{bmatrix} \begin{bmatrix} 2 & -1 & 0 \\ 1 & 0 & 1 \end{bmatrix}$$
$$= \begin{bmatrix} 2 & 0 \\ 5 & -2 \\ 1 & -1 \end{bmatrix} \begin{bmatrix} 2 & -1 & 0 \\ 1 & 0 & 1 \end{bmatrix}$$
$$= \begin{bmatrix} 4+0 & -2+0 & 0+0 \\ 10-2 & -5+0 & 0-2 \\ 2-1 & -1+0 & 0-1 \end{bmatrix} = \boxed{\begin{bmatrix} \mathbf{4} & \mathbf{-2} & \mathbf{0} \\ \mathbf{8} & \mathbf{-5} & \mathbf{-2} \\ \mathbf{1} & \mathbf{-1} & \mathbf{-1} \end{bmatrix}}$$

5. Let $[A] = \begin{bmatrix} 2 & 1 \\ -\frac{3}{2} & -\frac{1}{2} \end{bmatrix}$.

Inverse matrices are discussed in Section 12.4, beginning on p. 841. We write the augmented matrix $\left[A \,\vdots\, I\right]$ and proceed to transform, if possible, the left-hand side A into the identity matrix $[I_2]$:

$$[B] = \left[\begin{array}{cc|cc} 2 & 1 & 1 & 0 \\ -\frac{3}{2} & -\frac{1}{2} & 0 & 1 \end{array}\right] \to \left[\begin{array}{cc|cc} 1 & \frac{1}{2} & \frac{1}{2} & 0 \\ -\frac{3}{2} & -\frac{1}{2} & 0 & 1 \end{array}\right] \to \left[\begin{array}{cc|cc} 1 & \frac{1}{2} & \frac{1}{2} & 0 \\ 0 & \frac{1}{4} & \frac{3}{4} & 1 \end{array}\right]$$

$*$Row($\frac{1}{2}$,[B],1) $*$Row+($\frac{3}{2}$,[Ans],1,2)

$$\to \left[\begin{array}{cc|cc} 1 & \frac{1}{2} & \frac{1}{2} & 0 \\ 0 & 1 & 3 & 4 \end{array}\right] \to \left[\begin{array}{cc|cc} 1 & 0 & -1 & -2 \\ 0 & 1 & 3 & 4 \end{array}\right]$$

$*$Row(4,[Ans],2) $*$Row+($-\frac{1}{2}$,[Ans],2,1)

So, the result in the right half of this augmented matrix is the inverse of $[A]$. That is,
$$[A]^{-1} = \boxed{\begin{bmatrix} \mathbf{-1} & \mathbf{-2} \\ \mathbf{3} & \mathbf{4} \end{bmatrix}}.$$

We encourage you to check this answer by confirming $[A][A]^{-1} = [A]^{-1}[A] = [I]$.

6. Let $[A]=\begin{bmatrix}1&3&3\\1&4&3\\1&3&4\end{bmatrix}$.

Inverse matrices are discussed in Section 12.4, beginning on p. 841. We write the augmented matrix $\left[A \;\vdots\; I\right]$ and proceed to transform, if possible, the left-hand side A into the corresponding identity matrix $[I_3]$:

$$[B]=\left[\begin{array}{ccc|ccc}1&3&3&1&0&0\\1&4&3&0&1&0\\1&3&4&0&0&1\end{array}\right]\rightarrow\left[\begin{array}{ccc|ccc}1&3&3&1&0&0\\0&1&0&-1&1&0\\0&0&1&-1&0&1\end{array}\right]$$

∗Row+(–1,[B],1,2)
∗Row+(–1,[Ans],1,3)

$$\rightarrow\left[\begin{array}{ccc|ccc}1&0&3&4&-3&0\\0&1&0&-1&1&0\\0&0&1&-1&0&1\end{array}\right]\rightarrow\left[\begin{array}{ccc|ccc}1&0&0&7&-3&-3\\0&1&0&-1&1&0\\0&0&1&-1&0&1\end{array}\right]$$

∗Row+(–3,[Ans],2,1) ∗Row+(–3,[Ans],3,1)

So, the result in the right half of this augmented matrix is the inverse of $[A]$. That is,

$$[A]^{-1}=\boxed{\begin{bmatrix}\mathbf{7}&\mathbf{-3}&\mathbf{-3}\\\mathbf{-1}&\mathbf{1}&\mathbf{0}\\\mathbf{-1}&\mathbf{0}&\mathbf{1}\end{bmatrix}}.$$

You should check the answer; in particular, does $[A][A]^{-1} = [A]^{-1}[A]=[I]$?

7. The ***graphing method*** for solving systems of equations is introduced in Section 12.1 (see pp. 808-809).
The graphs of $2x-y=2$ and $3x-2y=1$ are shown on the next page; the former has a slope of 2 with a y-intercept of -2, and the latter has a slope of $\frac{3}{2}$ with a y-intercept of $-\frac{1}{2}$. Looking at the point of intersection, we conclude that the solution appears to be $\boxed{\mathbf{(3,\ 4)}}$. Of course, you should always check the solution — especially when using the graphing method.

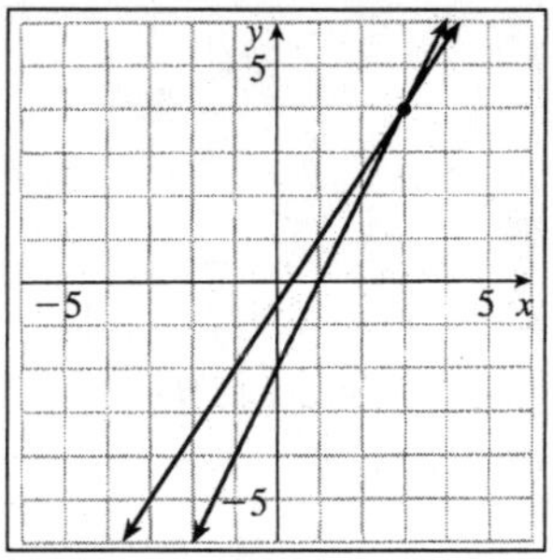

8. The ***addition method*** for solving systems of equations is introduced in Section 12.1 (see pp. 810-811).
Multiplying both sides of the first equation by 2 forces the coefficients of y to be opposites:

$$2\begin{cases} x+3y=3 \\ 4x-6y=-6 \end{cases} \rightarrow \quad +\begin{cases} 2x+6y=6 \\ 4x-6y=-6 \end{cases}$$

$$6x=0$$
$$x=0$$

Now, to obtain the value of y, we substitute the value of 0 for x in either of the given equations and proceed to solve as follows:

$$x+3y=3$$
$$(0)+3y=3$$
$$y=1$$

So, the solution is **(0, 1)**. You should then take the time to check your solution, using both of the given equations to do so.

9. Initially introduced in Section 12.1, the ***substitution method*** for solving systems of equations is discussed on textbook pp. 809-810.
Given the first equation $y=1-2x$, replace y in the second equation with $1-2x$:

$$5x+2y=1$$
$$5x+2(1-2x)=1$$
$$5x+2-4x=1$$
$$x=-1$$

Now, to find y, substitute the value of -1 for x in either of the given equations. In this problem, it happens to be more efficient to use the first equation:

$$y=1-2x=1-2(-1)=3$$

Therefore, the solution to this linear system is the ordered pair **(−1, 3)**. As always, you should take the time to check the solution, using both equations to do so.

10. The ***Gauss-Jordan elimination method*** is introduced in Section 12.3 (see pp. 830-832). We write the given system of equations in augmented matrix form, and then, by applying elementary row operations, we seek to transform it into row-reduced form.

$$[A]=\left[\begin{array}{ccc|c}1&1&1&2\\1&2&-2&1\\1&1&3&4\end{array}\right]\rightarrow\left[\begin{array}{ccc|c}1&1&1&2\\0&1&-3&-1\\0&0&2&2\end{array}\right]\rightarrow\left[\begin{array}{ccc|c}1&0&4&3\\0&1&-3&-1\\0&0&2&2\end{array}\right]$$

∗Row+(–1,[A],1,2) ∗Row+(–1,[Ans],2,1)
∗Row+(–1,[Ans],1,3)

$$\rightarrow\left[\begin{array}{ccc|c}1&0&4&3\\0&1&-3&-1\\0&0&1&1\end{array}\right]\rightarrow\left[\begin{array}{ccc|c}1&0&0&-1\\0&1&0&2\\0&0&1&1\end{array}\right]$$

∗Row($\frac{1}{2}$,[Ans],3) ∗Row+(–4,[Ans],3,1)
∗Row+(3,[Ans],3,2)

This final matrix represents the system $\begin{cases}1x+0y+0z=-1\\0x+1y+0z=2\\0x+0y+1z=1\end{cases}$.

Hence, the solution to the given system is **(–1, 2, 1)**. Finally, please make sure you check this answer, using all three original equations to do so.

11. Let $[A]=\begin{bmatrix}2&1\\-\frac{3}{2}&-\frac{1}{2}\end{bmatrix}$, $[X]=\begin{bmatrix}x\\y\end{bmatrix}$, and $[B]=\begin{bmatrix}13\\10\end{bmatrix}$.

The solution is $[X]=[A]^{-1}[B]$.

In Problem 5, we found $[A]^{-1}=\begin{bmatrix}-1&-2\\3&4\end{bmatrix}$.

So,

$$[X]=[A]^{-1}[B]=\begin{bmatrix}-1&-2\\3&4\end{bmatrix}\begin{bmatrix}13\\10\end{bmatrix}=\begin{bmatrix}-13-20\\39+40\end{bmatrix}=\begin{bmatrix}-33\\79\end{bmatrix}.$$

Therefore, the solution to the given system of equations is the ordered pair **(–33, 79)**. As always, you should check the answer, using both equations to do so.

12. Notice that the lines $2x - y + 2 = 0$ and $2x - y + 12 = 0$ share a slope of 2 (i.e., the lines are parallel). Similarly, $3x + 2y + 10 = 0$ and $3x + 2y - 18 = 0$ are parallel, with a slope of $-\frac{3}{2}$. Using the skills practiced in Section 12.5 along with our prior knowledge of graphing lines and half-planes (from Sections 11.1 and 11.2, respectively), we obtain:

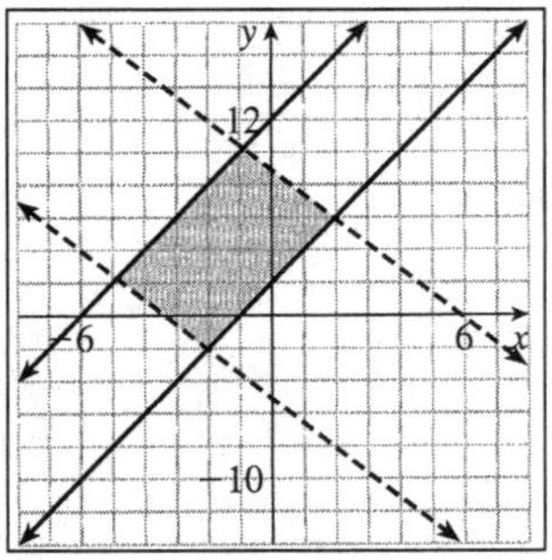

13. This is a mixture problem comparable to those in Section 12.2 (review, as necessary, textbook pp. 817-821).

Let $x =$ NUMBER OF BARS OF PRODUCT I

and $y =$ NUMBER OF BARS OF PRODUCT II.

Given that each bar of Product I contains 3 oz of metal A and 5 oz of metal B, it follows that x bars of Product I contain $3x$ oz of metal A and $5x$ oz of metal B. Similarly, since each bar of Product II contains 4 oz of metal A and 7 oz of metal B, it follows that y bars of Product II contain $4y$ oz of metal A and $7y$ oz of metal B. Additionally, recall that the desired alloy requires 33 oz of metal A and 56 oz of metal B. The following table summarizes all of this information:

	Metal A	***Metal B***
Product I	$3x$ oz	$5x$ oz
Product II	$4y$ oz	$7y$ oz
Totals	33 oz	56 oz

Thus, we have a system of linear equations:

$$\begin{cases} 3x + 4y = 33 \\ 5x + 7y = 56 \end{cases}$$

Using the addition method to solve it, we obtain:

$$\begin{matrix} 5\times \\ -3\times \end{matrix}\begin{cases} 3x + 4y = 33 \\ 5x + 7y = 56 \end{cases} \quad \rightarrow \quad +\begin{cases} 15x + 20y = 165 \\ -15x - 21y = -168 \end{cases}$$

$$-y = -3$$

$$y = 3$$

To obtain the value of x, we substitute the value of 3 for y in either of the two equations. We have chosen arbitrarily to use the first equation:

$$3x + 4y = 33$$
$$3x + 4(3) = 33$$
$$3x + 12 = 33$$
$$3x = 21$$
$$x = 7$$

So, the solution to the system is $(7, 3)$.

This means **she should use 7 bars of Product I and 3 bars of Product II**.

14. This problem is largely an extension of Problem 13 (see solution immediately above).

Let x = NUMBER OF UNITS OF PRODUCT I,

y = NUMBER OF UNITS OF PRODUCT II,

z = NUMBER OF UNITS OF PRODUCT III.

Like Problem 13, the given information gives rise to a system of equations. In this case, we obtain:

$$\begin{cases} 2x + 2y + 3z = 1{,}250 \\ x + 2y + 2z = 900 \\ x + y + 2z = 750 \end{cases}$$

This system can be solved in a variety of ways (e.g., addition, substitution, Gauss-Jordan, or the inverse matrix method). Its solution is $(x, y, z) = (100, 150, 250)$, which means **100 units of Product I, 150 of Product II, and 250 of Product III** may be manufactured (if all of the inventory is to be used).

15. This problem is almost identical to text Example 2 in Section 12.6. For additional reinforcement, we highly recommend you revisit this example's detailed solution (see textbook pp. 853-856).

Let x = number of acres to be planted in corn

and y = number of acres to be planted in wheat.

The farmer's goal ("objective") is to maximize profit, so the objective function must describe total profit. Since each acre of corn yields 100 bushels and each acre of wheat yields 40 bushels, it follows that x acres of corn yield $100x$ bushels and y acres of wheat yield $40y$ bushels. Given that the net profit is $\$2.10$ per bushel of corn and $\$2.50$ per bushel of wheat, the total profit P is:

$P = (100x$ bushels of corn) × $2.10/bushel + ($40y$ bushels of wheat) × $2.50/bushel

So, the objective function is $P = 210x + 100y$.

As for the constraints, we have:

- A maximum of 500 acres are available for planting corn and/or wheat, so $x + y \le 500$.
- The total yield, $(100x + 40y)$ bushels, cannot exceed the storage capacity of $18{,}000$ bushels, so $100x + 40y \le 18{,}000$.
- The costs associated with farming corn and wheat are respectively $\$120$ per acre and $\$60$ per acre. So, x acres of corn cost $\$(120x)$, and y acres of wheat cost $\$(60y)$. Assuming that the farmer's total costs cannot exceed the available capital, it follows that $120x + 60y \le 24{,}000$.
- Acreage cannot be a negative quantity, so $x \ge 0$ and $y \ge 0$.

Now, we have formulated this problem as a linear programming model:

<u>**Maximize**</u>: $P = 210x + 100y$

<u>**Subject to**</u>: $$\begin{cases} x \ge 0 \\ y \ge 0 \\ x + y \le 500 \\ 100x + 40y \le 18{,}000 \\ 120x + 60y \le 24{,}000 \end{cases}$$

The graph of these five constraints form the following region of feasible solutions:

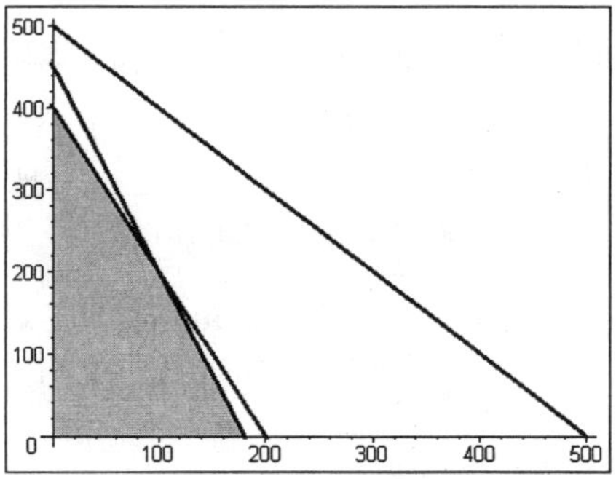

Like text Example 2 (see p. 856), notice that $x + y \le 500$ is a superfluous constraint.

To find the coordinates of this feasible region's four corner points, we solve the following systems of equations:

Corner Point	System to Solve	Suggested Method	Solution
Lower left	$\begin{cases} x=0 \\ y=0 \end{cases}$	n/a	$(0,0)$
Upper left	$\begin{cases} x=0 \\ 120x+60y=24{,}000 \end{cases}$	Substitution	$(0,400)$
Lower Right	$\begin{cases} y=0 \\ 100x+40y=18{,}000 \end{cases}$	Substitution	$(180,0)$
Upper Right	$\begin{cases} 100x+40y=18{,}000 \\ 120x+60y=24{,}000 \end{cases}$	Addition	$(100,200)$

By the linear programming theorem (textbook p. 852), we know the objective function's optimum value (maximum or minimum) will occur at one of these corner points. Obviously, it is not necessary to check the point $(0,0)$, which would yield no profit. So, to find the acres of corn and of wheat that should be planted to maximize profit, we evaluate the objective (profit) function $P=210x+100y$ at each of the remaining three corner points:

Corner Point	$P=210x+100y$
$(0,400)$	$P=210(0)+100(400)=40{,}000$
$(180,0)$	$P=210(180)+100(0)=37{,}800$
$(100,200)$	$P=210(100)+100(200)=41{,}000$

We conclude:

the maximum profit *P* = \$41,000 is achieved by planting 100 acres in corn, 200 acres of wheat, and leaving the remaining 200 acres unplanted.

APPENDICES

Appendix A

1. The worst-case scenario is to have obtained one card from each suit after the first four draws (i.e., 1 heart, 1 spade, 1 diamond, and 1 club). Therefore, in order to be absolutely certain that at least two cards are from the same suit, one **must draw at least 5 cards**.

3. This problem requires that we find a pattern and then use it to predict an answer. This very important method of thought is known as ***inductive reasoning***; it was first introduced in textbook Section 1.2 and was subsequently used to solve various problems throughout the text.

Let us consider the units digits of 3^{2001} and 2^{2001} separately.

We begin by listing powers of 3:

$$3^0 = 1,\ 3^1 = 3,\ 3^2 = 9,\ 3^3 = 2\underline{\mathbf{7}},\ 3^4 = 8\underline{\mathbf{1}},\ 3^5 = 24\underline{\mathbf{3}},\ 3^6 = 72\underline{\mathbf{9}},\ 3^7 = 2{,}18\underline{\mathbf{7}},\ \ldots.$$

Notice that the units digits seem to form a repeating pattern $(1, 3, 9, 7, 1, 3, 9, 7, \ldots)$. In particular,

$3^0, 3^4, 3^8, 3^{12}, \cdots$ have a units digit of 1

$3^1, 3^5, 3^9, 3^{13}, \cdots$ have a units digit of 3

$3^2, 3^6, 3^{10}, 3^{14}, \cdots$ have a units digit of 9

$3^3, 3^7, 3^{11}, 3^{15}, \cdots$ have a units digit of 7

Since 2000 is a multiple of 4 (just like $0, 4, 8, 12, \ldots$), 3^{2000} must have a units digit of 1 (just like $3^0, 3^4, 3^8, 3^{12}, \ldots$). Consequently, 3^{2001} has a units digit of 3.

Similarly, we consider the powers of 2:

$$2^0 = 1,\ 2^1 = 2,\ 2^2 = 4,\ 2^3 = 8,\ 2^4 = 1\underline{\mathbf{6}},\ 2^5 = 3\underline{\mathbf{2}},\ 2^6 = 6\underline{\mathbf{4}},\ 2^7 = 12\underline{\mathbf{8}},\ \ldots.$$

Beginning with 2^1, there is a repeating pattern $(2, 4, 8, 6, 2, 4, 8, 6, \ldots)$, namely:

$2^1, 2^5, 2^9, 2^{13}, \cdots$ have a units digit of 2

$2^2, 2^6, 2^{10}, 2^{14}, \cdots$ have a units digit of 4

$2^3, 2^7, 2^{11}, 2^{15}, \cdots$ have a units digit of 8

$2^4, 2^8, 2^{12}, 2^{16}, \cdots$ have a units digit of 6

Since 2000 is a multiple of 4 (just like $0, 4, 8, 12, \ldots$), 2^{2000} must have a units digit of 6 (just like $2^4, 2^8, 2^{12}, 2^{16}, \ldots$), and so 2^{2001} has a units digit of 2 (just like 2^1, 2^5,

2^9, 2^{13}, ...). Therefore, because 3^{2001} has a units digit of 3 and 2^{2001} has a units digit of 2, the difference $3^{2001} - ^{2001}$ must have a units digit of $3-2=\boxed{1}$.

5. **All 12 months** have 28 days (some months have more than 28 days).

7. Section 5.6 (see p. 277) introduced a very useful problem-solving step, one that was used frequently in subsequent chapters. In particular, this process involved translating ("evolving") problems into verbal equations. This problem — like so many before it — presents us with yet another opportunity to apply this powerful problem-solving tool. Translating the given information, we obtain:

ANN'S AGE = BRITTANY'S AGE + 3; BRITTANY'S AGE = CHELSEA'S AGE + 2;

DEIDRE'S AGE = ELYSSE'S AGE + 5; ELYSSE'S AGE = FAWN'S AGE + 1

Notice that there are relationships among Ann's age, Brittany's age, and Chelsea's age as well as relationships among Deidre's age, Elysse's age, and Fawn's age; however, **nothing can be said** about Deidre's age as compared to Chelsea's age.

9. To review similar measurement-conversion problems, please revisit textbook Section 7.4.

Since 1 megamile is equivalent to $1{,}000{,}000$ miles, 2.376 megamiles is $2{,}376{,}000$ miles. Because there are $1{,}000$ miles per kilomile, it follows that

$$2.376 \text{ megamiles} = 2{,}376{,}000 \text{ miles} = \frac{2{,}376{,}000 \text{ mi}}{1{,}000 \text{ mi/kilomile}} = \boxed{\textbf{2,376 kilomiles}}.$$

11. Problems similar to this were encountered in textbook Section 4.6. In fact, with the exception of the notation used, Problem 29 in the Chapter 4 Review (p. 233) and this particular problem are identical.

Upon recalling the text's extended order-of-operations agreement (on p. 29), we take note of the give expression's parentheses. The parentheses suggest we first compute $(1,2)$ and $(3,4)$ and then, afterwards, use their respective values to compute the final answer. That is,

$$(1,2) = 1\times 2+1+2=5 \text{ and } (3,4)=3\times 4+3+4=19.$$

Hence,

$$((1,2),(3,4)) = (5,19) = 5\times 19+5+19=\boxed{119}.$$

13. An understanding of ***arithmetic sequences*** can prove quite helpful in solving problems like this. Accordingly, please revisit Section 8.3 as necessary.

The set of odd numbers $\{1, 3, 5, 7, \ldots\}$ form an arithmetic sequence with first term $a_1 = 1$ and common difference $d = 2$. By applying the knowledge acquired on textbook p. 527, we can represent the nth odd number by the following (underlined) formula:

$$a_n = a_1 + (n-1)d = 1 + (n-1)(2) = 1 + 2n - 2 = \underline{\underline{2n-1}}.$$

Before proceeding, let us take a moment to confirm this formula's accuracy:

- For $n = 1$, we have $2(1) - 1 = 1$, the 1st odd number;
- For $n = 2$, we have $2(2) - 1 = 3$, the 2nd odd number;
- For $n = 3$, we have $2(3) - 1 = 5$, the 3rd odd number;

And so on...

Therefore, to find the 473rd odd number, we let $n = 473$ and obtain:

$$2n - 1 = 2(473) - 1 = \boxed{\mathbf{945}}.$$

15. This problem is identical to Problem 55 (Section 1.1) and similar to the Chapter 6 Review's Problem 1 (stated on textbook p. 418 and solved in this manual).
Consider this depiction of the described cube:

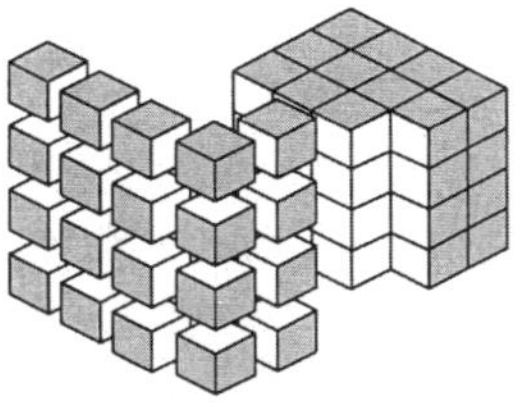

As shown, after cutting, there are $4^3 = \underline{64\text{ one-inch cubes}}$. The eight corners are painted on exactly three sides; the cubes positioned along each edge of the original 4-in cube (but not on a corner) will be painted on two sides.
On each of the original 4-in cube's six faces, the four middle cubes will be painted on one side. So, with 6 faces and 4 cubes per face painted on one side only, there must be a total of $6 \times 4 = \underline{24\text{ cubes painted on one side only}}$.
Therefore, the fraction of all the one-inch cubes painted on one side only is

$$\tfrac{24}{64} = \boxed{\mathbf{\tfrac{3}{8}}}.$$

17. Once again, as mentioned in our solution to Problem 11 on the previous page, we remind you that comparable problems were addressed in Section 4.6. In fact, this problem is nearly identical to Problem 30 in the Chapter 4 Review (p. 233).

We perform the operations in a step-by-step manner (beginning within each pair of parentheses) and obtain the following:

$$M(m(1,2), m(2,3)) = M(1,2) = \boxed{\mathbf{2}}.$$

19. Illustrations of some common polyhedra are provided on textbook p. 465 (see Figure 7.23).

a) **tetrahedron** b) **cube (hexahedron)** c) **octahedron**

d) **icosahedron** e) **dodecahedron**

21. Consider a pattern where we cross out multiples of 3:

$$1, 2, \not{3}, 4, 5, \not{6}, 7, 8, \not{9}, 10, 11, \not{12}, 13, 14, \not{15}, 16, 17, \not{18}, \ldots$$

We then explore this pattern, seeking to devise a general formula for the nth positive integer not divisible by 3. With n representing the nth positive integer that is not divisible by 3, we discover:

$$\begin{cases} \text{If } n \text{ is even, the } n^{\text{th}} \text{ number that is not divisible by 3 is } \tfrac{3}{2}n-1. \\ \text{If } n \text{ is odd, the } n^{\text{th}} \text{ number that is not divisible by 3 is } \tfrac{3}{2}n-\tfrac{1}{2}. \end{cases}$$

Before proceeding, let us take a moment to confirm this generalization:

- If $n=1$, then $\frac{3}{2}(1)-\frac{1}{2}=1$ is the 1st number not divisible by 3;
- If $n=2$, then $\frac{3}{2}(2)-1=2$ is the 2nd number not divisible by 3;
- If $n=3$, then $\frac{3}{2}(3)-\frac{1}{2}=4$ is the 3rd number not divisible by 3;
- If $n=4$, then $\frac{3}{2}(4)-1=5$ is the 4th number not divisible by 3;
- And so on…

Therefore, we let $n=1{,}000$ and find:

$\frac{3}{2}(1{,}000)-1=$ **1,499** is the 1,000th number not divisible by 3.

23. [Note: If you happen to check the calculation regarding 450 dollar bills per pound, you will find that 1 lb is actually about 454 g. However, there will be some waste in the printing process, so the textbook author's assumption that 1 lb $\approx$ 450 g is indeed acceptable.]

We begin by calculating the total weight of paper required to print a trillion one-dollar bills:

$$\frac{1\times10^{12}\text{ bills}}{450\text{ bills per lb}}=2.\overline{2}\times10^{9}\text{ lb}$$

Next, we calculate the volume of a tree, which we assume is cylindrical in shape. From Section 7.4 (p. 456), recall that the volume formula for a cylinder is $V=Bh$, where B is the area of the cylinder's circular base (i.e., $B=\pi r^2$) and h is the height of the tree. Given that an average tree has a height of 50 feet and a diameter of 12 inches (i.e., 1 ft), it follows that $h=50$ ft, $r=0.5$ ft, and the volume is:

$$V=Bh=\pi r^2 h=\pi(0.5)^2(50)\approx 39.2697\text{ ft}^3\text{ per tree.}$$

Assuming that each cubic foot of wood yields about 50 pounds of paper, it follows that $V\approx 1{,}963.5$ lb. Thus, the total number of trees required to print the money is

$$\frac{\text{total weight}}{\text{weight per tree}}=\frac{2.\overline{2}\times10^{9}}{V}\approx 1{,}131{,}768.$$

So, **more than a million trees** must be cut down to print a trillion $1 bills.

25. [**Reminder**: The concept of **area** and its associated formulas are discussed in Section 7.2.]
Of course, the two semicircles form one complete circle. So, instead of attempting (largely in vain) to calculate the area of the shaded region directly, notice that the desired area is the area of the square less the area of the circle. Letting s represent the length of the square's sides and r represent the radius of the circle, we obtain:

$$\text{AREA OF SHADED REGION} = \underbrace{\text{AREA OF SQUARE}} - \underbrace{\text{AREA OF CIRCLE}}$$
$$= s^2 - \pi r^2$$
$$= (8)^2 - \pi(4)^2$$

This area, $(64 - 16\pi)$ square inches, is approximately equal to **13.7 in.²**.

27. [**Reminder**: The concept of **area** and its associated formulas are discussed in Section 7.2.]
As shown below, since the two smaller circles have radii of 2 and 3 units, the largest circle has a diameter of 10 units and, consequently, a radius of 5 units.

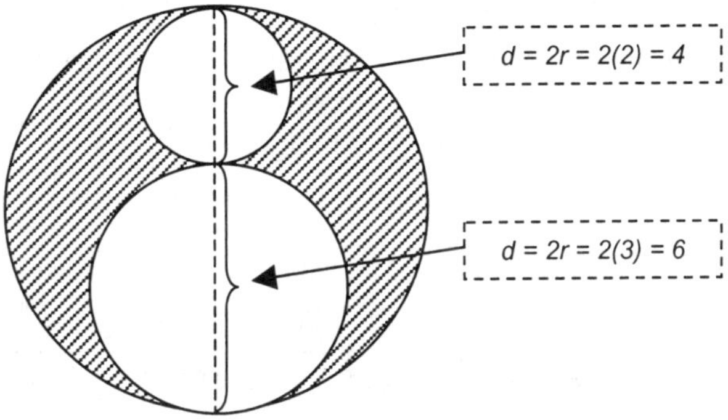

We then calculate the areas of the three circles:

$A = \pi(2)^2 = 4\pi$ (smallest area)

$B = \pi(3)^2 = 9\pi$ (middle area)

$C = \pi(5)^2 = 25\pi$ (largest area)

To find the area of the shaded region, we apply the same reasoning as employed in Problem 25 (see previous solution). In particular, the area of the shaded region is the area of the largest circle less the sum of the areas of the two smaller circles:

$$\text{AREA OF SHADED REGION} = C - (A + B) = 25\pi - (4\pi + 9\pi) = 12\pi$$

So, the ratio of the area of the smallest circle to the area of the shaded region is

$$\frac{\text{Area of Smallest Circle}}{\text{Area of Shaded Region}} = \frac{4\pi}{12\pi} = \boxed{\frac{1}{3}}.$$

29. Let x be the amount of money the gambler started with. Since she doubled her money on the first day while spending $\$30$, she had $\$(2x-30)$ at the end of the first day. Then, since she tripled her money on the second day while spending $\$20$, she had $3(2x-30)-20=6x-90-20=\$(6x-110)$ after the second day. The fact that the amount she had after the second day was the same as the amount she started with the first day, we write (and solve) the following equation:

$$6x-110=x$$
$$5x=110$$
$$x=22$$

So, we conclude that she had **\$22** at the start of the first day.

31. Let x be the cost per traveler if 100 persons travel. It then follows that $100x$ represents the cost of the charter. Similarly, since each person saves $\$78$ if 125 persons make the trip, the cost per traveler (for 125) travelers is $(x-78)$, and accordingly the cost of the charter would be $125(x-78)$.

Assuming that the cost of the charter remains the same regardless of the number of travelers, it then follows that:

$$100x=125(x-78)$$
$$100x=125x-9{,}750$$
$$25x=9{,}750$$
$$x=390$$

Thus, **the cost is \$390 if 100 persons make the trip**.

33. By employing the "numbered paths" approach (as encountered in textbook Section 1.1, pp. 6-10), we find there to be **33 paths** from point A to point B.

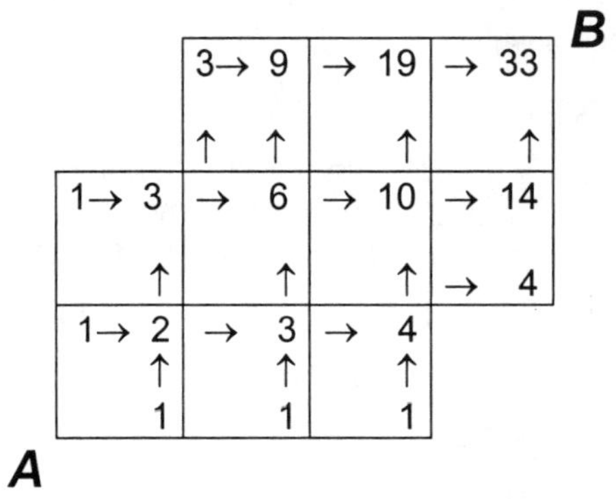

35. This problem is an extension of Problem 17 (of the Chapter 1 Review Question, on textbook p. 59), which requires an understanding of ***scientific notation***. Review, if necessary, Section 1.3.

Given that the national debt was $\$5.7$ trillion (on June 1, 2000) and the U.S. population at that time was 274.9 million, each person's share of the debt on that date was:

$$\frac{5.7\times 10^{12}}{2.749\times 10^{8}} \approx \$20{,}734.81.$$

If each person pays $\$1$ per day, it will take approximately $20{,}734.81$ days for the entire debt to be paid off. In term of years, it will take approximately

$$\left(\frac{5.7\times 10^{12}}{2.749\times 10^{8}}\right) \div 365 \approx 56.80770592 \text{ years of repayment.}$$

Therefore, **it will take until March of 2057** to repay the $\$5.7$ trillion debt. [**Disclaimer**: Realize that this is the date on which the current (6/1/00) debt would be repaid excluding all the accrued interest during this nearly 57-year span.]

37. BASIC programming is the topic of Section 5.9; the Fibonacci sequence is defined in Section 8.3 (and its relationship to the golden ratio is discussed in Section 6.6).

```
10  REM FIBONACCI SEQUENCE
20  PRINT "This program calculates the first 20 numbers in a Fibonacci Sequence"
30  PRINT
40  PRINT "NUMBER", "FIBONACCI SEQUENCE"
45  REM SET THE VALUES FOR THE FIRST TWO NUMBERS
60  LET FIRST = 1
70  LET SECOND = 1
80  PRINT "1", FIRST
90  PRINT "2", SECOND
95  REM RESET VALUES TO USE IN THE FORMULA
100 LET NEXTLAST = FIRST
110 LET LAST = SECOND
120 FOR J = 3 TO 20
125     REM CALCULATE THE NEXT FIBONACCI NUMBER IN THE SEQUENCE
130     NEW = LAST + NEXTLAST
140     PRINT J, NEW
145     REM RESET VALUES FOR THE NEXT STEP
150     LET NEXTLAST = LAST
160     LET LAST = NEW
170 NEXT J
180 END
```

39. **At least three** configurations can be seen:

(i) a little cube nestled in the corner of a larger cube;
(ii) a big cube with a cubical chunk removed from one corner;
(iii) two cubes meeting externally at a corner.